Undergraduate Texts in Mathematics

For other titles in this series, go to
http://www.springer.com/series/666

Sudhir R. Ghorpade
Balmohan V. Limaye

A Course in Multivariable Calculus and Analysis

With 79 figures

Springer

Sudhir R. Ghorpade
Department of Mathematics
Indian Institute of Technology, Bombay
Powai, Mumbai 400076
India
srg@math.iitb.ac.in

Balmohan V. Limaye
Department of Mathematics
Indian Institute of Technology, Bombay
Powai, Mumbai 400076
India
bvl@math.iitb.ac.in

ISSN 0172-6056
ISBN 978-1-4419-1620-4 e-ISBN 978-1-4419-1621-1
DOI 10.1007/978-1-4419-1621-1
Springer New York Dordrecht Heidelberg London

Library of Congress Control Number:2009940318

Mathematics Subject Classification (2000): 26-01, 26Bxx, 40-01, 40Bxx

Printed on acid-free paper

Springer is part of Springer Science+Business Media (www.springer.com)

Preface

Calculus of real-valued functions of several real variables, also known as multivariable calculus, is a rich and fascinating subject. On the one hand, it seeks to extend eminently useful and immensely successful notions in one-variable calculus such as limit, continuity, derivative, and integral to "higher dimensions." On the other hand, the fact that there is much more room to move about in the n-space $\mathbb{R}^n$ than on the real line $\mathbb{R}$ brings to the fore deeper geometric and topological notions that play a significant role in the study of functions of two or more variables.

Courses in multivariable calculus at an undergraduate level and even at an advanced level are often faced with the unenviable task of conveying the multifarious and multifaceted aspects of multivariable calculus to a student in the span of just about a semester or two. Ambitious courses and teachers would try to give some idea of the general Stokes's theorem for differential forms on manifolds as a grand generalization of the fundamental theorem of calculus, and prove the change of variables formula in all its glory. They would also try to do justice to important results such as the implicit function theorem, which really have no counterpart in one-variable calculus. Most courses would require the student to develop a passing acquaintance with the theorems of Green, Gauss, and Stokes, never mind the tricky questions about orientability, simple connectedness, etc. Forgotten somewhere is the initial promise that we shall do unto functions of several variables whatever we did in the previous course to functions of one variable. Also forgotten is a reasonable expectation that new and general concepts introduced in multivariable calculus should be neatly tied up with their relics in one-variable calculus. For example, the area of a bounded region in the plane, defined via double integrals, should be related to formulas for the areas of planar regions between two curves (given by equations in rectangular coordinates or in polar coordinates). Likewise, the volume of a solid in 3-space, defined via triple integrals, should be related to methods for computing volumes of solids of revolution, thereby resolving the mystery that the washer method and the shell method always give the same answer. Indeed, a conscientious student is likely to face a myriad of questions if the promise of extending one-variable calculus to "higher dimensions" is taken seriously. For instance: Why aren't we talking of monotonicity, which was such a big deal in one-variable calculus? Do Rolle's theorem and the mean value theorem, which were considered very important, have genuine analogues? Why is there no L'Hôpital's rule now? Can't we talk of convexity and concavity of functions of several variables, and in that case, shouldn't it have something to do with derivatives? Is it still true that the processes of differentiation and integration are inverses of each other, and if so, then how? Aren't there any numerical methods for approximating double integrals and triple integrals? Whatever happened to infinite series and improper integrals?

We thought and believed that questions and concerns such as those above are perfectly legitimate and should be addressed in a book on multivariable calculus. Thus, about a decade ago, when we taught together a course at IIT Bombay that combined one-variable calculus and multivariable calculus, we looked for books that addressed these questions and could be easily read by undergraduate students. There were a number of excellent books available, most notably, the two volumes of Apostol's *Calculus* and the two-volume *Introduction to Calculus and Analysis* by Courant and John. Besides, a wealth of material was available in classics of older genre such as the books of Bromwich and Hobson. However, we were mildly dissatisfied with some aspect or the other of the various books we consulted. As a first attempt to help our students, we prepared a set of notes, written in a telegraphic style, with detailed explanations given during the lectures. Subsequently, these notes and problem sets were put together into a booklet that has been in private circulation at IIT Bombay since March 1998. Goaded by the positive feedback received from colleagues and students, we decided to convert this booklet into a book. To begin with, we were no less ambitious. We wanted a self-contained and rigorous book of a reasonable size that covered one-variable as well as multivariable calculus, and adequately answered all the concerns expressed above. As years went by, and the size of our manuscript grew, we developed a better appreciation for the fraternity of authors of books, especially of serious books on calculus and real analysis. It was clear that choices had to be made. Along the way, we decided to separate out one-variable calculus and multivariable calculus. Our treatment of the former is contained in *A Course in Calculus and Real Analysis*, hereinafter referred to as ACICARA, published by Springer, New York, in its *Undergraduate Texts in Mathematics* series in 2006.

The present book may be viewed as a sequel to ACICARA, and it caters to theoretical as well as practical aspects of multivariable calculus. The table of contents should give a general idea of the topics covered in this book. It will be seen that we have made certain choices, some quite standard and some rather unusual. As is common with introductory books on multivariable calculus, we have mainly restricted ourselves to functions of two variables. We have also briefly indicated how the theory extends to functions of more than two variables. Wherever it seemed appropriate, we have worked out the generalizations to functions of three variables. Indeed, as explained in the first chapter, there is a striking change as we pass from the one-dimensional world of $\mathbb{R}$ and functions on $\mathbb{R}$ to the two-dimensional space $\mathbb{R}^2$ and functions on $\mathbb{R}^2$. On the other hand, the work needed to extend calculus on $\mathbb{R}^2$ to calculus on the n-dimensional space $\mathbb{R}^n$ for $n > 2$ is often relatively routine. Among the unusual choices that we have made is the noninclusion of line integrals, surface integrals, and the related theorems of Green, Gauss, and Stokes. Of course, we do realize that these topics are very important. However, a thorough treatment of them would have substantially increased the size of the book or diverted us from doing justice to the promise of developing, wherever possible, notions and results analogous to those in one-variable calculus. For readers interested in these important theorems, we have suggested a number of books in the *Notes and Comments* on Chapter 5.

The subject matter of this book is quite classical, and therefore the novelty, if any, lies mainly in the selection of topics and in the overall treatment. With this in view, we list here some of the topics discussed in this book that are normally not covered in texts at this level on multivariable calculus: monotonicity and bimonotonicity of functions of two variables and their relationship with partial differentiation; functions of bounded variation and bounded bivariation; rectangular Rolle's and mean value theorems; higher-order directional derivatives and their use in Taylor's theorem; convexity and its relation with the monotonicity of the gradient and the nonnegative definiteness of the Hessian; an exact analogue of the fundamental theorem of calculus for real-valued functions defined on a rectangle; cubature rules based on products and on triangulation for approximate evaluations of double integrals; conditional and unconditional convergence of double series and of improper double integrals.

Basic guiding principles and the organizational aspects of this book are similar to those in ACICARA. We have always striven for clarity and precision. We continue to distinguish between the intrinsic definition of a geometric notion and its analytic counterpart. A case in point is the notion of a saddle point of a surface, where we adopt a nonstandard definition that seems more geometric and intuitive. Complete proofs of all the results stated in the text, except the change of variables formula, are included, and as a rule, these do not depend on any of the exercises. Each chapter is divided into several sections that are numbered serially in that chapter. A section is often divided into several subsections, which are not numbered, but appear in the table of contents. When a new term is defined, it appears in boldface. Definitions are not numbered, but can be located using the index. Lemmas, propositions, examples, and remarks are numbered serially in each chapter. Moreover, for the convenience of readers, we have often included the statements of certain basic results in one-variable calculus. Each of these appears as a "Fact," and is also serially numbered in each chapter. Each such fact is accompanied by a reference, usually to ACICARA, where a proof can be found. The end of a proof of a lemma or a proposition is marked by the symbol $\Box$, while the symbol $\diamond$ marks the end of an example or a remark. Bibliographic details about the books and articles mentioned in the text and in this preface can be found in the list of references. Citations appear in square brackets. Each chapter concludes with *Notes and Comments*, where distinctive features of exposition are highlighted and pointers to relevant literature are provided. These *Notes and Comments* may be collectively viewed as an extended version of the preface, and a reader wishing to get a quick idea of what is new and different in this book might find it useful to browse through them. The exercises are divided into two parts: Part A, consisting of relatively routine problems, and Part B, containing those that are of a theoretical nature or are particularly challenging. Except for the first section of the first chapter, we have avoided using the more abstract vector notation and opted for classical notation involving explicit coordinates. We hope that this will seem more friendly to undergraduate students, while relatively advanced readers will have no difficulty in passing to vector notation and working out analogues of the notions and results in this book in the general setting of $\mathbb{R}^n$.

Although we view this book as a sequel to ACICARA, it should be emphasized that this is an independent book and can be read without having studied ACICARA. The formal prerequisite for reading this book is familiarity with one-variable calculus and occasionally, a nodding acquaintance with 2×2 and 3×3 matrices and their determinants. It would be useful if the reader has some mathematical maturity and an aptitude for mathematical proofs. This book can be used as a textbook for an undergraduate course in multivariable calculus. Parts of the book could be useful for advanced undergraduate and graduate courses in real analysis, or for self-study by students interested in the subject. For teachers and researchers, this may be a useful reference for topics that are skipped or cursorily treated in standard texts.

We thank our parent institution, IIT Bombay, and in particular its Department of Mathematics for providing excellent infrastructure. Financial assistance received from the Curriculum Development Cell at IIT Bombay is gratefully acknowledged. We are indebted to Rafikul Alam, Aldric Brown, Dinesh Karia, Swanand Khare, Rekha Kulkarni, Shobhan Mandal, Thamban Nair, S. H. Patil, P. Shunmugaraj, and especially R. R. Simha, for a critical reading of parts of this book and many useful suggestions. We thank Maria Zeltser for reviewing the entire book. We are especially thankful to Arunkumar Patil, who is mainly responsible for drawing the figures in this book. Thanks are also due to Ann Kostant and her colleagues at Springer, New York, for excellent cooperation, to C. L. Anthony for typing a substantial part of the manuscript, and to David Kramer for his thorough copyediting. Last, but not least, we express our gratitude toward Sharmila Ghorpade and Nirmala Limaye for their continued support. We would appreciate receiving comments, suggestions, and corrections. A dynamic errata together with relevant information about this book will be posted at `http://www.math.iitb.ac.in/~srg/acimc`, and we encourage the reader to visit this website for updates concerning this book.

Mumbai, India
August 2009

Sudhir Ghorpade
Balmohan Limaye

Contents

1

Vectors and Functions

Typically, a first course in calculus comprises of the study of real-valued functions of one real variable, that is, functions $f : D \to \mathbb{R}$, where D is a subset of the set $\mathbb{R}$ of all real numbers. We shall assume that the reader has had a first course in calculus and is familiar with basic properties of real numbers and functions of one real variable. For a ready reference, one may refer to [22], which is abbreviated throughout the text as ACICARA. However, for the convenience of the reader, relevant facts from one-variable calculus will be recalled whenever needed.

The basic object of our study will be the **n-dimensional (Euclidean) space** $\mathbb{R}^n$ consisting of n-tuples of real numbers, namely,

$$\mathbb{R}^n := \{(x_1, \ldots, x_n) : x_1, \ldots, x_n \in \mathbb{R}\},$$

and real-valued functions on subsets of $\mathbb{R}^n$. Whenever we write $\mathbb{R}^n$, it will be tacitly assumed that $n \in \mathbb{N}$, that is, n is a positive integer. Elements of $\mathbb{R}^n$ are sometimes referred to as **vectors** in n-space when $n > 1$. In contrast, the elements of $\mathbb{R}$ are referred to as **scalars**. Given a vector $\mathbf{x} = (x_1, \ldots, x_n)$ in $\mathbb{R}^n$ and $1 \le i \le n$, the scalar x_i is called the ith **coordinate** of $\mathbf{x}$.

The algebraic operations on $\mathbb{R}$ can be easily extended to $\mathbb{R}^n$ in a componentwise manner. Thus, we define the **sum** of $\mathbf{x} = (x_1, \ldots, x_n)$ and $\mathbf{y} = (y_1, \ldots, y_n)$ to be $\mathbf{x} + \mathbf{y} := (x_1 + y_1, \ldots, x_n + y_n)$. It is easily seen that addition defined in this way satisfies properties analogous to those in $\mathbb{R}$. In particular, the **zero vector** $\mathbf{0} := (0, \ldots, 0)$ plays a role similar to the number 0 in $\mathbb{R}$. We might wish to define the **product** of $(x_1, \ldots, x_n)$ and $(y_1, \ldots, y_n)$ to be $(x_1y_1, \ldots, x_ny_n)$. However, this kind of componentwise multiplication is not well behaved. For example, the componentwise product of the nonzero vectors $(1, 0)$ and $(0, 1)$ in $\mathbb{R}^2$ is the zero vector $(0, 0)$, and consequently, the reciprocals of these nonzero vectors cannot be defined. As a matter of fact, there is no reasonable notion whatsoever of division in $\mathbb{R}^n$, in general. (See the Notes and Comments at the end of this chapter.) Moreover, as explained later, the order relation on $\mathbb{R}$ extends only partially to $\mathbb{R}^n$ when $n > 1$.

S.R. Ghorpade and B.V. Limaye, *A Course in Multivariable Calculus and Analysis*,
Undergraduate Texts in Mathematics, DOI 10.1007/978-1-4419-1621-1_1,

For these reasons, the theory of functions of several variables differs significantly from that of functions of one-variable. However, once $n > 1$, there is not a great deal of difference between the smaller values of n and the larger values of n. This is particularly true with the basic aspects of the theory of functions of several variables that are developed here. With this in view and for the sake of simplicity, we shall almost exclusively restrict ourselves to the case $n = 2$. In this case, the space $\mathbb{R}^n$ can be effectively visualized as the plane. Also, graphs of real-valued functions of two variables may be viewed as surfaces in 3-space. More generally, a surface in 3-space can be given by (the zeros of) a function of three variables. With this in mind, we shall also occasionally allude to $\mathbb{R}^3$ and to real-valued functions of three variables.

In the first section below we discuss a number of preliminary notions concerning vectors in $\mathbb{R}^n$ and some important types of subsets of $\mathbb{R}^n$. Next, in Section 1.2, we develop some basic aspects of (real-valued) functions of two variables. Finally, in Section 1.3, we discuss some useful transformations or coordinate changes of the 3-space $\mathbb{R}^3$.

1.1 Preliminaries

We begin with a discussion of basic facts concerning algebraic operations, order properties, elementary inequalities, important types of subsets, etc. In these matters, there is hardly any simplification possible by restricting to $\mathbb{R}^2$, and thus we will work here with $\mathbb{R}^n$ for arbitrary $n \in \mathbb{N}$.

Algebraic Operations

We have already discussed the notion of addition of points in $\mathbb{R}^n$ and the fact that the corresponding analogue of algebraic properties in $\mathbb{R}$ holds in $\mathbb{R}^n$. More precisely, this means that the following properties hold. Note that each of these is an immediate consequence of the corresponding properties of real numbers. (See, for example, Section 1.1 of ACICARA.)

A1. $\mathbf{x} + (\mathbf{y} + \mathbf{z}) = (\mathbf{x} + \mathbf{y}) + \mathbf{z}$ *for all* $\mathbf{x}, \mathbf{y}, \mathbf{z} \in \mathbb{R}^n$.
A2. $\mathbf{x} + \mathbf{y} = \mathbf{y} + \mathbf{x}$ *for all* $\mathbf{x}, \mathbf{y} \in \mathbb{R}^n$.
A3. $\mathbf{x} + \mathbf{0} = \mathbf{x}$ *for all* $\mathbf{x} \in \mathbb{R}^n$.
A4. *Given any* $\mathbf{x} \in \mathbb{R}^n$, *there is* $\mathbf{x}' \in \mathbb{R}^n$ *such that* $\mathbf{x} + \mathbf{x}' = \mathbf{0}$.

These properties may be used tacitly in the sequel. As indicated earlier, we do not have a good notion of multiplication of points in $\mathbb{R}^n$ when $n > 2$. But we have useful notions of *scalar multiplication* and *dot product* that are defined as follows.

Given any $c \in \mathbb{R}$ and $\mathbf{x} = (x_1, \ldots, x_n) \in \mathbb{R}^n$, we define

$$c\mathbf{x} := (cx_1, \ldots, cx_n).$$

This is referred to as the **scalar multiplication** of the vector $\mathbf{x}$ by the scalar c. Geometrically speaking, the scalar multiple $c\mathbf{x}$ corresponds to stretching or contracting the vector $\mathbf{x}$ according as $c > 1$ or $0 < c < 1$, whereas if $c < 0$, then $c\mathbf{x}$ corresponds to the reflection of $\mathbf{x}$ about the origin followed by stretching or contracting.

Given any $\mathbf{x} = (x_1, \dots, x_n)$ and $\mathbf{y} = (y_1, \dots, y_n)$ in $\mathbb{R}^n$, the **dot product** (also known as the **inner product** or the **scalar product**) of $\mathbf{x}$ and $\mathbf{y}$ is the real number denoted by $\mathbf{x} \cdot \mathbf{y}$ and defined by

$$\mathbf{x} \cdot \mathbf{y} := x_1 y_1 + \cdots + x_n y_n.$$

The dot product permits us to talk about the "angle" between two vectors. We shall explain this in greater detail a little later.

We also have an analogue of the notion of the absolute value of a real number, which is defined as follows. Given any $\mathbf{x} = (x_1, \dots, x_n) \in \mathbb{R}^n$, the **norm** (also known as the **magnitude** or the **length**) of $\mathbf{x}$ is the nonnegative real number denoted by $|\mathbf{x}|$ and defined by

$$|\mathbf{x}| := \sqrt{\mathbf{x} \cdot \mathbf{x}} = \sqrt{x_1^2 + \cdots + x_n^2}.$$

Geometrically speaking, the norm $|\mathbf{x}|$ represents the distance between $\mathbf{x}$ and the origin $\mathbf{0} := (0, \dots, 0)$. More generally, for any $\mathbf{x}, \mathbf{y} \in \mathbb{R}^n$, the norm of their difference, that is, $|\mathbf{x} - \mathbf{y}|$, represents the distance between $\mathbf{x}$ and $\mathbf{y}$. A vector $\mathbf{u}$ in $\mathbb{R}^n$ for which $|\mathbf{u}| = 1$ is called a **unit vector** in $\mathbb{R}^n$. For example, in $\mathbb{R}^2$ the vectors $\mathbf{i} := (1, 0)$ and $\mathbf{j} := (0, 1)$ are unit vectors.

Elementary properties of scalar multiplication, dot product, and the norm are given in the following proposition. It may be remarked that the inequality in (iv) below is a restatement of the **Cauchy–Schwarz inequality** as given in Proposition 1.12 of ACICARA. But the proof given here is somewhat different. The inequality in (v) is referred to as the **triangle inequality**.

Proposition 1.1. *Given any $r, s \in \mathbb{R}$ and $\mathbf{x}, \mathbf{y}, \mathbf{z} \in \mathbb{R}^n$, we have*

(i) $(rs)\mathbf{x} = r(s\mathbf{x})$, $r(\mathbf{x} + \mathbf{y}) = r\mathbf{x} + r\mathbf{y}$ *and* $(r + s)\mathbf{x} = r\mathbf{x} + s\mathbf{x}$,
(ii) $\mathbf{x} \cdot \mathbf{y} = \mathbf{y} \cdot \mathbf{x}$, $(\mathbf{x} + \mathbf{y}) \cdot \mathbf{z} = \mathbf{x} \cdot \mathbf{z} + \mathbf{y} \cdot \mathbf{z}$ *and* $r(\mathbf{x} \cdot \mathbf{y}) = (r\mathbf{x}) \cdot \mathbf{y} = \mathbf{x} \cdot (r\mathbf{y})$,
(iii) $|\mathbf{x}| \geq 0$; *moreover,* $|\mathbf{x}| = 0 \iff \mathbf{x} = \mathbf{0}$,
(iv) $|\mathbf{x} \cdot \mathbf{y}| \leq |\mathbf{x}||\mathbf{y}|$,
(v) $|\mathbf{x} + \mathbf{y}| \leq |\mathbf{x}| + |\mathbf{y}|$,
(vi) $|r\mathbf{x}| = |r||\mathbf{x}|$.

Proof. Properties listed in (i), (ii), and (iii) are obvious. The inequality in (iv) is obvious if $\mathbf{x} = \mathbf{0}$. Assume that $\mathbf{x} \neq \mathbf{0}$. Let $a := \mathbf{x} \cdot \mathbf{x}$, $b := \mathbf{x} \cdot \mathbf{y}$, and $c := \mathbf{y} \cdot \mathbf{y}$. Then by (iii), $a > 0$. Given any $t \in \mathbb{R}$, consider $q(t) := at^2 + 2bt + c$. In view of (i) and (ii), we have $q(t) = (t\mathbf{x} + \mathbf{y}) \cdot (t\mathbf{x} + \mathbf{y})$, and hence by (iii), $q(t) \geq 0$ for all $t \in \mathbb{R}$. In particular, upon putting $t = -b/a$ and multiplying throughout by a, we obtain $ac - b^2 \geq 0$, that is, $b^2 \leq ac$. Hence $|b| \leq \sqrt{a}\sqrt{c}$, which proves (iv). The inequality in (v) follows from (ii) and (iv), since

$$|\mathbf{x}+\mathbf{y}|^2 = (\mathbf{x}+\mathbf{y})\cdot(\mathbf{x}+\mathbf{y}) = \mathbf{x}\cdot\mathbf{x}+2\mathbf{x}\cdot\mathbf{y}+\mathbf{y}\cdot\mathbf{y} \leq |\mathbf{x}|^2+2|\mathbf{x}||\mathbf{y}|+|\mathbf{y}|^2 = (|\mathbf{x}|+|\mathbf{y}|)^2.$$

The equality in (vi) is obvious. □

From parts (iii) and (iv) of Proposition 1.1, we see that if $\mathbf{x}, \mathbf{y} \in \mathbb{R}^n$ are any nonzero vectors, then $(\mathbf{x}\cdot\mathbf{y}/|\mathbf{x}||\mathbf{y}|)$ is in the closed interval $[-1, 1]$, that is, the subset $\{r \in \mathbb{R} : -1 \leq r \leq 1\}$ of $\mathbb{R}$. We define the **angle** between $\mathbf{x}$ and $\mathbf{y}$ to be $\cos^{-1}(\mathbf{x}\cdot\mathbf{y}/|\mathbf{x}||\mathbf{y}|)$. In other words, the angle between $\mathbf{x}$ and $\mathbf{y}$ is the unique real number $\theta \in [0, \pi]$ such that $\mathbf{x}\cdot\mathbf{y} = |\mathbf{x}||\mathbf{y}|\cos\theta$.

Order Properties

On $\mathbb{R}$, there is a natural order relation $\leq$ that permits us to compare any two real numbers. In fact, given any $a, b \in \mathbb{R}$, we write $a \leq b$ if $a = b$ or if a is to the left of b on the real line. As is well known, this order relation plays a crucial role in the study of functions of one-variable. In $\mathbb{R}^n$, the idea of one vector being to the left of another does not seem to make sense, and it is natural to ask whether there is an analogous order relation. To understand this better, let us first give precise definitions of what is meant by an order relation.

Let S be a set. A **relation** on S is a subset R of $S \times S$; for $x, y \in S$, if the pair (x, y) is in R, then we write xRy and say that x is **related to** y by the relation R. A relation $\leq$ on S is called a **partial order** on S if for any $x, y, z \in S$, the following three properties are satisfied: (i) [**reflexivity**] $x \leq x$, (ii) [**transitivity**] if $x \leq y$ and $y \leq z$, then $x \leq z$, and (iii) [**antisymmetry**] if $x \leq y$ and $y \leq x$, then $x = y$. If $\leq$ is a partial order on S, then S together with the relation $\leq$ is called a **partially ordered set**, or simply a **poset**. If S is a poset, then for any $x, y \in S$, it is customary to write $x \geq y$ as an equivalent form of $y \leq x$. A **total order**, or simply an **order**, on S is a partial order $\leq$ on S such that $x \leq y$ or $y \leq x$ for every $x, y \in S$. If $\leq$ is a total order on S, then the set S together with the relation $\leq$ is called a **totally ordered set** or a **linearly ordered set**.

For example, the natural order on $\mathbb{R}$ is a total order. It is clear that a total order is a partial order, but the converse need not be true. For instance, suppose on the set $\mathbb{N}$ of positive integers, we define $m \leq n$ if $m \mid n$, that is, if m divides n. Then it is easily seen that $\leq$ is a partial order on $\mathbb{N}$, but not a total order on $\mathbb{N}$. What makes the natural order on $\mathbb{R}$ particularly useful in the study of functions of one variable is not only the fact that it is a total order, but also that it has many useful properties. First, it is compatible with the algebraic operations, that is, for any $x, y, z \in \mathbb{R}$ and $c \in \mathbb{R}$, we have

$$x \leq y \Longrightarrow x + z \leq y + z \quad \text{and} \quad cx \leq cy \text{ if } c \geq 0, \text{ while } cx \geq cy \text{ if } c \leq 0.$$

Further, the natural order on $\mathbb{R}$ has the **archimedean property**, which can be stated as follows:[1]

For any $x, y \in \mathbb{R}$ with $x \geq 0$ and $x \neq 0$, there is $k \in \mathbb{N}$ such that $kx \geq y$.

Last, but not least, the natural order on $\mathbb{R}$ has the least upper bound property. To explain this, let us first note that if S is a poset, then a subset D of S is said to be **bounded above** if there is $\alpha \in S$ such that $x \leq \alpha$ for all $x \in D$; any such α is called an **upper bound** of S. If a subset D of S is bounded above, then an element M of S is called a **least upper bound** or a **supremum** of D if M is an upper bound of S and $M \leq \alpha$ for every upper bound α of D. It is clear that if $D \subseteq S$ and D has a least upper bound M, then M is unique, and we shall denote it by $\sup S$. Similarly, a subset D of a poset S is said to be **bounded below** if there is $\beta \in S$ such that $\beta \leq x$ for all $x \in D$; any such β is called a **lower bound** of S. If a subset D of S is bounded below, then an element m of S is called a **greatest lower bound** or an **infimum** of D if m is a lower bound of S and $\beta \leq m$ for every lower bound β of D. Again it is clear that if $D \subseteq S$ and D has a least upper bound m, then m is unique, and we shall denote it by $\inf S$.

Now the **least upper bound property** for a poset S can be stated as follows: Every nonempty subset of S that is bounded above has a supremum in S. A key fact about the real numbers is that the set $\mathbb{R}$ together with its natural total order has the least upper bound property. In the case of $\mathbb{R}^n$, it is possible to define a total order, known as the lexicographic order, which is compatible with the algebraic operations of addition and scalar multiplication (Exercise 1), but it satisfies neither the archimedean property nor the least upper bound property. (See Exercise 32 as well as the Notes and Comments at the end of this chapter.) However, there is a more natural partial order on $\mathbb{R}^n$, described below, which not only is compatible with the algebraic operations on $\mathbb{R}^n$, but also satisfies the least upper bound property. Given any $\mathbf{x} = (x_1, \ldots, x_n)$ and $\mathbf{y} = (y_1, \ldots, y_n)$ in $\mathbb{R}^n$, we define

$$\mathbf{x} \leq \mathbf{y} \iff x_i \leq y_i \text{ for all } i = 1, \ldots, n.$$

Clearly, this is a partial order on $\mathbb{R}^n$, and it may be called the **product order** or the **componentwise order** on $\mathbb{R}^n$. If $n > 1$, then the product order on $\mathbb{R}^n$ is not a total order; for example, if $\mathbf{x} := (1, 0, 0, \ldots, 0)$ and $\mathbf{y} := (0, 1, 0, \ldots, 0)$, then neither $\mathbf{x} \leq \mathbf{y}$ nor $\mathbf{y} \leq \mathbf{x}$. However, the product order has a number of nice properties listed in the proposition below.

Proposition 1.2. *Let $\leq$ denote the product order on $\mathbb{R}^n$. Then for $r, s \in \mathbb{R}$ and $\mathbf{x}, \mathbf{y}, \mathbf{z} \in \mathbb{R}^n$, we have the following:*

[1] The archimedean property for $\mathbb{R}$ is often stated as follows: Given any $x \in \mathbb{R}$ with $x > 0$, there is $k \in \mathbb{N}$ such that $k > x$. The formulation given here is slightly different, but obviously equivalent. It has the advantage that it makes sense in the context of $\mathbb{R}^n$ as well.

(i) *Given any* $\mathbf{x}, \mathbf{y} \in \mathbb{R}^n$ *with* $\mathbf{x} \leq \mathbf{y}$, *we have* $\mathbf{x} + \mathbf{z} \leq \mathbf{y} + \mathbf{z}$ *for all* $\mathbf{z} \in \mathbb{R}^n$; *also, for any* $c \in \mathbb{R}$, *we have* $c\mathbf{x} \leq c\mathbf{y}$ *if* $c \geq 0$, *and* $c\mathbf{x} \geq c\mathbf{y}$ *if* $c \leq 0$.
(ii) *For any* $\mathbf{x} = (x_1, \ldots, x_n)$ *and* $\mathbf{y} = (y_1, \ldots, y_n)$ *in* $\mathbb{R}^n$ *such that* $x_i \geq 0$ *and* $x_i \neq 0$ *for each* $i = 1, \ldots, n$, *there is* $k \in \mathbb{N}$ *such that* $k\mathbf{x} \geq \mathbf{y}$.
(iii) *Every nonempty subset of* $\mathbb{R}^n$ *that is bounded above has a supremum in* $\mathbb{R}^n$. *Likewise, every nonempty subset of* $\mathbb{R}^n$ *that is bounded below has an infimum in* $\mathbb{R}^n$.

Proof. The properties listed in (i) are obvious. To prove (ii), use the archimedean property on $\mathbb{R}$ to find $k_i \in \mathbb{N}$ such that $k_i x_i \geq y_i$ for each $i = 1, \ldots, n$. Now $k := \max\{k_1, \ldots, k_n\}$ satisfies $kx_i \geq y_i$ for all $i = 1, \ldots, n$, and in particular, $k\mathbf{x} \geq \mathbf{y}$. Finally, let D be a nonempty subset of $\mathbb{R}^n$. Then for $1 \leq i \leq n$, the set D_i consisting of the ith coordinates of the elements of D is a nonempty subset of $\mathbb{R}$. If D is bounded above, then so is each D_i, and if $M_i := \sup D_i$, then $\mathbf{M} := (M_1, \ldots, M_n)$ is clearly the supremum of D in $\mathbb{R}^n$. Likewise, if D is bounded below, then so is each D_i, and if $m_i := \inf D_i$, then $\mathbf{m} := (m_1, \ldots, m_n)$ is clearly the infimum of D in $\mathbb{R}^n$. □

Remark 1.3. If D is a nonempty bounded subset of $\mathbb{R}^n$, then by part (iii) of Proposition 1.2, supremum and infimum (with respect to the product order) of D exist, and it is clear that these are unique. However, if $n > 1$, then the supremum and infimum can be far away from the elements of D. For example, if $D := \{\mathbf{x} = (x_1, x_2) \in \mathbb{R}^2 : 0 \leq x_1 \leq 1 \text{ and } x_2 = 1 - x_1\}$, then $\mathbf{M} := (1, 1)$ is the supremum of D and $\mathbf{m} := (0, 0)$ is the infimum of D, and we have $|\mathbf{x} - \mathbf{M}| = |\mathbf{x} - \mathbf{m}| \geq 1/\sqrt{2}$ for all $\mathbf{x} \in D$. ◇

Intervals, Disks, and Bounded Sets

Let us begin by recalling the general notion of an interval in $\mathbb{R}$. To this end, given any $a, b \in \mathbb{R}$, let $I_{a,b}$ denote the closed interval between a and b. In other words, $I_{a,b} = [a, b]$ if $a \leq b$, while $I_{a,b} = [b, a]$ if $b \leq a$; equivalently, $I_{a,b} := [\min\{a, b\}, \max\{a, b\}]$. A subset I of $\mathbb{R}$ is said to be an **interval** if $I_{a,b} \subseteq I$ for every $a, b \in I$. It is well known and elementary to see that a subset of I is an interval if and only if it is one among the familiar types of intervals, namely an open interval or a closed interval or a semiopen interval or semi-infinite interval or the doubly infinite interval $\mathbb{R}$. (See, for example, Proposition 1.7 of ACICARA.) We are now ready to define an analogous notion in $\mathbb{R}^n$.

Given any $\mathbf{a} = (a_1, \ldots, a_n)$ and $\mathbf{b} = (b_1, \ldots, b_n)$ in $\mathbb{R}^n$, let

$$I_{\mathbf{a},\mathbf{b}} := I_{a_1,b_1} \times \cdots \times I_{a_n,b_n}.$$

A subset I of $\mathbb{R}^n$ is said to be an n**-interval** if $I_{\mathbf{a},\mathbf{b}} \subseteq I$ for every $\mathbf{a}, \mathbf{b} \in I$. For example, if $I_1, \ldots, I_n$ are intervals in $\mathbb{R}$, then $I_1 \times \cdots \times I_n$ is an n-interval. It turns out that every n-interval is of this form.

Proposition 1.4. *Let $I \subseteq \mathbb{R}^n$ be an n-interval. Then $I = I_1 \times \cdots \times I_n$ for some intervals $I_1, \ldots, I_n$ in $\mathbb{R}$.*

Proof. For $1 \le j \le n$, let I_j denote the set of all possible jth coordinates of elements of I, that is, let

$$I_j := \{a \in \mathbb{R} : \text{there is } \mathbf{a} = (a_1, \ldots, a_n) \in I \text{ such that } a_j = a\}.$$

Let $a, b \in I_j$. Then there are $\mathbf{a} = (a_1, \ldots, a_n)$ and $\mathbf{b} = (b_1, \ldots, b_n)$ in I such that $a_j = a$ and $b_j = b$. Now, if $x \in I_{a,b}$, then $(a_1, \ldots, a_{j-1}, x, a_{j+1}, \ldots, a_n)$ is in $I_{\mathbf{a},\mathbf{b}}$ and hence in I; thus $x \in I_j$. This shows that each I_j is an interval. Next, it is clear that $I \subseteq I_1 \times \cdots \times I_n$. To prove the other inclusion, let $\mathbf{x} := (x_1, \ldots, x_n) \in I_1 \times \cdots \times I_n$. Then there are $\mathbf{a}_j = (a_{j,1}, \ldots, a_{j,n}) \in I$ such that $a_{j,j} = x_j$ for $j = 1, \ldots, n$. Now let $\mathbf{u}_j = (x_1, \ldots, x_j, a_{j,j+1}, \ldots, a_{j,n})$ for $j = 1, \ldots, n$. Observe that $\mathbf{u}_1 = (x_1, a_{1,2}, \ldots, a_{1,n}) = \mathbf{a}_1 \in I$. Since $\mathbf{a}_2 = (a_{2,1}, x_2, a_{2,3}, \ldots, a_{2,n})$, we see that $\mathbf{u}_2 = (x_1, x_2, a_{2,3}, \ldots, a_{2,n}) \in I_{\mathbf{u}_1, \mathbf{a}_2} \subseteq I$. Next, since $\mathbf{a}_3 = (a_{3,1}, a_{3,2}, x_3, a_{3,4}, \ldots, a_{3,n}) \in I$, we see that $\mathbf{u}_3 \in I_{\mathbf{u}_2, \mathbf{a}_3} \subseteq I$. Continuing in this manner, we see that $\mathbf{u}_j \in I_{\mathbf{u}_{j-1}, \mathbf{a}_j} \subseteq I$ for $j = 2, \ldots, n$. In particular, $\mathbf{x} = \mathbf{u}_n \in I$. Thus $I = I_1 \times \cdots \times I_n$. □

As a special case of the above proposition, we note that every 2-interval in $\mathbb{R}^2$ is of the form

$$I \times J := \left\{(x, y) \in \mathbb{R}^2 : x \in I \text{ and } y \in J\right\},$$

where I and J are intervals in $\mathbb{R}$. This fact will be used tacitly in the sequel.

An n-interval of the form $I_1 \times \cdots \times I_n$, where each of the $I_1, \ldots, I_n$ is a closed and bounded interval in $\mathbb{R}$, is called a **hypercuboid** in $\mathbb{R}^n$. In other words, a hypercuboid is a subset of $\mathbb{R}^n$ of the form $\{\mathbf{x} \in \mathbb{R}^n : \mathbf{a} \le \mathbf{x} \le \mathbf{b}\}$ for some $\mathbf{a}, \mathbf{b} \in \mathbb{R}^n$. Note that this is nonempty if and only if $\mathbf{a} \le \mathbf{b}$. A hypercuboid in $\mathbb{R}^n$ shall be referred to as a **cuboid** when $n = 3$ and as a **rectangle** when $n = 2$.

In the local study of a function of one variable near a point $c \in \mathbb{R}$, it is often helpful to consider symmetric open intervals about c, that is, open intervals of the form $(c - r, c + r)$, where r is a positive real number. In the case of $\mathbb{R}^n$, the corresponding role is played by sets that look like open disks or open squares when $n = 2$. General definitions are given below. These may be easier to follow if one notices that the symmetric open interval $(c-r, c+r)$ can be viewed as the set $\{x \in \mathbb{R} : |x - c| < r\}$.

Given any $\mathbf{c} = (c_1, \ldots, c_n) \in \mathbb{R}^n$ and any $r > 0$, we define

$$\mathbb{S}_r(\mathbf{c}) := \{\mathbf{x} = (x_1, \ldots, x_n) \in \mathbb{R}^n : |x_i - c_i| < r \text{ for } i = 1, \ldots, n\}$$

and

$$\mathbb{B}_r(\mathbf{c}) := \{\mathbf{x} = (x_1, \ldots, x_n) \in \mathbb{R}^n : |\mathbf{x} - \mathbf{c}| < r\}.$$

For example, if $n = 2$, then $\mathbb{S}_r(\mathbf{c})$ looks like a square (with its boundary excluded) with $\mathbf{c} = (c_1, c_2)$ at its "center" and each side having length $2r$

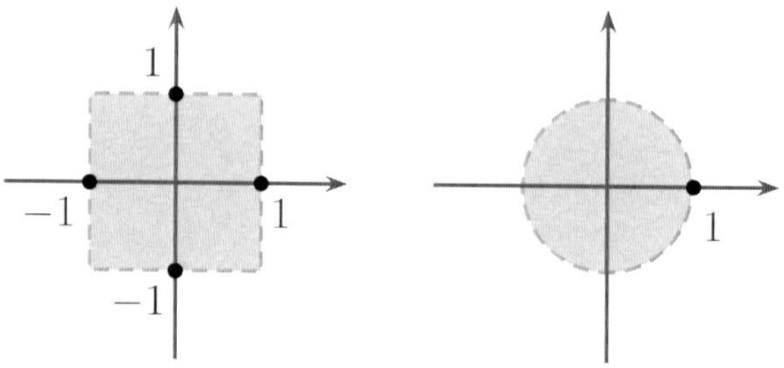

Fig. 1.1. The open square $\mathbb{S}_1(0,0)$ and the open disk $\mathbb{B}_1(0,0)$.

(Figure 1.1), whereas $\mathbb{B}_r(\mathbf{c})$ looks like a disk (with its boundary excluded) centered at $\mathbf{c} = (c_1, c_2)$ and of diameter $2r$ (Figure 1.1). Thus when $n = 2$, we shall refer to $\mathbb{S}_r(\mathbf{c})$ as the **open square centered at c of radius** r and to $\mathbb{B}_r(\mathbf{c})$ as the **open disk centered at c of radius** r. When $n = 2$ and $\mathbf{c} = (c_1, c_2)$, we will often write $\mathbb{S}_r(c_1, c_2)$ in place of $\mathbb{S}_r(\mathbf{c})$.

A subset D of $\mathbb{R}^n$ is said to be **bounded** if it is bounded above as well as bounded below, that is, if there are $\mathbf{a}, \mathbf{b} \in \mathbb{R}^n$ such that $\mathbf{a} \leq \mathbf{x} \leq \mathbf{b}$ for all $\mathbf{x} \in D$. Equivalently, D is bounded if there is $r > 0$ such that $D \subseteq \mathbb{S}_r(\mathbf{0})$. For example, if $D := \{\mathbf{x} \in \mathbb{R}^2 : |\mathbf{x}| < 1\}$, then D is bounded, whereas its complement $\mathbb{R}^2 \setminus D = \{\mathbf{x} \in \mathbb{R}^2 : |\mathbf{x}| \geq 1\}$ is not bounded.

If a subset D of $\mathbb{R}^n$ is nonempty and bounded, then we define the **diameter** of D, denoted by $\operatorname{diam}(D)$, to be the real number

$$\operatorname{diam}(D) := \sup\{|\mathbf{x} - \mathbf{y}| : \mathbf{x}, \mathbf{y} \in D\}.$$

Roughly speaking, the diameter of a nonempty bounded set is the largest possible distance between any two of its points. For example, if R is a rectangle of side lengths a, b, then $\operatorname{diam}(R) = \sqrt{a^2 + b^2}$.

Line Segments and Paths

Let $\mathbf{c}, \mathbf{d} \in \mathbb{R}^n$. The **line segment** joining $\mathbf{c}$ and $\mathbf{d}$ is defined to be the subset

$$\{(1-t)\mathbf{c} + t\mathbf{d} : t \in [0,1]\}$$

of $\mathbb{R}^n$. Note that the endpoints $\mathbf{c}$ and $\mathbf{d}$ correspond to the parameter values $t = 0$ and $t = 1$, respectively. Given any $D \subseteq \mathbb{R}^n$, we say that the line segment joining $\mathbf{c}$ and $\mathbf{d}$ lies in D if $(1-t)\mathbf{c} + t\mathbf{d} \in D$ for all $t \in [0,1]$.

If $n = 1$, a line segment is essentially the only way of joining $\mathbf{c}$ and $\mathbf{d}$. But if $n > 1$, then there is more room to move about, and the points $\mathbf{c}$ and $\mathbf{d}$ can be joined by many different paths. Formally, a **path** in $\mathbb{R}^n$ is an n-tuple $(x_1, \ldots, x_n)$ of continuous functions $x_1, \ldots, x_n : [\alpha, \beta] \to \mathbb{R}$, where $\alpha, \beta \in \mathbb{R}$ with $\alpha < \beta$. Often we will simply say that the path is given by

$(x_1(t), \ldots, x_n(t))$, $t \in [\alpha, \beta]$. It is then understood that $\alpha, \beta \in \mathbb{R}$ with $\alpha < \beta$ and $x_1, \ldots, x_n$ are continuous functions from $[\alpha, \beta]$ to $\mathbb{R}$.

Let Γ be a path in $\mathbb{R}^n$ given by $(x_1(t), \ldots, x_n(t))$, $t \in [\alpha, \beta]$. The endpoints $(x_1(\alpha), \ldots, x_n(\alpha))$ and $(x_1(\beta), \ldots, x_n(\beta))$ are called the **initial point** and the **terminal point** of Γ, respectively. If we let $\mathbf{c} := (x_1(\alpha), \ldots, x_n(\alpha))$ and $\mathbf{d} := (x_1(\beta), \ldots, x_n(\beta))$, then Γ is said to be a path from $\mathbf{c}$ to $\mathbf{d}$, or a path joining $\mathbf{c}$ to $\mathbf{d}$. Given any $D \subseteq \mathbb{R}^n$, we say that the path Γ lies in D if $(x_1(t), \ldots, x_n(t)) \in D$ for all $t \in [\alpha, \beta]$.

The two ways of connecting points in $\mathbb{R}^n$ (by a line segment or by a path) lead to the following definitions. A subset D of $\mathbb{R}^n$ is said to be

1. **convex** if the line segment joining any two points of D lies in D,
2. **path-connected** if any two points of D can be joined by a path that lies in D.

Examples 1.5. (i) Given any $r \in \mathbb{R}$ and $\mathbf{c} \in \mathbb{R}^n$, the sets $\mathbb{S}_r(\mathbf{c})$ and $\mathbb{B}_r(\mathbf{c})$ are convex. This follows from the definition of a line segment and the triangle inequality (part (v) of Proposition 1.1).

(ii) If $D \subseteq \mathbb{R}^n$ is convex, then for any $\mathbf{c} = (c_1, \ldots, c_n)$ and $\mathbf{d} = (d_1, \ldots, d_n)$ in D, the path corresponding to the line segment joining $\mathbf{c}$ to $\mathbf{d}$, that is, the path given by $((1-t)c_1 + td_1, \ldots, (1-t)c_n + td_n)$, $t \in [0, 1]$, lies in D. Hence D is path-connected. Thus every convex set is path-connected. However, the converse is not true. For example, if $D := \mathbb{R}^2 \setminus \mathbb{S}_1(0, 0)$ is the complement of the open square of radius 1 centered at the origin in $\mathbb{R}^2$, then D is path-connected, but D is not convex. (See Exercise 5.)

(iii) On $\mathbb{R}$, the notions of a convex set and a path-connected set coincide. Indeed, if $D \subseteq \mathbb{R}$, then

$$D \text{ is convex } \iff D \text{ is path-connected } \iff D \text{ is an interval in } \mathbb{R}.$$

The equivalence of the three notions follows from the definitions and the intermediate value theorem in $\mathbb{R}$, which implies that if $x : [\alpha, \beta] \to \mathbb{R}$ is continuous, then $x([\alpha, \beta])$ is an interval in $\mathbb{R}$.

(iv) If I is an n-interval, then using Proposition 1.4, we see that I is a convex subset of $\mathbb{R}^n$; in particular, by (ii) above, I is path-connected.

(v) Denote, as usual, by $\mathbb{Q}$ the set of all rational numbers, and consider $D := \mathbb{Q}^n = \{(r_1, \ldots, r_n) \in \mathbb{R}^n : r_1, \ldots, r_n \in \mathbb{Q}\}$. Then D is not path-connected. Indeed, if D were path-connected, then by restricting to one of the coordinates, we find that $\mathbb{Q}$ is path-connected, and hence by (iii) above, $\mathbb{Q}$ would be an interval in $\mathbb{R}$. But of course, $\mathbb{Q}$ is not an interval in $\mathbb{R}$ because we know (for example, from Proposition 1.6 of ACICARA) that between any two rational numbers, there is an irrational number. ◇

For more examples, see Exercises 5, 6, and 8.

1.2 Functions and Their Geometric Properties

In this section we shall restrict to $\mathbb{R}^n$ with $n = 2$ and develop a number of basic notions concerning the central object of our study, namely, a function of two variables. We give basic examples of functions, and note that there are two broad types: algebraic functions and transcendental functions. Prominent among the former are the polynomial functions and rational functions. Next, we consider a number of geometric properties of functions. Most of these properties are intimately related to notions such as continuity and differentiability that are studied in later chapters. What is of the essence here is to understand the intrinsic and geometric nature of these properties, and to realize that basic aspects can be studied without recourse to continuity and differentiability.

Basic Notions

A (real-valued) function of two variables is simply a function $f : D \to \mathbb{R}$, where D is a subset of $\mathbb{R}^2$. For instance, if $D := \mathbb{B}_1(0,0)$, then $f(x,y) := \sqrt{1 - x^2 - y^2}$ for $(x,y) \in D$ defines a function $f : D \to \mathbb{R}$. Composites of a function such as $f : D \to \mathbb{R}$ can be formed in at least three ways. For example, given any $E \subseteq \mathbb{R}$ such that $f(D) \subseteq E$ and a function $g : E \to \mathbb{R}$, we can form the composite $g \circ f : D \to \mathbb{R}$. Moreover, given any $E \subseteq \mathbb{R}$ and two functions $x, y : E \to \mathbb{R}$ such that $(x(t), y(t)) \in D$, we can form the composite $F : E \to \mathbb{R}$ defined by $F(t) := f(x(t), y(t))$ for $t \in E$. Similarly, if $E \subseteq \mathbb{R}^2$ and $x, y : E \to \mathbb{R}$ are such that $(x(u,v), y(u,v)) \in D$, then we can form the composite $F : E \to \mathbb{R}$ defined by $F(u,v) := f(x(u,v), y(u,v))$ for $(u,v) \in E$.

Sums, products, and scalar multiples of real-valued functions of two variables are given by pointwise addition, multiplication, and scalar multiplication, respectively. Thus, given any $D \subseteq \mathbb{R}^2$ and $f, g : D \to \mathbb{R}$ and $r \in \mathbb{R}$, we let $f + g$, fg, and rf be the functions from D to $\mathbb{R}$ defined by $(f+g)(x,y) := f(x,y) + g(x,y)$, $(fg)(x,y) := f(x,y)g(x,y)$, and $(rf)(x,y) := rf(x,y)$ for $(x,y) \in D$. Further, if g is such that $g(x,y) \neq 0$ for all $(x,y) \in D$, then the quotient f/g is the function from D to $\mathbb{R}$ defined by $(f/g)(x,y) := f(x,y)/g(x,y)$ for $(x,y) \in D$. Moreover, if f is such that $f(x,y) \geq 0$ for all $(x,y) \in D$, then for any $r \in \mathbb{R}$, the rth power f^r is the function from D to $\mathbb{R}$ defined by $f^r(x,y) := f(x,y)^r$ for $(x,y) \in D$. Sometimes, we write $f \leq g$ to mean that $f(x,y) \leq g(x,y)$ for all $(x,y) \in D$.

Given any $D \subseteq \mathbb{R}^2$ and $f : D \to \mathbb{R}$, the **graph** of f is defined to be the subset $\{(x, y, f(x,y)) : (x,y) \in D\}$ of $\mathbb{R}^3$; in other words, the graph of f is the *surface* in $\mathbb{R}^3$ given by $z = f(x,y)$, $(x,y) \in D$. For example, the graph of $f : \mathbb{B}_1(0,0) \to \mathbb{R}$ defined by $f(x,y) := \sqrt{1 - x^2 - y^2}$ is an upper hemisphere, whereas the graph of $f : \mathbb{R}^2 \to \mathbb{R}$ defined by $f(x,y) = x^2 + y^2$ is a paraboloid. (See Figure 1.2.)

In general, graphs of functions of two variables are difficult to draw, but we can get some idea of the graph by looking at certain curves associated with

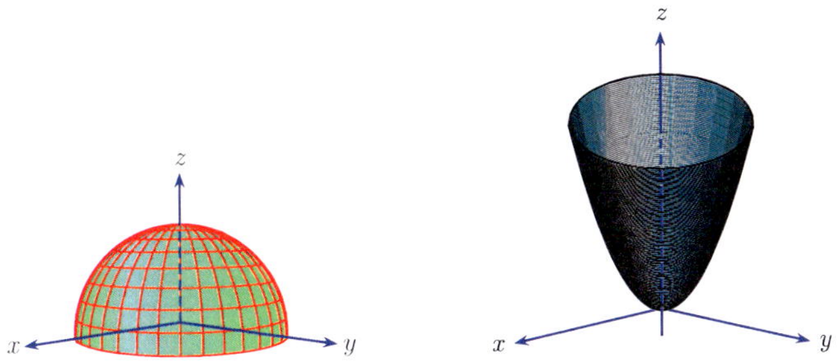

Fig. 1.2. The upper hemisphere $z = \sqrt{1 - x^2 - y^2}$ and the paraboloid $z = x^2 + y^2$.

the function. For $D \subseteq \mathbb{R}^2$ and $f : D \to \mathbb{R}$, the **level curve** of f corresponding to any $c \in \mathbb{R}$ is the curve in $\mathbb{R}^2$ given by $f(x, y) = c$, $(x, y) \in D$, that is, the subset $\{(x, y) \in D : f(x, y) = c\}$ of $\mathbb{R}^2$; the **contour line** of f corresponding to any $c \in \mathbb{R}$ is the curve in $\mathbb{R}^3$ obtained by intersecting the surface given by $z = f(x, y)$, $(x, y) \in D$, with the horizontal plane given by $z = c$, that is, the subset $\{(x, y, f(x, y)) \in \mathbb{R}^3 : (x, y) \in D \text{ and } f(x, y) = c\}$ of $\mathbb{R}^3$.

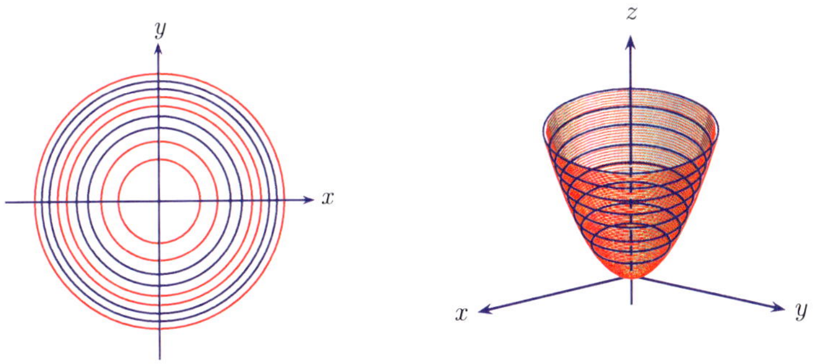

Fig. 1.3. The level curves and the contour lines for the function $f(x, y) := x^2 + y^2$.

For example, if $f : \mathbb{B}_1(0, 0) \to \mathbb{R}$ is defined by $f(x, y) := \sqrt{1 - x^2 - y^2}$ for $(x, y) \in \mathbb{B}_1(0, 0)$, then the level curves of f are concentric circles centered at the origin of radius ≤ 1, while the contour lines of f are the circles on the upper hemisphere obtained by intersecting it with planes given by $z = c$ as c varies over $[0, 1]$. The level curves of $f : \mathbb{R}^2 \to \mathbb{R}$ defined by $f(x, y) = x^2 + y^2$ for $(x, y) \in \mathbb{R}^2$ are also concentric circles centered at origin (of any radius), but for equally spaced values of c, the level curves $f(x, y) = c$ for $(x, y) \in \mathbb{R}^2$ are not as evenly placed as in the case of the hemisphere. (See Figure 1.3.)

We shall now discuss some basic examples of functions. The most basic among these are polynomial functions. Before discussing these, let us first recall a few basic notions from algebra.

A **polynomial** in two variables x and y (with real coefficients) is a finite sum of terms of the form cx^iy^j, where i,j are nonnegative integers and $c \in \mathbb{R}$; here c is called the **coefficient** of the term and $i+j$ is called its **total degree**, provided it is a nonzero term, that is, $c \neq 0$. Two polynomials are equal if they have identical nonzero terms. The **zero polynomial** is the polynomial having all of its coefficients equal to zero. The **total degree**, or simply the **degree**, of a nonzero polynomial is the maximum of the total degrees of its nonzero terms. A nonzero polynomial is said to be **homogeneous** if all its nonzero terms have the same total degree. For instance, $x^5y + 2x^4 + y^2 + 1$ and $x^3 + x^2y + 6xy^2$ are polynomials of total degree 6 and 3 respectively; the latter is homogeneous, while the former is not. Notice that we can evaluate a polynomial at points of $\mathbb{R}^2$. Thus, if $p(x,y)$ is a polynomial in the variables x and y, then for any $(x_0,y_0) \in \mathbb{R}^2$, by substituting x_0 for x and y_0 for y in $p(x,y)$, we obtain a real number, denoted by $p(x_0,y_0)$. Observe that if $p(x,y)$ is a homogeneous polynomial of degree d, then $p(\lambda x_0, \lambda y_0) = \lambda^d p(x_0,y_0)$ for all $(x_0,y_0) \in \mathbb{R}^2$ and $\lambda \in \mathbb{R}$.

Let $D \subseteq \mathbb{R}^2$ and let $f : D \to \mathbb{R}$ be a function. If there is a polynomial $p(x,y)$ in two variables such that $f(x_0,y_0) = p(x_0,y_0)$ for all $(x_0,y_0) \in D$, then f is said to be a **polynomial function** on D. In case $D = \mathbb{R}^2$ or more generally, when $D = I \times J$, where both I and J are intervals containing more than one point in $\mathbb{R}$, the polynomial $p(x,y)$ is uniquely determined by the function f. (See Exercise 36.) So in this case we may talk about the total degree or the coefficients of a polynomial function f. We say that f is a **rational function** on D if there are polynomials $p(x,y)$ and $q(x,y)$ in two variables such that $q(x_0,y_0) \neq 0$ for any $(x_0,y_0) \in D$ and $f(x_0,y_0) = p(x_0,y_0)/q(x_0,y_0)$ for all $(x_0,y_0) \in D$. The polynomials $p(x,y)$ and $q(x,y)$ are not uniquely determined by the rational function f even when $D = \mathbb{R}^2$. For example, $(x^3-x^2+x-1)/(x^2y^2+2x^2+y^2+2)$ and $(xy^2+x-y^2-1)/(y^4+3y^2+2)$ define the same rational function on $\mathbb{R}^2$, but the corresponding numerators and denominators are not the same. We say that f is an **algebraic function** on D if $z = f(x,y)$ satisfies an equation of the form

$$p_n(x,y)z^n + p_{n-1}(x,y)z^{n-1} + \cdots + p_1(x,y)z + p_0(x,y) = 0 \quad \text{for } (x,y) \in D,$$

where $n \in \mathbb{N}$ and $p_0(x,y), p_1(x,y), \ldots, p_n(x,y)$ are polynomials in two variables such that $p_n(x,y)$ is a nonzero polynomial. For example, $f : \mathbb{B}_1(0,0) \to \mathbb{R}$ defined by $f(x,y) := \sqrt{1-x^2-y^2}$ for $(x,y) \in \mathbb{B}_1(0,0)$, is an algebraic function, since $z = f(x,y)$ satisfies the equation $z^2 - (x^2+y^2-1) = 0$ for $(x,y) \in \mathbb{B}_1(0,0)$. Finally, if f is not an algebraic function, then it is said to be a **transcendental function**. For example, $f : \mathbb{R}^2 \to \mathbb{R}$ defined by $f(x,y) = \sin(xy)$ is a transcendental function because if $z = f(x,y)$ were to satisfy an equation of the kind above, then by substituting a suitable value for y we would find that the sine function (of one variable) was algebraic, but we know from one-variable calculus (for example, Proposition 7.29 of ACICARA) that the sine function is transcendental. (See Exercise 11.)

Another useful way to generate examples of functions is by combining or piecing together known functions. For example, if I and J are intervals in $\mathbb{R}$ and $\phi : I \to \mathbb{R}$ and $\psi : J \to \mathbb{R}$ are functions of one variable, then $f, g : I \times J \to \mathbb{R}$ defined by

$$f(x, y) := \phi(x) + \psi(y) \quad \text{and} \quad g(x, y) := \phi(x)\psi(y)$$

are functions of two variables on $I \times J$. If D_1 and D_2 are subsets of $\mathbb{R}^2$ and $f_1 : D_1 \to \mathbb{R}$ and $f_2 : D_2 \to \mathbb{R}$ are functions of two variables such that $f_1(x, y) = f_2(x, y)$ for all $(x, y) \in D_1 \cap D_2$, then $f : D_1 \cup D_2 \to \mathbb{R}$ defined by

$$f(x, y) := \begin{cases} f_1(x, y) & \text{if } (x, y) \in D_1, \\ f_2(x, y) & \text{if } (x, y) \in D_2, \end{cases}$$

gives a function of two variables on $D_1 \cup D_2$, which may be referred to as the function obtained by piecing together f_1 and f_2.

Bounded Functions

To talk about bounded functions, we use, in effect, the terminology applicable to the range of a given function. The definitions given below are analogous to the corresponding notions for functions of one variable. (See, for example, page 22 of ACICARA.)

Let $D \subseteq \mathbb{R}^2$ and let $f : D \to \mathbb{R}$ be any function. We say that f is **bounded above** on D if there is $\alpha \in \mathbb{R}$ such that $f(x, y) \leq \alpha$ for all $(x, y) \in D$; in this case, we say that f **attains its upper bound** on D if there is $(x_0, y_0) \in D$ such that $\sup\{f(x, y) : (x, y) \in D\} = f(x_0, y_0)$. Likewise, we say that f is **bounded below** on D if there is $\beta \in \mathbb{R}$ such that $f(x, y) \geq \beta$ for all $(x, y) \in D$; in this case, we say that f **attains its lower bound** on D if there is $(x_0, y_0) \in D$ such that $\inf\{f(x, y) : (x, y) \in D\} = f(x_0, y_0)$. Finally, we say that f is **bounded** on D if it is bounded above on D as well as bounded below on D; in this case, we say that f **attains its bounds** on D if it attains its upper bound on D and also attains its lower bound on D.

For example, $f : \mathbb{R}^2 \to \mathbb{R}$ defined by $f(x, y) := (x^2 + y^2)$ is bounded below (but not bounded above) on $\mathbb{R}^2$ and attains its lower bound (which is 0), while $f : \mathbb{R}^2 \to \mathbb{R}$ defined by $f(x, y) := -(x^2 + y^2)$ is bounded above (but not bounded below) on $\mathbb{R}^2$ and attains its upper bound (which is 0). The function $f : \mathbb{R}^2 \to \mathbb{R}$ defined by $f(x, y) := xy$ is neither bounded below nor bounded above on $\mathbb{R}^2$. Thus, each of these three functions fails to be bounded on $\mathbb{R}^2$. (See Figures 1.2 and 1.9.) On the other hand, $f : \mathbb{R}^2 \to \mathbb{R}$ defined by $f(x, y) := (x^2 + y^2)/(x^2 + y^2 + 1)$ is bounded on $\mathbb{R}^2$ and it attains its lower bound (which is 0), but does not attain its upper bound (which is 1). (See Figure 1.4.)

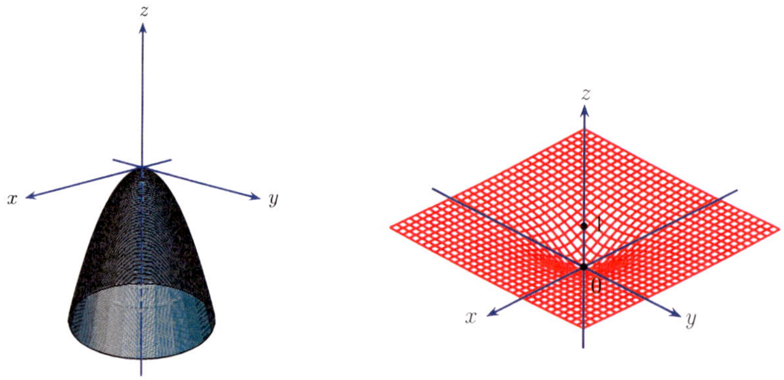

Fig. 1.4. Graphs of $f(x, y) := -(x^2 + y^2)$ and $f(x, y) := (x^2 + y^2)/(x^2 + y^2 + 1)$.

Monotonicity and Bimonotonicity

The notion of product order on $\mathbb{R}^2$ enables us to talk about monotonicity of a function of two variables. There is also a related but distinct notion for functions of two variables, called bimonotonicity, which will be discussed here.

Let $D \subseteq \mathbb{R}^2$ and let $f : D \to \mathbb{R}$ be any function. Also, let I and J be intervals in $\mathbb{R}$ such that $I \times J \subseteq D$. We say that

1. f is **monotonically increasing** on $I \times J$ if for all $(x_1, y_1), (x_2, y_2)$ in $I \times J$, we have
$$(x_1, y_1) \le (x_2, y_2) \implies f(x_1, y_1) \le f(x_2, y_2),$$
2. f is **monotonically decreasing** on $I \times J$ if for all $(x_1, y_1), (x_2, y_2)$ in $I \times J$, we have
$$(x_1, y_1) \le (x_2, y_2) \implies f(x_1, y_1) \ge f(x_2, y_2),$$
3. f is **monotonic** on $I \times J$ if f is monotonically increasing on $I \times J$ or monotonically decreasing on $I \times J$,
4. f is **bimonotonically increasing** on $I \times J$ if for all $(x_1, y_1), (x_2, y_2)$ in $I \times J$, we have
$$(x_1, y_1) \le (x_2, y_2) \implies f(x_1, y_2) + f(x_2, y_1) \le f(x_1, y_1) + f(x_2, y_2),$$
5. f is **bimonotonically decreasing** on $I \times J$ if for all $(x_1, y_1), (x_2, y_2)$ in $I \times J$, we have
$$(x_1, y_1) \le (x_2, y_2) \implies f(x_1, y_2) + f(x_2, y_1) \ge f(x_1, y_1) + f(x_2, y_2),$$
6. f is **bimonotonic** on $I \times J$ if f is bimonotonically increasing on $I \times J$ or bimonotonically decreasing on $I \times J$.

It may be noted that f is monotonically increasing on $I \times J$ if and only if it is (monotonically) increasing in each of the two variables, that is, for every fixed $x \in I$, the function $y \longmapsto f(x, y)$ from J to $\mathbb{R}$ is increasing on J, and for every fixed $y \in J$, the function $x \longmapsto f(x, y)$ from I to $\mathbb{R}$ is increasing on I. Likewise for monotonically decreasing functions. The following result gives conditions under which an increasing function in the variable x and an increasing function in the variable y can be added or multiplied to obtain a monotonic and/or bimonotonic function of two variables.

Proposition 1.6. *Let I, J be nonempty intervals in $\mathbb{R}$. Given any $\phi : I \to \mathbb{R}$ and $\psi : J \to \mathbb{R}$, consider $f : I \times J \to \mathbb{R}$ and $g : I \times J \to \mathbb{R}$ defined by*

$$f(x, y) := \phi(x) + \psi(y) \quad \text{and} \quad g(x, y) := \phi(x)\psi(y) \quad \text{for } (x, y) \in I \times J.$$

Then we have the following.

(i) *f is monotonically increasing on $I \times J$ if and only if ϕ is increasing on I and ψ is increasing on J.*
(ii) *Assume that $\phi(x) \geq 0$ and $\psi(y) \geq 0$ for all $x \in I$, $y \in J$, and also that $\phi(x_0) > 0$ and $\psi(y_0) > 0$ for some $x_0 \in I$ and some $y_0 \in J$. Then g is monotonically increasing on $I \times J$ if and only if ϕ is increasing on I and ψ is increasing on J.*
(iii) *f is always bimonotonically increasing and also bimonotonically decreasing on $I \times J$.*
(iv) *If ϕ is monotonic on I and ψ is monotonic on J, then g is bimonotonic on $I \times J$. More specifically, if ϕ and ψ are both increasing or both decreasing, then g is bimonotonically increasing, whereas if ϕ is increasing and ψ is decreasing, or vice-versa, then g is bimonotonically decreasing.*

Proof. Both (i) and (ii) are straightforward consequences of the definitions. Next, (iii) follows by noting that for all $(x_1, y_1), (x_2, y_2) \in I \times J$, we have $f(x_1, y_2) + f(x_2, y_1) = f(x_1, y_1) + f(x_2, y_2)$. Finally, the identity

$$g(x_2, y_2) + g(x_1, y_1) - g(x_1, y_2) - g(x_2, y_1) = (\phi(x_2) - \phi(x_1))\,(\psi(y_2) - \psi(y_1)),$$

valid for all $(x_1, y_1), (x_2, y_2) \in I \times J$, implies the assertions in (iv). □

Results similar to parts (i) and (ii) of Proposition 1.6 hold for monotonically decreasing functions. (See Exercise 15.) Also, the converse of part (iii) holds. (See Exercise 38.) The above proposition as well as the one below can be used to generate several examples of monotonic and bimonotonic functions.

Proposition 1.7. *Let I, J be nonempty intervals in $\mathbb{R}$. The the set*

$$I + J := \{x + y : x \in I \text{ and } y \in J\}$$

is an interval in $\mathbb{R}$. Further, let $\phi : I + J \to \mathbb{R}$ be any function and consider $f : I \times J \to \mathbb{R}$ defined by

$$f(x, y) := \phi(x + y) \quad \text{for } (x, y) \in I \times J.$$

Then we have the following:

(i) ϕ *is increasing on* $I + J \Longrightarrow f$ *is monotonically increasing on* $I \times J$.
(ii) ϕ *is decreasing on* $I + J \Longrightarrow f$ *is monotonically decreasing on* $I \times J$.
(iii) ϕ *is convex on* $I + J \Longrightarrow f$ *is bimonotonically increasing on* $I \times J$.
(iv) ϕ *is concave on* $I + J \Longrightarrow f$ *is bimonotonically decreasing on* $I \times J$.

Proof. Let $x_1, x_2 \in I$ and $y_1, y_2 \in J$ be such that $x_1+y_1 < x_2+y_2$ and consider $r \in \mathbb{R}$ such that $x_1 + y_1 < r < x_2 + y_2$. Then there is $t \in \mathbb{R}$ with $0 < t < 1$ [in fact, $t = (x_2+y_2-r)/(x_2+y_2-x_1-y_1)$] such that $r = t(x_1+y_1)+(1-t)(x_2+y_2)$. Hence if $x := tx_1 + (1-t)x_2$ and $y := ty_1 + (1-t)y_2$, then $x \in I$, $y \in J$ and $r = x + y$. Thus $I + J$ is an interval in $\mathbb{R}$.

Both (i) and (ii) are straightforward consequences of the definitions. Next, suppose ϕ is convex on $I + J$. Consider any $(x_1, y_1), (x_2, y_2) \in I \times J$ with $(x_1, y_1) \le (x_2, y_2)$ and $(x_1, y_1) \ne (x_2, y_2)$. Observe that

$$x_1+y_2 = \lambda(x_1+y_1)+(1-\lambda)(x_2+y_2) \quad \text{and} \quad x_2+y_1 = (1-\lambda)(x_1+y_1)+\lambda(x_2+y_2),$$

where $\lambda = (x_2 - x_1)/(x_2 - x_1 + y_2 - y_1)$. Hence $\phi(x_1 + y_2) \le \lambda\phi(x_1 + y_1) + (1-\lambda)\phi(x_2 + y_2)$ and $\phi(x_2 + y_1) \le (1-\lambda)\phi(x_1 + y_1) + \lambda\phi(x_2 + y_2)$. Consequently, $\phi(x_1 + y_2)+\phi(x_2 + y_1) \le \phi(x_1 + y_1)+\phi(x_2 + y_2)$, that is, $f(x_1, y_2)+f(x_2, y_1) \le f(x_1, y_1)+f(x_2, y_2)$. Thus f is bimonotonically increasing on $I\times J$. This proves (iii). The proof of (iv) is similar. □

Examples 1.8. (i) Consider $f : [-1, 1] \times [-1, 1] \to \mathbb{R}$ defined by

$$f(x, y) := \begin{cases} (x + 1)(y + 1) & \text{if } x + y < 0, \\ (x + 2)(y + 2) & \text{if } x + y \ge 0. \end{cases}$$

If we fix $y_0 \in [-1, 1]$ and consider the function $\phi : [-1, 1] \to \mathbb{R}$ defined by

$$\phi(x) := \begin{cases} (y_0 + 1)(x + 1) & \text{if } -1 \le x < -y_0, \\ (y_0 + 2)(x + 2) & \text{if } -y_0 \le x \le 1, \end{cases}$$

then it is easy to see that ϕ is increasing on $[-1, 1]$. Also, if we fix $x_0 \in [-1, 1]$ and consider $\psi : [-1, 1] \to \mathbb{R}$ defined by

$$\psi(y) := \begin{cases} (x_0 + 1)(y + 1) & \text{if } -1 \le y < -x_0, \\ (x_0 + 2)(y + 2) & \text{if } -x_0 \le y \le 1, \end{cases}$$

then it is easy to see that ψ is increasing on $[-1, 1]$. It follows that f is monotonically increasing on $[-1, 1] \times [-1, 1]$. However, f is not bimonotonic on $[-1, 1] \times [-1, 1]$. To see this, note that $(0, 0) \le (1, 1)$ and $f(0, 1)+f(1, 0) = 6+6 < 4+9 = f(0, 0)+f(1, 1)$, whereas $(-1, 0) \le (0, 1)$ and $f(-1, 1) + f(0, 0) = 3 + 4 > 0 + 6 = f(-1, 0) + f(0, 1)$.

(ii) Consider $f : \mathbb{R}^2 \to \mathbb{R}$ defined by $f(x, y) := \cos x + \sin y$. Using Proposition 1.6, we readily see that f is bimonotonic, but not monotonic.

(iii) Let $p \in \mathbb{R}$ and let $f : (0, \infty) \times (0, \infty) \to \mathbb{R}$ be defined by $f(x, y) := (x+y)^p$. Consider $\phi : (0, \infty) \to \mathbb{R}$ defined by $\phi(t) := t^p$. Clearly, ϕ is twice differentiable and $\phi'(t) = pt^{p-1}$, while $\phi''(t) = p(p-1)t^{p-2}$ for $t \in (0, \infty)$. It follows that ϕ is decreasing if $p \leq 0$, increasing if $p \geq 0$, convex if either $p \leq 0$ or $p \geq 1$, and concave if $0 \leq p \leq 1$. Thus, using Proposition 1.7, we see that f is monotonically decreasing and bimonotonically increasing if $p \leq 0$, monotonically increasing and bimonotonically decreasing if $0 \leq p \leq 1$, and both monotonically and bimonotonically increasing if $p \geq 1$. For another example of this kind, see Exercise 17. ◇

Remark 1.9. In Chapter 3, we shall define and study the notion of partial derivatives of a function of two variables. It will be shown that a function f of two variables is bimonotonically increasing if and only if the second-order mixed partial derivative f_{xy} is nonnegative, while f is bimonotonically decreasing if and only if f_{xy} is nonpositive. (See Proposition 3.55.) ◇

Functions of Bounded Variation

In general, the sum of two monotonic functions need not be monotonic. For example, $f : [0, 1] \times [0, 1] \to \mathbb{R}$ defined by $f(x, y) := x - y$ is a sum of monotonic functions (given by $(x, y) \longmapsto x$ and $(x, y) \longmapsto -y$), but it is neither increasing nor decreasing. On the other hand, since a monotonic function on a (closed) rectangle is bounded ($f : [a, b] \times [c, d] \to \mathbb{R}$ monotonic $\Longrightarrow$ the values of f lie between $f(a, c)$ and $f(b, d)$), sums of monotonic functions are bounded. In fact, they satisfy a stronger property defined below.

Let $a, b, c, d \in \mathbb{R}$ with $a \leq b$ and $c \leq d$, and let $f : [a, b] \times [c, d] \to \mathbb{R}$ be any function. Denote by S_f the subset of $\mathbb{R}$ consisting of finite sums of the form

$$\sum_{i=1}^{n} |f(x_i, y_i) - f(x_{i-1}, y_{i-1})|,$$

where $n \in \mathbb{N}$ and $(x_0, y_0), \ldots, (x_n, y_n)$ are any points in $\mathbb{R}^2$ satisfying

$$(a, c) = (x_0, y_0) \leq (x_1, y_1) \leq \cdots \leq (x_{n-1}, y_{n-1}) \leq (x_n, y_n) = (b, d).$$

If the set S_f is bounded above, then f is said to be of **bounded variation**. In this case, we denote the supremum of S_f by $V(f)$, and call it the **total variation** of f on $[a, b] \times [c, d]$.

We record below some elementary properties of functions of bounded variation. It may be noted that parts (ii) and (iii) of the proposition below can be readily used to produce several examples of functions of bounded variation. Henceforth, when we consider a rectangle of the form $[a, b] \times [c, d]$, it will be tacitly assumed that $a, b, c, d \in \mathbb{R}$ with $a \leq b$ and $c \leq d$.

Proposition 1.10. *Let* $f, g : [a, b] \times [c, d] \to \mathbb{R}$ *and* $r \in \mathbb{R}$. *Then*

(i) f *is of bounded variation* $\Longrightarrow$ f *is bounded,*
(ii) f *is monotonic* $\Longrightarrow$ f *is of bounded variation,*
(iii) f, g *are of bounded variation* $\Longrightarrow$ $f + g$, rf, fg *are of bounded variation.*

Proof. (i) Assume that f is of bounded variation. Given any $(x, y) \in [a, b] \times [c, d]$, we have $(a, c) \le (x, y) \le (b, d)$, and so $|f(x, y) - f(a, c)| + |f(b, d) - f(x, y)| \le V(f)$. This implies that $2|f(x, y)| \le |f(a, c)| + |f(b, d)| + V(f)$. Thus, f is bounded.

(ii) If f is monotonic, then for any $n \in \mathbb{N}$ and $(x_0, y_0), \ldots, (x_n, y_n) \in \mathbb{R}^2$ with $(a, c) = (x_0, y_0) \le (x_1, y_1) \le \cdots \le (x_n, y_n) = (b, d)$, we have

$$\sum_{i=1}^{n} |f(x_i, y_i) - f(x_{i-1}, y_{i-1})| = \left| \sum_{i=1}^{n} (f(x_i, y_i) - f(x_{i-1}, y_{i-1})) \right|,$$

which reduces to $|f(x_n, y_n) - f(x_0, y_0)|$; hence f is of bounded variation and moreover, $V(f) = |f(b, d) - f(a, c)|$.

(iii) Suppose f and g are of bounded variation. Using elementary properties of the absolute value (parts (v) and (vi) of Proposition 1.1 with $n = 1$), we see that $V(f + g) \le V(f) + V(g)$ and $V(rf) = |r|V(f)$. Moreover, if we let $M(f) := \sup\{|f(x, y)| : (x, y) \in [a, b] \times [c, d]\}$ and $M(g) := \sup\{|g(x, y)| : (x, y) \in [a, b] \times [c, d]\}$, then adding and subtracting appropriate quantities, we obtain $V(fg) \le M(f)V(g) + M(g)V(f)$. This proves (iii). □

From parts (ii) and (iii) of Proposition 1.10, we see that sums of monotonic functions are of bounded variation. We shall soon show that the converse is also true. For this purpose, we need the following auxiliary result.

Lemma 1.11. *If* $f : [a, b] \times [c, d] \to \mathbb{R}$ *is of bounded variation, then for any* $(x, y) \in [a, b] \times [c, d]$, *the restriction* $f|_{[a,x]\times[c,y]}$ *is of bounded variation and* $V\left(f|_{[a,x]\times[c,y]}\right) + |f(b, d) - f(x, y)| \le V(f)$.

Proof. Given any $n \in \mathbb{N}$ and any $(x_0, y_0), (x_1, y_1), \ldots, (x_n, y_n) \in \mathbb{R}^2$ satisfying $(a, c) = (x_0, y_0) \le (x_1, y_1) \le \cdots \le (x_n, y_n) = (x, y)$, we have

$$\sum_{i=1}^{n} |f(x_i, y_i) - f(x_{i-1}, y_{i-1})| + |f(b, d) - f(x, y)| \le V(f).$$

Hence $f|_{[a,x]\times[c,y]}$ is of bounded variation and its total variation is at most $V(f) - |f(b, d) - f(x, y)|$. □

If $f : [a, b] \times [c, d] \to \mathbb{R}$ is of bounded variation, then we define the corresponding **total variation function** $v_f : [a, b] \times [c, d] \to \mathbb{R}$ by $v_f(x, y) := V\left(f|_{[a,x]\times[c,y]}\right)$. The following result gives the so-called **Jordan decomposition** of a function of bounded variation.

Proposition 1.12. *If* $f : [a,b] \times [c,d] \to \mathbb{R}$ *is of bounded variation, then there are unique functions* $g, h : [a,b] \times [c,d] \to \mathbb{R}$ *such that* g *and* h *are monotonically increasing,* $f = g - h$*, and* $v_f = g + h$*.*

Proof. Define $g, h : [a,b] \times [c,d] \to \mathbb{R}$ by $g = \frac{1}{2}(v_f + f)$ and $h = \frac{1}{2}(v_f - f)$. Clearly, $f = g - h$ and $v_f = g + h$. Let $(x_1, y_1), (x_2, y_2) \in [a,b] \times [c,d]$ with $(x_1, y_1) \le (x_2, y_2)$. Applying Lemma 1.11 to the restriction $f|_{[a,x_2]\times[c,y_2]}$, we see that $v_f(x_1, y_1) + |f(x_2, y_2) - f(x_1, y_1)| \le v_f(x_2, y_2)$, and hence

$$g(x_2, y_2) - g(x_1, y_1) = \frac{1}{2}\left[v_f(x_2, y_2) - v_f(x_1, y_1) + f(x_2, y_2) - f(x_1, y_1)\right] \ge 0$$

as well as

$$h(x_2, y_2) - h(x_1, y_1) = \frac{1}{2}\left[v_f(x_2, y_2) - v_f(x_1, y_1) - f(x_2, y_2) + f(x_1, y_1)\right] \ge 0.$$

Thus g and h are monotonically increasing. The uniqueness of g and h is obvious from the conditions $f = g - h$ and $v_f = g + h$. □

As remarked earlier, Proposition 1.12 gives a characterization of functions of bounded variation. We have also seen that a function of bounded variation need not be monotonic, that is, the converse of part (ii) of Proposition 1.10 is not true. We shall now give an example to show that a bounded function need not be of bounded variation, that is, the converse of part (i) of Proposition 1.10 is not true. The same example shows that a bivariate function that is monotonically increasing in one variable and monotonically decreasing in the other need not be of bounded variation.

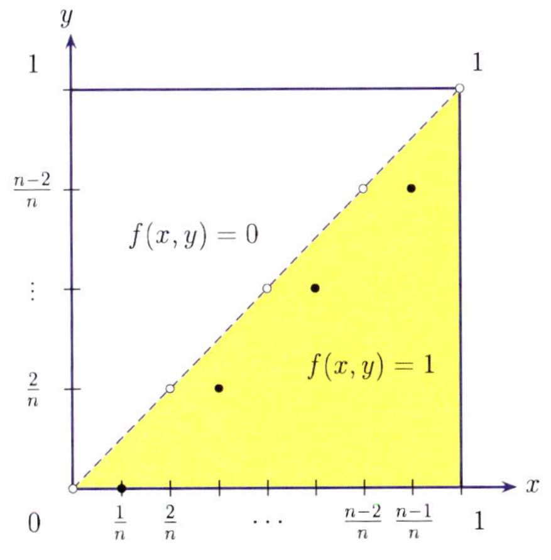

Fig. 1.5. Illustration of the function in Example 1.13 and the points (x_i, y_i) of the rectangle $[0,1] \times [0,1]$ that straddle the diagonal line $y = x$.

Example 1.13. Consider $f : [0,1] \times [0,1] \to \mathbb{R}$ defined by

$$f(x,y) := \begin{cases} 0 & \text{if } x \leq y, \\ 1 & \text{if } x > y. \end{cases}$$

Clearly, f is a bounded function. But if we consider points of the rectangle $[0,1] \times [0,1]$ that straddle the diagonal line $y = x$, then it is seen that there is too much variation in the values of f. (See Figure 1.5.) For example, if $n \in \mathbb{N}$ is even and for $i = 1, \ldots, n$, we let $(x_i, y_i) := (i/n, i/n)$ if i is even and $(x_i, y_i) := (i/n, (i-1)/n)$ if i is odd, then we clearly have $(0,0) = (x_0, y_0) \leq (x_1, y_1) \leq \cdots \leq (x_n, y_n) = (1,1)$ and

$$\sum_{i=1}^{n} |f(x_i, y_i) - f(x_{i-1}, y_{i-1})| = |1-0| + |0-1| + \cdots + |1-0| + |0-1| = n.$$

It follows that f is not of bounded variation on $[0,1] \times [0,1]$. ◇

Remark 1.14. For further results on functions of bounded variation, see Exercises 43–46, 48, and 49. ◇

Functions of Bounded Bivariation

Just as a sum of monotonic functions need not be monotonic, the sum of bimonotonic functions need not be bimonotonic. Accordingly, we are led to the following analogue of a function of bounded variation.

Let $a, b, c, d \in \mathbb{R}$ with $a \leq b$ and $c \leq d$, and let $f : [a,b] \times [c,d] \to \mathbb{R}$ be a function. Denote by T_f the set of finite double sums of the form

$$\sum_{i=1}^{n} \sum_{j=1}^{m} |f(x_i, y_j) + f(x_{i-1}, y_{j-1}) - f(x_i, y_{j-1}) - f(x_{i-1}, y_j)|,$$

where $n, m \in \mathbb{N}$ and $(x_0, y_0), \ldots, (x_n, y_m)$ are any points in $\mathbb{R}^2$ satisfying

$$a = x_0 \leq x_1 \leq \cdots \leq x_{n-1} \leq x_n = b \text{ and } c = y_0 \leq y_1 \leq \cdots \leq y_{m-1} \leq y_m = d.$$

If the set T_f is bounded above, then f is said to be of **bounded bivariation**. In this case, we denote the supremum of T_f by $W(f)$, and call it the **total bivariation** of f on $[a,b] \times [c,d]$.

The properties below are analogous to those in Proposition 1.10, except that part (i) needs an additional hypothesis (Example 1.19 (i)) and in part (iii), one has to exclude products of bimonotonic functions (Example 1.19 (ii)).

Proposition 1.15. *Let $f, g : [a,b] \times [c,d] \to \mathbb{R}$ and $r \in \mathbb{R}$. Then:*

(i) *If f is of bounded bivariation and, in addition, f is bounded on any two adjacent sides of the rectangle $[a,b] \times [c,d]$, then f is bounded.*

(ii) *If f is bimonotonic, then f is of bounded bivariation.*
(iii) *If f and g are of bounded bivariation, then so are $f+g$ and rf.*

Proof. (i) Assume that f is of bounded variation and f is bounded on the two sides $[a,b]\times\{c\}$ and $\{a\}\times[c,d]$. Given any $(x,y)\in[a,b]\times[c,d]$, we have $a\le x$ and $c\le y$. Hence $|f(x,y)+f(a,c)-f(x,c)-f(a,y)|\le W(f)$, and so $|f(x,y)|\le|f(a,c)|+|f(x,c)|+|f(a,y)|+W(f)$. This implies that f is bounded. A similar argument applies if f is bounded on any of the other two adjacent sides of $[a,b]\times[c,d]$.

(ii) If f is bimonotonic, then for any $m,n\in\mathbb{N}$ and $(x_0,y_0),\dots,(x_n,y_n)$ in $\mathbb{R}^2$ with $a=x_0\le x_1\le\cdots\le x_n=b$ and $c=y_0\le y_1\le\cdots\le y_m=d$, the double sum

$$\sum_{i=1}^{n}\sum_{j=1}^{m}|f(x_i,y_j)+f(x_{i-1},y_{j-1})-f(x_i,y_{j-1})-f(x_{i-1},y_j)|$$

is equal to $|f(x_n,y_n)+f(x_0,y_0)-f(x_0,y_n)-f(x_n,y_0)|$; so f is of bounded bivariation and moreover, $W(f)=|f(b,d)+f(a,c)-f(a,d)-f(b,c)|$.

(iii) Suppose f and g are of bounded bivariation. Using elementary properties of the absolute value, we see that $W(f+g)\le W(f)+W(g)$ and $W(rf)=|r|W(f)$. This proves (iii). □

From parts (ii) and (iii) of Proposition 1.15, we see that sums of bimonotonic functions are of bounded bivariation. We shall show that the converse is also true. For this purpose, we need the following auxiliary result, which is analogous to but a little more subtle than Lemma 1.11. We employ the following notation and terminology.

Let $a,b,c,d\in\mathbb{R}$ with $a\le b$ and $c\le d$. A collection $(x_0,y_0),\dots,(x_n,y_m)$ of points in $\mathbb{R}^2$ satisfying

$$a=x_0\le x_1\le\cdots\le x_n=b \quad\text{and}\quad c=y_0\le y_1\le\cdots\le y_m=d$$

will be referred to as a collection of **grid points** of $[a,b]\times[c,d]$. If P is such a collection of grid points of $[a,b]\times[c,d]$ and $f:[a,b]\times[c,d]\to\mathbb{R}$ is any function, then we will denote by $W(P,f)$ the following double sum

$$W(P,f):=\sum_{i=1}^{n}\sum_{j=1}^{m}|f(x_i,y_j)+f(x_{i-1},y_{j-1})-f(x_i,y_{j-1})-f(x_{i-1},y_j)|.$$

Note that the total bivariation $W(f)$ of f is the supremum of $W(P,f)$ as P varies over all possible collections of grid points of $[a,b]\times[c,d]$.

Lemma 1.16. *If $f:[a,b]\times[c,d]\to\mathbb{R}$ is of bounded bivariation, then for any $(x,y)\in[a,b]\times[c,d]$, each of the restrictions f_1,f_2,f_3,f_4 defined by*

$$f_1:=f|_{[a,x]\times[c,y]},\quad f_2:=f|_{[a,x]\times[y,d]},\quad f_3:=f|_{[x,b]\times[c,y]},\quad f_4:=f|_{[x,b]\times[y,d]}$$

is of bounded bivariation and $W(f)=W(f_1)+W(f_2)+W(f_3)+W(f_4)$.

Proof. Assume that f is of bounded bivariation and fix $(x,y) \in [a,b] \times [c,d]$. First, observe that if P_1, P_2, P_3, P_4 are any collections of grid points of the rectangles $[a,x] \times [c,y]$, $[a,x] \times [y,d]$, $[x,b] \times [c,y]$, $[x,b] \times [y,d]$, respectively, then by collating the grid points of P_1, P_2, P_3, P_4, we obtain a collection P of grid points of $[a,b]\times[c,d]$ with $W(P,f) = W(P_1,f_1)+W(P_2,f_2)+W(P_3,f_3)+W(P_4,f_4)$. In particular, $0 \le W(P_i,f_i) \le W(P,f) \le W(f)$ for $i = 1,2,3,4$. This shows that each f_i is of bounded bivariation and $W(f_i) \le W(f)$ for $i = 1,2,3,4$. Moreover, $W(P,f) \le W(f_1) + W(f_2) + W(f_3) + W(f_4)$.

Next, observe that if Q is any collection of grid points of $[a,b]\times[c,d]$ and if P is obtained by adjoining (x,y) to Q, then $W(Q,f) \le W(P,f)$. Now P can be regarded as a collection of grid points of $[a,b] \times [c,d]$ obtained by collating certain collections of grid points of the rectangles $[a,x] \times [c,y]$, $[a,x] \times [y,d]$, $[x,b] \times [c,y]$, $[x,b] \times [y,d]$, and hence

$$W(Q,f) \le W(P,f) \le W(f_1) + W(f_2) + W(f_3) + W(f_4).$$

Since Q is an arbitrary collection of grid points of $[a,b] \times [c,d]$, it follows that $W(f) \le W(f_1)+W(f_2)+W(f_3)+W(f_4)$. On the other hand, given any $\epsilon > 0$, we can find collections P_1, P_2, P_3, P_4 of grid points of $[a,x]\times[c,y]$, $[a,x]\times[y,d]$, $[x,b] \times [c,y]$, $[x,b] \times [y,d]$, respectively, such that $W(f_i) - \frac{\epsilon}{4} < W(P_i,f_i)$ for $i = 1,2,3,4$. Hence, if P denotes the collection of grid points of $[a,b] \times [c,d]$ obtained by collating P_1, P_2, P_3, P_4, then we have

$$\begin{aligned} &W(f_1) + W(f_2) + W(f_3) + W(f_4) - \epsilon \\ &\quad < W(P_1,f_1) + W(P_2,f_2) + W(P_3,f_3) + W(P_4,f_4) = W(P,f) \le W(f). \end{aligned}$$

Since $\epsilon > 0$ is arbitrary, $W(f) = W(f_1) + W(f_2) + W(f_3) + W(f_4)$. □

If $f : [a,b] \times [c,d] \to \mathbb{R}$ is of bounded bivariation, then we define the corresponding **total bivariation function** $w_f : [a,b] \times [c,d] \to \mathbb{R}$ by $w_f(x,y) := W\big(f|_{[a,x]\times[c,y]}\big)$. The following result gives the so-called **Jordan decomposition** of a function of bounded bivariation.

Proposition 1.17. *If $f : [a,b] \times [c,d] \to \mathbb{R}$ is of bounded bivariation, then there are unique functions $g, h : [a,b] \times [c,d] \to \mathbb{R}$ such that g and h are bimonotonically increasing, $f = g - h$, and $w_f = g + h$.*

Proof. Define $g,h : [a,b] \times [c,d] \to \mathbb{R}$ by $g = \frac{1}{2}(w_f + f)$ and $h = \frac{1}{2}(w_f - f)$. Clearly, $f = g - h$ and $w_f = g + h$. Consider any $(x_1,y_1),(x_2,y_2)$ in $[a,b] \times [c,d]$ with $(x_1,y_1) \le (x_2,y_2)$. Then the rectangle $R := [a,x_2] \times [c,y_2]$ has four subrectangles $R_1 := [a,x_1] \times [c,y_1]$, $R_2 := [a,x_1] \times [y_1,y_2]$, $R_3 := [x_1,x_2]\times[c,y_1]$, and $R_4 := [x_1,x_2]\times[y_1,y_2]$. (See Figure 1.6.) Applying Lemma 1.16 to $f|_R$, that is, to the restriction of f to R, we see that $w_f(x_2,y_2) = W_1 + W_2 + W_3 + W_4$, where $W_i := W(f|_{R_i})$ for $i = 1,2,3,4$. Moreover, applying Lemma 1.16 to $f|_{[a,x_2]\times[c,y_1]}$ as well as to $f|_{[a,x_1]\times[c,y_2]}$, we also see that $w_f(x_2,y_1) = W_1 + W_3$ and $w_f(x_1,y_2) = W_1 + W_2$. Now let us write

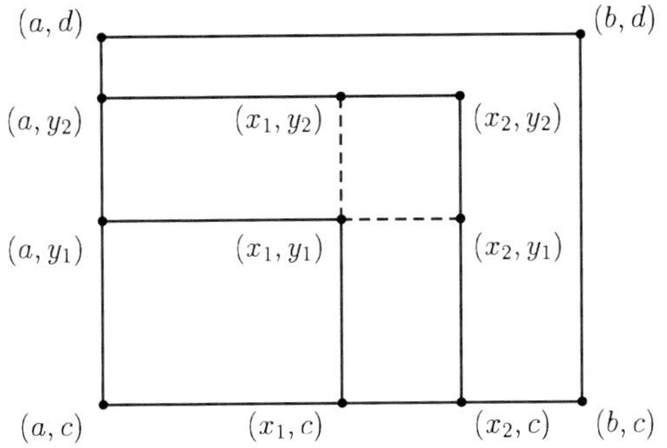

Fig. 1.6. Typical positions of the points (x_1, y_1) and (x_2, y_2) in the proof of Proposition 1.17 and the corresponding subrectangles.

$$g(x_2, y_2) + g(x_1, y_1) - g(x_2, y_1) - g(x_1, y_2) = \frac{1}{2}(A + B)$$

and

$$h(x_2, y_2) + h(x_1, y_1) - h(x_2, y_1) - h(x_1, y_2) = \frac{1}{2}(A - B),$$

where

$$A := w_f(x_2, y_2) + w_f(x_1, y_1) - w_f(x_2, y_1) - w_f(x_1, y_2)$$

and

$$B := f(x_2, y_2) + f(x_1, y_1) - f(x_2, y_1) - f(x_1, y_2).$$

Since $A = (W_1 + W_2 + W_3 + W_4) + W_1 - (W_1 + W_3) - (W_1 + W_2) = W_4$ and $|B| \le W_4$, it follows that g and h are bimonotonically increasing. The uniqueness of g and h is obvious, since $f = g - h$ and $w_f = g + h$. □

Remark 1.18. For further results on functions of bounded bivariation, see Exercises 22, 43, 44, and 46–49. ◇

Examples 1.19. (i) Consider $f : [0, 1] \times [0, 1] \to \mathbb{R}$ defined by

$$f(x, y) := \begin{cases} (1/x) + (1/y) & \text{if } x \neq 0 \text{ and } y \neq 0, \\ 1/x & \text{if } x \neq 0 \text{ and } y = 0, \\ 1/y & \text{if } x = 0 \text{ and } y \neq 0, \\ 0 & \text{if } x = 0 \text{ and } y = 0. \end{cases}$$

Then f is of the form $\phi(x) + \psi(y)$, and hence (by part (iii) of Proposition 1.6), f is bimonotonic. In particular, f is of bounded bivariation. But clearly, f is not bounded on any of the four sides of $[0, 1] \times [0, 1]$. (As a consequence, f is not of bounded variation on $[0, 1] \times [0, 1]$.) This shows that the additional hypothesis in part (i) of Proposition 1.15 about boundedness on two adjacent sides of the rectangle cannot be dropped.

(ii) Consider $f, g : [0,1] \times [0,1] \to \mathbb{R}$ defined by

$$f(x,y) := \begin{cases} 1/x & \text{if } x \neq 0, \\ 0 & \text{if } x = 0, \end{cases} \quad \text{and} \quad g(x,y) := y.$$

Clearly f and g are bimonotonic on $[0,1] \times [0,1]$; indeed, each is a function of the form $\phi(x) + \psi(y)$. In particular, f and g are of bounded bivariation. However, fg is not of bounded bivariation. Indeed, if fg were of bounded bivariation, then by considering grids of the form $(x_0, y_0), \ldots, (x_n, y_m)$, where $m = 1$, $y_0 = 0$, and $y_1 = 1$, we see that f would be of bounded variation. But then by part (i) of Proposition 1.10, f would have to be bounded, which is not the case. Thus, we could not have included products in the statement of part (iii) of Proposition 1.15.

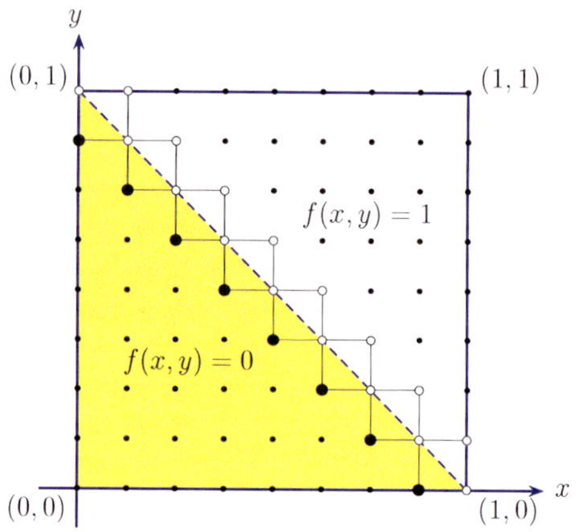

Fig. 1.7. Illustration of the function in Example 1.19 (iii) and the points (x_i, y_i) of the rectangle $[0,1] \times [0,1]$ that straddle the diagonal line $y = 1 - x$.

(iii) Consider $f : [0,1] \times [0,1] \to \mathbb{R}$ defined by

$$f(x,y) := \begin{cases} 0 & \text{if } x + y \leq 1, \\ 1 & \text{if } x + y > 1. \end{cases}$$

Then f is monotonically increasing in $[0,1] \times [0,1]$. Indeed, given any $(x_1, y_1), (x_2, y_2) \in [0,1] \times [0,1]$ with $(x_1, y_1) \leq (x_2, y_2)$, we have $x_1 + y_1 \leq x_2 + y_2$, and hence $x_1 + y_1 > 1$ implies $x_2 + y_2 > 1$. So either $f(x_1, y_1) = 0 \leq f(x_2, y_2)$ or $f(x_1, y_1) = 1$, in which case $f(x_2, y_2) = 1$. Being monotonically increasing, f is of bounded variation on $[0,1] \times [0,1]$. But if we consider points of the rectangle $[0,1] \times [0,1]$ that straddle the diagonal line $y = 1 - x$, then it is seen that there is too much bivariation in the values of f. (See Figure 1.7.) For example, if $n \in \mathbb{N}$ and we let $x_i := i/n$ and $y_j := j/n$ for $0 \le i, j \le n$, then

$$\sum_{i=1}^{n}\sum_{j=1}^{n} |f(x_i, y_j) + f(x_{i-1}, y_{j-1}) - f(x_i, y_{j-1}) - f(x_{i-1}, y_j)|$$
$$\ge \sum_{k=1}^{n} |f(x_k, y_{n-k+1}) + f(x_{k-1}, y_{n-k}) - f(x_k, y_{n-k}) - f(x_{k-1}, y_{n-k+1})|$$
$$= \sum_{k=1}^{n} |1 + 0 - 0 - 0| = n.$$

It follows that f is not of bounded bivariation on $[0,1] \times [0,1]$. ◇

Remark 1.20. The concepts of bimonotonicity and bounded bivariation introduced in this chapter for functions of two variables can be extended to n-fold monotonicity and bounded n-fold variation for functions of n variables. To this end, it is useful to consider the difference operator $\triangle$ defined as follows. Given any $\mathbf{a} = (a_1, \ldots, a_n)$ and $\mathbf{b} = (b_1, \ldots, b_n)$ in $\mathbb{R}^n$ with $\mathbf{a} \le \mathbf{b}$, that is, $a_i \le b_i$ for $i = 1, \ldots, n$, and any $f : [\mathbf{a}, \mathbf{b}] \to \mathbb{R}$, define

$$\triangle_{\mathbf{a}}^{\mathbf{b}} f := \sum_{\mathbf{c}} k(\mathbf{c}) f(\mathbf{c}),$$

where the summation is over all $\mathbf{c} = (c_1, \ldots, c_n) \in \mathbb{R}^n$ such that $c_i \in \{a_i, b_i\}$ for $i = 1, \ldots, n$, and for any such $\mathbf{c}$,

$$k(\mathbf{c}) := k_1 \cdots k_n, \quad \text{where for } 1 \le i \le n, \quad k_i = \begin{cases} 1 & \text{if } c_i = b_i, \\ -1 & \text{if } c_i = a_i. \end{cases}$$

For example, if $n = 1$, then $\triangle_{a_1}^{b_1} f = f(b_1) - f(a_1)$, while if $n = 2$, then

$$\triangle_{(a_1,a_2)}^{(b_1,b_2)} f = f(b_1, b_2) + f(a_1, a_2) - f(b_1, a_2) - f(a_1, b_2),$$

and if $n = 3$, then

$$\triangle_{(a_1,a_2,a_3)}^{(b_1,b_2,b_3)} f = f(b_1, b_2, b_3) + f(b_1, a_2, a_3) + f(a_1, b_2, a_3) + f(a_1, a_2, b_3)$$
$$- f(b_1, b_2, a_3) - f(a_1, b_2, b_3) - f(b_1, a_2, b_3) - f(a_1, a_2, a_3).$$

Now, f is said to be **n-fold monotonically increasing** if $\triangle_{\mathbf{x}}^{\mathbf{y}} f \ge 0$ for all $\mathbf{x}, \mathbf{y} \in [\mathbf{a}, \mathbf{b}]$ with $\mathbf{x} \le \mathbf{y}$. The remaining concepts are defined analogously. ◇

Convexity and Concavity

The notions of convex and concave functions from one-variable calculus admit a straightforward analogue to functions of several variables, provided we discuss convexity and concavity of a function on convex subsets of its domain.

Let $D \subseteq \mathbb{R}^2$ and $f : D \to \mathbb{R}$ be any function. Also, let A be a convex subset of D. We say that

1. f is **convex** on A if for all $(x_1, y_1), (x_2, y_2) \in A$ and $t \in (0,1)$, we have

$$f((1-t)(x_1, y_1) + t(x_2, y_2)) \leq (1-t)f(x_1, y_1) + tf(x_2, y_2),$$

2. f is **concave** on A if for all $(x_1, y_1), (x_2, y_2) \in A$ and $t \in (0,1)$, we have

$$f((1-t)(x_1, y_1) + t(x_2, y_2)) \geq (1-t)f(x_1, y_1) + tf(x_2, y_2).$$

Changing the inequalities $\leq$ and $\geq$ to strict inequalities $<$ and $>$, respectively, in 1 and 2 above, we obtain the notions of **strictly convex** and **strictly concave** functions.

Geometrically speaking, convex functions are those whose graph lies below the triangle in the plane determined by three points on the graph. More precisely, if $D \subseteq \mathbb{R}^2$ is convex and not a line segment in $\mathbb{R}^2$, then $f : D \to \mathbb{R}$ is convex on D if and only if for any noncollinear points $P_i := (x_i, y_i)$, $i = 1, 2, 3$, in D and any (x, y) in the triangle with P_1, P_2, P_3 as its vertices, we have $f(x, y) \leq g(x, y)$, where $z = g(x, y)$ is the equation of the plane passing through $(x_i, y_i, f(x_i, y_i))$ for $i = 1, 2, 3$. Similarly for concave functions. (See Exercise 50.)

Examples 1.21. (i) If $f : \mathbb{R}^2 \to \mathbb{R}$ is the **norm function** on $\mathbb{R}^2$ given by $f(x, y) := \sqrt{x^2 + y^2}$ for $(x, y) \in \mathbb{R}^2$, then by part (v) of Proposition 1.1 we see that f is convex on $\mathbb{R}^2$.

(ii) Let I, J be intervals in $\mathbb{R}$. Then $I \times J$ is a convex set in $\mathbb{R}^2$. Further, if $\phi : I \to \mathbb{R}$ is convex on I and $\psi : J \to \mathbb{R}$ is convex on J, then the function $f : I \times J \to \mathbb{R}$ defined by $f(x, y) := \phi(x) + \psi(y)$ is convex on $I \times J$. For instance, $f : \mathbb{R}^2 \to \mathbb{R}$ defined by $f(x, y) := e^x + |y|$ for $(x, y) \in \mathbb{R}^2$ is convex on $\mathbb{R}^2$.

(iii) If $D \subseteq \mathbb{R}^2$ is convex and $f : D \to \mathbb{R}$ is any function, then f is concave on D if and only if $-f$ is convex on D. Using this, (i) and (ii) above give rise to examples of concave functions. ◇

Local Extrema and Saddle Points

The notions of local maxima and local minima for functions of one variable extend easily to functions of two (or more) variables. Geometrically, the local extrema correspond to the peaks and dips on the graph. Moreover, a new and interesting phenomenon emerges, namely, that of a *saddle point*. To define the latter, we first introduce some terminology concerning paths in $\mathbb{R}^2$.

Let Γ be a path in $\mathbb{R}^2$ given by $(x(t), y(t))$, $t \in [\alpha, \beta]$. The path Γ is said to **pass through** a point $(x_0, y_0) \in \mathbb{R}^2$ if there is $t_0 \in (\alpha, \beta)$ such that $(x(t_0), y(t_0)) = (x_0, y_0)$. As in one-variable calculus (for example, Section 4.1 of ACICARA), we say that the **tangent** to Γ at a point $(x(t_0), y(t_0))$, where $t_0 \in (\alpha, \beta)$, is **defined** if x, y are differentiable at t_0 and $(x'(t_0), y'(t_0)) \neq (0, 0)$. In this case, we will refer to the pair $(x'(t_0), y'(t_0))$ as the **tangent vector** to Γ at $(x(t_0), y(t_0))$. In general, we will say that Γ is a **regular path** if the tangent is always defined, that is, if the functions x, y are differentiable on (α, β) and $(x'(t), y'(t)) \neq (0,0)$ for all $t \in (\alpha, \beta)$.

Let $D \subseteq \mathbb{R}^2$ and let (x_0, y_0) be an interior point of D. Suppose Γ lies in D, that is, $(x(t), y(t)) \in D$ for all $t \in [\alpha, \beta]$, and Γ passes through (x_0, y_0), that is, $(x_0, y_0) = (x(t_0), y(t_0))$ for some $t_0 \in (\alpha, \beta)$. Given any $f : D \to \mathbb{R}$, the function $F : [\alpha, \beta] \to \mathbb{R}$ given by $F(t) := f(x(t), y(t))$ is sometimes referred to as the **restriction** of f to the path Γ. We shall say that f has a **local maximum at** (x_0, y_0) **along** Γ if F has a local maximum at t_0. Likewise, we say that f has a **local minimum at** (x_0, y_0) **along** Γ if F has a local minimum at t_0.

Suppose Γ_1 and Γ_2 are regular paths in $\mathbb{R}^2$ given by $(x_1(t), y_1(t))$, $t \in [\alpha_1, \beta_1]$, and by $(x_2(t), y_2(t))$, $t \in [\alpha_2, \beta_2]$, respectively. Also, suppose both Γ_1 and Γ_2 pass through a point $(x_0, y_0) \in \mathbb{R}^2$, so that there are $t_i \in (\alpha_i, \beta_i)$ with $(x(t_i), y(t_i)) = (x_0, y_0)$ for $i = 1, 2$. Then Γ_1 and Γ_2 are said to **intersect transversally** at (x_0, y_0) if the tangent vectors at (x_0, y_0) are defined and are not multiples of each other, that is, $(x_1'(t_1), y_1'(t_1))$ and $(x_2'(t_2), y_2'(t_2))$ are both different from $(0,0)$ and there is no $\lambda \in \mathbb{R}$ such that $(x_1'(t_1), y_1'(t_1)) = \lambda(x_2'(t_2), y_2'(t_2))$.

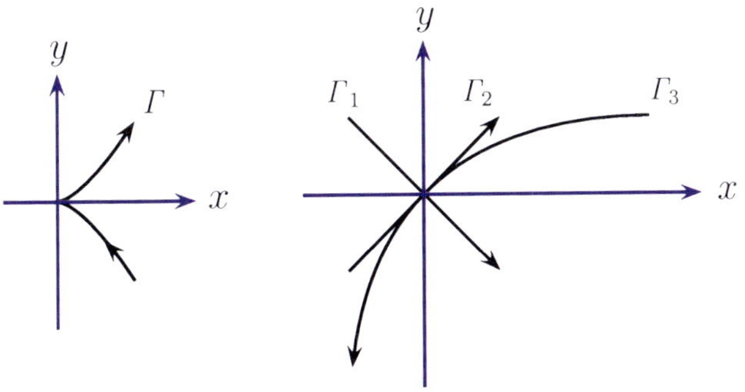

Fig. 1.8. A nonregular path Γ and regular paths Γ_1, Γ_2, Γ_3; note that Γ_1 and Γ_2 as well as Γ_1 and Γ_3 intersect transversally, but Γ_2 and Γ_3 do not.

For example, the path Γ given by (t^2, t^3), $t \in [-1, 1]$, is not a regular path. (See Figure 1.8.) On the other hand, consider the paths Γ_1, Γ_2, and Γ_3 given by $(t, -t)$, (t, t), and $(2t + t^2, 2t - t^2)$, respectively, with $t \in [-1, 1]$ in each case. Each of these paths is regular and passes through the origin $(0, 0)$. The tangent vectors to Γ_1, Γ_2, and Γ_3 at the origin are $(1, -1)$, $(1, 1)$, and $(2, 2)$, respectively. Hence Γ_1 and Γ_2 intersect transversally at $(0, 0)$; also, Γ_1 and Γ_3 intersect transversally at $(0, 0)$, but Γ_2 and Γ_3 do not intersect transversally at $(0, 0)$. (See Figure 1.8.)

Let $D \subseteq \mathbb{R}^2$ and let (x_0, y_0) be an interior point of D. We say that a function $f : D \to \mathbb{R}$ has

1. a **local maximum** at (x_0, y_0) if there is $\delta > 0$ such that $\mathbb{S}_\delta(x_0, y_0) \subseteq D$ and $f(x, y) \leq f(x_0, y_0)$ for all $(x, y) \in \mathbb{S}_\delta(x_0, y_0)$,
2. a **local minimum** at (x_0, y_0) if there is $\delta > 0$ such that $\mathbb{S}_\delta(x_0, y_0) \subseteq D$ and $f(x, y) \geq f(x_0, y_0)$ for all $(x, y) \in \mathbb{S}_\delta(x_0, y_0)$,
3. a **saddle point** at (x_0, y_0) if there are regular paths Γ_1 and Γ_2 lying in D and intersecting transversally at (x_0, y_0) such that f has a local maximum at (x_0, y_0) along Γ_1, while f has a local minimum at (x_0, y_0) along Γ_2.

As in the case of functions of one variable, we can define stronger versions of the above notions in which the adjective *strict* is added. Thus, $f : D \to \mathbb{R}$ has a **strict local maximum** at (x_0, y_0) of D if there is $\delta > 0$ such that $\mathbb{S}_\delta(x_0, y_0) \subseteq D$ and $f(x, y) > f(x_0, y_0)$ for all $(x, y) \in \mathbb{S}_\delta(x_0, y_0)$ with $(x, y) \neq (x_0, y_0)$. A **strict local minimum** is defined similarly. Further, f is said to have a **strict saddle point** at (x_0, y_0) if there are regular paths Γ_1 and Γ_2 lying in D and intersecting transversally at (x_0, y_0) such that f has a strict local maximum at (x_0, y_0) along Γ_1, while f has a strict local minimum at (x_0, y_0) along Γ_2.

Examples 1.22. (i) The function $f : \mathbb{R}^2 \to \mathbb{R}$ defined by $f(x, y) := -(x^2 + y^2)$ has a local maximum at $(0, 0)$. (See Figure 1.4.)

(ii) The function $f : \mathbb{R}^2 \to \mathbb{R}$ defined by $f(x, y) := x^2+y^2$ has a local minimum at $(0, 0)$. (See Figure 1.2.)

(iii) The function $f : \mathbb{R}^2 \to \mathbb{R}$ defined by $f(x, y) := xy$ has a saddle point at $(0, 0)$. (See Figure 1.9.) To see this, consider the paths Γ_1 and Γ_2 given by $(t, -t)$, $t \in [-1, 1]$, and by (t, t), $t \in [-1, 1]$, respectively. We have seen that these are regular paths in $\mathbb{R}^2$ that intersect transversally at $(0, 0)$. Moreover, we have $f(t, -t) = -t^2$ and $f(t, t) = t^2$ for $t \in [-1, 1]$. Hence f has a local maximum at $(0, 0)$ along Γ_1, and a local minimum at $(0, 0)$ along Γ_2.

(iv) The function $f : \mathbb{R}^2 \to \mathbb{R}$ defined by $f(x, y) := x^4 + y^3$ has neither a local maximum nor a local minimum at $(0, 0)$. To see this, note that $f(0, 0) = 0$ and f takes both positive as well as negative values in any open square centered at the origin. [For example, $f(r, 0) = r^4 > 0$ and $f(0, -r) = -r^3 < 0$ for any $r > 0$.] It turns out that f does not have a saddle point at $(0, 0)$. (See Figure 1.9.) In fact, if Γ_1 and Γ_2 are any regular paths in $\mathbb{R}^2$ such that f has a local maximum at $(0, 0)$ along Γ_1 and a local minimum at $(0, 0)$ along Γ_2, then it can be shown that Γ_1 and Γ_2 do not intersect transversally at the origin. A proof of this will be given later, in Example 4.13 (v) of Chapter 4. ◇

In the first three examples above, a stronger conclusion is valid. Namely, in Examples 1.22 (i), (ii), and (iii), f has a strict local maximum, a strict local minimum, and a strict saddle point, respectively, at $(0, 0)$.

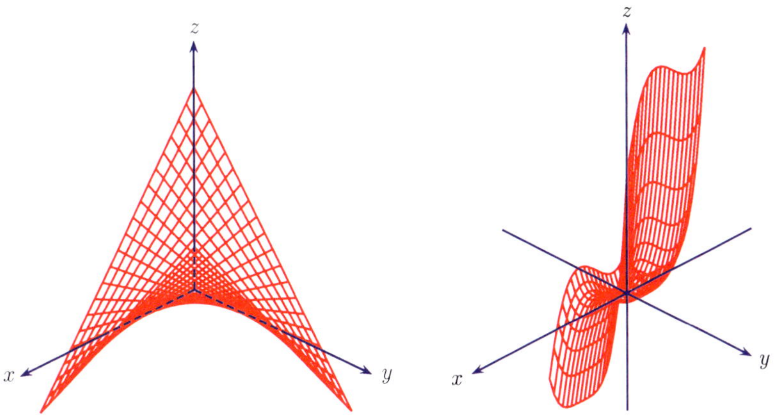

g

Fig. 1.9. Graphs of $f(x, y) := xy$ and $f(x, y) := x^4 + y^3$.

Intermediate Value Property

Let us begin by recalling from one-variable calculus that if $D \subseteq \mathbb{R}$, then a function $\phi : D \to \mathbb{R}$ is said to have the Intermediate Value Property (IVP) on an interval $I \subseteq D$ if for any $a, b \in I$ and any $r \in \mathbb{R}$ between $\phi(a)$ and $\phi(b)$, there is $c \in I_{a,b}$ such that $r = \phi(c)$. This notion has the following straightforward analogue for functions of several variables.

Let $D \subseteq \mathbb{R}^2$. A function $f : D \to \mathbb{R}$ is said to have the **Intermediate Value Property**, or in short, the **IVP**, on a 2-interval $I \times J \subseteq D$ if for any $(x_1, y_1), (x_2, y_2) \in I \times J$ and any $r \in \mathbb{R}$ between $f(x_1, y_1)$ and $f(x_2, y_2)$, there is $(x_0, y_0) \in I_{(x_1,y_1),(x_2,y_2)}$ such that $r = f(x_0, y_0)$.

Proposition 1.23. *Let $D \subseteq \mathbb{R}^2$ and let $f : D \to \mathbb{R}$ be a function. Then for any 2-interval $I \times J \subseteq D$,*

$$f \text{ has the IVP on } I \times J \Longrightarrow f(I \times J) \text{ is an interval in } \mathbb{R}.$$

Proof. Let $a, b \in f(I \times J)$. Then $a = f(x_1, y_1)$ and $b = f(x_2, y_2)$ for some $(x_1, y_1), (x_2, y_2) \in I \times J$. If $r \in I_{a,b}$, then by the IVP of f on $I \times J$, there is $(x_0, y_0) \in I_{(x_1,y_1),(x_2,y_2)}$ such that $f(x_0, y_0) = r$. Since $I_{(x_1,y_1),(x_2,y_2)} \subseteq I \times J$, we see that $I_{a,b} \subseteq f(I \times J)$. This proves that $f(I \times J)$ is an interval. □

Remark 1.24. It is easy to see that the converse of Proposition 1.23 is not true. In fact, in contrast to one-variable calculus, the converse is not true even for monotonic functions. (Compare Remark 1.21 in ACICARA.) For example, consider $I = J = [0, 1]$ and $f : I \times J \to \mathbb{R}$ defined by $f(x, y) := [x] + y$, where $[x]$ denotes the integer part of x. Clearly, f is monotonic as well as bimonotonic. Moreover, $f(I \times J) = [0, 2]$ is an interval in $\mathbb{R}$. Note, however, that the real number $\frac{3}{4}$ lies between $f(0, 0) = 0$ and $f(1, \frac{1}{2}) = \frac{3}{2}$, but $\frac{3}{4}$ is not the value of f at any point on the 2-interval $I_{(0,0),(1,\frac{1}{2})} = [0,1] \times [0, \frac{1}{2}]$. Indeed, the image of this 2-interval is $[0, \frac{1}{2}] \cup [1, \frac{3}{2}]$, which is not an interval in $\mathbb{R}$. ◇

As indicated by the example in Remark 1.24, a simple modification of the result in Proposition 1.23 yields a characterization of the IVP.

Proposition 1.25. *Let $D \subseteq \mathbb{R}^2$ and let $f : D \to \mathbb{R}$ be a function. Then for any 2-interval $I \times J \subseteq D$,*

$$f \text{ has the IVP on } I \times J \iff f(E) \text{ is an interval in } \mathbb{R} \text{ for every 2-interval } E \subseteq I \times J.$$

Proof. The implication "$\Longrightarrow$" follows from Proposition 1.23. To prove the converse, let $(x_1, y_1), (x_2, y_2) \in I \times J$ and let r be a real number that lies between $f(x_1, y_1)$ and $f(x_2, y_2)$. Let E denote the 2-interval $I_{(x_1,y_1),(x_2,y_2)}$. Since $f(E)$ is an interval in $\mathbb{R}$, it follows that $r = f(x_0, y_0)$ for some $(x_0, y_0) \in E$. Since $E \subseteq I \times J$, it follows that f has the IVP on $I \times J$. □

1.3 Cylindrical and Spherical Coordinates

For points in $\mathbb{R}^2$, one has the familiar notion of polar coordinates. These provide an alternative and useful way to represent points in the plane other than the origin. Recall that the polar coordinates of $(x, y) \in \mathbb{R}^2 \setminus \{(0,0)\}$ are given by (r, θ), where r and θ are real numbers determined by the equations $x = r\cos\theta$ and $y = r\sin\theta$ and the conditions $r > 0$ and $\theta \in (-\pi, \pi]$. The precise relationship is stated below. For a proof of this, one may refer to Proposition 7.20 of ACICARA.

Fact 1.26. *If $x, y \in \mathbb{R}$ are such that $(x, y) \neq (0, 0)$, then r and θ defined by*

$$r := \sqrt{x^2 + y^2} \quad \text{and} \quad \theta := \begin{cases} \cos^{-1}\left(\dfrac{x}{r}\right) & \text{if } y \geq 0, \\[2ex] -\cos^{-1}\left(\dfrac{x}{r}\right) & \text{if } y < 0, \end{cases}$$

satisfy the following properties:

$$r, \theta \in \mathbb{R}, \quad r > 0, \quad \theta \in (-\pi, \pi], \quad x = r\cos\theta, \quad \text{and} \quad y = r\sin\theta.$$

Conversely, if $r, \theta \in \mathbb{R}$ are such that $r > 0$ and $\theta \in (-\pi, \pi]$, then $x := r\cos\theta$ and $y := r\sin\theta$ are real numbers such that $(x, y) \neq (0,0)$, $r = \sqrt{x^2 + y^2}$, and θ equals $\cos^{-1}(x/r)$ or $-\cos^{-1}(x/r)$ according as $y \geq 0$ or $y < 0$.

In the 3-space $\mathbb{R}^3$, there are at least two important and useful representations of points, and these are known as cylindrical coordinates and spherical coordinates. Of these, the former is a straightforward extension of the notion of polar coordinates, and we describe it first.

Cylindrical Coordinates

The **cylindrical coordinates** of a point that is not on the z-axis, that is, a point (x, y, z) in $\mathbb{R}^3$ for which $(x, y) \neq (0, 0)$, are defined to be the triple (r, θ, z), where (r, θ) are the polar coordinates of (x, y). Thus the cylindrical coordinates are related to the rectangular coordinates (x, y, z) by the equations

$$x = r\cos\theta, \quad y = r\sin\theta, \quad \text{and} \quad z = z,$$

where the real numbers r and θ satisfy the conditions

$$r > 0 \quad \text{and} \quad \theta \in (-\pi, \pi].$$

As an immediate consequence of Fact 1.26, we obtain a one-to-one correspondence between the sets

$$\{(x, y, z) \in \mathbb{R}^3 : (x, y) \neq (0, 0)\} \quad \text{and} \quad \{(r, \theta, z) \in \mathbb{R}^3 : r > 0,\ \theta \in (-\pi, \pi]\}.$$

The equations for r, θ, and z in terms of the rectangular coordinates are

$$r = \sqrt{x^2 + y^2}, \quad \theta = \begin{cases} \cos^{-1}\left(\dfrac{x}{r}\right) & \text{if } y \geq 0, \\ -\cos^{-1}\left(\dfrac{x}{r}\right) & \text{if } y < 0, \end{cases} \quad \text{and} \quad z = z.$$

Notice that if we fix $r_0 > 0$, then the points whose cylindrical coordinates (r, θ, z) satisfy $r = r_0$ constitute a cylinder of radius r_0 with its axis along the z-axis. See, for instance, the picture on the left in Figure 1.10.

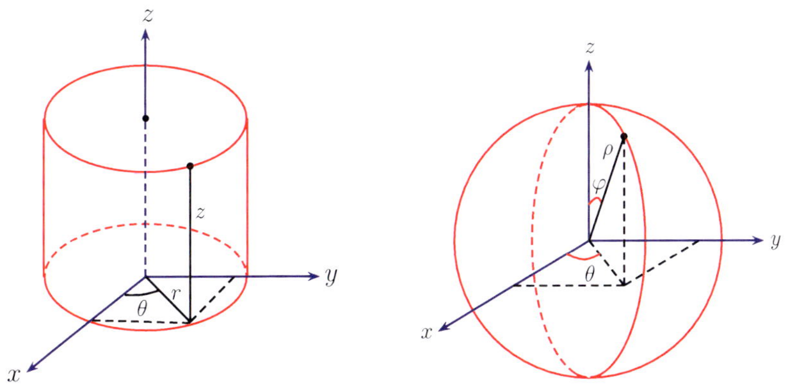

Fig. 1.10. Illustrations of cylindrical and spherical coordinates.

Spherical Coordinates

The **spherical coordinates** of a point that is not on the z-axis, that is, a point (x, y, z) in $\mathbb{R}^3$ for which $(x, y) \neq (0, 0)$, are defined to be the triple (r, θ, φ) in $\mathbb{R}^3$ determined by the equations

$$x = \rho \sin\varphi \cos\theta, \quad y = \rho \sin\varphi \sin\theta, \quad z = \rho \cos\varphi,$$

and the conditions

$$\rho, \varphi, \theta \in \mathbb{R}, \quad \rho > 0, \quad \varphi \in (0, \pi), \quad \theta \in (-\pi, \pi].$$

Geometrically speaking, ρ is the distance from the vector (x, y, z) to the origin $(0, 0, 0)$, while φ is the angle made by the vector (x, y, z) with the vector $(0, 0, 1)$ on positive z-axis, and θ is the angle made in the xy-plane by the vector $(x, y, 0)$ with the vector $(1, 0, 0)$ on the positive x-axis. See, for instance, the picture on the right in Figure 1.10. The following proposition justifies the above definition and describes the precise relationship between rectangular coordinates and spherical coordinates.

Proposition 1.27. *If $x, y, z \in \mathbb{R}$ with $(x, y) \neq (0, 0)$, then ρ, φ, θ defined by*

$$\rho := \sqrt{x^2 + y^2 + z^2}, \quad \varphi := \cos^{-1}\frac{z}{\rho}, \quad \theta := \begin{cases} \cos^{-1}\left(\dfrac{x}{\rho\sin\varphi}\right) & \text{if } y \geq 0, \\[2ex] -\cos^{-1}\left(\dfrac{x}{\rho\sin\varphi}\right) & \text{if } y < 0, \end{cases}$$

satisfy the conditions

$$\rho, \varphi, \theta \in \mathbb{R}, \quad \rho > 0, \quad \varphi \in (0, \pi), \quad \theta \in (-\pi, \pi],$$

and the equations

$$x = \rho \sin\varphi \cos\theta, \quad y = \rho \sin\varphi \sin\theta, \quad z = \rho \cos\varphi.$$

Conversely, if $\rho, \varphi, \theta \in \mathbb{R}$ are such that $\rho > 0$, $\varphi \in (0, \pi)$, and $\theta \in (-\pi, \pi]$, then the real numbers x, y, z defined by

$$x := \rho \sin\varphi \cos\theta, \quad y := \rho \sin\varphi \sin\theta, \quad z := \rho \cos\varphi,$$

are such that $(x, y) \neq (0, 0)$, $\rho = \sqrt{x^2 + y^2 + z^2}$, $\varphi = \cos^{-1}(z/\rho)$, and θ equals $\cos^{-1}(x/\rho\sin\varphi)$ or $-\cos^{-1}(x/\rho\sin\varphi)$ according as $y \geq 0$ or $y < 0$.

Proof. Suppose $x, y, z \in \mathbb{R}$ with $(x, y) \neq (0, 0)$ are given. Define ρ, φ, and θ by the formulas displayed above. Since $(x, y) \neq (0, 0)$, we see that $\rho > 0$ and $|z/\rho| < 1$. Consequently, $\varphi := \cos^{-1}(z/\rho) \in (0, \pi)$. Clearly, $z = \rho\cos\varphi$, and by Fact 1.26, we see that $(\rho\sin\varphi, \theta)$ are the polar coordinates of (x, y). Hence $\theta \in (-\pi, \pi]$ and moreover, $x = \rho\sin\varphi\cos\theta$ and $y = \rho\sin\varphi\sin\theta$.

Conversely, suppose $\rho, \varphi, \theta \in \mathbb{R}$ are such that $\rho > 0$, $\varphi \in (0, \pi)$, and $\theta \in (-\pi, \pi]$. Define $x := \rho \sin\varphi \cos\theta$, $y := \rho \sin\varphi \sin\theta$, and $z := \rho \cos\varphi$. Then $x^2 + y^2 = \rho^2 \sin^2\varphi > 0$, and hence $(x, y) \neq (0, 0)$. Also, it is clear that

$$\rho = \sqrt{x^2 + y^2 + z^2} \quad \text{and} \quad \varphi = \cos^{-1}(z/\rho).$$

Finally, applying Fact 1.26 with $r := \rho \sin\varphi$, we readily see that θ equals $\cos^{-1}(x/\rho \sin\varphi)$ or $-\cos^{-1}(x/\rho \sin\varphi)$ according as $y \geq 0$ or $y < 0$. □

As an immediate consequence of Proposition 1.27, we obtain a one-to-one correspondence between the sets $\{(x, y, z) \in \mathbb{R}^3 : (x, y) \neq (0, 0)\}$ and $\{(\rho, \varphi, \theta) \in \mathbb{R}^3 : \rho > 0,\ \varphi \in (0, \pi),\ \theta \in (\pi, \pi]\}$.

Remark 1.28. The arguments used in the proof of Proposition 1.27 also give the following relation between the cylindrical and the spherical coordinates of a point. Let P be a point of $\mathbb{R}^3$ that is not on the z-axis. If (ρ, φ, θ) are the spherical coordinates of P, then the cylindrical coordinates of P are given by (r, θ, z), where θ is common to both sets of coordinates, while the other coordinates are determined by the relations

$$r = \rho \sin\varphi \quad \text{and} \quad z = \rho \cos\varphi.$$

On the other hand, if the cylindrical coordinates of P are (r, θ, z), then the spherical coordinates are given by (ρ, φ, θ), where θ is common to both sets of coordinates, while the other coordinates are determined by the relations

$$\rho = \sqrt{r^2 + z^2} \quad \text{and} \quad \varphi = \cos^{-1}\left(z/\sqrt{r^2 + z^2}\right).$$

These formulas can also be verified directly. ◇

Notes and Comments

A course in multivariable calculus generally proceeds along the lines of a course in one-variable calculus. Several of the notions and results in the setting of $\mathbb{R}$ *have a natural analogue in the context of* $\mathbb{R}^n$*. However, there are some fundamental differences. For example, as mentioned in the text, there is, in general, no reasonable notion whatsoever of division in* $\mathbb{R}^n$*. There have been attempts to understand this phenomenon and sometimes to overcome the obstacles. For example, if* $n = 2$*, then by defining multiplication suitably, one can ensure that division by nonzero vectors is possible. This leads to the so-called complex numbers, and in turn to an important and beautiful subject known as complex analysis. A famous theorem that goes back to Frobenius (1878) asserts that for* $n > 2$*, a reasonable notion of multiplication in* $\mathbb{R}^n$ *is possible only* *when $n = 4$ and $n = 8$, and this leads to the "quaternions" and "octonions," respectively. However, when $n = 4$, the multiplication is not commutative and when $n = 8$, it is neither commutative nor associative. For an in-depth look at numbers in general, and complex numbers in particular, one can consult the book [16] by Ebbinghaus et al. For more on quaternions, octonions, and the theorem of Frobenius, one can consult the book of Kantor and Solodovnikov [33] and the expository article of Baez [4].*

Yet another difference between $\mathbb{R}$ and $\mathbb{R}^n$ is the apparent absence of a natural order on $\mathbb{R}^n$ for $n > 1$. There are, of course, total orders on $\mathbb{R}^n$, such as the lexicographic order on $\mathbb{R}^n$ (Exercise 1), that are compatible with the algebraic operations, but they fail to satisfy the archimedean property and the least upper bound property. In fact, Hölder showed in 1901 that there cannot be an archimedean total order compatible with addition on $\mathbb{R}^n$ if $n > 1$. More precisely, he proved that a totally ordered group is archimedean if and only if it is order-isomorphic to a subgroup of $\mathbb{R}$. For more details and a proof of this result, we refer to the book of Fuchs [20]. We have argued in this text that the product order or the componentwise order on $\mathbb{R}^n$ is a suitable extension of the natural order on $\mathbb{R}$. The product order is only a partial order, but it is compatible with the addition and scalar multiplication on $\mathbb{R}^n$ and satisfies the least upper bound property and a weak form of the archimedean property. It is more suitable for analysis on $\mathbb{R}^n$ because unlike the lexicographic order, which gives progressively less importance to later components, the product order treats all the components equitably. We have used the product order on $\mathbb{R}^n$ to discuss monotonicity for functions of several variables. Moreover, we have discussed an interesting variant of monotonicity for functions of two variables, called here bimonotonicity. Continuing on this theme, we have also considered functions of bounded variation and the so-called functions of bounded bivariation. These notions go back to Arzelà (1905) and Vitali (1908). The interested reader may consult Sections 254–256 of Hobson's treatise [32], the survey papers [9, 10] of Clarkson and Adams, and Section 4 in Chapter III of Hildebrand's book [31].

As in ACICARA*, we have defined geometric notions such as local extrema before derivatives enter the picture. The notion of a saddle point is also treated in the same vein, and our definition does not involve partial derivatives or the discriminant. In fact, this definition differs from the definitions found in most texts. But arguably it is more natural and geometric. We shall revisit saddle points and explain this point further in Chapter 4. Cylindrical and spherical coordinates are introduced in this chapter and are handled them with some care and precision, just as we treated polar coordinates in Chapter 7 of* ACICARA.

Exercises

Part A

1. The **lexicographic order** or the **dictionary order** on $\mathbb{R}^n$ is defined as follows. For $\mathbf{x} = (x_1, \ldots, x_n)$ and $\mathbf{y} = (y_1, \ldots, y_n)$ in $\mathbb{R}^n$, we define $\mathbf{x} \preceq \mathbf{y}$ if

either $\mathbf{x} = \mathbf{y}$ or if the first nonzero coordinate in $\mathbf{y} - \mathbf{x}$ is positive. We may write $\mathbf{x} \succeq \mathbf{y}$ as an equivalent form of $\mathbf{y} \preceq \mathbf{x}$. Show that $\preceq$ is a total order on $\mathbb{R}^n$. Further, show that $\preceq$ is compatible with the algebraic operations in the sense that for any $\mathbf{x}, \mathbf{y} \in \mathbb{R}^n$ with $\mathbf{x} \preceq \mathbf{y}$, we have $\mathbf{x} + \mathbf{z} \preceq \mathbf{y} + \mathbf{z}$ for all $\mathbf{z} \in \mathbb{R}^n$, and also, $c\mathbf{x} \preceq c\mathbf{y}$ or $c\mathbf{x} \succeq c\mathbf{y}$ according as $c \geq 0$ or $c \leq 0$.

2. **(Parallelogram Law)** Show that $|\mathbf{x} + \mathbf{y}|^2 + |\mathbf{x} - \mathbf{y}|^2 = 2|\mathbf{x}|^2 + 2|\mathbf{y}|^2$ for all $\mathbf{x}, \mathbf{y} \in \mathbb{R}^n$.
3. Let $\mathbf{c} = (c_1, \ldots, c_n) \in \mathbb{R}^n$ and $r \in \mathbb{R}$ with $r > 0$.
   (i) Show that $\mathbb{B}_r(\mathbf{c}) \subseteq \mathbb{S}_r(\mathbf{c}) \subseteq \mathbb{B}_{r\sqrt{n}}(\mathbf{c})$.
   (ii) If we let $\mathbb{M}_r(\mathbf{c}) := \{\mathbf{x} \in \mathbb{R}^n : |x_1 - c_1| + \cdots + |x_n - c_n| < r\}$, then show that $\mathbb{M}_r(\mathbf{c}) \subseteq \mathbb{B}_r(\mathbf{c}) \subseteq \mathbb{M}_{rn}(\mathbf{c})$. (Hint: $(a_1 + \cdots + a_n)^2 \leq n(a_1^2 + \cdots + a_n^2)$ for any $a_1, \ldots, a_n \in [0, \infty)$; see, for example, page 34 of ACICARA.)
   (iii) Sketch the sets in (i) and (ii) above when $n = 2$, $r = 1$, and $\mathbf{c} = (0, 0)$.
4. Give an example of a subset D of $\mathbb{R}^2$ such that D is not a 2-interval, but D has the property that $I_{\mathbf{a},\mathbf{b}} \subseteq D$ for every $\mathbf{a}, \mathbf{b} \in D$ with $\mathbf{a} \leq \mathbf{b}$.
5. Show that if I and J are intervals in $\mathbb{R}$, then the complement of their product, that is, the set $D := \mathbb{R}^2 \setminus (I \times J) = \{(x, y) \in \mathbb{R}^2 : (x, y) \notin I \times J\}$, is path-connected. Further, show that if I and J are nonempty bounded intervals in $\mathbb{R}$, then D is not convex.
6. Let $r, s \in \mathbb{R}$ with $0 < r < s$, and let $\mathbf{c} \in \mathbb{R}^n$. Show that the sets $\mathbb{B}_s(\mathbf{c}) \setminus \mathbb{B}_r(\mathbf{c})$ and $\mathbb{S}_s(\mathbf{c}) \setminus \mathbb{S}_r(\mathbf{c})$ are path-connected, but not convex.
7. Let $k \in \mathbb{N}$ and $P_1, \ldots, P_k \in \mathbb{R}^n$. A **convex combination** of $P_1, \ldots, P_k$ is an element in $\mathbb{R}^n$ of the form $\lambda_1 P_1 + \cdots + \lambda_k P_k$, where $\lambda_1, \ldots, \lambda_k \in [0, 1]$ and $\lambda_1 + \cdots + \lambda_k = 1$. Show that if $D \subseteq \mathbb{R}^n$ is convex, then the convex combination of any k points in D is in D.
8. Let $S \subseteq \mathbb{R}^n$. The **convex hull** of S in $\mathbb{R}^n$ is defined as the set of all convex combinations of finitely many elements of S. Show that the convex hull of S is a convex, and hence a path-connected, subset of $\mathbb{R}^n$. Deduce that a line segment in $\mathbb{R}^n$ as well as a triangle in $\mathbb{R}^n$, being the convex hull of two distinct points or three noncollinear points, is a convex subset of $\mathbb{R}^n$.
9. Describe the level curves and contour lines for $f : D \to \mathbb{R}$ corresponding to the values $c = -3, -2, -1, 0, 1, 2, 3, 4$, where $f(x, y)$ is given by
   (i) $x - y$, (ii) xy, (iii) $x^2 + y^2$, (iv) y/x, (v) $\sqrt{5 - x^2 - y^2}$,
   and where $D := \mathbb{R}^2$ in (i), (ii), and (iii), while $D := \{(x, y) \in \mathbb{R}^2 : x \neq 0\}$ in (iv) and $D := \{(x, y) \in \mathbb{R}^2 : x^2 + y^2 \leq 5\}$ in (v).
10. Let $D \subseteq \mathbb{R}^2$ and let $f : D \to \mathbb{R}$ be a function. Fix $(x_0, y_0) \in D$ and let $D_1 := \{x \in \mathbb{R} : (x, y_0) \in D\}$ and $D_2 := \{y \in \mathbb{R} : (x_0, y) \in D\}$. Consider $\phi : D_1 \to \mathbb{R}$ and $\psi : D_2 \to \mathbb{R}$ defined by $\phi(x) := f(x, y_0)$ for $x \in D_1$ and $\psi(y) := f(x_0, y)$ for $y \in D_2$. Show that if f is a rational function, then both ϕ and ψ are rational functions. Also show that if f is an algebraic function, then both ϕ and ψ are algebraic functions.
11. Show that $f : \mathbb{R}^2 \to \mathbb{R}$ defined by $f(x, y) := |xy|$ is not a rational function, and $g : \mathbb{R}^2 \to \mathbb{R}$ defined by $g(x, y) := \sin(xy)$ is a transcendental function.
12. Consider $D \subseteq \mathbb{R}^2$ and $f : D \to \mathbb{R}$ defined by either of the following. Determine in each case whether f is bounded above. If it is, then find an

upper bound. Also determine whether f is bounded below. If it is, then find a lower bound. Further, determine whether f attains its upper bound or lower bound.

(i) $D := \mathbb{S}_1(0,0)$ and $f(x,y) := x^2 + y^2 - 1$,
(ii) $D := \mathbb{S}_\pi(0,0)$ and $f(x,y) := \sin(xy)$,
(iii) $D := \mathbb{S}_{\pi/4}(0,0)$ and $f(x,y) := \tan(x+y)$.

13. Let I, J be nonempty intervals in $\mathbb{R}$. Given any $\phi : I \to \mathbb{R}$ and $\psi : J \to \mathbb{R}$, define $f, g : I \times J \to \mathbb{R}$ by $f(x,y) := \phi(x) + \psi(y)$ and $g(x,y) := \phi(x)\psi(y)$.
    (i) Show that f is bounded on $I \times J$ if and only if ϕ is bounded on I and ψ is bounded on J.
    (ii) Assume that ϕ and ψ are not identically zero, that is, $\phi(x_0) \neq 0$ and $\psi(y_0) \neq 0$ for some $x_0 \in I$ and some $y_0 \in J$. Show that g is bounded on $I \times J$ if and only if ϕ is bounded on I and ψ is bounded on J.
14. Let I and J be nonempty intervals in $\mathbb{R}$ and let $f : I \times J \to \mathbb{R}$ be a function that is monotonically increasing as well as monotonically decreasing on $I \times J$. Show that f is a constant function.
15. Let I, J be nonempty intervals in $\mathbb{R}$. Given any $\phi : I \to \mathbb{R}$ and $\psi : J \to \mathbb{R}$, define $f, g : I \times J \to \mathbb{R}$ by $f(x,y) := \phi(x) + \psi(y)$ and $g(x,y) := \phi(x)\psi(y)$.
    (i) Show that f is monotonically decreasing on $I \times J$ if and only if ϕ is decreasing on I and ψ is decreasing on J.
    (ii) Show that if ϕ is decreasing on I and ψ is decreasing on J, and if $\phi(x) \geq 0$ and $\psi(y) \geq 0$ for all $x \in I$ and $y \in J$, then g is monotonically decreasing on $I \times J$. Conversely, show that if g is monotonically decreasing on $I \times J$ and if $\phi(x) > 0$ and $\psi(y) > 0$ for all $x \in I$ and $y \in J$, then ϕ is decreasing on I and ψ is decreasing on J. Give an example to show that this converse may not be true if ϕ and ψ are nonnegative but not strictly positive.
16. Consider $f : \mathbb{R}^2 \to \mathbb{R}$ defined by $f(x,y) := x - y$ for $(x,y) \in \mathbb{R}^2$. Is f monotonic on $\mathbb{R} \times \mathbb{R}$? Is f bimonotonic on $\mathbb{R} \times \mathbb{R}$? Justify your answers.
17. Let $p \in \mathbb{R}$ and let $g : (1,\infty) \times (1,\infty) \to \mathbb{R}$ be defined by $g(x,y) := [\ln(x+y)]^p$ for $(x,y) \in (1,\infty) \times (1,\infty)$. Show that g is monotonically decreasing and bimonotonically increasing if $p \leq 0$, whereas g is monotonically increasing and bimonotonically decreasing if $0 \leq p \leq 1$. What can be said about g if $p > 1$? (Hint: Proposition 1.7.)
18. Consider $f : [0,2] \times [0,2] \to \mathbb{R}$ defined by

$$f(x,y) := \begin{cases} (x+1)(y+1) & \text{if } x+y \geq 2, \\ xy & \text{if } x+y < 2. \end{cases}$$

    Show that f is monotonically increasing on $[0,2] \times [0,2]$, but f is not bimonotonic on $[0,2] \times [0,2]$.
19. Let I and J be closed and bounded intervals in $\mathbb{R}$. Show that a monotonic function on the rectangle $I \times J$ is bounded. Give an example to show that a bimonotonic function on $I \times J$ need not be bounded.

20. Let $f : [a,b] \times [c,d] \to \mathbb{R}$ be a function on the rectangle $[a,b] \times [c,d]$. For a fixed $y \in [c,d]$, let $\phi_y : [a,b] \to \mathbb{R}$ denote the function defined by $\phi_y(x) := f(x,y)$. Also, for a fixed $x \in [a,b]$, let $\psi_x : [c,d] \to \mathbb{R}$ denote the function defined by $\psi_x(y) := f(x,y)$. Assume that f is bimonotonically increasing and prove the following.
   (i) If ϕ_c is increasing on $[a,b]$, then so is ϕ_y for every $y \in [c,d]$.
   (ii) If ψ_a is increasing on $[c,d]$, then so is ψ_x for every $x \in [a,b]$.
   (iii) If ϕ_c is increasing on $[a,b]$ and ψ_a is increasing on $[c,d]$, then f is monotonically increasing on $[a,b] \times [c,d]$.
21. Consider $f : [0,1] \times [0,1] \to \mathbb{R}$ defined by $f(x,y) := 0$ if $x \le y$ and $f(x,y) := 1$ if $x > y$. We have seen in Example 1.13 that f is not of bounded variation. Show that f is not of bounded bivariation. (Hint: Example 1.19 (iii))
22. Let $f : [a,b] \times [c,d] \to \mathbb{R}$ be of bounded bivariation. Show that the corresponding total variation function w_f is monotonically as well as bimonotonically increasing on $[a,b] \times [c,d]$.
23. Let $D := \{(x,y) \in \mathbb{R}^2 : x > 0 \text{ and } y > 0\}$ and $f : D \to \mathbb{R}$ be defined by $f(x,y) := \ln(1/xy)$. Show that D is a convex subset of $\mathbb{R}^2$ and f is a convex function on D. (Hint: $f(x,y) := \ln(1/x) + \ln(1/y)$.)
24. Let $D \subseteq \mathbb{R}^2$ be a convex set containing $(0,0)$ and let $f : D \to \mathbb{R}$ be a convex function. Show that for any $(x,y) \in D$, we have $(-x,-y) \in D$ and $f(-x,-y) \ge -f(x,y)$.
25. **(Jensen's inequality)** Let $D \subseteq \mathbb{R}^2$ be convex and let $f : D \to \mathbb{R}$ be any function. Given any $k \in \mathbb{N}$ with $k > 1$, show that f is convex on D if and only if $f(\lambda_1 P_1 + \cdots + \lambda_k P_k) \le \lambda_1 f(P_1) + \cdots + \lambda_k f(P_k)$ for all $\lambda_1, \ldots, \lambda_k \in [0,1]$ with $\lambda_1 + \cdots + \lambda_k = 1$.
26. Show that $f : \mathbb{R}^2 \to \mathbb{R}$ defined by $f(x,y) := x^2 - y^3$ has neither a local maximum nor a local minimum at $(0,0)$.
27. Show that $f : \mathbb{R}^2 \to \mathbb{R}$ defined by $f(x,y) := x^2 - y^2$ has a saddle point at $(0,0)$.
28. Recall that for any $a \in \mathbb{R}$, the integer part of a is denoted by $[a]$. Consider $f : \mathbb{R}^2 \to \mathbb{R}$ defined by one of the following. In each case, determine whether f has the IVP on $\mathbb{R}^2$.
   (i) $f(x,y) := x + y$, (ii) $f(x,y) := [x] + [y]$, (iii) $f(x,y) = x + [y]$.
29. Let $R := [0,2] \times [0,2]$ and let $f : R \to \mathbb{R}$ be defined by

$$f(x,y) := \begin{cases} 0 & \text{if } 0 \le x \le 1 \text{ and } 0 \le y \le 1, \\ x-1 & \text{if } 1 < x \le 2 \text{ and } 0 \le y \le 1, \\ 1 & \text{if } 0 \le x \le 2 \text{ and } 1 < y \le 2. \end{cases}$$

   Show that f is monotonic, $f(R)$ is an interval, but f does not have the IVP on R.
30. Find the cylindrical coordinates as well as the spherical coordinates of the points in $\mathbb{R}^3$ whose Cartesian coordinates are as follows:
   (i) $(1,0,0)$, (ii) $(0,1,0)$, (iii) $(1,1,0)$, (iv) $(1,-1,0)$.

31. Let $c \in \mathbb{R}$ be given. Describe geometrically the surface defined by the following equations in cylindrical coordinates:
(i) $r = c$, (ii) $\theta = c$, (iii) $z = c$,
and by the following equations in spherical coordinates:
(i) $\rho = c$, (ii) $\varphi = c$, (iii) $\theta = c$.

Part B

32. Consider the lexicographic order $\preceq$ on $\mathbb{R}^n$ defined in Exercise 1. Given any $\mathbf{x}, \mathbf{y} \in \mathbb{R}^n$, we will write $\mathbf{x} \prec \mathbf{y}$ if $\mathbf{x} \preceq \mathbf{y}$ and $\mathbf{x} \neq \mathbf{y}$; likewise, we will write $\mathbf{x} \succ \mathbf{y}$ if $\mathbf{x} \succeq \mathbf{y}$ and $\mathbf{x} \neq \mathbf{y}$. Show that if $n > 1$, then $\preceq$ is not archimedean, that is, show that there are $\mathbf{x}, \mathbf{y} \in \mathbb{R}^n$ with $\mathbf{x} \succ \mathbf{0}$ and $\mathbf{y} \succ \mathbf{0}$ such that $k\mathbf{x} \prec \mathbf{y}$ for all $k \in \mathbb{N}$. Show, however, that the following weak version holds. For any $\mathbf{x} = (x_1, \ldots, x_n)$ and $\mathbf{y} = (y_1, \ldots, y_n)$ in $\mathbb{R}^n$ with $x_1 > 0$, there is $k \in \mathbb{N}$ such that $k\mathbf{x} \succ \mathbf{y}$. Further, show that if $n > 1$, then $\mathbb{R}^n$ does not satisfy the least upper bound property with respect to $\preceq$, that is, there is a nonempty subset S of $\mathbb{R}^n$ such that S is bounded above but S does not have a supremum with respect to $\preceq$.
33. Given any $p \in \mathbb{R}$ with $p \geq 1$ and $\mathbf{x} = (x_1, \ldots, x_n) \in \mathbb{R}^n$, define the p-norm of $\mathbf{x}$ by
$$\|\mathbf{x}\|_p := \left(|x_1|^p + \cdots + |x_n|^p\right)^{1/p}.$$
Show that $\|\mathbf{x}\|_p \geq 0$, and moreover, $\|\mathbf{x}\|_p = 0 \iff \mathbf{x} = 0$. Also, show that for any $\mathbf{x}, \mathbf{y} \in \mathbb{R}^n$ and $r \in \mathbb{R}$, we have
$$\|\mathbf{x} + \mathbf{y}\|_p \leq \|\mathbf{x}\|_p + \|\mathbf{y}\|_p \quad \text{and} \quad \|r\mathbf{x}\|_p = |r|\,\|\mathbf{x}\|_p.$$
(Hint: The first assertion is essentially the Minkowski inequality for sums; see, for example, page 281 of ACICARA.)
34. Extend the p-norm defined in Exercise 33 to the case $p = \infty$ as follows.
$$\|\mathbf{x}\|_\infty := \max\left(|x_1|, \ldots, |x_n|\right) \quad \text{for } \mathbf{x} = (x_1, \ldots, x_n) \in \mathbb{R}^n.$$
Given any p, q with $1 \leq p \leq q \leq \infty$, prove that
$$\|\mathbf{x}\|_q \leq \|\mathbf{x}\|_p \leq n^\lambda \|\mathbf{x}\|_q \quad \text{for all } \mathbf{x} \in \mathbb{R}^n,$$
where $\lambda := 1/p$ if $p < q = \infty$ and $\lambda := (q-p)/pq$ if $p \leq q < \infty$, while $\lambda := 0$ if $p = q = \infty$. (Hint: For the first inequality, reduce to the case $\|\mathbf{x}\|_p = 1$, and note that $p \leq q$ and $|x_i| \leq 1$ implies $|x_i|^q \leq |x_i|^p$. For the second inequality, use the power mean inequality; see, for example, page 286 of ACICARA.)
35. Let p be such that $1 \leq p \leq \infty$. For $\mathbf{c} \in \mathbb{R}^n$ and $r > 0$, let
$$\mathbb{B}_r^{(p)}(\mathbf{c}) := \{\mathbf{x} \in \mathbb{R}^n : \|\mathbf{x} - \mathbf{c}\|_p < r\}.$$
Show that if q is such that $p \leq q \leq \infty$ and λ is as in Exercise 34 above, then $\mathbb{B}_r^{(p)}(\mathbf{c}) \subseteq \mathbb{B}_r^{(q)}(\mathbf{c}) \subseteq \mathbb{B}_{rn^\lambda}^{(p)}(\mathbf{c})$. Show that the inclusions in parts (i) and (ii) of Exercise 3 follow as particular cases.

36. Let $p(x_1, \ldots, x_n)$ be a polynomial in n variables $x_1, \ldots, x_n$ with coefficients in $\mathbb{R}$. This means that $p(x_1, \ldots, x_n)$ is a finite sum of terms of the form $cx_1^{i_1}x_2^{i_2}\cdots x_n^{i_n}$, where $c \in \mathbb{R}$ and $i_1, \ldots, i_n$ are nonnegative integers; here c is called the **coefficient** of the term and in case $c \neq 0$, the sum $i_1 + \cdots + i_n$ is called the **total degree** of the term. By a **zero** or a **root** of $p(x_1, \ldots, x_n)$ in $\mathbb{R}^n$ we mean a point $\mathbf{a} = (a_1, \ldots, a_n) \in \mathbb{R}^n$ such that $p(a_1, \ldots, a_n) = 0$, that is, by substituting a_i in place of x_i for each $i = 1, \ldots, n$ in $p(x_1, \ldots, x_n)$, we obtain the value 0.
    (i) Show that if $n = 1$ and $p(x_1)$ is a nonzero polynomial (that is, not all of its coefficients are zero), then it has at most finitely many zeros.
    (ii) Give an example to show that if $n > 1$, then there can be a nonzero polynomial in n variables with infinitely many zeros in $\mathbb{R}^n$.
    (iii) If there are subsets $E_1, \ldots, E_n$ of $\mathbb{R}$ such that E_j is an infinite set for each $j = 1, \ldots, n$ and we have $p(a_1, \ldots, a_n) = 0$ for all $(a_1, \ldots, a_n) \in E_1 \times \cdots \times E_n$, then show that $p(x_1, \ldots, x_n)$ must be the zero polynomial, that is, all its coefficients are zero.

    Now suppose f is a polynomial function on a subset D of $\mathbb{R}^n$, that is, suppose there is a polynomial $p(x_1, \ldots, x_n)$ in n variables with coefficients in $\mathbb{R}$ such that $f(a_1, \ldots, a_n) = p(a_1, \ldots, a_n)$ for all $(a_1, \ldots, a_n) \in D$. Show that if $D = E_1 \times \cdots \times E_n$, where E_j is an interval containing more than one point in $\mathbb{R}$ for each $j = 1, \ldots, n$, then the polynomial $p(x_1, \ldots, x_n)$ is uniquely determined by the function f.
37. Let (x_0, y_0) and (x_1, y_1) be any two points in $\mathbb{R}^2$ and let Γ be a path joining them, that is, Γ is the path given by $(x(t), y(t))$, $t \in [\alpha, \beta]$, where $x, y : [\alpha, \beta] \to \mathbb{R}$ are continuous functions with $(x(\alpha), y(\alpha)) = (x_0, y_0)$ and $(x(\beta), y(\beta)) = (x_1, y_1)$. Prove that the image of Γ, that is, the set $\{(x(t), y(t)) : t \in [\alpha, \beta]\}$, is a closed and path-connected subset of $\mathbb{R}^2$.
38. Let I and J be nonempty intervals in $\mathbb{R}$ and let $f : I \times J \to \mathbb{R}$ be any function that is bimonotonically increasing as well as bimonotonically decreasing. Show that there exist functions $\phi : I \to \mathbb{R}$ and $\psi : J \to \mathbb{R}$ such that

$$f(x, y) = \phi(x) + \psi(y) \quad \text{for all } (x, y) \in I \times J.$$

    Further, show that in this case the functions ϕ and ψ are unique up to a constant, that is, if there are functions $\phi_1 : I \to \mathbb{R}$ and $\psi_1 : J \to \mathbb{R}$ such that $f(x, y) = \phi_1(x) + \psi_1(y)$ for all $(x, y) \in I \times J$, then $\phi_1 = \phi + c$ and $\psi_1 = \psi - c$ for some $c \in \mathbb{R}$.
39. Let a and c be any real numbers. Given a function $f : [a, \infty) \times [c, \infty) \to \mathbb{R}$, let $F : [a, \infty) \times [c, \infty) \to \mathbb{R}$ be the function defined by

$$F(x, y) := f(x, y) - f(x, c) - f(a, y) + f(a, c) \quad \text{for } (x, y) \in [a, \infty) \times [c, \infty).$$

    Show that if f is bimonotonic, then F is monotonic.
40. Define a relation $\leq^*$ on $\mathbb{R}^2$ as follows: For (x_1, y_1) and (x_2, y_2) in $\mathbb{R}^2$,

$$(x_1, y_1) \leq^* (x_2, y_2) \quad \text{if} \quad x_1 \leq x_2 \text{ and } y_1 \geq y_2.$$

Show that $\leq^*$ is a partial order on $\mathbb{R}^2$. Next, let I, J be intervals in $\mathbb{R}$ and let $f : I \times J \to \mathbb{R}$ be a function. Define f to be **antimonotonically increasing** on $I \times J$ if

$$(x_1, y_1), (x_2, y_2) \in I \times J \text{ and } (x_1, y_1) \leq^* (x_2, y_2) \Longrightarrow f(x_1, y_1) \leq f(x_2, y_2),$$

and **antimonotonically decreasing** on $I \times J$ if

$$(x_1, y_1), (x_2, y_2) \in I \times J \text{ and } (x_1, y_1) \leq^* (x_2, y_2) \Longrightarrow f(x_1, y_1) \geq f(x_2, y_2).$$

We say that f is **antimonotonic** on $I \times J$ if f is antimonotonically increasing on $I \times J$ or antimonotonically decreasing on $I \times J$. Let $I^* := \{x \in \mathbb{R} : -x \in I\}$ and $J^* := \{y \in \mathbb{R} : -y \in J\}$, and define $f_* : I^* \times J \to \mathbb{R}$ and $f^* : I \times J^* \to \mathbb{R}$ by $f_*(x, y) := f(-x, y)$ and $f^*(x, y) = f(x, -y)$. Show that

$$\begin{aligned} f \text{ is antimonotonic on } I \times J &\Longleftrightarrow f_* \text{ is monotonic on } I^* \times J \\ &\Longleftrightarrow f^* \text{ is monotonic on } I \times J^*. \end{aligned}$$

41. Let $a, b \in \mathbb{R}$ with $a \leq b$. A function $\phi : [a, b] \to \mathbb{R}$ of one variable is said to be of **bounded variation** on $[a, b]$ if the set $S := \{\sum_{i=1}^{n} |\phi(x_i) - \phi(x_{i-1})| : n \in \mathbb{N} \text{ and } 1 = x_0 \leq x_1 \leq \cdots \leq x_n = b\}$ is bounded above in $\mathbb{R}$. In this case, we denote the supremum of S by $V(\phi)$ and call this the **total variation** of ϕ on $[a, b]$. Prove that $\phi : [a, b] \to \mathbb{R}$ is of bounded variation if and only if ϕ is a difference of two monotonically increasing functions.
42. Given any $n \geq 0$, let $\phi_n : [0, 1] \to \mathbb{R}$ be defined by $\phi_n(0) := 0$ and $\phi_n(x) := x^n \sin(1/x)$ for $0 < x \leq 1$. Show that ϕ_0 and ϕ_1 are not of bounded variation on $[0, 1]$, whereas ϕ_2 is of bounded variation on $[0, 1]$.
43. Let $\phi : [a, b] \to \mathbb{R}$ and $\psi : [c, d] \to \mathbb{R}$ be functions of one variable. Define $f, g : [a, b] \times [c, d] \to \mathbb{R}$ by $f(x, y) := \phi(x) + \psi(y)$ and $g(x, y) := \phi(x)\psi(y)$.
   (i) Show that f is of bounded variation on $[a, b] \times [c, d]$ if and only if ϕ is of bounded variation on $[a, b]$ and ψ is of bounded variation on $[c, d]$.
   (ii) Assume that ϕ and ψ are not identically zero, that is, $\phi(x_0) \neq 0$ and $\psi(y_0) \neq 0$ for some $x_0 \in [a, b]$ and $y_0 \in [c, d]$. Show that g is of bounded variation on $[a, b] \times [c, d]$ if and only if ϕ is of bounded variation on $[a, b]$ and ψ is of bounded variation on $[c, d]$.
   (iii) Show that f is always of bounded bivariation on $[a, b] \times [c, d]$.
   (iv) Assume that ϕ and ψ are not constant functions, that is, $\phi(x_*) \neq \phi(x^*)$ and $\psi(y_*) \neq \psi(y^*)$ for some $x_*, x^* \in [a, b]$ with $x_* < x^*$ and $y_*, y^* \in [c, d]$ with $y_* < y^*$. Show that g is of bounded bivariation on $[a, b] \times [c, d]$ if and only if ϕ is of bounded variation on $[a, b]$ and ψ is of bounded variation on $[c, d]$.
44. Let $f : [a, b] \times [c, d] \to \mathbb{R}$ be a function. Show that $V(f) = 0$ if and only if f is a constant function, whereas $W(f) = 0$ if and only if there are functions $\phi : [a, b] \to \mathbb{R}$ and $\psi : [c, d] \to \mathbb{R}$ such that $f(x, y) = \phi(x) + \psi(y)$ for all $(x, y) \in [a, b] \times [c, d]$.

45. Let $f : [a,b] \times [c,d] \to \mathbb{R}$ be of bounded variation. Fix $x^* \in (a,b)$ and let $f_1 = f_{[a,x^*]\times[c,d]}$ and $f_2 = f_{[x^*,b]\times[c,d]}$. Show that $V(f) \le V(f_1) + V(f_2)$. Give an example to show that the inequality can be strict, that is, we can have $V(f) < V(f_1) + V(f_2)$.
46. Let $f : [a,b] \times [c,d] \to \mathbb{R}$ be any function, and let $\phi_y : [a,b] \to \mathbb{R}$ and $\psi_x : [c,d] \to \mathbb{R}$ be as in Exercise 20.
    (i) Show that if f is of bounded variation on $[a,b] \times [c,d]$, then ϕ_y is of bounded variation on $[a,b]$ for each $y \in [c,d]$, and ψ_x is of bounded variation on $[c,d]$ for each $x \in [a,b]$.
    (ii) Use Example 1.13 to show that the converse of the assertion in (i) above is not true.
    (iii) Suppose f is of bounded bivariation on $[a,b] \times [c,d]$, and, in addition, ϕ_c is of bounded variation on $[a,b]$ and ψ_a is of bounded variation on $[c,d]$. Show that f is of bounded variation on $[a,b] \times [c,d]$, and $V(f) \le 2W(f) + V(\phi_c) + V(\psi_a)$. Also, show that there are unique functions $p, q : [a,b] \times [c,d] \to \mathbb{R}$ such that both p and q are monotonically as well as bimonotonically increasing, $f = p - q$, and $w_f = p + q - v_{\phi_c} - v_{\psi_a}$. (Hint: Let g, h be determined by $g + h = w_f$ and $g - h = f - \phi_c - \psi_a$, and let $\alpha_c, \beta_c, \gamma_a, \delta_a$ be determined by $\phi_c = \alpha_c - \beta_c$, $v_{\phi_c} = \alpha_c + \beta_c$, $\psi_a = \gamma_a - \delta_a$, and $v_{\psi_a} = \gamma_a + \delta_a$. Consider $p := g + \alpha_c + \gamma_a$ and $q := h + \beta_c + \delta_a$.)
    (iv) Use Example 1.19 (iii) to show that the converse of the assertion in (iii) above is not true.
47. Let $f : [a,b] \times [c,d] \to \mathbb{R}$ be a function of bounded bivariation. Define $\tilde{f} : [a,b] \times [c,d] \to \mathbb{R}$ by $\tilde{f}(x,y) := f(x,y) - f(a,y) - f(x,c) + f(a,c)$. Show that there are unique functions $\tilde{p}, \tilde{q} : [a,b] \times [c,d] \to \mathbb{R}$ such that $\tilde{p}$ and $\tilde{q}$ are monotonically as well as bimonotonically increasing, $\tilde{f} = \tilde{p} - \tilde{q}$, and $w_{\tilde{f}} = \tilde{p} + \tilde{q}$.
48. Let $f : [a,b] \times [c,d] \to \mathbb{R}$ be a function with the property that there is $\delta > 0$ such that $|f(x,y)| \ge \delta$ for all $(x,y) \in [a,b] \times [c,d]$.
    (i) If f is of bounded variation, then show that so is $1/f$.
    (ii) Give an example in which f is of bounded bivariation, but $1/f$ is not so.
49. Let $f : [a,b] \times [c,d] \to \mathbb{R}$ be a function.
    (i) Suppose there is $K > 0$ such that

$$|f(x_2,y_2) - f(x_1,y_1)| \le K\,(x_2 - x_1 + y_2 - y_1)$$

    for all $(x_1,y_1), (x_2,y_2) \in [a,b] \times [c,d]$ with $(x_1,y_1) \le (x_2,y_2)$. Then show that f is of bounded variation and $V(f) \le K(b - a + d - c)$.
    (ii) Suppose there is $M > 0$ such that

$$|f(x_2,y_2) + f(x_1,y_1) - f(x_2,y_1) - f(x_1,y_2)| \le M\,|x_2 - x_1|\,|y_2 - y_1|$$

    for all $(x_1,y_1), (x_2,y_2) \in [a,b] \times [c,d]$ with $(x_1,y_1) \le (x_2,y_2)$. Then show that f is of bounded bivariation and $W(f) \le M(b-a)(d-c)$.

50. Let $D \subseteq \mathbb{R}^2$ be convex and let $f : D \to \mathbb{R}$ be any function. Assume that D is not a line segment, that is, given any $P_1, P_2 \in D$, there is some $P \in D$ such that $P \neq (1-t)P_1 + tP_2$ for all $t \in \mathbb{R}$, that is, P is not on the line joining P_1 and P_2.
   (i) Let P_1, P_2, P_3 be any noncollinear points in D. Write $P_i := (x_i, y_i)$, $z_i := f(x_i, y_i)$, and $Q_i = (x_i, y_i, z_i)$ for $i = 1, 2, 3$. Show that Q_1, Q_2, Q_3 lie on a plane in $\mathbb{R}^3$ given by $z = Ax + By + C$ for uniquely determined $A, B, C \in \mathbb{R}$. (Hint: Those familiar with determinants may note that P_1, P_2, P_3 are noncollinear if and only if the 3×3 matrix having $(x_i, y_i, 1)$ as its ith row, for $i = 1, 2, 3$, has a nonzero determinant. Further, the equation of the plane in $\mathbb{R}^3$ passing through (x_i, y_i, z_i) is given by $\Delta(x, y, z) = 0$, where $\Delta(x, y, z)$ is the determinant of the 4×4 matrix whose first row is $(x, y, z, 1)$ and $(i+1)$th row is $(x_i, y_i, z_i, 1)$ for $i = 1, 2, 3$.)
   (ii) Show that f is convex on D if and only if for any noncollinear points P_1, P_2, P_3 in D and any (x, y) in the triangle with P_1, P_2, P_3 as its vertices, we have $f(x, y) \leq g(x, y)$, where $z = g(x, y)$ is the equation of the plane passing through $(x_i, y_i, f(x_i, y_i))$ for $i = 1, 2, 3$.
51. Consider $R := [-1, 1] \times [-1, 1]$ and $f, g : R \to \mathbb{R}$ defined by $f(0, 0) := 0$, $g(0, 0) := 0$, while $f(x, y) := 2|x|y/(x^2 + y^2)$ and $g(x, y) := 2|y|y/(x^2 + y^2)$ for $(x, y) \neq (0, 0)$. Show that f has the IVP on R, whereas g does not have the IVP on R. (Hint: Let $E := I_{(x_1, y_1), (x_2, y_2)}$ be a 2-interval in R. If $(0, 0) \notin E$, then consider the restriction of f to the line segment joining (x_1, y_1) and (x_2, y_2) and use the intermediate value theorem of one-variable calculus (Proposition 3.13 of ACICARA). If $(0, 0) \in E$, then observe that $f(\{t\} \times [0, t]) = [0, 1]$ and $f([-t, 0] \times \{-t\}) = [-1, 0]$ for any $t \in (0, 1]$. As for g, consider the image of a line segment on the y-axis.)
52. Let $D \subseteq \mathbb{R}^2$ and let $f : D \to \mathbb{R}$ be a function. Also let C be a convex subset of D. Let us say that f has the **Strong Intermediate Value Property**, or in short, the **SIVP**, on C if for any $(x_1, y_1), (x_2, y_2) \in C$ and any $r \in \mathbb{R}$ between $f(x_1, y_1)$ and $f(x_2, y_2)$, there is a point on the line joining (x_1, y_1) and (x_2, y_2) whose image under f is r, that is, there is $t \in [0, 1]$ such that $r = f((1-t)x_1 + tx_2, (1-t)y_1 + ty_2)$.
   (i) Show that if f has the SIVP on C, then $f(C)$ is an interval in $\mathbb{R}$. Give an example to show that the converse is not true.
   (ii) Show that f has the SIVP on C if and only if $f(E)$ is an interval for every convex subset E of C.
   (iii) In case C is a 2-interval and f has the SIVP on C, then show that f has the IVP on C. Give an example to show that the converse is not true. (Hint: Exercise 51.)
53. Let a be a positive real number. Determine the equation in cylindrical coordinates of a helix on the cylinder $x^2 + y^2 = a^2$ and the equation in spherical coordinates of a great circle on the sphere $x^2 + y^2 + z^2 = a^2$.

2

Sequences, Continuity, and Limits

In this chapter, we introduce the fundamental notions of continuity and limit of a real-valued function of two variables. As in ACICARA, the definitions as well as proofs of basic results will be given using sequences. There are, actually, two possible generalizations of real sequences that seem natural in the setting of two variables. First, functions defined on $\mathbb{N}$ with values in $\mathbb{R}^2$, and second, functions defined on $\mathbb{N}^2$ with values in $\mathbb{R}$. As we shall see, for developing the notions of continuity and limit of a function of two variables, only the former is relevant, and it is studied in this chapter. The study of the latter will be taken up in Chapter 7.

This chapter is organized as follows. Sequences in $\mathbb{R}^2$ are introduced in Section 2.1 below and their fundamental properties, including the Bolzano–Weierstrass Theorem and the Cauchy Criterion, are derived from the corresponding results for sequences in $\mathbb{R}$. We also use the notion of sequence to introduce basic topological notions of closed and open sets, boundary points, and interior points, and also the closure and the interior of subsets of $\mathbb{R}^2$. Section 2.2 deals with the notion of continuity, and it is shown here that continuous functions on path-connected subsets of $\mathbb{R}^2$ or on closed and bounded subsets of $\mathbb{R}^2$ possess several nice properties. An important result known as the Implicit Function Theorem is also proved in this section. Finally, in Section 2.3 we introduce limits of functions of two variables. The definition is given using sequences, while most of the basic properties are proved using a simple observation that the existence of limit of a function at a point is equivalent to the continuity of an associated function at that point.

2.1 Sequences in $\mathbb{R}^2$

A **sequence** in $\mathbb{R}^2$ is a function from $\mathbb{N}$ to $\mathbb{R}^2$. Typically, a sequence in $\mathbb{R}^2$ is denoted by $\big((x_n, y_n)\big)$, $\big((u_n, v_n)\big)$, etc. The value of a sequence $\big((x_n, y_n)\big)$ at $n \in \mathbb{N}$ is given by the element (x_n, y_n) of $\mathbb{R}^2$, and this element is called the

S.R. Ghorpade and B.V. Limaye, *A Course in Multivariable Calculus and Analysis*, Undergraduate Texts in Mathematics, DOI 10.1007/978-1-4419-1621-1_2,

nth **term** of that sequence. In case the terms of a sequence $\big((x_n, y_n)\big)$ lie in a subset D of $\mathbb{R}^2$, then we say that $\big((x_n, y_n)\big)$ is a sequence in D.

The notions of boundedness and convergence extend readily from the setting of sequences in $\mathbb{R}$ to sequences in $\mathbb{R}^2$. Let $\big((x_n, y_n)\big)$ be a sequence in $\mathbb{R}^2$. We say that $\big((x_n, y_n)\big)$ is **bounded** if there is $\alpha \in \mathbb{R}$ such that $|(x_n, y_n)| \le \alpha$ for all $n \in \mathbb{N}$. The sequence $\big((x_n, y_n)\big)$ is said to be **convergent** if there is $(x_0, y_0) \in \mathbb{R}^2$ that satisfies the following condition: For every $\epsilon > 0$ there is $n_0 \in \mathbb{N}$ such that $(x_n, y_n) \in \mathbb{S}_\epsilon(x_0, y_0)$ for all $n \ge n_0$, that is,

$$|x_n - x_0| < \epsilon \text{ and } |y_n - y_0| < \epsilon \quad \text{for all } n \ge n_0.$$

In this case, we say that $\big((x_n, y_n)\big)$ **converges** to (x_0, y_0) or that (x_0, y_0) is a **limit** of $\big((x_n, y_n)\big)$, and write $(x_n, y_n) \to (x_0, y_0)$. If $\big((x_n, y_n)\big)$ does not converge to (x_0, y_0), then we write $(x_n, y_n) \not\to (x_0, y_0)$; if $\big((x_n, y_n)\big)$ is not convergent, then it is said to be **divergent**.

A sequence $\big((x_n, y_n)\big)$ in $\mathbb{R}^2$ gives rise to two sequences (x_n) and (y_n) in $\mathbb{R}$, and vice versa. It turns out that the properties of $\big((x_n, y_n)\big)$ can be completely understood in terms of the properties of the sequences (x_n) and (y_n) in $\mathbb{R}$.

Proposition 2.1. *Given a sequence $\big((x_n, y_n)\big)$ in $\mathbb{R}^2$, we have the following.*

(i) *If $\big((x_n, y_n)\big)$ is convergent, then it has a unique limit.*
(ii) *$\big((x_n, y_n)\big)$ is bounded $\iff$ both (x_n) and (y_n) are bounded.*
(iii) *$\big((x_n, y_n)\big)$ is convergent $\iff$ both (x_n) and (y_n) are convergent. In fact, for $(x_0, y_0) \in \mathbb{R}^2$, we have $(x_n, y_n) \to (x_0, y_0) \iff x_n \to x_0$ and $y_n \to y_0$.*

Proof. Each of (i), (ii), and (iii) is immediate from the definitions. □

As noted in part (i) of Proposition 2.1, if $\big((x_n, y_n)\big)$ is a convergent sequence in $\mathbb{R}^2$, then it has a unique limit in $\mathbb{R}^2$. The limit of $\big((x_n, y_n)\big)$ is sometimes written as $\lim_{n\to\infty}(x_n, y_n)$ or as $\displaystyle\lim_{n\to\infty}(x_n, y_n)$.

Examples 2.2. (i) If $\big((x_n, y_n)\big)$ is a **constant sequence** in $\mathbb{R}^2$, that is, if there is $(x_0, y_0) \in \mathbb{R}^2$ such that $(x_n, y_n) = (x_0, y_0)$ for all $n \in \mathbb{N}$, then clearly, $\big((x_n, y_n)\big)$ is convergent and $(x_n, y_n) \to (x_0, y_0)$.

(ii) If $\big((x_n, y_n)\big)$ is the sequence in $\mathbb{R}^2$ defined by $(x_n, y_n) := (1/n, -1/n)$ for all $n \in \mathbb{N}$, then clearly, $\big((x_n, y_n)\big)$ is convergent and $(x_n, y_n) \to (0, 0)$.

(iii) The sequence $\big((x_n, y_n)\big)$ in $\mathbb{R}^2$ defined by $(x_n, y_n) := (1/n, (-1)^n)$ for all $n \in \mathbb{N}$ is divergent, since the sequence $((-1)^n)$ in $\mathbb{R}$ is divergent. ◇

Basic properties of sequences in $\mathbb{R}^2$ readily follow from the corresponding properties of sequences in $\mathbb{R}$. For ease of reference, we recall the relevant results for sequences in $\mathbb{R}$. For proofs, one may refer to pages 45–47 of ACICARA.

Fact 2.3. *Let (a_n) and (b_n) be sequences in $\mathbb{R}$, and let $a, b, \alpha, \beta \in \mathbb{R}$.*

(i) *If $a_n \to a$ and $b_n \to b$, then $a_n + b_n \to a + b$ and $a_n b_n \to ab$.*

(ii) *If* $a_n \to a$, *then for any* $r \in \mathbb{R}$, *we have* $ra_n \to ra$.
(iii) *If* $a \neq 0$ *and* $a_n \neq 0$ *for all* $n \in \mathbb{N}$, *then* $(1/a_n) \to (1/a)$.
(iv) *Let* $a_n \to a$. *If there is* $\ell \in \mathbb{N}$ *such that* $a_n \geq \alpha$ *for all* $n \geq \ell$, *then* $a \geq \alpha$. *Likewise, if there is* $m \in \mathbb{N}$ *such that* $a_n \leq \beta$ *for all* $n \geq m$, *then* $a \leq \beta$.
(v) *If* $a_n \to a$ *and* $a_n \geq 0$ *for all* $n \in \mathbb{N}$, *then* $a_n^{1/k} \to a^{1/k}$ *for any* $k \in \mathbb{N}$.
(vi) **(Sandwich Theorem in** $\mathbb{R}$**)** *If* (b_n) *and* (c_n) *are sequences such that* $b_n \to a$ *and* $c_n \to a$, *and if there is* $m \in \mathbb{N}$ *such that* $b_n \leq a_n \leq c_n$ *for all* $n \geq m$, *then* $a_n \to a$.

A few of these facts yield the result that sums, dot products, and scalar multiples of sequences in $\mathbb{R}^2$ converge, respectively, to the sums, dot products, and scalar multiples of the corresponding limits.

Proposition 2.4. *Let* $\big((x_n, y_n)\big)$ *and* $\big((u_n, v_n)\big)$ *be sequences in* $\mathbb{R}^2$, *and let* $(x_0, y_0), (u_0, v_0) \in \mathbb{R}^2$.

(i) *If* $(x_n, y_n) \to (x_0, y_0)$ *and* $(u_n, v_n) \to (u_0, v_0)$, *then* $(x_n, y_n) + (u_n, v_n) \to (x_0, y_0) + (u_0, v_0)$ *and* $(x_n, y_n) \cdot (u_n, v_n) \to (x_0, y_0) \cdot (u_0, v_0)$.
(ii) *If* $(x_n, y_n) \to (x_0, y_0)$, *then for any* $r \in \mathbb{R}$, $r(x_n, y_n) \to r(x_0, y_0)$.

Proof. Immediate consequence of part (iii) of Proposition 2.1 together with parts (i) and (ii) of Fact 2.3. □

Analogues of properties of sequences in $\mathbb{R}$ that depend on order relations, are considered in Exercise 2.

Subsequences and Cauchy Sequences

Let $\big((x_n, y_n)\big)$ be a sequence in $\mathbb{R}^2$. If $n_1, n_2, \ldots$ are positive integers such that $n_k < n_{k+1}$ for each $k \in \mathbb{N}$, then the sequence $\big((x_{n_k}, y_{n_k})\big)$, whose terms are $(x_{n_1}, y_{n_1}), (x_{n_2}, y_{n_2}), \ldots$, is called a **subsequence** of $\big((x_n, y_n)\big)$. The sequence $\big((x_n, y_n)\big)$ is said to be **Cauchy** if for every $\epsilon > 0$ there is $n_0 \in \mathbb{N}$ such that $|x_n - x_m| < \epsilon$ and $|y_n - y_m| < \epsilon$ for all $n, m \geq n_0$. It is clear that $\big((x_n, y_n)\big)$ is Cauchy if and only if both (x_n) and (y_n) are Cauchy sequences in $\mathbb{R}$.

Let us recall the following basic facts about sequences in $\mathbb{R}$. For proofs, one may refer to pages 45, 56, and 58 of ACICARA.

Fact 2.5. *Let* (a_n) *be a sequence in* $\mathbb{R}$. *Then we have the following.*

(i) (a_n) *is convergent* $\Longrightarrow$ (a_n) *is bounded.*
(ii) **(Bolzano–Weierstrass Theorem in** $\mathbb{R}$**)** *If* (a_n) *is bounded, then* (a_n) *has a convergent subsequence.*
(iii) (a_n) *is convergent* $\Longleftrightarrow$ (a_n) *is bounded and every convergent subsequence of* (a_n) *has the same limit.*
(iv) **(Cauchy Criterion in** $\mathbb{R}$**)** (a_n) *is Cauchy* $\Longleftrightarrow$ (a_n) *is convergent.*

These facts, in turn, lead to the following results.

Proposition 2.6. *Given a sequence* $\big((x_n, y_n)\big)$ *in* $\mathbb{R}^2$, *we have the following.*

(i) $\big((x_n, y_n)\big)$ *is convergent* $\Longrightarrow$ $\big((x_n, y_n)\big)$ *is bounded.*
(ii) **(Bolzano–Weierstrass Theorem)** *If* $\big((x_n, y_n)\big)$ *is a bounded sequence, then* $\big((x_n, y_n)\big)$ *has a convergent subsequence.*
(iii) $\big((x_n, y_n)\big)$ *is convergent* $\Longleftrightarrow$ $\big((x_n, y_n)\big)$ *is bounded and every convergent subsequence of* $\big((x_n, y_n)\big)$ *has the same limit.*
(iv) **(Cauchy Criterion)** $\big((x_n, y_n)\big)$ *is Cauchy* $\Longleftrightarrow$ $\big((x_n, y_n)\big)$ *is convergent.*

Proof. Clearly, (i) is an immediate consequence of parts (ii) and (iii) of Proposition 2.1 and part (i) of Fact 2.5. To prove (ii), suppose $\big((x_n, y_n)\big)$ is bounded. Then (x_n) is a bounded sequence in $\mathbb{R}$ and hence by part (ii) of Fact 2.5, (x_n) has a convergent subsequence, say (x_{n_k}). Now, (y_n) is a bounded sequence in $\mathbb{R}$ and hence so is (y_{n_k}). So, by part (ii) of Fact 2.5, (y_{n_k}) has a convergent subsequence, say $\big(y_{n_{k_j}}\big)$. Clearly, $\big((x_{n_{k_j}}, y_{n_{k_j}})\big)$ is a convergent subsequence of $\big((x_n, y_n)\big)$. This proves (ii). Next, if $\big((x_n, y_n)\big)$ is convergent, then it is clear that it is bounded and every convergent subsequence of $\big((x_n, y_n)\big)$ has the same limit. To prove the converse, suppose $\big((x_n, y_n)\big)$ is bounded. By (ii), $\big((x_n, y_n)\big)$ has a convergent subsequence. Suppose (x_0, y_0) is the (same) limit for every convergent subsequence of $\big((x_n, y_n)\big)$. If $(x_n, y_n) \not\to (x_0, y_0)$, then there are $\epsilon > 0$ and positive integers $n_1 < n_2 < \cdots$ such that $\max\{|x_{n_k} - x_0|, |y_{n_k} - y_0|\} \geq \epsilon$ for all $k \in \mathbb{N}$. Now, $\big((x_{n_k}, y_{n_k})\big)$ is bounded and hence by (ii), it has a convergent subsequence. Moreover, this subsequence must converge to (x_0, y_0). This is a contradiction. Thus (iii) is proved. Finally, (iv) follows from part (iii) of Proposition 2.1, part (iv) of Fact 2.5, and our earlier observation that $\big((x_n, y_n)\big)$ is Cauchy if and only if both (x_n) and (y_n) are Cauchy sequences in $\mathbb{R}$. □

The result in part (iv) of Proposition 2.6 is sometimes referred to as the **Cauchy completeness** of $\mathbb{R}^2$. A similar result holds for $\mathbb{R}^n$.

Closure, Boundary, and Interior

Let $D \subseteq \mathbb{R}^2$. We say that D is **closed** if every convergent sequence in D converges to a point of D. The set of all points in $\mathbb{R}^2$ that are limits of convergent sequences in D is called the **closure** of D and is denoted by $\overline{D}$. It is clear that D is closed if and only if $\overline{D} = D$. A point of $\mathbb{R}^2$ is said to be a **boundary point** of D if there is a sequence in D that converges to it and also a sequence in $\mathbb{R}^2 \setminus D$ that converges to it. The set of all boundary points of D in $\mathbb{R}^2$ is called the **boundary** of D (in $\mathbb{R}^2$), and is denoted by ∂D. It is easy to see that $\partial D = \partial(\mathbb{R}^2 \setminus D)$, that is, the boundary of a set coincides with the boundary of its complement. A relation between the closure and the boundary is described by the following.

Proposition 2.7. *Given any* $D \subseteq \mathbb{R}^2$, *we have* $\overline{D} = D \cup \partial D$.

Proof. Let $(x, y) \in \overline{D}$. Then there is a sequence in D converging to (x, y). Further, if $(x, y) \notin D$, then the constant sequence $\big((x_n, y_n)\big)$ defined by $(x_n, y_n) = (x, y)$ for all $n \in \mathbb{N}$ gives a sequence in $\mathbb{R}^2 \setminus D$ converging to (x, y), and so in this case, $(x, y) \in \partial D$. It follows that $\overline{D} \subseteq D \cup \partial D$. On the other hand, if $(x, y) \in D \cup \partial D$, then it is clear, using either a constant sequence or the definition of ∂D, that $(x, y) \in \overline{D}$, and so $D \cup \partial D \subseteq \overline{D}$. □

Proposition 2.8. *Let D be a nonempty subset of $\mathbb{R}^2$ such that $D \neq \mathbb{R}^2$. Then ∂D is nonempty.*

Proof. Since D is nonempty, there is some $(x_0, y_0) \in D$, and since $D \neq \mathbb{R}^2$, there is some $(x_1, y_1) \in \mathbb{R}^2 \setminus D$. Consider the line segment joining these two points, that is, consider $L := \{t \in [0, 1] : (1 - t)(x_0, y_0) + t(x_1, y_1) \in D\}$. Then L is a nonempty subset of $\mathbb{R}$ bounded above by 1. Let $t^* := \sup L$ and $(x^*, y^*) := (1 - t^*)(x_0, y_0) + t^*(x_1, y_1)$. We claim that (x^*, y^*) is a boundary point of D. To see this, let (t_n) be a sequence in L such that $t_n \to t^*$. Let $(x_n, y_n) := (1 - t_n)(x_0, y_0) + t_n(x_1, y_1)$ for $n \in \mathbb{N}$. Clearly $\big((x_n, y_n)\big)$ is a sequence in D that converges to (x^*, y^*). Further, if $t^* < 1$, then we can find $s_n \in \mathbb{R}$ for $n \in \mathbb{N}$ such that $s_n \to t^*$ and $t^* < s_n \leq 1$, and we let $(u_n, v_n) := (1 - s_n)(x_0, y_0) + s_n(x_1, y_1)$ for $n \in \mathbb{N}$, whereas if $t^* = 1$, then we let $(u_n, v_n) := (x_1, y_1)$ for $n \in \mathbb{N}$. In any case, we see that $\big((u_n, v_n)\big)$ is a sequence in $\mathbb{R}^2 \setminus D$ that converges to (x^*, y^*). This proves the claim. □

Let D be a subset of $\mathbb{R}^2$ and let (x_0, y_0) be any point of $\mathbb{R}^2$. We say that (x_0, y_0) is an **interior point** of D if $(x_0, y_0) \in D$ and (x_0, y_0) is not a boundary point of D. It is easy to see that (x_0, y_0) is an interior point of D if and only if there is $r > 0$ such that $\mathbb{S}_r(x_0, y_0) \subseteq D$. The **interior** of D is defined to be the set of all interior points of D. Clearly, the interior of D is a subset of D. We say that D is **open** if every point of D is an interior point of D. The following proposition shows the connection between the notions of an open set and a closed set.

Proposition 2.9. *Let $D \subseteq \mathbb{R}^2$. Then D is closed if and only if $\mathbb{R}^2 \setminus D$ is open.*

Proof. First, suppose D is a closed set. Let $(x_0, y_0) \in \mathbb{R}^2 \setminus D$. If (x_0, y_0) is not an interior point of $\mathbb{R}^2 \setminus D$, then there is a sequence $\big((x_n, y_n)\big)$ in the complement of $\mathbb{R}^2 \setminus D$, that is, in D, such that $(x_n, y_n) \to (x_0, y_0)$, and so $(x_0, y_0) \in \overline{D} = D$, which is a contradiction. This proves that $\mathbb{R}^2 \setminus D$ is an open set. Conversely, suppose $\mathbb{R}^2 \setminus D$ is open. Let $\big((x_n, y_n)\big)$ be any sequence in D such that $(x_n, y_n) \to (x_0, y_0)$ for some $(x_0, y_0) \in \mathbb{R}^2$. Then (x_0, y_0) cannot be an interior point of $\mathbb{R}^2 \setminus D$. But since $\mathbb{R}^2 \setminus D$ is open, it follows that $(x_0, y_0) \notin \mathbb{R}^2 \setminus D$, that is, $(x_0, y_0) \in D$. This proves that D is closed. □

Example 2.10. Let $\alpha, \beta \in \mathbb{R}$ with $\alpha > 0$ and $\beta > 0$. Consider the sets $D_1 := \{(x, y) \in \mathbb{R}^2 : |x| \leq \alpha \text{ and } |y| \leq \beta\}$, $D_2 := \{(x, y) \in \mathbb{R}^2 : |x| < \alpha \text{ and } |y| \leq \beta\}$, $D_3 := \{(x, y) \in \mathbb{R}^2 : |x| \leq \alpha \text{ and } |y| < \beta\}$, and $D_4 := \{(x, y) \in \mathbb{R}^2 : |x| < \alpha \text{ and } |y| < \beta\}$. In view of part (iv) of Fact 2.3, we readily see that D_1 is closed, D_4 is open, whereas D_2 and D_3 are neither closed nor open. Further, for each $i = 1, 2, 3, 4$, the closure of D_i is D_1, the interior of D_i is D_4, and the boundary of D_i is the set $\{(x, y) \in \mathbb{R}^2 : |x| = \alpha \text{ and } |y| = \beta\}$. ◇

Remark 2.11. The notions discussed in this section concerning sequences in $\mathbb{R}^2$, closed sets, closure, boundary points, boundary, interior points, interior, and open sets admit a straightforward extension to $\mathbb{R}^3$ and more generally, to $\mathbb{R}^n$ for any $n \in \mathbb{N}$. To avoid a notational conflict, one may denote a sequence in $\mathbb{R}^n$ by $(\mathbf{x}_k)$, where the parameter k runs through $\mathbb{N}$ and $\mathbf{x}_k \in \mathbb{R}^n$ for each $k \in \mathbb{N}$. It may be instructive to formulate precise analogues of the notions and results in this section for $\mathbb{R}^n$ and write down proofs of analogous results in the general case. This may also be a good opportunity to review the results in this section. ◇

2.2 Continuity

Let D be a subset of $\mathbb{R}^2$ and let (x_0, y_0) be any point in D. A function $f : D \to \mathbb{R}$ is said to be **continuous** at (x_0, y_0) if for every sequence $\big((x_n, y_n)\big)$ in D such that $(x_n, y_n) \to (x_0, y_0)$, we have $f(x_n, y_n) \to f(x_0, y_0)$. If f is not continuous at (x_0, y_0), then we say that f is **discontinuous** at (x_0, y_0). When f is continuous at every $(x_0, y_0) \in D$, we say that f is **continuous** on D.

Examples 2.12. (i) If D is any subset of $\mathbb{R}^2$ and $f : D \to \mathbb{R}$ is a constant function on D, that is, if there is $c \in \mathbb{R}$ such that $f(x, y) = c$ for all $(x, y) \in D$, then clearly, f is continuous on D.

(ii) If $f : \mathbb{R}^2 \to \mathbb{R}$ is the norm function given by $f(x, y) := \sqrt{x^2 + y^2}$ for $(x, y) \in \mathbb{R}^2$, then f is continuous on $\mathbb{R}^2$. To see this, let $(x_0, y_0) \in \mathbb{R}$ be any point and let $\big((x_n, y_n)\big)$ be a sequence in $\mathbb{R}^2$ such that $(x_n, y_n) \to (x_0, y_0)$. Then by part (iii) of Proposition 2.1, the sequences (x_n) and (y_n) in $\mathbb{R}$ are such that $x_n \to x_0$ and $y_n \to y_0$. Hence, by parts (i) and (v) of Fact 2.3, we see that $\sqrt{x_n^2 + y_n^2} \to \sqrt{x_0^2 + y_0^2}$. Thus f is continuous on $\mathbb{R}^2$.

(iii) Consider the **coordinate functions** $p_1, p_2 : \mathbb{R}^2 \to \mathbb{R}$ defined by $p_1(x, y) := x$ and $p_2(x, y) := y$ for $(x, y) \in \mathbb{R}^2$. Then by part (iii) of Proposition 2.1, we immediately see that p_1 and p_2 are continuous on $\mathbb{R}^2$.

(iv) Let $D \subseteq \mathbb{R}^2$ and let us fix $(x_0, y_0) \in D$. Consider

$$D_1 := \{x \in \mathbb{R} : (x, y_0) \in D\} \quad \text{and} \quad D_2 := \{y \in \mathbb{R} : (x_0, y) \in D\}.$$

Notice that the set D_1 depends on y_0, whereas D_2 depends on x_0. Given any $f : D \to \mathbb{R}$, let $\phi : D_1 \to \mathbb{R}$ and $\psi : D_2 \to \mathbb{R}$ be functions of one variable defined by

$$\phi(x) := f(x, y_0) \quad \text{for } x \in D_1 \quad \text{and} \quad \psi(y) := f(x_0, y) \quad \text{for } y \in D_2.$$

These functions will play a useful role in the study of the function f of two variables around the point (x_0, y_0). If f is continuous at (x_0, y_0), then

ϕ is continuous at x_0 and ψ is continuous at y_0. To see this, let (x_n) be a sequence in D_1 such that $x_n \to x_0$. Then $(x_n, y_0) \to (x_0, y_0)$ and hence $f(x_n, y_0) \to f(x_0, y_0)$, that is, $\phi(x_n) \to \phi(x_0)$. Thus ϕ is continuous at x_0. Similarly, ψ is continuous at y_0. ◇

Let us recall that the sign of a continuous function of one variable is preserved in a neighborhood of that point. More precisely, we have the following. For a proof, one may refer to page 68 of ACICARA.

Fact 2.13. *Let $E \subseteq \mathbb{R}$, $c \in E$, and let $\phi : E \to \mathbb{R}$ be continuous at c. If $\phi(c) > 0$, then there is $\delta > 0$ such that $\phi(x) > 0$ for all $x \in E \cap (c - \delta, c + \delta)$. Likewise, if $\phi(c) < 0$, then there is $\delta > 0$ such that $\phi(x) < 0$ for all $x \in E \cap (c - \delta, c + \delta)$.*

A similar result holds for functions of two variables.

Lemma 2.14. *Let $D \subseteq \mathbb{R}^2$, $(x_0, y_0) \in D$, and let $f : D \to \mathbb{R}$ be a function that is continuous at (x_0, y_0). If $f(x_0, y_0) > 0$, then there is $\delta > 0$ such that $f(x, y) > 0$ for all $(x, y) \in D \cap \mathbb{S}_\delta(x_0, y_0)$. Likewise, if $f(x_0, y_0) < 0$, then there is $\delta > 0$ such that $f(x, y) < 0$ for all $(x, y) \in D \cap \mathbb{S}_\delta(x_0, y_0)$.*

Proof. First, suppose $f(x_0, y_0) > 0$. If there is no $\delta > 0$ with the desired property, then for each $n \in \mathbb{N}$, we can find $(x_n, y_n) \in D \cap \mathbb{S}_{1/n}(x_0, y_0)$ such that $f(x_n, y_n) \leq 0$. Now $(x_n, y_n) \to (x_0, y_0)$, and since f is continuous at (x_0, y_0), we have $f(x_n, y_n) \to f(x_0, y_0)$. Hence, by part (iv) of Fact 2.3, $f(x_0, y_0) \leq 0$, which is a contradiction. The proof when $f(x_0, y_0) < 0$ is similar. □

Proposition 2.15. *Let $D \subseteq \mathbb{R}^2$, $(x_0, y_0) \in D$, $r \in \mathbb{R}$, and let $f, g : D \to \mathbb{R}$ be continuous at (x_0, y_0). Then $f + g$, rf, and fg are continuous at (x_0, y_0). In case $f(x_0, y_0) \neq 0$, there is $\delta > 0$ such that $f(x, y) \neq 0$ for all $(x, y) \in D \cap \mathbb{S}_\delta(x_0, y_0)$, and the function $1/f : D \cap \mathbb{S}_\delta(x_0, y_0) \to \mathbb{R}$ is continuous at (x_0, y_0). In case there is $\delta > 0$ such that $f(x, y) \geq 0$ for all $(x, y) \in D \cap \mathbb{S}_\delta(x_0, y_0)$, the function $f^{1/k} : D \cap \mathbb{S}_\delta(x_0, y_0) \to \mathbb{R}$ is continuous at (x_0, y_0) for every $k \in \mathbb{N}$.*

Proof. The continuity of $f + g$, rf, and fg at (x_0, y_0) follows readily from parts (i) and (ii) of Fact 2.3. In case $f(x_0, y_0) \neq 0$, we have either $f(x_0, y_0) > 0$ or $f(x_0, y_0) < 0$. Thus, by Lemma 2.14, there is $\delta > 0$ such that $f(x, y) \neq 0$ for all $(x, y) \in D \cap \mathbb{S}_\delta(x_0, y_0)$. Now, by part (iii) of Fact 2.3, we see that the function $1/f : D \cap \mathbb{S}_\delta(x_0, y_0) \to \mathbb{R}$ is continuous at (x_0, y_0). Finally, the assertion about the continuity of $f^{1/k}$ at (x_0, y_0) is a direct consequence of part (v) of Fact 2.3. □

As in the case of functions of one variable, we can easily deduce from Proposition 2.15 the following. Suppose $D \subseteq \mathbb{R}^2$ and $f, g : D \to \mathbb{R}$ are continuous at $(x_0, y_0) \in D$. Then the difference $f - g$ is continuous at (x_0, y_0). Also, if $g(x_0, y_0) \neq 0$, then the quotient f/g is continuous at (x_0, y_0). Further, if there is $\delta > 0$ such that $f(x) \geq 0$ for all $x \in D \cap \mathbb{S}_\delta(x_0, y_0)$, then for every positive rational number r, the function f^r is continuous at (x_0, y_0). Similarly, if $f(x_0, y_0) > 0$, then for every negative rational number r the function f^r is continuous at (x_0, y_0).

Examples 2.16. (i) Using Proposition 2.15 and the above remarks, we see that every polynomial function on $\mathbb{R}^2$ is continuous and every rational function is continuous wherever it is defined, that is, if $p(x, y)$ and $q(x, y)$ are polynomials in two variables and if $D := \{(x, y) \in \mathbb{R}^2 : q(x, y) \neq 0\}$, then the rational function $f : D \to \mathbb{R}$ defined by $f(x, y) := p(x, y)/q(x, y)$ for $(x, y) \in D$ is continuous on D. Moreover, if $E = \{(x, y) \in \mathbb{R}^2 : p(x, y) \geq 0 \text{ and } q(x, y) > 0\}$, then for any $m, n \in \mathbb{N}$, the algebraic function $g : E \to \mathbb{R}$ defined by $g(x, y) := p(x, y)^{1/m}/q(x, y)^{1/n}$ for $(x, y) \in E$, is continuous on E.

(ii) Consider $f : \mathbb{R}^2 \to \mathbb{R}$ defined as follows.

$$f(x, y) := \begin{cases} \dfrac{xy}{x^2 + y^2} & \text{if } (x, y) \neq (0, 0), \\ 0 & \text{if } (x, y) = (0, 0). \end{cases}$$

Then f is not continuous at $(0, 0)$. To see this, consider a sequence in $\mathbb{R}^2$ approaching $(0, 0)$ along the line $y = x$; for example, the sequence $((1/n, 1/n))$. Then $(1/n, 1/n) \to (0, 0)$, but $f(1/n, 1/n) \to 1/2 \neq f(0, 0)$.

(iii) Consider a variant of the function in (ii), namely, $f : \mathbb{R}^2 \to \mathbb{R}$ given by

$$f(x, y) := \begin{cases} \dfrac{x^2 y}{x^2 + y^2} & \text{if } (x, y) \neq (0, 0), \\ 0 & \text{if } (x, y) = (0, 0). \end{cases}$$

Then f is continuous at $(0, 0)$. To see this, note that for any $(x, y) \in \mathbb{R}^2$, we have $x^2 \leq x^2 + y^2$ and consequently, $|f(x, y)| \leq |y|$. Hence if $((x_n, y_n))$ is any sequence in $\mathbb{R}^2$ with $(x_n, y_n) \to (0, 0)$, then $y_n \to 0$, and as a result, $f(x_n, y_n) \to 0 = f(0, 0)$.

(iv) Consider a variant of the function in (iii), namely, $f : \mathbb{R}^2 \to \mathbb{R}$ given by

$$f(x, y) := \begin{cases} \dfrac{x^2 y}{x^4 + y^2} & \text{if } (x, y) \neq (0, 0), \\ 0 & \text{if } (x, y) = (0, 0). \end{cases}$$

Then $f(x, y)$ approaches 0 along every line passing through the origin [indeed, $f(0, y) = 0$ and $f(x, mx) = mx/(x^2 + m^2) \to 0$ as $x \to 0$]. However, f is not continuous at $(0, 0)$. To see this, consider a sequence in $\mathbb{R}^2$ approaching $(0, 0)$ along the parabola $y = x^2$; for example, the sequence $((1/n, 1/n^2))$. Then $(1/n, 1/n^2) \to (0, 0)$, but $f(1/n, 1/n^2) \to 1/2 \neq f(0, 0)$.

(v) Consider a variant of the function in (iv), namely, $f : \mathbb{R}^2 \to \mathbb{R}$ given by

$$f(x,y) := \begin{cases} \dfrac{x^3y}{x^4+y^2} & \text{if } (x,y) \neq (0,0), \\ 0 & \text{if } (x,y) = (0,0). \end{cases}$$

Then f is continuous at $(0,0)$. To see this, use the A.M.-G.M. Inequality (given, for example, on page 12 of ACICARA) to obtain $2|x^2y| \le x^4 + y^2$, and hence $|f(x,y)| \le |x|/2$ for all $(x,y) \in \mathbb{R}^2$. Thus, if $\big((x_n,y_n)\big)$ is any sequence in $\mathbb{R}^2$ with $(x_n,y_n) \to (0,0)$, then we see that $x_n \to 0$ and as a result, $f(x_n,y_n) \to 0 = f(0,0)$. ◇

Composition of Continuous Functions

We now show that the composition of continuous functions is continuous. It may be noted that for functions of two variables, three types of composites are possible. Thus, the following result is stated in three parts.

Proposition 2.17. *Let $D \subseteq \mathbb{R}^2$, $(x_0,y_0) \in D$, and let $f : D \to \mathbb{R}$ be continuous at (x_0,y_0).*

(i) *Suppose $E \subseteq \mathbb{R}$ is such that $f(D) \subseteq E$. If $g : E \to \mathbb{R}$ is continuous at $f(x_0,y_0)$, then $g \circ f : D \to \mathbb{R}$ is continuous at (x_0,y_0).*

(ii) *Suppose $E \subseteq \mathbb{R}$, $t_0 \in E$, and $x, y : E \to \mathbb{R}$ are such that $(x(t),y(t)) \in D$ for all $t \in E$ and $(x(t_0),y(t_0)) = (x_0,y_0)$. If x, y are continuous at t_0, then $F : E \to \mathbb{R}$ defined by $F(t) := f(x(t),y(t))$ is continuous at t_0.*

(iii) *Suppose $E \subseteq \mathbb{R}^2$, $(u_0,v_0) \in E$, and $x, y : E \to \mathbb{R}$ are such that $(x(u,v),y(u,v)) \in D$ for all $(u,v) \in E$ and $(x(u_0,v_0),y(u_0,v_0)) = (x_0,y_0)$. If x, y are continuous at (u_0,v_0), then $F : E \to \mathbb{R}$ defined by $F(u,v) := f(x(u,v),y(u,v))$ is continuous at (u_0,v_0).*

Proof. (i) Suppose E and g satisfy the hypotheses in (i). Let $\big((x_n,y_n)\big)$ be a sequence in D such that $(x_n,y_n) \to (x_0,y_0)$. By the continuity of f at (x_0,y_0), we obtain $f(x_n,y_n) \to f(x_0,y_0)$. Now $(f(x_n,y_n))$ is a sequence in $f(D)$, and hence by the continuity of g at $f(x_0,y_0)$, we obtain $g\left(f(x_n,y_n)\right) \to g\left(f(x_0,y_0)\right)$. So $g \circ f : D \to \mathbb{R}$ is continuous at (x_0,y_0).

(ii) Suppose E, t_0, and the functions x, y satisfy the hypotheses in (ii), and F is as defined in (ii). Let (t_n) be a sequence in E such that $t_n \to t_0$. By the continuity of x and y at t_0, we obtain $x(t_n) \to x(t_0)$ and $y(t_n) \to y(t_0)$. Thus, by part (iii) of Proposition 2.1, $\big(x(t_n),y(t_n)\big)$ is a sequence in D that converges to (x_0,y_0). Hence by the continuity of f at (x_0,y_0), we obtain $f(x(t_n),y(t_n)) \to f(x_0,y_0)$, that is, $F(t_n) \to F(t_0)$. So F is continuous at t_0.

(iii) Suppose E, (u_0,v_0), and the functions x, y satisfy the hypotheses in (iii), and F is as defined in (iii). Let (u_n,v_n) be a sequence in E such that $(u_n,v_n) \to (u_0,v_0)$. By the continuity of x and y at (u_0,v_0), we obtain $x(u_n,v_n) \to x(u_0,v_0)$ and $y(u_n,v_n) \to y(u_0,v_0)$. Thus, by part (iii) of Proposition 2.1, $\big(x(u_n,v_n),y(u_n,v_n)\big)$ is a sequence in D that converges to (x_0,y_0). Hence by the continuity of f at (x_0,y_0), we obtain $f(x(u_n,v_n),y(u_n,v_n)) \to f(x_0,y_0)$, that is, $F(u_n,v_n) \to F(u_0,v_0)$. So F is continuous at (u_0,v_0). □

Examples 2.18. (i) By part (i) of Proposition 2.17, $f : \mathbb{R}^2 \to \mathbb{R}$ defined by $f(x, y) := \sin(xy)$ is continuous at each $(x_0, y_0) \in \mathbb{R}^2$, and $g : \mathbb{R}^2 \to \mathbb{R}$ defined by $g(x, y) := \cos(x + y)$ is continuous at each $(x_0, y_0) \in \mathbb{R}^2$.

(ii) By part (ii) of Proposition 2.17, if $f(x, y)$ is any polynomial in two variables, then $F : \mathbb{R} \to \mathbb{R}$ defined by $F(t) := f(e^t, \sin t)$ for $t \in \mathbb{R}$ is continuous at every $t_0 \in \mathbb{R}$.

(iii) By part (iii) of Proposition 2.17, if $f(x, y)$ is any polynomial in two variables, then $F : \mathbb{R}^2 \to \mathbb{R}$ defined by $F(u, v) := f(\sin(uv), \cos(u + v))$ for $(u, v) \in \mathbb{R}^2$ is continuous at every $(u_0, v_0) \in \mathbb{R}^2$.

(iv) Consider the functions that give the polar coordinates of a point in $\mathbb{R}^2$ other than the origin. (See Section 1.3 and, in particular, Fact 1.26.) More precisely, consider $r : \mathbb{R}^2 \to \mathbb{R}$ and $\theta : \mathbb{R}^2 \setminus \{(0, 0)\} \to \mathbb{R}$ defined by

$$r(x, y) := \sqrt{x^2 + y^2} \qquad \text{and} \quad \theta(x, y) := \begin{cases} \cos^{-1}\left(\dfrac{x}{r(x, y)}\right) & \text{if } y \geq 0, \\[2ex] -\cos^{-1}\left(\dfrac{x}{r(x, y)}\right) & \text{if } y < 0. \end{cases}$$

Then, as seen already in Example 2.12 (ii), the function r is continuous on $\mathbb{R}^2$. Also, we know that $\cos^{-1} : [-1, 1] \to \mathbb{R}$ is a continuous function of one variable. (See, for example, page 252 of ACICARA.) Consequently, by Proposition 2.15 and part (i) of Proposition 2.17, we see that the function θ is continuous at every $(x_0, y_0) \in \mathbb{R}^2$ for which $y_0 \neq 0$. Also, θ is continuous on the positive x-axis. To see this, note that if $(x_0, 0) \in \mathbb{R}^2$ with $x_0 > 0$ and if $((x_n, y_n))$ is any sequence in $\mathbb{R}^2 \setminus \{(0, 0)\}$ converging to $(x_0, 0)$, then

$$|\theta(x_n, y_n)| = \left|\cos^{-1}\left(\frac{x_n}{\sqrt{x_n^2 + y_n^2}}\right)\right| \to \left|\cos^{-1}\left(\frac{x_0}{|x_0|}\right)\right| = |\cos^{-1}(1)| = 0,$$

and hence $\theta(x_n, y_n) \to 0$. However, at points on the negative x-axis, the function θ is discontinuous. To see this, fix $(x_0, 0) \in \mathbb{R}^2$ with $x_0 < 0$. Clearly, we can find sequences $((x_n, y_n))$ and $((u_n, v_n))$ in $\mathbb{R}^2 \setminus \{(0, 0)\}$ converging to $(x_0, 0)$ such that $y_n \geq 0$ and $v_n < 0$ for all $n \in \mathbb{N}$. Now,

$$\theta(x_n, y_n) = \cos^{-1}\left(\frac{x_n}{\sqrt{x_n^2 + y_n^2}}\right) \to \cos^{-1}\left(\frac{x_0}{|x_0|}\right) = \cos^{-1}(-1) = \pi,$$

whereas

$$\theta(u_n, v_n) = -\cos^{-1}\left(\frac{u_n}{\sqrt{u_n^2 + v_n^2}}\right) \to -\cos^{-1}\left(\frac{x_0}{|x_0|}\right) = -\cos^{-1}(-1) = -\pi.$$

Thus, θ is discontinuous at every point of $\{(x, y) \in \mathbb{R}^2 : x < 0 \text{ and } y = 0\}$. In fact, given any $x_0 < 0$, we can take $x_n = u_n = x_0$ for all $n \in \mathbb{N}$ in the above argument, and this shows that the function from $(-\infty, 0]$ to $\mathbb{R}$ given by $y \mapsto \theta(x_0, y)$ is discontinuous at 0. On the other hand, the functions that give the rectangular coordinates of a point in the (polar) plane are continuous. More precisely, the functions $x, y : \mathbb{R}^2 \to \mathbb{R}$ defined by $x(r, \theta) := r\cos\theta$ and $y(r, \theta) := r\sin\theta$ are continuous on $\mathbb{R}^2$. $\diamond$

Piecing Continuous Functions on Overlapping Subsets

An effective way to construct continuous functions of one variable is to piece together two continuous function defined on overlapping subsets that intersect at a single point, provided their values agree at the common point of intersection. (See, for example, Proposition 3.5 of ACICARA.) We now obtain a similar result for functions of two variables. A precise statement is given below, and the key hypothesis in this result is illustrated in Figure 2.1.

Proposition 2.19. *Let D_1 and D_2 be subsets of $\mathbb{R}^2$ and let $f_1 : D_1 \to \mathbb{R}$ and $f_2 : D_2 \to \mathbb{R}$ be continuous functions such that $f_1(x, y) = f_2(x, y)$ for all $(x, y) \in D_1 \cap D_2$. Let $D := D_1 \cup D_2$ and let $f : D \to \mathbb{R}$ be defined by*

$$f(x, y) := \begin{cases} f_1(x, y) & \text{if } (x, y) \in D_1, \\ f_2(x, y) & \text{if } (x, y) \in D_2. \end{cases}$$

If D_i is closed in D, that is, $\overline{D_i} \cap D = D_i$ for $i = 1, 2$, then f is continuous.

Proof. Since f_1 and f_2 agree on $D_1 \cap D_2$, it is clear that f is well defined. Assume now that each D_i is closed in D for $i = 1, 2$. Fix $(x_0, y_0) \in D$. Let $\big((x_n, y_n)\big)$ be a sequence in D such that $(x_n, y_n) \to (x_0, y_0)$. In case there is $n_1 \in \mathbb{N}$ such that $(x_n, y_n) \in D_1$ for all $n \geq n_1$, then $(x_0, y_0) \in D_1$ since D_1 is closed in D; further, by the continuity of f_1 on D_1, we obtain $f(x_n, y_n) = f_1(x_n, y_n) \to f_1(x_0, y_0) = f(x_0, y_0)$. Similarly, in case there is $n_2 \in \mathbb{N}$ such that $(x_n, y_n) \in D_2$ for all $n \geq n_2$, then $(x_0, y_0) \in D_2$ and $f(x_n, y_n) \to f(x_0, y_0)$. In the remaining case, there are two subsequences $\big((x_{\ell_k}, y_{\ell_k})\big)$ and $\big((x_{m_k}, y_{m_k})\big)$ of $\big((x_n, y_n)\big)$ such that $(x_{\ell_k}, y_{\ell_k}) \in D_1$ and $(x_{m_k}, y_{m_k}) \in D_2$ for all $k \in \mathbb{N}$, and moreover, $\mathbb{N} = \{\ell_1, \ell_2, \ldots\} \cup \{m_1, m_2, \ldots\}$. Clearly, $(x_{\ell_k}, y_{\ell_k}) \to (x_0, y_0)$ and $(x_{m_k}, y_{m_k}) \to (x_0, y_0)$. Now, since each D_i is closed in D, we have $(x_0, y_0) \in D_1 \cap D_2$; further, since each f_i is continuous at (x_0, y_0), we have $f(x_{\ell_k}, y_{\ell_k}) = f_1(x_{\ell_k}, y_{\ell_k}) \to f_1(x_0, y_0) = f(x_0, y_0)$ and $f(x_{m_k}, y_{m_k}) = f_2(x_{m_k}, y_{m_k}) \to f_2(x_0, y_0) = f(x_0, y_0)$. Since $\mathbb{N} = \{\ell_1, \ell_2, \ldots\} \cup \{m_1, m_2, \ldots\}$, it follows that $f(x_n, y_n) \to f(x_0, y_0)$. This proves that f is continuous at (x_0, y_0). $\square$

Examples 2.20. (i) Consider the semiopen rectangles $D_1 := (0, 1] \times [-1, 1]$ and $D_2 := [1, 2) \times [-1, 1]$. (See Figure 2.1.) Note that neither D_1 nor D_2 is closed in $\mathbb{R}^2$, but each D_i is closed in $D := D_1 \cup D_2$ for $i = 1, 2$. Thus the hypothesis of Proposition 2.19 is satisfied, and continuous functions on D_1 and D_2 that agree on $D_1 \cap D_2 = \{1\} \times [-1, 1]$ extend to a continuous function on D.

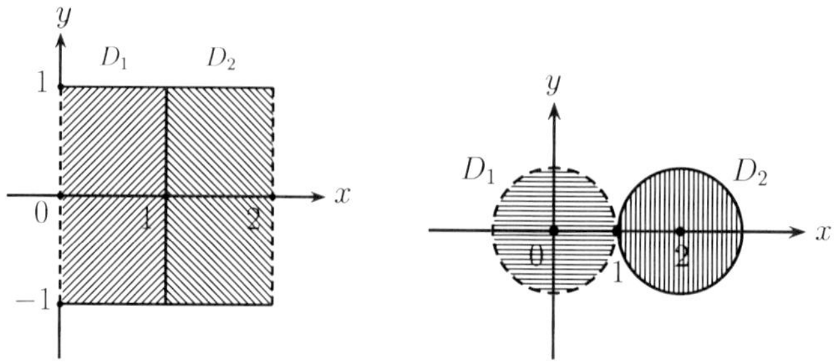

Fig. 2.1. Illustration of the conditions $\overline{D_1} \cap D = D_1$ and $\overline{D_2} \cap D = D_2$ in Proposition 2.19 that are satisfied in Example 2.20(i) and violated in Example 2.20(ii).

(ii) Let D_1 be the open disk $\mathbb{B}_1(0,0)$ and let D_2 the closure of the disk $\mathbb{B}_1(2,0)$, that is, $D_1 := \{(x,y) \in \mathbb{R}^2 : x^2 + y^2 < 1\}$ and $D_2 := \{(x,y) \in \mathbb{R}^2 : (x-2)^2 + y^2 \le 1\}$. (See Figure 2.1 (ii).) Consider $f_1 : D_1 \to \mathbb{R}$ and $f_2 : D_2 \to \mathbb{R}$ defined by $f_1(x,y) := 0$ for all $(x,y) \in D_1$ and $f_2(x,y) := 1$ for all $(x,y) \in D_2$. Clearly, f_1 and f_2 are continuous. Moreover, $D_1 \cap D_2 = \emptyset$ and hence $f : D_1 \cup D_2 \to \mathbb{R}$ as given in Proposition 2.19 is well defined. But f is not continuous at $(1,0)$, since $(x_n, y_n) := \left(1 - \frac{1}{n}, 0\right) \to (1,0)$, whereas $f(x_n, y_n) = f_1(x_n, y_n) = 0$ for all $n \in \mathbb{N}$, and thus $f(x_n, y_n) \not\to 1 = f(1,0)$. This shows that the hypothesis $\overline{D_i} \cap D = D_i$ for $i = 1, 2$ in Proposition 2.19 cannot be dropped. ◇

An easy inductive argument shows that the result in Proposition 2.19 can be extended to piece together continuous functions not just on two overlapping sets, but on any finite number of sets, provided they agree on all pairwise intersections and each of the sets is closed in the union of all the sets. For our purpose, it will suffice to record the following special case of partitioning a set into four quadrants at a given point.

Corollary 2.21. *Let $D \subseteq \mathbb{R}$ and let $f : D \to \mathbb{R}$ be a function. Given any $(x_0, y_0) \in D$, let $D_1 := \{(x,y) \in D : x \ge x_0$ and $y \ge y_0\}$, $D_2 := \{(x,y) \in D : x \le x_0$ and $y \ge y_0\}$, $D_3 := \{(x,y) \in D : x \le x_0$ and $y \le y_0\}$, $D_4 := \{(x,y) \in D : x \ge x_0$ and $y \le y_0\}$, and $f_i = f_{|D_i}$ for $i = 1, \ldots, 4$. Then f is continuous if and only if f_i is continuous for each $i = 1, \ldots, 4$.*

Proof. If f is continuous, then clearly f_i is continuous for each $i = 1, \ldots, 4$. To prove the converse, consider $E_1 := D_1 \cup D_2$ and $E_2 := D_3 \cup D_4$, and also $g_i := f_{|E_i}$ for $i = 1, 2$. Using Proposition 2.19, we see that the continuity of f_1 and f_2 implies the continuity of g_1, while the continuity of f_3 and f_4 implies the continuity of g_2. Further, the continuity of f follows from the continuity of g_1 and g_2 using Proposition 2.19 again. □

Characterizations of Continuity

We have chosen to define continuity of a function at a point using sequences. Alternative definitions are possible, as is shown by the result below.

Proposition 2.22. *Let $D \subseteq \mathbb{R}^2$, $(x_0, y_0) \in D$, and let $f : D \to \mathbb{R}$ be any function. Then the following are equivalent.*

(i) *f is continuous at (x_0, y_0), that is, for every sequence $\big((x_n, y_n)\big)$ in D such that $(x_n, y_n) \to (x_0, y_0)$, we have $f(x_n, y_n) \to f(x_0, y_0)$.*
(ii) *For every $\epsilon > 0$, there is $\delta > 0$ such that $|f(x, y) - f(x_0, y_0)| < \epsilon$ for all $(x, y) \in D \cap \mathbb{S}_\delta(x_0, y_0)$.*
(iii) *For every open subset V of $\mathbb{R}$ containing $f(x_0, y_0)$, there is an open subset U of $\mathbb{R}^2$ containing (x_0, y_0) such that $f(U \cap D) \subseteq V$, that is, $f(x, y) \in V$ for all $(x, y) \in U \cap D$.*

Proof. Assume that (i) holds. If (ii) does not hold, then there is $\epsilon > 0$ such that for every $\delta > 0$, there is (x, y) in $D \cap \mathbb{S}_\delta(x_0, y_0)$ with the property that $|f(x, y) - f(x_0, y_0)| \geq \epsilon$. Consequently, for each $n \in \mathbb{N}$, there is (x_n, y_n) in $D \cap \mathbb{S}_{1/n}(x_0, y_0)$ such that $|f(x_n, y_n) - f(x_0, y_0)| \geq \epsilon$. But then $(x_n, y_n) \to (x_0, y_0)$ and $f(x_n, y_n) \not\to f(x_0, y_0)$. This contradicts (i). Thus, (i) $\Rightarrow$ (ii).

Next, assume that (ii) holds. Let V be an open subset of $\mathbb{R}$ containing $f(x_0, y_0)$. Then there is $\epsilon > 0$ such that $(f(x_0, y_0) - \epsilon, f(x_0, y_0) + \epsilon) \subseteq V$. By (ii), we can find $\delta > 0$ such that $|f(x, y) - f(x_0, y_0)| < \epsilon$ for all $(x, y) \in D \cap \mathbb{S}_\delta(x_0, y_0)$. Thus, if we let $U = \mathbb{S}_\delta(x_0, y_0)$, then U is an open subset of $\mathbb{R}^2$ containing (x_0, y_0) such that $f(U \cap D) \subseteq V$. Thus, (ii) $\Rightarrow$ (iii).

Finally, assume that (iii) holds. Let $\big((x_n, y_n)\big)$ be any sequence in D such that $(x_n, y_n) \to (x_0, y_0)$. Given any $\epsilon > 0$, take V to be the open interval $(f(x_0, y_0) - \epsilon, f(x_0, y_0) + \epsilon)$ in $\mathbb{R}$. By (iii), there is an open subset U of $\mathbb{R}^2$ containing (x_0, y_0) such that $f(U \cap D) \subseteq V$. Since U is open, there is $\delta > 0$ such that $\mathbb{S}_\delta(x_0, y_0) \subseteq U$. Further, since $(x_n, y_n) \to (x_0, y_0)$, there is $n_0 \in \mathbb{N}$ such that $(x_n, y_n) \in \mathbb{S}_\delta(x_0, y_0)$ for all $n \geq n_0$. Consequently, $f(x_n, y_n)$ is in $(f(x_0, y_0) - \epsilon, f(x_0, y_0) + \epsilon)$, that is, $|f(x_n, y_n) - f(x_0, y_0)| < \epsilon$ for all $n \geq n_0$. Thus, $f(x_n, y_n) \to f(x_0, y_0)$, and so (iii) $\Rightarrow$ (i).

This proves the equivalence of (i), (ii), and (iii). □

Corollary 2.23. *Let $D \subseteq \mathbb{R}^2$ be open in $\mathbb{R}^2$ and let $f : D \to \mathbb{R}$ be any function. Then f is continuous on D if and only if for every open subset V of $\mathbb{R}$, the set $f^{-1}(V) := \{(x, y) \in D : f(x, y) \in V\}$ is open in $\mathbb{R}^2$.*

Proof. Follows easily from Proposition 2.22. □

Example 2.24. Clearly, the intervals $(0, \infty)$, $(-\infty, 0)$ and the set $\mathbb{R} \setminus \{0\}$ are open subsets of $\mathbb{R}$. Thus, as a consequence of Corollary 2.23, we see that if $f : \mathbb{R}^2 \to \mathbb{R}$ is continuous, then each of the sets $\{(x, y) \in \mathbb{R}^2 : f(x, y) > 0\}$, $\{(x, y) \in \mathbb{R}^2 : f(x, y) < 0\}$, and $\{(x, y) \in \mathbb{R}^2 : f(x, y) \neq 0\}$ is open in $\mathbb{R}^2$. ◇

Continuity and Boundedness

A bounded function need not be continuous. Consider, for example, the function $f : \mathbb{R}^2 \to \mathbb{R}$ defined by

$$f(x,y) := \begin{cases} 1 & \text{if both } x \text{ and } y \text{ are rational,} \\ 0 & \text{otherwise.} \end{cases}$$

Clearly, f is bounded but f is not continuous at any point of $\mathbb{R}^2$. Also, a continuous function need not be bounded. For example, $g : \mathbb{R}^2 \to \mathbb{R}$ and $h : (0,1) \times (0,1) \to \mathbb{R}$ defined by

$$g(x,y) := x + y \quad \text{and} \quad h(x,y) := 1/(x+y)$$

are both continuous, but neither g nor h is a bounded function. It may be noted that the domain of g is closed, but not bounded, whereas the domain of h is bounded, but not closed. The following result shows that the situation is nicer if the domain is closed as well as bounded.

Proposition 2.25. *Let $D \subseteq \mathbb{R}^2$ be closed and bounded, and let $f : D \to \mathbb{R}$ be continuous. Then f is bounded, that is, $f(D) := \{f(x,y) : (x,y) \in D\}$ is a bounded subset of $\mathbb{R}$. Also, $f(D)$ is a closed subset of $\mathbb{R}$. As a consequence, if D is nonempty, then f attains its bounds, that is, there are $(a,b),(c,d) \in D$ such that $f(a,b) = \sup f(D)$ and $f(c,d) = \inf f(D)$.*

Proof. Suppose f is not bounded above. Then for each $n \in \mathbb{N}$, there is $(x_n, y_n) \in D$ such that $f(x_n, y_n) > n$. Since D is bounded, by the Bolzano–Weierstrass Theorem (part (ii) of Proposition 2.6), the sequence $((x_n, y_n))$ has a convergent subsequence, say $((x_{n_k}, y_{n_k}))$. Suppose $(x_{n_k}, y_{n_k}) \to (x_0, y_0)$. Then $(x_0, y_0) \in D$, since D is closed, and $f(x_{n_k}, y_{n_k}) \to f(x_0, y_0)$, since f is continuous. On the other hand, $f(x_{n_k}, y_{n_k}) > n_k$ for each $k \in \mathbb{N}$, and $n_k \to \infty$ as $k \to \infty$, which leads to a contradiction. Hence f must be bounded above. Similarly, it can be seen that f is bounded below. Thus $f(D)$ is bounded. Next, suppose (z_n) is a sequence in $f(D)$ such that $z_n \to r$ for some $r \in \mathbb{R}$. Write $z_n = f(x_n, y_n)$, where $(x_n, y_n) \in D$ for $n \in \mathbb{N}$. As before, $((x_n, y_n))$ has a convergent subsequence, say $((x_{n_k}, y_{n_k}))$, which must converge to a point (x_0, y_0) of D. Since f is continuous at (x_0, y_0), $z_{n_k} = f(x_{n_k}, y_{n_k}) \to f(x_0, y_0)$, and hence $r = f(x_0, y_0)$, which shows that $r \in f(D)$. Thus $f(D)$ is closed. Finally, if D is nonempty, then $f(D)$ is a nonempty bounded subset of $\mathbb{R}$ and thus $M := \sup f(D)$ and $m := \inf f(D)$ are well defined. By the definition of supremum and infimum, for each $n \in \mathbb{N}$, we can find $(a_n, b_n), (c_n, d_n) \in D$ such that $M - \frac{1}{n} < f(a_n, b_n) \le M$ and $m \le f(c_n, d_n) < m + \frac{1}{n}$. Consequently, $f(a_n, b_n) \to M$ and $f(c_n, d_n) \to m$. Since $f(D)$ is closed, $M, m \in f(D)$, that is, $f(a,b) = \sup f(D)$ and $f(c,d) = \inf f(D)$ for some $(a,b),(c,d) \in D$. □

Remark 2.26. Subsets of $\mathbb{R}^2$ (and more generally, of $\mathbb{R}^n$) that are both closed and bounded are often referred to as compact sets. Thus, the above proposition says that the continuous image of a compact set is compact. For more on compactness, see Exercise 17. ◇

Continuity and Monotonicity

For functions of one variable, there is no direct relationship between continuity and monotonicity. Indeed, it suffices to consider the integer part function $x \longmapsto [x]$ and the absolute value function $x \longmapsto |x|$ to conclude that a monotonic function need not be continuous and a continuous function need not be monotonic. For functions of two variables, a similar situation prevails. In fact, using the product order on $\mathbb{R}^2$, we have introduced in Chapter 1 two distinct notions: monotonicity and bimonotonicity. We will show below that neither of these implies or is implied by continuity.

Examples 2.27. (i) Consider $f : [-1,1] \times [-1,1] \to \mathbb{R}$ defined by $f(x,y) := xy$. Clearly, f is continuous but not monotonic on $[-1,1] \times [-1,1]$. Note, however, that f is bimonotonically increasing on $[-1,1] \times [-1,1]$, since we have $x_2y_2 + x_1y_1 - x_2y_1 - x_1y_2 = (x_2 - x_1)(y_2 - y_1)$ for all $(x_1, y_1), (x_2, y_2) \in \mathbb{R}^2$.

(ii) Consider $f : [-1,1] \times [-1,1] \to \mathbb{R}$ defined by $f(x,y) := (x+y)^3$. Clearly, f is continuous. However, f is not bimonotonic on $[-1,1] \times [-1,1]$. To see this, observe that $(x_1, y_1) := (0,0)$ and $(x_2, y_2) := (1,1)$ are points of $[-1,1] \times [-1,1]$ satisfying $(x_1, y_1) \le (x_2, y_2)$ and

$$f(x_1, y_1) + f(x_2, y_2) - f(x_1, y_2) - f(x_2, y_1) = 0 + 8 - 1 - 1 = 6 > 0,$$

whereas $(u_1, v_1) := (-1,-1)$ and $(u_2, v_2) := (0,0)$ are points of $[-1,1] \times [-1,1]$ satisfying $(u_1, v_1) \le (u_2, v_2)$ and

$$f(u_1, v_1) + f(u_2, v_2) - f(u_1, v_2) - f(u_2, v_1) = -8 + 0 + 1 + 1 = -6 < 0.$$

(iii) Consider $f : [-1,1] \times [-1,1] \to \mathbb{R}$ defined by

$$f(x,y) := \begin{cases} 1 & \text{if } x > 0 \text{ and } y > 0, \\ 0 & \text{otherwise.} \end{cases}$$

It is easy to see that f is monotonically as well as bimonotonically increasing, but not continuous on $[-1,1] \times [-1,1]$. ◇

Continuity, Bounded Variation, and Bounded Bivariation

In general, a function of bounded variation need not be continuous. Likewise for a function of bounded bivariation. In fact, Example 2.27 (iii) provides a common counterexample. We have seen earlier that a continuous function need not be monotonic or bimonotonic. The following example shows that it need not even be of bounded variation or of bounded bivariation.

Example 2.28. Consider $f : [0,1] \times [0,1] \to \mathbb{R}$ defined by

$$f(x,y) := \begin{cases} xy \cos(\pi/2x) & \text{if } x \neq 0, \\ 0 & \text{if } x = 0. \end{cases}$$

Clearly, f is continuous on $(0,1] \times [0,1]$. Moreover, since $|f(x,y)| \leq |xy|$ for all $(x,y) \in [0,1] \times [0,1]$, it is readily seen that f is continuous at $(0,y)$ for each $y \in [0,1]$. Thus, f is continuous on $[0,1] \times [0,1]$. Next, given any even positive integer n, say $n = 2k$ for some $k \in \mathbb{N}$, if we consider the points

$$x_0 = 0 = y_0 \quad \text{and} \quad x_i := \frac{1}{n+1-i} \quad \text{and} \quad y_i = 1 \text{ for } i = 1, \ldots, n,$$

then we have $(0,0) = (x_0, y_0) \leq (x_1, y_1) \leq \cdots \leq (x_n, y_n) = (1,1)$ and moreover, $f(x_i, y_i) = 0$ if i is even and $f(x_i, y_i) = \pm x_i$ if i is odd. Thus

$$\sum_{i=1}^{n} |f(x_i, y_i) - f(x_{i-1}, y_{i-1})| = \frac{1}{n} + \frac{1}{n} + \frac{1}{n-2} + \frac{1}{n-2} + \cdots + \frac{1}{2} + \frac{1}{2} = \sum_{i=1}^{k} \frac{1}{i}.$$

Since the set $\{\sum_{i=1}^{k}(1/i) : k \in \mathbb{N}\}$ is not bounded above (as is shown, for example, on page 51 of ACICARA), it follows that f is not of bounded variation on $[0,1] \times [0,1]$.

Furthermore, if we let $n = 2k$ and $x_0, x_1, \ldots, x_n$ be as above, but take $m = 1$, $y_0 = 0$, and $y_1 = 1$, then $0 = x_0 \leq x_1 \leq \cdots \leq x_n = 1$ and $0 = y_0 \leq y_1 = 1$, and moreover, for any $i \geq 0$, we have $f(x_i, 0) = 0$, whereas $f(x_i, 1) = 0$ if i is even and $f(x_i, 1) = \pm x_i$ if i is odd, and thus

$$\sum_{i=1}^{n} \sum_{j=1}^{m} |f(x_i, y_j) + f(x_{i-1}, y_{j-1}) - f(x_i, y_{j-1}) - f(x_{i-1}, y_j)| = \sum_{i=1}^{k} \frac{1}{i}.$$

It follows, therefore, that f is not of bounded bivariation on $[0,1] \times [0,1]$. $\diamond$

Remark 2.29. Using Exercise 38, a refined version of the Jordan decomposition (Propositions 1.12 and 1.17) can be obtained for continuous functions. Namely, a continuous function of bounded variation is a difference of continuous monotonic functions, whereas a continuous function of bounded bivariation is a difference of continuous bimonotonic functions. $\diamond$

Continuity and Convexity

In general, a continuous function is neither convex nor concave. For example, consider $D := [-1,1] \times [-1,1]$ and $f : D \to \mathbb{R}$ defined by $f(x,y) := x^3 + y^3$. Clearly, f is continuous. But f is neither convex nor concave. To see this, observe that $\left(-\frac{1}{2}, -\frac{1}{2}\right) = \frac{1}{2}(-1,-1) + \frac{1}{2}(0,0)$ and $\left(\frac{1}{2}, \frac{1}{2}\right) = \frac{1}{2}(1,1) + \frac{1}{2}(0,0)$, but $f\left(-\frac{1}{2}, -\frac{1}{2}\right) = -\frac{1}{4} > -1 = \frac{1}{2}f(-1,-1) + \frac{1}{2}f(0,0)$, and $f\left(\frac{1}{2}, \frac{1}{2}\right) = \frac{1}{4} < 1 = \frac{1}{2}f(1,1) + \frac{1}{2}f(0,0)$. Moreover, a convex function need not be continuous. For example, if $D := \{(x,y) \in \mathbb{R}^2 : x^2 + y^2 \leq 1\}$ is the closed unit disk and $f : D \to \mathbb{R}$ is a variant of the norm function defined by

$$f(x,y) := \begin{cases} \sqrt{x^2+y^2} & \text{if } x^2 + y^2 < 1, \\ 2 & \text{if } x^2 + y^2 = 1, \end{cases}$$

then f is convex on D, but not continuous on D. Here, the continuity of f fails precisely at the boundary points of D. In fact, we will show that a convex function is always continuous at the interior points of its domain. First, we prove a couple of auxiliary results, which may also be of independent interest.

Lemma 2.30. *Let $a, b, c, d \in \mathbb{R}$ with $a < b$ and $c < d$. Then every real-valued convex function on the closed rectangle $[a, b] \times [c, d]$ in $\mathbb{R}^2$ is bounded.*

Proof. Let $D := [a, b] \times [c, d]$ and let $f : D \to \mathbb{R}$ be any convex function. Define $M := \max\{f(a, c), f(a, d), f(b, c), f(b, d)\}$. Let $(x, y) \in D$. Then there is s in $[0, 1]$ such that $x = (1-s)a + sb$. Using the convexity of f on D, we see that $f(x, y) \le (1-s)f(a, y) + sf(b, y)$. Further, there is $t \in [0, 1]$ such that $y = (1-t)c + td$. Again, using the convexity of f on D, we obtain

$$\begin{aligned} f(x,y) &\le (1-s)\left[(1-t)f(a,c) + tf(a,d)\right] + s\left[(1-t)f(b,c) + tf(b,d)\right] \\ &\le (1-s)\left[(1-t)M + tM\right] + s\left[(1-t)M + tM\right] \;=\; M. \end{aligned}$$

It follows that M is an upper bound for f. Next, consider the center point $(p, q) := \left(\frac{a+b}{2}, \frac{c+d}{2}\right)$ of D and let $(u, v) := (a + b - x, c + d - y)$. Clearly, $(u, v) \in D$ and $(p, q) = \frac{1}{2}(x, y) + \frac{1}{2}(u, v)$. Hence using the convexity of f, we obtain $f(p, q) \le \frac{1}{2}f(x, y) + \frac{1}{2}f(u, v) \le \frac{1}{2}f(x, y) + M$, that is, $f(x, y) \ge m$, where $m := 2\left(f(p, q) - M\right)$. It follows that m is a lower bound for f. □

Lemma 2.31. *Let D be convex and open in $\mathbb{R}^2$, and let $f : D \to \mathbb{R}$ be convex. Also, let $[a, b] \times [c, d]$ be a closed rectangle contained in D, where $a, b, c, d \in \mathbb{R}$ with $a < b$ and $c < d$. Then there is $K \in \mathbb{R}$ such that*

$$|f(x,y) - f(u,v)| \le K\left(|x-u| + |y-v|\right) \quad \textit{for all } (x,y),(u,v) \in [a,b]\times[c,d].$$

Proof. Since D is open, there is $\delta > 0$ such that $[a-\delta, b+\delta]\times[c-\delta, d+\delta] \subseteq D$. By Lemma 2.30, there are $m, M \in \mathbb{R}$ such that $m \le f(z, w) \le M$ for all $(z, w) \in [a-\delta, b+\delta] \times [c-\delta, d+\delta]$. Now, fix any $(x, y), (u, v) \in [a, b] \times [c, d]$. The case $(x, y) = (u, v)$ is trivial, and so we will assume that $(x, y) \ne (u, v)$. Then $\ell := |x - u| + |y - v| > 0$, and we can consider $z := u + \frac{\delta}{\ell}(u - x)$ and $w := v + \frac{\delta}{\ell}(v - y)$. Since $|u - x| \le \ell$, that is, $-\ell \le u - x \le \ell$, we have $u - \delta \le z \le u + \delta$, and hence $z \in [a - \delta, b + \delta]$. Similarly, $w \in [c - \delta, d + \delta]$. In particular, $(z, w) \in D$. Moreover, it can be easily verified that

$$u = \frac{\delta}{\ell+\delta}\,x + \frac{\ell}{\ell+\delta}\,z \quad \text{and} \quad v = \frac{\delta}{\ell+\delta}\,y + \frac{\ell}{\ell+\delta}\,w.$$

Thus $(u, v) = (1 - t)(x, y) + t(z, w)$, where $t := \ell/(\ell + \delta)$. Since $0 < t < 1$, using the convexity of f on D, we obtain $f(u, v) \le (1 - t)f(x, y) + tf(z, w)$. Further, since $0 < t < \ell/\delta$, we see that

$$f(u,v) - f(x,y) \le t\left[f(z,w) - f(x,y)\right] \le \frac{\ell}{\delta}\left[M - m\right] = K\left(|x-u| + |y-v|\right),$$

where $K := (M - m)/\delta$. Similarly, $f(x, y) - f(u, v) \le K\left(|x - u| + |y - v|\right)$. This proves the desired inequality for $|f(u, v) - f(x, y)|$. □

We are now ready to show that a convex function is continuous at all the interior points of its domain. This is an immediate consequence of the above lemma. (See also Exercise 10.)

Proposition 2.32. *Let D be a convex subset of $\mathbb{R}^2$ and let $f : D \to \mathbb{R}$ be a convex function. Then f is continuous at every interior point of D. In particular, if D is also open in $\mathbb{R}^2$, then f is continuous on D.*

Proof. Let (x_0, y_0) be an interior point of D. Then there is $r > 0$ such that $R := [x_0 - r, x_0 + r] \times [y_0 - r, y_0 + r]$ is contained in D. By Lemma 2.31, there is $K \in \mathbb{R}$ such that $|f(x, y) - f(x_0, y_0)| \le K(|x - x_0| + |y - y_0|)$ for $(x, y) \in R$. This implies that if $((x_n, y_n))$ is a sequence in D such that $(x_n, y_n) \to (x_0, y_0)$, then $f(x_n, y_n) \to f(x_0, y_0)$. Thus, f is continuous at (x_0, y_0). □

Continuity and Intermediate Value Property

A result of fundamental importance in one-variable calculus is that continuous functions possess the intermediate value property (IVP). For ease of reference, we state this result below; see, for example, Proposition 3.13 of ACICARA.

Fact 2.33. (Intermediate Value Theorem) *Let D be a subset of $\mathbb{R}$ and let $\phi : D \to \mathbb{R}$ be a continuous function. Then ϕ has the IVP on every interval $I \subseteq D$, that is, if $a, b \in I$ with $a < b$ and $r \in \mathbb{R}$ is between $\phi(a)$ and $\phi(b)$, then there is $c \in [a, b]$ such that $\phi(c) = r$; in particular, $\phi(I)$ is an interval in $\mathbb{R}$.*

The following result may be viewed as an analogue of Fact 2.33 for real-valued continuous functions of two variables.

Proposition 2.34 (Bivariate Intermediate Value Theorem). *Let D be a subset of $\mathbb{R}^2$ and let $f : D \to \mathbb{R}$ be a continuous function. Then $f(E)$ is an interval in $\mathbb{R}$ for every path-connected subset E of D. In particular, f has the IVP on every 2-interval in D.*

Proof. Suppose $E \subseteq D$ is path-connected. Let $z_1, z_2 \in f(E)$ and let r be any real number between z_1 and z_2. Then $z_1 = f(x_1, y_1)$ and $z_2 = f(x_2, y_2)$ for some $(x_1, y_1), (x_2, y_2) \in E$. Since E is path-connected, there is a path Γ joining (x_1, y_1) to (x_2, y_2) that lies in E. Let $x, y : [\alpha, \beta] \to \mathbb{R}$ be continuous functions such that Γ is given by $(x(t), y(t))$, $t \in [\alpha, \beta]$. Consider $F : [\alpha, \beta] \to \mathbb{R}$ defined by $F(t) := f(x(t), y(t))$. By part (ii) of Proposition 2.17, F is continuous, and by Fact 2.33, F has the IVP on $[\alpha, \beta]$. It follows that $r = F(t_0)$ for some $t_0 \in [\alpha, \beta]$, and hence $r \in f(E)$. This proves that $f(E)$ is an interval in $\mathbb{R}$. Finally, every 2-interval is path-connected (Example 1.5 (iv)), and so in view of Proposition 1.25, we see that f has the IVP on every 2-interval in D. □

The following example shows that the converse of the above result is not true, that is, the IVP does not imply continuity.

Example 2.35. Consider $f : [0,1] \times [0,1] \to \mathbb{R}$ defined by

$$f(x,y) := \begin{cases} \cos(1/y) & \text{if } y \neq 0, \\ 0 & \text{if } y = 0. \end{cases}$$

Then f is not continuous on $[0,1] \times [0,1]$, because, for example, $(0,\, 1/n\pi) \to (0,0)$, but $f(0,\, 1/n\pi) = (-1)^n \not\to f(0,0) = 0$. Note, however, that f is continuous on $[0,1] \times (0,1]$. We show that f has the IVP on $[0,1] \times [0,1]$. Let $r \in \mathbb{R}$ be an intermediate value of f, that is, r is between $f(x_1,y_1)$ and $f(x_2,y_2)$ for some $(x_1,y_1),(x_2,y_2) \in [0,1] \times [0,1]$. If $y_1 > 0$ and $y_2 > 0$, then by the continuity of f on $[0,1] \times (0,1]$ and Proposition 2.34, we see that $r = f(x,y)$ for some $(x,y) \in I_{(x_1,y_1),(x_2,y_2)}$. If $y_1 = y_2 = 0$, then $f(x_1,y_1) = f(x_2,y_2) = 0$ and there is nothing to prove. Thus we may assume, without loss of generality, that $y_1 = 0$ and $y_2 > 0$. Choose $k \in \mathbb{N}$ such that $(1/k\pi) < y_2$. Now $y_1 < (1/(k+2)\pi) < (1/k\pi) < y_2$, and therefore $\cos(1/y)$ assumes every value between -1 and 1 as y varies from y_1 to y_2. It follows that $r = f(x_1,y)$ for some $y \in [y_1,y_2]$. Thus f has the IVP on $[0,1] \times [0,1]$. ◇

Corollary 2.36. *Let D be a nonempty, path-connected, closed, and bounded subset of $\mathbb{R}^2$ and let $f : D \to \mathbb{R}$ be a continuous function. Then the range $f(D)$ of f is a closed and bounded interval in $\mathbb{R}$.*

Proof. First, note that since D is nonempty, so is $f(D)$. By Proposition 2.25, $f(D)$ is bounded, and moreover, if $m := \inf f(D)$ and $M := \sup f(D)$, then $f(D) \subseteq [m,M]$ and $m, M \in f(D)$. Further, by Proposition 2.34, $f(D)$ is an interval in $\mathbb{R}$. It follows that $f(D) = [m,M]$. □

Uniform Continuity

The notion of uniform continuity for functions of one variable can be easily extended to functions of two variables. Let D be a subset of $\mathbb{R}^2$. A function $f : D \to \mathbb{R}$ is said to be **uniformly continuous** on D if for any sequences $\big((x_n,y_n)\big)$ and $\big((u_n,v_n)\big)$ in D such that $|(x_n,y_n) - (u_n,v_n)| \to 0$, we have $|f(x_n,y_n) - f(u_n,v_n)| \to 0$.

Specializing one of the two sequences to a constant sequence, we readily see that a uniformly continuous function is continuous. As in the case of functions of one variable, the converse is true if the domain is closed and bounded.

Proposition 2.37. *Let $D \subseteq \mathbb{R}^2$ be a closed and bounded set. Then every continuous function on D is uniformly continuous on D.*

Proof. Suppose $f : D \to \mathbb{R}$ is continuous but not uniformly continuous on D. Then there are sequences $\big((x_n,y_n)\big)$ and $\big((u_n,v_n)\big)$ in D such that $|(x_n,y_n) - (u_n,v_n)| \to 0$, but $|f(x_n,y_n) - f(u_n,v_n)| \not\to 0$. The latter implies that there are $\epsilon > 0$ and positive integers $n_1 < n_2 < \cdots$ such that $\big|f(x_{n_k},y_{n_k}) - f(u_{n_k},v_{n_k})\big| \geq \epsilon$ for all $k \in \mathbb{N}$. Now, by the Bolzano–Weierstrass

Theorem (part (ii) of Proposition 2.6), $\big((x_{n_k}, y_{n_k})\big)$ has a convergent subsequence, say $\big((x_{n_{k_j}}, y_{n_{k_j}})\big)$. If $(x_{n_{k_j}}, y_{n_{k_j}}) \to (x_0, y_0)$, then $(u_{n_{k_j}}, v_{n_{k_j}}) \to (x_0, y_0)$, because $|(x_n, y_n) - (u_n, v_n)| \to 0$. Since f is continuous on D, we see that $\left|f(x_{n_{k_j}}, y_{n_{k_j}}) - f(u_{n_{k_j}}, v_{n_{k_j}})\right| \to |f(x_0, y_0) - f(x_0, y_0)| = 0$. But this is a contradiction, since $\left|f(x_{n_{k_j}}, y_{n_{k_j}}) - f(u_{n_{k_j}}, v_{n_{k_j}})\right| \geq \epsilon$ for all $j \in \mathbb{N}$. □

Examples 2.38. (i) Consider $f : \mathbb{R}^2 \to \mathbb{R}$ defined by $f(x, y) := x + y$. Then it is clear that f is uniformly continuous on $\mathbb{R}^2$.

(ii) If $D \subseteq \mathbb{R}^2$ and $f : D \to \mathbb{R}$ is uniformly continuous, then for every fixed $(x_0, y_0) \in D$, the functions $\phi : D_1 \to \mathbb{R}$ and $\psi : D_2 \to \mathbb{R}$, defined as in Example 2.12(iv), are uniformly continuous. This follows from the definition of uniform continuity by specializing one of the coordinates in the two sequences to a constant sequence.

(iii) Consider $D \subseteq \mathbb{R}^2$ and $f : D \to \mathbb{R}$ given by

$$D := \{(x, y) \in \mathbb{R}^2 : x, y \in [0, 1] \text{ and } (x, y) \neq (0, 0)\} \quad \text{and} \quad f(x, y) := \frac{1}{x + y}.$$

Then f is continuous on D but not uniformly continuous on D. To see the latter, consider the sequences $\big((x_n, y_n)\big)$ and $\big((u_n, v_n)\big)$ in D given by $(x_n, y_n) := (1/n, 0)$ and $(u_n, v_n) := (1/(n + 1), 0)$ for $n \in \mathbb{N}$. We have $|(x_n, y_n) - (u_n, v_n)| = 1/n(n+1) \to 0$, but $|f(x_n, y_n) - f(u_n, v_n)| = |n - (n + 1)| = 1 \not\to 0$. Alternatively, we could use (ii) above and the fact that $\phi : (0, 1] \to \mathbb{R}$ defined by $\phi(x) = f(x, 0) = 1/x$ is not uniformly continuous on $(0, 1]$. (See Example 3.18 (ii) on page 80 of ACICARA.) It may be noted here that the domain of f is bounded but not closed.

(iv) Consider $f : \mathbb{R}^2 \to \mathbb{R}$ defined by $f(x, y) := x^2 + y^2$. Then f is continuous on $\mathbb{R}^2$, but not uniformly continuous on $\mathbb{R}^2$. To see the latter, consider the sequences $\big((x_n, y_n)\big)$ and $\big((u_n, v_n)\big)$ in D given by $(x_n, y_n) := (n, 0)$ and $(u_n, v_n) := (n-(1/n), 0)$ for $n \in \mathbb{N}$. We have $|(x_n, y_n) - (u_n, v_n)| = 1/n \to 0$, but $|f(x_n, y_n) - f(u_n, v_n)| = \left|n^2 - [n^2 - 2 + (1/n^2)]\right| = 2-(1/n^2) \not\to 0$. Alternatively, we could use (ii) above and the fact that $\phi : \mathbb{R} \to \mathbb{R}$ defined by $\phi(x) = f(x, 0) = x^2$ is not uniformly continuous on $\mathbb{R}$. (See Example 3.18 (iii) on page 80 of ACICARA.) It may be noted here that the domain of f is closed but not bounded. On the other hand, the restriction of f to any bounded subset of $\mathbb{R}^2$ is uniformly continuous. ◇

A criterion for the uniform continuity of a function of two variables that does not involve convergence of sequences can be given as follows. The result below may be compared with Proposition 2.22.

Proposition 2.39. *Let $D \subseteq \mathbb{R}^2$. Consider a function $f : D \to \mathbb{R}$. Then f is uniformly continuous on D if and only if it satisfies the following ϵ-δ condition: For every $\epsilon > 0$, there is $\delta > 0$ such that*

$$(x, y), (u, v) \in D \text{ and } |(x, y) - (u, v)| < \delta \Longrightarrow |f(x, y) - f(u, v)| < \epsilon.$$

Proof. Assume that f is uniformly continuous on D. Suppose the ϵ-δ condition does not hold. Then there is $\epsilon > 0$ such that for any $\delta > 0$, we can find $(x, y), (u, v) \in D$ for which $|(x, y) - (u, v)| < \delta$, but $|f(x, y) - f(u, v)| \geq \epsilon$. Considering $\delta := 1/n$ for $n \in \mathbb{N}$, we obtain sequences $\big((x_n, y_n)\big)$ and $\big((u_n, v_n)\big)$ in D such that $|(x_n, y_n) - (u_n, v_n)| < \frac{1}{n}$ and $|f(x_n, y_n) - f(u_n, v_n)| \geq \epsilon$ for all $n \in \mathbb{N}$. Consequently, $|(x_n, y_n) - (u_n, v_n)| \to 0$, but $|f(x_n, y_n) - f(u_n, v_n)| \not\to 0$. This contradicts the assumption that f is uniformly continuous on D.

Conversely, assume that the ϵ-δ condition is satisfied. Suppose $\big((x_n, y_n)\big)$ and $\big((u_n, v_n)\big)$ are any sequences in D such that $|(x_n, y_n) - (u_n, v_n)| \to 0$. Let $\epsilon > 0$ be given. Then there is $\delta > 0$ such that if $(x, y), (u, v) \in D$ satisfy $|(x, y) - (u, v)| < \delta$, then $|f(x, y) - f(u, v)| < \epsilon$. Now, for this $\delta > 0$, we can find $n_0 \in \mathbb{N}$ such that $|(x_n, y_n) - (u_n, v_n)| < \delta$ for all $n \geq n_0$. Consequently, $|f(x_n, y_n) - f(u_n, v_n)| < \epsilon$ for all $n \geq n_0$. Thus $|f(x_n, y_n) - f(u_n, v_n)| \to 0$. This proves the uniform continuity of f on D. □

Implicit Function Theorem

In the study of functions of one variable, one considers the so-called *implicitly defined curves*, that is, curves given by equations of the form $f(x, y) = 0$, $(x, y) \in D$, where $f : D \to \mathbb{R}$ is a real-valued function of two variables. Heuristically, such an equation defines one of the variables as a function of the other; for example, it may define y as a function of x. In other words, from the equation $f(x, y) = 0$, we may be able to solve for y in terms of x. In fact, this is tacitly assumed when one does implicit differentiation in calculus of functions of one variable. The following result asserts that it is possible to solve the equation $f(x, y) = 0$ locally, around a point (x_0, y_0) satisfying $f(x_0, y_0) = 0$, provided f is continuous in each variable and is either a strictly increasing or a strictly decreasing function of y, for each fixed x. Moreover, the solution $y = \eta(x)$ is unique and it is a continuous function of x.

Proposition 2.40 (Implicit Function Theorem). *Let $D \subseteq \mathbb{R}^2$ and (x_0, y_0) be an interior point of D, and let $f : D \to \mathbb{R}$ satisfy $f(x_0, y_0) = 0$. Assume that there is $r > 0$ with $\mathbb{S}_r(x_0, y_0) \subseteq D$ such that the following conditions hold.*

(a) *Given any $x \in (x_0 - r, x_0 + r)$, the function $\psi : (y_0 - r, y_0 + r) \to \mathbb{R}$ defined by $\psi(y) := f(x, y)$ is continuous. Also, given any $y \in (y_0 - r, y_0 + r)$, the function $\phi : (x_0 - r, x_0 + r) \to \mathbb{R}$ defined by $\phi(x) := f(x, y)$ is continuous.*
(b) *Given any $x \in (x_0 - r, x_0 + r)$, the function $\psi : (y_0 - r, y_0 + r) \to \mathbb{R}$ defined by $\psi(y) := f(x, y)$ is strictly monotonic.*

Then there are $\delta > 0$ and a unique continuous function $\eta : (x_0 - \delta, x_0 + \delta) \to \mathbb{R}$ with $\eta(x_0) = y_0$ such that $(x, \eta(x)) \in \mathbb{S}_r(x_0, y_0)$ and $f(x, \eta(x)) = 0$ for all $x \in (x_0 - \delta, x_0 + \delta)$.

Proof. In view of (b), let us first suppose that $\psi_0 : (y_0 - r, y_0 + r) \to \mathbb{R}$ defined by $\psi_0(y) := f(x_0, y)$ is strictly increasing on $(y_0 - r, y_0 + r)$.

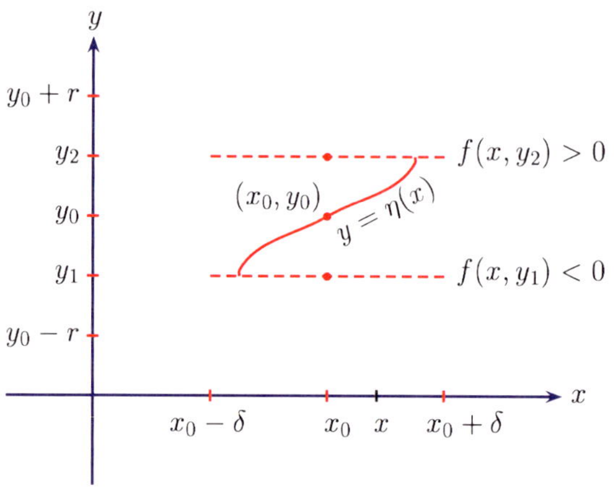

Fig. 2.2. Illustration of the proof of the Implicit Function Theorem.

Choose any $y_1 \in (y_0 - r, y_0)$ and $y_2 \in (y_0, y_0 + r)$. Since $f(x_0, y_0) = 0$ and the function ψ_0 is strictly increasing on $(y_0 - r, y_0 + r)$, we see that $f(x_0, y_1) < 0$ and $f(x_0, y_2) > 0$. By continuity, the sign of f is preserved on small horizontal segments of the lines $y = y_1$ and $y = y_2$. (See Figure 2.2.) More precisely, using (a), we see that the function defined by $x \longmapsto f(x, y_1)$ is continuous on $(x_0 - r, x_0 + r)$, and hence it follows from Fact 2.13 that there is $\delta_1 > 0$ with $\delta_1 \le r$ such that $f(x, y_1) < 0$ for all $x \in (x_0 - \delta_1, x_0 + \delta_1)$. Similarly, there is $\delta_2 > 0$ with $\delta_2 \le r$ such that $f(x, y_2) > 0$ for all $x \in (x_0 - \delta_2, x_0 + \delta_2)$. Let $\delta := \min\{\delta_1, \delta_2\}$. Then

$$f(x, y_1) < 0 < f(x, y_2) \quad \text{for all } x \in (x_0 - \delta, x_0 + \delta).$$

Thus, given any $x \in (x_0 - \delta, x_0 + \delta)$, the function $\psi : (y_0 - r, y_0 + r) \to \mathbb{R}$ defined by $\psi(y) := f(x, y)$ satisfies $\psi(y_1) < 0 < \psi(y_2)$. Also by (a), ψ is continuous. Hence by the IVP of ψ, there is $y \in (y_1, y_2)$ such that $\psi(y) = 0$, that is, $f(x, y) = 0$. Moreover, since $\psi(y_1) < \psi(y_2)$, it follows from (b) that ψ is strictly increasing on $(y_0 - r, y_0 + r)$, and hence y is uniquely determined by x. Thus if we write $y = \eta(x)$, then we obtain a unique function $\eta : (x_0 - \delta, x_0 + \delta) \to \mathbb{R}$ such that $\eta(x) \in (y_1, y_2)$ and $f(x, \eta(x)) = 0$ for all $x \in (x_0 - \delta, x_0 + \delta)$. In particular, since $f(x_0, y_0) = 0$ and $y_0 \in (y_1, y_2)$, we have $\eta(x_0) = y_0$.

To prove the continuity of η, fix any $x^\star \in (x_0 - \delta, x_0 + \delta)$ and let (x_n) be a sequence in $(x_0 - \delta, x_0 + \delta)$ such that $x_n \to x^\star$. We have seen above that for any $x \in (x_0 - \delta, x_0 + \delta)$, the function $\psi : (y_0 - r, y_0 + r) \to \mathbb{R}$ defined by $\psi(y) = f(x, y)$ is strictly increasing. Fix $y_1, y_2 \in (y_0 - r, y_0 + r)$ as above, so that $y_1 < \eta(x) < y_2$ for all $x \in (x_0 - \delta, x_0 + \delta)$. Let $\epsilon > 0$ be given and let us suppose ϵ is so small that $y_1 < \eta(x^\star) - \epsilon < \eta(x^\star) + \epsilon < y_2$, that is, $0 < \epsilon < \min\{\eta(x^\star) - y_1,\ y_2 - \eta(x^\star)\}$. Using (a) and (b), we see that

$$f(x_n, \eta(x^\star) - \epsilon) \to f(x^\star, \eta(x^\star) - \epsilon) \quad \text{and} \quad f(x^\star, \eta(x^\star) - \epsilon) < f(x^\star, \eta(x^\star)) = 0.$$

Hence there is $n_1 \in \mathbb{N}$ such that $f(x_n, \eta(x^\star) - \epsilon) < 0$ for all $n \geq n_1$. Similarly, $f(x_n, \eta(x^\star) + \epsilon) \to f(x^\star, \eta(x^\star) + \epsilon) > f(x^\star, \eta(x^\star)) = 0$, and hence there is $n_2 \in \mathbb{N}$ such that $f(x_n, \eta(x^\star) + \epsilon) > 0$ for all $n \geq n_2$. Let $n_0 = \max\{n_1, n_2\}$. Then $f(x_n, \eta(x^\star) - \epsilon) < 0 < f(x_n, \eta(x^\star) + \epsilon)$ for all $n \geq n_0$. But since $f(x_n, \eta(x_n)) = 0$, it follows from (*b*) that $\eta(x^\star) - \epsilon < \eta(x_n) < \eta(x^\star) + \epsilon$, that is, $|\eta(x_n) - \eta(x^\star)| < \epsilon$ for all $n \geq n_0$. Thus, $\eta(x_n) \to \eta(x^\star)$. This proves that η is continuous on $(x_0 - \delta, x_0 + \delta)$.

The case in which $\psi_0 : (y_0 - r, y_0 + r) \to \mathbb{R}$ defined by $\psi_0(y) := f(x_0, y)$ is strictly decreasing on $(y_0 - r, y_0 + r)$ is proved similarly. Alternatively, it follows from applying the result proved above to $-f$. □

Example 2.41. Consider $f : \mathbb{R}^2 \to \mathbb{R}$ defined by $f(x, y) = x^2 + y^2 - 1$. Then $C := \{(x, y) \in \mathbb{R}^2 : f(x, y) = 0\}$ is the unit circle in $\mathbb{R}^2$. If $(x_0, y_0) \in C$ and $y_0 \neq 0$, then we can easily see that the hypotheses of the Implicit Function Theorem are satisfied, and the "solution" is given by $\eta(x) := \sqrt{1 - x^2}$ or by $\eta(x) := -\sqrt{1 - x^2}$ according as $y_0 > 0$ or $y_0 < 0$. ◇

Remark 2.42. We have a straightforward analogue of the Implicit Function Theorem for solving $f(x, y) = 0$ for x in terms of y. In this situation, condition (a) in Proposition 2.40 remains the same, while (b) is replaced by the condition that for any $y \in (y_0 - r, y_0 + r)$, the function $\phi : (x_0 - r, x_0 + r) \to \mathbb{R}$ defined by $\phi(x) := f(x, y)$ is strictly monotonic. The conclusion would be that there are $\delta > 0$ and a unique continuous function $\xi : (y_0 - \delta, y_0 + \delta) \to \mathbb{R}$ with $\xi(y_0) = x_0$ such that $(\xi(y), y) \in \mathbb{S}_r(x_0, y_0)$ and $f(\xi(y), y) = 0$ for all $y \in (y_0 - \delta, y_0 + \delta)$. This can be proved in a manner similar to the proof of Proposition 2.40. Alternatively, it follows from applying Proposition 2.40 to the function $(x, y) \mapsto f(y, x)$ and the point (y_0, x_0). ◇

An important consequence of the Implicit Function Theorem is that a continuous real-valued function of one variable that is strictly monotonic in an interval about a point admits a continuous (and strictly monotonic) inverse, locally. A more precise statement appears below. This result may be viewed as a special case of the so-called **Inverse Function Theorem**.

Proposition 2.43. *Let I be an interval in $\mathbb{R}$ and $x_0 \in I$. Suppose $f : I \to \mathbb{R}$ is continuous and strictly monotonic on $I_1 := (x_0 - r, x_0 + r) \cap I$ for some $r > 0$. Let $y_0 := f(x_0)$, $J := f(I)$, and $J_1 := f(I_1)$. Then there are $\delta > 0$ and a unique continuous function $\xi : (y_0 - \delta, y_0 + \delta) \cap J \to \mathbb{R}$ such that $\xi(y_0) = x_0$ and $f(\xi(y)) = y$ for all $y \in (y_0 - \delta, y_0 + \delta) \cap J$. In particular, $f^{-1} : J_1 \to \mathbb{R}$ is continuous at y_0.*

Proof. First, let us consider the case in which x_0 is an interior point of I. Then we may choose $r > 0$ such that $(x_0 - r, x_0 + r) \subseteq I$, and therefore $I_1 = (x_0 - r, x_0 + r)$. Consider $h : \mathbb{S}_r(x_0, y_0) \to \mathbb{R}$ defined by $h(x, y) := f(x) - y$. Then h is continuous, $h(x_0, y_0) = 0$, and given any $y \in (y_0 - r, y_0 + r)$, the function from I_1 to $\mathbb{R}$ given by $x \longmapsto h(x, y)$ is strictly monotonic. Hence by the Implicit Function Theorem (Proposition 2.40 and Remark 2.42), there are $\delta > 0$ and a unique continuous function $\xi : (y_0-\delta, y_0+\delta) \to \mathbb{R}$ with $\xi(y_0) = x_0$ such that $(\xi(y), y) \in \mathbb{S}_r(x_0, y_0)$ and $h(\xi(y), y) = 0$ for all $y \in (y_0 - \delta, y_0 + \delta)$. Consequently, $f(\xi(y)) = y$ for all $y \in (y_0 - \delta, y_0 + \delta) \cap J$ and, in particular, $(y_0 - \delta, y_0 + \delta) \subseteq J$. Since f is continuous and strictly monotonic on $I_1 = (x_0 - r, x_0 + r) \subseteq I$, it follows that y_0 is an interior point of $J_1 := f(I_1)$ and $f^{-1} = \xi$ on J_1. Hence $f^{-1} : J_1 \to \mathbb{R}$ is continuous at y_0.

In case x_0 is an endpoint of I, we can extend f to a continuous, strictly monotonic function f^* on a larger interval I^* such that x_0 is an interior point of I^*. For example, if f is strictly increasing and $I = [x_0, b)$, then we may take $I^* := [x_0 - 1, b)$ and $f^*(x) := f(x)$ if $x \in [x_0, b)$ and $f^*(x) := (x - x_0) + y_0$ if $x \in [x_0 - 1, x_0)$. Applying the arguments in the previous paragraph to f^*, we obtain the desired result. □

As an immediate corollary of Proposition 2.43, we obtain an alternative proof of the Continuous Inverse Theorem for functions of one variable (given, for example, on page 78 of ACICARA), which asserts that a continuous one-one function defined on an interval has a continuous inverse. To this end, we shall use the following fact from the theory of functions of one variable, which is completely elementary in the sense that neither the statement nor the proof involves the notions of continuity or limits. For a proof of this fact and also for some related results, one may refer to page 29 of ACICARA.

Fact 2.44. *Let I be an interval in $\mathbb{R}$. If $f : I \to \mathbb{R}$ is one-one and has the IVP on I, then f is strictly monotonic on I.*

Corollary 2.45. *Let I be an interval in $\mathbb{R}$ and let $f : I \to \mathbb{R}$ be a one-one continuous function. Then the inverse function $f^{-1} : f(I) \to \mathbb{R}$ is continuous.*

Proof. By part (i) of Fact 2.33, f has the IVP on I. So, by Fact 2.44, f is strictly monotonic on I. Hence by Proposition 2.43, f^{-1} is continuous. □

The notion of continuity can be extended to functions of three or more variables in a completely analogous manner. Most results extend to this case in a straightforward way. A result for which the extension to functions of three variables may not be immediate is the Implicit Function Theorem (Proposition 2.40). Recall that the latter may be roughly stated by saying that if around a point, $f(x, y)$ is continuous in x as well as in y and strictly monotonic in y, then we can solve the equation $f(x, y) = 0$ for y in terms of x around that point. It turns out that for functions of three variables, in order to solve $f(x, y, z) = 0$ for z in terms of x and y around a point, what we need apart from the strict monotonicity in z is not just the continuity in each of the three variables, but the continuity in the variable z and the (bivariate) continuity in x and y. In effect, the statement as well as the proof of Proposition 2.40 generalize easily if the variable x is replaced by two (or more) variables. For ease of reference, we record below a precise statement of this result. Formulation of analogues as in Remark 2.42 and a general result in the case of functions of n variables is left to the reader.

Proposition 2.46 (Trivariate Implicit Function Theorem). *Let $D \subseteq \mathbb{R}^3$, $(x_0, y_0, z_0) \in D$, and $f : D \to \mathbb{R}$ be such that $f(x_0, y_0, z_0) = 0$. Assume that there is $r > 0$ with $\mathbb{S}_r(x_0, y_0, z_0) \subseteq D$ and the following conditions hold:*

(a) *Given any $(x, y) \in \mathbb{S}_r(x_0, y_0)$, the function $\psi : (z_0 - r, z_0 + r) \to \mathbb{R}$ defined by $\psi(z) = f(x, y, z)$ is continuous. Also, given any $z \in (z_0 - r, z_0 + r)$, the function $\phi : \mathbb{S}_r(x_0, y_0) \to \mathbb{R}$ defined by $\phi(x, y) = f(x, y, z)$ is continuous.*
(b) *Given any $(x, y) \in \mathbb{S}_r(x_0, y_0)$, the function $\psi : (z_0 - r, z_0 + r) \to \mathbb{R}$ defined by $\psi(z) = f(x, y, z)$ is strictly monotonic.*

Then there are $\delta > 0$ and a unique continuous function $\zeta : \mathbb{S}_\delta(x_0, y_0) \to \mathbb{R}$ with $\zeta(x_0, y_0) = z_0$ such that $(x, y, \zeta(x, y)) \in \mathbb{S}_r(x_0, y_0, z_0)$ and $f(x, y, \zeta(x, y)) = 0$ for all $(x, y) \in \mathbb{S}_\delta(x_0, y_0)$.

Proof. The proof is similar to that of Proposition 2.40 if we make appropriate notational changes. □

2.3 Limits

Let $D \subseteq \mathbb{R}^2$ and $(x_0, y_0) \in \mathbb{R}^2$. Assume that an open square of positive radius centered at (x_0, y_0), except possibly the center, is contained in D, that is, $\mathbb{S}_r(x_0, y_0) \setminus \{(x_0, y_0)\} \subseteq D$ for some $r > 0$. Let $f : D \to \mathbb{R}$ be any function. We say that a **limit** of f as (x, y) tends to (x_0, y_0) exists if there is a real number ℓ such that whenever a sequence $\big((x_n, y_n)\big)$ in $D \setminus \{(x_0, y_0)\}$ converges to (x_0, y_0), we have $f(x_n, y_n) \to \ell$. We then write $f(x, y) \to \ell$ as $(x, y) \to (x_0, y_0)$. It may be noted that there do exist sequences in $D \setminus \{(x_0, y_0)\}$ that converge to (x_0, y_0). For example,

$$(x_n, y_n) := \left(x_0 - \frac{r}{n+1}, \; y_0 - \frac{r}{n+1}\right) \quad \text{for } n \in \mathbb{N}$$

defines one such sequence. Using this and the fact that the limit of a sequence in $\mathbb{R}^2$ is unique (part (i) of Proposition 2.1), we readily see that if a limit of f as (x, y) tends to (x_0, y_0) exists, then it is unique. With this in view, if $f(x, y) \to \ell$ as $(x, y) \to (x_0, y_0)$, then we may refer to ℓ as *the* limit of $f(x, y)$ as (x, y) tends to (x_0, y_0), and write

$$\lim_{(x,y)\to(x_0,y_0)} f(x, y) = \ell.$$

Examples 2.47. (i) Consider $f : \mathbb{R}^2 \to \mathbb{R}$ defined by $f(0, 0) := 1$ and $f(x, y) := \sin(xy)$ for $(x, y) \in \mathbb{R}^2 \setminus \{(0, 0)\}$. Then the limit of f as (x, y) tends to $(0, 0)$ exists and is equal to 0. Indeed, if $\big((x_n, y_n)\big)$ is a sequence in $\mathbb{R}^2 \setminus \{(0, 0)\}$ such that $(x_n, y_n) \to (0, 0)$, then $x_n y_n \to 0$, and by the continuity of the sine function, $\sin(x_n y_n) \to \sin 0 = 0$, that is, $f(x_n, y_n) \to 0$.

(ii) Consider $f : \mathbb{R}^2 \to \mathbb{R}$ defined by

$$f(x, y) = \begin{cases} x + y & \text{if } x \neq y, \\ 1 & \text{if } x = y. \end{cases}$$

Then the limit of f as (x, y) tends to $(0, 0)$ does not exist. This can be seen by considering two sequences approaching $(0, 0)$, one along the line $y = x$ and another staying away from this line. For example, if $(x_n, y_n) := (1/n, 1/n)$ and $(u_n, v_n) := (-1/n, 1/n)$ for $n \in \mathbb{N}$, then $((x_n, y_n))$ and $((u_n, v_n))$ are sequences in $\mathbb{R}^2 \setminus \{(0, 0)\}$ converging to $(0, 0)$, but $f(x_n, y_n) \to 1$ and $f(u_n, v_n) \to 0$.

(iii) Consider $f : \mathbb{R}^2 \setminus \{(0, 0)\} \to \mathbb{R}$ given by $f(x, y) = xy/(x^2 + y^2)$ for $(x, y) \in \mathbb{R}^2$, $(x, y) \neq (0, 0)$. Then the limit of f as (x, y) tends to $(0, 0)$ does not exist. This can also be seen by considering two sequences approaching $(0, 0)$, along different lines through the origin. For example, if $(x_n, y_n) := (1/n, 1/n)$ and $(u_n, v_n) := (1/n, 2/n)$ for $n \in \mathbb{N}$, then $((x_n, y_n))$ and $((u_n, v_n))$ are sequences in $\mathbb{R}^2 \setminus \{(0, 0)\}$ converging to $(0, 0)$, but $f(x_n, y_n) \to \frac{1}{2}$ and $f(u_n, v_n) \to \frac{2}{5}$. ◇

Limits and Continuity

The concepts of continuity and limit are related in a similar way as in the case of functions of one variable.

Proposition 2.48. *Let $D \subseteq \mathbb{R}^2$ and let $(x_0, y_0) \in \mathbb{R}^2$ be an interior point of D, that is, $\mathbb{S}_r(x_0, y_0) \subseteq D$ for some $r > 0$. Let $f : D \to \mathbb{R}$ be any function. Then f is continuous at (x_0, y_0) if and only if the limit of f as (x, y) tends to (x_0, y_0) exists and is equal to $f(x_0, y_0)$.*

Proof. Assume that f is continuous at (x_0, y_0). Let $((x_n, y_n))$ be any sequence in D such that $(x_n, y_n) \to (x_0, y_0)$. By the continuity of f at (x_0, y_0), we see that $f(x_n, y_n) \to f(x_0, y_0)$. It follows that the limit of f as (x, y) tends to (x_0, y_0) exists and is equal to $f(x_0, y_0)$.

To prove the converse, assume that the limit of f as (x, y) tends to (x_0, y_0) exists and is equal to $f(x_0, y_0)$. Let $((x_n, y_n))$ be any sequence in D such that $(x_n, y_n) \to (x_0, y_0)$. If there is $n_0 \in \mathbb{N}$ such that $(x_n, y_n) = (x_0, y_0)$ for all $n \geq n_0$, then it is clear that $f(x_n, y_n) \to f(x_0, y_0)$. Otherwise, there are positive integers $n_1, n_2, \ldots$ such that $n_1 < n_2 < \cdots$ and $\{n \in \mathbb{N} : (x_n, y_n) \neq (x_0, y_0)\} = \{n_k : k \in \mathbb{N}\}$. Now, $((x_{n_k}, y_{n_k}))$ is a sequence in $D \setminus \{(x_0, y_0)\}$ that converges to (x_0, y_0), and therefore $f(x_{n_k}, y_{n_k}) \to f(x_0, y_0)$. Since $f(x_n, y_n) = f(x_0, y_0)$ for all $n \in \mathbb{N} \setminus \{n_k : k \in \mathbb{N}\}$, it follows that $f(x_n, y_n) \to f(x_0, y_0)$. Hence f is continuous at (x_0, y_0). □

As a consequence, we obtain a useful characterization for the existence of the limit of a function in terms of the continuity of an associated function.

Corollary 2.49. *Let $D \subseteq \mathbb{R}^2$ and $(x_0, y_0) \in \mathbb{R}^2$ be such that D contains $\mathbb{S}_r(x_0, y_0) \setminus \{(x_0, y_0)\}$ for some $r > 0$. Given a function $f : D \to \mathbb{R}$ and $\ell \in \mathbb{R}$, let $F : D \cup \{(x_0, y_0)\} \to \mathbb{R}$ be the function defined by*

$$F(x, y) := \begin{cases} f(x, y) & \text{if } (x, y) \in D \setminus \{(x_0, y_0)\}, \\ \ell & \text{if } (x, y) = (x_0, y_0). \end{cases}$$

Then

$$\lim_{(x,y)\to(x_0,y_0)} f(x, y) \text{ exists and is equal to } \ell \iff F \text{ is continuous at } (x_0, y_0).$$

Proof. Since $f(x, y) = F(x, y)$ for $(x, y) \in D \setminus \{(x_0, y_0)\}$, it is clear that $\lim_{(x,y)\to(x_0,y_0)} f(x, y)$ exists if and only if $\lim_{(x,y)\to(x_0,y_0)} F(x, y)$ exists, and in this case the two limits are equal. Further, since (x_0, y_0) is an interior point of $D \cup \{(x_0, y_0)\}$ and $F(x_0, y_0) = \ell$, the desired result follows from applying Proposition 2.48 to F. □

Examples 2.50. (i) In view of Proposition 2.48 and Example 2.16 (i), we see that every rational function has a limit wherever it is defined, that is, if $p(x, y)$ and $q(x, y)$ are polynomials in two variables and if $(x_0, y_0) \in \mathbb{R}^2$ is such that $q(x_0, y_0) \neq 0$, then

$$\lim_{(x,y)\to(x_0,y_0)} \frac{p(x, y)}{q(x, y)} = \frac{p(x_0, y_0)}{q(x_0, y_0)}.$$

On the other hand, if $q(x_0, y_0) = 0$, then the limit of $p(x, y)/q(x, y)$ may not exist, in general. For example, for any $m, k \in \mathbb{N}$, the rational function $f(x, y) := x^m/y^k$ does not have a limit as (x, y) tends to $(0, 0)$. To see this, it suffices to approach $(0, 0)$ along the parametric curve given by $(x(t), y(t)) = (\alpha t^k, \beta t^m)$, $t \in [-1, 1]$, where α, β are any nonzero constants. For example, if $(x_n, y_n) := (1/n^k, 1/n^m)$ and $(u_n, v_n) := (2/n^k, 1/n^m)$ for $n \in \mathbb{N}$, then $((x_n, y_n))$ and $((u_n, v_n))$ are sequences in $\mathbb{R}^2 \setminus \{(0, 0)\}$ converging to $(0, 0)$, but $f(x_n, y_n) \to 1$ and $f(u_n, v_n) \to 2^m$.

(ii) Consider $f : \mathbb{R}^2 \setminus \{(0, 0)\} \to \mathbb{R}$ defined by $f(x, y) = x^2y/(x^2 + y^2)$. Then in view of Proposition 2.48 and Example 2.16 (i), we see that the limit of $f(x, y)$ as (x, y) tends to $(0, 0)$ exists and is equal to 0. ◇

Thanks to Corollary 2.49, basic properties of limits of real-valued functions of two variables can be deduced from the corresponding properties of continuous functions.

Proposition 2.51. *Let $D \subseteq \mathbb{R}^2$ and $(x_0, y_0) \in \mathbb{R}^2$ be such that D contains $\mathbb{S}_t(x_0, y_0) \setminus \{(x_0, y_0)\}$ for some $t > 0$. Let $f, g : D \to \mathbb{R}$, and let $\ell, m \in \mathbb{R}$ be such that*

$$\lim_{(x,y)\to(x_0,y_0)} f(x, y) = \ell \quad \text{and} \quad \lim_{(x,y)\to(x_0,y_0)} g(x, y) = m.$$

Then for any $r \in \mathbb{R}$, the limits of $f+g$, rf, and fg as (x, y) tends to (x_0, y_0) exist, and are equal to $\ell + m$, $r\ell$, and ℓm respectively. Moreover, if $\ell \neq 0$, then there is $\delta > 0$ such that $f(x, y) \neq 0$ for all $(x, y) \in D \cap \mathbb{S}_\delta(x_0, y_0) \setminus \{(x_0, y_0)\}$, and the limit of $1/f : D \cap \mathbb{S}_\delta(x_0, y_0) \setminus \{(x_0, y_0)\} \to \mathbb{R}$ as (x, y) tends to (x_0, y_0) exists, and is equal to $1/\ell$.

Proof. Let $F, G : D \cup \{(x_0, y_0)\} \to \mathbb{R}$ be the functions defined by letting $F(x, y) := f(x, y)$ and $G(x, y) := g(x, y)$ for $(x, y) \in D \setminus \{(x_0, y_0)\}$ and setting $F(x_0, y_0) := \ell$ and $G(x_0, y_0) := m$. By Corollary 2.49, F and G are continuous at (x_0, y_0). So the assertion concerning the limits of $f+g$, rf, and fg follow from Proposition 2.15 and Corollary 2.49. If $\ell \neq 0$, then the desired existence of δ and the limit of $1/f$ follow from Lemma 2.14, Proposition 2.15, and Corollary 2.49. □

As in the case of functions of one variable, if there are certain inequalities among the values of real-valued functions of two variables, then the same prevail when we pass to limits, provided the limits exist. But of course, strict inequalities such as $<$ can change to weak inequalities such as $\leq$ when we pass to the limit. (See Exercise 11.) On the other hand, strict inequalities on limits yield strict inequalities on the values of the corresponding function around the point where the limit is taken. (See Exercise 12.) Moreover, for nonnegative functions, extraction of roots is preserved by passing to limits.

Proposition 2.52. *Let $D, (x_0, y_0), r, f, g, \ell$, and m be as in Proposition 2.51.*

(i) *If there is $\delta > 0$ with $\delta \leq r$ such that $f(x, y) \leq g(x, y)$ for all (x, y) in $\mathbb{S}_\delta(x_0, y_0) \setminus \{(x_0, y_0)\}$, then $\ell \leq m$. Conversely, if $\ell < m$, then there is $\delta > 0$ such that $\delta \leq r$ and $f(x, y) < g(x, y)$ for all $(x, y) \in \mathbb{S}_\delta(x_0, y_0) \setminus \{(x_0, y_0)\}$.*

(ii) *If $f(x, y) \geq 0$ for all $(x, y) \in D$, then $\ell \geq 0$ and for each $k \in \mathbb{N}$, the limit of $f^{1/k} : D \to \mathbb{R}$ as (x, y) tends to (x_0, y_0) exists, and is equal to $\ell^{1/k}$.*

(iii) **[Sandwich Theorem]** *If $\ell = m$ and if there is $h : D \to \mathbb{R}$ such that $f(x, y) \leq h(x, y) \leq g(x, y)$ for all $(x, y) \in D$, then the limit of h as (x, y) tends to (x_0, y_0) exists, and is equal to ℓ.*

Proof. Consider $H : D \cup \{(x_0, y_0)\} \to \mathbb{R}$ defined by $H(x, y) := g(x, y) - f(x, y)$ for $(x, y) \in D \setminus \{(x_0, y_0)\}$ and $H(x_0, y_0) := m - \ell$. By Corollary 2.49 and Proposition 2.51, H is continuous at (x_0, y_0). If $\ell > m$, then $H(x_0, y_0) < 0$ and hence by Lemma 2.14, there is $\eta > 0$ such that $H(x, y) < 0$, that is, $f(x, y) > g(x, y)$ for all $(x, y) \in D \cap \mathbb{S}_\eta(x_0, y_0)$. This contradicts the assumption on f and g. Hence $\ell \leq m$. Conversely, suppose $\ell < m$. Then $H(x_0, y_0) > 0$, and hence by Lemma 2.14, there is $\delta > 0$ such that $H(x, y) > 0$ for all $(x, y) \in D \cap \mathbb{S}_\delta(x_0, y_0)$, and so $f(x, y) < g(x, y)$ for all $(x, y) \in D \cap \mathbb{S}_\delta(x_0, y_0)$. This proves (i). Next, if $f(x, y) \geq 0$ for all $(x, y) \in D$, then by (i), we obtain $\ell \geq 0$. Further, given any $k \in \mathbb{N}$, the assertion about the limit of $f^{1/k}$ follows from Proposition 2.15 and Corollary 2.49. Finally, (iii) is an immediate consequence of part (vi) of Fact 2.3. □

As in the case of functions of one variable, a criterion for the existence of the limit of a real-valued function of two variables that does not involve convergence of sequences can be given as follows.

Proposition 2.53. *Let $D \subseteq \mathbb{R}^2$ and $(x_0, y_0) \in \mathbb{R}^2$ be such that D contains $\mathbb{S}_r(x_0, y_0) \setminus \{(x_0, y_0)\}$ for some $r > 0$ and let $f : D \to \mathbb{R}$ be a function. Then the limit of $f(x, y)$ as (x, y) tends to (x_0, y_0) exists if and only if there is $\ell \in \mathbb{R}$ satisfying the following ϵ-δ condition: For every $\epsilon > 0$, there is $\delta > 0$ such that*

$$(x, y) \in D \cap \mathbb{S}_\delta(x_0, y_0) \text{ and } (x, y) \neq (x_0, y_0) \Longrightarrow |f(x, y) - \ell| < \epsilon.$$

Proof. Given $\ell \in \mathbb{R}$, let $F : D \cup \{(x_0, y_0)\} \to \mathbb{R}$ be the function associated with f and ℓ as in Corollary 2.49. Using the equivalence of (i) and (ii) in Proposition 2.22 together with Corollary 2.49, we obtain the desired result. □

The above characterization yields the following analogue of the Cauchy Criterion for sequences in $\mathbb{R}^2$ (part (iv) of Proposition 2.6).

Proposition 2.54 (Cauchy Criterion for Limits of Functions). *Suppose $D \subseteq \mathbb{R}^2$ and $(x_0, y_0) \in \mathbb{R}^2$ are such that D contains $\mathbb{S}_r(x_0, y_0) \setminus \{(x_0, y_0)\}$ for some $r > 0$. Let $f : D \to \mathbb{R}$ be a function. Then $\lim_{(x,y)\to(x_0,y_0)} f(x, y)$ exists if and only if for every $\epsilon > 0$, there is $\delta > 0$ such that*

$$(x, y), (u, v) \in D \cap \mathbb{S}_\delta(x_0, y_0) \setminus \{(x_0, y_0)\} \Longrightarrow |f(x, y) - f(u, v)| < \epsilon.$$

Proof. Assume that $\ell := \lim_{(x,y)\to(x_0,y_0)} f(x, y)$ exists. Let $\epsilon > 0$ be given. By Proposition 2.53, there is $\delta > 0$ such that $|f(x, y) - \ell| < \epsilon/2$ for all (x, y) in $D \cap \mathbb{S}_\delta(x_0, y_0) \setminus \{(x_0, y_0)\}$. Hence for $(x, y), (u, v) \in D \cap \mathbb{S}_\delta(x_0, y_0) \setminus \{(x_0, y_0)\}$, we obtain $|f(x, y) - f(u, v)| \le |f(x, y) - \ell| + |\ell - f(u, v)| < (\epsilon/2) + (\epsilon/2) = \epsilon$, as desired. The converse follows readily from the Cauchy Criterion for limits of sequences in $\mathbb{R}$ (part (iv) of Fact 2.5). □

Limit from a Quadrant

An analogue of the notion of left(-hand) or right(-hand) limits for functions of one variable is given by limits from any one of the four quadrants for functions of two variables. These may be defined as follows.

Let $D \subseteq \mathbb{R}^2$ and $(x_0, y_0) \in \mathbb{R}^2$ be such that $(x_0, x_0 + r) \times (y_0, y_0 + r) \subseteq D$ for some $r > 0$. Given a function $f : D \to \mathbb{R}$, we say that a **limit of f from the first quadrant** as (x, y) tends to (x_0, y_0) exists if there is a real number ℓ such that whenever $\big((x_n, y_n)\big)$ is a sequence in $D \setminus \{(x_0, y_0)\}$ satisfying $(x_n, y_n) \ge (x_0, y_0)$ for all $n \in \mathbb{N}$ and $(x_n, y_n) \to (x_0, y_0)$, we have $f(x_n, y_n) \to \ell$. It is easy to see that if such a limit exists, then it is unique. In this case, we write

$$f(x, y) \to \ell \text{ as } (x, y) \to \left(x_0^+, y_0^+\right) \quad \text{or} \quad \lim_{(x,y)\to\left(x_0^+, y_0^+\right)} f(x, y) = \ell.$$

Similarly, we can define limits of f from the second, the third, and the fourth quadrants. Obvious analogues of the above notation are then used.

Remark 2.55. For limits from a quadrant, Corollary 2.49 admits a straightforward analogue. More precisely, let $D \subseteq \mathbb{R}^2$ and $(x_0, y_0) \in \mathbb{R}^2$ be such that $(x_0, x_0 + r) \times (y_0, y_0 + r) \subseteq D$ for some $r > 0$. Consider $D_1 := \{(x, y) \in D : x \geq x_0 \text{ and } y \geq y_0\}$ and $F_1 : D_1 \cup \{(x_0, y_0)\} \to \mathbb{R}$ defined by

$$F_1(x, y) := \begin{cases} f(x, y) & \text{if } (x, y) \in D_1 \setminus \{(x_0, y_0)\}, \\ \ell & \text{if } (x, y) = (x_0, y_0). \end{cases}$$

Then

$$\lim_{(x,y)\to(x_0^+, y_0^+)} f(x, y) \text{ exists and is equal to } \ell \iff F_1 \text{ is continuous at } (x_0, y_0).$$

This can be proved by a similar argument as in Corollary 2.49. Moreover, analogous results for limits from the second, the third, and the fourth quadrants can be readily obtained. $\diamond$

Proposition 2.56. *Let $D \subseteq \mathbb{R}^2$ and $(x_0, y_0) \in \mathbb{R}^2$ be such that D contains $\mathbb{S}_r(x_0, y_0) \setminus \{(x_0, y_0)\}$ for some $r > 0$. Let $f : D \to \mathbb{R}$ be a function and let $\ell \in \mathbb{R}$. Then $\lim_{(x,y)\to(x_0,y_0)} f(x, y) = \ell$ if and only if $\lim_{(x,y)\to(x_0^+, y_0^+)} f(x, y)$, $\lim_{(x,y)\to(x_0^-, y_0^+)} f(x, y)$, $\lim_{(x,y)\to(x_0^-, y_0^-)} f(x, y)$, and $\lim_{(x,y)\to(x_0^+, y_0^-)} f(x, y)$ exist and are all equal to ℓ. If, in addition, $(x_0, y_0) \in D$, then f is continuous at (x_0, y_0) if and only if the limit of f from each of the four quadrants as (x, y) tends to (x_0, y_0) exists and they are all equal to $f(x_0, y_0)$.*

Proof. If $\lim_{(x,y)\to(x_0,y_0)} f(x, y) = \ell$, then it is clear that the limit of f from each of the four quadrants as (x, y) tends to (x_0, y_0) exists and they are all equal to ℓ. To prove the converse, suppose the limit of f from each of the four quadrants as (x, y) tends to (x_0, y_0) exists and they are all equal to ℓ. Consider $F : D \cup \{(x_0, y_0)\} \to \mathbb{R}$ defined by $F(x_0, y_0) := \ell$ and $F(x, y) := f(x, y)$ for $(x, y) \in D$ with $(x, y) \neq (x_0, y_0)$. Let $D_1 := \{(x, y) \in D : x \geq x_0 \text{ and } y \geq y_0\}$, $D_2 := \{(x, y) \in D : x \leq x_0 \text{ and } y \geq y_0\}$, $D_3 := \{(x, y) \in D : x \leq x_0 \text{ and } y \leq y_0\}$, and $D_4 := \{(x, y) \in D : x \geq x_0 \text{ and } y \leq y_0\}$. Also, let $\widetilde{D}_i := D_i \cup \{(x_0, y_0)\}$ and $F_i := F_{|\widetilde{D}_i}$ for $i = 1, 2, 3, 4$. In view of Remark 2.55, we see that F_i is continuous at (x_0, y_0) for $i = 1, 2, 3, 4$. Hence by Corollary 2.21, F is continuous at (x_0, y_0), and therefore by Corollary 2.49, $\lim_{(x,y)\to(x_0,y_0)} f(x, y) = \ell$.

In case $(x_0, y_0) \in D$, the assertion about the continuity of f at (x_0, y_0) follows from what is proved above and Proposition 2.48. $\square$

Approaching Infinity

Let $D \subseteq \mathbb{R}^2$ be such that D contains a product of semi-infinite open intervals of the form $(a, \infty) \times (c, \infty)$, where $a, c \in \mathbb{R}$. Given a function $f : D \to \mathbb{R}$, we say that a **limit** of f as (x, y) tends to (∞, ∞) exists if there is a real number ℓ satisfying the following property:

$$\big((x_n, y_n)\big) \text{ any sequence in } D \text{ with } x_n \to \infty \text{ and } y_n \to \infty \implies f(x_n, y_n) \to \ell.$$

In this case the real number ℓ is unique and it is sometimes denoted by $\lim_{(x,y)\to(\infty,\infty)} f(x, y)$. Similarly, we can define a limit of f as $(x, y) \to (-\infty, \infty)$, or as $(x, y) \to (-\infty, -\infty)$, or as $(x, y) \to (\infty, -\infty)$, provided of course the domain D of f contains a product of semi-infinite open intervals of the form $(-\infty, b) \times (c, \infty)$, $(-\infty, b) \times (-\infty, d)$, or $(a, \infty) \times (-\infty, d)$, as the case may be, for some $a, b, c, d \in \mathbb{R}$. An alternative definition that is analogous to the ϵ-δ characterization (Proposition 2.53) can be given for such limits. It should suffice to consider the case of limits as $(x, y) \to (\infty, \infty)$. We leave a formulation of the statement and proofs in the other three cases as an exercise.

Proposition 2.57. *Let $D \subseteq \mathbb{R}^2$ be such that $D \supseteq (a, \infty) \times (c, \infty)$ for some $a, c \in \mathbb{R}$, and let $f : D \to \mathbb{R}$ be a function. Then $\lim_{(x,y)\to(\infty,\infty)} f(x, y)$ exists if and only if there is $\ell \in \mathbb{R}$ satisfying the following ϵ-(α, β) condition: For every $\epsilon > 0$, there are $\alpha, \beta \in \mathbb{R}$ such that*

$$(x, y) \in D \text{ with } (x, y) \geq (\alpha, \beta) \implies |f(x, y) - \ell| < \epsilon.$$

Proof. Assume that $\lim_{(x,y)\to(\infty,\infty)} f(x, y)$ exists and is equal to a real number ℓ. Suppose the ϵ-(α, β) condition is not satisfied. Then there is $\epsilon > 0$ such that for every $\alpha, \beta \in \mathbb{R}$, we can find $(x, y) \in D$ with $(x, y) \geq (\alpha, \beta)$, but $|f(x, y) - \ell| \geq \epsilon$. Taking $(\alpha, \beta) = (n, n)$, as n varies over $\mathbb{N}$, we obtain a sequence $\big((x_n, y_n)\big)$ in D such that $x_n \to \infty$ and $y_n \to \infty$, but $f(x_n, y_n) \not\to \ell$. This contradicts $\lim_{(x,y)\to(\infty,\infty)} f(x, y) = \ell$.

Conversely, assume the ϵ-(α, β) condition. Let $\big((x_n, y_n)\big)$ be a sequence in D such that $x_n \to \infty$ and $y_n \to \infty$. Given any $\epsilon > 0$, find $\alpha, \beta \in \mathbb{R}$ for which $\alpha > a$ and $\beta > c$. Now, there is $n_0 \in \mathbb{N}$ such that $(x_n, y_n) \geq (\alpha, \beta)$ for all $n \geq n_0$, and hence $|f(x_n, y_n) - \ell| < \epsilon$ for all $n \geq n_0$. Thus $f(x_n, y_n) \to \ell$, and so $\lim_{(x,y)\to(\infty,\infty)} f(x, y) = \ell$. □

As in the case of functions of one variable, in some cases ∞ or $-\infty$ can be regarded as a "limit" of a function of two variables. Let $D \subseteq \mathbb{R}^2$ and $(x_0, y_0) \in \mathbb{R}^2$ be such that D contains $\mathbb{S}_r(x_0, y_0) \setminus \{(x_0, y_0)\}$ for some $r > 0$ and let $f : D \to \mathbb{R}$ be any function. We say that $f(x, y)$ tends to ∞ as (x, y) tends to (x_0, y_0) if for every sequence $\big((x_n, y_n)\big)$ in $D\setminus\{(x_0, y_0)\}$ that converges to (x_0, y_0), we have $f(x_n, y_n) \to \infty$. We then write

$$f(x, y) \to \infty \quad \text{as} \quad (x, y) \to (x_0, y_0).$$

Likewise, we say that $f(x, y)$ tends to $-\infty$ as (x, y) tends to (x_0, y_0) if for every sequence $\big((x_n, y_n)\big)$ in $D \setminus \{(x_0, y_0)\}$ that converges to (x_0, y_0), we have $f(x_n, y_n) \to -\infty$. We then write

$$f(x,y) \to -\infty \text{ as } (x,y) \to (x_0,y_0).$$

For example,

$$\frac{1}{x^2+y^2} \to \infty \text{ as } (x,y) \to (0,0) \quad \text{and} \quad -\frac{1}{x^2+y^2} \to -\infty \text{ as } (x,y) \to (0,0).$$

We now give an analogue of Proposition 2.53 for a real-valued function of two variables that tends to ∞ or to $-\infty$.

Proposition 2.58. *Let $D \subseteq \mathbb{R}^2$ and $(x_0,y_0) \in \mathbb{R}^2$ be such that D contains $\mathbb{S}_r(x_0,y_0)\setminus\{(x_0,y_0)\}$ for some $r > 0$ and let $f : D \to \mathbb{R}$ be any function. Then $f(x,y) \to \infty$ as $(x,y) \to (x_0,y_0)$ if and only if the following α-δ condition holds: For every $\alpha \in \mathbb{R}$, there is $\delta > 0$ such that*

$$(x,y) \in D \cap \mathbb{S}_\delta(x_0,y_0) \text{ and } (x,y) \neq (x_0,y_0) \Longrightarrow f(x,y) > \alpha.$$

Likewise, $f(x,y) \to -\infty$ as $(x,y) \to (x_0,y_0)$ if and only if the following β-δ condition holds: For every $\beta \in \mathbb{R}$, there is $\delta > 0$ such that

$$(x,y) \in D \cap \mathbb{S}_\delta(x_0,y_0) \text{ and } (x,y) \neq (x_0,y_0) \Longrightarrow f(x,y) < \beta.$$

Proof. Assume that $f(x,y) \to \infty$ as $(x,y) \to (x_0,y_0)$. If the α-δ condition does not hold, then there exists $\alpha \in \mathbb{R}$ such that for every $\delta > 0$, there is $(x,y) \in D \cap \mathbb{S}_\delta(x_0,y_0)$ with $(x,y) \neq (x_0,y_0)$ and $f(x,y) \leq \alpha$. Taking $\delta = 1/n$ as n varies over $\mathbb{N}$, we obtain a sequence $\big((x_n,y_n)\big)$ in $D\setminus\{(x_0,y_0)\}$ such that $(x_n,y_n) \to (x_0,y_0)$, but $f(x_n,y_n) \not\to \infty$. This contradicts the assumption.

Conversely, assume the α-δ condition. Let $\big((x_n,y_n)\big)$ be a sequence in $D\setminus\{(x_0,y_0)\}$ such that $(x_n,y_n) \to (x_0,y_0)$, and let $\alpha > 0$ be given. Then there is $\delta > 0$ such that $f(x,y) > \alpha$ for all $(x,y) \in D \cap \mathbb{S}_\delta(x_0,y_0)$ with $(x,y) \neq (x_0,y_0)$. Further, there is $n_0 \in \mathbb{N}$ such that $(x_n,y_n) \in \mathbb{S}_\delta(x_0,y_0)$ for $n \geq n_0$. Hence $f(x_n,y_n) > \alpha$ for $n \geq n_0$. Thus $f(x,y) \to \infty$ as $(x,y) \to (x_0,y_0)$.

The equivalence of the condition $f(x,y) \to -\infty$ as $(x,y) \to (x_0,y_0)$ with the β-δ condition is proved similarly. □

Recall that we have defined the notion of a monotonically increasing function of two variables using the product order on $\mathbb{R}^2$. We show below that for such functions, existence of a limit from the first or the third quadrant is equivalent to boundedness properties.

Proposition 2.59. *Let $a,b,c,d \in \mathbb{R}\cup\{-\infty,\infty\}$ with $a < b$ and $c < d$ be such that either $a,c \in \mathbb{R}$ or $a = c = -\infty$, and either $b,d \in \mathbb{R}$ or $b = d = \infty$. Let $f : (a,b)\times(c,d) \to \mathbb{R}$ be a monotonically increasing function. Then*

(i) $\lim_{(x,y)\to(b^-,d^-)} f(x,y)$ *exists if and only if f is bounded above; in this case, $\lim_{(x,y)\to(b^-,d^-)} f(x,y) = \sup\{f(x,y) : (x,y) \in (a,b)\times(c,d)\}$. If f is not bounded above, then $f(x,y) \to \infty$ as $(x,y) \to (b^-,d^-)$.*

(ii) $\lim_{(x,y)\to(a^+,c^+)} f(x,y)$ *exists if and only if* f *is bounded below; in this case,* $\lim_{(x,y)\to(a^+,c^+)} f(x,y) = \inf\{f(x,y) : (x,y) \in (a,b) \times (c,d)\}$. *If* f *is not bounded below, then* $f(x,y) \to -\infty$ *as* $(x,y) \to (a^+,c^+)$.

Proof. (i) Suppose f is bounded above. Let $M := \sup\{f(x,y) : (x,y) \in (a,b) \times (c,d)\}$. Given any $\epsilon > 0$, there is $(b_0,d_0) \in (a,b) \times (c,d)$ such that $M - \epsilon < f(b_0,d_0) \le M$. Now, if $((x_n,y_n))$ is any sequence in $(a,b)\times(c,d)$ such that $(x_n,y_n) \to (b,d)$, then there is $n_0 \in \mathbb{N}$ such that $(b_0,d_0) \le (x_n,y_n)$ for $n \ge n_0$. Since f is monotonically increasing, we obtain $M-\epsilon < f(x_n,y_n) \le M$ for $n \ge n_0$. It follows that $\lim_{(x,y)\to(b^-,d^-)} f(x,y)$ exists and is equal to M.

On the other hand, suppose f is not bounded above. Let $\alpha \in \mathbb{R}$. Then there is $(b_0,d_0) \in (a,b) \times (c,d)$ such that $f(b_0,d_0) > \alpha$. Since f is monotonically increasing, we see that $f(x,y) > \alpha$ for all $(x,y) \in (b_0,b) \times (d_0,d)$. Now, if $((x_n,y_n))$ is any sequence in $(a,b) \times (c,d)$ such that $(x_n,y_n) \to (b,d)$, then there is $n_0 \in \mathbb{N}$ such that $(b_0,d_0) \le (x_n,y_n)$ for $n \ge n_0$, and hence $f(x_n,y_n) > \alpha$ for $n \ge n_0$. Thus $f(x_n,y_n) \to \infty$ as $(x,y) \to (b^-,d^-)$. It follows that $f(x,y) \to \infty$ as $(x,y) \to (b^-,d^-)$. This proves (i).

(ii) The proof of this part is similar to the proof of part (i) above. □

A result similar to the one above holds for monotonically decreasing functions. (See Exercise 31.) Consequently, we see that if $f : (a,b) \times (c,d) \to \mathbb{R}$ is a monotonic function, then

$$\lim_{(x,y)\to(b^-,d^-)} f(x,y) \text{ and } \lim_{(x,y)\to(a^+,c^+)} f(x,y) \text{ exist} \iff f \text{ is bounded.}$$

However, for a bounded monotonic function, limits along the other two quadrants may not exist. For example, consider $f : [-1,1] \times [-1,1] \to \mathbb{R}$ defined by

$$f(x,y) := \begin{cases} (x+2)(y+2) & \text{if } x+y \ge 0, \\ (x+1)(y+1) & \text{if } x+y < 0. \end{cases}$$

We have noted in Example 1.8 (i) that f is monotonically increasing. Also, it is clear that f is bounded and (consequently, or otherwise) the limits of f from the first and the third quadrants as (x,y) tends to $(0,0)$ exist. But the limits of f from the second and the fourth quadrants as (x,y) tends to $(0,0)$ do not exist. To see this, consider the sequences in $\mathbb{R}^2$ defined by $(x_n,y_n) := \left(-\frac{1}{n}, \frac{2}{n}\right)$ and $(x_n',y_n') := \left(-\frac{2}{n}, \frac{1}{n}\right)$ for $n \in \mathbb{N}$. Then

$$(x_n,y_n) \to 0 \text{ and } (x_n',y_n') \to 0, \quad \text{but} \quad f(x_n,y_n) \to 4 \text{ and } f(x_n',y_n') \to 1.$$

Likewise, if $(x_n,y_n) := \left(\frac{2}{n}, -\frac{1}{n}\right)$ and $(x_n',y_n') := \left(\frac{1}{n}, -\frac{2}{n}\right)$ for $n \in \mathbb{N}$, then

$$(x_n,y_n) \to 0 \text{ and } (x_n',y_n') \to 0, \quad \text{but} \quad f(x_n,y_n) \to 4 \text{ and } f(x_n',y_n') \to 1.$$

Thus $\lim_{(x,y)\to(0^-,0^+)} f(x,y)$ and $\lim_{(x,y)\to(0^+,0^-)} f(x,y)$ do not exist.

In Exercise 40 of Chapter 1, we introduced the notion of an antimonotonic function. It can be seen that if $f : (a,b) \times (c,d) \to \mathbb{R}$ is antimonotonic, then

$\lim_{(x,y)\to(a^+,d^-)} f(x,y)$ and $\lim_{(x,y)\to(b^-,c^+)} f(x,y)$ exist $\iff$ f is bounded.

(See Exercise 35.)

Remark 2.60. The notion of limit of a real-valued function of two variables admits a straightforward extension to real-valued functions of three or more variables. Moreover, analogues of all the results in Section 2.3 concerning limits can be easily formulated and proved in this case. ◇

Notes and Comments

For the local study around a point in $\mathbb{R}^2$ (and more generally, in $\mathbb{R}^n$), there are at least two natural analogues of the notion of an interval around a point in $\mathbb{R}$: open disks and open squares. These two are essentially equivalent, in the sense that an open disk can be inscribed in an open square with the same center, and vice versa. (See Exercise 3 of Chapter 1). In this book, we have preferred to use open squares instead of open disks. This approach is slightly unusual, but it pays off in several proofs that appear subsequently.

The development of topics discussed in this chapter proceeds along similar lines as in ACICARA. *Sequences in $\mathbb{R}^2$ are introduced first and their basic properties are derived quickly from the corresponding properties of sequences in $\mathbb{R}$. The notion of continuity is defined using convergence of sequences, and basic properties of continuous functions are proved using properties of sequences in $\mathbb{R}^2$. These include a result on piecing together continuous functions on overlapping domains, which does not seem easy to locate in the literature. Standard results about continuous functions on connected domains and on compact domains are included, except that for pedagogical reasons, we have preferred the terminology of path-connected sets and of closed and bounded sets. It may be remarked that the more general notions of connectedness and compactness are of fundamental importance in analysis and topology; for an introduction, we refer to Exercises 17, 18, 19, 20 21, and also the books of Rudin [48] and Munkres [40]. For a convex function of one variable, continuity at an interior point was relegated to an exercise in* ACICARA. *A similar result holds for convex functions of several variables, but proving it is a little more involved, and we have chosen to give a detailed proof for functions of two variables, using arguments similar to those in the book of Roberts and Varberg [47]. For an alternative proof, one may consult the book of Fleming [19].*

Following Hardy [29], we state and prove the Implicit Function Theorem under a weak hypothesis of continuity in each of the two variables and strict monotonicity in one of the variables. That this is possible appears to have been first observed by Besicovitch. (See the footnote on p. 203 of [29].) This version of the Implicit Function Theorem can be used to give an alternative proof of the Continuous Inverse Theorem. Also, it will pave the way for proving the classical version of the Implicit Function Theorem in Chapter 3.

Limits of functions of two variables are defined using sequences. We have deduced basic properties of limits from the corresponding properties of continuous functions. Perhaps the only nonstandard notion introduced here is that of a limit from a quadrant. This provides an interesting analogue of the notion in one-variable calculus of left(-hand) and right(-hand) limits. In general, for functions of n variables, the notion will have to deal with 2^n hyperoctants.

Exercises

Part A

1. Consider the sequence in $\mathbb{R}^2$ whose nth term is defined by one of the following. Determine whether it is convergent. If it is, then find its limit. (i) $(1/n, n^2)$, (ii) $(n, 1/n^2)$, (iii) $(1/n, 1/n^2)$, (iv) $(1/n, (-1)^n/n)$, (v) $\left(1 + \frac{1}{1!} + \frac{1}{2!} + \cdots + \frac{1}{n!}, \ln n\right)$, (vi) $\left(\left(1+\frac{1}{n}\right)^n, \left(1-\frac{1}{n}\right)^n\right)$.
2. A sequence $((x_n, y_n))$ in $\mathbb{R}^2$ is said to be
   bounded above if there is $(\alpha_1, \alpha_2) \in \mathbb{R}^2$ such that $(x_n, y_n) \le (\alpha_1, \alpha_2)$, that is, $x_n \le \alpha_1$ and $y_n \le \alpha_2$ for all $n \in \mathbb{N}$,
   bounded below if there is $(\beta_1, \beta_2) \in \mathbb{R}^2$ such that $(\beta_1, \beta_2) \le (x_n, y_n)$, that is, $\beta_1 \le x_n$ and $\beta_2 \le y_n$ for all $n \in \mathbb{N}$,
   monotonically increasing if $(x_n, y_n) \le (x_{n+1}, y_{n+1})$ for all $n \in \mathbb{N}$,
   monotonically decreasing if $(x_n, y_n) \ge (x_{n+1}, y_{n+1})$ for all $n \in \mathbb{N}$,
   monotonic if it is monotonically increasing or decreasing.

   Prove the following.
   (i) A monotonically increasing sequence in $\mathbb{R}^2$ is bounded above if and only if it is convergent. Also, if $((x_n, y_n))$ is monotonically increasing and bounded above, then $\lim_{n\to\infty}(x_n, y_n) = \sup\{(x_n, y_n) : n \in \mathbb{N}\}$.
   (ii) A monotonically decreasing sequence in $\mathbb{R}^2$ is bounded below if and only if it is convergent. Also, if $((x_n, y_n))$ is monotonically decreasing and bounded below, then $\lim_{n\to\infty}(x_n, y_n) = \inf\{(x_n, y_n) : n \in \mathbb{N}\}$.
   (iii) A monotonic sequence in $\mathbb{R}^2$ is convergent if and only if it is bounded.
3. Is it true that every sequence in $\mathbb{R}^2$ has a monotonic subsequence? Justify your answer. [Note: It may be remarked that every sequence in $\mathbb{R}$ has a monotonic subsequence; see page 55 of ACICARA.]
4. Let $(x_0, y_0) \in \mathbb{R}^2$. We say that (x_0, y_0) is a **cluster point** of a sequence $((x_n, y_n))$ in $\mathbb{R}^2$ if there is a subsequence $((x_{n_k}, y_{n_k}))$ of $((x_n, y_n))$ such that $(x_{n_k}, y_{n_k}) \to (x_0, y_0)$. Show that if $(x_n, y_n) \to (x_0, y_0)$, then (x_0, y_0) is the only cluster point of $((x_n, y_n))$. Also, show that the converse is not true, that is, there is a sequence $((x_n, y_n))$ in $\mathbb{R}^2$ that has a unique cluster point but is not convergent.
5. If a subset D of $\mathbb{R}^2$ is bounded, then show that its closure $\overline{D}$ is also a bounded subset of $\mathbb{R}^2$.

6. Find the closure, the boundary, and the interior of the following subsets of $\mathbb{R}^2$. Also, determine whether these subsets are closed or open.
(i) $\{(x, y) \in \mathbb{R}^2 : 0 \le x < 1 \text{ and } 0 < y \le 2\}$, (ii) $\{(x, x^2) : x \in \mathbb{R}\}$,
(iii) any finite subset of $\mathbb{R}^2$, (iv) $\{(m, n) : m, n \in \mathbb{N}\}$,
(v) $\{(1/m, 1/n) : m, n \in \mathbb{N}\}$, (vi) $\{(r, s) : r, s \in \mathbb{Q}\}$.
7. Let $D \subseteq \mathbb{R}^2$. Show that the closure of D is the smallest closed subset of $\mathbb{R}^2$ containing D and the interior of D is the largest open subset of D.
8. Let $f, g : [-1, 1] \times [-1, 1] \to \mathbb{R}$ be the functions defined by
$$f(x, y) := (x + y)^2 \quad \text{and} \quad g(x, y) := \begin{cases} (x + y)^2 & \text{if } x + y \ge 0, \\ -(x + y)^2 & \text{if } x + y < 0. \end{cases}$$
Show that both f and g are continuous on $[-1, 1] \times [-1, 1]$. Further show that f is bimonotonic but g is not bimonotonic on $[-1, 1] \times [-1, 1]$.
9. Consider $f : \mathbb{R}^2 \to \mathbb{R}$ defined by $f(0, 0) := 0$ and for $(x, y) \neq (0, 0)$, by one of the following. In each case, determine whether f is continuous.
(i) $\dfrac{xy^2}{x^2 + y^2}$, (ii) $\dfrac{xy^2}{x^2 + y^4}$, (iii) $\dfrac{x^3 y}{x^6 + y^2}$, (iv) $\dfrac{x^2}{x^2 + y^2}$,
(v) $xy \ln(x^2 + y^2)$, (vi) $\dfrac{x^3}{x^2 + y^2}$, (vii) $\dfrac{x^4 y}{x^2 + y^2}$,
(viii) $\dfrac{x^3 y - xy^3}{x^2 + y^2}$, (ix) $\dfrac{\sin(x + y)}{|x| + |y|}$, (x) $\dfrac{\sin^2(x + y)}{|x| + |y|}$.
10. Let D be convex and open in $\mathbb{R}^2$, and let $f : D \to \mathbb{R}$ be a convex function. If $[a, b] \times [c, d]$ is a closed rectangle contained in D, where $a, b, c, d \in \mathbb{R}$ with $a < b$ and $c < d$, then show that f satisfies a **Lipschitz condition** on $[a, b] \times [c, d]$, that is, there is $L \in \mathbb{R}$ such that
$$|f(x, y) - f(u, v)| \le L\, |(x, y) - (u, v)| \quad \text{for all } (x, y), (u, v) \in [a, b] \times [c, d].$$
(Hint: Use Lemma 2.31, or give a proof similar to that of Lemma 2.31.)
11. Let $D := \mathbb{S}_1(0, 0) \setminus \{(0, 0)\}$ and let $f, g : D \to \mathbb{R}$ be defined by $f(x, y) := |x| + |y|$ and $g(x, y) := \frac{1}{2}(|x| + |y|)$. Show that $f(x, y) < g(x, y)$ for all $(x, y) \in D$, but $\lim_{(x,y) \to (0,0)} f(x, y) = \lim_{(x,y) \to (0,0)} g(x, y)$.
12. Show that there is $\delta > 0$ such that $\sin(xy) < \cos(xy)$ for all $(x, y) \in \mathbb{S}_\delta(0, 0)$. (Hint: Proposition 2.52.)
13. Consider $f : \mathbb{R}^2 \to \mathbb{R}$ defined by one of (i)–(iv) below. Determine whether the two-variable limit $\lim_{(x,y) \to (0,0)} f(x, y)$ and the two iterated limits $\lim_{x \to 0} \left[\lim_{y \to 0} f(x, y)\right]$ and $\lim_{y \to 0} \left[\lim_{x \to 0} f(x, y)\right]$ exist. If they do, then find them.
$$\text{(i) } f(x, y) := x + y, \quad \text{(ii) } f(x, y) := \begin{cases} \dfrac{x^2 y^2}{x^2 y^2 + (x - y)^2} & \text{if } (x, y) \neq (0, 0), \\ 0 & \text{if } (x, y) = (0, 0), \end{cases}$$
$$\text{(iii) } f(x, y) := \begin{cases} \dfrac{x + y}{x - y} & \text{if } x \neq y, \\ 0 & \text{if } x = y, \end{cases} \quad \text{(iv) } f(x, y) := \begin{cases} x \sin \dfrac{1}{y} & \text{if } y \neq 0, \\ 0 & \text{if } y = 0. \end{cases}$$

Part B

14. Show that a sequence in $\mathbb{R}^2$ is convergent if and only if it is bounded and all its convergent subsequences have the same limit. (Hint: Bolzano–Weierstrass Theorem.)
15. Let m, n be nonnegative integers and let $i, j \in \mathbb{N}$ be even. Let $f : \mathbb{R}^2 \to \mathbb{R}$ be defined by $f(0,0) := 0$ and $f(x,y) := x^m y^n/(x^i + y^j)$ for $(x,y) \neq (0,0)$. Show that f is continuous at $(0,0)$ if and only if $mj + ni > ij$.
16. Let $E \subseteq \mathbb{R}$ be open in $\mathbb{R}$ and let $\Phi = (x, y)$ be a pair of real-valued functions $x, y : E \to \mathbb{R}$. Show that both x and y are continuous on E if and only if the set $\Phi^{-1}(V) := \{t \in E : (x(t), y(t)) \in V\}$ is open in $\mathbb{R}$ for every open subset V of $\mathbb{R}^2$.
17. Let $D \subseteq \mathbb{R}^2$. A family $\{U_\alpha : \alpha \in A\}$ indexed by an arbitrary set A is called an **open cover** of D if each U_α is open in $\mathbb{R}^2$ and D is contained in the union of U_α as α varies over A. Such an open cover is said to have a **finite subcover** if there are finitely many indices $\alpha_1, \ldots, \alpha_n \in A$ such that $D \subseteq U_{\alpha_1} \cup \cdots \cup U_{\alpha_n}$. The set D is said to be **compact** if every open cover of D has a finite subcover. Prove the following.
    (i) If D is finite, then D is compact.
    (ii) If D is compact and $E \subseteq D$ is closed, then E is compact. (Hint: If $\{U_\alpha : \alpha \in A\}$ is an open cover of D, then consider $\{U_\alpha : \alpha \in A\} \cup \{D \setminus E\}$.)
    (iii) If D is compact, then D is closed. (Hint: If $(x_0, y_0) \in D \setminus \partial D$, then the set of open squares centered at (x, y) and of radius $|(x,y) - (x_0, y_0)|/2$, as (x, y) varies over D, is an open cover of D.)
    (iv) If D is compact, then D is bounded.
    (v) If $D = [a, b] \times [c, d]$ is a closed rectangle, then D is compact. (Hint: Use the midpoints $(a + b)/2$ and $(c + d)/2$ to subdivide D into four smaller rectangles. If an open cover of D has no finite subcover, then the same holds for one of the smaller rectangles. Continue this process and look at the limiting situation.)
    (vi) **(Heine–Borel Theorem)** D is compact $\iff$ D is closed and bounded.

    Generalize the definition and the properties above to subsets of $\mathbb{R}^n$.
18. Let $D \subseteq \mathbb{R}^2$ and $E \subseteq \mathbb{R}$. Prove the following.
    (i) If D is compact and $f : D \to \mathbb{R}$ is continuous, then the range $f(D)$ is closed and bounded.
    (ii) If E is closed and bounded and $x, y : E \to \mathbb{R}$ are continuous, then the subset $\{(x(t), y(t)) : t \in E\}$ of $\mathbb{R}^2$ is compact.
19. If $D \subseteq \mathbb{R}^2$ is path-connected and $f : D \to \mathbb{R}$ is a continuous function such that the image $f(D)$ is a finite set, then show that f is a constant function. Is the conclusion valid if D is not path-connected? Justify your answer. (Hint: If D has two points, take a path $(x(t), y(t))$ joining them. Consider $t \mapsto f(x(t), y(t))$ and use Fact 2.33.)
20. If $D \subseteq \mathbb{R}^2$ is path-connected, then show that D cannot be written as a union of two disjoint, nonempty open subsets of D. (Hint: If it could, then there would be a continuous function $f : D \to \{0, 1\}$. Use Exercise 19.)

21. Let D be an open subset of $\mathbb{R}^2$. If D cannot be written as a union of two disjoint, nonempty open subsets of D, then show that D is path-connected.
22. Let D be a bounded subset of $\mathbb{R}^2$ and let $\overline{D}$ denote its closure. Suppose $f : D \to \mathbb{R}$ be a continuous function. Prove that f is uniformly continuous on D if and only if there is a continuous function $\bar{f} : \overline{D} \to \mathbb{R}$ such that $\bar{f}|_D = f$.
23. Let $f : [0,1] \times [0,1] \to \mathbb{R}$ be the **bivariate Thomae function** defined by

$$f(x,y) := \begin{cases} 1 & \text{if } x = 0 \text{ and } y \in \mathbb{Q} \cap [0,1], \\ 1/q & \text{if } x, y \in \mathbb{Q} \cap [0,1] \text{ and } x = p/q \text{ for some} \\ & \text{relatively prime positive integers } p \text{ and } q, \\ 0 & \text{otherwise.} \end{cases}$$

Show that the set of discontinuities of f is $\{(x,y) \in [0,1] \times [0,1] : x, y \in \mathbb{Q}\}$.
24. **(Duhamel's Theorem)** Let $a, b \in \mathbb{R}$ with $a < b$ and $D := [a,b] \times [a,b]$. If $f : D \to \mathbb{R}$ is continuous and $\phi : [a,b] \to \mathbb{R}$ is defined by $\phi(x) := f(x,x)$ for $x \in [a,b]$, then show that ϕ is Riemann integrable on $[a,b]$. Further, show that given any $\epsilon > 0$, there is $\delta > 0$ such that for every partition $P := \{x_0, x_1, \ldots, x_n\}$ of $[a,b]$ with $\mu(P) < \delta$, and every $c_i, \tilde{c}_i \in [x_{i-1}, x_i]$, for $i = 1, \ldots, n$, we have

$$\left| \int_a^b \phi(x)\, dx - \sum_{i=1}^{n} f(c_i, \tilde{c}_i)\ (x_i - x_{i-1}) \right| < \epsilon.$$

25. **(Bliss's Theorem)** If $\phi, \psi : [a,b] \to \mathbb{R}$ are continuous, then show that given any $\epsilon > 0$, there is $\delta > 0$ such that for every partition $P := \{x_0, x_1, \ldots, x_n\}$ of $[a,b]$ with $\mu(P) < \delta$, and every $c_i, \tilde{c}_i \in [x_{i-1}, x_i]$, for $i = 1, \ldots, n$, we have

$$\left| \int_a^b \phi(x)\psi(x)\, dx - \sum_{i=1}^{n} \phi(c_i)\psi(\tilde{c}_i)\ (x_i - x_{i-1}) \right| < \epsilon.$$

26. Let $D \subseteq \mathbb{R}$ and $t_0 \in \mathbb{R}$ be such that D contains $(t_0 - r, t_0) \cup (t_0, t_0 + r)$ for some $r > 0$. For each $t \in D$, let $f_t : [a,b] \to \mathbb{R}$ be a Riemann integrable function. Suppose $f(x) := \lim_{t \to t_0} f_t(x)$ for $x \in [a,b]$, and $f_t \to f$ uniformly in the sense that for every $\epsilon > 0$, there is $\delta > 0$ such that

$$t \in D,\ 0 < |t - t_0| < \delta,\ x \in [a,b] \Longrightarrow |f_t(x) - f(x)| < \epsilon.$$

Show that $f : [a,b] \to \mathbb{R}$ is Riemann integrable. Further, show that $\lim_{t \to t_0} \int_a^b f_t(x)dx$ exists and is equal to $\int_a^b f(x)dx$. Deduce that if $F : [\alpha, \beta] \times [a,b] \to \mathbb{R}$ is continuous, then for each $t_0 \in [\alpha, \beta]$, we have

$$\lim_{t \to t_0} \int_a^b F(t,x)dx = \int_a^b \lim_{t \to t_0} F(t,x)dx = \int_a^b F(t_0, x)dx.$$

Conclude that $\phi : [\alpha, \beta] \to \mathbb{R}$ defined by $\phi(t) := \int_a^b F(t,x)dx$ is continuous.

27. Let $D \subseteq \mathbb{R}$ be such that D contains $[c, \infty)$ for some $c \in \mathbb{R}$. For each $t \in D$, let $f_t : [a, b] \to \mathbb{R}$ be a Riemann integrable function. Suppose $f(x) := \lim_{t\to\infty} f_t(x)$ for $x \in [a, b]$, and $f_t \to f$ uniformly in the sense that for every $\epsilon > 0$, there is $s \in D$ such that $|f_t(x) - f(x)| < \epsilon$ for all $t \in D$ with $t \geq s$ and all $x \in [a, b]$. Show that $f : [a, b] \to \mathbb{R}$ is Riemann integrable. Further, show that $\lim_{t\to\infty} \int_a^b f_t(x)dx$ exists and is equal to $\int_a^b f(x)dx$.
28. Let $D \subseteq \mathbb{R}^2$ and $(x_0, y_0) \in \mathbb{R}^2$ be such that D contains a punctured square $\mathbb{S}_r(x_0, y_0)\backslash\{(x_0, y_0)\}$ for some $r > 0$. Suppose $f : D \to \mathbb{R}$ is such that $\lim_{(x,y)\to(x_0,y_0)} f(x, y)$ exists and is equal to ℓ. Prove the following.
    (i) If $\lim_{y\to y_0} f(x, y)$ exists for every fixed $x \in (x_0 - r, x_0) \cup (x_0, x_0 + r)$, then the **iterated limit** $\lim_{x\to x_0} \left[\lim_{y\to y_0} f(x, y)\right]$ exists and is equal to ℓ.
    (ii) If $\lim_{x\to x_0} f(x, y)$ exists for every fixed $y \in (y_0 - r, y_0) \cup (y_0, y_0 + r)$, then the **iterated limit** $\lim_{y\to y_0} \left[\lim_{x\to x_0} f(x, y)\right]$ exists and is equal to ℓ.
29. Use Exercise 13 (ii) to show that even when both the iterated limits in (i) and (ii) of Exercise 28 exist, they may not be equal. Also, use Exercise 13 (iv) to show that the existence of the two-variable limit does not imply that the one-variable limits in (i) and (ii) of Exercise 28 exist.
30. Let $D \subseteq \mathbb{R}^2$ be such that D contains $(a, \infty) \times (c, \infty)$ for some $a, c \in \mathbb{R}$. Suppose $f : D \to \mathbb{R}$ is such that $\lim_{(x,y)\to(\infty,\infty)} f(x, y)$ exists and is equal to ℓ.
    (i) If $\lim_{y\to\infty} f(x, y)$ exists for every fixed $x \geq a$, then prove that the **iterated limit** $\lim_{x\to\infty} \left[\lim_{y\to\infty} f(x, y)\right]$ exists and is equal to ℓ.
    (ii) If $\lim_{x\to\infty} f(x, y)$ exists for every fixed $y \geq c$, then prove that the **iterated limit** $\lim_{y\to\infty} \left[\lim_{x\to\infty} f(x, y)\right]$ exists and is equal to ℓ.
31. Let $a, b, c, d \in \mathbb{R}$ with $a < b$ and $c < d$, and let $f : (a, b) \times (c, d) \to \mathbb{R}$ be a monotonically decreasing function. Prove the following.
    (i) $\lim_{(x,y)\to(b^-, d^-)} f(x, y)$ exists if and only if f is bounded below; in this case, $\lim_{(x,y)\to(b^-, d^-)} f(x, y) = \inf\{f(x, y) : (x, y) \in (a, b) \times (c, d)\}$. If f is not bounded below, then $f(x, y) \to -\infty$ as $(x, y) \to (b^-, d^-)$.
    (ii) $\lim_{(x,y)\to(a^+, c^+)} f(x, y)$ exists if and only if f is bounded above; in this case, $\lim_{(x,y)\to(a^+, c^+)} f(x, y) = \sup\{f(x, y) : (x, y) \in (a, b) \times (c, d)\}$. If f is not bounded above, then $f(x, y) \to \infty$ as $(x, y) \to (a^+, c^+)$.
32. Let $a, b, c, d \in \mathbb{R}$ with $a < b$ and $c < d$, and let $f : (a, b) \times (c, d) \to \mathbb{R}$ be a monotonically increasing function. Show that for every $(x_0, y_0) \in (a, b) \times (c, d)$, both $\lim_{(x,y)\to(x_0^-, y_0^-)} f(x, y)$ and $\lim_{(x,y)\to(x_0^+, y_0^+)} f(x, y)$ exist, and $\lim_{(x,y)\to(x_0^-, y_0^-)} f(x, y) \leq f(x_0, y_0) \leq \lim_{(x,y)\to(x_0^+, y_0^+)} f(x, y)$. Also, show that if $(x_1, y_1) \in (a, b) \times (c, d)$ with $x_0 < x_1$ and $y_0 < y_1$, then $\lim_{(x,y)\to(x_0^+, y_0^+)} f(x, y) \leq \lim_{(x,y)\to(x_1^-, y_1^-)} f(x, y)$. Formulate and prove an analogue of these properties for monotonically decreasing functions.

33. Let $D \subseteq \mathbb{R}^2$ and (x_0, y_0) be any point of $\mathbb{R}^2$. If there is a sequence $((x_n, y_n))$ in $D \setminus \{(x_0, y_0)\}$ such that $(x_n, y_n) \to (x_0, y_0)$, then (x_0, y_0) is called a **limit point** (or an **accumulation point**) of D.
   (i) Show that (x_0, y_0) is a limit point of D if and only if for every $r > 0$, there is $(x, y) \in D$ such that $0 < |(x, y) - (x_0, y_0)| < r$.
   (ii) If (x_0, y_0) is a limit point of D, then show that for every $r > 0$, the open disk $\mathbb{B}_r(x_0, y_0)$ as well as the open square $\mathbb{S}_r(x_0, y_0)$ contain infinitely many points of the set D.
   (iii) If D is a finite subset of $\mathbb{R}^2$, show that D has no limit point.
   (iv) Determine all the limit points of D if $D := \mathbb{N} \times \mathbb{N}$, or $D := \mathbb{Q} \times \mathbb{Q}$, or $D := \{(\frac{1}{n}, \frac{1}{m}) : n, m \in \mathbb{N}\}$, or $D := (a, b) \times (c, d)$, or $D := [a, b) \times (c, d]$, where $a, b, c, d \in \mathbb{R}$ with $a < b$ and $c < d$.
   (v) Let $((x_n, y_n))$ be a sequence in $\mathbb{R}^2$ and suppose $D = \{(x_n, y_n) : n \in \mathbb{N}\}$ is the set of all its terms. Show that a limit point of D is a cluster point of the sequence $((x_n, y_n))$. Give an example to show that a cluster point of $((x_n, y_n))$ need not be a limit point of D.
34. Let $D \subseteq \mathbb{R}^2$ and let (x_0, y_0) be a limit point of D. We say that a **limit** of a function $f : D \to \mathbb{R}$ as (x, y) tends to (x_0, y_0) exists if there is a real number ℓ such that whenever $((x_n, y_n))$ is any sequence in $D \setminus \{(x_0, y_0)\}$ that converges to (x_0, y_0), we have $f(x_n, y_n) \to \ell$; in this case ℓ is called a **limit** of f as (x, y) tends to (x_0, y_0). Show that if a limit of f as (x, y) tends to (x_0, y_0) exists, then it must be unique. Also, prove analogues of Propositions 2.48, 2.51, 2.52, 2.53, 2.54 and Corollary 2.49.
35. Let $a, b, c, d \in \mathbb{R}$ with $a < b$ and $c < d$, and let $f : (a, b) \times (c, d) \to \mathbb{R}$ be an antimonotonic function. Show that both $\lim_{(x,y)\to(a^+, d^-)} f(x, y)$ and $\lim_{(x,y)\to(b^-, c^+)} f(x, y)$ exist if and only if f is bounded. (Hint: Exercise 40 of Chapter 1)
36. Let $a, b, c, d \in \mathbb{R}$ with $a < b$ and $c < d$, and let $D := (a, b] \times (c, d]$ and $f : D \to \mathbb{R}$ be a bimonotonic function.
   (i) Define $F : D \to \mathbb{R}$ by $F(x, y) := f(x, y) - f(x, d) - f(b, y) + f(b, d)$. Show that either F is monotonically increasing and bounded below, or F is monotonically decreasing and bounded above.
   (ii) If the one-variable limits $\lim_{x\to b^-} f(x, d)$ and $\lim_{y\to d^-} f(b, y)$ exist, then show that $\lim_{(x,y)\to(b^-, d^-)} f(x, y)$ exists.
37. Let $a, b, c, d \in \mathbb{R}$ with $a < b$ and $c < d$. State and prove results analogous to those in Exercise 36 above for functions defined on $[a, b) \times [c, d)$, $[a, b) \times (c, d]$, and $(a, b] \times [c, d)$. (Hint: For $[a, b) \times (c, d]$ and $(a, b] \times [c, d)$, consider the notion of antimonotonicity.)
38. Let $a, b, c, d \in \mathbb{R}$ with $a < b$ and $c < d$, and let $f : [a, b] \times [c, d] \to \mathbb{R}$ be any function. Show that if f is of bounded variation and v_f is continuous, then f is continuous. On the other hand, give an example to show that if f is of bounded bivariation and w_f is continuous, then f need not be continuous.

3

Partial and Total Differentiation

The notion of derivative of a function of one-variable does not really have a solitary analogue for functions of several variables. Indeed, for a function of two (or more) variables, there is a plethora of derivatives depending on whether we choose to become partial to one of the variables, or opt to move about in a specific direction, or prefer to take the total picture in consideration. The first two viewpoints lead to the notions of partial derivatives and directional derivatives, while the last leads to a somewhat more abstract notion of differentiability and, in turn, to the notion of total derivative. We define partial and directional derivatives in Section 3.1, and prove a number of basic properties including two distinct analogues of the mean value theorem and a version of Taylor's theorem using higher-order directional derivatives. In Section 3.2, we study the notion of differentiability and prove the classical version of the Implicit Function Theorem. It may be remarked that those wishing to bypass the abstract notion of differentiability can always replace it, wherever invoked, by a slightly stronger but more pragmatic condition on the existence and continuity of partial derivatives. (See Proposition 3.33.) These readers can, therefore, skip all of Section 3.2 except perhaps the classical version of the Implicit Function Theorem. Some key results regarding differentiable functions of two variables such as the classical version of Taylor's theorem and the chain rule are discussed in Section 3.3. Next, in Section 3.4, we revisit the notions of monotonicity, bimonotonicity, convexity, and concavity introduced in Chapter 1, and relate these to partial derivatives. Finally, in Section 3.5, we briefly outline how some of the results discussed in previous sections extend to functions of three variables, and also discuss the notions of tangent plane and normal line, which can be better understood in the context of surfaces defined (implicitly) by functions of three variables.

S.R. Ghorpade and B.V. Limaye, *A Course in Multivariable Calculus and Analysis*,
Undergraduate Texts in Mathematics, DOI 10.1007/978-1-4419-1621-1_3,

3.1 Partial and Directional Derivatives

Let us first recall the notion of derivative for a function of one variable. Let $D \subseteq \mathbb{R}$ and let c be an interior point of D, that is, $(c-r, c+r) \subseteq D$ for some $r > 0$. A function $f : D \to \mathbb{R}$ is said to be differentiable at c if the limit

$$\lim_{h \to 0} \frac{f(c+h) - f(c)}{h}$$

exists; in this case the value of the limit is denoted by $f'(c)$ and is called the derivative of f at c. Now suppose $D \subseteq \mathbb{R}^2$ and (x_0, y_0) is an interior point of D, that is, $\mathbb{S}_r(x_0, y_0) \subseteq D$ for some $r > 0$. For a function $f : D \to \mathbb{R}$, it might seem natural to consider a limit such as

$$\lim_{(h,k) \to (0,0)} \frac{f(x_0 + h, y_0 + k) - f(x_0, y_0)}{(h, k)}.$$

But this doesn't make sense for the simple reason that division of a real number by a point in $\mathbb{R}^2$ has not been defined. There are ways to get around this problem but they are not particularly easy, and we defer a discussion of the notion of differentiability for functions of two (or more) variables to a later section. For the moment, we shall see that choosing to become partial to one of the variables makes things easier and leads to a useful notion.

Partial Derivatives

Let $D \subseteq \mathbb{R}^2$ and let $f : D \to \mathbb{R}$ be any function. Fix $(x_0, y_0) \in D$ and define $D_1, D_2 \subseteq \mathbb{R}$ by $D_1 := \{x \in \mathbb{R} : (x, y_0) \in D\}$ and $D_2 := \{y \in \mathbb{R} : (x_0, y) \in D\}$. If x_0 is an interior point of D_1, we define the **partial derivative** of f with respect to x at (x_0, y_0) to be the limit

$$\lim_{h \to 0} \frac{f(x_0 + h, y_0) - f(x_0, y_0)}{h},$$

provided this limit exists. It is denoted by $f_x(x_0, y_0)$. Likewise, if y_0 is an interior point of D_2, we define the **partial derivative** of f with respect to y at (x_0, y_0) to be the limit

$$\lim_{k \to 0} \frac{f(x_0, y_0 + k) - f(x_0, y_0)}{k},$$

provided this limit exists. It is denoted by $f_y(x_0, y_0)$. These partial derivatives are also called the **first-order partial derivatives** or simply the **first partials** of f at (x_0, y_0). They are sometimes denoted by

$$\frac{\partial f}{\partial x}(x_0, y_0) \quad \text{and} \quad \frac{\partial f}{\partial y}(x_0, y_0)$$

instead of $f_x(x_0, y_0)$ and $f_y(x_0, y_0)$, respectively. If these partial derivatives exist, then the pair $(f_x(x_0, y_0), f_y(x_0, y_0))$ is called the **gradient** of f at (x_0, y_0) and is denoted by $\nabla f(x_0, y_0)$. Thus

$$\nabla f(x_0, y_0) = \left(\frac{\partial f}{\partial x}(x_0, y_0), \frac{\partial f}{\partial y}(x_0, y_0) \right).$$

The partial derivative $f_x(x_0, y_0)$ gives the rate of change in f at (x_0, y_0) along the x-axis, whereas $f_y(x_0, y_0)$ gives the rate of change in f at (x_0, y_0) along the y-axis. In practice, finding the partial derivative of f with respect to x amounts to taking the derivative of $f(x, y)$ as a function of x, treating y as a constant. Indeed, if $\phi : D_1 \to \mathbb{R}$ is the function of one variable defined by $\phi(x) := f(x, y_0)$, then ϕ is differentiable at x_0 if and only if the partial derivative of f with respect to x at (x_0, y_0) exists; in this case $f_x(x_0, y_0) = \phi'(x_0)$. Similarly, if $\psi : D_2 \to \mathbb{R}$ is defined by $\psi(y) = f(x_0, y)$ for $y \in D_2$, then ψ is differentiable at y_0 if and only if the partial derivative of f with respect to y at (x_0, y_0) exists; in this case $f_y(x_0, y_0) = \psi'(y_0)$. As a consequence, we see that partial derivatives of sums, scalar multiples, products, reciprocals, and radicals possess exactly the same properties as derivatives of functions of one variable. Moreover, since differentiability implies continuity for functions of one variable, we see that if the partial derivatives of f at (x_0, y_0) exist, then ϕ is continuous at x_0 and ψ is continuous at y_0. However, as Example 3.1 (iii) below shows, existence of both the partial derivatives at a point does not imply continuity at that point.

Analogous to the left(-hand) and the right(-hand) derivatives in one-variable calculus, we have the concepts of left(-hand) and right(-hand) partial derivatives at points that are akin to endpoints of an interval in $\mathbb{R}$. Let, as before, $D \subseteq \mathbb{R}^2$ and let $f : D \to \mathbb{R}$ be a function. Fix $(x_0, y_0) \in D$, and let $D_1 := \{x \in \mathbb{R} : (x, y_0) \in D\}$ and $D_2 := \{y \in \mathbb{R} : (x_0, y) \in D\}$. If there is $r > 0$ such that $(x_0 - r, x_0] \subseteq D_1$, then we define the **left(-hand) partial derivative** of f with respect to x at (x_0, y_0) to be the limit

$$\lim_{h \to 0^-} \frac{f(x_0 + h, y_0) - f(x_0, y_0)}{h},$$

provided this limit exists. It is denoted by $(f_x)_-(x_0, y_0)$. On the other hand, if there is $r > 0$ such that $[x_0, x_0 + r) \subseteq D$, then the **right(-hand) partial derivative** of f with respect to x at (x_0, y_0) is defined to be the above limit with $h \to 0^-$ replaced by $h \to 0^+$. It is denoted by $(f_x)_+(x_0, y_0)$. Likewise, we define the left(-hand) and right(-hand) partial derivatives of f with respect to y at (x_0, y_0). These are denoted by $(f_y)_-(x_0, y_0)$ and $(f_y)_-(x_0, y_0)$ respectively.

In case D is the rectangle $[a, b] \times [c, d]$, then for each $(x_0, y_0) \in D$, we have $D_1 = [a, b]$ and $D_2 = [c, d]$. If $a < x_0 < b$, then x_0 is an interior point of D_1 and it is clear that the partial derivative of f with respect to x at (x_0, y_0) exists if and only if both the left(-hand) and the right(-hand) partial derivatives of f with respect to x at (x_0, y_0) exist and are equal. Likewise for partial derivatives with respect to y when $c < y_0 < d$. If $x_0 = a$ or $x_0 = b$, then x_0 is not an interior point of D_1, but the right(-hand) partial derivative of f with respect to x at (x_0, y_0) can still be defined when $x_0 = a$, while the left(-hand) derivative of f with respect to x at (x_0, y_0) can be defined when $x_0 = b$. Likewise if $y_0 = c$ or $y_0 = d$. With this in view, we shall say that the **partial derivative** f_x of f exists on $[a, b] \times [c, d]$ if f_x exists at each point of $(a, b) \times [c, d]$, $(f_x)_+$ exists at each point of $\{a\} \times [c, d]$, and $(f_x)_-$ exists at each point of $\{b\} \times [c, d]$. In this case, we will simply write $f_x(a, y_0)$ to denote $(f_x)_+(a, y_0)$ and $f_x(b, y_0)$ to denote $(f_x)_-(b, y_0)$ for every $y_0 \in [c, d]$. In this way, we obtain a function $f_x : [a, b] \times [c, d] \to \mathbb{R}$. A similar convention holds for f_y.

Examples 3.1. (i) Let $f : \mathbb{R}^2 \to \mathbb{R}$ be given by $f(x, y) := x^2 + y^2$. Then both the partial derivatives of f exist at every point of $\mathbb{R}^2$; in fact, $f_x(x_0, y_0) = 2x_0$ and $f_y(x_0, y_0) = 2y_0$ for any $(x_0, y_0) \in \mathbb{R}^2$.

(ii) Let $f : \mathbb{R}^2 \to \mathbb{R}$ be the norm function given by $f(x, y) := \sqrt{x^2 + y^2}$. Then both the partial derivatives of f exist at every point of $\mathbb{R}^2$ except the origin; in fact, for any $(x_0, y_0) \in \mathbb{R}^2$ with $(x_0, y_0) \neq (0, 0)$,

$$f_x(x_0, y_0) = \frac{x_0}{\sqrt{x_0^2 + y_0^2}} \quad \text{and} \quad f_y(x_0, y_0) = \frac{y_0}{\sqrt{x_0^2 + y_0^2}}.$$

To examine whether any of the partial derivatives exist at $(0, 0)$, we have to resort to the definition. This leads to a limit of the quotient $h/|h|$ as h approaches 0. Clearly, such a limit does not exist. It follows that $f_x(0, 0)$ and $f_y(0, 0)$ do not exist.

(iii) Let $f : \mathbb{R}^2 \to \mathbb{R}$ be given by $f(0, 0) := 0$ and $f(x, y) := xy/(x^2 + y^2)$ for $(x, y) \neq (0, 0)$. Then for any $h, k \in \mathbb{R}$ with $h \neq 0$ and $k \neq 0$, we have

$$\frac{f(0 + h, 0) - f(0, 0)}{h} = 0 \quad \text{and} \quad \frac{f(0, 0 + k) - f(0, 0)}{k} = 0.$$

Hence $f_x(0, 0)$ and $f_y(0, 0)$ exist and are both equal to 0. However, as seen already in Example 2.16 (ii), f is not continuous at $(0, 0)$.

(iv) Let $f : \mathbb{R}^2 \to \mathbb{R}$ be given by $f(x, y) = |x| + |y|$ for $(x, y) \in \mathbb{R}^2$. Clearly, f is continuous at $(0, 0)$. But for any $h, k \in \mathbb{R}$ with $h \neq 0$ and $k \neq 0$, we have

$$\frac{f(0 + h, 0) - f(0, 0)}{h} = \frac{|h|}{h} \quad \text{and} \quad \frac{f(0, 0 + k) - f(0, 0)}{k} = \frac{|k|}{k}.$$

Hence $f_x(0, 0)$ and $f_y(0, 0)$ do not exist. However, the left(-hand) and the right(-hand) partial derivatives of f at $(0, 0)$ do exist. Indeed, $(f_x)_+(0, 0) = 1 = (f_y)_+(0, 0)$, while $(f_x)_-(0, 0) = -1 = (f_y)_-(0, 0)$. On the other hand, if we let g and h denote the restrictions of f to the rectangles $[-1, 1] \times [0, 1]$ and $[0, 1] \times [-1, 1]$ respectively, then in accordance with our conventions, $g_y(0, 0)$ and $h_x(0, 0)$ do exist and are both equal to 1.

(v) Let $\theta : \mathbb{R}^2 \setminus \{(0,0)\} \to \mathbb{R}$ be the polar angle function defined by $\theta(x,y) := \cos^{-1}\left(x/\sqrt{x^2+y^2}\right)$ if $y \geq 0$ and $\theta(x,y) := -\cos^{-1}\left(x/\sqrt{x^2+y^2}\right)$ if $y < 0$. We have seen in Example 2.18 (iv) that θ is continuous on all of $\mathbb{R}^2$ except on the negative x-axis. Now let us examine the existence of partial derivatives of θ. First, consider $(x_0, y_0) \in \mathbb{R}^2$ with $y_0 \neq 0$. Then there is $r > 0$ such that θ is given by only one of the two expressions above throughout $\mathbb{S}_r(x_0, y_0)$. So, if we remember that the derivative of $\cos^{-1} t$ is $-1/\sqrt{1-t^2}$ for $t \in (-1,1)$, and apply standard rules of differentiation of functions of one variable, then we see that both the partial derivatives of θ exist and

$$\theta_x(x_0, y_0) = \frac{-y_0}{x_0^2 + y_0^2} \quad \text{and} \quad \theta_y(x_0, y_0) = \frac{x_0}{x_0^2 + y_0^2} \quad \text{provided } y_0 \neq 0.$$

Next, let us consider points on the x-axis. We have $\theta(x, 0) = 0$ if $x > 0$ and $\theta(x, 0) = \pi$ if $x < 0$. Hence if $x_0 \in \mathbb{R}$ with $x_0 \neq 0$, then $\theta_x(x_0, 0)$ exists and is equal to 0. Moreover, if $x_0 > 0$, then in view of Example 2.18 (iv), L'Hôpital's rule for $\frac{0}{0}$ indeterminate forms (Proposition 4.37 of ACICARA), and the expression above for $\theta_y(x_0, y_0)$, we see that

$$\begin{aligned}\theta_y(x_0, 0) &:= \lim_{k \to 0} \frac{\theta(x_0, k) - \theta(x_0, 0)}{k} \\ &= \lim_{k \to 0} \frac{\theta(x_0, k)}{k} = \lim_{k \to 0} \frac{x_0/(x_0^2 + k^2)}{1} = \frac{1}{x_0}.\end{aligned}$$

Finally, if $x_0 < 0$, then from Example 2.18 (iv), we know that the function $\theta_0 : (-\infty, 0] \to \mathbb{R}$ defined by $\theta_0(y) := \theta(x_0, y)$ is not continuous at $y = 0$. Hence θ_0 cannot be differentiable at 0. In other words, $\theta_y(x_0, 0)$ does not exist if $x_0 < 0$. $\diamond$

We have seen in Example 3.1 (iii) above that existence of both the partial derivatives does not imply continuity. However, it is easy to show that it does imply the continuity in each of the two variables, and also, bivariate continuity in case one or both of the partial derivatives are bounded. For the latter, we need to use a basic result in one-variable calculus known as the mean value theorem, or, in short, the MVT. Let us first recall the statement. A proof can be found, for example, on page 120 of ACICARA.

Fact 3.2 (MVT). *Let $a, b \in \mathbb{R}$ with $a < b$. If $\phi : [a,b] \to \mathbb{R}$ is continuous on $[a,b]$ and differentiable on (a,b), then there is $c \in (a,b)$ such that*

$$\phi(b) - \phi(a) = \phi'(c)(b-a).$$

Proposition 3.3. *For any $f : [a,b] \times [c,d] \to \mathbb{R}$, we have the following:*

(i) *If f_x exists on $[a,b] \times [c,d]$, then for each fixed $y_0 \in [c,d]$, the function from $[a,b]$ to $\mathbb{R}$ given by $x \longmapsto f(x, y_0)$ is continuous.*

(ii) *If f_y exists on $[a,b] \times [c,d]$, then for each fixed $x_0 \in [a,b]$, the function from $[c,d]$ to $\mathbb{R}$ given by $y \longmapsto f(x_0, y)$ is continuous.*

(iii) *If both f_x and f_y exist, and if one of them is bounded on $[a,b] \times [c,d]$, then f is continuous on $[a,b] \times [c,d]$.*

Proof. (i) Fix $y_0 \in [c,d]$. The existence of $f_x(x_0, y_0)$ for every $x_0 \in [c,d]$ readily implies that the function of one variable given by $x \longmapsto f(x, y_0)$ is differentiable, and hence continuous, on $[a,b]$.

(ii) Proof of (ii) is similar to that of (i) above.

(iii) Assume that both f_x and f_y exist, and f_x is bounded on $[a,b] \times [c,d]$. Then there is $\alpha \in \mathbb{R}$ such that $|f_x(u,v)| \leq \alpha$ for all $(u,v) \in [a,b] \times [c,d]$. Fix $(x_0, y_0) \in [a,b] \times [c,d]$. Given any $(x,y) \in [a,b] \times [c,d]$ with $x \neq x_0$, by the MVT (Fact 3.2), we see that there is $u \in \mathbb{R}$ between x and x_0 such that

$$f(x,y) - f(x_0,y) = f_x(u,y)(x - x_0) \quad \text{and so} \quad |f(x,y) - f(x_0,y)| \leq \alpha |x - x_0|.$$

Consequently,

$$\begin{aligned} |f(x,y) - f(x_0,y_0)| &\leq |f(x,y) - f(x_0,y)| + |f(x_0,y) - f(x_0,y_0)| \\ &\leq \alpha |x - x_0| + |f(x_0,y) - f(x_0,y_0)| \,. \end{aligned}$$

Moreover, these inequalities are clearly valid if $x = x_0$. Thus, in view of (ii), $f(x,y) \to f(x_0,y_0)$ as $(x,y) \to (x_0,y_0)$. So f is continuous at (x_0,y_0). □

Directional Derivatives

The notion of partial derivatives can be easily generalized to that of a *directional derivative*, which measures the rate of change of a function at a point along a given direction. We specify a direction by specifying a unit vector. Let $\mathbf{u} = (u_1, u_2)$ be a unit vector in $\mathbb{R}^2$, so that $u_1^2 + u_2^2 = 1$. Also let $D \subseteq \mathbb{R}^2$ and $f : D \to \mathbb{R}$ be any function. Let $(x_0, y_0) \in D$ be such that D contains a segment of the line passing through (x_0, y_0) in the direction of $\mathbf{u}$, that is, 0 is an interior point of $D_0 := \{t \in \mathbb{R} : (x_0 + tu_1, y_0 + tu_2) \in D\}$. We define the **directional derivative** of f at (x_0, y_0) along $\mathbf{u}$ to be the limit

$$\lim_{t \to 0} \frac{f(x_0 + tu_1, y_0 + tu_2) - f(x_0, y_0)}{t},$$

provided this limit exists. It is denoted by $\mathbf{D_u} f(x_0, y_0)$. Note that if $\mathbf{v} = -\mathbf{u}$, then $\mathbf{D_v} f(x_0, y_0) = -\mathbf{D_u} f(x_0, y_0)$. Note also that if $\mathbf{i} := (1,0)$ and $\mathbf{j} := (0,1)$, then $\mathbf{D_i} f(x_0, y_0) = f_x(x_0, y_0)$ and $\mathbf{D_j} f(x_0, y_0) = f_y(x_0, y_0)$.

Examples 3.4. (i) Let $f : \mathbb{R}^2 \to \mathbb{R}$ be given by $f(x,y) := x^2 + y^2$. Given any unit vector $\mathbf{u} = (u_1, u_2)$ in $\mathbb{R}^2$ and any $(x_0, y_0) \in \mathbb{R}^2$, for every $t \in \mathbb{R}$ with $t \neq 0$, the quotient $[f(x_0 + tu_1, y_0 + tu_2) - f(x_0, y_0)]/t$ is equal to

$$\frac{(x_0+tu_1)^2+(y_0+tu_2)^2-(x_0^2+y_0^2)}{t} = \frac{2tx_0u_1+t^2u_1^2+2ty_0u_2+t^2u_2^2}{t}$$
$$= 2x_0u_1+2y_0u_2+t.$$

It follows that $\mathbf{D_u}f(x_0, y_0)$ exists and is equal to $2x_0u_1 + 2y_0u_2$. Thus, in view of Example 3.1 (i), we see that $\mathbf{D_u}f(x_0, y_0) = f_x(x_0, y_0)u_1 + f_y(x_0, y_0)u_2$.

(ii) Let $f : \mathbb{R}^2 \to \mathbb{R}$ be given by $f(x, y) := \sqrt{x^2 + y^2}$. Given any unit vector $\mathbf{u} = (u_1, u_2)$ in $\mathbb{R}^2$ and any $(x_0, y_0) \in \mathbb{R}^2$, for every $t \in \mathbb{R}$ with $t \neq 0$, the quotient $[f(x_0 + tu_1, y_0 + tu_2) - f(x_0, y_0)]/t$ is equal to

$$\frac{\sqrt{(x_0+tu_1)^2+(y_0+tu_2)^2}-\sqrt{x_0^2+y_0^2}}{t}$$
$$= \frac{2tx_0u_1+2ty_0u_2+t^2}{t\left(\sqrt{(x_0+tu_1)^2+(y_0+tu_2)^2}+\sqrt{x_0^2+y_0^2}\right)}.$$

It follows that if $(x_0, y_0) \neq (0, 0)$, then $\mathbf{D_u}f(x_0, y_0)$ exists and

$$\mathbf{D_u}f(x_0, y_0) = \frac{x_0u_1 + y_0u_2}{\sqrt{x_0^2+y_0^2}}.$$

Thus, in view of Example 3.1 (ii), we see once again that $\mathbf{D_u}f(x_0, y_0) = f_x(x_0, y_0)u_1 + f_y(x_0, y_0)u_2$ for $(x_0, y_0) \neq (0, 0)$. On the other hand, $\mathbf{D_u}f(0, 0)$ does not exist, since the quotient $t/|t|$ does not have a limit as $t \to 0$.

(iii) Let $f : \mathbb{R}^2 \to \mathbb{R}$ be given by $f(0, 0) := 0$ and $f(x, y) := x^2y/(x^4 + y^2)$ for $(x, y) \neq (0, 0)$. Given any unit vector $\mathbf{u} = (u_1, u_2)$ in $\mathbb{R}^2$ and any $t \in \mathbb{R}$ with $t \neq 0$, the quotient $[f(0 + tu_1, 0 + tu_2) - f(0, 0)]/t$ is equal to $u_1^2u_2/(u_1^4t^2 + u_2^2)$. It follows that $\mathbf{D_u}f(0, 0)$ exists and

$$\mathbf{D_u}f(0,0) = \begin{cases} \dfrac{u_1^2}{u_2} & \text{if } u_2 \neq 0, \\ 0 & \text{if } u_2 = 0. \end{cases}$$

In particular, $f_x(0, 0) = 0 = f_y(0, 0)$. Thus, we see this time that $\mathbf{D_u}f(0, 0) \neq f_x(0, 0)u_1 + f_y(0, 0)u_2$, unless $u_1 = 0$ or $u_2 = 0$. Notice that in view of Example 2.16 (iv), f is not continuous at $(0, 0)$ even though all the directional derivatives of f at $(0, 0)$ exist.

(iv) Let $f : \mathbb{R}^2 \to \mathbb{R}$ be given by $f(x, y) = |x| + |y|$ for $(x, y) \in \mathbb{R}^2$. Let $\mathbf{u} = (u_1, u_2)$ be any unit vector. Then $|u_1| + |u_2| \neq 0$ and for any $t \in \mathbb{R}$ with $t \neq 0$, we have

$$\frac{f(0+tu_1, 0+tu_2) - f(0,0)}{t} = \frac{|t|}{t}\left(|u_1| + |u_2|\right).$$

Hence $\mathbf{D_u}f(0, 0)$ does not exist. Notice that here, f is clearly continuous at $(0, 0)$, but none of the directional derivatives of f at $(0, 0)$ exist. ◇

We shall now use the notion of a directional derivative to derive an analogue of the MVT (Fact 3.2) in the case of functions of two variables. Clearly, such an analogue should, roughly speaking, say that the difference between the values of a function at two distinct points is equal to the product of the distance between the two points and the value of the "derivative" at a point "lying between" them. To make the idea of "lying between" precise, we simply restrict to the line segment joining the two points, and then it becomes clear that the appropriate notion of "derivative" to consider here is that of the directional derivative in the direction of this line segment. More precisely, we have the following.

Proposition 3.5 (Bivariate Mean Value Theorem). *Let $D \subseteq \mathbb{R}^2$ and let D° denote the interior of D. Suppose $(x_0, y_0), (x_1, y_1)$ are distinct points of D such that $L := \{(x(t), y(t)) \in \mathbb{R}^2 : t \in (0,1)\} \subseteq D^\circ$, where*

$$x(t) := x_0 + t(x_1 - x_0) \quad \text{and} \quad y(t) := y_0 + t(y_1 - y_0).$$

Let $\mathbf{u} = (u_1, u_2)$ be the unit vector given by

$$\mathbf{u} = (u_1, u_2) := \frac{1}{r}(x_1 - x_0, y_1 - y_0), \quad \text{where} \quad r := \sqrt{(x_1 - x_0)^2 + (y_1 - y_0)^2},$$

and let $f : D \to \mathbb{R}$ be a continuous function such that $\mathbf{D_u} f$ exists at each point of L. Then there is $(c, d) \in L$ such that

$$f(x_1, y_1) - f(x_0, y_0) = r\mathbf{D_u} f(c, d).$$

Proof. Let $F : [0,1] \to \mathbb{R}$ be defined by $F(t) := f(x(t), y(t))$ for $t \in [0,1]$. By part (ii) of Proposition 2.17, F is continuous on $[0,1]$. Moreover, given any $t_0 \in (0,1)$ and $t \in [0,1]$ with $t \neq t_0$, we have

$$x(t) = x_0 + t(x_1 - x_0) = x_0 + t_0(x_1 - x_0) + (t - t_0)(x_1 - x_0) = x(t_0) + (t - t_0)ru_1,$$

and similarly, $y(t) = y(t_0) + (t - t_0)ru_2$, and hence

$$\frac{F(t) - F(t_0)}{t - t_0} = \frac{f(x(t_0) + (t - t_0)ru_1, y(t_0) + (t - t_0)ru_2) - f(x(t_0), y(t_0))}{t - t_0}.$$

Thus, multiplying the numerator and the denominator of the expression on the right by r, we see that F is differentiable at t_0 and

$$F'(t_0) = r\mathbf{D_u} f(x(t_0), y(t_0)).$$

Hence by the MVT (Fact 3.2) applied to F, there is $\theta \in (0,1)$ such that $F(1) - F(0) = (1 - 0)F'(\theta)$. Consequently, $(c, d) := (x(\theta), y(\theta))$ is a point of L, and we have $f(x_1, y_1) - f(x_0, y_0) = r\mathbf{D_u} f(c, d)$. □

We have seen in Examples 3.4 that in many (but not all) situations, the directional derivative $\mathbf{D_u} f$ equals $\nabla f \cdot \mathbf{u}$. In this case an alternative version of the Bivariate Mean Value Theorem can be given as in the corollary below.

Corollary 3.6. *Let $D \subseteq \mathbb{R}^2$ and let D° denote the interior of D. Suppose $(x_0, y_0), (x_1, y_1)$ are distinct points of D such that $L := \{(x(t), y(t)) \in \mathbb{R}^2 : t \in (0,1)\} \subseteq D^\circ$, where $x(t) := x_0 + t(x_1 - x_0)$* and *$y(t) := y_0 + t(y_1 - y_0)$. Let $f : D \to \mathbb{R}$ be a continuous function such that $\mathbf{D_u}f$ exists at each point of L for all unit vectors $\mathbf{u} \in \mathbb{R}^2$, and moreover $\mathbf{D_u}f(x(t), y(t)) = \nabla f(x(t), y(t)) \cdot \mathbf{u}$ for all unit vectors $\mathbf{u} \in \mathbb{R}^2$ and $t \in (0,1)$. Then there is $(c, d) \in L$ such that*

$$f(x_1, y_1) - f(x_0, y_0) = (x_1 - x_0) f_x(c, d) + (y_1 - y_0) f_y(c, d).$$

In particular, if there are $m, M \in \mathbb{R}$ such that $m \le f_x(u, v) \le M$ and also $m \le f_y(u, v) \le M$ for all $(u, v) \in L$, then the following **Bivariate Mean Value Inequality** *holds:*

$$m\,(x_1 - x_0 + y_1 - y_0) \le f(x_1, y_1) - f(x_0, y_0) \le M\,(x_1 - x_0 + y_1 - y_0).$$

Proof. Let $h := x_1 - x_0$, $k := y_1 - y_0$, and $r := \sqrt{h^2 + k^2}$. Define

$$\mathbf{u} := (u_1, u_2) := \frac{1}{r}(h, k) = \frac{1}{r}\,(x_1 - x_0, y_1 - y_0).$$

Then $\mathbf{u}$ is a unit vector in $\mathbb{R}^2$ with $r\mathbf{u} = (h, k)$. Thus, in view of the assumption on $\mathbf{D_u}f$, we see that for any $(c, d) \in L$,

$$r\mathbf{D_u}f(c, d) = r\nabla f(c, d) \cdot \mathbf{u} = h f_x(c, d) + k f_y(c, d).$$

Thus, the desired result follows from Proposition 3.5. □

Higher-Order Partial Derivatives

Let $D \subseteq \mathbb{R}^2$ be an open set and let us fix $(x_0, y_0) \in D$. It is clear that x_0 is an interior point of $D_1 := \{x \in \mathbb{R} : (x, y_0) \in D\}$ and y_0 is an interior point of $D_2 := \{y \in \mathbb{R} : (x_0, y) \in D\}$. Let $f : D \to \mathbb{R}$ be any function.

If $f_x(x_0, y_0)$ exists at every $(x_0, y_0) \in D$, then we obtain a function from D to $\mathbb{R}$ given by $(x, y) \mapsto f_x(x, y)$. It is denoted by f_x and called the **partial derivative** of f with respect to x on D. The **partial derivative** of f with respect to y on D, denoted by f_y, is defined in a similar way. Sometimes these partial derivatives on D are denoted by $\frac{\partial f}{\partial x}$ and $\frac{\partial f}{\partial y}$ instead of f_x and f_y, respectively. In case both f_x and f_y are defined on D, then we define the **gradient** of f on D to be the transformation $\nabla f : D \to \mathbb{R}^2$ given by $\nabla f(x, y) = (f_x(x, y), f_y(x, y))$ for $(x, y) \in D$.

In case f_x is defined on D, we can consider its partial derivatives at any point of D. Let $(x_0, y_0) \in D$. The partial derivative of $f_x : D \to \mathbb{R}$ with respect to x at (x_0, y_0), if it exists, is denoted by $f_{xx}(x_0, y_0)$. Also, the partial derivative of f_x with respect to y at (x_0, y_0), if it exists, is denoted by $f_{xy}(x_0, y_0)$. In case f_y is defined on D, we can similarly define $f_{yx}(x_0, y_0)$ and $f_{yy}(x_0, y_0)$. These partial derivatives are sometimes denoted by

$$\frac{\partial^2 f}{\partial x^2}(x_0, y_0), \quad \frac{\partial^2 f}{\partial y \partial x}(x_0, y_0), \quad \frac{\partial^2 f}{\partial x \partial y}(x_0, y_0), \quad \text{and} \quad \frac{\partial^2 f}{\partial y^2}(x_0, y_0)$$

instead of $f_{xx}(x_0, y_0)$, $f_{xy}(x_0, y_0)$, $f_{yx}(x_0, y_0)$, and $f_{yy}(x_0, y_0)$, respectively. Collectively, these are referred to as the **second-order partial derivatives** or simply the **second partials** of f at (x_0, y_0). Among these, the middle two, namely,

$$f_{xy}(x_0, y_0) = \frac{\partial^2 f}{\partial y \partial x}(x_0, y_0) \quad \text{and} \quad f_{yx}(x_0, y_0) = \frac{\partial^2 f}{\partial x \partial y}(x_0, y_0),$$

are called the **mixed (second-order) partial derivatives** of f, or simply the **mixed partials** of f. The order in which x and y appear in mixed partial derivatives can sometimes be a matter of confusion. The order may be easier to remember if one notes that

$$f_{xy} = (f_x)_y = \frac{\partial}{\partial y}\left(\frac{\partial f}{\partial x}\right) = \frac{\partial^2 f}{\partial y \partial x} \quad \text{and} \quad f_{yx} = (f_y)_x = \frac{\partial}{\partial x}\left(\frac{\partial f}{\partial y}\right) = \frac{\partial^2 f}{\partial x \partial y}.$$

Finally, we remark that in light of our conventions in respect of left(-hand) and right(-hand) partial derivatives, the higher-order partial derivatives of any $f : [a, b] \times [c, d] \to \mathbb{R}$ are defined in a similar manner.

Examples 3.7. (i) Let $f : \mathbb{R}^2 \to \mathbb{R}$ be given by $f(x, y) := x^2 y + xy^2$. Given any $(x_0, y_0) \in \mathbb{R}^2$, we have $f_x(x_0, y_0) = 2x_0 y_0 + y_0^2$ and $f_y(x_0, y_0) = x_0^2 + 2x_0 y_0$, and consequently, $f_{xx}(x_0, y_0) = 2y_0$ and $f_{yy}(x_0, y_0) = 2x_0$, while

$$f_{xy}(x_0, y_0) = 2x_0 + 2y_0 = f_{yx}(x_0, y_0).$$

(ii) Let $f : \mathbb{R}^2 \to \mathbb{R}$ be given by $f(x, y) := \sin xy$. Given any $(x_0, y_0) \in \mathbb{R}^2$, we have $f_x(x_0, y_0) = y_0 \cos(x_0 y_0)$ and $f_y(x_0, y_0) = x_0 \cos(x_0 y_0)$, and consequently, $f_{xx}(x_0, y_0) = -y_0^2 \sin(x_0 y_0)$ and $f_{yy}(x_0, y_0) = -x_0^2 \sin(x_0 y_0)$, while

$$f_{xy}(x_0, y_0) = \cos(x_0 y_0) - x_0 y_0 \sin(x_0 y_0) = f_{yx}(x_0, y_0). \qquad \diamond$$

In the examples above, the mixed partials turned out to be equal. Presently, we will show that this is always the case whenever the first-order partial derivatives exist and one of the mixed second-order partial derivatives is continuous. To this end, we will use yet another version of the MVT (Fact 3.2) for bivariate functions defined on a rectangle, which, in turn, will be deduced from a version of Rolle's theorem. To begin with, let us recall the statement of Rolle's theorem from one-variable calculus. It may be noted that this is an immediate consequence of the MVT (Fact 3.2). A direct proof can be found, for example, on page 119 of ACICARA.

Fact 3.8 (Rolle's Theorem). *Let $a, b \in \mathbb{R}$ with $a < b$. If $f : [a, b] \to \mathbb{R}$ is continuous on $[a, b]$, differentiable on (a, b), and satisfies $f(a) = f(b)$, then there is $c \in (a, b)$ such that $f'(c) = 0$.*

In the two-variable analogue below, an interval is replaced by a rectangle, that is, a product of intervals, and the equality of function values on the boundary of an interval is replaced by the equality of sums of function values on the opposite endpoints of a rectangle. Alternatively, the alternating sum of function values at the corner points is required to be zero. The hypothesis concerning continuity and differentiability is also analogous.

Proposition 3.9 (Rectangular Rolle's Theorem). *Let $a, b, c, d \in \mathbb{R}$ with $a < b$ and $c < d$, and let $f : [a, b] \times [c, d] \to \mathbb{R}$ satisfy the following.*

- *For each fixed $y_0 \in [c, d]$, the function given by $x \longmapsto f(x, y_0)$ is continuous on $[a, b]$ and differentiable on (a, b).*
- *For each fixed $x_0 \in (a, b)$, the function given by $y \longmapsto f_x(x_0, y)$ is continuous on $[c, d]$ and differentiable on (c, d).*
- *$f(a, c) + f(b, d) = f(b, c) + f(a, d)$.*

Then there is $(x_0, y_0) \in (a, b) \times (c, d)$ such that $f_{xy}(x_0, y_0) = 0$.

Proof. Consider $\phi : [a, b] \to \mathbb{R}$ defined by $\phi(x) := f(x, d) - f(x, c)$. Then ϕ is continuous on $[a, b]$, differentiable on (a, b), and $\phi(a) = f(a, d) - f(a, c) = f(b, d) - f(b, c) = \phi(b)$. Hence by Rolle's Theorem (Fact 3.8), there is x_0 in (a, b) such that $\phi'(x_0) = 0$, that is, $f_x(x_0, c) = f_x(x_0, d)$. Next, consider $\psi : [c, d] \to \mathbb{R}$ defined by $\psi(y) := f_x(x_0, y)$. Then ψ is continuous on $[c, d]$, differentiable on (c, d), and $\psi(c) = f_x(x_0, c) = f_x(x_0, d) = \psi(d)$. Hence by Rolle's Theorem (Fact 3.8), there is $y_0 \in (c, d)$ such that $\psi'(y_0) = 0$, that is, $f_{xy}(x_0, y_0) = 0$. □

Remark 3.10. Another version of Rolle's Theorem (Fact 3.8) is given in Exercise 26 of Chapter 4. ◇

Proposition 3.11 (Rectangular Mean Value Theorem). *Let $a, b, c, d \in \mathbb{R}$ with $a < b$ and $c < d$, and let $f : [a, b] \times [c, d] \to \mathbb{R}$ satisfy the following.*

- *For each fixed $y_0 \in [c, d]$, the function given by $x \longmapsto f(x, y_0)$ is continuous on $[a, b]$ and differentiable on (a, b).*
- *For each fixed $x_0 \in (a, b)$, the function given by $y \longmapsto f_x(x_0, y)$ is continuous on $[c, d]$ and differentiable on (c, d).*

Then there is $(x_0, y_0) \in (a, b) \times (c, d)$ such that

$$f(b, d) + f(a, c) - f(b, c) - f(a, d) = (b - a)(d - c) f_{xy}(x_0, y_0).$$

Proof. Consider $s \in \mathbb{R}$ and $F : [a, b] \times [c, d] \to \mathbb{R}$ defined by

$$F(x, y) := f(x, y) + f(a, c) - f(x, c) - f(a, y) - s(x - a)(y - c).$$

Observe that $F(a,c) = F(a,d) = F(b,c) = 0$. Thus, if we choose s such that $F(b,d) = 0$, that is, if we take

$$s := \frac{f(b,d) + f(a,c) - f(b,c) - f(a,d)}{(b-a)(d-c)},$$

then we have $F(a,c) + F(b,d) = F(b,c) + F(a,d)$. So by the Rectangular Rolle's Theorem, there is $(x_0, y_0) \in (a,b) \times (c,d)$ such that $F_{xy}(x_0, y_0) = 0$, that is, $f_{xy}(x_0, y_0) = s$. This yields the desired result. □

The Rectangular Mean Value Theorem can be used to estimate the value of a function at one of the corner points of a rectangle, provided its values at the remaining corner points are known and we have bounds for one of its second-order mixed partial derivatives at the interior points of the rectangle.

Corollary 3.12 (Rectangular Mean Value Inequality). *Let $a, b, c, d \in \mathbb{R}$ and let $f : [a,b] \times [c,d] \to \mathbb{R}$ satisfy the following.*

- *For each fixed $y_0 \in [c,d]$, the function given by $x \longmapsto f(x, y_0)$ is continuous on $[a,b]$ and differentiable on (a,b).*
- *For each fixed $x_0 \in (a,b)$, the function given by $y \longmapsto f_x(x_0, y)$ is continuous on $[c,d]$ and differentiable on (c,d), and moreover, there are $m, M \in \mathbb{R}$ such that $m \le f_{xy}(x_0, y_0) \le M$ for all $(x_0, y_0) \in (a,b) \times (c,d)$.*

Then

$$m(b-a)(d-c) \le f(b,d) + f(a,c) - f(b,c) - f(a,d) \le M(b-a)(d-c).$$

Proof. If $a < b$ and $c < d$, then the desired inequalities are an immediate consequence of Proposition 3.11, whereas if $a = b$ or $c = d$, then each of the three expressions in the above inequalities is zero. □

Remark 3.13. In each of the last three results, namely, the Rectangular Rolle's Theorem, the Rectangular Mean Value Theorem, and the Rectangular Mean Value Inequality, the hypothesis that for each $y_0 \in [c,d]$, the function given by $x \longmapsto f(x, y_0)$ is continuous on $[a,b]$ can be weakened. Indeed, as the proofs show, it suffices to assume that the function $\phi : [a,b] \to \mathbb{R}$ defined by $\phi(x) := f(x,d) - f(x,c)$ is continuous. Moreover, each of these three results admit a straightforward analogue with f_x and f_{xy} replaced by f_y and f_{yx}. ◇

We are now ready to prove the equality of mixed second-order partial derivatives provided one of them is continuous.

Proposition 3.14 (Mixed Partials Theorem). *Let $D \subseteq \mathbb{R}^2$ be an open set and let (x_0, y_0) be any point of D. Let $f : D \to \mathbb{R}$ be such that both f_x and f_y exist on D. If f_{xy} or f_{yx} exists on D and is continuous at (x_0, y_0), then both $f_{xy}(x_0, y_0)$ and $f_{yx}(x_0, y_0)$ exist and*

$$f_{xy}(x_0, y_0) = f_{yx}(x_0, y_0).$$

Proof. Assume that f_{xy} exists on D and is continuous at (x_0, y_0). Let $\epsilon > 0$ be given. Since D is an open subset of $\mathbb{R}^2$ and f_{xy} is continuous at (x_0, y_0), by Proposition 2.22, there is $\delta > 0$ such that $\mathbb{S}_\delta(x_0, y_0) \subseteq D$ and

$$(u, v) \in \mathbb{S}_\delta(x_0, y_0) \Longrightarrow |f_{xy}(u, v) - f_{xy}(x_0, y_0)| < \epsilon.$$

Fix $(h, k) \in \mathbb{S}_\delta(0, 0)$ with $h \neq 0$ and $k \neq 0$. By the Rectangular Mean Value Theorem (Proposition 3.11), there is $(c, d) \in \mathbb{S}_\delta(x_0, y_0)$ such that

$$f(x_0 + h, y_0 + k) - f(x_0 + h, y_0) - f(x_0, y_0 + k) + f(x_0, y_0) = hk f_{xy}(c, d).$$

The left-hand side of the above equation can be written as $G(y_0 + k) - G(y_0)$, where $G : (y_0 - \delta, y_0 + \delta) \to \mathbb{R}$ is defined by $G(y) := f(x_0 + h, y) - f(x_0, y)$. Consequently,

$$\left| \frac{G(y_0 + k) - G(y_0)}{hk} - f_{xy}(x_0, y_0) \right| = |f_{xy}(c, d) - f_{xy}(x_0, y_0)| < \epsilon.$$

Since f_y exists on D, the function G is differentiable at y_0 and $G'(y_0) = f_y(x_0 + h, y_0) - f_y(x_0, y_0)$. Hence, taking the limit as $k \to 0$ (with h fixed), we see that for $0 < |h| < \delta$,

$$\left| \frac{f_y(x_0 + h, y_0) - f_y(x_0, y_0)}{h} - f_{xy}(x_0, y_0) \right| = \left| \frac{G'(y_0)}{h} - f_{xy}(x_0, y_0) \right| \leq \epsilon.$$

Since $\epsilon > 0$ is arbitrary, we conclude that $f_{yx}(x_0, y_0)$ exists and is equal to $f_{xy}(x_0, y_0)$. The case in which f_{yx} exists on D and is continuous at (x_0, y_0) is proved similarly in view of the last statement in Remark 3.13. □

Remark 3.15. In light of our conventions in respect of left(-hand) as well as right(-hand) partial derivatives, it is easily seen that a result analogous to the Mixed Partials Theorem (Proposition 3.14) holds when $D = [a, b] \times [c, d]$ and (x_0, y_0) is any point of D. ◇

Example 3.16. Consider $f : \mathbb{R}^2 \longrightarrow \mathbb{R}$ given by

$$f(x, y) := \begin{cases} xy\dfrac{x^2 - y^2}{x^2 + y^2} & \text{if } (x, y) \neq (0, 0), \\ 0 & \text{if } (x, y) = (0, 0). \end{cases}$$

It is clear that f_x, f_y, f_{xy}, and f_{yx} exist on $\mathbb{R}^2 \setminus \{(0, 0)\}$. Also, it is easy to see that $f_x(0, y_0) = -y_0$ for any $y_0 \in \mathbb{R}$ and $f_y(x_0, 0) = x_0$ for any $x_0 \in \mathbb{R}$. Hence $f_{xy}(0, 0) = -1 \neq 1 = f_{yx}(0, 0)$. Thus from Proposition 3.14, it follows that neither f_{xy} nor f_{yx} can be continuous at $(0, 0)$. ◇

Let $D \subseteq \mathbb{R}^2$ be an open set and $f : D \to \mathbb{R}$ a function on D. In case a second-order partial derivative of f is defined at every point of D, we can consider partial derivatives of the second-order partial derivatives with respect to x or with respect to y. This leads to the **third-order partial derivatives** $f_{xxx}, f_{xxy}, f_{xyx}, f_{xyy}, f_{yxx}, f_{yxy}, f_{yyx}$, and f_{yyy}. Applying the Mixed Partials Theorem to f_x and to f_y, we see that $f_{xxy} = f_{xyx}$ and $f_{yyx} = f_{yxy}$ at each point of D at which these mixed third-order partial derivatives are continuous. Moreover, if $f_{xy} = f_{yx}$ on D, then taking partial derivatives with respect to x and with respect to y, we obtain $f_{xyx} = f_{yxx}$ and $f_{yxy} = f_{xyy}$. Thus, if the second-order partial derivatives exist and are continuous on D, and the third-order partial derivatives exist on D, then we have $f_{xxy} = f_{xyx} = f_{yxx}$ and $f_{xyy} = f_{yxy} = f_{yyx}$ at each point (x_0, y_0) of D where these third-order partial derivatives are continuous. At such a point, instead of eight possible third-order partial derivatives, it suffice to consider only four, and these may be denoted by

$$\frac{\partial^3 f}{\partial x^3}(x_0, y_0), \quad \frac{\partial^3 f}{\partial x^2 \partial y}(x_0, y_0), \quad \frac{\partial^3 f}{\partial x \partial y^2}(x_0, y_0), \quad \text{and} \quad \frac{\partial^3 f}{\partial y^3}(x_0, y_0),$$

instead of $f_{xxx}(x_0, y_0)$, $f_{yxx}(x_0, y_0)$, $f_{yyx}(x_0, y_0)$, and $f_{yyy}(x_0, y_0)$, respectively. Continuing in this way, for each $n \in \mathbb{N}$, we can consider the **nth-order partial derivatives** of f at any point (x_0, y_0) of D. As such, there are 2^n possibilities, but if the partial derivatives of order $< n$ exist and are continuous on D and those of nth-order exist on D and are continuous at (x_0, y_0), then it suffices to consider only $n + 1$ of them, namely,

$$\frac{\partial^n f}{\partial x^n}(x_0, y_0), \quad \frac{\partial^n f}{\partial x^{n-1} \partial y}(x_0, y_0), \quad \ldots, \quad \frac{\partial^n f}{\partial x \partial y^{n-1}}(x_0, y_0), \quad \frac{\partial^n f}{\partial y^n}(x_0, y_0),$$

or, in short,

$$\frac{\partial^n f}{\partial x^{n-m} \partial y^m}(x_0, y_0) \quad \text{for } m = 0, 1, \ldots, n.$$

Examples 3.17. (i) Let I and J be nonempty open intervals in $\mathbb{R}$, and let $\phi : I \to \mathbb{R}$ and $\psi : J \to \mathbb{R}$ be infinitely differentiable functions of one variable. Consider $f, g : I \times J \to \mathbb{R}$ defined by

$$f(x, y) = \phi(x) + \psi(y) \quad \text{and} \quad g(x, y) = \phi(x)\psi(y).$$

Using an easy induction on $\ell + m$, where ℓ, m are nonnegative integers, we see that all the higher-order partial derivatives of f and g exist on $I \times J$, and are given, at any $(x_0, y_0) \in I \times J$, by

$$\frac{\partial^{\ell+m} f}{\partial x^\ell \partial y^m}(x_0, y_0) = \begin{cases} \phi(x_0) + \psi(y_0) & \text{if } \ell = 0 = m, \\ \phi^{(\ell)}(x_0) & \text{if } \ell \geq 1 \text{ and } m = 0, \\ \psi^{(m)}(y_0) & \text{if } \ell = 0 \text{ and } m \geq 1, \\ 0 & \text{if } \ell \geq 1 \text{ and } m \geq 1. \end{cases}$$

Likewise, the higher-order partial derivatives of g are given by

$$\frac{\partial^{\ell+m} g}{\partial x^\ell \partial y^m}(x_0, y_0) = \phi^{(\ell)}(x_0)\psi^{(m)}(y_0) \text{ for } \ell \geq 0, \ m \geq 0, \text{ and } (x_0, y_0) \in I \times J.$$

(ii) Given any nonnegative integers i, j, let $g_{i,j} : \mathbb{R}^2 \to \mathbb{R}$ denote the function defined by $g_{i,j}(x,y) = x^i y^j$. Using (i) above, we see that for any nonnegative integers ℓ, m and any $(x_0, y_0) \in \mathbb{R}^2$, we have

$$\frac{\partial^{\ell+m} f_{i,j}}{\partial x^\ell \partial y^m}(x_0, y_0) = \begin{cases} \dfrac{i!\,j!}{(i-\ell)!\,(j-m)!}\, x_0^{i-\ell} y_0^{j-m} & \text{if } \ell \le i \text{ and } m \le j, \\ 0 & \text{if } \ell > i \text{ or } m > j. \end{cases}$$

Next, let $f : \mathbb{R}^2 \longrightarrow \mathbb{R}$ be a polynomial function, so that

$$f(x,y) := \sum_{\substack{i \ge 0 \\ i+j \le n}} \sum_{j \ge 0} c_{i,j} x^i y^j \quad \text{for all } (x,y) \in \mathbb{R}^2,$$

where n is a nonnegative integer and $c_{i,j} \in \mathbb{R}$ for $i, j \ge 0$ with $i + j \le n$. Let us set $c_{i,j} := 0$ if $i, j \ge 0$ with $i + j > n$. Also, as usual, let $a^0 := 1$ for any $a \in \mathbb{R}$ and set any empty sum equal to 0. Then in view of the above formula for the partial derivatives of $g_{i,j}$, we see that for any nonnegative integers ℓ, m and any $(x_0, y_0) \in \mathbb{R}^2$, we have

$$\frac{\partial^{\ell+m} f}{\partial x^\ell \partial y^m}(x_0, y_0) = \sum_{\substack{i \ge \ell \\ i+j \le n}} \sum_{j \ge m} c_{i,j} \frac{i!\,j!}{(i-\ell)!\,(j-m)!}\, x_0^{i-\ell} y_0^{j-m}.$$

Thus, the partial derivatives of f of any order exist at every $(x_0, y_0) \in \mathbb{R}^2$ and are given by the above formula.

(iii) As remarked earlier, taking partial derivatives is tantamount to differentiating a function of one variable, and thus usual rules of differentiation are applicable. Consider, in particular, the effect of the chain rule on a composite function of the form $f(x,y) := g(u(x,y))$, where $g : E \to \mathbb{R}$ and $u : D \to \mathbb{R}^2$ are such that $u(D) \subseteq E$, and where E is an open subset of $\mathbb{R}$ and D is an open subset of $\mathbb{R}^2$. Assume that all the partial derivatives of u exist on D and g is infinitely differentiable on E. Then all the higher-order partial derivatives of f exist on D. However, finding an explicit formula for them in terms of the partial derivatives of u and the derivatives of f does not seem easy. In the special case in which u is linear, that is, $u(x,y) := ax + by$ for $(x,y) \in D$, where $a, b \in \mathbb{R}$, it is easy to see that at any $(x_0, y_0) \in D$, upon letting $u_0 := u(x_0, y_0) = ax_0 + by_0$, we have

$$\frac{\partial^{\ell+m} f}{\partial x^\ell \partial y^m}(x_0, y_0) = g^{(\ell+m)}(u_0) a^\ell b^m \quad \text{for } \ell \ge 0,\ m \ge 0.$$

Another special case is treated in Exercise 39. Formal statements of the chain rule are discussed later, in Section 3.3. ◇

Given any $h, k \in \mathbb{R}$, we define the **partial differential operator** $\mathcal{D}_{h,k}$ as follows:

$$\mathcal{D}_{h,k} := h\frac{\partial}{\partial x} + k\frac{\partial}{\partial y}.$$

$\mathcal{D}_{h,k}$ transforms a real-valued function of two variables to another real-valued function of two variables. Thus, if $D \subseteq \mathbb{R}^2$ is open and $f : D \to \mathbb{R}$ is such that both the partial derivatives of f exist at every point of D, then $\mathcal{D}_{h,k}f : D \to \mathbb{R}$ is the function defined by

$$(\mathcal{D}_{h,k}f)(x_0, y_0) := h\frac{\partial f}{\partial x}(x_0, y_0) + k\frac{\partial f}{\partial y}(x_0, y_0) = hf_x(x_0, y_0) + kf_y(x_0, y_0),$$

where (x_0, y_0) varies over D. The operator notation $\mathcal{D}_{h,k}$ has the advantage that we can consider formal powers (or successive composites) of $\mathcal{D}_{h,k}$, and these allow us to consider a combination of the nth-order partial derivatives at once. Thus, for any $n \in \mathbb{N}$, we define

$$\mathcal{D}^n_{h,k} := \left(h\frac{\partial}{\partial x} + k\frac{\partial}{\partial y}\right)^n := \sum_{m=0}^{n} \binom{n}{m} h^{n-m}k^m \frac{\partial^n}{\partial x^{n-m}\partial y^m}.$$

In other words, if $D \subseteq \mathbb{R}^2$ is open and $f : D \to \mathbb{R}$ has continuous partial derivatives of order $\le n$ at every point of D, then $\mathcal{D}^n_{h,k}f : D \to \mathbb{R}$ is the function defined by

$$(\mathcal{D}^n_{h,k}f)(x_0, y_0) := \sum_{m=0}^{n} \binom{n}{m} h^{n-m}k^m \frac{\partial^n f}{\partial x^{n-m}\partial y^m}(x_0, y_0),$$

where (x_0, y_0) varies over D. For example,

$$(\mathcal{D}^2_{h,k}f)(x_0, y_0) = h^2\frac{\partial^2 f}{\partial x^2}(x_0, y_0) + 2hk\frac{\partial^2 f}{\partial x \partial y}(x_0, y_0) + k^2\frac{\partial^2 f}{\partial y^2}(x_0, y_0).$$

Example 3.18. Consider, as before, a polynomial function $f : \mathbb{R}^2 \to \mathbb{R}$ given by

$$f(x, y) := \sum_{\substack{i \ge 0 \\ i+j \le n}} \sum_{j \ge 0} c_{i,j}x^i y^j \quad \text{for all } (x, y) \in \mathbb{R}^2.$$

Putting $(x_0, y_0) = (0, 0)$ in the formula for higher-order partial derivatives of f given in Example 3.17 (ii), we see that

$$\frac{\partial^{\ell+m} f}{\partial x^\ell \partial y^m}(0, 0) = \ell!\, m!\, c_{\ell,m} \quad \text{for any nonnegative integers } \ell, m.$$

Consequently, given any nonnegative integer p, we have

$$\left(\mathcal{D}^p_{h,k}f\right)(0, 0) = p! \sum_{m=0}^{p} c_{p-m,m}\, h^{p-m}k^m \quad \text{for any } (h, k) \in \mathbb{R}^2.$$

Dividing both sides by $p!$ and summing as p varies from 0 to n, we see that

$$f(h,k) = \sum_{p=0}^{n} \frac{1}{p!} \left(\mathcal{D}^p_{h,k} f\right)(0,0) \quad \text{for any } (h,k) \in \mathbb{R}^2.$$

Thus, we obtain an alternative expression for $f(x,y)$ in terms of its higher-order partial derivatives at $(0,0)$. ◇

Higher-Order Directional Derivatives

Let $D \subseteq \mathbb{R}^2$ be an open subset of $\mathbb{R}^2$ and let $\mathbf{u} = (u_1, u_2)$ be a unit vector in $\mathbb{R}^2$. Then for every $(x_0, y_0) \in D$, the real number 0 is clearly an interior point of $D_0 := \{t \in \mathbb{R} : (x_0 + tu_1, y_0 + tu_2) \in D\}$. Let $f : D \to \mathbb{R}$ be any function.

If $\mathbf{D_u}f(x_0, y_0)$ exists for every $(x_0, y_0) \in D$, then we obtain a function from D to $\mathbb{R}$ given by $(x,y) \longmapsto \mathbf{D_u}f(x,y)$. It is denoted by $\mathbf{D_u}f$ and called the **directional derivative** of f on D along $\mathbf{u}$. Since $\mathbf{D_u}f$ is a real-valued function of two variables, we can further consider its directional derivatives along $\mathbf{u}$ at points of D. If these exist at every point of D, then we obtain a function from D to $\mathbb{R}$ given by $(x,y) \longmapsto \mathbf{D_u}(\mathbf{D_u}f)(x,y)$. It is denoted by $\mathbf{D}^2_\mathbf{u}f$ and called the **second-order directional derivative** of f on D along $\mathbf{u}$. In general, we can make a recursive definition as follows. Let $f_0 := \mathbf{D}^0_\mathbf{u}f := f$ and suppose $f_i := \mathbf{D}^i_\mathbf{u}f$ has been defined for $i = 0, 1, \ldots, n-1$, where $n \in \mathbb{N}$. If the directional derivative of f_{n-1} on D along $\mathbf{u}$ exists, then we denote it by $\mathbf{D}^n_\mathbf{u}f$ and call it the ***n*th-order directional derivative** of f on D along $\mathbf{u}$. Note that $\mathbf{D}^1_\mathbf{u}f := \mathbf{D_u}f$.

Remark 3.19. Suppose $f : D \to \mathbb{R}$ and $\mathbf{u}$ are as above, and the directional derivative $\mathbf{D_u}f$ of f on D along $\mathbf{u}$ exists. If $\mathbf{v}$ is any unit vector in $\mathbb{R}^2$ and if the directional derivative of $\mathbf{D_u}f$ on D along $\mathbf{v}$ exists, then we obtain a function from D to $\mathbb{R}$ given by $(x,y) \longmapsto \mathbf{D_v}(\mathbf{D_u}f)(x,y)$. It may be denoted by $D^2_{\mathbf{uv}}f$. Proceeding in this manner, one can formulate a notion of nth-order directional derivatives along an ordered n-tuple of unit vectors in $\mathbb{R}^2$. However, we shall not have any occasion to use such notions in their full generality, and hence we refrain from discussing it further. ◇

We have seen earlier that a bivariate analogue of the MVT can be formulated using directional derivatives. In a similar vein, we can formulate and prove a bivariate analogue of Taylor's theorem using higher-order directional derivatives. First, let us recall the statement of Taylor's theorem from one-variable calculus. For a proof, one may refer to page 122 of ACICARA.

Fact 3.20. *Let $a, b \in \mathbb{R}$ with $a < b$ and n a nonnegative integer. If $f : [a,b] \to \mathbb{R}$ is such that $f', f'', \ldots, f^{(n)}$ exist on $[a,b]$ and further, $f^{(n)}$ is continuous on $[a,b]$ and differentiable on (a,b), then there is $c \in (a,b)$ satisfying*

$$f(b) = f(a) + f'(a)(b-a) + \cdots + \frac{f^{(n)}(a)}{n!}(b-a)^n + \frac{f^{(n+1)}(c)}{(n+1)!}(b-a)^{n+1}.$$

Here is an analogous result for functions of two variables.

Proposition 3.21 (Bivariate Taylor Theorem). *Let $D \subseteq \mathbb{R}^2$ be an open set and let (x_0, y_0) and (x_1, y_1) be distinct points in D such that the line segment joining them is in D, that is, $L := \{(x(t), y(t)) \in \mathbb{R}^2 : t \in [0,1]\} \subseteq D$, where*

$$x(t) := x_0 + t(x_1 - x_0) \text{ and } y(t) := y_0 + t(y_1 - y_0).$$

Let $\mathbf{u} = (u_1, u_2)$ be the unit vector given by

$$\mathbf{u} = (u_1, u_2) := \frac{1}{r}(x_1 - x_0, y_1 - y_0), \quad \text{where} \quad r := \sqrt{(x_1 - x_0)^2 + (y_1 - y_0)^2}.$$

Let n be a nonnegative integer and let $f : D \to \mathbb{R}$ be such that $\mathbf{D}^i_{\mathbf{u}} f$ exists and is continuous on D for $i = 0, \ldots, n$, and moreover, $\mathbf{D}^{n+1}_{\mathbf{u}} f$ exists at $(x(t), y(t))$ for each $t \in (0,1)$. Then there is $(c, d) \in L \setminus \{(x_0, y_0), (x_1, y_1)\}$ such that

$$f(x_1, y_1) = \sum_{i=0}^{n} \frac{r^i}{i!} \left(\mathbf{D}^i_{\mathbf{u}} f\right)(x_0, y_0) + \frac{r^{n+1}}{(n+1)!} \left(\mathbf{D}^{n+1}_{\mathbf{u}} f\right)(c, d).$$

Proof. For $i = 0, \ldots, n$, define $f_i : D \to \mathbb{R}$ by $f_i := \mathbf{D}^i_{\mathbf{u}} f$ and $F_i : [0,1] \to \mathbb{R}$ by $F_i(t) := f_i(x(t), y(t))$ for $t \in [0,1]$. Define $f_{n+1} : L \setminus \{(x_0, y_0), (x_1, y_1)\} \to \mathbb{R}$ by $f_{n+1} := \mathbf{D}^{n+1}_{\mathbf{u}} f$ and $F_{n+1} : (0,1) \to \mathbb{R}$ by $F_{n+1}(t) := f_{n+1}(x(t), y(t))$ for $t \in (0,1)$.

Using the definition of directional derivative, we see, as in the proof of the Bivariate Mean Value Theorem (Proposition 3.5) that if $n \geq 1$, then $F := F_0$ is differentiable on $[0,1]$ and

$$F'(t) = r\left(\mathbf{D}_{\mathbf{u}} f\right)(x(t), y(t)) = r f_1(x(t), y(t)) = r F_1(t) \quad \text{for all } t \in [0,1].$$

Similarly, if $n \geq 2$, then F_1 is differentiable on $[0,1]$ and $F_1'(t) = rF_2(t)$ for all $t \in [0,1]$. Hence $F := F_0$ is twice differentiable on $[0,1]$ and

$$F''(t) = rF_1'(t) = r^2 F_2(t) \quad \text{for all } t \in [0,1].$$

Continuing in this way, we see that for $i = 0, \ldots, n$, the ith-order derivative of F exists on $[0,1]$ and $F^{(i)}(t) = r^i F_i(t)$ for all $t \in [0,1]$. Moreover, the $(n+1)$th-order derivative of F exists on $(0,1)$ and $F^{(n+1)}(t) = r^{n+1} F_{n+1}(t)$ for all $t \in (0,1)$. Hence by Taylor's theorem of one-variable calculus (Fact 3.20) applied to F, there is $\theta \in (0,1)$ such that

$$F(1) = \sum_{i=0}^{n} \frac{F^{(i)}(0)}{i!} + \frac{F^{(n+1)}(\theta)}{(n+1)!}.$$

Consequently, $(c, d) := (x(\theta), y(\theta)) \in L \setminus \{(x_0, y_0), (x_1, y_1)\}$, and we have

$$f(x_1, y_1) = \sum_{i=0}^{n} \frac{r^i}{i!} \left(\mathbf{D}^i_{\mathbf{u}} f\right)(x_0, y_0) + \frac{r^{n+1}}{(n+1)!} \left(\mathbf{D}^{n+1}_{\mathbf{u}} f\right)(c, d),$$

as desired. □

In case the identity $\mathbf{D_u} f = \nabla f \cdot \mathbf{u}$ is satisfied by f as well as its higher-order directional derivatives, we obtain the following alternative version of the Bivariate Taylor Theorem that is analogous to Corollary 3.6.

Corollary 3.22. *Let $D \subseteq \mathbb{R}^2$ be an open set and n a nonnegative integer. Let (x_0, y_0) and (x_1, y_1) be distinct points in D and let L be the line segment joining them. Assume that $L \subseteq D$. Let $f : D \to \mathbb{R}$ be such that f has continuous partial derivatives of order $\le n+1$ at every point of D and moreover, the higher-order directional derivatives $\mathbf{D}^i_{\mathbf{u}} f$ exist at every point of D for all unit vectors $\mathbf{u}$ in $\mathbb{R}^2$ and $i = 0, 1, \ldots, n+1$. Assume further that for any unit vector $\mathbf{u}$ in $\mathbb{R}^2$, the functions $f_i := \mathbf{D}^i_{\mathbf{u}} f$ satisfy*

$$(\mathbf{D_u} f_i)(x, y) = \nabla f_i(x, y) \cdot \mathbf{u} \quad \text{for all } (x, y) \in D \text{ and } i = 0, \ldots, n,$$

where $f_0 := f$. Then there is $(c, d) \in L \setminus \{(x_0, y_0), (x_1, y_1)\}$ such that

$$f(x_1, y_1) = \sum_{i=0}^{n} \frac{1}{i!} \left(\mathcal{D}^i_{h,k} f\right)(x_0, y_0) + \frac{1}{(n+1)!} \left(\mathcal{D}^{n+1}_{h,k} f\right)(c, d),$$

where $h := x_1 - x_0$ and $k := y_1 - y_0$.

Proof. With h and k as above, let $r := \sqrt{h^2 + k^2}$ and $\mathbf{u} := (h/r,\, k/r)$. Then $\mathbf{u}$ is a unit vector in $\mathbb{R}^2$ with $r\mathbf{u} = (h, k)$. As in the proof of Corollary 3.6, for any $(x, y) \in D$ and $i = 0, 1, \ldots, n$, we have

$$r f_{i+1}(x, y) = r\mathbf{D_u} f_i(x, y) = \left(h\frac{\partial f_i}{\partial x} + k\frac{\partial f_i}{\partial y}\right)(x, y) = (\mathcal{D}_{h,k} f_i)(x, y).$$

Successively using the above identity, we see that $r^i \left(\mathbf{D}^i_{\mathbf{u}} f\right) = r^i f_i = (\mathcal{D}_{h,k} f_i)^i$ for $i = 0, \ldots, n+1$. So the desired result follows from Proposition 3.21. □

3.2 Differentiability

The difficulties involved in generalizing the notion of differentiability from functions of one variable to functions of two (or more) variables were discussed at the beginning of Section 3.1. We shall show in this section how to overcome these difficulties. The key idea here is twofold: (i) a realization that the derivative of a real-valued function of two variables may not be a single number but possibly a pair of real numbers, and (ii) an observation that the problem of division by a point (h, k) in $\mathbb{R}^2$ can be solved by replacing (h, k) with its norm $|(h, k)| := \sqrt{h^2 + k^2}$. To understand this better, let us first note that if $D \subseteq \mathbb{R}$ and c is an interior point of D, then a function $f : D \to \mathbb{R}$ is differentiable at c if and only if there is $\alpha \in \mathbb{R}$ such that

$$\lim_{h \to 0} \frac{f(c+h) - f(c) - \alpha h}{|h|} = 0.$$

In this case, α is the derivative of f at c. Now suppose $D \subseteq \mathbb{R}^2$ and (x_0, y_0) is an interior point of D. A function $f : D \to \mathbb{R}$ is said to be **differentiable** at (x_0, y_0) if there is $(\alpha_1, \alpha_2) \in \mathbb{R}^2$ such that

$$\lim_{(h,k)\to(0,0)} \frac{f(x_0 + h, y_0 + k) - f(x_0, y_0) - \alpha_1 h - \alpha_2 k}{\sqrt{h^2 + k^2}} = 0.$$

In this case, we call the pair (α_1, α_2) the **total derivative**[1] of f at (x_0, y_0).

Let us note that if f is differentiable at (x_0, y_0) and if (α_1, α_2) is the total derivative of f at (x_0, y_0), then letting (h, k) approach $(0, 0)$ along the x-axis or the y-axis, we see that

$$\lim_{h\to 0} \frac{f(x_0 + h, y_0) - f(x_0, y_0) - \alpha_1 h}{|h|} = 0 = \lim_{k\to 0} \frac{f(x_0, y_0 + k) - f(x_0, y_0) - \alpha_2 k}{|k|}.$$

Consequently, both $f_x(x_0, y_0)$ and $f_y(x_0, y_0)$ exist and are equal to α_1 and α_2, respectively. In other words, if f is differentiable, then the gradient of f at (x_0, y_0) exists and

$$\text{the total derivative of } f \text{ at } (x_0, y_0) = \nabla f(x_0, y_0).$$

Thus in checking the differentiability of f at (x_0, y_0), it is clear which values of α_1 and α_2 can possibly work, and the task reduces to checking whether the corresponding two-variable limit exists and is equal to zero. Also, if either of the partial derivatives does not exist at a point, then we can be sure that f is not differentiable at that point. This is illustrated by Example 3.23 (iii). On the other hand, existence of both the partial derivatives at (x_0, y_0) is not sufficient for f to be differentiable at (x_0, y_0), and this will be seen later in Example 3.29 (i).

Examples 3.23. (i) Let $f : \mathbb{R}^2 \to \mathbb{R}$ be the constant function given by $f(x, y) := 1$ for all $(x, y) \in \mathbb{R}^2$. It is clear that f is differentiable at any $(x_0, y_0) \in \mathbb{R}^2$ and the total derivative at (x_0, y_0) is $(0, 0)$.

(ii) Let $f : \mathbb{R}^2 \to \mathbb{R}$ be given by $f(x, y) := x^2 + y^2$. Given any $(x_0, y_0) \in \mathbb{R}^2$, we have $f_x(x_0, y_0) = 2x_0$ and $f_y(x_0, y_0) = 2y_0$, and moreover,

$$\lim_{(h,k)\to(0,0)} \frac{f(x_0 + h, y_0 + k) - f(x_0, y_0) - 2x_0 h - 2y_0 k}{\sqrt{h^2 + k^2}}$$
$$= \lim_{(h,k)\to(0,0)} \frac{h^2 + k^2}{\sqrt{h^2 + k^2}} = \lim_{(h,k)\to(0,0)} \sqrt{h^2 + k^2} = 0.$$

It follows that f is differentiable at (x_0, y_0) and $\nabla f(x_0, y_0) = (2x_0, 2y_0)$.

[1] In modern treatments of multivariable calculus, instead of the pair (α_1, α_2), the linear map from $\mathbb{R}^2$ to $\mathbb{R}$ given by $(h, k) \mapsto \alpha_1 h + \alpha_2 k$ is called the (total) derivative of f at (x_0, y_0). However, the pair (α_1, α_2) and the corresponding linear map determine each other uniquely. For this reason and for the sake of simplicity, we have chosen to call the pair (α_1, α_2) the (total) derivative of f at (x_0, y_0).

(iii) Let $f : \mathbb{R}^2 \to \mathbb{R}$ be the norm function given by $f(x,y) := \sqrt{x^2+y^2}$. We have seen in Example 3.1 (ii) that both $f_x(0,0)$ and $f_y(0,0)$ do not exist. Hence f is not differentiable at $(0,0)$.

(iv) Let $f : \mathbb{R}^2 \to \mathbb{R}$ defined by $f(x,y) = |xy|$. It is easily seen that $f_x(0,0) = 0 = f_y(0,0)$. Moreover, for any $(h,k) \in \mathbb{R}^2$, we have $|h| \le \sqrt{h^2+k^2}$ and thus if $(h,k) \neq (0,0)$, then

$$\frac{f(h,k) - f(0,0) - 0 \cdot h - 0 \cdot k}{\sqrt{h^2+k^2}} = \frac{|hk|}{\sqrt{h^2+k^2}} \le |k| \to 0 \text{ as } (h,k) \to (0,0).$$

Hence, f is differentiable at $(0,0)$ and $\nabla f(0,0) = (0,0)$. ◇

For functions of one variable, one has a useful characterization of differentiability given by Carathéodory's lemma. Let us recall its statement; for a proof we refer to page 107 of ACICARA.

Fact 3.24 (Carathéodory's Lemma). *Let $D \subseteq \mathbb{R}$ and let c be an interior point of D. Then $f : D \to \mathbb{R}$ is differentiable at c if and only if there exists a function $f_1 : D \to \mathbb{R}$ such that $f(x) - f(c) = (x-c)f_1(x)$ for all $x \in D$, and f_1 is continuous at c. Moreover, if these conditions hold, then $f'(c) = f_1(c)$.*

If the conditions of Fact 3.24 hold, then the function f_1 is uniquely determined by f and c, and f_1 is called the **increment function** associated with f and c. For functions of two variables, there is an analogous characterization of differentiability, and it will play an important role in the sequel.

Proposition 3.25 (Increment Lemma). *Let $D \subseteq \mathbb{R}^2$ and let (x_0, y_0) be an interior point of D. Then $f : D \to \mathbb{R}$ is differentiable at (x_0, y_0) if and only if there exist functions $f_1, f_2 : D \to \mathbb{R}$ such that f_1 and f_2 are continuous at (x_0, y_0) and*

$$f(x,y) - f(x_0,y_0) = (x-x_0)f_1(x,y) + (y-y_0)f_2(x,y) \quad \text{for all } (x,y) \in D.$$

Moreover, if these conditions hold, then $\nabla f(x_0,y_0) = (f_1(x_0,y_0), f_2(x_0,y_0))$.

Proof. Assume that $f : D \to \mathbb{R}$ is differentiable at (x_0, y_0). Then there is $(\alpha_1, \alpha_2) \in \mathbb{R}^2$ such that

$$\lim_{(x,y)\to(x_0,y_0)} \frac{F(x,y)}{\sqrt{(x-x_0)^2 + (y-y_0)^2}} = 0,$$

where

$$F(x,y) := f(x,y) - f(x_0,y_0) - \alpha_1(x-x_0) - \alpha_2(y-y_0) \quad \text{for } (x,y) \in D.$$

Define $f_1, f_2 : D \to \mathbb{R}$ by $f_i(x_0,y_0) := \alpha_i$ for $i = 1, 2$, and for $(x,y) \neq (x_0,y_0)$,

$$f_1(x,y) := \alpha_1 + \frac{(x-x_0)F(x,y)}{(x-x_0)^2 + (y-y_0)^2}, \quad f_2(x,y) := \alpha_2 + \frac{(y-y_0)F(x,y)}{(x-x_0)^2 + (y-y_0)^2}.$$

Then for all $(x, y) \in D$ with $(x, y) \neq (x_0, y_0)$, we have

$$\begin{aligned}
&(x - x_0)f_1(x, y) + (y - y_0)f_2(x, y)\\
&= \alpha_1(x - x_0) + \frac{(x - x_0)^2 F(x, y)}{(x - x_0)^2 + (y - y_0)^2} + \alpha_2(y - y_0) + \frac{(y - y_0)^2 F(x, y)}{(x - x_0)^2 + (y - y_0)^2}\\
&= \alpha_1(x - x_0) + \alpha_2(y - y_0) + F(x, y).
\end{aligned}$$

Thus, using the definition of $F(x, y)$ and making a direct verification when $(x, y) = (x_0, y_0)$, we obtain

$$f(x, y) - f(x_0, y_0) = (x - x_0)f_1(x, y) + (y - y_0)f_2(x, y) \quad \text{for all } (x, y) \in D.$$

Moreover, given any $(x, y) \in \mathbb{R}^2$ with $(x, y) \neq (x_0, y_0)$, we clearly have

$$\frac{|x - x_0|}{\sqrt{(x - x_0)^2 + (y - y_0)^2}} \leq 1 \quad \text{and} \quad \frac{|y - y_0|}{\sqrt{(x - x_0)^2 + (y - y_0)^2}} \leq 1,$$

and hence for $(x, y) \in D$ with $(x, y) \neq (x_0, y_0)$, and for $i = 1, 2$, we have

$$|f_i(x, y) - \alpha_i| \leq \left| \frac{F(x, y)}{\sqrt{(x - x_0)^2 + (y - y_0)^2}} \right| \to 0 \quad \text{as } (x, y) \to (x_0, y_0).$$

Thus, in view of Proposition 2.48, we see that f_1 and f_2 are continuous at (x_0, y_0).

Conversely, assume that there are $f_1, f_2 : D \to \mathbb{R}$ that are continuous at (x_0, y_0) and satisfy

$$f(x, y) - f(x_0, y_0) = (x - x_0)f_1(x, y) + (y - y_0)f_2(x, y) \quad \text{for all } (x, y) \in D.$$

Define $\alpha_1 := f_1(x_0, y_0)$ and $\alpha_2 := f_2(x_0, y_0)$, and for $(x, y) \in D$, let $F(x, y)$ be as before. Then

$$F(x, y) = (x - x_0)(f_1(x, y) - \alpha_1) + (y - y_0)(f_2(x, y) - \alpha_2) \quad \text{for } (x, y) \in D.$$

Consequently, for any $(x, y) \in D$ with $(x, y) \neq (x_0, y_0)$, we have

$$\left| \frac{F(x, y)}{\sqrt{(x - x_0)^2 + (y - y_0)^2}} \right| \leq |f_1(x, y) - \alpha_1| + |f_2(x, y) - \alpha_2|.$$

Since both f_1 and f_2 are continuous at (x_0, y_0), we have $|f_1(x, y) - \alpha_1| \to 0$ and $|f_2(x, y) - \alpha_2| \to 0$ as $(x, y) \to (x_0, y_0)$. Hence,

$$\lim_{(x,y)\to(x_0,y_0)} \frac{F(x, y)}{\sqrt{(x - x_0)^2 + (y - y_0)^2}} = 0.$$

It follows that f is differentiable at (x_0, y_0) and also that $\nabla f(x_0, y_0) = (f_1(x_0, y_0), f_2(x_0, y_0))$. □

A pair (f_1, f_2) of functions satisfying the conditions in the Increment Lemma will be called a **pair of increment functions** associated with the function f and the point (x_0, y_0). Thus the Increment Lemma may be paraphrased by saying that the differentiability of f at (x_0, y_0) is equivalent to the existence of a pair of increment functions associated with f and (x_0, y_0).

Remarks 3.26. (i) As in the case of functions of one variable, differentiability of a function of two variables is a local condition. In other words, if $D \subseteq \mathbb{R}^2$ and (x_0, y_0) is an interior point of D, then $f : D \to \mathbb{R}$ is differentiable at (x_0, y_0) if and only if there is $\delta > 0$ with $\mathbb{S}_\delta(x_0, y_0) \subseteq D$ such that the restriction $f_{|\mathbb{S}_\delta(x_0,y_0)}$ of f to $\mathbb{S}_\delta(x_0, y_0)$ is differentiable at (x_0, y_0). In particular, the Increment Lemma can be applied to such a restriction of f, and to check differentiability of f at (x_0, y_0) it suffices to find a pair of increment functions on $\mathbb{S}_\delta(x_0, y_0)$ for some $\delta > 0$.

(ii) In contrast to the case of functions of one variable, a pair of increment functions associated with a function $f : D \to \mathbb{R}$ of two variables and an interior point (x_0, y_0) of D may not be unique. Indeed, let (f_1, f_2) be a pair of increment functions associated with f and (x_0, y_0), and let $h : D \to \mathbb{R}$ be any function that is continuous at (x_0, y_0). Define $g_1, g_2 : D \to \mathbb{R}$ by

$$g_1(x, y) = f_1(x, y) + (y - y_0)h(x, y) \quad \text{and} \quad g_2(x, y) = f_2(x, y) - (x - x_0)h(x, y).$$

Then (g_1, g_2) is also a pair of increment functions associated with f and (x_0, y_0). Thus, when f is differentiable at $(x_0, y_))$, there are infinitely many pairs of increment functions associated with f and (x_0, y_0). ◇

Example 3.27. Consider $f : \mathbb{R}^2 \to \mathbb{R}$ given by $f(x, y) := x$, and let (x_0, y_0) be any point of $\mathbb{R}^2$. Define $f_1, f_2 : \mathbb{R}^2 \to \mathbb{R}$ by $f_1(x, y) := 1$ and $f_2(x, y) := 0$. Then it is clear that (f_1, f_2) is a pair of increment functions associated with f and (x_0, y_0). Thus, f is differentiable at (x_0, y_0) and $\nabla f(x_0, y_0) = (1, 0)$. Similarly, $g : \mathbb{R}^2 \to \mathbb{R}$ given by $g(x, y) := y$ is differentiable at any $(x_0, y_0) \in \mathbb{R}^2$ and $\nabla g(x_0, y_0) = (0, 1)$. ◇

An immediate consequence of the Increment Lemma is the following.

Proposition 3.28. *Let $D \subseteq \mathbb{R}^2$ and let (x_0, y_0) be an interior point in D. If $f : D \to \mathbb{R}$ is differentiable at (x_0, y_0), then f is continuous at (x_0, y_0).*

Proof. If (f_1, f_2) is a pair of increment functions associated with f and (x_0, y_0), then

$$f(x, y) = f(x_0, y_0) + (x - x_0)f_1(x, y) + (y - y_0)f_2(x, y) \qquad \text{for all } (x, y) \in D.$$

Consequently, the continuity of f at (x_0, y_0) follows from the continuity of f_1 and f_2 at (x_0, y_0). □

Examples 3.29. (i) Let $f : \mathbb{R}^2 \to \mathbb{R}$ be given by $f(0,0) := 0$ and $f(x,y) := x^2y/(x^4+y^2)$ for $(x,y) \neq (0,0)$. We have seen in Example 2.16 (iv) that f is not continuous at $(0,0)$. Hence, f is not differentiable at $(0,0)$. On the other hand, it may be recalled from Example 3.4 (iii) that all the directional derivatives of f at $(0,0)$ exist.

(ii) Let $f : \mathbb{R}^2 \to \mathbb{R}$ be the norm function given by $f(x,y) = \sqrt{x^2+y^2}$. We have seen in Example 2.12 (ii) that f is continuous at $(0,0)$, but we have also seen in Example 3.23 (iii) that f is not differentiable at $(0,0)$. Thus, the converse of Proposition 3.28 is not true. ◇

An easy application of the Increment Lemma shows that (total) derivatives of sums, scalar multiples, products, reciprocals, and radicals of real-valued functions of two variables behave in the same way as in the one-variable case.

Proposition 3.30. *Let $D \subseteq \mathbb{R}^2$ and let (x_0, y_0) be an interior point of D. Suppose $r \in \mathbb{R}$ and $f, g : D \to \mathbb{R}$ are functions that are differentiable at (x_0, y_0). Then $f+g$, rf, and fg are differentiable at (x_0, y_0); moreover,*

$$\nabla(f+g)(x_0,y_0) = \nabla f(x_0,y_0) + \nabla g(x_0,y_0), \quad \nabla(rf)(x_0,y_0) = r\nabla f(x_0,y_0),$$

and

$$\nabla(fg)(x_0,y_0) = g(x_0,y_0)\nabla f(x_0,y_0) + f(x_0,y_0)\nabla g(x_0,y_0).$$

In case $f(x_0,y_0) \neq 0$, then there is $\delta > 0$ such that $\mathbb{S}_\delta(x_0,y_0) \subseteq D$ and $f(x,y) \neq 0$ for all $(x,y) \in \mathbb{S}_\delta(x_0,y_0)$; moreover, $1/f : D \cap \mathbb{S}_\delta(x_0,y_0) \to \mathbb{R}$ is differentiable at (x_0,y_0) and

$$\nabla\left(\frac{1}{f}\right)(x_0,y_0) = -\frac{1}{f(x_0,y_0)^2}\nabla f(x_0,y_0).$$

In case there is $\delta > 0$ such that $\mathbb{S}_\delta(x_0,y_0) \subseteq D$ and $f(x,y) > 0$ for all $(x,y) \in \mathbb{S}_\delta(x_0,y_0)$, then for every $k \in \mathbb{N}$, the function $f^{1/k} : \mathbb{S}_\delta(x_0,y_0) \to \mathbb{R}$ is differentiable at (x_0,y_0) and

$$\nabla\left(f^{1/k}\right)(x_0,y_0) = \frac{1}{k}f(x_0,y_0)^{(1/k)-1}\nabla f(x_0,y_0).$$

Proof. Let (f_1, f_2) and (g_1, g_2) denote, respectively, pairs of increment functions associated with f and g and the point (x_0, y_0). Using Propositions 2.15 and 3.28, we readily see that (f_1+g_1, f_2+g_2), (rf_1, rf_2), $(f_1 g + f(x_0,y_0)g_1,\ f_2 g + f(x_0,y_0)g_2)$ are, respectively, pairs of increment functions associated with $f+g$, rf, fg and the point (x_0,y_0). In case $f(x_0,y_0) \neq 0$, in view of Proposition 3.28 and Lemma 2.14, we see that there is $\delta > 0$ such that $\mathbb{S}_\delta(x_0,y_0) \subseteq D$ and for all $(x,y) \in \mathbb{S}_\delta(x_0,y_0)$, we have $f(x,y) \neq 0$ and

$$\frac{1}{f(x,y)} - \frac{1}{f(x_0,y_0)} = (x-x_0)\left[\frac{-f_1(x,y)}{f(x,y)f(x_0,y_0)}\right] + (y-y_0)\left[\frac{-f_2(x,y)}{f(x,y)f(x_0,y_0)}\right].$$

Thus, in view of Proposition 2.15 and Proposition 3.28, it follows that $(-f_1/f(x_0, y_0)f, -f_2/f(x_0, y_0)f)$ is a pair of increment functions associated with $1/f$ and (x_0, y_0). Finally, if $f(x_0, y_0) > 0$, then by Proposition 3.28 and Lemma 2.14, there is $\delta > 0$ such that $\mathbb{S}_\delta(x_0, y_0) \subseteq D$ and $f(x, y) > 0$ for all $(x, y) \in \mathbb{S}_\delta(x_0, y_0)$. Now fix any $(x, y) \in \mathbb{S}_\delta(x_0, y_0)$. For simplicity, write $F(x, y) := f(x, y)^{1/k}$. Then

$$f(x, y) - f(x_0, y_0) = F(x, y)^k - F(x_0, y_0)^k = [F(x, y) - F(x_0, y_0)]\, G(x, y),$$

where $G(x, y) := F(x, y)^{k-1} + F(x_0, y_0)F(x, y)^{k-2} + \cdots + F(x_0, y_0)^{k-1}$. Hence

$$F(x, y) - F(x_0, y_0) = (x - x_0)\frac{f_1(x, y)}{G(x, y)} + (y - y_0)\frac{f_2(x, y)}{G(x, y)}.$$

By Proposition 2.15 and Proposition 3.28, F is continuous at (x_0, y_0) and therefore so is G. It follows that $(f_1/G, f_2/G)$ is a pair of increment functions for $f^{1/k}$. Now apply the Increment Lemma. □

Remark 3.31. With notation and hypotheses as in the above proposition, we can deduce, using the results for sums and scalar multiples, that the difference $f-g$ is differentiable at (x_0, y_0) and $\nabla(f-g)(x_0, y_0) = \nabla f(x_0, y_0) - \nabla g(x_0, y_0)$. Also, using the results for products and reciprocals, we see that if $g(x_0, y_0) \neq 0$, then the quotient f/g is differentiable at (x_0, y_0) and

$$\nabla\left(\frac{f}{g}\right)(x_0, y_0) = \frac{g(x_0, y_0)\nabla f(x_0, y_0) - f(x_0, y_0)\nabla g(x_0, y_0)}{g(x_0, y_0)^2}.$$

Further, we can deduce, using the result for products, reciprocals, and radicals, that if $r \in \mathbb{Q}$, then the rational power f^r is differentiable at (x_0, y_0) and

$$\nabla\left(f^r\right)(x_0, y_0) = r f(x_0, y_0)^{r-1} \nabla f(x_0, y_0),$$

provided $f(x_0, y_0) \neq 0$ if r is a negative integer, and $f(x_0, y_0) > 0$ if r is not an integer. ◇

Example 3.32. Using Proposition 3.30 together with Example 3.23 (i) and Example 3.27, we see that every polynomial function in two variables is differentiable on $\mathbb{R}^2$. Moreover, in view of Remark 3.31, we also see that every rational function of two variables is differentiable at each point of $\mathbb{R}^2$ where it is defined. Also, if $f(x, y)$ is a rational function of two variables and r is a rational number, then the algebraic function f^r is differentiable at $(x_0, y_0) \in \mathbb{R}^2$, provided f is defined at (x_0, y_0) and $f(x_0, y_0) \neq 0$. ◇

While the Increment Lemma gives an alternative way to check differentiability of a function of two variables, in practice, neither the definition nor the Increment Lemma is particularly effective in ascertaining differentiability. Note, however, that Proposition 3.28 does give a necessary condition for differentiability, namely, continuity. This can sometimes be used to show that a function is not differentiable at a point. Likewise, if either of the partial derivatives does not exist at a point, then the function cannot be differentiable at that point. We have also seen in Examples 3.29 (i) and (ii) that neither continuity nor the existence of partial derivatives is sufficient to ascertain differentiability. But it turns out that the existence and continuity of partial derivatives imply differentiability. This result, proved below, gives a very useful set of sufficient conditions for differentiability.

Proposition 3.33. *Let $D \subseteq \mathbb{R}^2$ and let (x_0, y_0) be an interior point of D. Let $f : D \to \mathbb{R}$ be such that both f_x and f_y exist on $D \cap \mathbb{S}_\delta(x_0, y_0)$ for some $\delta > 0$. If one of them is continuous at (x_0, y_0), then f is differentiable at (x_0, y_0).*

Proof. Suppose f_x is continuous at (x_0, y_0). Since (x_0, y_0) is an interior point of D, we may assume without loss of generality that $\mathbb{S}_\delta(x_0, y_0) \subseteq D$. In view of Remark 3.26 (i), it suffices to find a pair of increment functions associated with $f_{|\mathbb{S}_\delta(x_0,y_0)}$ and (x_0, y_0). To this end, let us first observe that for any $(x, y) \in \mathbb{S}_\delta(x_0, y_0)$, we have $(x_0, y) \in \mathbb{S}_\delta(x_0, y_0)$, and moreover, we can decompose $f(x, y) - f(x_0, y_0)$ along the "hook" (Figure 3.1) linking (x, y) and (x_0, y_0), that is, we can write $f(x, y) - f(x_0, y_0) = A(x, y) + B(y)$, where

$$A(x, y) := f(x, y) - f(x_0, y) \quad \text{and} \quad B(y) := f(x_0, y) - f(x_0, y_0).$$

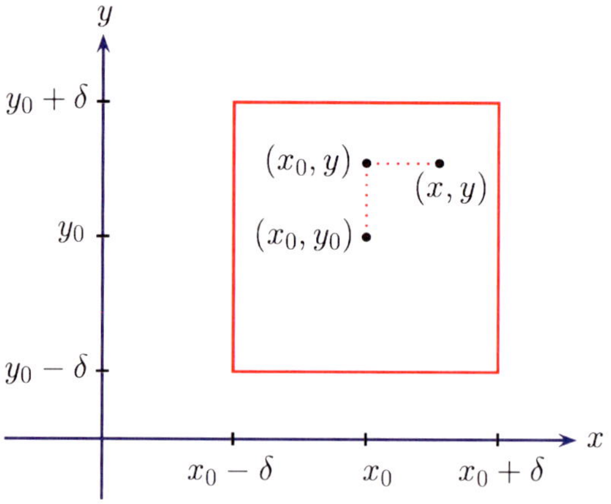

Fig. 3.1. The "hook" linking (x, y) and (x_0, y_0).

Let us define functions $f_1, f_2 : \mathbb{S}_\delta(x_0, y_0) \to \mathbb{R}$ by

$$f_1(x, y) := \begin{cases} \dfrac{A(x, y)}{x - x_0} & \text{if } x \neq x_0, \\ \\ f_x(x_0, y) & \text{if } x = x_0, \end{cases} \quad \text{and} \quad f_2(x, y) := \begin{cases} \dfrac{B(y)}{y - y_0} & \text{if } y \neq y_0, \\ \\ f_y(x_0, y_0) & \text{if } y = y_0. \end{cases}$$

Then it is easily seen that

$$f(x, y) - f(x_0, y_0) = (x - x_0)f_1(x, y) + (y - y_0)f_2(x, y) \text{ for all } (x, y) \in \mathbb{S}_\delta(x_0, y_0).$$

Moreover, since f_x exists on $\mathbb{S}_\delta(x_0, y_0)$, we see that if $x \in (x_0 - \delta, x_0 + \delta)$ with $x \neq x_0$ and $y \in (y_0 - \delta, y_0 + \delta)$, then by the MVT (Fact 3.2), there is $c \in \mathbb{R}$ between x_0 and x such that $A(x, y) = f(x, y) - f(x_0, y) = (x - x_0)f_x(c, y)$, and hence $f_1(x, y) = f_x(c, y)$. Now, since f_x continuous at (x_0, y_0), we see that f_1 is continuous at (x_0, y_0). Also, since $f_y(x_0, y_0)$ exists, we see that f_2 is continuous at (x_0, y_0). Thus it follows from the Increment Lemma that f is differentiable at (x_0, y_0). The case in which f_y is continuous at (x_0, y_0) is proved similarly. □

The first example below illustrates how Proposition 3.33 can be used to determine differentiability, while the second example shows that the converse of Proposition 3.33 is not true. Another example of a differentiable function whose partial derivatives exist but are not continuous is given in Exercise 29.

Examples 3.34. (i) Consider $f : \mathbb{R}^2 \to \mathbb{R}$ defined by $f(0, 0) := 0$ and $f(x, y) := x^2y^2/(x^4 + y^2)$ for $(x, y) \neq (0, 0)$. It is easily seen that $f_x(0, 0) = 0 = f_y(0, 0)$. Also, for $(x_0, y_0) \neq (0, 0)$,

$$f_x(x_0, y_0) = \frac{2x_0y_0^2(y_0^2 - x_0^4)}{(x_0^4 + y_0^2)^2} \quad \text{and} \quad f_y(x_0, y_0) = \frac{2x_0^6y_0}{(x_0^4 + y_0^2)^2}.$$

Moreover, since $(x_0^4 + y_0^2)^2 \geq y_0^4$ and $(x_0^4 + y_0^2)^2 \geq 2x_0^4y_0^2$, we see that $|f_x(x_0, y_0)| \leq 2|x_0| + |x_0| = 3|x_0|$ for $(x_0, y_0) \neq (0, 0)$, and therefore f_x is continuous at $(0, 0)$. Hence by Proposition 3.33, f is differentiable at $(0, 0)$. Note, however, that $f_y(x_0, x_0^2) = \frac{1}{2}$ for all $x_0 \neq 0$, and hence f_y is not continuous at $(0, 0)$.

(ii) Consider $f : \mathbb{R}^2 \to \mathbb{R}$ defined by $f(x, y) := |xy|$. We have seen in Example 3.23 (iv) that f is differentiable at $(0, 0)$. Let $(x_0, y_0) \in \mathbb{R}^2$. For any $h \in \mathbb{R}$ with $h \neq 0$,

$$\frac{f(x_0 + h, y_0) - f(x_0, y_0)}{h} = \frac{|y_0|\left(|x_0 + h| - |x_0|\right)}{h}.$$

Consequently, $f_x(x_0, 0) = 0$, whereas $f_x(0, y_0)$ does not exist if $y_0 \neq 0$. Similarly, it can be seen that $f_y(0, y_0) = 0$, whereas $f_y(x_0, 0)$ does not exist if $x_0 \neq 0$. Thus neither f_x nor f_y exists on $\mathbb{S}_\delta(0, 0)$ for any $\delta > 0$. ◇

Differentiability and Directional Derivatives

We shall now show that if a function f of two variables is differentiable, then all its directional derivatives exist and they can be computed by the simple formula $\mathbf{D_u}f = \nabla f \cdot \mathbf{u}$. More precisely, we have the following.

Proposition 3.35. *Let $D \subseteq \mathbb{R}^2$ and let (x_0, y_0) be an interior point of D. If $f : D \to \mathbb{R}$ is differentiable at (x_0, y_0), then for every unit vector $\mathbf{u} = (u_1, u_2)$ in $\mathbb{R}^2$, the directional derivative $\mathbf{D_u} f(x_0, y_0)$ exists and moreover,*

$$\mathbf{D_u} f(x_0, y_0) = \nabla f(x_0, y_0) \cdot \mathbf{u} = f_x(x_0, y_0)u_1 + f_y(x_0, y_0)u_2.$$

Proof. Let (f_1, f_2) be a pair of increment functions associated with f and (x_0, y_0). Then for any $t \in \mathbb{R}$ such that $(x_0 + tu_1, y_0 + tu_2) \in D$,

$$f(x_0+tu_1, y_0+tu_2) - f(x_0, y_0) = tu_1 f_1(x_0+tu_1, y_0+tu_2) + tu_2 f_2(x_0+tu_1, y_0+tu_2).$$

Thus, using Proposition 2.15 and the continuity of f_1 and f_2 at (x_0, y_0), we see that $\mathbf{D_u} f(x_0, y_0)$ exists and moreover,

$$\mathbf{D_u} f(x_0, y_0) = f_1(x_0, y_0)u_1 + f_2(x_0, y_0)u_2 = f_x(x_0, y_0)u_1 + f_y(x_0, y_0)u_2,$$

as desired. □

The above result suggests the following geometric interpretation of the gradient. Let $D \subseteq \mathbb{R}^2$ and let (x_0, y_0) be an interior point of D. Let $f : D \to \mathbb{R}$ be differentiable at (x_0, y_0) and suppose $\nabla f(x_0, y_0) \neq (0, 0)$. Given any unit vector $\mathbf{u} = (u_1, u_2)$,

$$\mathbf{D_u} f(x_0, y_0) = \nabla f(x_0, y_0) \cdot \mathbf{u} = |\nabla f(x_0, y_0)| \cos\theta\,,$$

where $\theta \in [0, \pi]$ is the angle between $\nabla f(x_0, y_0)$ and $\mathbf{u}$. Thus, if we keep in mind the fact that $\mathbf{D_u}(x_0, y_0)$ measures the rate of change in f in the direction of $\mathbf{u}$, then we can make the following observations.

1. $\mathbf{D_u} f(x_0, y_0)$ is maximum when $\cos\theta = 1$, that is, when $\theta = 0$. Thus near (x_0, y_0), $\mathbf{u} = \nabla f(x_0, y_0)/|\nabla f(x_0, y_0)|$ is the direction in which f increases most rapidly.
2. $\mathbf{D_u} f(x_0, y_0)$ is minimum when $\cos\theta = -1$, that is, when $\theta = \pi$. Thus near (x_0, y_0), $\mathbf{u} = -\nabla f(x_0, y_0)/|\nabla f(x_0, y_0)|$ is the direction in which f decreases most rapidly.
3. $\mathbf{D_u} f(x_0, y_0) = 0$ when $\cos\theta = 0$, that is, when $\theta = \pi/2$. Thus near (x_0, y_0), $\mathbf{u} = \pm\left(f_y(x_0, y_0), -f_x(x_0, y_0)\right)/|\nabla f(x_0, y_0)|$ are the directions of no change in f.

For example, consider $f : \mathbb{R}^2 \to \mathbb{R}$ defined by $f(x, y) = 4 - x^2 - y^2$. We have $f_x = -2x$ and $f_y = -2y$. So at $(x_0, y_0) = (1, 1)$, the gradient is given by $\nabla f(1, 1) = (-2, -2)$. Thus, near $(1, 1)$, the steepest ascent on the surface $z = f(x, y)$ is in the direction of $\nabla f(1, 1)/|\nabla f(1, 1)| = \left(-1/\sqrt{2}, -1/\sqrt{2}\right)$, while the steepest descent is in the reverse direction, namely, $\left(1/\sqrt{2}, 1/\sqrt{2}\right)$. The directions of no change are $\pm\left(1/\sqrt{2}, -1/\sqrt{2}\right)$.

Proposition 3.35 is also useful in showing that certain functions are not differentiable even though the gradient may exist. Indeed, it suffices to find a single unit vector $\mathbf{u}$ such that the identity $\mathbf{D_u} f = \nabla f \cdot \mathbf{u}$ fails to hold. On the other hand, even when this identity holds for all unit vectors $\mathbf{u}$, the function f may not be differentiable. These remarks are illustrated below.

Examples 3.36. (i) Let $f : \mathbb{R}^2 \to \mathbb{R}$ be given by $f(x,y) := \sqrt{|xy|}$. It is easy to see that f is continuous at $(0,0)$ and $f_x(0,0) = 0 = f_y(0,0)$. On the other hand, given a unit vector $\mathbf{u} = (u_1, u_2)$ in $\mathbb{R}^2$ and any $t \in \mathbb{R}$ with $t \neq 0$, we have

$$\frac{f(0+tu_1, 0+tu_2) - f(0,0)}{t} = \frac{|t|\sqrt{|u_1u_2|}}{t}.$$

It follows that the directional derivative $\mathbf{D_u}f(0,0)$ does not exist whenever u_1 and u_2 are nonzero, for example, if $\mathbf{u} = (1/\sqrt{2}, 1/\sqrt{2})$. Hence, by Proposition 3.35, we conclude that f is not differentiable at $(0,0)$.

(ii) Let $f : \mathbb{R}^2 \to \mathbb{R}$ be given by $f(0,0) := 0$ and $f(x,y) := x^2y/(x^2+y^2)$ for $(x,y) \neq (0,0)$. We have seen in Example 2.16 (iii) that f is continuous at $(0,0)$. Also, given a unit vector $\mathbf{u} = (u_1, u_2)$ in $\mathbb{R}^2$ and any $t \in \mathbb{R}$ with $t \neq 0$, we have

$$\frac{f(0+tu_1, 0+tu_2) - f(0,0)}{t} = \frac{t^3u_1^2u_2}{t(t^2u_1^2 + t^2u_2^2)} = u_1^2u_2.$$

It follows that the directional derivative $\mathbf{D_u}f(0,0)$ exists and is equal to $u_1^2u_2$. In particular, $f_x(0,0) = 0 = f_y(0,0)$. Consequently, $\mathbf{D_u}f(0,0) \neq \nabla f(0,0) \cdot \mathbf{u}$ whenever u_1 and u_2 are nonzero. Hence, by Proposition 3.35, we conclude that f is not differentiable at $(0,0)$.

(iii) Let $f : \mathbb{R}^2 \to \mathbb{R}$ be given by $f(0,0) := 0$ and $f(x,y) := x^3y/(x^4+y^2)$ for $(x,y) \neq (0,0)$. We have seen in Example 2.16 (v) that f is continuous at $(0,0)$. Moreover, for any unit vector $\mathbf{u} = (u_1, u_2)$ in $\mathbb{R}^2$ and any $t \in \mathbb{R}$ with $t \neq 0$, we have

$$\frac{f(0+tu_1, 0+tu_2) - f(0,0)}{t} = \frac{t^4u_1^3u_2}{t(t^4u_1^4 + t^2u_2^2)} = \frac{tu_1^3u_2}{t^2u_1^4 + u_2^2},$$

and so, considering separately the cases $u_2 = 0$ and $u_2 \neq 0$, we see that $\mathbf{D_u}f(0,0)$ exists and is equal to 0. In particular, $f_x(0,0) = 0 = f_y(0,0)$. Consequently, $\mathbf{D_u}f(0,0) = \nabla f(0,0) \cdot \mathbf{u}$ for all unit vectors $\mathbf{u}$. On the other hand, if we consider

$$Q(h,k) := \frac{f(0+h, 0+k) - f(0,0) - 0 \cdot h - 0 \cdot k}{\sqrt{h^2+k^2}} = \frac{h^3k}{(h^4+k^2)\sqrt{h^2+k^2}},$$

then $Q(h,k) \not\to 0$ as $(h,k) \to (0,0)$. To see this, consider a sequence in $\mathbb{R}^2 \setminus \{(0,0)\}$ approaching $(0,0)$ along the parabola $k = h^2$. For example, if $(a_n, b_n) := (1/n, 1/n^2)$ for $n \in \mathbb{N}$, then $Q(a_n, b_n) \to 1/2$. It follows that f is not differentiable at $(0,0)$. This shows that the converse of Proposition 3.35 is not true. In fact, it shows that a function can satisfy all the necessary conditions for differentiability given in Propositions 3.28 and 3.35, but still it may fail to be differentiable. ◇

It may be worthwhile to record the following consequence of the sufficient and the necessary conditions for differentiability proved in this section.

Corollary 3.37. *Let $D \subseteq \mathbb{R}^2$ and let (x_0, y_0) be an interior point of D. If $f : D \to \mathbb{R}$ is such that both f_x and f_y exist on $D \cap \mathbb{S}_\delta(x_0, y_0)$ for some $\delta > 0$, and one of them is continuous at (x_0, y_0), then*

(i) *f is continuous at (x_0, y_0),*
(ii) *for every unit vector $\mathbf{u} = (u_1, u_2)$, the directional derivative $\mathbf{D_u} f(x_0, y_0)$ exists and moreover,*

$$\mathbf{D_u} f(x_0, y_0) = \nabla f(x_0, y_0) \cdot \mathbf{u} = f_x(x_0, y_0)u_1 + f_y(x_0, y_0)u_2.$$

Proof. By Proposition 3.33, f is differentiable at (x_0, y_0). Hence (i) follows from Proposition 3.28, while (ii) follows from Proposition 3.35. □

Implicit Differentiation

In calculus of functions of a single variable, one encounters the process of implicit differentiation. Typically, this is applied to equations of the form $f(x, y) = 0$, which are "implicitly differentiated," treating y as a function of x, so as to obtain an equation such as

$$P(x, y) + Q(x, y)\frac{dy}{dx} = 0.$$

Using this, the derivative of y with respect to x is computed at points where $Q(x, y)$ does not vanish. To gain a proper perspective on this process and to put it on a firm footing, one has to take recourse to functions of two variables and an important result known as the Implicit Function Theorem. To begin with, note that $P(x, y)$ and $Q(x, y)$ are, in fact, the partial derivatives $f_x(x, y)$ and $f_y(x, y)$, and the process of differentiation at a point can be justified if the chain rule is applicable and if the equation $f(x, y) = 0$ does indeed define y as a function of x, at least around the point at which derivatives are taken. The Implicit Function Theorem, in the form given below, enables us to justify the latter. It may be recalled that we had already proved a version of the Implicit Function Theorem in the context of continuous functions. The following is the classical version and the one that is most often used in practice.

Proposition 3.38 (Classical Version of Implicit Function Theorem). *Let $D \subseteq \mathbb{R}^2$, $f : D \to \mathbb{R}$, and $(x_0, y_0) \in D$ be such that $f(x_0, y_0) = 0$. Assume that there is $r > 0$ with $\mathbb{S}_r(x_0, y_0) \subseteq D$ and the following conditions hold:*

(a) *f_x and f_y exist at every point of $\mathbb{S}_r(x_0, y_0)$,*
(b) *f_y is continuous at (x_0, y_0) and $f_y(x_0, y_0) \neq 0$.*

Then there are $\delta > 0$ *and a unique continuous function* $\eta : (x_0 - \delta, x_0 + \delta) \to \mathbb{R}$ *with* $\eta(x_0) = y_0$ *such that* $(x, \eta(x)) \in \mathbb{S}_r(x_0, y_0)$ *and* $f(x, \eta(x)) = 0$ *for all* $x \in (x_0 - \delta, x_0 + \delta)$. *Moreover,* η *is differentiable at* x_0 *and*

$$\eta'(x_0) = -\frac{f_x(x_0, y_0)}{f_y(x_0, y_0)}.$$

Further, if the condition (b) is replaced by the stronger condition

(b*) f_y *is continuous on* $\mathbb{S}_r(x_0, y_0)$ *and* $f_y(x_0, y_0) \neq 0$,

then η *is differentiable at every point of* $(x_0 - \delta, x_0 + \delta)$, *and moreover,*

$$f_y(x, \eta(x)) \neq 0 \quad \text{and} \quad \eta'(x) = -\frac{f_x(x, \eta(x))}{f_y(x, \eta(x))} \quad \text{for all } x \in (x_0 - \delta, x_0 + \delta).$$

Proof. Assume that $f_y(x_0, y_0) > 0$. Since f_y is continuous on $\mathbb{S}_r(x_0, y_0)$, by Lemma 2.14, we see that there is $t > 0$ with $t \le r$ such that $f_y(x, y) > 0$ for all $(x, y) \in \mathbb{S}_t(x_0, y_0)$. Thus from the first derivative test of one-variable calculus (for instance, part (iii) of Proposition 4.27 of ACICARA), we see that for each $x \in (x_0 - t, x_0 + t)$, the function given by $y \longmapsto f(x, y)$ is strictly increasing, that is, the condition (b) of the Implicit Function Theorem (Proposition 2.40) is satisfied. Also, by (a) above, we see that the condition (a) of Proposition 2.40 is satisfied. Hence there there are $\delta > 0$ with $\delta \le t \le r$ and a unique continuous function $\eta : (x_0 - \delta, x_0 + \delta) \to (y_0 - t, y_0 + t)$ such that $\eta(x_0) = y_0$ and $f(x, \eta(x)) = 0$ for all $x \in (x_0 - \delta, x_0 + \delta)$. Furthermore, by (a), (b), and Proposition 3.33, we see that f is differentiable at (x_0, y_0), and hence by the Increment Lemma (Proposition 3.25), there are $f_1, f_2 : \mathbb{S}_\delta(x_0, y_0) \to \mathbb{R}$ that are continuous at (x_0, y_0) and satisfy

$$f(x, y) - f(x_0, y_0) = (x - x_0) f_1(x, y) + (y - y_0) f_2(x, y) \quad \text{for } (x, y) \in \mathbb{S}_\delta(x_0, y_0).$$

Since f_2 is continuous at (x_0, y_0) and $f_2(x_0, y_0) = f_y(x_0, y_0) \neq 0$, we can find $\delta' > 0$ such that $\delta' \le \delta$ and $f_2(x, y) \neq 0$ for all $(x, y) \in \mathbb{S}_{\delta'}(x_0, y_0)$. Further, since η is continuous at x_0, we can find $\delta'' > 0$ with $\delta'' \le \delta'$ such that $|\eta(x) - y_0| < \delta'$ whenever $x \in (x_0 - \delta'', x_0 + \delta'')$. Putting $y = \eta(x)$, we obtain

$$\eta(x) - \eta(x_0) = \eta(x) - y_0 = -\frac{f_1(x, \eta(x))}{f_2(x, \eta(x))}(x - x_0) \quad \text{for } x \in (x_0 - \delta'', x_0 + \delta'').$$

Moreover, since f_1, f_2 are continuous at (x_0, y_0) with $f_1(x_0, y_0) = f_x(x_0, y_0)$ and $f_2(x_0, y_0) = f_y(x_0, y_0) \neq 0$, by Proposition 2.15 together with part (ii) of Proposition 2.17 and by Carathéodory's Lemma (Fact 3.24), we see that η is differentiable at x_0 and $\eta'(x_0)$ satisfies the desired formula.

Finally, suppose (b) is replaced by (b*). Let t, δ, and η be as above. Since $\delta \le t$, we have $f_y(x, y) \neq 0$ for all $(x, y) \in \mathbb{S}_t(x_0, y_0)$ and hence $f_y(x, \eta(x)) \neq 0$ for all $x \in (x_0 - \delta, x_0 + \delta)$. Fix any $x_1 \in (x_0 - \delta, x_0 + \delta)$ and put $y_1 := \eta(x_1)$. Let $r_1 > 0$ be such that $\mathbb{S}_{r_1}(x_1, y_1) \subseteq \mathbb{S}_r(x_0, y_0)$. Applying what we have proved so far to (x_1, y_1) instead of (x_0, y_0), we see that there are $\delta_1 > 0$ with $\delta_1 \le r_1$ and a unique continuous function $\eta_1 : (x_1 - \delta_1, x_1 + \delta_1) \to \mathbb{R}$ with $\eta_1(x_1) = y_1$ such that $(x, \eta(x)) \in \mathbb{S}_{r_1}(x_1, y_1)$ and $f(x, \eta_1(x)) = 0$ for all $x \in (x_1 - \delta_1, x_1 + \delta_1)$. Moreover, η_1 is differentiable at x_1 and $\eta_1'(x_1) = -f_x(x_1, y_1)/f_y(x_1, y_1)$. Without loss of generality, we may assume that $\delta_1 > 0$ is so small that $I_1 := (x_1 - \delta_1, x_1 + \delta_1) \subseteq (x_0 - \delta, x_0 + \delta)$. Now, by the uniqueness of η_1, it follows that $\eta|_{I_1} = \eta_1$. In particular, η is differentiable at x_1 and $\eta'(x_1)$ satisfies the desired formula. Since x_1 was an arbitrary element of $(x_0 - \delta, x_0 + \delta)$, the last assertion in the theorem is proved.

The case in which $f_y(x_0, y_0) < 0$ is proved similarly. □

Examples 3.39. (i) Let $m, n \in \mathbb{N}$ and consider $f : \mathbb{R}^2 \to \mathbb{R}$ given by $f(x, y) := x^m + y^n - 1$. Then $f_x = mx^{m-1}$ and $f_y = ny^{n-1}$ exist and are continuous on $\mathbb{R}^2$. Thus, by the Implicit Function Theorem (Proposition 3.38), for any $(x_0, y_0) \in \mathbb{R}^2$ with $f(x_0, y_0) = 0$ and $y_0 \ne 0$, the equation $f(x, y) = 0$ can be solved for y in terms of x, locally near (x_0, y_0). Likewise, for any $(x_0, y_0) \in \mathbb{R}^2$ with $f(x_0, y_0) = 0$ and $x_0 \ne 0$, we have $f_x(x_0, y_0) \ne 0$, and hence the equation $f(x, y) = 0$ can be solved for x in terms of y, locally near (x_0, y_0). Notice that in this example both $f_x(x_0, y_0)$ and $f_y(x_0, y_0)$ are zero when $(x_0, y_0) = (0, 0)$ but then $f(0, 0) \ne 0$. It follows that for every $(x_0, y_0) \in \mathbb{R}^2$ with $f(x_0, y_0) = 0$, there are $\delta > 0$ and a differentiable function $\eta : (x_0 - \delta, x_0 + \delta) \to \mathbb{R}$ with $f(x, \eta(x)) = 0$ for all $x \in (x_0 - \delta, x_0 + \delta)$, or a differentiable function $\xi : (y_0 - \delta, y_0 + \delta) \to \mathbb{R}$ with $f(\xi(y), y) = 0$ for all $y \in (y_0 - \delta, y_0 + \delta)$.

(ii) Consider $f : \mathbb{R}^2 \to \mathbb{R}$ defined by $f(x, y) = y^2 - x^3$. Then $f_x(0, 0) = f_y(0, 0) = 0$. So the Implicit Function Theorem is not applicable near $(0, 0)$. Indeed, the "solutions" for y in terms of x, or for x in terms of y, namely, $y = \pm\sqrt{x^3}$ or $x = \sqrt[3]{y^2}$, are not differentiable at the origin. ◇

Remark 3.40. As in Remark 2.42, we have a straightforward analogue of the Classical Version of the Implicit Function Theorem, which corresponds to solving $f(x, y) = 0$ for x in terms of y. In this situation, condition (a) in Proposition 3.38 remains the same, while in (b) and (b*), one has to replace f_y by f_x. The conclusion would be that there are $\delta > 0$ and $\xi : (y_0 - \delta, y_0 + \delta) \to \mathbb{R}$ as in Remark 2.42. Moreover, ξ would be differentiable and $\xi' = -f_y/f_x$. Combining either of the two situations, we can make a unified statement of the Classical Version of the Implicit Function Theorem as follows.

Let $D \subseteq \mathbb{R}^2$ and let (x_0, y_0) be an interior point of D. Let $f : D \to \mathbb{R}$ have continuous partial derivatives in $\mathbb{S}_r(x_0, y_0)$ for some $r > 0$ with $\mathbb{S}_r(x_0, y_0) \subseteq D$. Suppose $f(x_0, y_0) = 0$ and $\nabla f(x_0, y_0) \ne (0, 0)$. Then there are $\delta > 0$, $t_0 \in \mathbb{R}$, and differentiable functions $x, y : (t_0 - \delta, t_0 + \delta) \to \mathbb{R}$ such that $(x(t_0), y(t_0)) = (x_0, y_0)$, and for every $t \in (t_0 - \delta, t_0 + \delta)$, we have $(x(t), y(t)) \in \mathbb{S}_r(x_0, y_0)$ and $f(x(t), y(t)) = 0$. Moreover, $(x'(t), y'(t)) \ne (0, 0)$ and $f_x(x(t), y(t))x'(t) + f_y(x(t), y(t))y'(t) = 0$ for all $t \in (t_0 - \delta, t_0 + \delta)$. In other words, if $f(x_0, y_0) = 0$ and $\nabla f(x_0, y_0) \ne (0, 0)$, then there is a regular path passing through (x_0, y_0) and lying on the surface $z = f(x, y)$. ◇

As in Proposition 2.43, we can derive a classical version of the so-called **Inverse Function Theorem** for real-valued functions of one variable as a consequence of the Classical Version of the Implicit Function Theorem.

Proposition 3.41. *Let I be an interval in $\mathbb{R}$ and x_0 an interior point of I. Suppose $f : I \to \mathbb{R}$ is continuously differentiable on $(x_0 - r, x_0 + r)$ for some $r > 0$ with $(x_0 - r, x_0 + r) \subseteq I$ and $f'(x_0) \neq 0$. Let $y_0 := f(x_0)$ and $J := f(I)$. Then there are $\delta > 0$ with $(y_0 - \delta, y_0 + \delta) \subseteq J$ and a unique differentiable function $\xi : (y_0 - \delta, y_0 + \delta) \to \mathbb{R}$ such that $\xi(y_0) = x_0$ and $f(\xi(y)) = y$ for all $y \in (y_0 - \delta, y_0 + \delta)$. Moreover, there is $t > 0$ with $t < r$ such that f is one-one on $(x_0 - t, x_0 + t)$ and f^{-1} is differentiable at y_0 with $\left(f^{-1}\right)'(y_0) = 1/f'(x_0)$.*

Proof. Consider $h : \mathbb{S}_r(x_0, y_0) \to \mathbb{R}$ defined by $h(x, y) := f(x) - y$. Then $h(x_0, y_0) = 0$, both h_x and h_y exist and are continuous on $\mathbb{S}_r(x_0, y_0)$, and $h_x(x_0, y_0) = f'(x_0) \neq 0$. Hence by the Classical Version of the Implicit Function Theorem (Proposition 3.38 and Remark 3.40), there are $\delta > 0$ and a unique continuous function $\xi : (y_0 - \delta, y_0 + \delta) \to \mathbb{R}$ with $\xi(y_0) = x_0$ such that $(\xi(y), y) \in \mathbb{S}_r(x_0, y_0)$ and $h(\xi(y), y) = 0$ for all $y \in (y_0 - \delta, y_0 + \delta)$. Moreover, ξ is differentiable at y_0 and $\xi'(y_0) = -h_y(x_0, y_0)/h_x(x_0, y_0) = 1/f'(x_0)$. Consequently, $f(\xi(y)) = y$ for all $y \in (y_0 - \delta, y_0 + \delta)$, and in particular, $(y_0 - \delta, y_0 + \delta) \subseteq J$. Moreover, since $f'(x_0) \neq 0$ and f' is continuous on $(x_0 - r, x_0 + r)$, there is $t > 0$ with $t < r$ such that $f'(x) \neq 0$ for all $x \in (x_0 - t, x_0 + t)$. Hence by the IVP of f' and the first derivative test of one-variable calculus (or more specifically, by part (ii) of Corollary 4.28 of ACICARA), we see that f is strictly monotonic, and in particular, one-one, on $(x_0 - t, x_0 + t)$. It follows that $f^{-1} = \xi$ on $f\left((x_0 - t, x_0 + t)\right)$. Also, in view of the continuity and strict monotonicity of f on $(x_0 - t, x_0 + t)$, we see that y_0 is an interior point of $f(x_0 - t, x_0 + t)$. Hence f^{-1} is differentiable at y_0 with $\left(f^{-1}\right)'(y_0) = 1/f'(x_0)$. □

As an immediate corollary of Proposition 3.41, we obtain a version of the differentiable inverse theorem of one-variable calculus (given, for example, on page 112 of ACICARA).

Corollary 3.42. *Let I be an open interval in $\mathbb{R}$ and let $f : I \to \mathbb{R}$ be a continuously differentiable function such that $f'(x) \neq 0$ for all $x \in I$. Then the inverse function $f^{-1} : f(I) \to \mathbb{R}$ is continuously differentiable and $\left(f^{-1}\right)'(f(x)) = 1/f'(x)$ for all $x \in I$.*

Proof. Apply Proposition 3.41 at each point of I to obtain the differentiability of f^{-1} and the formula $\left(f^{-1}\right)'(f(x)) = 1/f'(x)$ for all $x \in I$. This formula implies the continuity of the derivative of f^{-1}, since f' is continuous and $f'(x) \neq 0$ for all $x \in I$. □

3.3 Taylor's Theorem and Chain Rule

In this section, we discuss two important results in multivariable calculus known as Taylor's theorem and the chain rule. We have already discussed Taylor's theorem using the notion of higher-order directional derivatives. We give here a classical version that is more widely used in practice. The chain rule for functions of two (or more) variables is an analogue of the chain rule for functions of one variable (given, for example, on page 111 of ACICARA).

Bivariate Taylor Theorem

We have already discussed in Section 3.1 analogues of the mean value theorem (MVT) and Taylor's theorem for functions of two variables using the notion of directional derivatives. We have also stated alternative versions of these results in case the directional derivatives satisfy an identity of the form $\mathbf{D_u}f = \nabla f \cdot \mathbf{u}$. Subsequently, in Section 3.2, we have seen that such an identity is a consequence of differentiability of the function, and in particular, a consequence of the continuity of the partial derivatives. We can now put these facts together and derive the classical version of the Bivariate Mean Value Theorem and the Bivariate Taylor Theorem.

Proposition 3.43 (Classical Version of Bivariate Mean Value Theorem). *Let D be a convex and open subset of $\mathbb{R}^2$, and let $f : D \to \mathbb{R}$ be any differentiable function. Given any distinct points (x_0, y_0) and (x_1, y_1) in D, there is $(c, d) \in D$ lying on the line segment joining (x_0, y_0) and (x_1, y_1), with $(c, d) \neq (x_i, y_i)$ for $i = 0, 1$, such that*

$$\begin{aligned} f(x_1, y_1) - f(x_0, y_0) &= (x_1 - x_0) f_x(c, d) + (y_1 - y_0) f_y(c, d) \\ &= (x_1 - x_0,\, y_1 - y_0) \cdot \nabla f(c, d). \end{aligned}$$

Proof. Since D is convex, the line segment joining (x_0, y_0) and (x_1, y_1) is contained in D, that is, $L := \{(x_0 + t(x_1 - x_0), y_0 + t(y_1 - y_0)) : t \in [0, 1]\} \subseteq D$. Also, since f is differentiable on D, by Proposition 3.35, we have $\mathbf{D_u}f(x, y) = \nabla f(x, y) \cdot \mathbf{u}$ for $(x, y) \in D$ and all unit vectors $\mathbf{u}$ in $\mathbb{R}^2$. Thus, the desired result follows from Corollary 3.6. □

Remark 3.44. With notation and hypotheses as in Proposition 3.43, if we write $h := x_1 - x_0$ and $k := y_1 - y_0$, then the conclusion of the Bivariate Mean Value Theorem may be paraphrased by saying that

$$f(x_0 + h, y_0 + k) = f(x_0, y_0) + h f_x(x_0 + \theta h, y_0 + \theta k) + k f_y(x_0 + \theta h, y_0 + \theta k)$$

for some $\theta \in (0, 1)$. ◇

Corollary 3.45. *Let $D \subseteq \mathbb{R}^2$ be nonempty, convex, and open in $\mathbb{R}^2$, and let $f : D \to \mathbb{R}$ be any function. Then f is a constant function on D if and only if f is differentiable and both f_x and f_y vanish identically on D.*

Proof. If f is a constant function, then it is obvious that f is differentiable and $f_x = f_y = 0$ on D. Conversely, suppose f is differentiable and both f_x and f_y vanish identically on D. Let (x_0, y_0) be any point of D. Since $f_x = f_y = 0$ on D, by Proposition 3.43, for any $(x_1, y_1) \in D$ with $(x_1, y_1) \neq (x_0, y_0)$ we have $f(x_1, y_1) - f(x_0, y_0) = 0$, that is, $f(x_1, y_1) = f(x_0, y_0)$. Thus, f is a constant function on D. □

Remark 3.46. If $D \subseteq \mathbb{R}^2$ is not convex, then there do exist differentiable functions on D whose gradient does not vanish identically on D. For example, if $D = \mathbb{S}_1(0,0) \cup \mathbb{S}_1(2,2)$ is a disjoint union of two open squares and $f : D \to \mathbb{R}$ is defined by $f(x, y) = 1$ if $x \in \mathbb{S}_1(0,0)$ and $f(x,y) = 2$ if $x \in \mathbb{S}_1(2,2)$, then clearly D is nonempty and open, and $f_x = f_y = 0$ on D, but f is not a constant function. This shows that the hypothesis that D is convex cannot be dropped from Corollary 3.45. However, it can be shown that a weaker hypothesis on D, namely, that D is path-connected, also suffices. (See Exercise 37.) ◇

Proposition 3.47 (Classical Version of Bivariate Taylor Theorem). *Let D be a convex and open subset of $\mathbb{R}^2$, and let n be a nonnegative integer. If $f : D \to \mathbb{R}$ is such that all the partial derivatives of f of order $\leq n+1$ exist and are continuous on D, then for any distinct points (x_0, y_0) and (x_1, y_1) in D, there is $(c, d) \in D$ lying on the line segment joining (x_0, y_0) and (x_1, y_1), with $(c, d) \neq (x_i, y_i)$ for $i = 0, 1$, such that*

$$f(x_1, y_1) = \sum_{i=0}^{n} \frac{1}{i!}\left(h\frac{\partial}{\partial x} + k\frac{\partial}{\partial y}\right)^i f(x_0, y_0) + \frac{1}{(n+1)!}\left(h\frac{\partial}{\partial x} + k\frac{\partial}{\partial y}\right)^{n+1} f(c, d),$$

where $h := x_1 - x_0$ and $k := y_1 - y_0$. Alternatively,

$$f(x_1, y_1) = P_n(x_1, y_1) + \sum_{\substack{\ell \geq 0 \\ \ell+m=n+1}} \sum_{m \geq 0} \frac{\partial^{n+1} f}{\partial x^\ell \partial y^m}(c, d) \frac{(x - x_0)^\ell}{\ell!} \frac{(y - y_0)^m}{m!},$$

where

$$P_n(x, y) := \sum_{\substack{\ell \geq 0 \\ \ell+m \leq n}} \sum_{m \geq 0} \frac{\partial^{\ell+m} f}{\partial x^\ell \partial y^m}(x_0, y_0) \frac{(x - x_0)^\ell}{\ell!} \frac{(y - y_0)^m}{m!} \quad \text{for any } (x, y) \in \mathbb{R}^2.$$

Proof. Since D is convex, the line segment joining (x_0, y_0) and (x_1, y_1) is contained in D, that is, $L := \{(x_0 + t(x_1 - x_0), y_0 + t(y_1 - y_0)) : t \in [0, 1]\} \subseteq D$. Let $\mathbf{u} = (u_1, u_2)$ be the unit vector given by

$$\mathbf{u} = (u_1, u_2) := \frac{1}{r}(x_1 - x_0, y_1 - y_0), \quad \text{where} \quad r := \sqrt{(x_1 - x_0)^2 + (y_1 - y_0)^2}.$$

Now, by Propositions 3.33 and 3.35, $f_0 := f$ is differentiable on D and the directional derivative $\mathbf{D_u} f$ exists at every point of D. More generally, using

Propositions 3.33 and 3.35 together with induction on i, we see that for $i = 1, \ldots, n$, the ith directional derivative $\mathbf{D}_{\mathbf{u}}^i f$ exists at every point of D and moreover, if $f_i : D \to \mathbb{R}$ is defined by $f_i := \mathbf{D}_{\mathbf{u}}^i f$, then f_i is differentiable on D. Consequently, by Propositions 3.28 and 3.35, for each $i = 0, \ldots, n$, the function f_i is continuous on D and satisfies $\mathbf{D}_{\mathbf{u}} f_i(x, y) = \nabla f_i(x, y) \cdot \mathbf{u}$ for every $(x, y) \in D$. Thus, by Corollary 3.22, there is $(c, d) \in L \setminus \{(x_0, y_0), (x_1, y_1)\}$ such that

$$f(x_1, y_1) = \sum_{i=0}^{n} \frac{1}{i!} \left(h\frac{\partial}{\partial x} + k\frac{\partial}{\partial y} \right)^i f(x_0, y_0) + \frac{1}{(n+1)!} \left(h\frac{\partial}{\partial x} + k\frac{\partial}{\partial y} \right)^{n+1} f(c, d),$$

where $h := x_1 - x_0$ and $k := y_1 - y_0$. This proves the first assertion. To prove the alternative expression for $f(x_1, y_1)$, note that for $i = 0, \ldots, n+1$, we have

$$\frac{1}{i!} \left(h\frac{\partial}{\partial x} + k\frac{\partial}{\partial y} \right)^i f(x, y) = \sum_{\substack{\ell \geq 0 \\ \ell + m = i}} \sum_{m \geq 0} \frac{\partial^i f}{\partial x^\ell \partial y^m}(x, y) \frac{h^\ell}{\ell!} \frac{k^m}{m!} \quad \text{for any } (x, y) \in D.$$

The last two displayed equations yield the desired result. □

Remarks 3.48. (i) The classical version of the Bivariate Mean Value Theorem corresponds to the case $n = 0$ of the Bivariate Taylor Theorem.

(ii) The case $n = 1$ of the classical version of the Bivariate Taylor Theorem is sometimes called the **Extended Bivariate Mean Value Theorem**. It can be stated as follows: *If $D \subseteq \mathbb{R}^2$ is convex and open, and $f : D \to \mathbb{R}$ has continuous first-order and second-order partial derivatives on D, then for any distinct points (x_0, y_0) and (x_1, y_1) in D, there is $(c, d) \in D$ on the line segment joining (x_0, y_0) and (x_1, y_1), with $(c, d) \neq (x_i, y_i)$ for $i = 0, 1$, such that*

$$f(x, y) = f(x_0, y_0) + \left(h\frac{\partial}{\partial x} + k\frac{\partial}{\partial x} \right) f(x_0, y_0) + \frac{1}{2} \left(h\frac{\partial}{\partial x} + k\frac{\partial}{\partial x} \right)^2 f(c, d).$$

(iii) The polynomial $P_n(x, y)$ defined in Proposition 3.47 above is called the nth **(bivariate) Taylor polynomial** of f around (x_0, y_0). The difference $R_n := f - P_n$ is called the **(bivariate) remainder** of order n. Thus, Proposition 3.47 relates the function f to its Taylor polynomial by saying that for every $(x, y) \in D$, there is $(c, d) \in D$ such that

$$f(x, y) = P_n(x, y) + \sum_{\substack{\ell \geq 0 \\ \ell + m = n+1}} \sum_{m \geq 0} \frac{\partial^{n+1} f}{\partial x^\ell \partial y^m}(c, d) \frac{(x - x_0)^\ell}{\ell!} \frac{(y - y_0)^m}{m!}.$$

The last expression is sometimes referred to as the **Bivariate Taylor Formula** for f around (x_0, y_0). This formula shows that the remainder R_n is given by

$$R_n(x,y) = \sum_{\substack{\ell\ge 0\ m\ge 0 \\ \ell+m=n+1}} \frac{\partial^{n+1} f}{\partial x^\ell \partial y^m}(c,d)\frac{(x-x_0)^\ell}{\ell!}\frac{(y-y_0)^m}{m!}$$

for some $(c,d) \in D$. ◇

The following corollary of the Bivariate Taylor Formula generalizes Corollary 3.45 and gives a characterization of polynomial functions on convex open subsets of $\mathbb{R}^2$.

Corollary 3.49. *Let D be a nonempty, convex, and open subset of $\mathbb{R}^2$, and let $f : D \to \mathbb{R}$ be any function. Let n be a nonnegative integer. Then f is a polynomial function on D of total degree $\le n$ if and only if all partial derivatives of f of order $\le n+1$ exist and are continuous on D, and moreover, all the $(n+1)$th-order partial derivatives of f vanish identically on D.*

Proof. Suppose f is a polynomial function on D of total degree $\le n$, that is, $f(x,y)$ is a finite sum of terms of the form $c_{i,j}x^i y^j$ where i,j are nonnegative integers with $i+j \le n$ and $c_{i,j} \in \mathbb{R}$. In view of Example 3.17 (ii), we see that the partial derivatives of f of every order exist and are continuous on D, and moreover, all the $(n+1)$th-order partial derivatives of f vanish identically on D. To prove the converse, it suffices to fix some $(x_0, y_0) \in D$ and apply the Bivariate Taylor Formula for f around (x_0, y_0). □

Examples 3.50. (i) Let I and J be nonempty open intervals in $\mathbb{R}$, and let $\phi : I \to \mathbb{R}$ and $\psi : J \to \mathbb{R}$ be infinitely differentiable functions of one variable. Consider $f, g : I \times J \to \mathbb{R}$ defined by

$$f(x,y) = \phi(x) + \psi(y) \quad \text{and} \quad g(x,y) = \phi(x)\psi(y).$$

Let $x_0 \in I$, $y_0 \in J$, and let n be a nonnegative integer. If $Q_n(\phi)$ and $Q_n(\psi)$ denote the nth Taylor polynomials of ϕ and ψ around x_0 and y_0 respectively, then using the formulas in Example 3.17 (i), we see that the nth Taylor polynomial of f around (x_0, y_0) is given by $Q_n(\phi)(x) + Q_n(\psi)(y)$, whereas the nth Taylor polynomial of g around (x_0, y_0) is given by the sum of terms of total degree $\le n$ in the product $Q_n(\phi)(x)Q_n(\psi)(y)$. For instance, the nth Taylor polynomials around $(0,0)$ of $f, g : \mathbb{R}^2 \to \mathbb{R}$ defined by $f(x,y) := e^x + e^y$ and $g(x,y) := e^{x+y}$ are given by

$$P_n(f)(x,y) = \sum_{\ell=0}^{n} \frac{x^\ell}{\ell!} + \sum_{m=0}^{n} \frac{y^m}{m!} \quad \text{and} \quad P_n(g)(x,y) = \sum_{\substack{\ell\ge 0\ m\ge 0 \\ \ell+m\le n}} \frac{x^\ell}{\ell!}\frac{y^m}{m!}.$$

(ii) Let E be an open subset of $\mathbb{R}$ and D an open subset of $\mathbb{R}^2$ such that $\{x+y : (x,y) \in D\} \subseteq E$. Suppose $g : E \to \mathbb{R}$ is infinitely differentiable and $f : D \to \mathbb{R}$ is defined by $f(x,y) := g(x+y)$. Given any nonnegative integer n and any $(x_0, y_0) \in D$, if $Q_n(g)$ is the nth Taylor polynomial of

g around $x_0 + y_0$, then using the formulas in Example 3.17 (iii), we see that the nth Taylor polynomial of f around (x_0, y_0) is given by

$$\begin{aligned} P_n(f)(x,y) &= \sum_{\substack{\ell \geq 0 \; m \geq 0 \\ \ell+m \leq n}} \sum g^{(\ell+m)}(x_0+y_0) \frac{(x-x_0)^\ell}{\ell!} \frac{(y-y_0)^m}{m!} \\ &= \sum_{j=0}^{n} \frac{g^{(j)}(x_0+y_0)}{j!} \sum_{\ell=0}^{j} \frac{j!}{\ell!(j-\ell)!}(x-x_0)^\ell (y-y_0)^{j-\ell} \\ &= \sum_{j=0}^{n} \frac{g^{(j)}(x_0+y_0)}{j!}(x+y-x_0-y_0)^j \\ &= Q_n(g)(x+y). \end{aligned}$$

For instance, the nth Taylor polynomial of $f : \mathbb{S}_{1/2}(0,0) \to \mathbb{R}$ defined by $f(x,y) := 1/(1-x-y)$ around $(0,0)$ is given by $\sum_{j=0}^{n}(x+y)^j$. Results similar to those above for functions of the form $f(x,y) := g(xy)$ are given in Exercise 40. ◇

Chain Rule

The Chain Rule for functions of two variables is a widely used technique for computing derivatives of composite functions. As in Proposition 2.17 concerning the continuity of composite functions, we state the Chain Rule in three parts, applicable to three ways of forming composites.

Proposition 3.51 (Chain Rule). *Let $D \subseteq \mathbb{R}^2$ and let (x_0, y_0) be an interior point of D and let $f : D \to \mathbb{R}$ be differentiable at (x_0, y_0).*

(i) *Let $E \subseteq \mathbb{R}$ be such that $f(D) \subseteq E$ and $f(x_0, y_0)$ is an interior point of E. If $g : E \to \mathbb{R}$ is differentiable at $f(x_0, y_0)$, then the function $F : D \to \mathbb{R}$ defined by $F := g \circ f$ is differentiable at (x_0, y_0) and*

$$\nabla F(x_0, y_0) = g'\left(f(x_0, y_0)\right) \nabla f(x_0, y_0).$$

(ii) *Let $E \subseteq \mathbb{R}$ and let t_0 be an interior point of E. If $x, y : E \to \mathbb{R}$ are differentiable at t_0 and if $(x(t), y(t)) \in D$ for all $t \in E$ and $(x(t_0), y(t_0)) = (x_0, y_0)$, then the function $F : E \to \mathbb{R}$ defined by $F(t) := f(x(t), y(t))$ for $t \in E$ is differentiable at t_0, and*

$$F'(t_0) = \nabla f(x_0, y_0) \cdot (x'(t_0), y'(t_0)) = f_x(x_0, y_0)x'(t_0) + f_y(x_0, y_0)y'(t_0).$$

(iii) *Let $E \subseteq \mathbb{R}^2$ and let (u_0, v_0) be an interior point of E. If $x, y : E \to \mathbb{R}$ are differentiable at (u_0, v_0) and if $(x(u,v), y(u,v)) \in D$ for all $(u,v) \in E$ and $(x(u_0, v_0), y(u_0, v_0)) = (x_0, y_0)$, then the function $F : E \to \mathbb{R}$ defined by $F(u,v) := f(x(u,v), y(u,v))$ for $(u,v) \in E$ is differentiable at (u_0, v_0), and $\nabla F(u_0, v_0)$ is equal to*

$$\left(\nabla f(x_0, y_0) \cdot (x_u(u_0, v_0), y_u(u_0, v_0))\,,\ \nabla f(x_0, y_0) \cdot (x_v(u_0, v_0), y_v(u_0, v_0))\right).$$

Proof. By the Increment Lemma, there is a pair (f_1, f_2) of increment functions associated with f and (x_0, y_0). Thus $f_1, f_2 : D \to \mathbb{R}$ are continuous at (x_0, y_0) with $(f_1(x_0, y_0), f_2(x_0, y_0)) = \nabla f(x_0, y_0)$, and

$$f(x,y) - f(x_0, y_0) = (x - x_0)f_1(x,y) + (y - y_0)f_2(x,y) \quad \text{for all } (x,y) \in D.$$

This result will play a crucial role in proving (i), (ii), and (iii).

(i) Suppose E and g satisfy the hypotheses in (i). By Carathéodory's Lemma (Fact 3.24), there is an increment function $g_1 : E \to \mathbb{R}$ associated with g and $z_0 := f(x_0, y_0)$. Thus g_1 is continuous at z_0 with $g_1(z_0) = g'(z_0)$ and

$$g(z) - g(z_0) = (z - z_0)g_1(z) \quad \text{for all } z \in E.$$

Since $f(D) \subseteq E$, it follows that for every $(x,y) \in D$,

$$F(x,y) - F(x_0, y_0) = (g_1 \circ f)(x,y)\left[(x - x_0)f_1(x,y) + (y - y_0)f_2(x,y)\right].$$

Moreover, by Propositions 3.28, 2.15 and part (i) of Proposition 2.17, we see that $(g_1 \circ f)f_1$ and $(g_1 \circ f)f_2$ are continuous at (x_0, y_0). Thus we conclude from the Increment Lemma that F is differentiable at (x_0, y_0) and $\nabla F(x_0, y_0) = (g_1 \circ f)(x_0, y_0)\nabla f(x_0, y_0) = g'\left(f(x_0, y_0)\right)\nabla f(x_0, y_0)$.

(ii) Suppose E, t_0 and x, y satisfy the hypotheses in (ii). By Carathéodory's Lemma (Fact 3.24), there are increment functions $x_1, y_1 : E \to \mathbb{R}$ associated, respectively, with x, y, and t_0. Thus x_1, y_1 are continuous at t_0 with $(x_1(t_0), y_1(t_0)) = (x'(t_0), y'(t_0))$, and for all $t \in E$,

$$x(t) - x(t_0) = (t - t_0)x_1(t) \quad \text{and} \quad y(t) - y(t_0) = (t - t_0)y_1(t).$$

Given any $t \in E$, we have $(x(t), y(t)) \in D$, and so $F(t) := f(x(t), y(t))$ satisfies

$$F(t) - F(t_0) = (x(t) - x(t_0))\, f_1\left(x(t), y(t)\right) + (y(t) - y(t_0))\, f_2\left(x(t), y(t)\right).$$

If $F_1 : E \to \mathbb{R}$ is defined by $F_1(t) := x_1(t)f_1\left(x(t), y(t)\right) + y_1(t)f_2\left(x(t), y(t)\right)$, then we have

$$F(t) - F(t_0) = (t - t_0)\, F_1(t) \quad \text{for all } t \in E.$$

Moreover, in view of part (ii) of Proposition 2.17, we see that F_1 is continuous at t_0. Thus, we conclude from Carathéodory's Lemma (Fact 3.24) that F is differentiable at t_0 and that $F'(t_0) = F_1(t_0) = \nabla f(x_0, y_0) \cdot (x'(t_0), y'(t_0))$.

(iii) Suppose E, (u_0, v_0), and x, y satisfy the hypotheses in (iii). Let

$$E_1 := \{u \in \mathbb{R} : (u, v_0) \in E\} \quad \text{and} \quad E_2 := \{v \in \mathbb{R} : (u_0, v) \in E\}.$$

Also, let $\phi, \xi : E_1 \to \mathbb{R}$ and $\psi, \eta : E_2 \to \mathbb{R}$ be functions defined by

$$\phi(u) := x(u, v_0), \quad \xi(u) := y(u, v_0), \quad \psi(v) := x(u_0, v), \quad \eta(v) := y(u_0, v).$$

Then E_1, u_0, and ϕ, ξ satisfy the hypotheses in (ii), and hence $F_1 : E_1 \to \mathbb{R}$ defined by $F_1(u) = f(\phi(u), \xi(u))$ for $u \in E_1$ is differentiable at u_0 and $F_1'(u_0) = \nabla f(x_0, y_0) \cdot (\phi'(u_0), \xi'(u_0))$. Similarly, E_2, v_0, and ψ, η satisfy the hypotheses in (ii), and hence $F_2 : E_2 \to \mathbb{R}$ defined by $F_2(v) = f(\psi(v), \eta(v))$ for $v \in E_2$ is differentiable at v_0 and $F_2'(v_0) = \nabla f(x_0, y_0) \cdot (\psi'(v_0), \eta'(v_0))$. Also, since $F_1(u) = f(x(u, v_0), y(u, v_0))$ for $u \in E_1$ and $F_2(v) = f(x(u_0, v), y(u_0, v))$ for $v \in E_2$, it follows that

$$F_u(u_0, v_0) = F_1'(u_0) = \nabla f(x_0, y_0) \cdot (x_u(u_0, v_0), y_u(u_0, v_0)),$$
$$F_v(u_0, v_0) = F_2'(v_0) = \nabla f(x_0, y_0) \cdot (x_v(u_0, v_0), y_v(u_0, v_0)).$$

Hence $\nabla F(u_0, v_0)$ satisfies the desired equality. □

Remark 3.52. It is often helpful to write the identities given by the Chain Rule in a slightly informal but suggestive notation as follows.

(i) If $z = f(x, y)$ and $w = g(z)$, then w is a function of (x, y), and

$$\frac{\partial w}{\partial x} = \frac{dw}{dz}\frac{\partial z}{\partial x} \quad \text{and} \quad \frac{\partial w}{\partial y} = \frac{dw}{dz}\frac{\partial z}{\partial y}.$$

(ii) If $z = f(x, y)$ and if $x = x(t)$, $y = y(t)$, then z is a function of t, and

$$\frac{dz}{dt} = \frac{\partial z}{\partial x}\frac{dx}{dt} + \frac{\partial z}{\partial y}\frac{dy}{dt}.$$

(iii) If $z = f(x, y)$ and if $x = x(u, v)$, $y = y(u, v)$, then z is a function of (u, v), and

$$\frac{\partial z}{\partial u} = \frac{\partial z}{\partial x}\frac{\partial x}{\partial u} + \frac{\partial z}{\partial y}\frac{\partial y}{\partial u} \quad \text{and} \quad \frac{\partial z}{\partial v} = \frac{\partial z}{\partial x}\frac{\partial x}{\partial v} + \frac{\partial z}{\partial y}\frac{\partial y}{\partial v}.$$

It should be noted that the identities in (i), (ii), and (iii) above are valid when the concerned (partial) derivatives are evaluated at appropriate points and when the hypothesis of Proposition 3.51 holds. In view of Proposition 3.33, the latter holds if each of the (partial) derivatives exists in an open square around the relevant point and is continuous at that point.

We also remark that the displayed identities in (ii) and (iii) above can be written in a unified form, using *matrix notation*, as follows:

$$\frac{dz}{dt} = \begin{bmatrix} \frac{\partial z}{\partial x} & \frac{\partial z}{\partial y} \end{bmatrix} \begin{bmatrix} \frac{\partial x}{\partial t} \\ \frac{\partial y}{\partial t} \end{bmatrix} \quad \text{and} \quad \begin{bmatrix} \frac{\partial z}{\partial u} & \frac{\partial z}{\partial v} \end{bmatrix} = \begin{bmatrix} \frac{\partial z}{\partial x} & \frac{\partial z}{\partial y} \end{bmatrix} \begin{bmatrix} \frac{\partial x}{\partial u} & \frac{\partial x}{\partial v} \\ \frac{\partial y}{\partial u} & \frac{\partial y}{\partial v} \end{bmatrix}.$$

The 2×2 matrix on the right-hand side of the second identity is called the **Jacobian matrix** of the functions x and y with respect to the variables u and v, or more precisely, the **Jacobian matrix** of the transformation $\Phi : E \to \mathbb{R}^2$ defined by $\Phi(u, v) := (x(u, v), y(u, v))$. The determinant of this matrix gives a function $J(\Phi) : E \to \mathbb{R}$, and it is called the **Jacobian** of Φ or the **Jacobian** of the functions x and y with respect to the variables u and v. Thus,

$$J(\Phi)(u,v) := \det \begin{bmatrix} \dfrac{\partial x}{\partial u} & \dfrac{\partial x}{\partial v} \\ \dfrac{\partial y}{\partial u} & \dfrac{\partial y}{\partial v} \end{bmatrix} := \frac{\partial x}{\partial u}\frac{\partial y}{\partial v} - \frac{\partial x}{\partial v}\frac{\partial y}{\partial u} \quad \text{for } (u,v) \in E.$$

The Jacobian of x and y with respect to u and v is sometimes denoted by

$$\frac{\partial(x,y)}{\partial(u,v)}.$$

Similar notation prevails in the case of n functions of n variables.

It will be useful to record the following consequence of the Chain Rule for Jacobian matrices of composite functions. Suppose E is an open subset of $\mathbb{R}^2$ and $\Phi : E \to \mathbb{R}^2$ is as above, that is, $\Phi(u,v) := (x(u,v), y(u,v))$, where the component functions $x, y : E \to \mathbb{R}$ are differentiable on E. Let $D \subseteq \mathbb{R}^2$ be an open subset with $\Phi(E) \subseteq D$. If $\Psi : D \to \mathbb{R}^2$ given by $\Psi(x,y) := (w(x,y), z(x,y))$ is such that the component functions $w, z : D \to \mathbb{R}$ are differentiable on D, then the component functions of $\Psi \circ \Phi : E \to \mathbb{R}^2$ are differentiable on E and

$$J\left(\Psi \circ \Phi\right)(u,v) = J\left(\Psi\right)\left(x(u,v), y(u,v)\right) J\left(\Phi\right)(u,v) \quad \text{for all } (u,v) \in E.$$

In other words, the Jacobian of the composite is equal to the product of the Jacobians. To see this, note that the above identity can be written as follows:

$$\begin{bmatrix} \dfrac{\partial w}{\partial u} & \dfrac{\partial w}{\partial v} \\ \dfrac{\partial z}{\partial u} & \dfrac{\partial z}{\partial v} \end{bmatrix} = \begin{bmatrix} \dfrac{\partial w}{\partial x} & \dfrac{\partial w}{\partial y} \\ \dfrac{\partial z}{\partial x} & \dfrac{\partial z}{\partial y} \end{bmatrix} \begin{bmatrix} \dfrac{\partial x}{\partial u} & \dfrac{\partial x}{\partial v} \\ \dfrac{\partial y}{\partial u} & \dfrac{\partial y}{\partial v} \end{bmatrix}.$$

The equality of each entry in the matrix on the left with the corresponding entry of the product of the two matrices on the right is a direct consequence of part (iii) of Proposition 3.51. ◇

Examples 3.53. (i) Consider $f : \mathbb{R}^2 \to \mathbb{R}$ and $g : \mathbb{R} \to \mathbb{R}$ defined by $f(x,y) := xy$ for $(x,y) \in \mathbb{R}^2$ and $g(z) = \sin z$ for $z \in \mathbb{R}$. By the Chain Rule, the composite function $F : \mathbb{R}^2 \to \mathbb{R}$ given by $F(x,y) := (g \circ f)(x,y) = \sin(xy)$ is differentiable at every point of $\mathbb{R}^2$ and

$$\frac{\partial F}{\partial x} = \frac{dF}{dz}\frac{\partial z}{\partial x} = (\cos xy)y \quad \text{and} \quad \frac{\partial F}{\partial y} = \frac{dF}{dz}\frac{\partial z}{\partial y} = (\cos xy)x.$$

(ii) Consider $f : \mathbb{R}^2 \to \mathbb{R}$ defined by $f(x,y) := x^2 + y^2$ for $(x,y) \in \mathbb{R}^2$. Further, let $x, y : \mathbb{R} \to \mathbb{R}$ be defined by $x(t) := e^t$ and $y(t) := \sin t$ for

$t \in \mathbb{R}$. By the Chain Rule, the composite function $F : \mathbb{R}^2 \to \mathbb{R}$ given by $F(t) := f(x(t), y(t)) = e^{2t} + \sin^2 t$ is differentiable at every point of $\mathbb{R}$ and

$$\frac{dF}{dt} = \frac{\partial f}{\partial x}\frac{dx}{dt} + \frac{\partial f}{\partial y}\frac{dy}{dt} = (2x(t))e^t + (2y(t))\cos t = 2e^{2t} + 2\sin t \cos t.$$

(iii) As in (ii) above, consider $f : \mathbb{R}^2 \to \mathbb{R}$ defined by $f(x, y) := x^2 + y^2$ for $(x, y) \in \mathbb{R}^2$. Let $x, y : \mathbb{R}^2 \to \mathbb{R}$ be defined by $x(u, v) := u^2 - v^2$ and $y(u, v) := 2uv$ for $(u, v) \in \mathbb{R}^2$. By the Chain Rule, the composite function $F : \mathbb{R}^2 \to \mathbb{R}$ given by

$$F(u, v) := f(x(u, v), y(u, v)) = (u^2 - v^2)^2 + (2uv)^2 = u^4 + 2u^2v^2 + v^4$$

is differentiable at every point of $\mathbb{R}^2$ and

$$\frac{\partial F}{\partial u} = \frac{\partial F}{\partial x}\frac{\partial x}{\partial u} + \frac{\partial F}{\partial y}\frac{\partial y}{\partial u} = 2(u^2 - v^2)(2u) + 2(2uv)(2v) = 4u(u^2 + v^2),$$
$$\frac{\partial F}{\partial v} = \frac{\partial F}{\partial x}\frac{\partial x}{\partial v} + \frac{\partial F}{\partial y}\frac{\partial y}{\partial v} = 2(u^2 - v^2)(-2v) + 2(2uv)(2u) = 4v(u^2 + v^2).$$

Thus, $\nabla F(u, v) = 4(u^2 + v^2)(u, v)$ for all $(u, v) \in \mathbb{R}^2$. Note that the Jacobian of the functions x and y with respect to the variables u and v is equal to $(2u)(2u) - (-2v)(2v) = 4(u^2 + v^2)$.

(iv) As in (iii) above, consider $f : \mathbb{R}^2 \to \mathbb{R}$ defined by $f(x, y) := x^2 + y^2$ for $(x, y) \in \mathbb{R}^2$. Let $x, y : \mathbb{R}^2 \to \mathbb{R}$ be defined by $x(r, \theta) := r\cos\theta$ and $y(r, \theta) := r\sin\theta$ for $(r, \theta) \in \mathbb{R}^2$. By the Chain Rule, the composite function $F : \mathbb{R}^2 \to \mathbb{R}$ given by

$$F(r, \theta) := f(x(r, \theta), y(r, \theta)) = (r\cos\theta)^2 + (r\sin\theta)^2 = r^2$$

is differentiable at every point of $\mathbb{R}^2$ and

$$\frac{\partial F}{\partial r} = \frac{\partial F}{\partial x}\frac{\partial x}{\partial r} + \frac{\partial F}{\partial y}\frac{\partial y}{\partial r} = (2r\cos\theta)(\cos\theta) + (2r\sin\theta)(\sin\theta) = 2r,$$
$$\frac{\partial F}{\partial \theta} = \frac{\partial F}{\partial x}\frac{\partial x}{\partial \theta} + \frac{\partial F}{\partial y}\frac{\partial y}{\partial \theta} = (2r\cos\theta)(-r\sin\theta) + (2r\sin\theta)(r\cos\theta) = 0.$$

Thus, $\nabla F(r, \theta) = (2r, 0)$ for all $(r, \theta) \in \mathbb{R}^2$. Note that the Jacobian of the functions x and y with respect to the variables r and θ is equal to r.

(v) Consider $f : \mathbb{R}^2 \to \mathbb{R}$ defined by $f(0, 0) := 0$ and $f(x, y) := xy/(x^2 + y^2)$ for $(x, y) \neq (0, 0)$. Also consider $x, y : \mathbb{R} \to \mathbb{R}$ defined by $x(t) = t$ and $y(t) := t^2$ for $t \in \mathbb{R}$. We have seen in Example 3.1 (iii) that f_x and f_y exist on $\mathbb{R}^2$ and that $f_x(0, 0) = 0 = f_y(0, 0)$. Also, both x and y are differentiable on $\mathbb{R}$ and $x'(t) = 1$ and $y'(t) = 2t$ for $t \in \mathbb{R}$. On the other hand, if $F : \mathbb{R}^2 \to \mathbb{R}$ is defined by $F(t) := f(x(t), y(t))$, then $F(t) = t/(1 + t^2)$ and $F'(t) = (1 - t^2)/(1 + t^2)^2$ for all $t \in \mathbb{R}$. Thus, we see that

$$F'(0) = 1, \quad \text{whereas} \quad f_x(0, 0)x'(0) + f_y(0, 0)y'(0) = 0.$$

This shows that the hypothesis in the Chain Rule that f is differentiable cannot be dropped. ◇

3.4 Monotonicity and Convexity

We have seen in Section 1.2 that the notion of monotonicity for functions of one variable admits two distinct analogues in the setting of real-valued functions of two variables. In this section, we will relate these to the sign of certain partial derivatives, thereby obtaining analogues of the first derivative test of one-variable calculus. Next, we will obtain useful criteria for functions of bounded variation as well as for functions of bounded bivariation in terms of partial derivatives. The gradient of a real-valued function of two (or more) variables is a vector-valued function of two (or more) variables. We will see that there is a natural notion of monotonicity for such functions. Subsequently, we will relate monotonicity of the gradient with the notions of convexity and concavity of functions of two (or more) variables.

Monotonicity and First Partials

Let us recall that a function $f : I \times J \to \mathbb{R}$ defined on a product of intervals I and J in $\mathbb{R}$ is said to be monotonically increasing if

$$(x_1, y_1), (x_2, y_2) \in I \times J \text{ and } (x_1, y_1) \le (x_2, y_2) \implies f(x_1, y_1) \le f(x_2, y_2).$$

If this condition holds with $\le$ replaced by $\ge$ in the last inequality, then f is said to be monotonically decreasing. Thus, in effect, monotonicity of a function of two variables is the same as monotonicity in each of the two variables. Likewise, taking the partial derivative of a function of two variables amounts to treating it as a function of one variable (regarding the other variable as constant) and taking the derivative as in one-variable calculus. Thus, it is natural to expect that the following characterization holds.

Proposition 3.54. *Let $D \subseteq \mathbb{R}^2$ and let I, J be any intervals in $\mathbb{R}$ such that $I \times J \subseteq D$. Let $f : D \to \mathbb{R}$ be such that both f_x and f_y exist on $I \times J$. Then*

(i) *f is monotonically increasing on $I \times J \iff f_x \ge 0$ and $f_y \ge 0$ on $I \times J$.*
(ii) *f is monotonically decreasing on $I \times J \iff f_x \le 0$ and $f_y \le 0$ on $I \times J$.*

Proof. (i) Suppose f is monotonically increasing on $I \times J$. Then given any $(x_0, y_0) \in I \times J$, we find $[f(x_0 + h, y_0) - f(x_0, y_0)]/h \ge 0$ for any $h \in \mathbb{R}$ such that $h \ne 0$ and $(x_0 + h, y_0) \in I \times J$. Taking the limit as $h \to 0$, it follows that $f_x(x_0, y_0) \ge 0$. In a similar manner, we see that $f_y(x_0, y_0) \ge 0$.

Conversely, suppose $f_x \ge 0$ and $f_y \ge 0$ on $I \times J$. Let $(x_1, y_1), (x_2, y_2) \in I \times J$ be such that $(x_1, y_1) \le (x_2, y_2)$. Let us first show that $f(x_1, y_1) \le f(x_2, y_1)$. This is trivial if $x_1 = x_2$. If $x_1 < x_2$, consider $\phi : [x_1, x_2] \to \mathbb{R}$ defined by $\phi(t) := f(t, y_1)$ for $t \in [x_1, x_2]$. Using the MVT (Fact 3.2), we see that $\phi(x_2) - \phi(x_1) = \phi'(c)(x_2 - x_1)$ for some $c \in \mathbb{R}$ with $x_1 < c < x_2$. Consequently, $f(x_2, y_1) - f(x_1, y_1) = f_x(c, y_1)(x_2 - x_1) \ge 0$, that is, $f(x_1, y_1) \le f(x_2, y_1)$. Next, we will show that $f(x_2, y_1) \le f(x_2, y_2)$. This is trivial if $y_1 = y_2$, whereas if $y_1 < y_2$, then this follows in a similar manner by applying the MVT (Fact 3.2) to $\psi : [y_1, y_2] \to \mathbb{R}$ defined by $\psi(s) := f(x_2, s)$. Combining the two inequalities, we obtain $f(x_1, y_1) \le f(x_2, y_2)$. Thus, f is monotonically increasing on $I \times J$.

(ii) A proof similar to that of (i) can be given. Alternatively, (ii) follows by applying (i) to $-f$. □

Bimonotonicity and Mixed Partials

Let us recall that a function $f : I \times J \to \mathbb{R}$ defined on a product of intervals I and J in $\mathbb{R}$ is said to be bimonotonically increasing if

$$(x_1, y_1) \le (x_2, y_2) \implies f(x_1, y_2) + f(x_2, y_1) \le f(x_1, y_1) + f(x_2, y_2).$$

If this condition holds with $\le$ replaced by $\ge$ in the last inequality, then f is said to be bimonotonically decreasing. It turns out that these notions admit a neat characterization in terms of a mixed second-order partial derivative.

Proposition 3.55. *Let $D \subseteq \mathbb{R}^2$ and let I, J be any intervals in $\mathbb{R}$ such that $I \times J \subseteq D$. Let $f : D \to \mathbb{R}$ be such that f_x and f_{xy} exist on $I \times J$. Then*

(i) *f is bimonotonically increasing on $I \times J \iff f_{xy} \ge 0$ on $I \times J$.*
(ii) *f is bimonotonically decreasing on $I \times J \iff f_{xy} \le 0$ on $I \times J$.*

Proof. (i) Suppose f is bimonotonically increasing on $I \times J$. Given any $(x_0, y_0) \in I \times J$ and nonzero $h, k \in \mathbb{R}$, we see that

$$\begin{aligned} 0 &\le \frac{f(x_0 + h, y_0 + k) + f(x_0, y_0) - f(x_0, y_0 + k) - f(x_0 + h, y_0)}{hk} \\ &= \frac{1}{k}\left(\frac{f(x_0 + h, y_0 + k) - f(x_0, y_0 + k)}{h} - \frac{f(x_0 + h, y_0) - f(x_0, y_0)}{h}\right) \end{aligned}$$

whenever $(x_0 + h, y_0 + k) \in I \times J$. Taking the limit first as $h \to 0$ and then as $k \to 0$, it follows that $f_{xy}(x_0, y_0) \ge 0$.

Conversely, suppose $f_{xy} \ge 0$ on $I \times J$. Let $(x_1, y_1), (x_2, y_2) \in I \times J$ be such that $(x_1, y_1) \le (x_2, y_2)$. If $x_1 < x_2$ and $y_1 < y_2$, then by the Rectangular Mean Value Theorem (Proposition 3.11), there is $(c, d) \in I \times J$ such that

$$f(x_1, y_1) + f(x_2, y_2) - f(x_1, y_2) - f(x_2, y_1) = f_{xy}(c, d)(y_2 - y_1)(x_2 - x_1) \ge 0.$$

Moreover, the expression on the left is equal to zero if $x_1 = x_2$ or if $y_1 = y_2$. It follows that f is bimonotonically increasing on $I \times J$.

(ii) A proof similar to that of (i) can be given. Alternatively, (ii) follows by applying (i) to $-f$. □

Remark 3.56. Characterizations similar to those in Proposition 3.55 hold with f_x and f_{xy} replaced throughout by f_y and f_{yx}. ◇

Bounded Variation and Boundedness of First Partials

Let us recall that a function $f : [a,b] \times [c,d] \to \mathbb{R}$ is said to be of bounded variation on $[a,b] \times [c,d]$ if the set of all finite sums of the form

$$\sum_{i=1}^{n} |f(x_i, y_i) - f(x_{i-1}, y_{i-1})|,$$

where $n \in \mathbb{N}$ and $(x_0, y_0), \ldots, (x_n, y_n)$ vary over points of $\mathbb{R}^2$ satisfying

$$(a,c) = (x_0, y_0) \le (x_1, y_1) \le \cdots \le (x_{n-1}, y_{n-1}) \le (x_n, y_n) = (b,d),$$

is bounded above. We have seen in Section 1.2 that a real-valued function on a rectangle is of bounded variation if and only if it is the difference of two real-valued monotonically increasing functions. In practice, it is not always easy to verify whether a given function is of bounded variation using the definition or this characterization. However, the following simple criterion is often helpful.

Proposition 3.57. *Let $f : [a,b] \times [c,d] \to \mathbb{R}$ be continuous on $[a,b] \times [c,d]$ and differentiable on $(a,b) \times (c,d)$. Assume that f_x exists and is bounded on $(a,b) \times [c,d]$, while f_y exists and is bounded on $[a,b] \times (c,d)$. Then f is of bounded variation on $[a,b] \times [c,d]$.*

Proof. Let $\alpha \in \mathbb{R}$ be such that $|f_x(u,v)| \le \alpha$ for all $(u,v) \in (a,b) \times [c,d]$ and $|f_y(u,v)| \le \alpha$ for all $(u,v) \in [a,b] \times (c,d)$. Suppose $n \in \mathbb{N}$ and $(x_0, y_0), \ldots, (x_n, y_n)$ are any points of $\mathbb{R}^2$ such that

$$(a,c) = (x_0, y_0) \le (x_1, y_1) \le \cdots \le (x_{n-1}, y_{n-1}) \le (x_n, y_n) = (b,d).$$

In view of the Bivariate Mean Value Inequality (Corollary 3.6) together with Proposition 3.35, we see that

$$|f(x_i, y_i) - f(x_{i-1}, y_{i-1})| \le \alpha\big(x_i - x_{i-1} + y_i - y_{i-1}\big) \quad \text{for } i = 1, \ldots, n,$$

and consequently,

$$\sum_{i=1}^{n} |f(x_i, y_i) - f(x_{i-1}, y_{i-1})| \le \alpha\,(b - a + d - c)\,.$$

Thus f is of bounded variation on $[a,b] \times [c,d]$. □

The above proof not only shows that the function f as in Proposition 3.57 is of bounded variation, but also gives an upper bound for the total variation of f on $[a,b] \times [c,d]$. Moreover, we also have the following useful corollary, which is perhaps more useful in practice.

Corollary 3.58. *If $f : [a,b] \times [c,d] \to \mathbb{R}$ has continuous partial derivatives on $[a,b] \times [c,d]$, then f is of bounded variation on $[a,b] \times [c,d]$.*

Proof. By Proposition 3.33, f is differentiable on $(a,b) \times (c,d)$. Moreover, by Proposition 2.25, the first partials f_x and f_y are bounded on $[a,b] \times [c,d]$. Hence the desired result follows from Proposition 3.57. □

Bounded Bivariation and Boundedness of Mixed Partials

Let us recall that a function $f : [a, b] \times [c, d] \to \mathbb{R}$ is said to be of bounded bivariation on $[a, b] \times [c, d]$ if the set of all finite double sums of the form

$$\sum_{i=1}^{n} \sum_{j=1}^{m} |f(x_i, y_j) + f(x_{i-1}, y_{j-1}) - f(x_i, y_{j-1}) - f(x_{i-1}, y_j)|,$$

where $n, m \in \mathbb{N}$ and $(x_0, y_0), \ldots, (x_n, y_m)$ vary over points of $\mathbb{R}^2$ satisfying

$$a = x_0 \le x_1 \le \cdots \le x_{n-1} \le x_n = b \text{ and } c = y_0 \le y_1 \le \cdots \le y_{m-1} \le y_m = d,$$

is bounded above. We have seen in Section 1.2 that a real-valued function on a rectangle is of bounded bivariation if and only if it is the difference of two real-valued bimonotonically increasing functions. Again, in practice, it is not always easy to verify whether a given function is of bounded bivariation using the definition or this characterization. However, the following simple criterion is often helpful.

Proposition 3.59. *Let $f : [a, b] \times [c, d] \to \mathbb{R}$ be continuous on $[a, b] \times [c, d]$. Assume that f_x exists and is continuous on $(a, b) \times [c, d]$, while f_{xy} exists and is bounded on $(a, b) \times (c, d)$. Then f is of bounded bivariation on $[a, b] \times [c, d]$.*

Proof. Let $\beta \in \mathbb{R}$ be such that $|f_{xy}(u, v)| \le \beta$ for all $(u, v) \in (a, b) \times (c, d)$. Suppose $m, n \in \mathbb{N}$ and $(x_0, y_0), \ldots, (x_n, y_m)$ are any points of $\mathbb{R}^2$ such that

$$a = x_0 \le x_1 \le \cdots \le x_{n-1} \le x_n = b \text{ and } c = y_0 \le y_1 \le \cdots \le y_{m-1} \le y_m = d.$$

For $i = 1, \ldots, n$ and $j = 1, \ldots, m$, using the Rectangular Mean Value Inequality (Corollary 3.12), we see that

$$|f(x_i, y_j) + f(x_{i-1}, y_{j-1}) - f(x_i, y_{j-1}) - f(x_{i-1}, y_j)| \le \beta A_{ij},$$

where $A_{ij} := (x_i - x_{i-1})(y_j - y_{j-1})$. Now $\sum_{i=1}^{n} \sum_{j=1}^{m} A_{ij} = (b - a)(d - c)$, and consequently,

$$\sum_{i=1}^{n} \sum_{j=1}^{m} |f(x_i, y_j) + f(x_{i-1}, y_{j-1}) - f(x_i, y_{j-1}) - f(x_{i-1}, y_j)| \le \beta(b-a)(d-c).$$

Thus f is of bounded bivariation on $[a, b] \times [c, d]$. □

Remark 3.60. In light of the last statement in Remark 3.13, it is readily seen that a result similar Proposition 3.59 holds with f_x and f_{xy} replaced throughout by f_y and f_{yx}. ◇

The proof of Proposition 3.59 not only shows that the function f therein is of bounded bivariation, but also gives an upper bound for the total bivariation of f on $[a, b] \times [c, d]$.

Convexity and Monotonicity of Gradient

The notions of convex functions and concave functions of two variables have been discussed in Section 1.2. As in one-variable calculus, we can relate these notions to the monotonicity of the (total) derivative, which we now define. To motivate this, let us first note that if $I \subseteq \mathbb{R}$ is an interval in $\mathbb{R}$, then $f : I \to \mathbb{R}$ is monotonically increasing if for every $x_1, x_2 \in I$ with $x_1 < x_2$, we have $f(x_1) \leq f(x_2)$. Equivalently, $f : I \to \mathbb{R}$ is monotonically increasing if and only if

$$\big(f(x_2) - f(x_1)\big) \cdot (x_2 - x_1) \geq 0 \quad \text{for all } x_1, x_2 \in I.$$

This condition generalizes easily if we replace the function of one variable by the gradient of a function of two (or more) variables and we replace the multiplication of real numbers by the dot product of vectors.

Let $D \subseteq \mathbb{R}^2$ be open and convex, and let $f : D \to \mathbb{R}$ be any differentiable function. We say that ∇f is **monotonically increasing** on D if

$$\big(\nabla f(x_2, y_2) - \nabla f(x_1, y_1)\big) \cdot (x_2 - x_1, y_2 - y_1) \geq 0 \quad \text{for all } (x_1, y_1), (x_2, y_2) \in D.$$

Likewise, we say that ∇f is **monotonically decreasing** on D if

$$\big(\nabla f(x_2, y_2) - \nabla f(x_1, y_1)\big) \cdot (x_2 - x_1, y_2 - y_1) \leq 0 \quad \text{for all } (x_1, y_1), (x_2, y_2) \in D.$$

We say that ∇f is **monotonic** on D if it is monotonically increasing or monotonically decreasing on D.

We shall now see that the convexity or the concavity of f can be characterized in terms of the monotonicity of ∇f in exactly the same manner as in the case of functions of one variable. (See, for example, Section 4.3 of ACICARA.) First, we require an auxiliary result.

Lemma 3.61. *Let $D \subseteq \mathbb{R}^2$ be open and convex, and let $f : D \to \mathbb{R}$ be any function. If f is differentiable at some $(x_0, y_0) \in D$ and f is convex on D, then*

$$f(x, y) - f(x_0, y_0) \geq \nabla f(x_0, y_0) \cdot (x - x_0, y - y_0) \quad \text{for all } (x, y) \in D.$$

Conversely, if f is differentiable on D and satisfies

$$f(x, y) - f(x_0, y_0) \geq \nabla f(x_0, y_0) \cdot (x - x_0, y - y_0) \quad \text{for all } (x_0, y_0), (x, y) \in D,$$

then f is convex on D.

Proof. Suppose f is differentiable at some $(x_0, y_0) \in D$ and f is convex on D. Let (f_1, f_2) be a pair of increment functions associated with f and (x_0, y_0). Then $f_1, f_2 : D \to \mathbb{R}$ are continuous at (x_0, y_0) and we have

$$f(x, y) - f(x_0, y_0) = (x - x_0) f_1(x, y) + (y - y_0) f_2(x, y) \quad \text{for all } (x, y) \in D.$$

Fix any $(x, y) \in D$. Let $h := x - x_0$ and $k := y - y_0$. Then for any $t \in (0, 1)$, the point $(x_0 + th, y_0 + tk) = t(x, y) + (1 - t)(x_0, y_0)$ is in D because D is convex; further, since f is convex, we have

$$f(x_0+th, y_0+tk) \le tf(x, y)+(1-t)f(x_0, y_0) = t\,[f(x, y)-f(x_0, y_0)]-f(x_0, y_0).$$

Consequently, for every $t \in (0, 1)$, we have

$$\begin{aligned} t\,[f(x, y) - f(x_0, y_0)] &\ge f(x_0 + th, y_0 + tk) - f(x_0, y_0) \\ &= t\,[hf_1(x_0 + th, y_0 + tk) + kf_2(x_0 + th, y_0 + tk)]\,. \end{aligned}$$

It follows that for every $t \in (0, 1)$, we have

$$f(x, y) - f(x_0, y_0) \ge hf_1(x_0 + th, y_0 + tk) + kf_2(x_0 + th, y_0 + tk).$$

Since $(f_1(x_0, y_0), f_2(x_0, y_0)) = (f_x(x_0, y_0), f_y(x_0, y_0)) = \nabla f(x_0, y_0)$, by taking the limit as $t \to 0^+$ and using the continuity of f_1, f_2 at (x_0, y_0), we obtain

$$f(x, y) - (x_0, y_0) \ge hf_x(x_0, y_0) + kf_y(x_0, y_0) = \nabla f(x_0, y_0) \cdot (x - x_0, y - y_0).$$

This proves the desired inequality.

Conversely, suppose f is differentiable on D and satisfies

$$f(x, y) - f(x_0, y_0) \ge \nabla f(x_0, y_0) \cdot (x - x_0, y - y_0) \quad \text{for all } (x_0, y_0), (x, y) \in D.$$

Consider any $(x_1, y_1), (x_2, y_2) \in D$ and $t \in (0, 1)$. Define

$$(x_0, y_0) := (1 - t)(x_1, y_1) + t(x_2, y_2) = (x_1, y_1) + t(x_2 - x_1, y_2 - y_1).$$

Since D is convex, $(x_0, y_0) \in D$. Moreover, by the hypothesis, we have

$$f(x_i, y_i) - f(x_0, y_0) \ge \nabla f(x_0, y_0) \cdot (x_i - x_0, y_i - y_0) \quad \text{for } i = 1, 2.$$

Multiplying the inequality corresponding to $i = 1$ by $(1 - t)$, the inequality corresponding to $i = 2$ by t, and then adding the two, we obtain

$$[(1 - t)f(x_1, y_1) + tf(x_2, y_2)] - f(x_0, y_0) \ge \nabla f(x_0, y_0) \cdot (x_3, y_3),$$

where $(x_3, y_3) := [(1 - t)(x_1 - x_0, y_1 - y_0) + t(x_2 - x_0, y_2 - y_0)]$. But since $(x_3, y_3) = (1 - t)(x_1, y_1) + t(x_2, y_2) - (x_0, y_0) = (0, 0)$, we obtain

$$f(x_0, y_0) \le (1 - t)f(x_1, y_1) + tf(x_2, y_2).$$

This proves that f is convex on D. □

Remark 3.62. It may be noted that the proof of the "converse" in Lemma 3.61 does not make an essential use of the hypothesis that f is differentiable, except for the existence of the gradient ∇f. In fact, the first assertion in this lemma can also be proved under the weaker hypothesis that $\nabla f(x_0, y_0)$ exists. This follows from a general property of convex functions, which states that if $D \subseteq \mathbb{R}^2$ is open and convex, $(x_0, y_0) \in D$, and $f : D \to \mathbb{R}$ is a convex function such that $\nabla f(x_0, y_0)$ exists, then f is differentiable at (x_0, y_0). A proof of this property is slightly technical, and will not be needed in the text; however, it is sketched in Exercise 42. ◇

The next lemma is an elementary characterization of the convexity of a function f of two variables in terms of the convexity of functions of one variable obtained by restricting f to various line segments in the domain of f.

Lemma 3.63. *Let $D \subseteq \mathbb{R}^2$ be convex, and let $f : D \to \mathbb{R}$ be any function. Then f is convex on D if and only if for every $(x_1, y_1), (x_2, y_2) \in D$, the function $F : [0,1] \to \mathbb{R}$ defined by $F(t) := f\left(x_1 + t(x_2 - x_1),\, y_1 + t(y_2 - y_1)\right)$ is convex on $[0,1]$.*

Proof. Suppose f is convex on D. Fix any $(x_1, y_1), (x_2, y_2) \in D$, and consider $F : [0,1] \to \mathbb{R}$ be defined by $F(t) := f\left(x_1 + t(x_2 - x_1),\, y_1 + t(y_2 - y_1)\right)$. Given any $t_1, t_2 \in [0,1]$ and $\lambda \in (0,1)$, let us take $t := (1-\lambda)t_1 + \lambda t_2$. Then $t \in (t_1, t_2) \subseteq (0,1)$ and

$$\begin{aligned} x_1 + t(x_2 - x_1) &= [(1-\lambda) + \lambda]x_1 + [(1-\lambda)t_1 + \lambda t_2](x_2 - x_1) \\ &= (1-\lambda)[(1-t_1)x_1 + t_1 x_2] + \lambda[(1-t_2)x_1 + t_2 x_2]. \end{aligned}$$

Thus, if we let $u_1 := (1-t_1)x_1 + t_1 x_2$ and $u_2 := (1-t_2)x_1 + t_2 x_2$, then $x_1 + t(x_2 - x_1) = (1-\lambda)u_1 + \lambda u_2$. Similarly, if we let $v_1 := (1-t_1)y_1 + t_1 y_2$ and $v_2 := (1-t_2)y_1 + t_2 y_2$, then $y_1 + t(y_2 - y_1) = (1-\lambda)v_1 + \lambda v_2$. Now since D is convex, $(u_1, v_1), (u_2, v_2) \in D$, and further, since f is convex,

$$F(t) = f\left((1-\lambda)(u_1, v_1) + \lambda(u_2, v_2)\right) \le (1-\lambda)f(u_1, v_1) + \lambda f(u_2, v_2).$$

It follows that $F((1-\lambda)t_1 + \lambda t_2) \le (1-\lambda)F(t_1) + \lambda F(t_2)$. Hence F is convex on $[0,1]$.

To prove the converse, let (x_1, y_1) and (x_2, y_2) be any two points of D, and suppose $F : [0,1] \to \mathbb{R}$ defined by $F(t) := f\left(x_1 + t(x_2 - x_1),\, y_1 + t(y_2 - y_1)\right)$ is convex on $[0,1]$. Then

$$F(t) = F\left((1-t)\cdot 0 + t \cdot 1\right) \le (1-t)F(0) + tF(1) \quad \text{for all } t \in (0,1),$$

and therefore,

$$f\left((1-t)(x_1, y_1) + t(x_2, y_2)\right) \le (1-t)f(x_1, y_1) + tf(x_2, y_2) \quad \text{for all } t \in (0,1).$$

This yields the desired result. □

Remark 3.64. The argument used in the first part of the above proof can also be used to prove a *local version* of Lemma 3.63, namely, if $D \subseteq \mathbb{R}^2$ is convex and $f : D \to \mathbb{R}$ is convex on D, and if $(x_0, y_0) \in D$ is such that $(x_0 + th,\, y_0 + tk) \in D$ for all $t \in I$ and $(h,k) \in \mathbb{R}^2$, where I is an interval in $\mathbb{R}$ containing 0, then $F : I \to \mathbb{R}$ defined by $F(t) := f(x_0 + th,\, y_0 + tk)$ for $t \in I$ is convex on I. To see this, note that for any $t_1, t_2 \in I$ and any $\lambda \in (0,1)$, we have $x_0 + [(1-\lambda)t_1 + \lambda t_2]h = (1-\lambda)(x_0 + t_1 h) + \lambda(x_0 + t_2 h)$, and $y_0 + [(1-\lambda)t_1 + \lambda t_2]k = (1-\lambda)(y_0 + t_1 k) + \lambda(y_0 + t_2 k)$; hence by the convexity of f on D, we obtain

$$F\left((1-\lambda)t_1+\lambda t_2\right) \le (1-\lambda)f(x_0+t_1h,\, y_0+t_1k)+\lambda f(y_0+t_2h,\, y_0+t_2k)$$
$$= (1-\lambda)F(t_1)+\lambda F(t_2),$$

which proves that F is convex on I. ◇

We are now ready to prove a characterization of convexity in terms of the monotonicity of the gradient.

Proposition 3.65. *Let $D \subseteq \mathbb{R}^2$ be open and convex, and let $f : D \to \mathbb{R}$ be any differentiable function on D. Then*

$$f \text{ is convex on } D \iff \nabla f \text{ is monotonically increasing on } D.$$

Proof. Suppose f is convex on D. Let (x_1, y_1) and (x_2, y_2) be any two points of D. By Lemma 3.61, we have

$$f(x_2, y_2) - f(x_1, y_1) \ge \nabla f(x_1, y_1) \cdot (x_2 - x_1,\, y_2 - y_1)$$

and

$$f(x_1, y_1) - f(x_2, y_2) \ge \nabla f(x_2, y_2) \cdot (x_1 - x_2,\, y_1 - y_2).$$

Adding the two inequalities above, we obtain

$$\left(\nabla f(x_2, y_2) - \nabla f(x_1, y_1)\right) \cdot (x_2 - x_1,\, y_2 - y_1) \ge 0.$$

Hence ∇f is monotonically increasing on D.

Conversely, suppose ∇f is monotonically increasing on D. Fix any (x_1, y_1) and (x_2, y_2) in D. Let $E := \{t \in \mathbb{R} : (x_1 + t(x_2 - x_1),\, y_1 + t(y_2 - y_1) \in D\}$. Note that since D is convex, E contains the interval $[0, 1]$. Let $x, y : E \to \mathbb{R}$ and let $F : E \to \mathbb{R}$ be defined by

$$x(t) := x_1 + t(x_2 - x_1), \quad y(t) := y_1 + t(y_2 - y_1), \quad \text{and} \quad F(t) := f(x(t), y(t)).$$

Since D is open in $\mathbb{R}^2$, by part (ii) of Proposition 2.23, we see that E is open in $\mathbb{R}$. By the Chain Rule (part (ii) of Proposition 3.51), F is differentiable and for any $t \in E$, we have

$$F'(t) = \nabla f(x(t), y(t)) \cdot (x'(t), y'(t)) = \nabla f(x(t), y(t)) \cdot (x_2 - x_1,\, y_2 - y_1).$$

In particular, for any $t_1, t_2 \in [0, 1]$, we have

$$F'(t_2) - F'(t_1) = \left[\nabla f(x(t_2), y(t_2)) - \nabla f(x(t_1), y(t_1))\right] \cdot (x_2 - x_1,\, y_2 - y_1).$$

Now, since $(x(t_2) - x(t_1),\, y(t_2) - y(t_1)) = (t_2 - t_1)(x_2 - x_1,\, y_2 - y_1)$, we see that $(t_2 - t_1)\left(F'(t_2) - F'(t_1)\right)$ equals

$$\left[\nabla f(x(t_2), y(t_2)) - \nabla f(x(t_1), y(t_1))\right] \cdot \left(x(t_2) - x(t_1),\, y(t_2) - y(t_1)\right).$$

Further, since ∇f is monotonic on D, we have $(t_2 - t_1)\left(F'(t_2) - F'(t_1)\right) \ge 0$. It follows that F' is monotonically increasing on $[0, 1]$. Hence by a standard result in one-variable calculus (for instance, part (i) of Proposition 4.31 of ACICARA), F is convex on $[0, 1]$. Thus, by Lemma 3.63, we conclude that f is convex on D. □

Remark 3.66. Let $D \subseteq \mathbb{R}^2$ be open and convex, and let $f : D \to \mathbb{R}$ be any function. It is clear that f is concave on D if and only if $-f$ is convex on D. Also, if f is differentiable on D, then ∇f is monotonically decreasing on D if and only if $\nabla(-f)$ is monotonically increasing on D. Thus, Proposition 3.65 readily implies the following characterization of concavity:

$$f \text{ is concave on } D \Longleftrightarrow \nabla f \text{ is monotonically decreasing on } D.$$

The notions of strict convexity and strict concavity of a real-valued function of two variables are defined by changing the inequalities $\le$ and $\ge$ in the definition of convexity and concavity to $<$ and $>$, respectively. Similarly, we can easily formulate the notion of the gradient being strictly increasing or strictly decreasing. Now, each of the results in this subsection has an analogue for strictly convex functions. In particular, if f is differentiable, then we have the following characterization:

$$f \text{ is strictly convex on } D \Longleftrightarrow \nabla f \text{ is strictly increasing on } D.$$

Applying this to $-f$, we see also that

$$f \text{ is strictly concave on } D \Longleftrightarrow \nabla f \text{ is strictly decreasing on } D.$$

Thus, we obtain analogues of the results in one-variable calculus for strictly convex and strictly concave functions (given, for example, on page 128 of ACICARA). ◇

Convexity and Nonnegativity of Hessian

We shall now proceed to investigate whether the characterization of convexity of a function of one variable in terms of the nonnegativity of its second derivative (given, for example, on page 129 of ACICARA) has an analogue for a function of two variables. To this end, we have first to reckon with the fact if f is a function of two variables, then there are four possible second-order partial derivatives, namely, f_{xx}, f_{xy}, f_{yx}, and f_{yy}. If these are well behaved, then we have $f_{xy} = f_{yx}$, and we need consider only three. Collectively, these are captured in the form

$$\left(h\frac{\partial}{\partial x} + k\frac{\partial}{\partial x}\right)^2 f = h^2\frac{\partial^2 f}{\partial x^2} + 2hk\frac{\partial^2 f}{\partial x \partial y} + k^2\frac{\partial^2 f}{\partial y^2}.$$

Treating h and k as variables and evaluating the partials at a particular point, we obtain a homogeneous polynomial of total degree 2 in (h, k) given by

$$Q(h, k) = ah^2 + 2bhk + ck^2, \qquad \text{where } a, b, c \in \mathbb{R}.$$

Such a polynomial is called a **binary quadratic form** in the variables h and k. If $D \subseteq \mathbb{R}^2$ is open and $f : D \to \mathbb{R}$ has the property that the first-order and second-order partial derivatives of f exist and are continuous on D, then for any $(x_0, y_0) \in D$, the associated binary quadratic form

$$h^2 \frac{\partial^2 f}{\partial x^2}(x_0, y_0) + 2hk \frac{\partial^2 f}{\partial x \partial y}(x_0, y_0) + k^2 \frac{\partial^2 f}{\partial y^2}(x_0, y_0)$$

is called the **Hessian form** of f at (x_0, y_0).

In general, a binary quadratic form $Q(h, k)$ is said to be **nonnegative definite** if it always takes nonnegative values, that is, $Q(s, t) \geq 0$ for every $(s, t) \in \mathbb{R}^2$. For $D \subseteq \mathbb{R}^2$ and $f : D \to \mathbb{R}$ as above, we shall say that the Hessian form of f is **nonnegative definite** on D if the Hessian form of f at (x_0, y_0) is nonnegative definite for every $(x_0, y_0) \in D$.

It turns out that the nonnegativity of the second derivative of a function of one variable is analogous to the nonnegative definiteness of the Hessian form of a function of two variables. This is the theme that we shall develop in the remainder of this section.

Proposition 3.67. *Let $D \subseteq \mathbb{R}^2$ be open and convex, and let $f : D \to \mathbb{R}$ be such that the first and second order partial derivatives of f exist and are continuous on D. Then*

$$f \text{ is convex on } D \iff \text{Hessian form of } f \text{ is nonnegative definite on } D.$$

Proof. Since the first-order partial derivatives of f exist and are continuous on D, by Proposition 3.33, we see that f is differentiable on D.

Suppose f is convex on D. Let $(x_0, y_0) \in D$ and $(h, k) \in \mathbb{R}^2$. Since D is open, there is $\delta > 0$ such that $(x_0+th, y_0+tk) \in D$ for all $t \in (-\delta, \delta)$. Consider $F : (-\delta, \delta) \to \mathbb{R}$ defined by $F(t) := f(x_0 + th, y_0 + tk)$. By Proposition 3.33 and the Chain Rule (part (ii) of Proposition 3.51), F is differentiable and for any $t \in (-\delta, \delta)$, we have

$$F'(t) = (hf_x + kf_y)(x_0 + th, y_0 + tk).$$

Applying similar reasoning to F', we see that F is twice differentiable and for any $t \in (-\delta, \delta)$, we have

$$F''(t) = (h[hf_{xx} + kf_{xy}] + k[hf_{yx} + kf_{yy}])(x_0 + th, y_0 + tk).$$

Thus, in view of the Mixed Partials Theorem (Proposition 3.14), we have

$$F''(t) = \left(h^2 \frac{\partial^2}{\partial x^2} + 2hk \frac{\partial^2}{\partial x \partial y} + k^2 \frac{\partial^2}{\partial y^2}\right) f(x_0 + th,\ y_0 + tk).$$

Now, in view of Remark 3.64, F is convex on $(-\delta, \delta)$, and hence by a standard result in one-variable calculus (for instance, part (i) of Proposition 4.32 of ACICARA), we have $F''(t) \geq 0$ for all $t \in (-\delta, \delta)$. In particular, $F''(0) \geq 0$, and hence

$$h^2 \frac{\partial^2 f}{\partial x^2}(x_0, y_0) + 2hk \frac{\partial^2 f}{\partial x \partial y}(x_0, y_0) + k^2 \frac{\partial^2 f}{\partial y^2}(x_0, y_0) \geq 0.$$

This proves that the Hessian form of f is nonnegative definite on D.

Conversely, suppose the Hessian form of f is nonnegative definite on D. Let $(x_0, y_0), (x, y) \in D$ and take $h := x - x_0$ and $k := y - y_0$. By the Extended Bivariate Mean Value Theorem (Remark 3.48 (ii)), there is $(c, d) \in D$ such that

$$f(x, y) = f(x_0, y_0) + \left(h\frac{\partial}{\partial x} + k\frac{\partial}{\partial x}\right) f(x_0, y_0) + \frac{1}{2}\left(h\frac{\partial}{\partial x} + k\frac{\partial}{\partial y}\right)^2 f(c, d).$$

Further, since the Hessian form of f at (c, d) is nonnegative definite, we obtain

$$f(x, y) - f(x_0, y_0) \geq \left(h\frac{\partial}{\partial x} + k\frac{\partial}{\partial y}\right) f(x_0, y_0) = \nabla f(x_0, y_0) \cdot (x - x_0, y - y_0).$$

Hence by Lemma 3.61, f is convex on D. □

Remark 3.68. A binary quadratic form $Q(h, k) := ah^2 + 2bhk + ck^2$ is said to be **positive definite** if $Q(s, t) > 0$ for all $(s, t) \in \mathbb{R}^2$ with $(s, t) \neq (0, 0)$. If $D \subseteq \mathbb{R}^2$ and $f : D \to \mathbb{R}$ are as in Proposition 3.67, then the Hessian form of f is **positive definite** on D if the Hessian form of f at (x_0, y_0) is positive definite for all $(x_0, y_0) \in D$. For such a function f,

the Hessian form of f is positive definite on $D \Longrightarrow f$ is strictly convex on D.

This can proved using arguments exactly similar to those in the second half of the proof of Proposition 3.67. However, as in the case of functions of one variable, the converse is not true. For example, if $f : \mathbb{R}^2 \to \mathbb{R}$ is defined by $f(x, y) = x^4 + y^4$, then f is strictly convex but the Hessian form of f at $(0, 0)$ is not positive definite. ◇

A priori, the nonnegative definiteness of the Hessian form of a function does not seem like a condition that is easy to check in practice. In general, to know that a binary quadratic form $Q(h, k)$ is nonnegative definite appears to require substitution of all possible pairs of real numbers (s, t) to check whether the resulting value $Q(s, t)$ is nonnegative. Interestingly, this can often be avoided because there is a simple and practical test to check whether a binary quadratic form is nonnegative definite. This test is given in the following proposition in a purely algebraic set-up, and later we use it to derive a simple criterion for convexity.

Proposition 3.69. *Let $Q(h, k) := ah^2 + 2bhk + ck^2$ be a binary quadratic form in the variables h and k with coefficients a, b, c in $\mathbb{R}$. Then*

$$Q(h, k) \text{ is nonnegative definite} \iff a \geq 0, \ c \geq 0 \text{ and } ac - b^2 \geq 0.$$

Proof. Suppose $Q(h,k)$ is nonnegative definite. Then $a = Q(1,0) \geq 0$ and $c = Q(0,1) \geq 0$. In case $a \neq 0$, consider

$$Q(b,-a) = ab^2 - 2ab^2 + ca^2 = ca^2 - ab^2 = a(ac - b^2).$$

Since $Q(b,-a) \geq 0$ and $a > 0$, we must have $ac - b^2 \geq 0$. Next, in case $a = 0$ and $c \neq 0$, consider

$$Q(c,-b) = ac^2 - 2cb^2 + cb^2 = ac^2 - cb^2 = c(ac - b^2).$$

Since $Q(c,-b) \geq 0$ and $c > 0$, we must have $ac - b^2 \geq 0$. Finally, in case $a = 0$ and $c = 0$, we have $2b = Q(1,1) \geq 0$ and $-2b = Q(1,-1) \geq 0$, which implies that $b = 0$; hence, in this case $ac - b^2 = 0$.

Conversely, suppose $a \geq 0$, $c \geq 0$ and $ac - b^2 \geq 0$. Let $\Delta := ac - b^2$. In case $a > 0$, the identity

$$aQ(h,k) = a^2h^2 + 2abhk + ack^2 = (ah + bk)^2 + \Delta k^2$$

implies that $Q(h,k) \geq 0$ for all $(h,k) \in \mathbb{R}^2$. In case $c > 0$, the identity

$$cQ(h,k) = ach^2 + 2bchk + c^2k^2 = (bh + ck)^2 + \Delta h^2$$

implies that $Q(s,t) \geq 0$ for all $(s,t) \in \mathbb{R}^2$. In case $a = c = 0$, the condition $ac - b^2 \geq 0$ implies that $b = 0$, and hence $Q(s,t) = 0$ for all $(s,t) \in \mathbb{R}^2$. Thus, in any case, $Q(h,k)$ is nonnegative definite. □

Remark 3.70. For the positive definiteness of a binary quadratic form $Q(h,k) := ah^2 + 2bhk + ck^2$, we have the following characterization:

$$Q(h,k) \text{ is positive definite } \iff a > 0 \text{ and } ac - b^2 > 0.$$

The proof is similar to that of Proposition 3.69. In fact, it is simpler. On the other hand, the example $Q(h,k) := -k^2$ shows that the conditions $a \geq 0$ and $ac - b^2 \geq 0$ are not sufficient to imply that $Q(h,k)$ is nonnegative definite. ◇

The real number $\Delta := ac - b^2$ will be called the **discriminant**[2] of the binary quadratic form $Q(h,k) := ah^2 + 2bhk + ck^2$. In case $D \subseteq \mathbb{R}^2$ is open and $f : D \to \mathbb{R}$ is such that the first-order and second-order partial derivatives of f exist and are continuous on D, then for any $(x_0, y_0) \in D$, we define the **discriminant** of f at (x_0, y_0) to be the discriminant of the Hessian form of f at (x_0, y_0), and denote it by $\Delta f(x_0, y_0)$; in other words,

[2] The binary quadratic form $ah^2 + 2bhk + ck^2$ corresponds to the quadratic polynomial $ax^2 + 2bx + c$, whose classical discriminant is $(2b)^2 - 4ac = 4(b^2 - ac)$. The reason we have ignored the constant factor 4 and, more significantly, reversed the sign while defining the discriminant of $Q(h,k)$ is because of its connection with the theory of matrices. This connection is explained in Remark 3.73. The notion of discriminant in the manner that we have defined generalizes easily to the case of quadratic forms in more than two variables.

$$\Delta f(x_0, y_0) := f_{xx}(x_0, y_0) f_{yy}(x_0, y_0) - [f_{xy}(x_0, y_0)]^2 .$$

An immediate consequence of Propositions 3.67 and 3.69 is the following practical criterion for convexity.

Proposition 3.71. *Let $D \subseteq \mathbb{R}^2$ be open and convex, and let $f : D \to \mathbb{R}$ be such that the first-order and second-order partial derivatives of f exist and are continuous on D. Then*

$$f \text{ is convex on } D \iff f_{xx}(x_0, y_0) \geq 0, \; f_{yy}(x_0, y_0) \geq 0 \text{ and } \Delta f(x_0, y_0) \geq 0 \text{ for all } (x_0, y_0) \in D.$$

Proof. By Proposition 3.69, we see that the Hessian form of f is nonnegative definite if and only if $f_{xx}(x_0, y_0) \geq 0$, $f_{yy}(x_0, y_0) \geq 0$, and $\Delta f(x_0, y_0) \geq 0$ for all $(x_0, y_0) \in D$. Hence the desired result follows from Proposition 3.67. □

Remark 3.72. In view of Remarks 3.68 and 3.70, we see that if $D \subseteq \mathbb{R}^2$ and $f : D \to \mathbb{R}$ are as in Proposition 3.71, then the positivity of $f_{xx}(x_0, y_0)$ and $\Delta f(x_0, y_0)$ for all $(x_0, y_0) \in D$ implies that f is strictly convex on D. However, the example discussed in Remark 3.68, namely, $f(x, y) := x^4 + y^4$, shows that the converse is not true in general. ◇

Remark 3.73. The study of binary quadratic forms is closely related to the study of 2×2 real symmetric matrices. Indeed, the binary quadratic form $Q(h, k) := ah^2 + 2bhk + ck^2$ can be expressed as the matrix product

$$\mathbf{h}^T A \mathbf{h}, \quad \text{where} \quad A := \begin{bmatrix} a & b \\ b & c \end{bmatrix} \quad \text{and} \quad \mathbf{h} := \begin{bmatrix} h \\ k \end{bmatrix},$$

where $\mathbf{h}^T$ denotes, as usual, the transpose $[h, k]$ of $\mathbf{h}$. In this set-up, the discriminant Δ of $Q(h, k)$ is precisely equal to the determinant of A. Note also that the condition for nonnegative definiteness in Proposition 3.69, namely, $a \geq 0$, $c \geq 0$, and $ac - b^2 \geq 0$, can be formulated by stating that the *principal minors*[3] of A are nonnegative. On the other hand, the condition for positive definiteness in Remark 3.70, namely, $a > 0$ and $ac - b^2 > 0$, can be formulated by stating that the *leading principal minors* of A are positive. In case $D \subseteq \mathbb{R}^2$ is open and $f : D \to \mathbb{R}$ is such that the first-order and second-order partial derivatives of f exist and are continuous on D, then the 2×2 real symmetric matrix corresponding to the Hessian form of f at $(x_0, y_0) \in D$ is given by

$$\begin{bmatrix} f_{xx}(x_0, y_0) & f_{xy}(x_0, y_0) \\ f_{xy}(x_0, y_0) & f_{yy}(x_0, y_0) \end{bmatrix} = \begin{bmatrix} f_{xx}(x_0, y_0) & f_{xy}(x_0, y_0) \\ f_{yx}(x_0, y_0) & f_{yy}(x_0, y_0) \end{bmatrix}.$$

[3] In general, for any matrix, a **minor** is the determinant of a square submatrix. If the rows and columns chosen to form the submatrix have the same indices, then it is called a **principal minor**; further, if these indices are consecutive, starting from 1, then it is called a **leading principal minor**.

Classically, the matrix on the right is called the **Hessian matrix** of f at (x_0, y_0), and this is, in fact, the reason why we have called the associated binary quadratic form the Hessian form of f at (x_0, y_0). Note also that the discriminant of f at (x_0, y_0) is exactly the determinant of the Hessian matrix of f at (x_0, y_0). ◇

3.5 Functions of Three Variables

In this section we briefly indicate how the theory developed so far in this chapter extends to functions of three (or more) variables. Details are provided only when the extension is not obvious. Along the way, we will also encounter some new concepts and results. For example, the notions of tangent planes and normal lines to surfaces will be introduced here. Graphs of functions of two variables are particular cases of surfaces in 3-space, and we could have discussed these notions in earlier sections. But as we shall see, tangent planes and normal lines can be better understood in the context of functions of three variables. Likewise, it is in the context of functions of three (or more) variables that the question of solving two equations for two of the variables becomes meaningful. We shall see that an answer can be given by an appropriate analogue of the Implicit Function Theorem.

Extensions and Analogues

The notion of partial derivatives of a function f of three variables is defined similarly. There are, of course, three possible partial derivatives, denoted by f_x, f_y, and f_z or by

$$\frac{\partial f}{\partial x}, \ \frac{\partial f}{\partial y}, \ \text{and} \ \frac{\partial f}{\partial z}.$$

There is also a similar notion of the directional derivative $\mathbf{D_u}f$ of f along a unit vector $\mathbf{u} := (u_1, u_2, u_3)$ in $\mathbb{R}^3$. The Bivariate Mean Value Theorem extends easily to the Trivariate Mean Value Theorem.

The higher-order partial derivatives of $f : D \to \mathbb{R}$, where $D \subseteq \mathbb{R}^3$, are defined in a similar manner as in the case of functions of two variables. This time around, we have $3^2 = 9$ possible second-order partial derivatives, namely, $f_{xx}, f_{xy}, f_{xz}, f_{yx}, f_{yy}, f_{yz}, f_{zx}, f_{zy}$, and f_{zz}. The equality of any two whose variables in the subscript differ only in their order holds at those points P_0 where at least one of them exists in $\mathbb{S}_r(P_0)$ for some $r > 0$ and is continuous at P_0. In other words, we have a Mixed Partials Theorem for functions of three variables analogous to the corresponding result for functions of two variables (Proposition 3.14). Moreover, the former can be proved as a consequence of the latter. For example, to prove that $f_{xy}(x_0, y_0, z_0) = f_{yx}(x_0, y_0, z_0)$, it suffices to consider the function of two variables defined by $(x, y) \longmapsto f(x, y, z_0)$ and apply the Mixed Partials Theorem (Proposition 3.14). As a result, if $D \subseteq \mathbb{R}^3$

is open in $\mathbb{R}^3$, then for any $f : D \to \mathbb{R}$, we have equality of any two nth-order partial derivatives of f whose variables in the subscript differ only in their order, provided all partial derivatives of f of order $\leq n$ exist and are continuous on D. In such a case, a typical nth-order derivative of f at $P_0 = (x_0, y_0, z_0)$ may be written as

$$\frac{\partial^n f}{\partial x^p \partial y^q \partial z^r}(P_0) = \frac{\partial^n f}{\partial x^p \partial y^q \partial z^r}(x_0, y_0, z_0),$$

where p, q, r are nonnegative integers with $p + q + r = n$.

An alternative approach to prove the Mixed Partials Theorem for functions of three variables is to use a "Cuboidal Mean Value Theorem." In other words, we first obtain analogues of the Rectangular Rolle's Theorem (Proposition 3.9) and the Rectangular Mean Value Theorem (Proposition 3.11). These analogues can be readily formulated and proved for functions of three or more variables. To this end, it is convenient to use the notation introduced in Remark 1.20 and notice that the quantity $f(b, d) + f(a, c) - f(b, c) - f(a, d)$ appearing in the statement of the Rectangular Mean Value Theorem (Proposition 3.11) can be succinctly expressed as $\triangle_{(a,c)}^{(b,d)} f$. A statement for functions of three variables is given in Exercise 43.

Successive directional derivatives (along the same unit vector) lead us to higher-order directional derivatives of a function of three variables. This notion behaves in the same way as in the case of functions of two variables. We can easily formulate and prove the Trivariate Taylor Theorem in the setting of higher-order directional derivatives, thus obtaining results analogous to Proposition 3.21 and Corollary 3.22.

The notion of differentiability extends easily. Thus, if $D \subseteq \mathbb{R}^3$ and (x_0, y_0, z_0) is an interior point of D, then $f : D \to \mathbb{R}$ is **differentiable** at (x_0, y_0, z_0) if there is $(\alpha_1, \alpha_2, \alpha_3) \in \mathbb{R}^3$ such that

$$\lim_{(h,k,\ell)\to(0,0,0)} \frac{f(x_0 + h, y_0 + k, z_0 + \ell) - f(x_0, y_0, z_0) - \alpha_1 h - \alpha_2 k - \alpha_3 \ell}{\sqrt{h^2 + k^2 + \ell^2}} = 0.$$

In this case, we call the triple $(\alpha_1, \alpha_2, \alpha_3)$ the **total derivative** of f at (x_0, y_0, z_0). If f is differentiable at (x_0, y_0, z_0), then f_x, f_y, and f_z exist at (x_0, y_0, z_0) and are equal to α_1, α_2, and α_3, respectively. In other words, the **gradient** ∇f of f exists at (x_0, y_0, z_0) and

$$\text{the total derivative of } f \text{ at } (x_0, y_0, z_0) \;=\; \nabla f(x_0, y_0, z_0).$$

An analogue of the Increment Lemma (Proposition 3.25) is readily obtained in the above situation. It shows that the differentiability of f at (x_0, y_0, z_0) is equivalent to the existence of a triple (f_1, f_2, f_3) of **increment functions** associated with f and (x_0, y_0, z_0), that is, the existence of functions $f_1, f_2, f_3 : D \to \mathbb{R}$ that are continuous at (x_0, y_0, z_0) and are such that for every $(x, y, z) \in D$, the difference $f(x, y, z) - f(x_0, y_0, z_0)$ is equal to

$$(x - x_0)f_1(x, y, z) + (y - y_0)f_2(x, y, z) + (z - z_0)f_3(x, y, z).$$

If these conditions hold, then we necessarily have $f_1(x_0, y_0, z_0) = f_x(x_0, y_0, z_0)$, $f_2(x_0, y_0, z_0) = f_y(x_0, y_0, z_0)$, and $f_3(x_0, y_0, z_0) = f_z(x_0, y_0, z_0)$. Existence of all the partial derivatives in $\mathbb{S}_r(x_0, y_0, z_0)$ for some $r > 0$ and the continuity of any two of them at (x_0, y_0, z_0) is sufficient for f to be differentiable at (x_0, y_0, z_0). The proof is similar to that of the corresponding result for functions of two variables (Proposition 3.33). Thus, if f_x and f_y are continuous at (x_0, y_0, z_0), then one writes

$$f(x, y, z) - f(x_0, y_0, z_0) = A(x, y, z) + B(y, z) + C(z),$$

where $A(x, y, z) := f(x, y, z) - f(x_0, y, z)$, $B(y, z) := f(x_0, y, z) - f(x_0, y_0, z)$, and $C(z) := f(x_0, y_0, z) - f(x_0, y_0, z_0)$. Next, $f_1, f_2, f_3 : D \to \mathbb{R}$ are defined using A, B, C, and one proceeds along the same lines as in the proof of Proposition 3.33. The necessary conditions for differentiability of f are exactly the same as in the case of functions of two variables. Thus, if $f : D \to \mathbb{R}$ is differentiable at (x_0, y_0, z_0), then f is continuous at (x_0, y_0, z_0), and for any unit vector $\mathbf{u} = (u_1, u_2, u_3)$ in $\mathbb{R}^3$, the directional derivative $\mathbf{D_u}f(x_0, y_0, z_0)$ exists and is equal to $\nabla f(x_0, y_0, z_0) \cdot \mathbf{u}$.

Important results such as the classical versions of the Implicit Function Theorem as well as of Taylor's Theorem, and the Chain Rule remain valid for functions of three (or more) variables, and play a useful role. We have already considered the Trivariate Implicit Function Theorem in Proposition 2.46. However, it is the classical version given below that is most easily remembered and widely used in practice.

Proposition 3.74 (Classical Version of Trivariate Implicit Function Theorem). *Let $D \subseteq \mathbb{R}^3$, $f : D \to \mathbb{R}$ and $(x_0, y_0, z_0) \in D$ be such that f has continuous partial derivatives in $\mathbb{S}_r(x_0, y_0, z_0)$ for some $r > 0$ with $\mathbb{S}_r(x_0, y_0, z_0) \subseteq D$, and $f(x_0, y_0, z_0) = 0$, while $f_z(x_0, y_0, z_0) \neq 0$. Then we can solve the equation $f(x, y, z) = 0$ for z in terms of x and y around (x_0, y_0), that is, there are $\delta > 0$ and a unique continuous function $\zeta : \mathbb{S}_\delta(x_0, y_0) \to \mathbb{R}$ with $\zeta(x_0, y_0) = z_0$ such that $(x, y, \zeta(x, y)) \in \mathbb{S}_r(x_0, y_0, z_0)$ and $f(x, y, \zeta(x, y)) = 0$ for all $(x, y) \in \mathbb{S}_\delta(x_0, y_0)$. Moreover, ζ is differentiable on $\mathbb{S}_\delta(x_0, y_0)$, and for any $(x, y) \in \mathbb{S}_\delta(x_0, y_0)$, we have $f_z(x, y, \zeta(x, y)) \neq 0$ and*

$$\nabla\zeta(x, y) = \left(-\frac{f_x(x, y, \zeta(x, y))}{f_z(x, y, \zeta(x, y))}, \; -\frac{f_y(x, y, \zeta(x, y))}{f_z(x, y, \zeta(x, y))} \right).$$

Proof. This is proved in the same way as in Proposition 3.38: the continuity of partial derivatives is used to verify that the hypothesis of Proposition 2.46 holds. This yields a continuous function ζ that has the desired properties. Moreover, if (f_1, f_2, f_3) is a triple of increment functions associated with f and (x_0, y_0, z_0) and we substitute $z = \zeta(x, y)$, then $(-f_1/f_3, -f_2/f_3)$ becomes a pair of increment functions associated to ζ and (x_0, y_0). Consequently, ζ

is differentiable at (x_0, y_0) and $\nabla\zeta(x_0, y_0)$ is given by the desired formula. Thanks to the continuity of f_z on $\mathbb{S}_r(x_0, y_0, z_0)$, the differentiability of ζ and the formula for $\nabla\zeta$ extends to all of $\mathbb{S}_\delta(x_0, y_0)$. □

A careful analysis of the proof of Proposition 3.74 shows that a slightly weaker hypothesis suffices, namely, instead of requiring the three partial derivatives of f to be continuous, it is enough if f_z and either of f_x and f_y is continuous.

Remark 3.75. As in Remark 3.40, there is a straightforward analogue of the classical version of the Trivariate Implicit Function Theorem, which corresponds to "solving" the equation $f(x, y, z) = 0$ for y in terms of x and z, or for x in terms of y and z, near $P_0 := (x_0, y_0, z_0)$. The key hypothesis would be $f_y(P_0) \neq 0$, or $f_x(P_0) \neq 0$, instead of $f_z(P_0) \neq 0$. One can combine these three results by stating the key hypothesis as $\nabla f(P_0) \neq 0$. Then the conclusion would be that there are $\delta > 0$, $(u_0, v_0) \in \mathbb{R}^2$, and differentiable functions $x, y, z : \mathbb{S}_\delta(u_0, v_0) \to \mathbb{R}$ with $(x(u_0, v_0), y(u_0, v_0), z(u_0, v_0)) = P_0$ such that $(x(u, v), y(u, v), z(u, v)) \in \mathbb{S}_r(x_0, y_0, z_0)$ and $f(x(u, v), y(u, v), z(u, v)) = 0$ for all $(u, v) \in \mathbb{S}_\delta(u_0, v_0)$. Moreover, at any $(u, v) \in \mathbb{S}_\delta(u_0, v_0)$, the vectors (x_u, y_u, z_u) and (x_v, y_v, z_v) are nonzero and are not multiplies of each other, and we have

$$f_x x_u + f_y y_u + f_z z_u = 0 \quad \text{and} \quad f_x x_v + f_y y_v + f_z z_v = 0,$$

where the partial derivatives of x and y are evaluated at (u, v), while the partial derivatives of f are evaluated at $(x(u, v), y(u, v), z(u, v))$. ◇

The classical version of the Trivariate Taylor Theorem is quite analogous to the Bivariate Taylor Theorem with polynomials in three variables entering the picture. The Chain Rule for real-valued functions of three variables can be formulated in various situations analogous to those in Proposition 3.51. For example, if we consider composite functions of the form $g(t) := f(x(t), y(t), z(t))$, then the Chain Rule would tell us that

$$\frac{dg}{dt} = \frac{\partial f}{\partial x}\frac{dx}{dt} + \frac{\partial f}{\partial y}\frac{dy}{dt} + \frac{\partial f}{\partial z}\frac{dz}{dt}.$$

Similarly, if $g(u, v) := f(x(u, v), y(u, v), z(u, v))$, then we will have

$$\frac{\partial g}{\partial u} = \frac{\partial f}{\partial x}\frac{\partial x}{\partial u} + \frac{\partial f}{\partial y}\frac{\partial y}{\partial u} + \frac{\partial f}{\partial z}\frac{\partial z}{\partial u}$$

and

$$\frac{\partial g}{\partial v} = \frac{\partial f}{\partial x}\frac{\partial x}{\partial v} + \frac{\partial f}{\partial y}\frac{\partial y}{\partial v} + \frac{\partial f}{\partial z}\frac{\partial z}{\partial v}.$$

All the equalities above are valid when the derivatives are evaluated at appropriate points that are assumed to be interior points of the domains of the concerned functions, and moreover, all the functions are assumed to be differentiable at these interior points. We leave it to the reader to make a more formal statement of the Chain Rule for functions of three variables, and also to prove it along the same lines as in Proposition 3.51.

As an application of the Chain Rule, we can obtain a natural extension of the classical version of the Trivariate Implicit Function Theorem, which permits us to "solve" two equations in three variables, say $f(x, y, z) = 0$ and $g(x, y, z) = 0$, for y and z in terms of x.

Proposition 3.76. *Let $D \subseteq \mathbb{R}^3$ and let (x_0, y_0, z_0) be an interior point of D. Let $f, g : D \to \mathbb{R}$ have continuous partial derivatives in $\mathbb{S}_r(x_0, y_0, z_0)$ for some $r > 0$ with $\mathbb{S}_r(x_0, y_0, z_0) \subseteq D$. If $f(x_0, y_0, z_0) = 0 = g(x_0, y_0, z_0)$ and*

$$f_y(x_0, y_0, z_0)g_z(x_0, y_0, z_0) - f_z(x_0, y_0, z_0)g_y(x_0, y_0, z_0) \neq 0,$$

then there are $\delta > 0$ and continuous functions $\phi, \psi : (x_0 - \delta, x_0 + \delta) \to \mathbb{R}$ with $\phi(x_0) = y_0$ and $\psi(x_0) = z_0$ such that $(x, \phi(x), \psi(x)) \in \mathbb{S}_r(x_0, y_0, z_0)$ and $f(x, \phi(x), \psi(x)) = 0 = g(x, \phi(x), \psi(x))$ for all $x \in (x_0 - \delta, x_0 + \delta)$. Moreover, ϕ and ψ are differentiable on $(x_0 - \delta, x_0 + \delta)$, and for any $x \in (x_0 - \delta, x_0 + \delta)$, we have $f_y g_z - g_z f_y \neq 0$ at $(x, \phi(x), \psi(x))$ and also

$$\phi'(x) = \frac{f_z g_x - f_x g_z}{f_y g_z - f_z g_y} \quad \text{and} \quad \psi'(x) = \frac{f_x g_y - f_y g_x}{f_y g_z - f_z g_y},$$

where the partial derivatives of f and g are evaluated at $(x, \phi(x), \psi(x))$.

Proof. Let $P_0 := (x_0, y_0, z_0)$. Since $f_y g_z - f_z g_y \neq 0$ at P_0, both $f_z(P_0)$ and $g_z(P_0)$ cannot be zero. Suppose $f_z(P_0) \neq 0$. Since the partial derivatives of f and g are continuous on $\mathbb{S}_r(x_0, y_0, z_0)$, replacing r by a smaller positive value, we may assume that $f_y g_z - f_z g_y \neq 0$ and $f_z \neq 0$ at every point of $\mathbb{S}_r(x_0, y_0, z_0)$. By the Classical Version of the Trivariate Implicit Function Theorem (Proposition 3.74), there are $t > 0$ and $\zeta : \mathbb{S}_t(x_0, y_0) \to \mathbb{R}$ with $\zeta(x_0, y_0) = z_0$ such that $(x, y, \zeta(x, y)) \in \mathbb{S}_r(x_0, y_0, z_0)$ and $f(x, y, \zeta(x, y)) = 0$ for all $(x, y) \in \mathbb{S}_t(x_0, y_0)$. Moreover, ζ is differentiable on $\mathbb{S}_t(x_0, y_0)$ and for any $(x, y) \in \mathbb{S}_t(x_0, y_0)$, we have $\nabla\zeta = (-f_x/f_z, -f_y/f_z)$, where the partial derivatives of f are evaluated at $(x, y, \zeta(x, y))$. Define $h : \mathbb{S}_t(x_0, y_0) \to \mathbb{R}$ by $h(x, y) := g(x, y, \zeta(x, y))$. Now $h(x_0, y_0) = g(P_0) = 0$, and by the Chain Rule for functions of three variables, h is differentiable in $\mathbb{S}_t(x_0, y_0)$ with

$$h_y = g_y + g_z\zeta_y = g_y + g_z(-f_y/f_z) = -(f_y g_z - f_z g_y)/f_z,$$

where the partial derivatives of f and g are evaluated at $(x, y, \zeta(x, y))$. In particular, h_y is continuous and $h_y \neq 0$ on $\mathbb{S}_t(x_0, y_0)$. Hence by the classical version of the Implicit Function Theorem (Proposition 3.38) applied to the function h, there are $\delta > 0$ and $\eta : (x_0 - \delta, x_0 + \delta) \to \mathbb{R}$ with $\eta(x_0) = y_0$ such that $(x, \eta(x)) \in \mathbb{S}_t(x_0, y_0)$ and $h(x, \eta(x)) = 0$ for all $x \in (x_0 - \delta, x_0 + \delta)$. Moreover, η is differentiable and $\eta'(x) = -h_x(x, \eta(x))/h_y(x, \eta(x))$ for all

$x \in (x_0 - \delta, x_0 + \delta)$. Now, let $\phi, \psi : (x_0 - \delta, x_0 + \delta) \to \mathbb{R}$ be defined by $\phi(x) := \eta(x)$ and $\psi(x) := \zeta(x, \eta(x))$. It is readily verified that ϕ and ψ satisfy the desired properties.

The case in which $g_z(P_0) \neq 0$ is proved similarly. □

Remark 3.77. As in Remark 3.75, there is a straightforward analogue of Proposition 3.76, which corresponds to "solving" two equations $f(x, y, z) = 0$ and $g(x, y, z) = 0$ for x and z in terms of y, or for x and y in terms of z, near $P_0 := (x_0, y_0, z_0)$. The key hypothesis would be $f_x g_z - f_z g_x \neq 0$ at P_0, or that $f_x g_y - f_y g_x \neq 0$ at P_0, instead of $f_y g_z - f_z g_y \neq 0$ at P_0. One can combine these three situations to arrive at the following version of Proposition 3.76.

Let $D \subseteq \mathbb{R}^3$ and let $P_0 := (x_0, y_0, z_0)$ be an interior point of D. Let $f, g : D \to \mathbb{R}$ have continuous partial derivatives in $\mathbb{S}_r(P_0)$ for some $r > 0$ with $\mathbb{S}_r(P_0) \subseteq D$. If $f(P_0) = 0 = g(P_0)$ and some 2×2 minor of the 2×3 matrix

$$\begin{pmatrix} f_x(P_0) & f_y(P_0) & f_z(P_0) \\ g_x(P_0) & g_y(P_0) & g_z(P_0) \end{pmatrix}$$

is nonzero [or equivalently, its rows, namely, $\nabla f(P_0)$ and $\nabla g(P_0)$, are not multiples of each other], then there are $\delta > 0$, $t_0 \in \mathbb{R}$, and differentiable functions $x, y, z : (t_0 - \delta, t_0 + \delta) \to \mathbb{R}$ with $(x(t_0), y(t_0), z(t_0)) = P_0$ such that $(x(t), y(t), z(t)) \in \mathbb{S}_r(P_0)$ and $f\,(x(t), y(t), z(t)) = 0 = g\,(x(t), y(t), z(t))$ for all $t \in (t_0 - \delta, t_0 + \delta)$. Moreover, $(x'(t), y'(t), z'(t)) \neq (0, 0, 0)$ and

$$f_x\,(x(t), y(t), z(t))\,x'(t) + f_y\,(x(t), y(t), z(t))\,y'(t) + f_z\,(x(t), y(t), z(t))\,z'(t) = 0$$

as well as

$$g_x\,(x(t), y(t), z(t))\,x'(t) + g_y\,(x(t), y(t), z(t))\,y'(t) + g_z\,(x(t), y(t), z(t))\,z'(t) = 0$$

for all $t \in (t_0 - \delta, t_0 + \delta)$. This unified version may be compared with the unified statement of the classical version of the Implicit Function Theorem for functions of two variables, given in Remark 3.40. In an analogous manner, we can formulate and prove a general version of the Implicit Function Theorem for solving m equations in n variables, where $m, n \in \mathbb{N}$ with $m < n$. The special case $n = 2m$ can be used to prove a general version of the so-called **Inverse Function Theorem** for inverting m functions in m variables. ◇

Tangent Planes and Normal Lines to Surfaces

Let us first review the notion of a tangent line to a curve as discussed in a course on one-variable calculus. Let $D \subseteq \mathbb{R}$ and let c be an interior point of D. If $f : D \to \mathbb{R}$ is differentiable at c, then the line given by

$$y - f(c) = f'(c)(x - c)$$

is called the **tangent line** to the curve $y = f(x)$, $x \in D$, at the point $(c, f(c))$. Often, one extends this notion to parametrically defined curves and implicitly defined curves as follows. (See, for example, pages 114–116 of ACICARA.)

Let C be a parametrically defined curve given by $(x(t), y(t))$, $t \in D$, and let t_0 be an interior point of D such that both x and y are differentiable at t_0 and $(x'(t_0), y'(t_0)) \neq (0, 0)$. Then the **tangent line** to C at the point $(x(t_0), y(t_0))$ is defined to be the line given by

$$\big(y - y(t_0)\big)x'(t_0) = \big(x - x(t_0)\big)y'(t_0).$$

Notice that this line can be represented parametrically by

$$\big(x(t_0),\ y(t_0)\big) + \lambda\big(x'(t_0),\ y'(t_0)\big), \quad \text{where } \lambda \text{ varies over } \mathbb{R}.$$

In other words, this is the line passing through $(x(t_0), y(t_0))$ and having the direction of the vector $(x'(t_0), y'(t_0))$. This is, in fact, the reason why in the context of paths, we called $(x'(t_0), y'(t_0))$ the **tangent vector** to C at $(x(t_0), y(t_0))$ when we introduced saddle points in Chapter 1. The line in $\mathbb{R}^2$ passing through $(x(t_0), y(t_0))$ and perpendicular to the tangent line to C at $(x(t_0), y(t_0))$ is called the **normal line** to C at $(x(t_0), y(t_0))$. It may be noted that this normal line is parametrically given by

$$\big(x(t_0),\ y(t_0)\big) + \lambda\big(y'(t_0),\ -x'(t_0)\big), \quad \text{where } \lambda \text{ varies over } \mathbb{R}.$$

Now consider an implicitly defined curve, that is, a curve in $\mathbb{R}^2$ defined by an equation of the form $F(x, y) = 0$, $(x, y) \in E$. The notion of a tangent line to such a curve is defined in one-variable calculus, if at all, in a rather ad hoc way. Indeed, one performs "implicit differentiation" with respect to x or with respect to y, so as to arrive at an equation of the form

$$P(x, y) + Q(x, y)\frac{dy}{dx} = 0 \quad \text{or} \quad R(x, y) + S(x, y)\frac{dx}{dy} = 0.$$

Now, if (x_0, y_0) is a point on the curve, so that $(x_0, y_0) \in E$ and $F(x_0, y_0) = 0$, then the **tangent line** to the curve at (x_0, y_0) is the line given by

$$y - y_0 = -\frac{P(x_0, y_0)}{Q(x_0, y_0)}(x - x_0) \quad \text{or by} \quad x - x_0 = -\frac{R(x_0, y_0)}{S(x_0, y_0)}(y - y_0)$$

according as $Q(x_0, y_0) \neq 0$ or $S(x_0, y_0) \neq 0$. This approach can now be streamlined in the light of the theory of functions of two variables. First, the process of "implicit differentiation" is readily justified by the classical version of the Implicit Function Theorem (Proposition 3.38). Next, the condition $Q(x_0, y_0) \neq 0$ or $S(x_0, y_0) \neq 0$ corresponds to the more intrinsic condition $F_y(x_0, y_0) \neq 0$ or $F_x(x_0, y_0) \neq 0$, or equivalently, $\nabla F(x_0, y_0) \neq (0, 0)$. This is, of course, the condition needed for the Implicit Function Theorem to be applicable. When this condition is satisfied, the **tangent line** to the curve $F(x, y) = 0$, $(x, y) \in E$, at the point (x_0, y_0) is the line in $\mathbb{R}^2$ given by

$$F_x(x_0, y_0)(x - x_0) + F_y(x_0, y_0)(y - y_0) = 0,$$

whereas the **normal line** to this curve at (x_0, y_0) is the line in $\mathbb{R}^2$ given by

$$F_y(x_0, y_0)(x - x_0) - F_x(x_0, y_0)(y - y_0) = 0.$$

In a similar manner, the notion of a tangent plane to the graph of a function of two variables can be obtained as a special case of the notion of tangent plane to an implicitly defined surface given by $f(x, y, z) = 0$, $(x, y, z) \in D$, where $D \subseteq \mathbb{R}^3$ and $f : D \to \mathbb{R}$. More precisely, if (x_0, y_0, z_0) is an interior point of D such that f is differentiable at (x_0, y_0, z_0) and $\nabla f(x_0, y_0, z_0) \neq (0, 0, 0)$, then we define the **tangent plane** to the surface $f(x, y, z) = 0$, $(x, y, z) \in D$, at the point (x_0, y_0, z_0) to be the plane in $\mathbb{R}^3$ given by

$$f_x(x_0, y_0, z_0)(x - x_0) + f_y(x_0, y_0, z_0)(y - y_0) + f_z(x_0, y_0, z_0)(z - z_0) = 0.$$

In particular, if $E \subseteq \mathbb{R}^2$ and $F : E \to \mathbb{R}$ is differentiable at an interior point (x_0, y_0) of E, then the **tangent plane** to the graph of F at the point $(x_0, y_0, F(x_0, y_0))$ is the tangent plane to the surface defined by $z - F(x, y) = 0$, $(x, y) \in E$, namely, the plane in $\mathbb{R}^3$ given by

$$z - F(x_0, y_0) = F_x(x_0, y_0)(x - x_0) + F_y(x_0, y_0)(y - y_0).$$

These notions extend easily to the case of functions of n variables for any $n > 1$. Thus, associated to a function of n variables and an interior point of the domain at which it is differentiable with nonzero gradient, there will be a **tangent hyperplane** in $\mathbb{R}^n$ given by the vanishing of one linear equation in the n variables.

The notion of a tangent vector to a parametrically defined curve in $\mathbb{R}^2$ readily extends to parametrically defined curves in $\mathbb{R}^3$. Thus, if C is a parametrically defined curve in $\mathbb{R}^3$ given by $(x(t), y(t), z(t))$, $t \in D$, where $D \subseteq \mathbb{R}$, and if t_0 is an interior point of D such that $x, y, z : D \to \mathbb{R}$ are differentiable at t_0 and if $(x'(t_0), y'(t_0), z'(t_0)) \neq (0, 0, 0)$, then $(x'(t_0), y'(t_0), z'(t_0))$ is called the **tangent vector** to C at the point $P_0 := (x(t_0), y(t_0), z(t_0))$. The line in $\mathbb{R}^3$ given parametrically by

$$P_0 + \lambda(x'(t_0), y'(t_0), z'(t_0)), \quad \text{where } \lambda \text{ varies over } \mathbb{R},$$

is called the **tangent line** to C at P_0.

It may be interesting to note that the tangent lines at a point to curves lying on a surface are contained in the tangent plane to the surface at that point. More precisely, suppose S is an implicitly defined surface in $\mathbb{R}^3$ given by $f(x, y, z) = 0$, $(x, y, z) \in D$, and C is a parametrically defined curve given by $(x(t), y(t), z(t))$, $t \in D$, such that C lies in D, that is, $(x(t), y(t), z(t)) \in D$ for all $t \in D$. Let t_0 be an interior point of D and let $(x_0, y_0, z_0) := (x(t_0), y(t_0), z(t_0))$. Assume that the tangent vector $\mathbf{v}_0 := (x'(t_0), y'(t_0), z'(t_0))$ to C at $P_0 := (x_0, y_0, z_0)$ is defined. Now, if P_0

is an interior point of D and f is differentiable at P_0, then the claim is that $P_0 + \lambda \mathbf{v}_0$ is on the tangent plane to S at P_0 for every $\lambda \in \mathbb{R}$. To see this, note that since C lies in D, we have $f(x(t), y(t), z(t)) = 0$ for all $t \in D$. Hence by the Chain Rule,

$$f_x(P_0)x'(t_0) + f_y(P_0)y'(t_0) + f_z(P_0)z'(t_0) = 0, \text{ that is, } \nabla f(P_0) \cdot \mathbf{v}_0 = 0.$$

It follows that if we write $P_0 + \lambda \mathbf{v}_0 = (x, y, z)$, then we have

$$f_x(x_0, y_0, z_0)(x - x_0) + f_y(x_0, y_0, z_0)(y - y_0) + f_z(x_0, y_0, z_0)(z - z_0) = 0.$$

This proves the claim. In case $\nabla f(P_0) \neq 0$, then the line in $\mathbb{R}^3$ given parametrically by

$$P_0 + \lambda \nabla f(P_0), \quad \text{where } \lambda \text{ varies over } \mathbb{R},$$

is called the **normal line** to S at the point P_0, and either of the unit vectors $\pm \nabla f(P_0)/|\nabla f(P_0)|$ is called a **normal vector** to S at P_0. Note that in case $f_x(P_0)$, $f_y(P_0)$, and $f_z(P_0)$ are all nonzero, then the normal line can also be described by the equations

$$\frac{x - x_0}{f_x(P_0)} = \frac{y - y_0}{f_y(P_0)} = \frac{z - z_0}{f_z(P_0)}.$$

Let S be a parametrically defined surface given by $\big(x(u,v),\, y(u,v),\, z(u,v)\big)$, $(u, v) \in E$, where $E \subseteq \mathbb{R}^2$ and $x, y, z : E \to \mathbb{R}$ are differentiable at an interior point $Q_0 = (u_0, v_0)$ of E. Assume that the vectors $(x_u(Q_0),\, y_u(Q_0),\, z_u(Q_0))$ and $(x_v(Q_0),\, y_v(Q_0),\, z_v(Q_0))$ are nonzero and are not multiples of each other. Equivalently, assume that $\mathbf{n}_0 := \big(y_u z_v - z_u y_v,\; z_u x_v - x_u z_v,\; x_u y_v - y_u x_v\big)$, where all the partials are evaluated at Q_0, is a nonzero vector in $\mathbb{R}^3$. Let $P_0 := (x(Q_0),\, y(Q_0),\, z(Q_0))$. Then the **tangent plane** to S at P_0 is defined to be the plane in $\mathbb{R}^3$ given parametrically by

$$P_0 + \lambda\big(x_u(Q_0),\, y_u(Q_0),\, z_u(Q_0)\big) + \mu\big(x_v(Q_0),\, y_v(Q_0),\, z_v(Q_0)\big),\ (\lambda, \mu) \in \mathbb{R}^2.$$

Moreover, the line in $\mathbb{R}^3$ given parametrically by

$$P_0 + \lambda \mathbf{n_0}, \quad \text{where } \lambda \text{ varies over } \mathbb{R},$$

is called the **normal line** to S at P_0.

Examples 3.78. (i) Consider $f : \mathbb{R}^3 \to \mathbb{R}$ defined by $f(x, y, z) := x^2 + y^2 + z^2 - 1$. Then the tangent plane to the surface given by $f(x, y, z) = 0$ at the point $(0, 0, 1)$ is given by $0(x - 0) + 0(y - 0) + 2(z - 1) = 0$, that is, the plane given by $z = 1$. The normal line at this point is given parametrically by $(0, 0, 1 + 2\lambda)$ as λ varies over $\mathbb{R}$, that is, by the z-axis. Notice that the surface here is the unit sphere and the point $(0, 0, 1)$ is its north pole.

(ii) Consider $f : \mathbb{R}^3 \to \mathbb{R}$ defined by $f(x, y, z) := e^x + \sin y - \cos z$. Then the surface given by $f(x, y, z) = 0$ passes through the origin, and the tangent plane at the origin is given by $x + y = 0$. The normal line at the origin is given parametrically by $(\lambda, \lambda, 0)$, as λ varies over $\mathbb{R}$, that is, by the intersection of the planes given by $x = y$ and $z = 0$.

(iii) Consider the cylinder S in $\mathbb{R}^3$ given parametrically by $(\cos u, \sin u, v)$, $(u, v) \in \mathbb{R}^2$. Let $(u_0, v_0) \in \mathbb{R}^2$ and let $P_0 := (\cos u_0, \sin u_0, v_0)$ be the corresponding point on S. Then the tangent plane to S at P_0 is given by $P_0 + \lambda(-\sin u_0, \cos u_0, 0) + \mu(0, 0, 1)$, $(\lambda, \mu) \in \mathbb{R}^2$, whereas the normal line to S at P_0 is given by $P_0 + \lambda(\cos u_0, \sin u_0, 0)$, $\lambda \in \mathbb{R}$. $\diamond$

Convexity and Ternary Quadratic Forms

The notions of convexity and concavity were already defined in the context of real-valued functions of n variables. There is no difficulty in extending the notion of monotonicity of the gradient and using it to characterize convexity or concavity in exactly the same way as in Section 3.3. On the other hand, the characterizations in terms of the second-order partial derivatives call for some explanation. To begin with, instead of a binary quadratic form one has to consider a **ternary quadratic form**, that is, a homogeneous polynomial of total degree 2 in three variables. If the variables are denoted by h, k, and ℓ, then a ternary quadratic form looks like

$$Q(h, k, \ell) := ah^2 + ck^2 + r\ell^2 + 2bhk + 2qk\ell + 2p\ell h,$$

where a, b, c, p, q, r are real numbers. In *matrix notation*, $Q(h, k, \ell)$ can be expressed as the matrix product

$$\mathbf{h}^T A\mathbf{h}, \quad \text{where} \quad A := \begin{bmatrix} a & b & p \\ b & c & q \\ p & q & r \end{bmatrix} \quad \text{and} \quad \mathbf{h} := \begin{bmatrix} h \\ k \\ \ell \end{bmatrix},$$

where $\mathbf{h}^T$ denotes the transpose $[h, k, \ell]$ of $\mathbf{h}$. We may refer to A as the 3×3 symmetric matrix corresponding to $Q(h, k, \ell)$.

If $D \subseteq \mathbb{R}^3$ is open and $f : D \to \mathbb{R}$ has the property that the first-order and second-order order partial derivatives of f exist and are continuous on D, then for any $(x_0, y_0, z_0) \in D$, the associated ternary quadratic form

$$h^2\frac{\partial^2 f}{\partial x^2} + k^2\frac{\partial^2 f}{\partial y^2} + \ell^2\frac{\partial^2 f}{\partial z^2} + 2hk\frac{\partial^2 f}{\partial x \partial y} + 2k\ell\frac{\partial^2 f}{\partial y \partial z} + 2\ell h\frac{\partial^2 f}{\partial z \partial x},$$

where all the second-order partial derivatives are evaluated at (x_0, y_0, z_0), is called the **Hessian form** of f at (x_0, y_0, z_0). The corresponding 3×3 symmetric matrix is called the **Hessian matrix** of f at (x_0, y_0, z_0).

In general, a ternary quadratic form $Q(h, k, \ell)$ or the corresponding 3×3 symmetric matrix A is said to be **nonnegative definite** if $Q(s, t, u) \geq 0$ for all $(s, t, u) \in \mathbb{R}^3$. For $D \subseteq \mathbb{R}^3$ and $f : D \to \mathbb{R}$ as above, we say that the Hessian form of f is **nonnegative definite** on D if the Hessian form of f at (x_0, y_0, z_0) is nonnegative definite for every $(x_0, y_0, z_0) \in D$. When D is convex (and open), the convexity of f on D is equivalent to the condition that the Hessian form of f is nonnegative definite on D. This is proved in exactly the same way as in Proposition 3.67.

The analogue of Proposition 3.69 that gives an algebraic characterization of nonnegative definiteness is far from obvious, and for this reason we give a complete statement and proof below.

Proposition 3.79. *Let $Q(h, k, \ell) := ah^2 + ck^2 + r\ell^2 + 2bhk + 2qk\ell + 2p\ell h$ be a ternary quadratic form in the variables h, k, and ℓ with coefficients a, b, c, p, q, r in $\mathbb{R}$. Let*

$$\Delta := \begin{vmatrix} a & b & p \\ b & c & q \\ p & q & r \end{vmatrix} := p(bq - cp) + q(bp - aq) + r(ac - b^2)$$

denote the determinant of the corresponding 3×3 matrix. Then

$$Q(h, k, \ell) \text{ is nonnegative definite}$$
$$\iff a \geq 0,\ c \geq 0,\ r \geq 0,\ ac - b^2 \geq 0,\ cr - q^2 \geq 0,\ ar - p^2 \geq 0, \text{ and } \Delta \geq 0.$$

Proof. Suppose $Q(h, k, \ell)$ is nonnegative definite. Then the binary quadratic forms $Q(h, k, 0)$, $Q(h, 0, \ell)$, and $Q(0, k, \ell)$ are nonnegative definite. Hence, by Proposition 3.69, each of a, c, r, $ac - b^2$, $cr - q^2$, and $ar - p^2$ is nonnegative. Further, we observe that $Q(bq - cp,\ bp - aq,\ ac - b^2) = (ac - b^2)\Delta$. Hence if $ac - b^2 \neq 0$, then $\Delta \geq 0$. By permuting the variables h, k, and ℓ cyclically, we obtain $Q(cr - q^2,\ pq - br,\ bq - cp) = (cr - q^2)\Delta$ and $Q(pq - br,\ ar - p^2,\ bp - aq) = (ar - p^2)\Delta$. Hence if $cr - q^2 \neq 0$ or if $ar - p^2 \neq 0$, then $\Delta \geq 0$. Finally, suppose $ac - b^2 = cr - q^2 = ar - p^2 = 0$. Now, if $a = 0$, then we must have $b = p = 0$. Similarly, if $c = 0$, then $b = q = 0$, while if $r = 0$, then $p = q = 0$. It follows that if $acr = 0$, then $\Delta = 0$. Next, suppose $acr \neq 0$. Consider $Q(b, -a, \ell) = 2\ell(bp - aq) + r\ell^2$. Since $Q(h, k, \ell)$ is nonnegative definite, we see that $2(bp - aq) + r\ell \geq 0$ if $\ell > 0$ and $2(bp - aq) + r\ell \leq 0$ if $\ell < 0$. Upon letting $\ell \to 0^+$, we see that $(bp - aq) \geq 0$, and upon letting $\ell \to 0^-$, we see that $(bp - aq) \leq 0$. Consequently, $bp - aq = 0$, that is, $aq = bp$. Hence $abq = b^2 p = acp$, and therefore $bq = cp$. Thus $bq - cp = 0$, $bp - aq = 0$, and $ac - b^2 = 0$. It follows that $\Delta = 0$.

Conversely, suppose each of a, c, r, $ac - b^2$, $ar - p^2$, $cr - q^2$, and Δ is nonnegative. In case $a = 0$, then the inequalities $ac - b^2 \geq 0$ and $ar - p^2 \geq 0$ imply that $b = 0$ and $p = 0$. Thus, in this case, $Q(h, k, \ell) = ck^2 + 2qk\ell + r\ell^2$, and this is nonnegative definite by Proposition 3.69. Similarly, if $c = 0$, then $b = q = 0$, while if $r = 0$, then $p = q = 0$, and in either of these cases, $Q(h, k, \ell)$ is nonnegative definite by Proposition 3.69. Suppose $a > 0$, $c > 0$, and $r > 0$. If $ac - b^2 > 0$, then the identity

$$a(ac-b^2)Q(h,k,\ell) = (ac-b^2)(ah+bk+p\ell)^2+[(ac-b^2)k+(aq-bp)\ell]^2+a\Delta\ell^2$$

implies that $Q(h,k,\ell)$ is nonnegative definite. Similarly, if $cr - q^2 > 0$, then

$$c(cr-q^2)Q(h,k,\ell) = (cr-q^2)(bh+ck+q\ell)^2+[(cr-q^2)\ell+(cp-bq)h]^2+c\Delta h^2$$

implies that $Q(h,k,\ell)$ is nonnegative definite, whereas if $ar - p^2 > 0$, then

$$r(ar-p^2)Q(h,k,\ell) = (ar-p^2)(ph+qk+r\ell)^2+[(ar-p^2)\ell+(br-pq)k]^2+r\Delta k^2$$

implies that $Q(h,k,\ell)$ is nonnegative definite. Finally, suppose $ac - b^2 = ar - p^2 = cr - q^2 = 0$. Then $bpq \neq 0$, because a, c, and r are positive. Moreover, $b^2p^2 = (ac)(ar) = a^2(cr) = a^2q^2$. Hence $bp = \pm aq$. On the other hand, $\Delta = 2q(bp - aq)$, and so if $bp = -aq$, then $\Delta = -4aq^2 < 0$, which is a contradiction. It follows that $bp = aq$, and as a consequence, $aQ(h,k,\ell) = (ah + bk + p\ell)^2$, which implies that $Q(h,k,\ell)$ is nonnegative definite. □

Remark 3.80. For the positive definiteness of a ternary quadratic form $Q(h,k,\ell) := ah^2 + ck^2 + r\ell^2 + 2bhk + 2qk\ell + 2p\ell h$, we have the following characterization:

$$Q(h,k,\ell) \text{ is positive definite } \iff a > 0, \ ac - b^2 > 0, \text{ and } \Delta > 0,$$

where Δ is the determinant of the 3×3 symmetric matrix A corresponding to $Q(h,k,\ell)$. The proof is similar to that of Proposition 3.79. In fact, it is much simpler. It may be noted that as in the case of binary quadratic forms, the nonnegative definiteness of $Q(h,k,\ell)$ is characterized by the nonnegativity of all the principal minors of A, while the positive definiteness of $Q(h,k,\ell)$ is characterized by the positivity of all the leading principal minors of A. This holds, in general, for quadratic forms in any number of variables. ◇

As a consequence of Proposition 3.79, we obtain a necessary and sufficient condition for the convexity of a function of three variables with continuous first-order and second-order partial derivatives. Also, we can obtain a sufficient condition for strict convexity. These results are analogous to those in Proposition 3.71 and Remark 3.72. We leave the precise formulation and proof to the reader.

Notes and Comments

The notion of differentiability of functions of two (or more) variables is rather subtle, and it is always a dilemma to decide how soon it should be introduced. We have chosen to work exclusively with partial derivatives and directional derivatives at an initial stage. Moreover, we have introduced higher-order directional derivatives in a manner analogous to higher-order partial derivatives.

This facilitates a compact formulation of Taylor's theorem that is very similar to the corresponding one-variable result. The proofs given here of the mean value theorem and Taylor's theorem do not use the Increment Lemma or the Chain Rule. Thus they are somewhat different from the standard proofs of these results. (See Exercise 38.) The notion of differentiability appears later, and the treatment here is elementary in so far as the total derivative is defined to be a vector (the gradient vector) rather than a linear map. The Increment Lemma (Proposition 3.25), which gives a useful characterization of differentiability, is used extensively in the sequel. All this is analogous to the Carathéodory Lemma (Fact 3.24) and its use in Chapter 4 of ACICARA.

Notions of monotonicity and bimonotonicity were defined in Chapter 1 for functions of two (or more) variables, and we have related them here to partial derivatives. We also obtain analogues of the criterion in one-variable calculus that shows that a function with a bounded derivative is of bounded variation. Here we use a "rectangular" version of the mean value theorem that appears, for example, in Rudin [48, Theorem 9.40]. The notions of convexity and concavity extend readily to functions of several variables, and one can obtain criteria in terms of derivatives in an analogous manner. Such material is often found in books on convex analysis but seldom in texts on multivariable calculus. Our treatment of this topic is partly influenced by the book of Roberts and Varberg [47]. In this context, we have also included an algebraic characterization of nonnegative definiteness of quadratic forms in two or three variables. This is usually proved in books on linear algebra with the help of eigenvalues by first proving a characterization of positive definiteness. We have avoided eigenvalues altogether and given a direct proof instead. In effect, we use here an explicit form of the so-called Lagrange–Beltrami identity. (See, for example, Section 6 in Chapter 2 of [6] and the article [23].) The case of binary quadratic forms is classical and goes back to Gauss's 1801 treatise [21].

Since we have restricted almost exclusively to real-valued functions, some important topics have not been covered. For example, we have not discussed higher-order differentiability and the notion of higher-order total derivatives. For a similar reason, we have not included details about the Inverse Function Theorem, except the special case of $n = 1$ *(Proposition 3.41) and a fleeting mention of the general case (Remark 3.77). For such topics and more, see, for example, the books of Courant and John [12] and Rudin [48]. For a more comprehensive treatment, see the little book of Spivak [54] and the classic of de Rham [14], which is now available in English.*

A notable exception to the development of multivariable calculus along the lines of one-variable calculus is the absence of any analogue of L'Hôpital's rule. This is mainly because the total derivative of a function of n *variables is not a real-valued function, unless* $n = 1$*. Thus quotients of real-valued functions of two or more variables and the quotients of their derivatives are birds of a different feather! Moreover, as we have pointed out in Chapter 1, there is no reasonable notion of division in* $\mathbb{R}^n$ *for* $n > 1$*.*

Exercises

Part A

1. Consider $f : \mathbb{R}^2 \to \mathbb{R}$ defined by $f(0,0) := 0$ and for $(x,y) \neq (0,0)$, by one of the following. In each case, determine whether the partial derivatives $f_x(0,0)$ and $f_y(0,0)$ exist. If they do, then find them.
(i) $\dfrac{xy^2}{x^2+y^2}$, (ii) $\dfrac{xy^2}{x^2+y^4}$, (iii) $\dfrac{x^3y}{x^6+y^2}$, (iv) $\dfrac{x^2}{x^2+y^2}$,
(v) $xy\ln(x^2+y^2)$, (vi) $\dfrac{x^3}{x^2+y^2}$, (vii) $\dfrac{x^4y}{x^2+y^2}$,
(viii) $\dfrac{x^3y-xy^3}{x^2+y^2}$, (ix) $\dfrac{\sin(x+y)}{|x|+|y|}$, (x) $\dfrac{\sin^2(x+y)}{|x|+|y|}$.
2. Consider $f : \mathbb{R}^2 \to \mathbb{R}$ defined by $f(x,y) := \big||x|-|y|\big| - |x| - |y|$. Determine whether (i) f is continuous at $(0,0)$, (ii) the partial derivatives $f_x(0,0)$ and $f_y(0,0)$ exist, and (iii) the directional derivative $\mathbf{D_u}f(0,0)$ exists. Is f differentiable at $(0,0)$? Justify your answer.
3. Let $D \subseteq \mathbb{R}^2$ be such that $(y,x) \in D$ whenever $(x,y) \in D$. Let $f,\ g : D \to \mathbb{R}$ satisfy $f(x,y) = g(y,x)$ for all $(x,y) \in D$. Given any $(x_0,y_0) \in D$, show that (i) if $f_x(x_0,y_0)$ exists, then $g_y(y_0,x_0)$ exists and equals $f_x(x_0,y_0)$, and (ii) if $f_y(x_0,y_0)$ exists, then $g_x(y_0,x_0)$ exists and equals $f_y(x_0,y_0)$.
4. Consider $f : \mathbb{R}^2 \to \mathbb{R}$ defined by $f(0,0) := 0$ and
$$f(x,y) := \begin{cases} x\sin(1/x) + y\sin(1/y) & \text{if } x \neq 0 \text{ and } y \neq 0, \\ x\sin 1/x & \text{if } x \neq 0 \text{ and } y = 0, \\ y \sin 1/y & \text{if } y \neq 0 \text{ and } x = 0. \end{cases}$$
Show that none of the partial derivatives of f exist at $(0,0)$ although f is continuous at $(0,0)$.
5. Let $f : \mathbb{R}^2 \to \mathbb{R}$ be defined by $f(x,y) := 0$ if $xy = 0$, and $f(x,y) := 1$ otherwise. Show that f is not continuous at $(0,0)$ although both the partial derivatives of f exist at $(0,0)$.
6. Let $f : \mathbb{R}^2 \to \mathbb{R}$ be defined by $f(x,y) := x^2 + y^2$ if x and y are both rational, and $f(x,y) := 0$ otherwise. Determine the points of $\mathbb{R}^2$ at which (i) f_x exists, (ii) f_y exists.
7. Consider $f : \mathbb{R}^2 \to \mathbb{R}$ defined by $f(0,0) := 0$ and for $(x,y) \neq (0,0)$, by one of the following. In each case, determine whether the directional derivative $\mathbf{D_u}f(0,0)$ exists for a unit vector $\mathbf{u}$ in $\mathbb{R}^2$. If it does, then check whether $\mathbf{D_u}f(0,0) = \nabla f(0,0) \cdot \mathbf{u}$ for a unit vector $\mathbf{u}$ in $\mathbb{R}^2$. Finally, determine whether f is differentiable at $(0,0)$.
(i) $\dfrac{x^2y}{x^2+y^2}$, (ii) $xy\dfrac{x^2-y^2}{x^2+y^2}$, (iii) $\dfrac{x^3}{x^2+y^2}$, (iv) $\dfrac{xy^2}{x^4+y^2}$,
(v) $\dfrac{x^5}{x^4+y^2}$, (vi) $\ln\left(x^2+y^2\right)$, (vii) $xy\ln\left(x^2+y^2\right)$.

8. Consider $f : \mathbb{R}^2 \to \mathbb{R}$ defined by $f(x, y) := (y/|y|)\sqrt{x^2 + y^2}$ if $y \neq 0$, and $f(x, y) := 0$ if $y = 0$. Show that f is continuous at $(0, 0)$, both $f_x(0, 0)$ and $f_y(0, 0)$ exist, $\mathbf{D_u}f(0, 0)$ exists for every unit vector $\mathbf{u}$ in $\mathbb{R}^2$, but f is not differentiable at $(0, 0)$.
9. Assume that $f : \mathbb{R}^2 \to \mathbb{R}$ is such that f_x and f_y exist in $\mathbb{S}_r(1, 2)$ for some $r > 0$ and are continuous at $(1, 2)$. If the directional derivative of f at $(1, 2)$ in the direction toward $(2, 3)$ is $2\sqrt{2}$ and in the direction toward $(1, 0)$ is -3, then find $f_x(1, 2)$, $f_y(1, 2)$, and the directional derivative of f at $(1, 2)$ in the direction toward $(4, 6)$.
10. Starting from $(1, 1)$, in which direction should one travel in order to obtain the most rapid rate of decrease of the function $f : \mathbb{R}^2 \to \mathbb{R}$ defined by $f(x, y) := (x + y - 2)^2 + (3x - y - 6)^2$?
11. About how much will the function $f(x, y) := \ln \sqrt{x^2 + y^2}$ change if the point (x, y) is moved from $(3, 4)$ a distance 0.1 unit straight toward $(3, 6)$?
12. Consider $f : \mathbb{R}^2 \to \mathbb{R}$ defined by $f(x, y) := (x + y)/\sqrt{2}$ if $x = y$, and $f(x, y) := 0$ otherwise. Show that $f_x(0, 0) = f_y(0, 0) = 0$ and $\mathbf{D_u}f(0, 0) = 1$, where $\mathbf{u} = \left(1/\sqrt{2},\ 1/\sqrt{2}\right)$. Deduce that f is not differentiable at $(0, 0)$.
13. Let m and n be positive integers. Consider $f : \mathbb{R}^2 \to \mathbb{R}$ defined by $f(x, y) := 0$ if $x = y$ and $f(x, y) := (x^m + y^n)/(x - y)$ if $x \neq y$. Show that f is discontinuous at $(0, 0)$. (Hint: Consider the equation $x^m + y^n - (x - y) = 0$, which defines y implicitly as a function of x.)
14. Let $f(x, y) := x^2 + 2xy$ for $(x, y) \in \mathbb{R}^2$ and $g(r, \theta) := f(r\cos\theta,\ r\sin\theta)$ for $(r, \theta) \in \mathbb{R}^2$. Determine the partial derivatives g_r and g_θ.
15. Let $f(x, y) := e^x + xy^2$ for $(x, y) \in \mathbb{R}^2$ and $g(u, v) := f(u + v,\ e^{u+v})$ for $(u, v) \in \mathbb{R}^2$. Determine the partial derivatives g_u and g_v.
16. Let D and E be open subsets of $\mathbb{R}^2$ and let $f : D \to \mathbb{R}$ and $x, y : E \to \mathbb{R}$ be such that $(x(u, v), y(u, v)) \in D$ for all $(u, v) \in E$. Define $g : D \to \mathbb{R}$ by $g(u, v) := f\left(x(u, v), y(u, v)\right)$. If the second-order partial derivatives of f as well as x and y exist and are continuous, then show that the same holds for g. Further, show that

$$g_{uu} = f_{xx}x_u^2 + 2f_{xy}x_u y_u + f_{yy}y_u^2 + f_x x_{uu} + f_y y_{uu},$$
$$g_{vv} = f_{xx}x_v^2 + 2f_{xy}x_v y_v + f_{yy}y_v^2 + f_x x_{vv} + f_y y_{vv},$$

and

$$g_{uv} = g_{vu} = f_{xx}\left(x_u x_v + y_u y_v\right) + f_{xy}\left(x_u y_v + x_v y_u\right) + f_x x_{uv} + f_y y_{uv}.$$

17. Given any nonnegative integer n, determine the nth Taylor polynomials around $(0, 0)$ of $f : \mathbb{S}_{1/2}(0, 0) \to \mathbb{R}$ defined by each of the following.

$$\frac{y - x}{(1 - x)(1 - y)}, \quad \sin x + \cos y, \quad \frac{1}{(1 - x)(1 - y)}, \quad (\sin x)(\cos y),$$
$$\sin(x + y), \quad \cos(x + y), \quad \ln(1 + xy), \quad (1 + xy)^r, \quad (1 - xy)^r,$$

where r denotes a rational number.

18. Let $f, g : [0,1] \times [0,1] \to \mathbb{R}$ be defined by $f(x,y) := \cos\left(\frac{\pi}{2}(x+y)\right)$ and $g(x,y) := (x+y-1)^3$. Show that f is monotonically decreasing, while g is monotonically increasing. Also show that neither f nor g is bimonotonic.
19. Find the equations of the tangent plane and the normal line to the surface $z = x^2 + y^2 - 2xy + 3y - x + 4$ at the point $(2, -3, 18)$.
20. Let $f : \mathbb{R}^2 \to \mathbb{R}$ be defined by $f(0,0) := 0$ and $f(x,y) := |x|y/\sqrt{x^2+y^2}$ for $(x,y) \neq (0,0)$. Let C_1, C_2, and C_3 be the curves in $\mathbb{R}^3$ given, respectively, by $(x_1(t),\, y_1(t),\, z_1(t)) := (t,\, 0,\, 0)$, $(x_2(t),\, y_2(t),\, z_2(t)) := (0,\, t,\, 0)$, and $(x_1(t),\, y_1(t),\, z_1(t)) := (t,\, t,\, t)$. Show that each of these three curves passes through $(0,0,0)$ and lies on the surface given by $z = f(x,y)$. Also, show that the tangent vector to C_i at $(0,0,0)$ is defined for each $i = 1,2,3$, but these tangent vectors are not coplanar.
21. Find equations of the line tangent to the curve of intersection of the two surfaces $x^2 + y^2 = 4$ and $z = x^2 + y^2$ at $(\sqrt{2}, \sqrt{2}, 4)$.
22. Let $F(x,y,z) := x^2 + 2xy - y^2 + z^2$ for $(x,y,z) \in \mathbb{R}^3$. Find the gradient of F at $(1,-1,3)$ and the equations of the tangent plane and the normal line to the surface $F(x,y,z) = 7$ at $(1,-1,3)$.
23. Find a constant c such that at any point of intersection of the two spheres $(x-c)^2 + y^2 + z^2 = 3$ and $x^2 + (y-1)^2 + z^2 = 1$, the corresponding tangent planes will be perpendicular to each other.
24. Find $\mathbf{D}_{\mathbf{u}}F(2,2,1)$, where $F(x,y,z) := 3x - 5y + 2z$ for $(x,y,z) \in \mathbb{R}^3$ and $\mathbf{u}$ is in the direction of the outward normal to the sphere $x^2 + y^2 + z^2 = 9$ at $(2,2,1)$.
25. Given $w := z\tan^{-1}(x/y)$ for $(x,y) \in \mathbb{R}^2$, $y \neq 0$, find $\frac{\partial^2 w}{\partial x^2} + \frac{\partial^2 w}{\partial y^2} + \frac{\partial^2 w}{\partial z^2}$.
26. Given $z = f(x,y)$, $x := u+v$, and $y := u-v$, show that $\frac{\partial^2 z}{\partial u \partial v} = \frac{\partial^2 z}{\partial x^2} - \frac{\partial^2 z}{\partial y^2}$.
27. Given $\sin(x+y) + \sin(y+z) = 1$ for $(x,y,z) \in \mathbb{R}^3$, find $\frac{\partial^2 z}{\partial x \partial y}$ at those $(x,y,z) \in \mathbb{R}^3$ for which $\cos(y+z) \neq 0$.
28. Let $z := f(x,y)$ have continuous second-order partial derivatives with respect to x and y. If $x := r\cos\theta$ and $y := r\sin\theta$, then show that (i) $f_x^2 + f_y^2 = z_r^2 + \dfrac{1}{r^2} z_\theta^2$ (ii) $f_{xx} + f_{yy} = z_{rr} + \dfrac{1}{r} z_r + \dfrac{1}{r^2} z_{\theta\theta}$.

Part B

29. Let $f : \mathbb{R} \to \mathbb{R}$ be defined by

$$f(x,y) := \begin{cases} (x^2+y^2)\sin\dfrac{1}{x^2+y^2} & \text{if } (x,y) \neq (0,0), \\ 0 & \text{if } (x,y) = (0,0). \end{cases}$$

Show that f is continuous at $(0,0)$, both the partial derivatives of f exist, but none of the partial derivatives of f is bounded in $\mathbb{S}_r(0,0)$ for any $r > 0$. [Note: This shows that the converse of the result in Proposition 3.3 is not true.] Also, show that f is differentiable at $(0,0)$, but none of the partial derivatives of f is continuous at $(0,0)$. [Note: This shows that the converse of the result in Proposition 3.33 is not true.]

30. **(Differentiation under the Integral)** Let $f : [a,b] \times [c,d] \to \mathbb{R}$ be continuous. If f_y exists and is continuous on $[a,b] \times [c,d]$, then show that the function $\psi : [c,d] \to \mathbb{R}$ defined by $\psi(y) := \int_a^b f(x,y)dx$ is differentiable and $\psi'(y_0) := \int_a^b f_y(x,y_0)dx$ for $y_0 \in [c,d]$. Moreover, ψ' is continuous on $[c,d]$. (Hint: Uniform continuity of f_y and Exercise 26 of Chapter 2)
31. **(Alternative Definition of Differentiability)** Let $D \subseteq \mathbb{R}^2$ and let (x_0,y_0) be an interior point of D. Prove that $f : D \to \mathbb{R}$ is differentiable at (x_0,y_0) if and only if there exist α, $\beta \in \mathbb{R}$ such that
$$\lim_{(h,k)\to(0,0)} \frac{|f(x_0+h,y_0+k) - f(x_0,y_0) - \alpha h - \beta k|}{\sqrt{h^2+k^2}} = 0.$$
32. **(Alternative Version of the Increment Lemma)** Let $D \subseteq \mathbb{R}^2$ and let (x_0,y_0) be an interior point of D. Prove that $f : D \to \mathbb{R}$ is differentiable at (x_0,y_0) if and only if there exist real numbers α, β, δ with $\delta > 0$ and functions ϵ_1, $\epsilon_2 : S_\delta(0,0) \to \mathbb{R}$ such that
$$f(x_0+h,y_0+k) = f(x_0,y_0) + \alpha h + \beta k + h\epsilon_1(h,k) + k\epsilon_2(h,k),$$
for all $(h,k) \in S_\delta(0,0)$, and
$$\lim_{(h,k)\to(0,0)} \epsilon_1(h,k) = 0 = \lim_{(h,k)\to(0,0)} \epsilon_2(h,k).$$
Moreover, if the above conditions hold, then $\nabla f(x_0,y_0) = (\alpha,\beta)$.
33. **(Young's Theorem)** Let $D \subseteq \mathbb{R}^2$, $(x_0,y_0) \in D$, and $f : D \to \mathbb{R}$ be such that both f_x and f_y exist and are differentiable in $\mathbb{S}_r(x_0,y_0)$ for some $r > 0$. Prove that the second-order partials of f exist on $\mathbb{S}_r(x_0,y_0)$ and $f_{xy}(x_0,y_0) = f_{yx}(x_0,y_0)$. (Compare Proposition 3.14.)
34. Let $D \subseteq \mathbb{R}^2$, $(x_0,y_0) \in D$, and $f : D \to \mathbb{R}$ be such that both f_x and f_y exist and are continuous in $\mathbb{S}_r(x_0,y_0)$ for some $r > 0$. Assume that $f(x_0,y_0) = 0$ and $f_y(x_0,y_0) \neq 0$. Then there are $\delta > 0$ and a differentiable function $\eta : (x_0-\delta, x_0+\delta) \to \mathbb{R}$ as given by the Implicit Function Theorem (Proposition 3.38). In case f_{xx}, f_{xy}, and f_{yy} exist and are continuous in $\mathbb{S}_r(x_0,y_0)$, show that η is twice differentiable on $(x_0-\delta, x_0+\delta)$ and
$$\frac{d^2\eta}{dx^2} = -\frac{(f_{xx}f_y^2 - 2f_xf_yf_{xy} + f_{yy}f_x^2)}{f_y^3},$$
where the partial derivatives on the right are evaluated at $(x,\eta(x))$.
35. Suppose the implicit equation $F(x,y,z) = 0$ determines each of the three variables as a function of the other two variables, that is, there exist functions ξ, η, ζ of two variables such that
$$F(\xi(y,z),y,z) = 0, \quad F(x,\eta(z,x),z) = 0, \quad \text{and} \quad F(x,y,\zeta(x,y)) = 0.$$
Assume that the partial derivatives of F, ξ, η, and ζ exist and are continuous. Show that $\left(\frac{\partial \xi}{\partial y}\right)\left(\frac{\partial \eta}{\partial z}\right)\left(\frac{\partial \zeta}{\partial x}\right) = -1.$

36. **(Euler's Theorem)** Suppose $F : \mathbb{R}^3 \to \mathbb{R}$ has the property that there exists $n \in \mathbb{N}$ such that $F(tx, ty, tz) = t^n F(x, y, z)$ for all $t \in \mathbb{R}$ and $(x, y, z) \in \mathbb{R}^3$. [Such a function is said to be homogeneous of degree n.] If the first-order partial derivatives of F exist and are continuous, then show that $x\frac{\partial F}{\partial x} + y\frac{\partial F}{\partial y} + z\frac{\partial F}{\partial z} = nF$.
37. Let $D \subseteq \mathbb{R}^2$ be a nonempty, path-connected, and open subset of $\mathbb{R}^2$, and let $f : D \to \mathbb{R}$ be any function. Show that f is a constant function on D if and only if both f_x and f_y exist and are identically zero on D. (Hint: Given any two points in D, there is a path in D joining them. Use Exercise 18 (ii), Corollary 3.45, and Exercise 19.)
38. Let D be a convex and open subset of $\mathbb{R}^2$, and let $(x_0, y_0), (x_1, y_1)$ be distinct points in D. Let $h := x_1 - x_0$, $k := y_1 - y_0$, and $x, y : \mathbb{R} \to \mathbb{R}$ be defined by $x(t) := x_0 + ht$ and $y(t) := y_0 + kt$. Suppose $f : D \to \mathbb{R}$ is any real-valued function and $F : [0, 1] \to \mathbb{R}$ is defined by $F(t) := f\left(x(t), y(t)\right)$.
   (i) If f is differentiable on D, then use the Chain Rule to show that F is continuous on $[0, 1]$, differentiable on $(0, 1)$, and satisfies

   $$F'(t) = hf_x\left(x(t), y(t)\right) + kf_y\left(x(t), y(t)\right) \qquad \text{for all } t \in (0, 1).$$

   Use this to deduce the classical version of Bivariate Mean Value Theorem as a consequence of the MVT for functions of one real variable.
   (ii) Let n be a nonnegative integer. If the partial derivatives of f of order $\le n+1$ exist and are continuous on D, then show that $F', \dots, F^{(n+1)}$ exist and are continuous on $[0, 1]$, and moreover,

   $$F^{(i)}(t) = \sum_{\substack{\ell \ge 0 \\ \ell+m=i}} \sum_{m \ge 0} \binom{i}{\ell} \frac{\partial^i f}{\partial x^\ell \partial y^m}\left(x(t), y(t)\right) h^\ell k^m \quad \text{for all } t \in [0, 1].$$

   Use this to deduce the classical version of the Bivariate Taylor Theorem as a consequence of Taylor's Theorem for functions of one real variable. (Hint: $E := \{t \in \mathbb{R} : (x(t), y(t)) \in D\}$ is open and contains $[0, 1]$. Use the Chain Rule and the Mixed Partials Theorem.)
39. Let E be an open subset of $\mathbb{R}$ and D be an open subset of $\mathbb{R}^2$ such that $\{xy : (x, y) \in D\} \subseteq E$. If $g : E \to \mathbb{R}$ is an infinitely differentiable function of one variable, then show that for the function $f : D \to \mathbb{R}$ defined by $f(x, y) := g(xy)$ for $(x, y) \in D$, all the higher-order partial derivatives exist. Further, show that given any nonnegative integers ℓ, m and any $(x_0, y_0) \in D$, the higher-order partial derivatives of f are given by

   $$\frac{\partial^{\ell+m} f}{\partial x^\ell \partial y^m}(x_0, y_0) = \sum_{i=0}^{\min\{\ell, m\}} \binom{\ell}{i}\binom{m}{i} i!\, g^{(\ell+m-i)}(x_0 y_0)\, x_0^{\ell-i}\, y_0^{m-i}.$$

   Deduce that if $(0, 0) \in D$, then $\frac{\partial^{\ell+m} f}{\partial x^\ell \partial y^m}(0, 0)$ is equal to 0 if $m \neq \ell$, and is equal to $\ell!\, g^{(\ell)}(0)$ if $m = \ell$.

40. Let E be the open interval $(-1, 1)$ in $\mathbb{R}$ and $D := E \times E$ the open square of radius 1 centered at $(0, 0)$. Suppose $g : E \to \mathbb{R}$ is infinitely differentiable and $f : D \to \mathbb{R}$ is defined by $f(x, y) := g(xy)$. Given a nonnegative integer n, let $Q_n(g)$ denote the nth Taylor polynomial of g at 0 and let $P_n(f)$ denote the nth Taylor polynomial of f around $(0, 0)$. Show that $P_n(f)(x, y) = Q_{[n/2]}(g)(xy)$ for $(x, y) \in \mathbb{R}^2$. Use this to determine the nth Taylor polynomial at $(0, 0)$ of the functions

$$e^{xy},\ \sin(xy),\ \cos(xy),\ \ln(1 + xy),\ (1 + xy)^r,\ (1 - xy)^r,$$

where r denotes a rational number. (Hint: Exercise 39)

41. Consider $f : \mathbb{R}^2 \to \mathbb{R}$ defined by $f(0, 0) := 0$ and $f(x, y) := e^{-1/(x^2+y^2)}$ for $(x, y) \neq (0, 0)$. Show that the partial derivatives of f at $(0, 0)$ of all orders exist, and moreover, they all have the value 0. Deduce that the nth Taylor polynomial of f around $(0, 0)$ is the zero polynomial for every nonnegative integer n.

42. Let $D \subseteq \mathbb{R}^2$ and let (x_0, y_0) be an interior point of D. Let $f : D \to \mathbb{R}$ be such that f is convex on $\mathbb{S}_\delta(x_0, y_0)$ for some $\delta > 0$ with $\mathbb{S}_\delta(x_0, y_0) \subseteq D$. Show that if both $f_x(x_0, y_0)$ and $f_y(x_0, y_0)$ exist, then f is differentiable at (x_0, y_0). (Hint: Consider $\phi(h, k) := f(x_0 + h, y_0 + k) - f(x_0, y_0) - hf_x(x_0, y_0) - kf_y(x_0, y_0)$ for $(h, k) \in \mathbb{S}_\delta(0, 0)$. Use the convexity of f to show that $\phi(h, k) \leq h\,(\phi(2h, 0)/2h) + k\,(\phi(0, 2k)/2k)$ for $0 < |h|, |k| < \delta$. Next, use Exercise 24 and the Cauchy–Schwarz inequality.)

43. **(Cuboidal Mean Value Theorem)** Let $a, b, c, d, p, q \in \mathbb{R}$ with $a < b$, $c < d$, and $p < q$, and let $f : [a, b] \times [c, d] \times [p, q] \to \mathbb{R}$ be a function satisfying the following three conditions: (i) For $y_0 \in [c, d]$ and $z_0 \in [p, q]$, the function given by $x \longmapsto f(x, y_0, z_0)$ is continuous on $[a, b]$ and differentiable on (a, b); (ii) for $x_0 \in (a, b)$ and $z_0 \in [p, q]$, the function given by $y \longmapsto f_x(x_0, y, z_0)$ is continuous on $[c, d]$ and differentiable on (c, d); and (iii) for $x_0 \in (a, b)$ and $y_0 \in (c, d)$, the function given by $z \longmapsto f_{xy}(x_0, y_0, z)$ is continuous on $[p, q]$ and differentiable on (p, q). Show that there is $(x_0, y_0, z_0) \in (a, b) \times (c, d) \times (p, q)$ such that

$$\triangle_{(a,c,p)}^{(b,d,q)} f = (b - a)(d - c)(q - p) f_{xyz}(x_0, y_0, z_0),$$

where, as in Remark 1.20,

$$\begin{aligned}\triangle_{(a,c,p)}^{(b,d,q)} f := {} & f(b, d, q) + f(b, c, p) + f(a, d, p) + f(a, c, q)\\ & - f(b, d, p) - f(a, d, q) - f(b, c, q) - f(a, c, p).\end{aligned}$$

(Hint: In case $\triangle_{(a,c,p)}^{(b,d,q)} f = 0$, consider $\phi : [a, b] \times [c, d] \to \mathbb{R}$ defined by $\phi(x, y) := f(x, y, q) - f(x, y, p)$ and use the Rectangular Rolle's Theorem (Proposition 3.9). In the general case, consider $F : [a, b] \times [c, d] \times [p, q] \to \mathbb{R}$ defined by $F(x, y, z) := f(x, y, z) + f(x, c, p) + f(a, y, p) + f(a, c, z) - f(x, y, p) - f(a, y, z) - f(x, c, z) - f(a, c, p) - s(x - a)(y - b)(z - c)$, where $s \in \mathbb{R}$ is so chosen that $F(b, d, q) = 0$.)

4

Applications of Partial Differentiation

In one-variable calculus, it is customary to apply the notion of differentiation to study local and global extrema of real-valued functions of one variable. (See, for example, Chapter 5 of ACICARA.) Here, we shall consider similar applications of the notion of differentiation to functions of two (or more) variables. As noted in Chapter 3, in multivariable calculus, the notion of differentiation manifests itself in several forms. The simplest among these are the partial derivatives, which together constitute the gradient. When the gradient exists, its vanishing turns out to be a necessary condition for a function to have a local extremum. We shall use this in Section 4.1 below to arrive at a useful recipe for determining the absolute (or global) extremum of a continuous real-valued function defined on a closed and bounded subset of $\mathbb{R}^2$. A variant of the optimization problem discussed in the first section will be considered in Section 4.2, where we try to determine the maximum or the minimum of a function subject to one or more constraints. Such problems are nicely and effectively handled by a technique known as the method of Lagrange multipliers. The theoretical as well as practical aspects of this method will be discussed here. In Section 4.3, we shall make a finer analysis involving the second-order partial derivatives to arrive at a sufficient condition for a function to have a local maximum or a local minimum or a saddle point. Finally, in Section 4.4, we revisit the Bivariate Taylor Theorem with a view toward approximating functions of two variables by linear or quadratic functions.

4.1 Absolute Extrema

We have seen in part (ii) of Proposition 2.25 that a continuous real-valued function defined on a closed and bounded subset of $\mathbb{R}^2$ is bounded and attains its bounds. In other words, if $D \subseteq \mathbb{R}^2$ is closed and bounded, and $f : D \to \mathbb{R}$ is continuous, then the **absolute minimum** and the **absolute maximum** of f on D, namely,

S.R. Ghorpade and B.V. Limaye, *A Course in Multivariable Calculus and Analysis*, Undergraduate Texts in Mathematics, DOI 10.1007/978-1-4419-1621-1_4,

$$m := \inf\{f(x,y) : (x,y) \in D\} \quad \text{and} \quad M := \sup\{f(x,y) : (x,y) \in D\}$$

exist, and moreover, there are $(x_1, y_1), (x_2, y_2) \in D$ such that

$$m = f(x_1, y_1) \quad \text{and} \quad M = f(x_2, y_2),$$

so that $m = \min\{f(x,y) : (x,y) \in D\}$ and $M = \max\{f(x,y) : (x,y) \in D\}$. The following question arises naturally. Knowing the function f, how does one find the **absolute extrema** m and M and points (x_1, y_1) and (x_2, y_2) where they are attained? As in one-variable calculus, it helps to consider the interior points of D at which the partial derivatives vanish or fail to exist, and also the boundary points of D.

Boundary Points and Critical Points

Recall that the notions of an interior point and a boundary point were defined in Section 2.1. It may suffice to remember that given any $D \subseteq \mathbb{R}^2$, a point $(x_0, y_0) \in \mathbb{R}^2$ is an **interior point** of D if and only if $\mathbb{S}_r(x_0, y_0) \subseteq D$ for some $r > 0$, whereas $(x_0, y_0) \in \mathbb{R}^2$ is a **boundary point** of D if and only if $\mathbb{S}_r(x_0, y_0)$ contains a point of D as well as of $\mathbb{R}^2 \setminus D$ for every $r > 0$. Observe that if $(x_0, y_0) \in D$, then (x_0, y_0) is either an interior point of D or a boundary point of D.

Given $D \subseteq \mathbb{R}^2$ and $f : D \to \mathbb{R}$, a point $(x_0, y_0) \in \mathbb{R}^2$ is called a **critical point** of f if (x_0, y_0) is an interior point of D such that either $\nabla f(x_0, y_0)$ does not exist, or $\nabla f(x_0, y_0)$ exists and $\nabla f(x_0, y_0) = (0, 0)$.

Let us recall a basic result from one-variable calculus that helps us answer a question similar to the one raised above. A proof is given, for example, on pages 117 and 118 of ACICARA.

Fact 4.1. *Let $E \subseteq \mathbb{R}$ and let t_0 be an interior point of E. If $\phi : D \to \mathbb{R}$ has a local extremum at t_0 and if ϕ is differentiable at t_0, then $\phi'(t_0) = 0$.*

Here is an analogous result for functions of two variables.

Lemma 4.2. *Let $D \subseteq \mathbb{R}^2$ and let (x_0, y_0) be an interior point of D. Suppose $f : D \to \mathbb{R}$ has a local extremum at (x_0, y_0). If $\mathbf{u} = (u_1, u_2)$ is a unit vector in $\mathbb{R}^2$ such that $\mathbf{D_u} f(x_0, y_0)$ exists, then $\mathbf{D_u} f(x_0, y_0) = 0$. In particular, if $\nabla f(x_0, y_0)$ exists, then $\nabla f(x_0, y_0) = (0, 0)$.*

Proof. Suppose f has a local minimum at (x_0, y_0). Then we can find $\delta > 0$ such that $\mathbb{S}_\delta(x_0, y_0) \subseteq D$ and $f(x,y) \geq f(x_0, y_0)$ for all $(x,y) \in \mathbb{S}_\delta(x_0, y_0)$. Consequently, if $A(t) := f(x_0 + tu_1, y_0 + tu_2) - f(x_0, y_0)$ for $t \in (-\delta, \delta)$, then

$$\frac{A(t)}{t} \leq 0 \quad \text{if } -\delta < t < 0 \quad \text{and} \quad \frac{A(t)}{t} \geq 0 \quad \text{if } 0 < t < \delta.$$

But $\mathbf{D_u} f(x_0, y_0)$ is, by definition, the limit of $A(t)/t$ as $t \to 0$. Hence if $\mathbf{D_u} f(x_0, y_0)$ exists, then $\mathbf{D_u} f(x_0, y_0) = 0$. The last assertion follows by taking $\mathbf{u} = (1, 0)$ and $\mathbf{u} = (0, 1)$. The case in which f has a local maximum at (x_0, y_0) is proved similarly. □

Examples 4.3. (i) Consider $f : \mathbb{R}^2 \to \mathbb{R}$ defined by $f(x, y) := -(x^2 + y^2)$. Then f is differentiable and $\nabla f(x, y) = (-2x, -2y)$ for $(x, y) \in \mathbb{R}^2$. Thus the only point where f can possibly have a local extremum is $(0, 0)$. Indeed, we have seen in Example 1.22 that f does have a local maximum at $(0, 0)$.

(ii) Consider $f : \mathbb{R}^2 \to \mathbb{R}$ defined by $f(x, y) := x^2 + y^2$. Then f is differentiable and $\nabla f(x, y) = (2x, 2y)$ for $(x, y) \in \mathbb{R}^2$. Thus the only point where f can possibly have a local extremum is $(0, 0)$. Indeed, we have seen in Example 1.22 that f does have a local minimum at $(0, 0)$.

(iii) Consider $f : \mathbb{R}^2 \to \mathbb{R}$ defined by $f(x, y) := xy$. Then f is differentiable and $\nabla f(x, y) = (y, x)$ for $(x, y) \in \mathbb{R}^2$. Thus the only point where f can possibly have a local extremum is $(0, 0)$. But $f(0, 0) = 0$ and for any $\delta > 0$, there are $(x_1, y_1), (x_2, y_2) \in \mathbb{S}_\delta(0, y_0)$ such that $f(x_1, y_1) < 0$ and $f(x_2, y_2) > 0$. For example, one can choose any $t \in (0, \delta)$ and let $(x_1, y_1) := (t, -t)$ and $(x_2, y_2) := (t, t)$. It follows that f has neither a local maximum nor a local minimum at $(0, 0)$. ◇

We are now in a position to identify the points at which an absolute extremum is attained.

Proposition 4.4. *Let $D \subseteq \mathbb{R}^2$ be closed and bounded, and let $f : D \to \mathbb{R}$ be a continuous function. Then the absolute minimum as well as the absolute maximum of f are attained either at a critical point of f or at a boundary point of D.*

Proof. By part (ii) of Proposition 2.25, f attains its absolute minimum as well as its absolute maximum on D. Let $(x_1, y_1) \in D$ be a point at which the absolute minimum of f is attained. Suppose (x_1, y_1) is an interior point of D. Then f has a local minimum at (x_1, y_1). If $\nabla f(x_1, y_1)$ exists, then by Lemma 4.2, $\nabla f(x_1, y_1) = 0$. It follows that (x_1, y_1) must be a critical point of D. Thus (x_1, y_1) is either a critical point of f or a boundary point of D.

A similar argument applies to a point at which the absolute maximum of f is attained. □

In practice, the critical points of a function are few in number, whereas the boundary points consist of "one-dimensional pieces." The function, when restricted to a "one-dimensional piece," is effectively a function of one variable. Thus, the methods of one-variable calculus (given, for example, in Section 5.1 of ACICARA) can be applied to determine the absolute extrema of the restrictions of the function to the "one-dimensional pieces." Thus, in view of Proposition 4.4, we have a plausible recipe to determine the absolute extrema and the points where they are attained:

First, determine the critical points of the function and the values of the function at these points. Next, determine the boundary of its domain. Restrict the function to the boundary components and determine the absolute extrema of the restricted function by one-variable methods. Compare the values of the function at all these points. The greatest value among them is the absolute maximum, while the least value is the absolute minimum.

This recipe is illustrated by the following examples.

Examples 4.5. (i) Let $D := [-2,2] \times [-2,2]$ and let $f : D \to \mathbb{R}$ be given by $f(x,y) := 4xy - 2x^2 - y^4$. Clearly, D is closed and bounded, and f is continuous. Thus the absolute extrema of f exist and are attained by f. To determine these, consider the partial derivatives of f. These exist at all interior points of D, and $f_x(x,y) = 4y - 4x$, while $f_y(x,y) = 4x - 4y^3$ for $(x,y) \in (-2,2) \times (-2,2)$. Thus,

$$\nabla f(x,y) = (0,0) \implies (x,y) = (0,0), (1,1), \text{ or } (-1,-1).$$

Also, $(x,y) \in D$ is a boundary point if and only if $x = \pm 2$ or $y = \pm 2$. The restrictions of f to its boundary components are the four functions from $[-2,2]$ to $\mathbb{R}$ given by $f(2,y)$, $f(-2,y)$, $f(x,-2)$, and $f(x,2)$. Due to symmetry $[f(-x,-y) = f(x,y)]$, it suffices to consider only the first and the last of these. So, let us determine the absolute maximum and minimum of $f(2,y)$ for $-2 \le y \le 2$ and of $f(x,2)$ for $-2 \le x \le 2$. As for $f(2,y) = 8y - 8 - y^4$, $y \in [-2,2]$, the only critical point is $y = \sqrt[3]{2}$, and the boundary points are $y = \pm 2$. Comparing the values of $f(2,y)$ at these three points, we see that the absolute maximum of $f(2,y)$ is at $y = \sqrt[3]{2}$ and the absolute minimum is at $y = -2$. Similarly, $f(x,2) = 8x - 2x^2 - 16$, $x \in [-2,2]$, and it is easily seen that the absolute maximum of $f(x,2)$ is at $x = 2$ and the absolute minimum is at $x = -2$. We can now tabulate all the relevant values as follows.

(x,y)	$(0,0)$	$(1,1)$	$(2,\sqrt[3]{2})$	$(2,-2)$	$(2,2)$
$f(x,y)$	0	1	$6\sqrt[3]{2} - 8$	-40	-8

Here we have disregarded the points $(-1,-1)$, $(-2,-\sqrt[3]{2})$, $(-2,2)$, and $(-2,-2)$ due to symmetry. It follows that the absolute maximum of f on D is 1, which is attained at $(1,1)$ as well as at $(-1,-1)$, and the absolute minimum of f on D is -40, which is attained at $(2,-2)$ as well as at $(-2,2)$.

(ii) Let us consider the problem of finding the triangle for which the product of the sines of its three angles is the largest. Since $\sin(\pi - x - y) = \sin(x+y)$ for all $x, y \in \mathbb{R}$, we may consider the function of two variables defined by

$$f(x,y) := \sin x \, \sin y \, \sin(x+y) \quad \text{for} \quad 0 \le x,\, y,\, x+y \le \pi,$$

and seek its absolute maximum. If x, y, or $x+y$ is 0 or π, then $f(x,y) = 0$. Thus f vanishes at each boundary point. For $0 < x,\, y,\, x+y < \pi$, the vanishing of the gradient means that

$$f_x(x,y) = \cos x \, \sin y \, \sin(x+y) + \sin x \, \sin y \, \cos(x+y) = 0,$$
$$f_y(x,y) = \sin x \, \cos y \, \sin(x+y) + \sin x \, \sin y \, \cos(x+y) = 0.$$

Since $0 < x,\, y < \pi$, we see that $\sin x$ and $\sin y$ are nonzero. Hence we obtain $\sin(2x+y) = 0 = \sin(x+2y)$. Also, since $0 < x+y < \pi$, we have

$0 < 2x + y,\ x + 2y < 2\pi$, and hence the only solution of $\sin(2x + y) = 0 = \sin(x + 2y)$ is given by $2x + y = \pi = x + 2y$, that is, $(x, y) = (\pi/3,\ \pi/3)$. Since f is positive at $(\pi/3, \pi/3)$, it follows that it has an absolute maximum at $(\pi/3, \pi/3)$. We thus conclude that the triangle for which the product of the sines of its three angles is the largest must be equilateral.

(iii) Let $a, b \in \mathbb{R}$ be positive and let $D := \{(x, y) \in \mathbb{R}^2 : b^2x^2 + a^2y^2 \leq a^2b^2\}$ be the region enclosed by the ellipse $(x^2/a^2) + (y^2/b^2) = 1$. Then D is closed and bounded, and $f : D \to \mathbb{R}$ defined by $f(x, y) := x^2 - y^2$ is continuous on D. Thus f has absolute extrema, and they are attained at points in D. Let us find them. To begin with, $\nabla f(x, y) = (2x, -2y)$ for $(x, y) \in \mathbb{R}^2$, and thus $(0, 0)$ is the only critical point of f in D. The boundary of D consists of the points on the ellipse, which is parametrically given by $(a \cos t, b \sin t)$, $0 \leq t \leq 2\pi$. The restriction of f to ∂D is essentially given by the function $g : [0, 2\pi] \to \mathbb{R}$ defined by $g(t) := a^2 \cos^2 t - b^2 \sin^2 t$. Clearly, g is differentiable on $(0, 2\pi)$ and $g'(t) = -(a^2 + b^2) \sin 2t$. Hence the critical points of g are $n\pi/2$ for $n = 1, 2, 3$. Also, the endpoints of $[0, 2\pi]$ are $n\pi/2$ for $n = 0, 4$. Now, $g(n\pi/2)$ equals a^2 if n is even and equals $-b^2$ if n is odd. It follows that the absolute maximum of g is a^2, which is attained at $t = 0, \pi, 2\pi$, while the absolute minimum of g is $-b^2$, which is attained at $t = \pi/2,\ 3\pi/2$. Since $f(0, 0) = 0$, we see that the absolute maximum of f is a^2, which is attained at $(\pm a, 0)$, while the absolute minimum of f is $-b^2$, which is attained at $(0, \pm b)$. ◇

Remark 4.6. The notion of a critical point readily extends to a function of n variables. Analogues of Lemma 4.2 and Proposition 4.4 can be easily formulated and proved along similar lines. Thus, if $D \subseteq \mathbb{R}^n$ and $f : D \to \mathbb{R}$ has absolute extrema (which would be the case if, for example, D is closed and bounded, and f is continuous), then they can be found by considering the values of f at its critical points together with the absolute extrema of f on the boundary of D. In general, if $D \subseteq \mathbb{R}^n$, then the boundary of D usually consists of several "$(n-1)$-dimensional pieces." Determination of absolute extrema of f on any one of them further gives rise to several $(n-2)$-dimensional pieces, and so on. At any rate, the method outlined in this section can be iteratively applied to determine absolute extrema of functions of n variables. In principle, it works for any $n \in \mathbb{N}$, but in practice, it is efficient when n is small. ◇

4.2 Constrained Extrema

In Section 4.1, we considered the problem of finding the absolute maximum or minimum of a function $f : D \to \mathbb{R}$, where $D \subseteq \mathbb{R}^2$. We showed that when D is closed and bounded and f is continuous, the absolute extrema exist and are attained at critical points of f or at boundary points of D. The boundary is usually given by the zero set of an equation such as $g(x, y) = 0$. For example, if D is the closed disk centered at (x_0, y_0) and of radius r, then we can let $g(x,y) := (x-x_0)^2 + (y-y_0)^2 - r^2$. In such a case, optimizing $f(x,y)$ on the boundary corresponds to finding the absolute extrema of f subject to the constraint $g(x,y) = 0$. In Section 4.1, we indicated how this could be done in some examples by solving $g(x,y) = 0$ for one of the variables, thereby reducing to a one-variable problem. We now provide an alternative by way of an elegant method to determine absolute extrema of a function of two (or more) variables, subject to the constraint given by the vanishing of another function or, more generally, by the vanishing of several other functions.

Lagrange Multiplier Method

The method of Lagrange for determining constrained extrema is based on the following result.

Proposition 4.7 (Lagrange Multiplier Theorem). *Let $D \subseteq \mathbb{R}^2$ and let (x_0, y_0) be an interior point of D. Suppose $f, g : D \to \mathbb{R}$ are such that the partial derivatives of f and g exist and are continuous in $\mathbb{S}_r(x_0, y_0)$ for some $r > 0$ with $\mathbb{S}_r(x_0, y_0) \subseteq D$. Let $C := \{(x,y) \in D : g(x,y) = 0\}$. Suppose the following three conditions are satisfied.*

(i) $(x_0, y_0) \in C$, *that is,* $g(x_0, y_0) = 0$,
(ii) $\nabla g(x_0, y_0) \neq (0,0)$, *and*
(iii) *the function f, when restricted to C, has a local extremum at (x_0, y_0).*

Then $\nabla f(x_0, y_0) = \lambda_0 \nabla g(x_0, y_0)$ for some $\lambda_0 \in \mathbb{R}$.

Proof. By (ii), $g_x(x_0, y_0) \neq 0$ or $g_y(x_0, y_0) \neq 0$. Suppose $g_y(x_0, y_0) \neq 0$. By the classical version of the Implicit Function Theorem (Proposition 3.38) applied to g, there are $\delta > 0$ with $\delta \leq r$ and $\eta : (x_0 - \delta, x_0 + \delta) \to \mathbb{R}$ with $\eta(x_0) = y_0$ such that $(x, \eta(x)) \in \mathbb{S}_r(x_0, y_0)$ and $g(x, \eta(x)) = 0$, that is, $(x, \eta(x)) \in C$, for all $x \in (x_0 - \delta, x_0 + \delta)$. Moreover, η is differentiable at x_0 and $\eta'(x_0) = -g_x(x_0, y_0)/g_y(x_0, y_0)$. Now consider $\phi : (x_0 - \delta, x_0 + \delta) \to \mathbb{R}$ defined by $\phi(x) := f(x, \eta(x))$. By the Chain Rule (part (ii) of Proposition 3.51), ϕ is differentiable at x_0 and

$$\phi'(x_0) = \nabla f(x_0, \eta(x_0)) \cdot (1,\ \eta'(x_0)) = f_x(x_0, y_0) - f_y(x_0, y_0) \frac{g_x(x_0, y_0)}{g_y(x_0, y_0)}.$$

On the other hand, by (iii), ϕ has a local extremum at x_0. Consequently, by Fact 4.1, $\phi'(x_0) = 0$, and so $f_y(x_0, y_0) g_x(x_0, y_0) = f_x(x_0, y_0) g_y(x_0, y_0)$. It follows that

$$\nabla f(x_0, y_0) = \lambda_0 \nabla g(x_0, y_0), \quad \text{where} \quad \lambda_0 := \frac{f_y(x_0, y_0)}{g_y(x_0, y_0)}.$$

The case in which $g_x(x_0, y_0) \neq 0$ is proved similarly. □

Thanks to the Lagrange Multiplier Theorem (Proposition 4.7), we have the following useful recipe to determine constrained extrema.

To determine the absolute extremum of a real-valued function f of two variables, subject to the constraint $g(x, y) = 0$, we consider a new variable λ, called an **undetermined multiplier**, and seek simultaneous solutions of

$$\nabla f(x, y) = \lambda \nabla g(x, y) \quad \text{and} \quad g(x, y) = 0.$$

If it can be ensured that f does have an absolute extremum on the zero set of g (which will certainly be the case if the zero set of g is closed and bounded, and f is continuous), then the absolute extremum of f is also a local extremum of f and it is necessarily attained either at a simultaneous solution (x_0, y_0) of the above two equations for which $\nabla g(x_0, y_0) \neq (0, 0)$ or at a point where the hypothesis of Proposition 4.7 does not hold. Thus the **Lagrange Multiplier Method** amounts to checking the values of f at such simultaneous solutions, and also at exceptional points such as points in the zero set of g at which ∇f or ∇g does not exist, or at which ∇g vanishes.

Examples 4.8. (i) Consider the problem of finding the maximum and the minimum of the function f given by $f(x, y) := xy$ on the unit circle, that is, subject to the constraint given by $x^2 + y^2 - 1 = 0$. Following the Lagrange Multiplier Method, we let $g(x, y) := x^2 + y^2 - 1$ for $(x, y) \in \mathbb{R}^2$ and consider the equations $\nabla f = \lambda \nabla g$ and $g(x, y) = 0$, that is,

$$y = 2\lambda x, \quad x = 2\lambda y, \quad \text{and} \quad x^2 + y^2 - 1 = 0.$$

These imply $4\lambda^2 = 1$, since $(x, y) = (0, 0)$ does not satisfy $x^2 + y^2 - 1 = 0$. Thus $\lambda = \pm 1/2$, and the simultaneous solutions of the above equations are given by $(x, y) = \left(\pm 1/\sqrt{2}, \pm 1/\sqrt{2}\right)$. Note that ∇g is nonzero at every solution of $g(x, y) = 0$. Also, the zero set of g, that is, the unit circle, is closed and bounded and f is continuous. Thus by the Lagrange Multiplier Theorem (Proposition 4.7), the maximum of f on the unit circle is $1/2$, which is attained at $\left(1/\sqrt{2}, 1/\sqrt{2}\right)$ and $\left(-1/\sqrt{2}, -1/\sqrt{2}\right)$, while the minimum is $-1/2$, which is attained at $\left(1/\sqrt{2}, -1/\sqrt{2}\right)$ and $\left(-1/\sqrt{2}, 1/\sqrt{2}\right)$.

(ii) Consider the problem of finding the shortest distance from the origin to a point on the cuspidal cubic given by $(x - 1)^3 = y^2$. This amounts to finding the minimum of the function $f : \mathbb{R}^2 \to \mathbb{R}$ given by $f(x, y) := x^2 + y^2$ subject to the constraint given by $g(x, y) := (x - 1)^3 - y^2 = 0$. Geometrically, it is obvious that the minimum is 1 and it is attained at $(1, 0)$. (See Figure 4.1.) But $\nabla f(1, 0) = (2, 0)$, while $\nabla g(1, 0) = (0, 0)$. Hence $\nabla f(1, 0) \neq \lambda \nabla g(1, 0)$ for any $\lambda \in \mathbb{R}$. This shows that the condition $\nabla g(x_0, y_0) \neq (0, 0)$ cannot be dropped from the Lagrange Multiplier Theorem. ◇

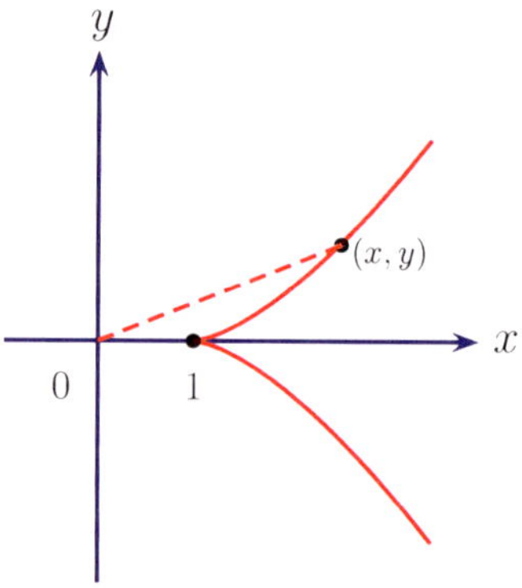

Fig. 4.1. Illustration of Example 4.8 (ii): Finding the minimum distance from the origin to points on the cuspidal cubic $(x-1)^3 = y^2$.

Case of Three Variables

We have seen in Section 3.5 that the classical version of the Implicit Function Theorem readily extends to functions of three (or more) variables. (See Proposition 3.74.) This, in turn, yields an extension of the Lagrange Multiplier Theorem (Proposition 4.7) for functions of three or more variables. Thus, the Lagrange Multiplier Method is applicable for such functions. Namely, to determine the maximum or the minimum of a function f subject to a constraint given by $g = 0$, we seek simultaneous solutions of $\nabla f = \lambda \nabla g$ and $g = 0$ at which $\nabla g \neq 0$. In case f and g have continuous partial derivatives and we know a priori that f does have absolute extrema on the zero set of g, then the extrema can be found by comparing the values of f at the simultaneous solutions of $\nabla f = \lambda \nabla g$ and $g = 0$ or at exceptional points such as those where $\nabla g = 0$ and $g = 0$.

Examples 4.9. (i) To find the maximum and the minimum of the function f given by $f(x, y, z) := x^2y^2z^2$ subject to the constraint that (x, y, z) lies on the unit sphere given by $\{(x, y, z) \in \mathbb{R}^3 : x^2 + y^2 + z^2 = 1\}$, we let $g(x, y, z) := x^2 + y^2 + z^2 - 1$ for $(x, y, z) \in \mathbb{R}^3$. Now $\nabla f = \lambda \nabla g$ implies

$$2xy^2z^2 - 2\lambda x = 0, \quad 2x^2yz^2 - 2\lambda y = 0, \quad \text{and} \quad 2x^2y^2z - 2\lambda z = 0.$$

Either a solution of this system of equations will have one of its coordinates 0 [and the value of f at such a point is 0], or else it must satisfy $x^2 = y^2 = z^2$ and $\lambda = x^4$. If, in addition, we require $g(x, y, z) = 0$, then we necessarily have $x = y = z = \pm 1/\sqrt{3}$. The value of f at each of the corresponding eight points is $1/27$. Also, f is continuous on the unit sphere, which is a closed and bounded set, and hence f attains its maximum as well as its minimum there. Therefore, subject to $g(x, y, z) = 0$, the maximum of f is $1/27$ [attained, for instance, at $(1/\sqrt{3}, 1/\sqrt{3}, 1/\sqrt{3})$] and the minimum of f is 0 [attained, for instance, at $(0,0,1)$]. Note that as a consequence, we obtain an alternative proof of the A.M.-G.M. inequality for three nonnegative real numbers as follows. Given any $a,b,c \in \mathbb{R}$ with $(a,b,c) \neq (0,0,0)$, let $r := \sqrt{a^2+b^2+c^2}$. Then $r \neq 0$ and the point $(x,y,z) := (a/r, b/r, c/r)$ is on the unit sphere in $\mathbb{R}^3$. Thus $f(x,y,z) \leq 1/27$, that is,

$$\frac{a^2b^2c^2}{r^6} = x^2y^2z^2 \leq \frac{1}{27} \quad \text{and hence} \quad \left(a^2b^2c^2\right)^{1/3} \leq \frac{(a^2+b^2+c^2)}{3}.$$

(ii) To find the points on the surface given by $z^2 = xy+4$ closest to the origin, we let $f(x,y,z) := x^2+y^2+z^2$ and $g(x,y,z) := xy+4-z^2$ for $(x,y,z) \in \mathbb{R}^3$. Now

$$\nabla f = \lambda \nabla g \Longrightarrow 2x = \lambda y, \quad 2y = \lambda x, \quad \text{and} \quad 2z = -2\lambda z.$$

Since $\lambda = 0$ implies $(x,y,z) = (0,0,0)$ and since $g(0,0,0) \neq 0$, we assume that $\lambda \neq 0$. Then $x = 0$ or $\lambda = \pm 2$. In case $x = 0$, we have $y = 0$, and if also $g(x,y,z) = 0$, then $z = \pm 2$. In case $x \neq 0$ and $\lambda = \pm 2$, we have $y = \pm x$ and $z = 0$, so that $g(x,y,z) = \pm x^2 + 4$. It follows that the only common solutions of $\nabla f = \lambda \nabla g$ and $g = 0$ are $(0,0,2)$, $(0,0,-2)$, $(2,-2,0)$, and $(-2,2,0)$. We have $f(0,0,\pm 2) = 4$ and $f(\pm 2, \mp 2, 0) = 8$. Now, f is continuous, and although the set $E := \{(x,y,z) \in \mathbb{R}^3 : g(x,y,z) = 0\}$ is not bounded, the set $E_1 = \{(x,y,z) \in E : x^2+y^2+z^2 \leq r\}$, where $r = x_0^2+y_0^2+z_0^2$ for some $(x_0,y_0,z_0) \in E$, is closed and bounded, and the minimum of f on E_1 equals the minimum of f on E. Note also that the only solution of $\nabla g = (0,0,0)$ is $(0,0,0)$ and this does not satisfy $g = 0$. Thus, by the Lagrange Multiplier Method, we conclude that $(0,0,\pm 2)$ are the points on the surface $z^2 = xy + 4$ closest to the origin. ◇

The Lagrange Multiplier Method can also be adapted to a situation in which there is more than one constraint. For example, suppose we want to find the absolute extremum of a function f of three variables x,y,z subject to the constraints given by $g(x,y,z) = 0$ and $h(x,y,z) = 0$. Then we can use the following variant of the Lagrange Multiplier Theorem.

Proposition 4.10 (A Variant of the Lagrange Multiplier Theorem). *Let $D \subseteq \mathbb{R}^3$ and let $P_0 := (x_0,y_0,z_0)$ be an interior point of D. Let $f,g,h : D \to \mathbb{R}$ have continuous partial derivatives in $\mathbb{S}_r(P_0)$ for some $r > 0$ with $\mathbb{S}_r(P_0) \subseteq D$. Let $C := \{(x,y,z) \in D : g(x,y,z) = 0 \text{ and } h(x,y,z) = 0\}$. Suppose the following three conditions are satisfied.*

(i) *$P_0 \in C$, that is, $g(P_0) = 0$ and $h(P_0) = 0$,*
(ii) *$\nabla g(P_0)$ and $\nabla h(P_0)$ are nonzero and are not multiples of each other,*
(iii) *the function f, when restricted to C, has a local extremum at P_0.*

Then $\nabla f(P_0) = \lambda_0 \nabla g(P_0) + \mu_0 \nabla h(P_0)$ for some $\lambda_0, \mu_0 \in \mathbb{R}$.

Proof. By the Implicit Function Theorem for solving two equations (Proposition 3.76 and Remark 3.77), there are $\delta > 0$, $t_0 \in \mathbb{R}$, and differentiable functions $x, y, z : (t_0 - \delta, t_0 + \delta) \to \mathbb{R}$ with $(x(t_0), y(t_0), z(t_0)) = P_0$ such that $(x(t), y(t), z(t)) \in \mathbb{S}_r(P_0)$ and $g(x(t), y(t), z(t)) = 0 = h(x(t), y(t), z(t))$ for all $t \in (t_0 - \delta, t_0 + \delta)$. Moreover, $\mathbf{v_0} := (x'(t_0), y'(t_0), z'(t_0)) \neq (0, 0, 0)$ and $\nabla g(P_0) \cdot \mathbf{v_0} = 0 = \nabla h(P_0) \cdot \mathbf{v_0}$. Consequently, $\mathbf{v_0}$ is orthogonal to the plane spanned by $\nabla g(P_0)$ and $\nabla h(P_0)$. This implies[1] that every vector $\mathbf{v} \in \mathbb{R}^3$ can be expressed as $\lambda \nabla g(P_0) + \mu \nabla h(P_0) + \nu \mathbf{v}_0$ for some $\lambda, \mu, \nu \in \mathbb{R}$. In particular, there are $\lambda_0, \mu_0, \nu_0 \in \mathbb{R}$ such that $\nabla f(P_0) = \lambda_0 \nabla g(P_0) + \mu_0 \nabla h(P_0) + \nu_0 \mathbf{v}_0$. Now consider $\phi : (t_0 - \delta, t_0 + \delta) \to \mathbb{R}$ defined by $\phi(t) := f(x(t), y(t), z(t))$. By the Chain Rule, ϕ is differentiable at t_0 and $\phi'(t_0) = \nabla f(P_0) \cdot (x'(t_0), y'(t_0), z'(t_0))$. On the other hand, by (iii), ϕ has a local extremum at t_0. Hence by Fact 4.1, $\phi'(t_0) = 0$, and so $\nabla f(P_0) \cdot \mathbf{v_0} = 0$. It follows that if $\nabla f(P_0) = \lambda_0 \nabla g(P_0) + \mu_0 \nabla h(P_0) + \nu_0 \mathbf{v}_0$ for some $\lambda_0, \mu_0, \nu_0 \in \mathbb{R}$, then $\nu_0 (\mathbf{v_0} \cdot \mathbf{v_0}) = (\nabla f(P_0) \cdot \mathbf{v_0}) = 0$. Since $\mathbf{v_0} \neq 0$, we have $\nu_0 = 0$ and so the desired assertion is proved. □

Thanks to Proposition 4.10, we have a **Lagrange Multiplier Method** for finding the absolute extrema of a function f of three variables x, y, z subject to two constraints given by $g(x, y, z) = 0$ and $h(x, y, z) = 0$. Namely, introduce new variables λ and μ, called **undetermined multipliers**, and seek simultaneous solutions of

$$\nabla f(x, y, z) = \lambda \nabla g(x, y, z) + \mu h(x, y, z) \quad \text{and} \quad g(x, y, z) = 0 = h(x, y, z).$$

If it can be ensured that f does have an absolute extremum on the intersection of the zero sets of g and h [which will certainly be the case if this intersection is closed and bounded, and f is continuous], then an absolute extremum is necessarily attained either at a simultaneous solution $P_0 := (x_0, y_0, z_0)$ of the above three equations for which $\nabla g(P_0)$ and $\nabla h(P_0)$ are nonzero and are not multiples of each other, or at a point where the hypothesis of Proposition 4.10 does not hold. Thus the absolute extrema of f can be determined by comparing the values of f at such simultaneous solutions and also at exceptional points such as those where ∇f, ∇g, or ∇h does not exist or where ∇g or ∇h vanishes or where they are multiples of each other.

Example 4.11. To find the point on the intersection of the two planes given by $x + y + z = 1$ and $3x + 2y + z = 6$ that is closest to the origin, we let $f(x, y, z) := x^2 + y^2 + z^2$, $g(x, y, z) := x + y + z - 1$ and $h(x, y, z) := 3x + 2y + z - 6$ for $(x, y, z) \in \mathbb{R}^3$. Now we have to find the absolute minimum of f subject to $g = 0$ and $h = 0$. Consider the equation $\nabla f = \lambda \nabla g + \mu \nabla h$. It yields

$$x = \frac{\lambda + 3\mu}{2}, \quad y = \frac{\lambda + 2\mu}{2}, \quad z = \frac{\lambda + \mu}{2}.$$

[1] This implication is an elementary fact in linear algebra or the study of vectors in 3-space. Its proof may be taken as an exercise or can be gleaned from the first few pages of any book on linear algebra.

Substituting these in the equations $g(x, y, z) = 0$ and $h(x, y, z) = 0$, we obtain

$$3\lambda + 6\mu = 2 \quad \text{and} \quad 3\lambda + 7\mu = 6.$$

This gives $\mu = 4$ and $\lambda = -22/3$, and therefore, $P_0 := (\frac{7}{3}, \frac{1}{3}, -\frac{5}{3})$ is the unique simultaneous solution of $\nabla f = \lambda \nabla g + \mu \nabla h$ and $g = h = 0$. Arguing as in Example 4.9 (ii), f must have an absolute minimum on the intersection of the zero sets of g and h. Also, ∇g and ∇h exist at every $P \in \mathbb{R}^3$, and we have $\nabla g(P) = (1, 1, 1)$ and $\nabla h(P) = (3, 2, 1)$; in particular, $\nabla g(P)$ and $\nabla h(P)$ are always nonzero and are not multiples of each other. Hence, we conclude that $P_0 = (\frac{7}{3}, \frac{1}{3}, -\frac{5}{3})$ is the desired point. ◇

4.3 Local Extrema and Saddle Points

The local analysis of a real-valued function f of two variables exhibits a phenomenon not encountered in the study of functions of one variable. Namely, apart from the local maxima and local minima, which are like peaks and dips on the surface $z = f(x, y)$, there can also be saddle points. As the name suggests, a saddle point is rather like the center point of a saddle that one puts on a horse. If one traverses along a certain path on the saddle, the center point appears as a peak, while along some other path, it appears as a dip. Our aim in this section is to describe analytic methods to locate the local extrema and saddle points.

In Section 1.2, we have given precise definitions of local extrema and saddle points. In Lemma 4.2, we showed that the gradient, if it exists, vanishes at points of local extrema. We will now see that a similar thing happens at a saddle point of a smooth function.

Proposition 4.12. *Let $D \subseteq \mathbb{R}^2$ and let (x_0, y_0) be an interior point of D. Suppose $f : D \to \mathbb{R}$ is differentiable at (x_0, y_0) and f has a saddle point at (x_0, y_0). Then $\nabla f(x_0, y_0) = (0, 0)$. Consequently, $\mathbf{D_u} f(x_0, y_0) = (0, 0)$ for every unit vector $\mathbf{u}$ in $\mathbb{R}^2$.*

Proof. Since f has a saddle point at (x_0, y_0), there are regular paths Γ_1 and Γ_2 lying in D and intersecting transversally at (x_0, y_0) such that f has a local maximum at (x_0, y_0) along Γ_1, while f has a local minimum at (x_0, y_0) along Γ_2. Let Γ_i be given by $(x_i(t), y_i(t))$, $t \in [\alpha_i, \beta_i]$, and let $t_i \in (\alpha_i, \beta_i)$ be such that $(x_i(t_i), y_i(t_i)) = (x_0, y_0)$ for $i = 1, 2$. Now $\phi_1 : [\alpha_1, \beta_1] \to \mathbb{R}$ defined by $\phi_1(t) := f(x_1(t), y_1(t))$ has a local maximum at t_1, while $\phi_2 : [\alpha_2, \beta_2] \to \mathbb{R}$ defined by $\phi_2(t) := f(x_2(t), y_2(t))$ has a local minimum at t_2. Since f is differentiable at (x_0, y_0) and x_i, y_i are differentiable at t_i, the Chain Rule (part (ii) of Proposition 3.51) shows that ϕ_i is differentiable at t_i and

$$\phi_i'(t_i) = \nabla f(x_0, y_0) \cdot (x_i'(t_i), y_i'(t_i)) \quad \text{for } i = 1, 2.$$

Further, since ϕ_i has a local extremum at t_i, by Fact 4.1 we have $\phi_i'(t_i) = 0$ for $i = 1, 2$. Thus

$$\nabla f(x_0, y_0) \cdot (x_1'(t_1), y_1'(t_1)) = (0, 0) = \nabla f(x_0, y_0) \cdot (x_2'(t_2), y_2'(t_2)).$$

Since the tangent vectors $(x_1'(t_1), y_1'(t_1))$ and $(x_2'(t_2), y_2'(t_2))$ are nonzero and are not multiples of each other, it follows that $\nabla f(x_0, y_0) = (0, 0)$. Hence by Proposition 3.35, $\mathbf{D_u} f(x_0, y_0) = (0, 0)$ for every unit vector $\mathbf{u}$ in $\mathbb{R}^2$. □

Examples 4.13. (i) Consider $f : \mathbb{R}^2 \to \mathbb{R}$ defined by $f(x, y) := xy$. Then f is differentiable and $\nabla f(x, y) = (y, x)$. Hence the only saddle point that f can possibly have is at $(0, 0)$. Indeed, we have seen in Example 1.22 (iii) that f does have a saddle point at $(0, 0)$. In fact, f has a strict saddle point at $(0, 0)$. (See Figure 1.4 on page 14.)

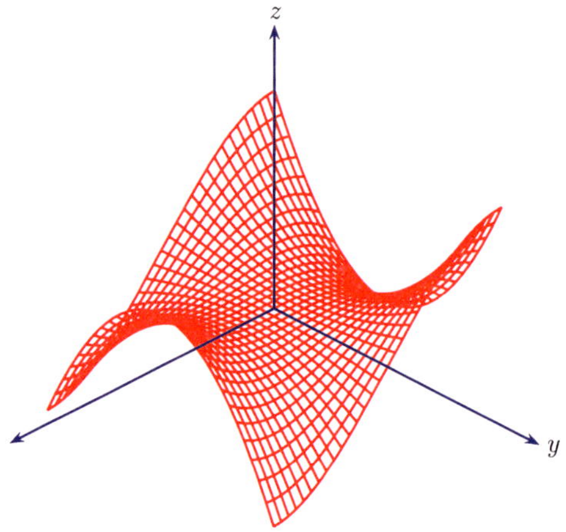

Fig. 4.2. Illustration of a monkey saddle: Graph of the function $f(x, y) := x^3 - 3xy^2$ in Example 4.13 (ii).

(ii) Consider $f : \mathbb{R}^2 \to \mathbb{R}$ defined by $f(x, y) := x^3 - 3xy^2$. Then f is differentiable and $\nabla f(x, y) = (3x^2 - 3y^2,\ -6xy)$. Hence the only saddle point that f can possibly have is at $(0, 0)$. Now, $f(t,\ \pm t/\sqrt{3}) = 0$ for all $t \in \mathbb{R}$. Thus it is trivial to see that if Γ_1 is the path given by $(t,\ t/\sqrt{3})$, $t \in [-1, 1]$, and Γ_2 is the path given by $(t,\ -t/\sqrt{3})$, $t \in [-1, 1]$, then Γ_1, Γ_2 are regular paths intersecting transversally at $(0, 0)$ such that f has a local maximum at $(0, 0)$ along Γ_1 and a local minimum at $(0, 0)$ along Γ_2. This does not, however, prove that f has a strict saddle point at $(0, 0)$ or that f does not have a local extremum at $(0, 0)$. To see this, let us observe that $f(x, y) = x(x - \sqrt{3}y)(x + \sqrt{3}y)$ and consider the parabolic paths $\widetilde{\Gamma}_1$ and $\widetilde{\Gamma}_2$ given by $(t\sqrt{3} - t^2,\ t + t^2\sqrt{3})$, $t \in [-1, 1]$, and by $(-t\sqrt{3} + t^2,\ t + t^2\sqrt{3})$, $t \in [-1, 1]$, respectively. Then $\widetilde{\Gamma}_1$ and $\widetilde{\Gamma}_2$ are regular paths intersecting transversally at $(0,0)$ such that f has a strict local maximum at $(0,0)$ along $\widetilde{\Gamma}_1$ and a strict local minimum at $(0,0)$ along $\widetilde{\Gamma}_2$. The graph of the function f looks like a saddle on which a two-legged animal with a long tail such as a monkey could ride. It is sometimes called a **monkey saddle**. (See Figure 4.2.)

(iii) Consider $f : \mathbb{R}^2 \to \mathbb{R}$ defined by $f(x,y) := x^2+y^2$. Then f is differentiable and $\nabla f(x,y) = (2x, 2y)$. Hence the only point where f can possibly have a saddle point is $(0,0)$. But as we have seen in Example 1.22 (ii), f has a local minimum at $(0,0)$, and in fact, $f(x,y) > 0 = f(0,0)$ for all $(x,y) \in \mathbb{R}^2$ with $(x,y) \neq (0,0)$. It follows that if Γ is any path in $\mathbb{R}^2$ passing through $(0,0)$ and f has a local maximum at $(0,0)$ along Γ, then Γ cannot be regular. Hence f does not have a saddle point at $(0,0)$. Similarly, one can see that for $f : \mathbb{R}^2 \to \mathbb{R}$ defined by $f(x,y) := -(x^2+y^2)$, the only possibility for a saddle point is $(0,0)$, but f does not have a saddle point at $(0,0)$. (See Figures 1.2 and 1.4 on pages 11 and 14.)

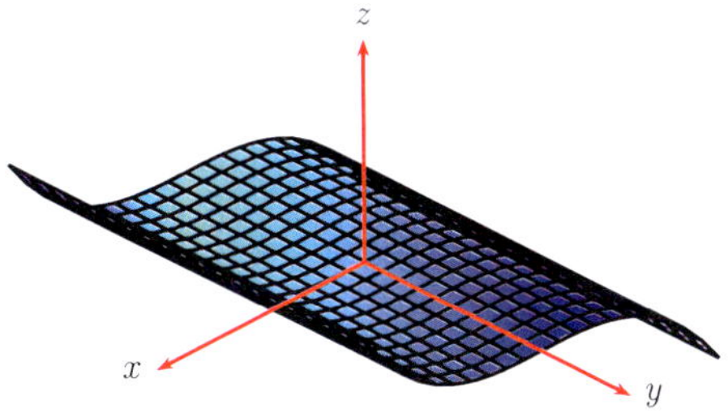

Fig. 4.3. Illustration of a fake saddle: Graph of $f(x,y) := x^3$ near the origin.

(iv) Consider $f : \mathbb{R}^2 \to \mathbb{R}$ defined by $f(x,y) := x^3$. Then f is differentiable and $\nabla f(x,y) = (3x^2, 0)$. Hence the only points at which f can possibly have a saddle point are points of the form $(0, y_0)$, where $y_0 \in \mathbb{R}$. Now let us suppose that $y_0 \in \mathbb{R}$ and f has a saddle point at $(0, y_0)$. Then there are regular paths Γ_1 and Γ_2 lying in $\mathbb{R}^2$ and intersecting transversally at $(0, y_0)$ such that f has a local maximum at $(0, y_0)$ along Γ_1, while f has a local minimum at $(0, y_0)$ along Γ_2. Let Γ_i be given by $(x_i(t), y_i(t))$, $t \in [\alpha_i, \beta_i]$, and $t_i \in (\alpha_i, \beta_i)$ be such that $(x_i(t_i), y_i(t_i)) = (0, y_0)$ for $i = 1, 2$. Note that a real number and its cube always have the same sign. With this in view, since $x_1(t_1) = 0$ and x_1^3 has a local maximum at t_1, we see that x_1 has a local maximum at t_1, and so $x_1'(t_1) = 0$. Similarly, x_2 has a local minimum at t_2, and so $x_2'(t_2) = 0$. Hence the two tangent vectors $(x_1'(t_1), y_1'(t_1))$ and $(x_2'(t_2), y_2'(t_2))$ are multiples of each other. It follows that f does not have a saddle point at $(0,0)$. (See Figure 4.3.)

(v) Consider $f : \mathbb{R}^2 \to \mathbb{R}$ defined by $f(x,y) := x^4 + y^3$. (See Figure 1.9.) Then f is differentiable and $\nabla f(x,y) = (4x^3, 3y^2)$. Hence the only point where f can possibly have a saddle point is $(0,0)$. Suppose Γ_i, given by

$(x_i(t), y_i(t))$, $t \in [\alpha_i, \beta_i]$, for $i = 1, 2$ are regular paths in $\mathbb{R}^2$ such that f has a local maximum at $(0,0)$ along Γ_1, and a local minimum at $(0,0)$ along Γ_2. Let $t_i \in (\alpha_i, \beta_i)$ be such that $(x_i(t_i), y_i(t_i)) = (0,0)$ for $i = 1, 2$. Now observe that given any $t \in (\alpha_1, \beta_1)$, if $x_1(t)^4 + y_1(t)^3 \leq 0$, then $y_1(t)^3 \leq -x_1(t)^4 \leq 0$ and hence $y_1(t) \leq 0$. On the other hand, given any $t \in (\alpha_2, \beta_2)$, if $x_2(t)^4 + y_2(t)^3 \geq 0$, then $y_2(t)^3 \geq -x_2(t)^4$ and hence $y_2(t) \geq -x_2(t)^{4/3}$, that is, $y_2(t) + x_2(t)^{4/3} \geq 0$. It follows, therefore, that $y_1(t)$ has a local maximum at $t = t_1$ and $y_2(t) + x_2(t)^{4/3}$ has a local minimum at $t = t_2$. Consequently, $y_1'(t_1) = 0$ and $y_2'(t_2) + \frac{4}{3}x_2(t_2)x_2'(t_2)^{1/3} = 0$. Since $x_2(t_2) = 0$, we find that $y_1'(t_1) = y_2'(t_2) = 0$, and so the two tangent vectors $(x_1'(t_1), y_1'(t_1))$ and $(x_2'(t_2), y_2'(t_2))$ are multiples of each other. In other words, Γ_1 and Γ_2 do not intersect transversally at $(0,0)$. It follows that f does not have a saddle point at $(0,0)$. $\diamond$

Discriminant Test

As we have seen thus far, the vanishing of the gradient is a necessary condition for a function to have a local extremum or a saddle point, but it is not a sufficient condition. To obtain a sufficient condition, we attempt to extend the first derivative test and the second derivative test of one-variable calculus (as given in Propositions 5.3 and 5.4 of ACICARA) to the case of functions of two variables. While it does not make sense to look at the sign of the gradient, we may consider the sign of the discriminant.

The basic idea is quite simple. Suppose the second-order partial derivatives of f exist and are continuous around (x_0, y_0). If $\nabla f(x_0, y_0) = (0,0)$, then by the Extended Mean Value Theorem, the difference $f(x, y) - f(x_0, y_0)$ can be expressed in terms of the Hessian form at some point near (x_0, y_0). It turns out that the behavior of the Hessian form at and around (x_0, y_0) is governed by its leading coefficient $f_{xx}(x_0, y_0)$ and the discriminant $\Delta f(x_0, y_0)$. Recall (from Section 3.4) that $\Delta f(x_0, y_0)$ is defined by

$$\Delta f(x_0, y_0) := f_{xx}(x_0, y_0) f_{yy}(x_0, y_0) - \left[f_{xy}(x_0, y_0)\right]^2.$$

More precisely, we have the so-called discriminant test, whose statement and proof will be given shortly. But first, we require the following algebraic result, which is a variant of Proposition 3.69.

Lemma 4.14. *Let $Q(h, k) := ah^2 + 2bhk + ck^2$ be a binary quadratic form in the variables h and k with $a, b, c \in \mathbb{R}$. Let $\Delta := ac - b^2$.*

(i) *If $a < 0$ and $\Delta > 0$, then $Q(x, y) < 0$ for all $(x, y) \in \mathbb{R}^2$, $(x, y) \neq (0,0)$.*
(ii) *If $a > 0$ and $\Delta > 0$, then $Q(x, y) > 0$ for all $(x, y) \in \mathbb{R}^2$, $(x, y) \neq (0,0)$.*
(iii) *If $\Delta < 0$, then there are two distinct lines L_1 and L_2 passing through the origin such that $Q(x, y) < 0$ for (x, y) on L_1 with $(x, y) \neq (0,0)$, while $Q(x, y) > 0$ for (x, y) on L_2 with $(x, y) \neq (0,0)$.*

Proof. Both (i) and (ii) follow from the identity

$$aQ(x,y) = a^2x^2 + 2abxy + acy^2 = (ax+by)^2 + \Delta y^2 \quad \text{for all } (x,y) \in \mathbb{R}^2.$$

To prove (iii), suppose $\Delta < 0$. We consider three cases.

Case 1: $a \neq 0$. In this case, a and $a\Delta$ have opposite signs,

$$Q(bt, -at) = \left(-ab^2 + ca^2\right) t^2 = a\Delta t^2, \quad \text{and} \quad Q(t,0) = at^2 \quad \text{for all } t \in \mathbb{R}.$$

So let $L_1 := \{(bt, -at) : t \in \mathbb{R}\}$ and $L_2 := \{(t,0) : t \in \mathbb{R}\}$ or vice versa according as $a > 0$ or $a < 0$.

Case 2: $a = 0$ and $c \neq 0$. In this case, c and $c\Delta$ have opposite signs,

$$Q(ct, -bt) = \left(ac^2 - cb^2\right) t^2 = c\Delta t^2, \quad \text{and} \quad Q(0,t) = ct^2 \quad \text{for all } t \in \mathbb{R}.$$

So let $L_1 := \{(ct, -bt) : t \in \mathbb{R}\}$ and $L_2 := \{(0,t) : t \in \mathbb{R}\}$ or vice versa according as $c > 0$ or $c < 0$.

Case 3: $a = c = 0$. In this case, $b \neq 0$,

$$Q(t,t) = 2bt^2, \quad \text{and} \quad Q(t,-t) = -2bt^2 \quad \text{for all } t \in \mathbb{R}.$$

So let $L_1 := \{(t,t) : t \in \mathbb{R}\}$ and $L_2 := \{(t,-t) : t \in \mathbb{R}\}$ or vice versa according as $b < 0$ or $b > 0$.

Thus, in all cases, we can find lines L_1 and L_2 as desired. □

Proposition 4.15 (Discriminant Test). *Let $D \subseteq \mathbb{R}^2$ and let (x_0, y_0) be an interior point of D. Suppose $f : D \to \mathbb{R}$ is such that the first-order and second-order partial derivatives of f exist and are continuous in $\mathbb{S}_r(x_0, y_0)$ for some $r > 0$ with $\mathbb{S}_r(x_0, y_0) \subseteq D$, and $\nabla f(x_0, y_0) = (0,0)$.*

(i) *If $\Delta f(x_0, y_0) > 0$ and $f_{xx}(x_0, y_0) < 0$, then f has a local maximum at (x_0, y_0).*
(ii) *If $\Delta f(x_0, y_0) > 0$ and $f_{xx}(x_0, y_0) > 0$, then f has a local minimum at (x_0, y_0).*
(iii) *If $\Delta f(x_0, y_0) < 0$, then f has a saddle point at (x_0, y_0).*

Proof. Since the second-order partial derivatives of f are continuous at (x_0, y_0), by Lemma 2.14, we can find some $\delta > 0$ with $\delta \leq r$ such that the signs of Δf and f_{xx} are preserved in $\mathbb{S}_\delta(x_0, y_0)$, that is, the following hold:

(a) If $\Delta f(x_0, y_0) \neq 0$, then $\Delta f(x,y)\Delta f(x_0, y_0) > 0$ for all $(x,y) \in \mathbb{S}_\delta(x_0, y_0)$.
(b) If $f_{xx}(x_0, y_0) \neq 0$, then $f_{xx}(x,y) f_{xx}(x_0, y_0) > 0$ for all $(x,y) \in \mathbb{S}_\delta(x_0, y_0)$.

Let $(x,y) \in \mathbb{S}_\delta(x_0, y_0)$ with $(x,y) \neq (x_0, y_0)$. Since $\nabla f(x_0, y_0) = (0,0)$, applying the Extended Bivariate Mean Value Theorem (Remark 3.48 (ii)) to f on $\mathbb{S}_\delta(x_0, y_0)$, we see that there is $\theta \in (0,1)$ such that upon letting $h := x - x_0$, $k := y - y_0$, and $(c,d) := (x_0 + \theta h,\ y_0 + \theta k)$, we have $(c,d) \in \mathbb{S}_\delta(x_0, y_0)$ and

$$f(x, y) - f(x_0, y_0) = \frac{1}{2}\left[f_{xx}(c, d)h^2 + 2f_{xy}(c, d)hk + f_{yy}(c, d)k^2\right].$$

Using (a) and (b) above and applying parts (i) and (ii) of Lemma 4.14 to the Hessian form of f at (c, d), we see that if $\Delta f(x_0, y_0) > 0$ and $f_{xx}(x_0, y_0) < 0$, then $f(x, y) < f(x_0, y_0)$, whereas if $\Delta f(x_0, y_0) > 0$ and $f_{xx}(x_0, y_0) > 0$, then $f(x, y) > f(x_0, y_0)$. Since (x, y) was an arbitrary point of $\mathbb{S}_\delta(x_0, y_0)$, assertions (i) and (ii) are proved.

Next, suppose $\Delta f(x_0, y_0) < 0$. Applying part (iii) of Lemma 4.14 to the Hessian form of f at (x_0, y_0), namely,

$$Q_0(h, k) := f_{xx}(x_0, y_0)h^2 + 2f_{xy}(x_0, y_0)hk + f_{yy}(x_0, y_0)k^2,$$

we see that there are $(h_1, k_1), (h_2, k_2) \in \mathbb{S}_1(0, 0)$ that are different from $(0, 0)$ and are not multiples of each other such that $Q_0(h_1, k_1) < 0$ and $Q_0(h_2, k_2) > 0$. Now for $i = 1, 2$, the second-order partial derivatives of f are continuous at (x_0, y_0), and hence by Lemma 2.14, there is $\delta_i > 0$, with $\delta_i < \delta$, such that for all $t \in (-\delta_i, \delta_i)$,

$$f_{xx}(x_0+th_i,\ y_0+tk_i)h_i^2+2f_{xy}(x_0+th_i,\ y_0+tk_i)h_ik_i+f_{yy}(x_0+th_i,\ y_0+tk_i)k_i^2$$

is nonzero and has the same sign as $Q_0(h_i, k_i)$. Thus, if for $i = 1, 2$, we let Γ_i denote the path given by $(x_0 + th_i,\ y_0 + tk_i)$, $t \in [-\delta_i, \delta_i]$, then we see that Γ_1 and Γ_2 are regular paths in D and they intersect transversally at (x_0, y_0). (See Figure 4.4.) Moreover, applying the Extended Bivariate Mean Value Theorem to f on $\mathbb{S}_{\delta_i}(x_0, y_0)$, we see, as in the proof of (i) and (ii) above, that for each $t \in (-\delta_i, \delta_i)$,

$$f(x_0 + th_i,\ y_0 + tk_i) - f(x_0, y_0) \text{ is } \leq 0 \text{ if } i = 1 \text{ and } \geq 0 \text{ if } i = 2.$$

It follows that $\phi_i : [-\delta_i, \delta_i] \to \mathbb{R}$ defined by $\phi_i(t) := f(x_0 + th_i,\ y_0 + tk_i)$ has a local maximum at $t = 0$ if $i = 1$ and a local minimum at $t = 0$ if $i = 2$. Thus f has a saddle point at (x_0, y_0), and assertion (iii) is proved. □

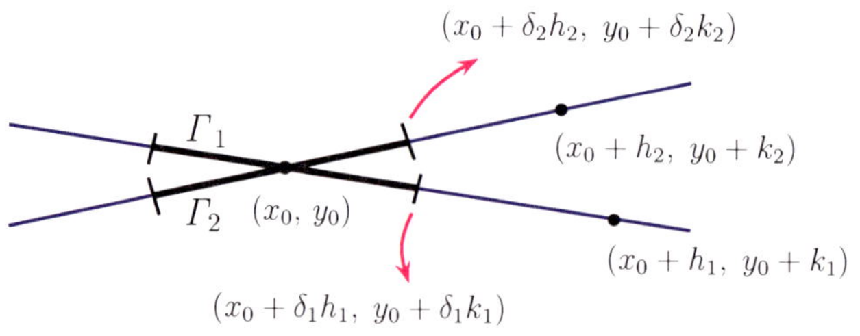

Fig. 4.4. Illustration of the paths Γ_1 and Γ_2 in the proof of Proposition 4.15.

Remarks 4.16. (i) If $\Delta f(x_0, y_0) := f_{xx}(x_0, y_0) f_{yy}(x_0, y_0) - [f_{xy}(x_0, y_0)]^2$ is positive, then we clearly have $f_{xx}(x_0, y_0) > 0 \iff f_{yy}(x_0, y_0) > 0$. Thus in parts (i) and (ii) of the Discriminant Test, the condition on the sign of f_{xx} can be replaced by an identical condition on the sign of f_{yy}.

(ii) Our proof of the Discriminant Test shows that the test can, in fact, be made stronger. Namely, with D, (x_0, y_0), and f as in Proposition 4.15, we have the following:

> *If $\Delta f(x_0, y_0) > 0$, then f has a strict local maximum or a strict local minimum at (x_0, y_0) according as $f_{xx}(x_0, y_0) < 0$ or $f_{xx}(x_0, y_0) > 0$, whereas if $\Delta f(x_0, y_0) < 0$, then f has a strict saddle point at (x_0, y_0).*

Hence assuming that $\Delta f(x_0, y_0) \neq 0$ (but not necessarily that $\nabla f(x_0, y_0) = (0, 0)$), we can deduce from the above stronger version of the Discriminant Test and Proposition 4.12 the following:

$$f \text{ has a saddle point at } (x_0, y_0) \iff \nabla f(x_0, y_0) = (0, 0) \text{ and } \Delta f(x_0, y_0) < 0.$$

In particular, if $\Delta f(x_0, y_0) \neq 0$, then f has a saddle point at (x_0, y_0) if and only if f has a strict saddle point at (x_0, y_0). ◇

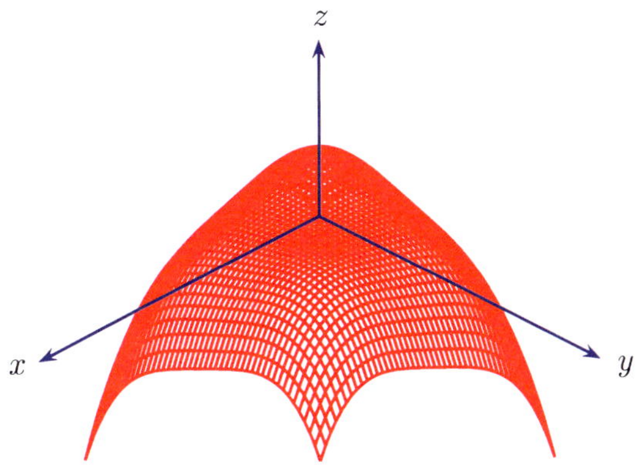

Fig. 4.5. Graph of the function $f(x, y) := 4xy - x^4 - y^4$ in Example 4.17 (ii).

Examples 4.17. (i) Consider an example that we have seen earlier, namely, $f : \mathbb{R}^2 \to \mathbb{R}$ defined by $f(x, y) := xy$. (See the graph on the left in Figure 1.9.) We have $\nabla f(x, y) = (y, x)$, so that $\nabla f(x, y) = (0, 0) \iff (x, y) = (0, 0)$. Moreover, $f_{xx} = f_{yy} = 0$, while $f_{xy} = 1$, and thus $\Delta f(x, y) := f_{xx} f_{yy} - f_{xy}^2 = -1 < 0$. So, it follows from the Discriminant Test that f has a saddle point at $(0, 0)$. In a similar manner, we can see that $f : \mathbb{R}^2 \to \mathbb{R}$ defined by $f(x, y) := x^2 - y^2$ has a saddle point at $(0, 0)$. More generally, if f is given by product of "distinct" linear forms, then a similar conclusion holds. (See Exercise 16.)

(ii) Consider $f : \mathbb{R}^2 \to \mathbb{R}$ defined by $f(x,y) := 4xy - x^4 - y^4$. Then f has continuous partial derivatives of all orders. Also, $f_x = 4(y - x^3)$ and $f_y = 4(x - y^3)$, and so

$$\nabla f(x,y) = (0,0) \iff (x,y) = (0,0),\ (1,1),\ \text{or}\ (-1,-1).$$

Further, $f_{xx} = -12x^2$, $f_{xy} = 4$, and $f_{yy} = -12y^2$, and so the discriminant is given by $\Delta f(x,y) := f_{xx}f_{yy} - f_{xy}^2 = 16(9x^2y^2 - 1)$. In particular, $\Delta f(0,0) = -16 < 0$ and $\Delta f(1,1) = \Delta f(-1,-1) = 128 > 0$. Also $f_{xx}(1,1) = f_{xx}(-1,-1) = -12 < 0$. By the Discriminant Test, f has a saddle point at $(0,0)$ and a local maximum at $(1,1)$ as well as at $(-1,-1)$. Compare these with the graph of f depicted in Figure 4.5.

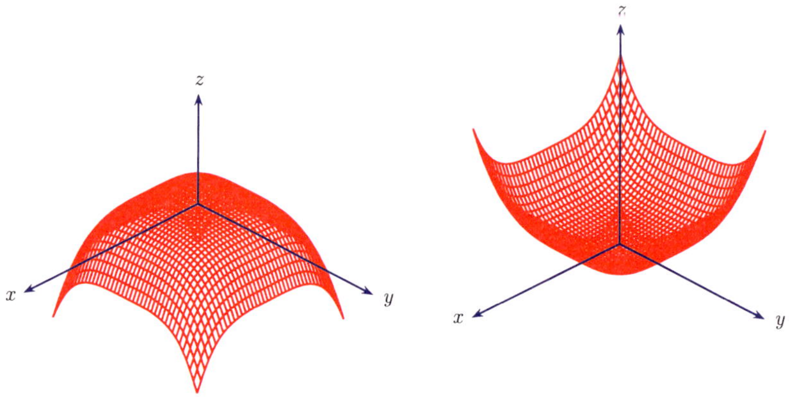

Fig. 4.6. Illustration of a local maximum and a local minimum: graphs of the functions $f(x,y) := -(x^4 + y^4)$ and $g(x,y) := x^4 + y^4$ in Example 4.17 (iii).

(iii) Consider $f : \mathbb{R}^2 \to \mathbb{R}$ defined by $f(x,y) := -(x^4 + y^4)$. Then $\nabla f(x,y) = (-4x^3, -4y^3)$ is $(0,0)$ only when $(x,y) = (0,0)$. But $\Delta f(0,0) = 0$, and so the Discriminant Test is not applicable to f at $(0,0)$. In fact, we can see directly that f has a local maximum at $(0,0)$. Similarly, if $g : \mathbb{R}^2 \to \mathbb{R}$ is defined by $g(x,y) := x^4 + y^4$, then $\nabla g(x,y) = (4x^3, 4y^3)$ is $(0,0)$ only when $(x,y) = (0,0)$. But $\Delta g(0,0) = 0$, and so the Discriminant Test is not applicable. In fact, we can see directly that g has a local minimum at $(0,0)$. (See Figure 4.6.)

(iv) Consider $f : \mathbb{R}^2 \to \mathbb{R}$ defined by $f(x,y) := x^3y - xy^3$. Then $\Delta f(0,0) = 0$, and so the Discriminant Test is not applicable to f at $(0,0)$. In fact, we can see directly that f has a saddle point at the origin. Indeed, it suffices to consider the paths Γ_1 and Γ_2 given by $(t, -t/2)$, $t \in [-1,1]$, and by $(t, t/2)$, $t \in [-1,1]$, respectively. The graph of the function f looks like a saddle on which a four-legged animal such as a dog could ride. It is sometimes called a **dog saddle**. (See Figure 4.7.)

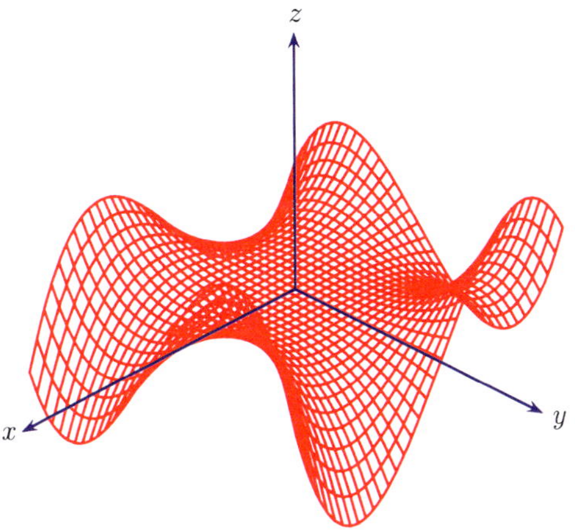

Fig. 4.7. Illustration of a dog saddle: graph of the function $f(x, y) := x^3y - xy^3$ in Example 4.17 (iv).

(v) Consider $f : \mathbb{R}^2 \to \mathbb{R}$ defined by $f(x, y) := x^3$. Then $\nabla f(0, 0) = (0, 0)$ and $\Delta f(0, 0) = 0$. Thus the Discriminant Test is not applicable to f at $(0, 0)$. We have seen in Example 4.13 (iv) that f has neither a local extremum nor a saddle point at $(0, 0)$. (See Figure 4.3.) ◇

4.4 Linear and Quadratic Approximations

In one-variable calculus (for example, Section 5.4 of ACICARA), we encounter the concept of a tangent line approximation or a linear approximation. In this way, if we know the value of a function of one variable and the value of its derivative at a particular point, then we can approximately determine the values of the function at all nearby points. Better approximations, known as quadratic, cubic, or in general, nth-degree approximations, can be obtained if we also know the values of the higher-order derivatives at that particular point. In effect, the given function is replaced, locally, by a linear or a quadratic or in general, a polynomial function. The key result that enables us to do this is Taylor's theorem. A similar situation prevails for functions of two (or more) variables. Let us first discuss the simplest of such approximations, namely the linear approximation of a function of two variables.

Linear Approximation

Let $D \subseteq \mathbb{R}^2$ and let (x_0, y_0) be an interior point of D. Suppose $f : D \to \mathbb{R}$ is such that the first-order partial derivatives of f at (x_0, y_0) exist. The function $L : \mathbb{R}^2 \to \mathbb{R}$ defined by

$$L(x,y) = f(x_0,y_0) + f_x(x_0,y_0)(x-x_0) + f_y(x_0,y_0)(y-y_0) \quad \text{for } (x,y) \in \mathbb{R}^2$$

is called the **linear approximation** to f around (x_0, y_0). Note that $L(x, y)$ is the first Taylor polynomial of f around (x_0, y_0). Geometrically speaking, $z = L(x, y)$ represents a plane that is precisely the tangent plane to the surface $z = f(x, y)$ at the point (x_0, y_0, z_0), where $z_0 := f(x_0, y_0)$. For this reason, L is also called the **tangent plane approximation** to f around (x_0, y_0). (See Figure 4.8.) The difference

$$e_1(x,y) := f(x,y) - L(x,y) \quad \text{for } (x,y) \in D$$

is called the **error** at (x, y) in the linear approximation to f around (x_0, y_0).

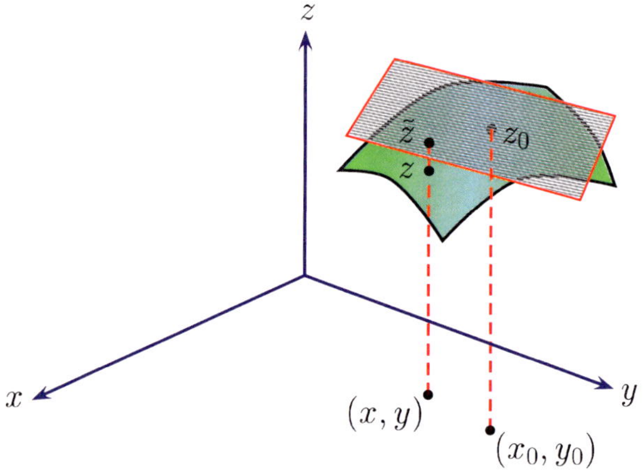

Fig. 4.8. Illustration of linear approximation, or the tangent plane approximation: for (x, y) near (x_0, y_0), the value of $z = f(x, y)$ is approximated by $\tilde{z} = L(x, y)$.

Proposition 4.18. *Let $D \subseteq \mathbb{R}^2$ and let (x_0, y_0) be an interior point of D. If $f : D \to \mathbb{R}$ is such that both $f_x(x_0, y_0)$ and $f_y(x_0, y_0)$ exist, then the linear approximation L to f around (x_0, y_0) is indeed an approximation to f around (x_0, y_0), that is,*

$$\lim_{(x,y)\to(x_0,y_0)} L(x,y) = f(x_0,y_0), \quad \textit{or equivalently,} \quad \lim_{(x,y)\to(x_0,y_0)} e_1(x,y) = 0.$$

Further, if f is differentiable at (x_0, y_0), then $e_1(x, y)$ rapidly approaches zero as $(x, y) \to (x_0, y_0)$ in the sense that

$$\lim_{(x,y)\to(x_0,y_0)} \frac{e_1(x,y)}{\sqrt{(x-x_0)^2 + (y-y_0)^2}} = 0.$$

Moreover, if the first-order and second-order partial derivatives of f exist and are continuous in $\mathbb{S}_r(x_0, y_0)$ for some $r > 0$ with $\mathbb{S}_r(x_0, y_0) \subseteq D$, then for any $(x_1, y_1) \in \mathbb{S}_r(x_0, y_0)$ with $(x_1, y_1) \neq (x_0, y_0)$, we have the error bound

$$|e_1(x_1, y_1)| \le \frac{M_2(x_1, y_1)}{2} \left(|x_1 - x_0| + |y_1 - y_0|\right)^2,$$

where $M_2(x_1, y_1)$ is an upper bound for $|f_{xx}|$, $|f_{yy}|$ and $|f_{xy}|$ on the open line segment $\{(x_0 + t(x_1 - x_0),\, y_0 + t(y_1 - y_0)) : t \in (0,1)\}$ joining (x_0, y_0) and (x_1, y_1).

Proof. It is obvious from the definition of L that $L(x, y) \to f(x_0, y_0)$, or equivalently, $e_1(x, y) \to 0$, as $(x, y) \to (x_0, y_0)$. The assertions about $e_1(x, y)$ rapidly approaching zero follow from the definition of differentiability of functions of two variables.

Suppose the first-order and second-order partial derivatives of f exist and are continuous in $\mathbb{S}_r(x_0, y_0)$ for some $r > 0$ with $\mathbb{S}_r(x_0, y_0) \subseteq D$. Now, $\mathbb{S}_r(x_0, y_0)$ is convex and open, and so the Extended Bivariate Mean Value Theorem (Remark 3.48 (ii)) is applicable to the restriction of f to $\mathbb{S}_r(x_0, y_0)$. Thus, given any $(x_1, y_1) \in \mathbb{S}_r(x_0, y_0)$ with $(x_1, y_1) \ne (x_0, y_0)$, there is some (c, d) on the line segment joining (x_0, y_0) and (x_1, y_1), with $(c, d) \ne (x_i, y_i)$ for $i = 1, 2$, such that

$$f(x_1, y_1) = L(x_1, y_1) + \frac{1}{2}\left(h^2 f_{xx}(c, d) + 2hk f_{xy}(c, d) + k^2 f_{yy}(c, d)\right),$$

where $h := x_1 - x_0$ and $k := y_1 - y_0$. This implies the desired error bound for $|e_1(x_1, y_1)| = |f(x_1, y_1) - L(x_1, y_1)|$. □

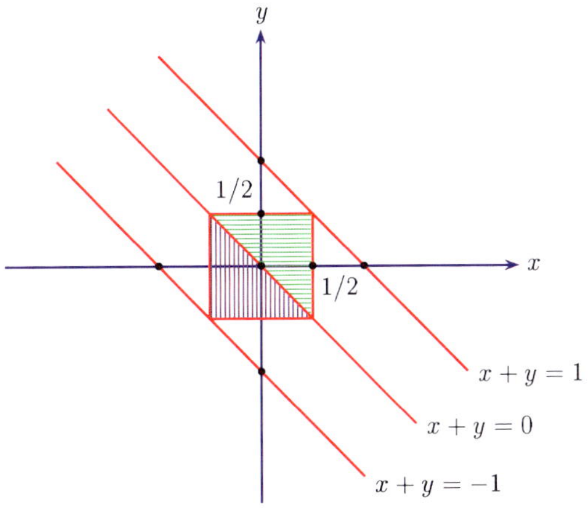

Fig. 4.9. Illustration of Example 4.19: the open square $\mathbb{S}_{1/2}(0, 0)$ subdivided into regions above and below the line given by $x + y = 0$.

Example 4.19. Let $D := \{(x, y) \in \mathbb{R}^2 : x + y \ne 1\}$ and consider $f : D \to \mathbb{R}$ defined by $f(x, y) := 1/(1 - x - y)$. Then $f_x = f_y = 1/(1 - x - y)^2$, and so the linear approximation to f for (x, y) near $(0,0)$ is given by $L(x,y) := 1 + x + y$ for $(x,y) \in \mathbb{R}^2$. Further, $f_{xx} = f_{xy} = f_{yy} = 2/(1-x-y)^3$. To find an upper bound for $|f_{xx}|$, $|f_{yy}|$, and $|f_{xy}|$ on an open line segment joining $(0,0)$ to a point nearby, say in the open square $\mathbb{S}_{1/2}(0,0)$, it is convenient to consider separately the points in the square that are above and below the line given by $x+y=0$. (See Figure 4.9.) Thus for any $(x_1,y_1) \in \mathbb{S}_{1/2}(0,0)$ with $x_1+y_1 \neq 1$ and $(x_1,y_1) \neq (0,0)$, we can take $M_2(x_1,y_1) = 2/(1-x_1-y_1)^3$ if $x_1+y_1>0$ and $M_2(x_1,y_1) = 2$ if $x_1+y_1 \le 0$. In particular, if $|x_1| < 0.1$ and $|y_1| < 0.1$, then we obtain $|e_1(x_1,y_1)| < 0.0782$ if $x_1+y_1>0$ and $|e_1(x_1,y_1)| < 0.04$ if $x_1 + y_1 \le 0$. ◇

Quadratic Approximation

As in the case of functions of one variable, quadratic approximations yield better estimates than linear approximations. Here are the basic definitions.

Let $D \subseteq \mathbb{R}^2$ and let (x_0,y_0) be an interior point of D. Suppose $f : D \to \mathbb{R}$ is such that the first-order and second-order partial derivatives of f at (x_0,y_0) exist. Define $Q : \mathbb{R}^2 \to \mathbb{R}$ by

$$Q(x,y) := L(x,y) + \frac{1}{2}\left[h^2 f_{xx}(x_0,y_0) + 2hk f_{xy}(x_0,y_0) + k^2 f_{yy}(x_0,y_0)\right],$$

where $h := x - x_0$, $k := y - y_0$, and $L(x,y) := f(x_0,y_0) + hf_x(x_0,y_0) + kf_y(x_0,y_0)$ for $(x,y) \in \mathbb{R}^2$. We call Q the **quadratic approximation** to f around (x_0,y_0). Note that $Q(x,y)$ is the second Taylor polynomial of f around (x_0,y_0). Geometrically speaking, $z = Q(x,y)$ represents a paraboloid passing through $(x_0,y_0,f(x_0,y_0))$ such that the surface given by $z = f(x,y)$ and this paraboloid have a common tangent at $(x_0,y_0,f(x_0,y_0))$. The difference

$$e_2(x,y) := f(x,y) - Q(x,y) \quad \text{for } (x,y) \in D$$

is called the **error** at (x,y) in the quadratic approximation to f around (x_0,y_0). In this situation, an analogue of Proposition 4.18 is the following.

Proposition 4.20. *Let $D \subseteq \mathbb{R}^2$ and let (x_0,y_0) be an interior point of D. Suppose $f : D \to \mathbb{R}$ is such that the first-order and second-order partial derivatives of f at (x_0,y_0) exist. Then the quadratic approximation Q to f around (x_0,y_0) is indeed an approximation to f around (x_0,y_0), that is,*

$$\lim_{(x,y)\to(x_0,y_0)} Q(x,y) = f(x_0,y_0), \quad \textit{or equivalently,} \quad \lim_{(x,y)\to(x_0,y_0)} e_2(x,y) = 0.$$

Further, if the first-order and second-order partial derivatives of f exist and are continuous in $\mathbb{S}_r(x_0,y_0)$ for some $r>0$ with $\mathbb{S}_r(x_0,y_0) \subseteq D$, then $e_2(x,y)$ approaches zero as $(x,y) \to (x_0,y_0)$ doubly rapidly in the sense that

$$\lim_{(x,y)\to(x_0,y_0)} \frac{e_2(x,y)}{(x-x_0)^2 + (y-y_0)^2} = 0.$$

Moreover, if the first-, second-, and third-order partial derivatives of f exist and are continuous in $\mathbb{S}_r(x_0, y_0)$ for some $r > 0$ with $\mathbb{S}_r(x_0, y_0) \subseteq D$, then for any $(x_1, y_1) \in \mathbb{S}_r(x_0, y_0)$ with $(x_1, y_1) \neq (x_0, y_0)$, we have the error bound

$$|e_2(x_1, y_1)| \leq \frac{M_3(x_1, y_1)}{3!} \left(|x_1 - x_0| + |y_1 - y_0|\right)^3,$$

where $M_3(x_1, y_1)$ is an upper bound for $|f_{xxx}|$, $|f_{xxy}|$, $|f_{xyy}|$, and $|f_{yyy}|$ on the open line segment $\{(x_0 + t(x_1 - x_0),\ y_0 + t(y_1 - y_0)) : t \in (0, 1)\}$ joining (x_0, y_0) and (x_1, y_1).

Proof. It is obvious from the definition of Q that $Q(x, y) \to f(x_0, y_0)$, or equivalently, $e_2(x, y) \to 0$, as $(x, y) \to (x_0, y_0)$. Further, if the first-order and second-order partial derivatives of f exist and are continuous in $\mathbb{S}_r(x_0, y_0)$ for some $r > 0$ with $\mathbb{S}_r(x_0, y_0) \subseteq D$, then by the Extended Bivariate Mean Value Theorem (Remark 3.48 (ii)), we see that given any $(x, y) \in \mathbb{S}_r(x_0, y_0)$ with $(x, y) \neq (x_0, y_0)$, there is some (c, d) on the line segment joining (x_0, y_0) and (x, y) such that

$$f(x, y) = L(x, y) + \frac{1}{2}\left[h^2 f_{xx}(c, d) + 2hk f_{xy}(c, d) + k^2 f_{yy}(c, d)\right],$$

where $h := x - x_0$, $k := y - y_0$, and $L(x, y) := f(x_0, y_0) + h f_x(x_0, y_0) + k f_y(x_0, y_0)$. Thus, upon letting $A := f_{xx}(c, d) - f_{xx}(x_0, y_0)$, $B := f_{xy}(c, d) - f_{xy}(x_0, y_0)$, and $C := f_{yy}(c, d) - f_{yy}(x_0, y_0)$, we obtain

$$|e_2(x, y)| = |f(x, y) - Q(x, y)| = \frac{1}{2}\left|Ah^2 + 2Bhk + Ck^2\right|.$$

Now $(x, y) \neq (x_0, y_0)$ implies that $(h, k) \neq (0, 0)$, and thus we have

$$\left|\frac{e_2(x, y)}{h^2 + k^2}\right| \leq \frac{|A|}{2}\frac{|h|^2}{|h^2 + k^2|} + \frac{|B|}{2}\frac{|2hk|}{|h^2 + k^2|} + \frac{|C|}{2}\frac{|k|^2}{|h^2 + k^2|} \leq \frac{|A| + |B| + |C|}{2}.$$

It is clear that if $(x, y) \to (x_0, y_0)$, then $(c, d) \to (x_0, y_0)$, and so by the continuity of the second-order partial derivatives of f at (x_0, y_0), it follows that $A \to 0$, $B \to 0$, and $C \to 0$ as $(x, y) \to (x_0, y_0)$. This implies the assertion about $e_2(x, y)$ approaching zero doubly rapidly.

Finally, suppose the first-, second-, and third-order partial derivatives of f exist and are continuous in $\mathbb{S}_r(x_0, y_0)$ for some $r > 0$ with $\mathbb{S}_r(x_0, y_0) \subseteq D$. Let $(x_1, y_1) \in \mathbb{S}_r(x_0, y_0)$ with $(x_1, y_1) \neq (x_0, y_0)$. By the classical version of the Bivariate Taylor Theorem (Proposition 3.47) with $n = 2$, there is some (c, d) on the line segment joining (x_0, y_0) and (x_1, y_1), with $(c, d) \neq (x_i, y_i)$ for $i = 1, 2$, such that $f(x_1, y_1)$ is equal to

$$Q(x_1, y_1) + \frac{1}{3!}\left(h^3 f_{xxx}(c, d) + 3h^2 k f_{xxy}(c, d) + 3hk^2 f_{xyy}(c, d) + k^3 f_{yyy}(c, d)\right),$$

where $h := x_1 - x_0$ and $k := y_1 - y_0$. This implies the desired error bound for $|e_2(x_1, y_1)| = |f(x_1, y_1) - Q(x_1, y_1)|$. □

Example 4.21. Let $D := \{(x,y) \in \mathbb{R}^2 : x+y \neq 1\}$ and consider $f : D \to \mathbb{R}$ defined by $f(x,y) := 1/(1-x-y)$. Then $f_x = f_y = 1/(1-x-y)^2$ and $f_{xx} = f_{xy} = f_{yy} = 2/(1-x-y)^3$. Thus, the quadratic approximation to f for (x,y) near $(0,0)$ is given by $Q(x,y) := 1+x+y+x^2+2xy+y^2$ for $(x,y) \in \mathbb{R}^2$. As in Example 4.19, it is convenient to consider separately the points in the open square $\mathbb{S}_{1/2}(0,0)$ that are above and below the line given by $x+y=0$. (See Figure 4.9.) Further, $f_{xxx} = f_{xxy} = f_{xyy} = f_{yyy} = 6/(1-x-y)^4$. Thus for any $(x_1,y_1) \in \mathbb{S}_{1/2}(0,0)$ with $(x_1,y_1) \neq (0,0)$, we can take $M_3(x_1,y_1) = 6/(1-x_1-y_1)^4$ if $x_1+y_1 > 0$ and $M_3(x_1,y_1) = 6$ if $x_1+y_1 \leq 0$. In particular, if $|x_1| < 0.1$ and $|y_1| < 0.1$, then we obtain $|e_2(x_1,y_1)| < 0.0196$ if $x_1+y_1 > 0$ and $|e_1(x_1,y_1)| < 0.008$ if $x_1+y_1 \leq 0$. Compare Example 4.19. ◇

Notes and Comments

In sequencing the topics in this chapter, we have followed the same principles as in Chapter 5 of ACICARA. *Tests for local extrema have no bearing on the determination of absolute extrema nor on the study of constrained extrema and the Lagrange multiplier method. Thus, absolute extrema and constrained extrema of functions of two (or more) variables are treated before considering local extrema and saddle points.*

As noted in Chapter 1 already, our definition of a saddle point differs from that found in most texts. Usually, a saddle point is defined as a critical point (that is, an interior point of the domain at which the gradient vanishes) where the function has neither a local maximum nor a local minimum. Some books define a saddle point as a critical point where the discriminant is negative. These definitions do ensure that the Discriminant Test can be proved fairly easily. But they do not seem to be completely in tune with the geometric idea of a saddle point. Indeed, functions such as $f(x,y) := x^3$ *or* $f(x,y) := x^4 + y^3$ *whose graph near the origin scarcely looks like a saddle would end up having a saddle point at the origin. Anomalies such as these are avoided in our definition. Moreover, to talk about a function having a saddle point, we do not have to presuppose that the function is differentiable at that point, let alone require that the point be a critical point. This is consistent with the tenet we have followed throughout this text: geometric notions come first and have an intrinsic definition; criteria involving derivatives and such come later, provided suitable differentiability conditions are satisfied.*

We have shown that if a real-valued function of two variables merely has first-order partial derivatives at a point, then it is approximated by a "linear" function around that point. Further, if the function is differentiable at that point, then the error in this approximation rapidly approaches zero. Further still, if the function has continuous first-order and second-order partial derivatives in a neighborhood of that point, then explicit error bounds for this error can be given. A similar situation is brought out for "quadratic" approximation. This graded approach seems noteworthy.

As an important application of differentiation of functions of one variable, one obtains sufficient conditions, usually ascribed to Picard, for the existence and uniqueness of a fixed point of a map from a closed and bounded interval in $\mathbb{R}$ *into itself. Another important application is a method of Newton for determining approximate solutions of equations of the form* $f(x) = 0$. *Analogues of these for functions of several variables are possible. But it is clear that these will necessarily involve vector-valued functions and their derivatives. Since we have restricted ourselves to only* real-valued *functions of several variables in this book, we have skipped applications such as these. However, for those interested in these applications, we suggest Chapter 54 (especially Sections 5, 6, 8, 15, 16, 19) of Vol. 3 of [18].*

Exercises

Part A

1. Find the absolute minimum and the absolute maximum of the function f given by $f(x, y) := 2x^2 - 4x + y^2 - 4y + 1$ on the closed triangular region bounded by the lines given by $x = 0$, $y = 2$, and $y = 2x$.
2. Find the absolute maximum and the absolute minimum of the function f given by $f(x, y) := (x^2 - 4x)\cos y$ over the rectangular region given by $1 \le x \le 3$, $-\pi/4 \le y \le \pi/4$.
3. Determine constants a and b such that the integral
$$\int_0^1 [ax + b - f(x)]^2 dx$$
is minimal if (i) $f(x) := x^2$, (ii) $f(x) := (x^2 + 1)^{-1}$.
4. The temperature at a point (x, y, z) in 3-space is given by $T(x, y, z) = 400xyz^2$. Find the highest temperature on the unit sphere $x^2 + y^2 + z^2 = 1$.
5. Consider the surface in $\mathbb{R}^3$ given by $z = xy + 1$. Find the point on the surface that is nearest to the origin.
6. Let $a, b, c, d \in \mathbb{R}$ with a, b, c not all zero and let $(x_0, y_0, z_0) \in \mathbb{R}^3$. Find the shortest distance between the plane given by $ax + by + cz = d$ and the point (x_0, y_0, z_0).
7. Let a, b, and c be nonzero real numbers. Find the minimum volume bounded by the planes given by $x = 0$, $y = 0$, $z = 0$, and a plane that is tangent to the ellipsoid given by $(x^2/a^2) + (y^2/b^2) + (z^2/c^2) = 1$.
8. Let a, b, and c be positive real numbers. Find the maximum value of $f(x, y, z) = x^a y^b z^c$ subject to the constraint given by $x + y + z = 1$. Deduce that if u, v, w are any positive real numbers, then
$$\left(\frac{u}{a}\right)^a \left(\frac{v}{b}\right)^b \left(\frac{w}{c}\right)^c \le \left(\frac{u+v+w}{a+b+c}\right)^{a+b+c}.$$

9. Let $s \in \mathbb{R}$ with $s > 0$ and let E be the set of all $(x, y, z) \in \mathbb{R}^3$ satisfying the inequalities $0 \le x,\ y,\ z \le s$, $x + y \ge z$, $y + z \ge x$, and $z + x \ge y$. For the function $f : E \to \mathbb{R}$ defined by $f(x, y, z) := s(s-x)(s-y)(s-z)$, show that $(2s/3, 2s/3, 2s/3)$ is the unique point in $\mathbb{R}^3$ where f has a maximum subject to the constraint given by $x + y + z = 2s$. Deduce that a triangle with a given perimeter $2s$ and maximum possible area is equilateral.
10. A space probe in the shape of the ellipsoid $4x^2 + y^2 + 4z^2 = 16$ enters the earth's atmosphere and the surface of the probe begins to heat. After one hour, the temperature at the point (x, y, z) on the surface of the probe is given by $T(x, y, z) = 8x^2 + 4yz - 16z + 600$. Find the hottest point on the surface of the probe.
11. Find the maximum value of the function $f : \mathbb{R}^3 \to \mathbb{R}$ given by $f(x, y, z) := xyz$ subject to the constraints given by $x + y + z = 40$ and $x + y = z$.
12. Find the minimum value of the function $f : \mathbb{R}^3 \to \mathbb{R}$ given by $f(x, y, z) := x^2 + y^2 + z^2$ subject to the constraints given by $x + 2y + 3z = 6$ and $x + 3y + 9z = 9$.
13. Find the maximum value of the function $f : \mathbb{R}^3 \to \mathbb{R}$ given by $f(x, y, z) := x^2 + 2y - z^2$ subject to the constraints given by $2x - y = 0$ and $y + z = 0$.
14. Show that the following functions have local minima at the indicated points.
(i) $f(x, y) := x^4 + y^4 + 4x - 32y - 7, \quad (x_0, y_0) = (-1, 2)$,
(ii) $f(x, y) := x^3 + 3x^2 - 2xy + 5y^2 - 4y^3, \quad (x_0, y_0) = (0, 0)$.
15. Consider $f : \mathbb{R}^2 \to \mathbb{R}$ defined by $f(x, y) := 3x^4 - 4x^2y + y^2$. Show that f has a local minimum at $(0, 0)$ on every line through $(0, 0)$. Does f have a local minimum at $(0, 0)$? Does f have a saddle point at $(0, 0)$?
16. Let a, b, c, d be any real numbers. Show that $f : \mathbb{R}^2 \to \mathbb{R}$ defined by $f(x, y) := (ax + by)(cx + dy)$ has a saddle point at $(0, 0)$ if $ad - bc \ne 0$ and has a local extremum at $(0, 0)$ if $ad - bc = 0$.
17. Consider $f : \mathbb{R}^2 \to \mathbb{R}$ defined by one of the following. In each case, find the points at which f has a local maximum, a local minimum, or a saddle point.
(i) $f(x, y) := (x^2 - y^2)e^{-(x^2+y^2)/2}$, (ii) $f(x, y) := x^3y^5$,
(iii) $f(x, y) = x^2 - y^2$, (iv) $f(x, y) := 6x^2 - 2x^3 + 3y^2 + 6xy$,
(v) $f(x, y) := x^3 + y^3 - 3xy + 15$, (vi) $f(x, y) := x^m$, where $m \in \mathbb{N}$.
18. Consider $f : \mathbb{R}^2 \to \mathbb{R}$ defined by

$$f(x, y) := \begin{cases} \min\{|x|, |y|\}, & \text{if } xy \ge 0, \\ -\min\{|x|, |y|\}, & \text{if } xy < 0. \end{cases}$$

Does $\nabla f(0, 0)$ exist? Is f differentiable at $(0, 0)$? Does f have a local extremum at $(0, 0)$? Does f have a saddle point at $(0, 0)$? Justify your answers.
19. Let $c_1, c_2 \in \mathbb{R}$ with $c_1 < c_2$. Show that $f : \mathbb{R}^2 \to \mathbb{R}$ defined by $f(x, y) := (y - c_1x^2)(y - c_2x^2)$ has a strict local minimum at $(0, 0)$ on

every line passing through the origin. Further, show that f has a saddle point at $(0,0)$. (Hint: Consider the segment of a line and a parabola passing through the origin.)

20. Find the linear approximation of $f(x,y) := \sqrt{x^2+y^2}$ near $(3,4)$.
21. Find the linear approximation of $f(x,y)$ near $(2,2)$ if $f(x,y) := 2x^2 + 4xy + y^2 + 12x - 12y + 16$. Also, find an upper bound for the error $e_1(x,y)$ that is valid for all $(x,y) \in \mathbb{R}^2$ such that $|x-2| < 0.1$ and $|y-2| < 0.1$.
22. Find the linear approximation of $f(x,y)$ near $(0,0)$ if
(i) $f(x,y) := \sqrt{1+x+y}$ for $(x,y) \in \mathbb{R}^2$ with $x+y \geq -1$,
(ii) $f(x,y) := 1/\sqrt{1-x-y}$ for $(x,y) \in \mathbb{R}^2$ with $x+y < 1$.
Find an estimate for the error $e_1(x,y)$ for $(x,y) \in \mathbb{R}^2$ with $x+y > 0$ and for $(x,y) \in \mathbb{R}^2$ with $x+y \leq 0$. Find an upper bound for the error $e_1(x,y)$ that is valid (a) for all $(x,y) \in \mathbb{R}^2$ with $|x| < 0.1$, $|y| < 0.1$, and $x+y > 0$ and (b) for all $(x,y) \in \mathbb{R}^2$ with $|x| < 0.1$, $|y| < 0.1$, and $x+y \leq 0$.
23. Find an approximate value of $[(0.99)e^{0.02}]^8$.
24. The dimensions of a cylindrical tin are known to change as follows. The radius changes from 3 inches to 2.9 inches, while the height changes from 4 inches to 4.2 inches. Estimate the change in the volume of the tin.

Part B

25. Let D denote the closed triangular region in $\mathbb{R}^2$ with vertices $(0,0)$, $(1,1)$, and $(1,-1)$. If $f : D \to \mathbb{R}$ is defined by $f(x,y) := \sqrt{x^2-y^2}$ for $(x,y) \in D$, then prove the following.
(i) f is continuous on D and $f(0,0) = f(1,1) = f(1,-1) = 0$.
(ii) At every interior point (x_0,y_0) of D, both f_x and f_y exist, but $\nabla f(x_0,y_0) \neq (0,0)$.
[Note: From the MVT (Fact 3.2), we know that the line joining any two points on a curve C given by $y = f(x)$, $x \in E$, is parallel to the tangent line to C at some interior point of E. The above example shows that there can be three points on a surface S given by $z = f(x,y)$, $(x,y) \in D$, such that no tangent plane to S at an interior point of D is parallel to the plane passing through the three points on S.]
26. Let D be a closed and bounded subset of $\mathbb{R}^2$ and let $f : D \to \mathbb{R}$ be a continuous function such that $f|_{\partial D}$ is constant. If the interior of D is nonempty and both f_x and f_y exist at every interior point of D, then show that there exists some interior point (x_0,y_0) of D such that $\nabla f(x_0,y_0) = 0$.
[Note: This result may be viewed a version of **Rolle's Theorem** (Fact 3.8) for functions of two variables.]
27. Given the conic section $Ax^2+2Bxy+Cy^2 = 1$, where $A > 0$ and $B^2 < AC$, let m and M denote respectively the distances from the origin to the nearest and the farthest points of the conic. Show that

$$M^2 = \frac{(A+C) + \sqrt{(A-C)^2 + 4B^2}}{2(AC-B^2)},$$

and find a companion formula for m^2.

28. Let p, q be positive real numbers such that $(1/p) + (1/q) = 1$. Determine the minimum of the function $f(x, y) := (x^p/p) + (y^q/q)$, $x, y \geq 0$, subject to the constraint given by $xy = 1$. Deduce **Hölder's inequality**, that is,

$$\sum_{i=1}^{n} a_i b_i \leq \left(\sum_{i=1}^{n} a_i^p\right)^{1/p} \left(\sum_{i=1}^{n} b_i^q\right)^{1/q},$$

for any nonnegative real numbers $a_1, \ldots, a_n, b_1, \ldots, b_n$.

29. Let $A, B, C, D, E, F \in \mathbb{R}$ with $A > 0$ and $B^2 < AC$. Consider $f : \mathbb{R}^2 \to \mathbb{R}$ defined by $f(x, y) := Ax^2 + 2Bxy + Cy^2 + 2Dx + 2Ey + F$. Show that there is a unique point (x_1, y_1) in $\mathbb{R}^2$ at which f has a local minimum, and further,

$$f(x_1, y_1) = Dx_1 + Ey_1 + F = \frac{1}{AC - B^2} \begin{vmatrix} A & B & D \\ B & C & E \\ D & E & F \end{vmatrix}.$$

(Hint: Transform the quadratic form to a sum of squares.)

30. Let $m, n \in \mathbb{N}$ and let $f : \mathbb{R}^2 \to \mathbb{R}$ be defined by $f(x, y) := x^m - y^n$. Show that f has a saddle point at $(0, 0)$ if and only if both m and n are even.

31. Let $m, n \in \mathbb{N}$ and let $f : \mathbb{R}^2 \to \mathbb{R}$ be defined by $f(x, y) := x^m + y^n$. Show that f does not have a saddle point at $(0, 0)$.

32. Let $n \in \mathbb{N}$ and let $f, g : \mathbb{R}^2 \to \mathbb{R}$ be defined by $f(x, y) := \text{Re}\,(x + iy)^n$ and $g(x, y) := \text{Im}\,(x + iy)^n$, where $x + iy$ is the complex number corresponding to $(x, y) \in \mathbb{R}^2$ and Re denotes the real part, while Im denotes the imaginary part. Show that if $n \geq 2$, then both f and g have a saddle point at $(0, 0)$.
[Note: The surface $z = f(x, y)$ is parametrically given by $x = r\cos\theta$, $y = r\sin\theta$, and $z = r^n \cos n\theta$, where $r \geq 0$ and $-\pi < \theta \leq \pi$. It is known as a **generalized monkey saddle**. Compare with Example 4.13 (ii) when $n = 3$.]

33. Let $m, n \in \mathbb{N}$ and let $f : \mathbb{R}^2 \to \mathbb{R}$ be defined by $f(x, y) := x^m y^n$. Show that f has a strict saddle point at $(0, 0)$ if and only if both m and n are odd.

5

Multiple Integration

In one-variable calculus, we study the theory of Riemann integration. (See, for example, Chapter 6 of ACICARA.) In this chapter, we will extend this theory to functions of several variables. As in the previous chapters, we shall mainly restrict to functions of two variables and briefly show how things work for functions of three variables. Further extension to the case of functions of n variables, where $n \geq 4$, is similar.

In Section 5.1 we consider the relatively simpler case of double integrals of functions defined on rectangles in $\mathbb{R}^2$. The general case of double integrals of functions defined on bounded subsets of $\mathbb{R}^2$ is developed in Section 5.2. This will lead, in particular, to the general concept of area of a bounded region in $\mathbb{R}^2$. Next, in Section 5.3, we discuss the change of variables formula for double integrals and prove it in an important special case. Finally, in Section 5.4, we will indicate how the theory of double integrals extends to triple integrals of functions defined on bounded subsets of $\mathbb{R}^3$, and discuss the general concept of volume of such subsets.

5.1 Double Integrals on Rectangles

In this chapter, by a **rectangle** we shall mean a nonempty closed rectangle in $\mathbb{R}^2$. In other words, a rectangle is a subset of $\mathbb{R}^2$ of the form

$$[a, b] \times [c, d] := \{(x, y) \in \mathbb{R}^2 : a \leq x \leq b \text{ and } c \leq y \leq d\},$$

where $a, b, c, d \in \mathbb{R}$ with $a < b$ and $c < d$. Likewise, by a **cuboid** we shall mean a nonempty closed cuboid in $\mathbb{R}^3$. Henceforth whenever we consider a rectangle of the form $[x_1, x_2] \times [y_1, y_2]$, it will be tacitly assumed that x_1, x_2, y_1, y_2 are real numbers with $x_1 < x_2$ and $y_1 < y_2$. Likewise, whenever we consider a cuboid of the form $[x_1, x_2] \times [y_1, y_2] \times [z_1, z_2]$, it will be tacitly assumed that $x_1, x_2, y_1, y_2, z_1, z_2$ are real numbers with $x_1 < x_2$, $y_1 < y_2$, and $z_1 < z_2$. Given

S.R. Ghorpade and B.V. Limaye, *A Course in Multivariable Calculus and Analysis*, Undergraduate Texts in Mathematics, DOI 10.1007/978-1-4419-1621-1_5,

a rectangle $R := [a,b] \times [c,d]$ and a cuboid $C := [x_1, x_2] \times [y_1, y_2] \times [z_1, z_2]$, it is usual to set

$$\text{Area}\,(R) := (b-a)(d-c) \quad \text{and} \quad \text{Vol}\,(C) := (x_2 - x_1)(y_2 - y_1)(z_2 - z_1),$$

and call these the **area** of R and the **volume** of C, respectively.

With the above understanding of the volume of a cuboid, let us investigate whether we can assign a meaning to what can reasonably be called the "volume" of a solid under a surface in $\mathbb{R}^3$. More precisely, consider a nonnegative bounded function defined on a rectangle $[a,b] \times [c,d]$, and the solid lying under its graph, above the xy-plane, and bounded by the planes given by $x = a$, $x = b$, $y = c$, and $y = d$. The problem of determining the "volume" of such a solid can be approached by subdividing the rectangle $[a,b] \times [c,d]$ into a finite number of subrectangles and then finding the sum of the volumes of the cuboids inscribed within the solid and also the sum of the volumes of cuboids that circumscribe the solid. (See Figure 5.1.) This leads to the notion of "double integral" of a bounded function, which, in turn, yields the desired notion of "volume" when the function is nonnegative. To arrive at these, we first formalize certain preliminary notions such as subdivisions of rectangles, volumes of inscribed and circumscribing cuboids.

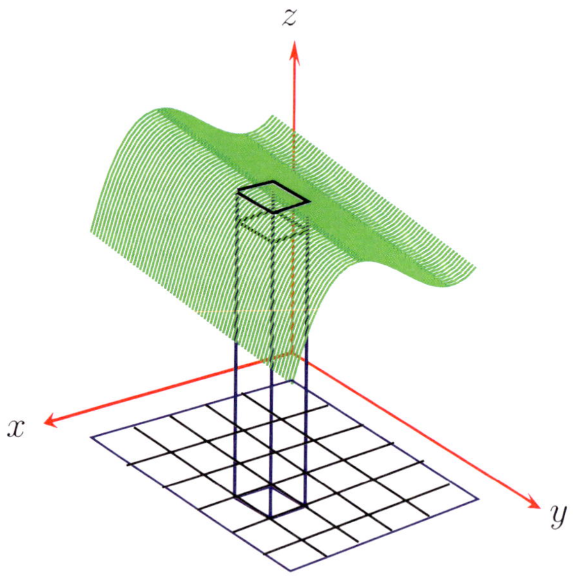

Fig. 5.1. Inscribed and circumscribing cuboids for a solid lying below a surface.

By a **partition** of a rectangle $[a,b] \times [c,d]$, we mean a finite set

$$P := \{(x_i, y_j) : i = 0, 1, \ldots, n \text{ and } j = 0, 1, \ldots, k\}$$

of points in $[a,b] \times [c,d]$ such that

$$a = x_0 < x_1 < \cdots < x_{n-1} < x_n = b \text{ and } c = y_0 < y_1 < \cdots < y_{k-1} < y_k = d.$$

The points (x_i, y_j), where $0 \le i \le n$ and $0 \le j \le k$, are sometimes called the **grid points** of P. For $1 \le i \le n$ and $1 \le j \le k$, the rectangle $[x_{i-1}, x_i] \times [y_{j-1}, y_j]$ is called the (i,j)th **subrectangle induced by** P. Further, we define the **mesh** of P to be

$$\mu(P) := \max\{x_1 - x_0, \ldots, x_n - x_{n-1}, y_1 - y_0, \ldots, y_k - y_{k-1}\}.$$

Note that the mesh of P is the maximum of the lengths of sides of the subrectangles induced by P. The mesh of P can be thought of as a measure of how finely the partition P subdivides the rectangle $[a,b] \times [c,d]$. The reason why $\mu(P)$ has been defined the way we have will be clear from the proof of Lemma 5.2. Further justification is given later, in Remark 5.33.

Example 5.1. The simplest partition of $[a,b] \times [c,d]$ is the one with only the corner points as its grid points, namely,

$$\mathsf{P}_{1,1} := \{(a,c), (a,d), (b,c), (b,d)\}.$$

More generally, for any $n, k \in \mathbb{N}$, the partition

$$\mathsf{P}_{n,k} := \{(x_i, y_j) : i = 0, 1, \ldots, n \text{ and } j = 0, 1, \ldots, k\},$$

where

$$x_i = a + \frac{i(b-a)}{n}, \quad i = 0, 1, \ldots, n, \quad \text{and} \quad y_j = c + \frac{j(d-c)}{k}, \quad j = 0, 1, \ldots, k,$$

divides the rectangle $[a,b] \times [c,d]$ into nk subrectangles of equal area, namely, $(b-a)(d-c)/nk$. We shall refer to $\mathsf{P}_{n,k}$ as the partition of $[a,b] \times [c,d]$ into $n \times k$ equal parts. Note that $\mu(\mathsf{P}_{n,k}) = \max\{1/n, 1/k\}$. It is clear that as n and k become large, $\mu(\mathsf{P}_{n,k})$ tends to zero and the subdivision of $[a,b] \times [c,d]$ corresponding to $\mathsf{P}_{n,k}$ becomes uniformly finer. ◇

Let $R := [a,b] \times [c,d]$ be a rectangle in $\mathbb{R}^2$ and let $f : R \to \mathbb{R}$ be a bounded function. Let us define

$$m(f) := \inf\{f(x,y) : (x,y) \in R\} \quad \text{and} \quad M(f) := \sup\{f(x,y) : (x,y) \in R\}.$$

Given a partition $P = \{(x_i, y_j) : i = 0, 1, \ldots, n \text{ and } j = 0, 1, \ldots, k\}$ of R, let

$$m_{i,j}(f) := \inf\{f(x,y) : (x,y) \in [x_{i-1}, x_i] \times [y_{j-1}, y_j]\},$$
$$M_{i,j}(f) := \sup\{f(x,y) : (x,y) \in [x_{i-1}, x_i] \times [y_{j-1}, y_j]\},$$

for $i = 1, \ldots, n$ and $j = 1, \ldots, k$. Clearly,

$$m(f) \le m_{i,j}(f) \le M_{i,j}(f) \le M(f) \quad \text{for all } i = 1, \ldots, n \text{ and } j = 1, \ldots, k.$$

We define the **lower double sum** and the **upper double sum** for the function f with respect to the partition P as follows:

$$L(P, f) := \sum_{i=1}^{n} \sum_{j=1}^{k} m_{i,j}(f)(x_i - x_{i-1})(y_j - y_{j-1}),$$

$$U(P, f) := \sum_{i=1}^{n} \sum_{j=1}^{k} M_{i,j}(f)(x_i - x_{i-1})(y_j - y_{j-1}).$$

Since

$$\sum_{i=1}^{n} \sum_{j=1}^{k} (x_i - x_{i-1})(y_j - y_{j-1}) = \sum_{i=1}^{n} (x_i - x_{i-1}) \sum_{j=1}^{k} (y_j - y_{j-1}) = (b-a)(d-c),$$

it follows that

$$m(f)(b-a)(d-c) \le L(P, f) \le U(P, f) \le M(f)(b-a)(d-c).$$

We now define the **lower double integral** of a bounded function f by

$$L(f) := \sup\{L(P, f) : P \text{ is a partition of } [a, b] \times [c, d]\}$$

and the **upper double integral** of f by

$$U(f) := \inf\{U(P, f) : P \text{ is a partition of } [a, b] \times [c, d]\}.$$

Given a partition P of $[a, b]\times[c, d]$, we say that a partition P^* of $[a, b]\times[c, d]$ is a **refinement** of P if every grid point of P is also a grid point of P^*. Given partitions P_1 and P_2 of $[a, b] \times [c, d]$, we say that a partition P^* of $[a, b] \times [c, d]$ is their **common refinement** if the grid points of P^* consist entirely of the grid points of P_1 and the grid points of P_2.

It turns out, as in the case of the Riemann integral, that as we refine a partition, the lower double sums can only become larger, whereas the upper double sums can only become smaller. To prove this, we will first analyze the effect of inserting one additional point, say (x^*, y^*) in a subrectangle induced by a partition $P = \{(x_i, y_j) : i = 0, 1, \ldots, n \text{ and } j = 0, 1, \ldots, k\}$. Note, however, that the resulting partition can have several additional points, namely, (x_i, y^*) for $i = 0, 1, \ldots, n$ and (x^*, y_j) for $j = 0, 1, \ldots, k$. We shall refer to such a refinement as a **one-step refinement** of P by the point (x^*, y^*).

Lemma 5.2. *Let $f : [a, b] \times [c, d] \to \mathbb{R}$ be a bounded function and let $\alpha \in \mathbb{R}$ be such that $|f(x, y)| \le \alpha$ for all $(x, y) \in [a, b]\times[c, d]$. Also, let P be a partition of $[a, b]\times[c, d]$ and P^* a one-step refinement of P by a point (x^*, y^*) of $[a, b]\times[c, d]$ not in P. Then*

$$0 \le U(P, f) - U(P^*, f) \le \alpha\mu\ell \quad \textit{and} \quad 0 \le L(P^*, f) - L(P, f) \le \alpha\mu\ell,$$

where $\mu := \mu(P)$ *denotes the mesh of* P *and* $\ell := 2\,(b - a + d - c)$ *denotes the perimeter of* $[a, b] \times [c, d]$.

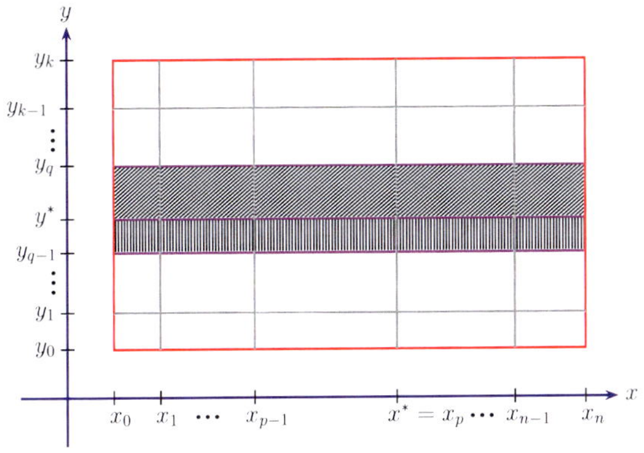

Fig. 5.2. Illustration of Case 1 in the proof of Lemma 5.2, where P^* splits the (i, q)th subrectangle induced by P into the top and the bottom subrectangles.

Proof. Let $P := \{(x_i, y_j) : i = 0, 1, \ldots, n \text{ and } j = 0, 1, \ldots, k\}$. Since (x^*, y^*) is not a grid point of P, it suffices to consider the following three cases.

Case 1. $x^* \in \{x_0, x_1, \ldots, x_n\}$, but $y^* \notin \{y_0, y_1, \ldots, y_k\}$.

In this case there are unique integers p, q with $0 \le p \le n$ and $1 \le q \le k$ such that $x^* = x_p$ and $y_{q-1} < y^* < y_q$. Now, for any $i = 1, \ldots, n$ and $j = 1, \ldots, k$, the (i, j)th subrectangle induced by P is also a subrectangle induced by P^*, provided $j \ne q$, while for $i = 1, \ldots, n$, the (i, q)th subrectangle induced by P splits into the bottom subrectangle $[x_{i-1}, x_i] \times [y_{q-1}, y^*]$ and the top subrectangle $[x_{i-1}, x_i] \times [y^*, y_q]$, which are among the subrectangles induced by P^*. (See Figure 5.2.) Thus, if for $i = 1, \ldots, n$, we let $M^B_{i,q}(f)$ and $M^T_{i,q}(f)$ denote the supremum of f on the bottom subrectangle and the top subrectangle respectively, then the difference $U(P, f) - U(P^*, f)$ reduces to

$$\sum_{i=1}^{n} (x_i - x_{i-1}) \left[M_{i,q}(f)(y_q - y_{q-1}) - M^T_{i,q}(f)(y_q - y^*) - M^B_{i,q}(f)(y^* - y_{q-1})\right].$$

Writing $(y_q - y_{q-1}) = (y_q - y^*) + (y^* - y_{q-1})$ in the first term of the above summands, we see that

$$U(P, f) - U(P^*, f) = \sum_{i=1}^{n} (x_i - x_{i-1}) \ \Big[\left(M_{i,q}(f) - M^T_{i,q}(f)\right)(y_q - y^*) + \left(M_{i,q}(f) - M^B_{i,q}(f)\right)(y^* - y_{q-1})\Big].$$

Now, $0 \le \left[M_{i,q}(f) - M_{i,q}^T(f)\right] \le 2\alpha$ and $0 \le \left[M_{i,q}(f) - M_{i,q}^B(f)\right] \le 2\alpha$ for each $i = 1, \ldots, n$, and so it follows that

$$\begin{aligned} 0 \le U(P, f) - U(P^*, f) &\le 2\alpha(y_q - y_{q-1}) \sum_{i=1}^{n} (x_i - x_{i-1}) \\ &= \alpha(y_q - y_{q-1})\,[2(b-a)] \le \alpha\mu\ell. \end{aligned}$$

In a similar way, we see that $0 \le L(P^*, f) - L(P, f) \le \alpha\mu\ell$.

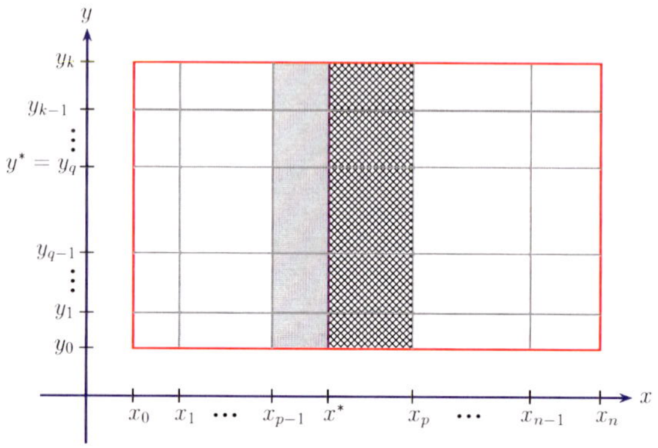

Fig. 5.3. Illustration of Case 2 in the proof of Lemma 5.2, where P^* splits the (p, j)th subrectangle induced by P into the left and the right subrectangles.

Case 2. $x^* \notin \{x_0, x_1, \ldots, x_n\}$, but $y^* \in \{y_0, y_1, \ldots, y_k\}$.
In this case there are unique integers p, q with $1 \le p \le n$ and $0 \le q \le k$ such that $x_{p-1} < x^* < x_p$ and $y^* = y_q$. We can then consider the left subrectangle $[x_{p-1}, x^*] \times [y_{j-1}, y_j]$ and the right subrectangle $[x^*, x_p] \times [y_{j-1}, y_j]$ for each $j = 1, \ldots, k$ (see Figure 5.3), and proceed as in Case 1 above to obtain

$$\begin{aligned} 0 \le U(P, f) - U(P^*, f) &\le 2\alpha(x_p - x_{p-1}) \sum_{j=1}^{k} (y_j - y_{j-1}) \\ &= \alpha(x_p - x_{p-1})\,[2(d-c)] \le \alpha\mu\ell, \end{aligned}$$

and also that $0 \le L(P^*, f) - L(P, f) \le \alpha\mu\ell$.

Case 3. $x^* \notin \{x_0, x_1, \ldots, x_n\}$ and $y^* \notin \{y_0, y_1, \ldots, y_k\}$.
In this case there are unique integers p, q with $1 \le p \le n$ and $1 \le q \le k$ such that $x_{p-1} < x^* < x_p$ and $y_{q-1} < y^* < y_q$. As in Case 1, P^* has the effect of splitting the (i, q)th subrectangle induced by P into the bottom subrectangle $[x_{i-1}, x_i] \times [y_{q-1}, y^*]$ and the top subrectangle $[x_{i-1}, x_i] \times [y^*, y_q]$ for each $i = 1, \ldots, n$ except $i = p$, and as in Case 2, splitting the (p, j)th subrectangle induced by P into the left subrectangle $[x_{p-1}, x^*] \times [y_{j-1}, y_j]$

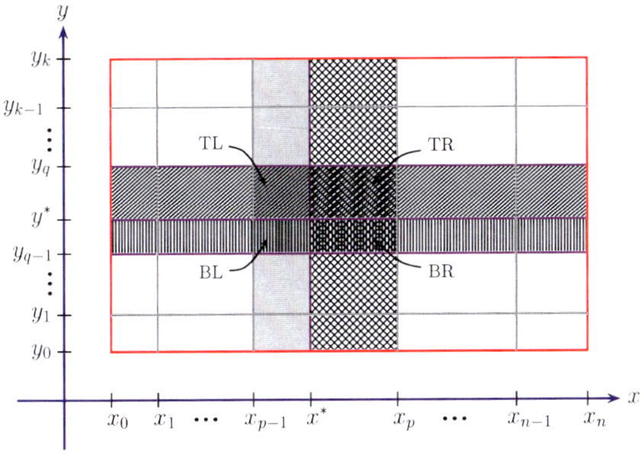

Fig. 5.4. Illustration of Case 3 in the proof of Lemma 5.2, where the refinement P^* splits the (p,q)th subrectangle induced by P into the top left (TL), the top right (TR), the bottom left (BL), and the bottom right (BR) subrectangles.

and the right subrectangle $[x^*, x_p] \times [y_{j-1}, y_j]$ for each $j = 1, \ldots, k$ except $j = q$. However, the (p,q)th subrectangle splits into four subrectangles, namely the bottom left subrectangle $[x_{p-1}, x^*] \times [y_{q-1}, y^*]$, the top left subrectangle $[x_{p-1}, x^*] \times [y^*, y_q]$, the bottom right subrectangle $[x^*, x_p] \times [y_{q-1}, y^*]$, and the top right subrectangle $[x^*, x_p] \times [y^*, y_q]$. (See Figure 5.4.) Let $M_{p,q}^{BL}(f)$, $M_{p,q}^{TL}(f)$, $M_{p,q}^{BR}(f)$, and $M_{p,q}^{TR}(f)$ denote, respectively, the supremum of f over these four subrectangles, and let

$$\begin{aligned} U_{p,q}^* := {} & M_{p,q}^{BL}(f)(x^* - x_{p-1})(y^* - y_{q-1}) + M_{p,q}^{TL}(f)(x^* - x_{p-1})(y_q - y^*) \\ & + M_{p,q}^{BR}(f)(x_p - x^*)(y^* - y_{q-1}) + M_{p,q}^{TR}(f)(x_p - x^*)(y_q - y^*). \end{aligned}$$

Also, let $U_{p,q} := M_{p,q}(f)(x_p - x_{p-1})(y_q - y_{q-1})$ denote the corresponding contribution to $U(P, f)$ from the (p,q)th subrectangle induced by P. Writing $(y_q - y_{q-1}) = (y_q - y^*) + (y^* - y_{q-1})$ and $(x_p - x_{p-1}) = (x_p - x^*) + (x^* - x_{p-1})$, we see that

$$\begin{aligned} U_{p,q} - U_{p,q}^* = {} & \left[M_{p,q}(f) - M_{p,q}^{BL}(f)\right](x^* - x_{p-1})(y^* - y_{q-1}) \\ & + \left[M_{p,q}(f) - M_{p,q}^{TL}(f)\right](x^* - x_{p-1})(y_q - y^*) \\ & + \left[M_{p,q}(f) - M_{p,q}^{BR}(f)\right](x_p - x^*)(y^* - y_{q-1}) \\ & + \left[M_{p,q}(f) - M_{p,q}^{TR}(f)\right](x_p - x^*)(y_q - y^*). \end{aligned}$$

Since the differences in the square brackets are clearly nonnegative and bounded above by 2α, we see that

$$0 \le U_{p,q} - U_{p,q}^* \le 2\alpha(x_p - x_{p-1})(y_q - y_{q-1}).$$

Combining this with the arguments in Case 1 as well as Case 2, we see that

$$
\begin{aligned}
0 \le U(P,f) - U(P^*,f) \le 2\alpha(y_q - y_{q-1}) &\sum_{i\neq p}(x_i - x_{i-1}) \\
+ 2\alpha(x_p - x_{p-1}) &\sum_{j\neq q}(y_j - y_{j-1}) \\
+ 2\alpha(x_p - x_{p-1})&(y_q - y_{q-1}).
\end{aligned}
$$

Since it is clear that $\sum_{i\neq p}(x_i - x_{i-1}) = (b-a) - (x_p - x_{p-1})$ and also that $\sum_{j\neq q}(y_j - y_{j-1}) = (d-c) - (y_q - y_{q-1})$, we obtain

$$
\begin{aligned}
0 &\le U(P,f) - U(P^*,f) \\
&\le 2\alpha\left[(y_q - y_{q-1})(b-a) + (x_p - x_{p-1})(d-c) - (x_p - x_{p-1})(y_q - y_{q-1})\right] \\
&\le \alpha\mu\left[2(b-a) + 2(d-c)\right] = \alpha\mu\ell.
\end{aligned}
$$

In a similar way, we see that $0 \le L(P^*,f) - L(P,f) \le \alpha\mu\ell$. □

We shall see below that the lower bound, namely 0, on the differences $U(P,f) - U(P^*,f)$ and $L(P^*,f) - L(P,f)$ proved in Lemma 5.2 has a number of nice consequences. It may be remarked that the upper bounds on these differences will be used only toward the end of this section, where we discuss Riemann double sums.

Proposition 5.3. *Let $f : [a,b] \times [c,d] \to \mathbb{R}$ be a bounded function.*

(i) *If P is partition of $[a,b] \times [c,d]$, and P^* is a refinement of P, then*

$$L(P,f) \le L(P^*,f) \quad \text{and} \quad U(P^*,f) \le U(P,f),$$

and consequently,

$$U(P^*,f) - L(P^*,f) \le U(P,f) - L(P,f).$$

(ii) *If P_1 and P_2 are partitions of $[a,b] \times [c,d]$, then $L(P_1,f) \le U(P_2,f)$.*
(iii) $L(f) \le U(f)$.

Proof. (i) Any refinement P^* of P can be obtained by a finite succession of one-step refinements starting with P. Hence by successively applying Lemma 5.2, we see that $0 \le L(P^*,f) - L(P,f)$ and $0 \le U(P,f) - U(P^*,f)$, that is, $L(P,f) \le L(P^*,f)$ and $U(P^*,f) \le U(P,f)$. As a consequence, we also have $U(P^*,f) - L(P^*,f) \le U(P,f) - L(P,f)$.

(ii) Let P^* denote the common refinement of partitions P_1 and P_2. Then in view of (i) above, $L(P_1,f) \le L(P^*,f) \le U(P^*,f) \le U(P_2,f)$.

(iii) Fix a partition P_0 of $[a,b] \times [c,d]$. By (ii) above, we have $L(P_0,f) \le U(P,f)$ for any partition P of $[a,b] \times [c,d]$. Hence $L(P_0,f) \le U(f)$. Now, since P_0 is an arbitrary partition of $[a,b] \times [c,d]$, we see that $L(f) \le U(f)$. □

Let $f : [a,b] \times [c,d] \to \mathbb{R}$ be a bounded function. Then f is said to be **integrable** on $[a,b] \times [c,d]$ if $L(f) = U(f)$. In other words, f is integrable if its lower double integral is equal to its upper double integral. In case f is integrable, the common value $L(f) = U(f)$ is called the **double integral**, or simply the **integral**, of f (on $[a,b] \times [c,d]$), and it is denoted by

$$\iint_{[a,b]\times[c,d]} f(x,y)d(x,y) \quad \text{or simply by} \quad \iint_{[a,b]\times[c,d]} f.$$

If f is integrable and also nonnegative, then the **volume** of the solid under the surface given by $z = f(x,y)$ and above the rectangle $[a,b] \times [c,d]$ is defined to be the double integral of f on $[a,b] \times [c,d]$. In other words,

$$\mathrm{Vol}(E_f) := \iint_{[a,b]\times[c,d]} f(x,y)d(x,y),$$

where

$$E_f := \{(x,y,z) \in \mathbb{R}^3 : a \le x \le b,\ c \le y \le d \text{ and } 0 \le z \le f(x,y)\}.$$

Basic Inequality and Criterion for Integrability

The following result gives an elementary but useful estimate for the absolute value of a double integral.

Proposition 5.4 (Basic Inequality). *Let $f : [a,b] \times [c,d] \to \mathbb{R}$ be an integrable function. If there are $\alpha, \beta \in \mathbb{R}$ such that $\beta \le f \le \alpha$, then we have*

$$\beta(b-a)(d-c) \le \iint_{[a,b]\times[c,d]} f(x,y)d(x,y) \le \alpha(b-a)(d-c).$$

In particular, if $|f| \le \alpha$, then we have

$$\left| \iint_{[a,b]\times[c,d]} f(x,y)d(x,y) \right| \le \alpha(b-a)(d-c).$$

Proof. As noted earlier, for every partition P of $[a,b] \times [c,d]$, we have

$$m(f)(b-a)(d-c) \le L(P,f) \le U(P,f) \le M(f)(b-a)(d-c).$$

This implies that $m(f)(b-a)(d-c) \le L(f) \le U(f) \le M(f)(b-a)(d-c)$. Now, since $\beta \le f(x) \le \alpha$ for all $x \in \mathbb{R}$, we have $\beta \le m(f)$ and $M(f) \le \alpha$. Also, since f is integrable, we have $L(f) = U(f)$. Using these facts, we readily obtain the desired inequalities. □

Examples 5.5. (i) Let $f(x,y) := 1$ for all $(x,y) \in [a,b] \times [c,d]$. Then for every partition $P := \{(x_i, y_j) : i = 0, 1, \ldots, n \text{ and } j = 0, 1, \ldots, k\}$, we have $m_{i,j}(f) = 1 = M_{i,j}(f)$ for all $i = 1, \ldots, n$ and $j = 1, \ldots, k$. Since

$$L(P, f) = U(P, f) = \sum_{i=1}^{n} \sum_{j=1}^{n} (x_i - x_{i-1})(y_j - y_{j-1}) = (b-a)(d-c),$$

we see that $L(f) = U(f) = (b-a)(d-c)$. Thus f is integrable and its double integral is equal to $(b-a)(d-c)$. In a similar manner, we see that if $r \in \mathbb{R}$ and f is the constant function on $[a,b] \times [c,d]$ defined by $f(x,y) := r$ for $(x,y) \in [a,b] \times [c,d]$, then f is integrable and

$$\iint_{[a,b]\times[c,d]} f(x,y) d(x,y) = \iint_{[a,b]\times[c,d]} r\, d(x,y) = r(b-a)(d-c).$$

(ii) Suppose $\phi : [a,b] \to \mathbb{R}$ is a bounded function of one variable. Let us regard ϕ as a function of two variables or in other words, consider $f : [a,b] \times [c,d] \to \mathbb{R}$ defined by $f(x,y) := \phi(x)$ for all $(x,y) \in [a,b] \times [c,d]$. Clearly, f is a bounded function. Now, if $P = \{(x_i, y_j) : i = 0, 1, \ldots, n \text{ and } j = 0, 1, \ldots, k\}$ is any partition of $[a,b] \times [c,d]$, then $P_1 := \{x_0, x_1, \ldots, x_n\}$ is a partition of $[a,b]$, whereas if Q is any partition of $[a,b]$, then $Q = P_1$ for some partition P of $[a,b] \times [c,d]$. (For example, if $Q = \{s_0, s_1, \ldots, s_m\}$, then we can take $P = \{(s_i, t_j) : i = 0, 1, \ldots, m \text{ and } j = 0, 1\}$, where $t_0 := c$ and $t_1 := d$.) Moreover, for any partition $P := \{(x_i, y_j) : i = 0, 1, \ldots, n \text{ and } j = 0, 1, \ldots, k\}$ of $[a,b] \times [c,d]$, we have $m_{i,j}(f) = m_i(\phi)$, where $m_i(\phi)$ denotes the infimum of ϕ on $[x_{i-1}, x_i]$, and therefore

$$\sum_{i=1}^{n} \sum_{j=1}^{n} (x_i - x_{i-1})(y_j - y_{j-1}) m_{i,j}(f) = (d-c) \sum_{i=1}^{n} (x_i - x_{i-1}) m_i(\phi),$$

that is, $L(P, f) = (d-c)L(P_1, \phi)$. Similarly, $U(P, f) = (d-c)U(P_1, \phi)$. Consequently, $L(f) = (d-c)L(\phi)$ and $U(f) = (d-c)U(\phi)$. Hence

$$f \text{ is integrable on } [a,b] \times [c,d] \iff \phi \text{ is Riemann integrable on } [a,b],$$

and in this case,

$$\iint_{[a,b]\times[c,d]} f(x,y) d(x,y) = (d-c) \int_a^b \phi(x) dx.$$

A similar conclusion holds in case $\psi : [c,d] \to \mathbb{R}$ is a bounded function of one variable and we define $f : [a,b] \times [c,d] \to \mathbb{R}$ by $f(x,y) := \psi(y)$ for all $(x,y) \in [a,b] \times [c,d]$.

(iii) It may be recalled that the **Dirichlet function** $\phi : [a,b] \to \mathbb{R}$ defined by

$$\phi(x) = \begin{cases} 1 & \text{if } x \text{ is a rational number,} \\ 0 & \text{if } x \text{ is an irrational number,} \end{cases}$$

provides a standard example of a function of one variable that is not Riemann integrable. (See, for instance, Example 6.4 (ii) of ACICARA.) Consider a variant of ϕ, namely, the function $f : [a,b] \times [c,d] \to \mathbb{R}$ defined by

$$f(x,y) = \begin{cases} 1 & \text{if } x \text{ and } y \text{ are rational numbers,} \\ 0 & \text{if } x \text{ or } y \text{ is an irrational number.} \end{cases}$$

We shall refer to f as the **bivariate Dirichlet function**. Let $P = \{(x_i, y_j) : i = 0, 1, \ldots, n \text{ and } j = 0, 1, \ldots, k\}$ be any partition of $[a,b] \times [c,d]$. Since both $[x_{i-1}, x_i]$ and $[y_{i-1}, y_j]$ contain a rational number as well as an irrational number, we see that $m_{i,j}(f) = 0$ and $M_{i,j}(f) = 1$ for all $i = 1, \ldots, n$ and $j = 1, \ldots, k$. Thus

$$L(P, f) = \sum_{i=1}^{n} \sum_{j=1}^{k} 0 \cdot (x_i - x_{i-1})(y_j - y_{j-1}) = 0,$$

whereas

$$U(P, f) = \sum_{i=1}^{n} \sum_{j=1}^{k} 1 \cdot (x_i - x_{i-1})(y_j - y_{j-1}) = (b-a)(d-c).$$

Consequently, $L(f) = 0$, whereas $U(f) = (b-a)(d-c)$. Since $a < b$ and $c < d$, we have $L(f) \neq U(f)$, that is, f is not integrable. ◇

The following result gives a useful criterion to determine whether a bounded function defined on $[a,b] \times [c,d]$ is integrable. It is exactly analogous to the corresponding criterion, also known as the Riemann Condition, for functions of one variable. (See, for example, Proposition 6.5 of ACICARA.)

Proposition 5.6 (Riemann Condition). *Let $f : [a,b] \times [c,d] \to \mathbb{R}$ be a bounded function. Then f is integrable if and only if for every $\epsilon > 0$, there is a partition P_ϵ of $[a,b] \times [c,d]$ such that*

$$U(P_\epsilon, f) - L(P_\epsilon, f) < \epsilon.$$

Proof. Suppose the stated condition is satisfied. Then for every $\epsilon > 0$, we have

$$0 \leq U(f) - L(f) \leq U(P_\epsilon, f) - L(P_\epsilon, f) < \epsilon.$$

Hence $L(f) = U(f)$, that is, f is integrable.

Conversely, suppose f is integrable. Let $\epsilon > 0$ be given. By the definitions of $U(f)$ and $L(f)$, there are partitions Q_ϵ and $\widetilde{Q}_\epsilon$ of $[a,b] \times [c,d]$ such that

$$U(Q_\epsilon, f) < U(f) + \frac{\epsilon}{2} \quad \text{and} \quad L(\widetilde{Q}_\epsilon, f) > L(f) - \frac{\epsilon}{2}.$$

Let P_ϵ denote the common refinement of Q_ϵ and $\widetilde{Q}_\epsilon$. Then by part (i) of Proposition 5.3,

$$L(f) - \frac{\epsilon}{2} < L(\widetilde{Q}_\epsilon, f) \le L(P_\epsilon, f) \le U(P_\epsilon, f) \le U(Q_\epsilon, f) < U(f) + \frac{\epsilon}{2}.$$

Since $L(f) = U(f)$, it follows that $U(P_\epsilon, f) - L(P_\epsilon, f) < \epsilon$, as desired □

Example 5.7. Let $f : [a, b] \times [c, d] \to \mathbb{R}$ be a bounded function such that $f(x, y) = 0$ for all $(x, y) \in (a, b) \times (c, d)$. Although the values of the function f on the sides of the rectangle $[a, b] \times [c, d]$ can be arbitrary (though bounded), we show that f is integrable and its double integral is equal to zero.

Let $\epsilon > 0$ satisfy $\epsilon < \min\{b - a, d - c\}/2$, and consider the partition $P_\epsilon = \{(x_i, y_j) : i, j = 0, 1, 2, 3\}$ of $[a, b] \times [c, d]$, where $x_0 := a$, $x_1 := a + \epsilon$, $x_2 := b - \epsilon$, $x_3 := b$ and $y_0 := c$, $y_1 := c + \epsilon$, $y_2 := d - \epsilon$, $y_3 := d$. (See Figure 5.5.)

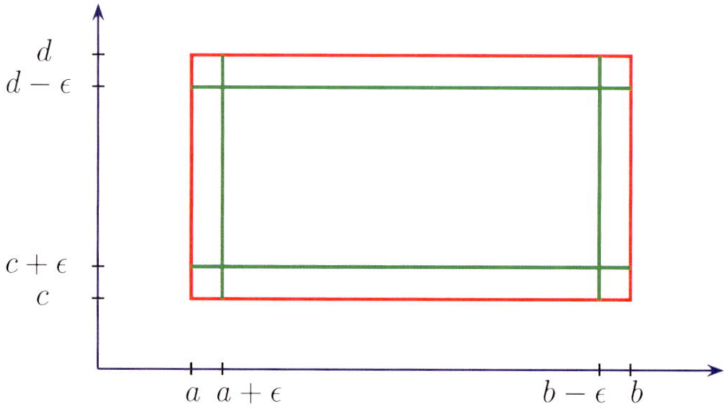

Fig. 5.5. The partition P_ϵ of $[a, b] \times [c, d]$ as in Example 5.7.

Since f is bounded, there is $\alpha > 0$ such that $-\alpha \le f(x, y) \le \alpha$ for all $(x, y) \in [a, b] \times [c, d]$. Further, since $f(x, y) = 0$ for all $(x, y) \in [a + \epsilon, b - \epsilon] \times [c + \epsilon, d - \epsilon]$, we have

$$U(P_\epsilon, f) \le \alpha\,[2\epsilon(b - a) + 2\epsilon(d - c)] = 2\alpha\epsilon(b - a + d - c)$$

and

$$L(P_\epsilon, f) \ge -\alpha\,[2\epsilon(b - a) + 2\epsilon(d - c)] = -2\alpha\epsilon(b - a + d - c).$$

Thus $U(P_\epsilon, f) - L(P_\epsilon, f) \le 4\alpha\epsilon(b - a + d - c)$. Since $\epsilon > 0$ can be taken arbitrarily small, the Riemann Condition (Proposition 5.6) shows that f is integrable. Moreover,

$$-2\alpha(b - a + d - c)\epsilon \le \iint_{[a,b]\times[c,d]} f(x, y)d(x, y) \le 2\alpha(b - a + d - c)\epsilon,$$

and since $\epsilon > 0$ is arbitrary, it follows that the double integral of f on $[a, b] \times [c, d]$ is equal to 0. ◇

Domain Additivity on Rectangles

A basic property of Riemann integrals is domain additivity, which is recalled below. A proof can be found, for example, on page 187 of ACICARA.

Fact 5.8 (Domain Additivity of Riemann Integrals). *Let $\phi : [a, b] \to \mathbb{R}$ be a bounded function and let $c \in (a, b)$. Then ϕ is integrable on $[a, b]$ if and only if ϕ is integrable on $[a, c]$ as well as on $[c, b]$. In this case,*

$$\int_a^b \phi(x)dx = \int_a^c \phi(x)dx + \int_c^b \phi(x)dx.$$

We shall now state and prove an analogous result for double integrals on rectangles.

Proposition 5.9 (Domain Additivity on Rectangles). *Let $f : [a, b] \times [c, d] \to \mathbb{R}$ be a bounded function.*

(i) *Let $s \in (a, b)$. Then f is integrable on $[a, b] \times [c, d]$ if and only if f is integrable on $[a, s] \times [c, d]$ as well as on $[s, b] \times [c, d]$. In this case,*

$$\iint_{[a,b]\times[c,d]} f(x, y)d(x, y) = \iint_{[a,s]\times[c,d]} f(x, y)d(x, y) + \iint_{[s,b]\times[c,d]} f(x, y)d(x, y).$$

(ii) *Let $t \in (c, d)$. Then f is integrable on $[a, b] \times [c, d]$ if and only if f is integrable on $[a, b] \times [c, t]$ as well as on $[a, b] \times [t, d]$. In this case,*

$$\iint_{[a,b]\times[c,d]} f(x, y)d(x, y) = \iint_{[a,b]\times[c,t]} f(x, y)d(x, y) + \iint_{[a,b]\times[t,d]} f(x, y)d(x, y).$$

Proof. (i) Assume that f is integrable on $[a, b] \times [c, d]$. Let $\epsilon > 0$ be given. Then, by the Riemann Condition (Proposition 5.6), there is a partition $P_\epsilon := \{(x_i, y_j) : i = 0, 1, \ldots, n \text{ and } j = 0, 1, \ldots, k\}$ of $[a, b] \times [c, d]$ such that $U(P_\epsilon, f) - L(P_\epsilon, f) < \epsilon$. Adjoining the points $(s, y_0), \ldots, (s, y_k)$ to the points of P_ϵ, if these are not already points of P_ϵ, we obtain a refinement $P_\epsilon^* = \{(x_i^*, y_j) : i = 0, 1, \ldots, n^* \text{ and } j = 0, 1, \ldots, k\}$ of P_ϵ, where $n^* \in \{n, n + 1\}$ and $\{x_i^* : i = 0, 1, \ldots, n^*\} = \{x_i : i = 0, 1, \ldots, n\} \cup \{s\}$. In particular, there is a unique $p \in \{1, \ldots, n\}$ such that $x_p^* = s$. Part (i) of Proposition 5.3 shows that

$$0 \le U(P_\epsilon^*, f) - L(P_\epsilon^*, f) \le U(P_\epsilon, f) - L(P_\epsilon, f) < \epsilon.$$

Now let g denote the restriction of f to $[a, s] \times [c, d]$, and let $Q_\epsilon^* := \{(x_i^*, y_j) : i = 0, 1, \ldots, p \text{ and } j = 0, 1, \ldots, k\}$. Then Q_ϵ^* is a partition of $[a, s] \times [c, d]$ and

$$\begin{aligned}U(Q_\epsilon^*, g) - L(Q_\epsilon^*, g) &= \sum_{i=1}^{p}\sum_{j=1}^{k}[M_{i,j}(g) - m_{i,j}(g)](x_i^* - x_{i-1}^*)(y_j - y_{j-1})\\ &= \sum_{i=1}^{p}\sum_{j=1}^{k}[M_{i,j}(f) - m_{i,j}(f)](x_i^* - x_{i-1}^*)(y_j - y_{j-1})\\ &\le \sum_{i=1}^{n}\sum_{j=1}^{k}[M_{i,j}(f) - m_{i,j}(f)](x_i^* - x_{i-1}^*)(y_j - y_{j-1})\\ &= U(P_\epsilon^*, f) - L(P_\epsilon^*, f),\end{aligned}$$

which is less than ϵ. Hence the Riemann Condition shows that g is integrable, that is, f is integrable on $[a, s] \times [c, d]$. Similarly, it can be seen that f is integrable on $[s, b] \times [c, d]$.

Conversely, suppose f is integrable on $[a, s]\times[c, d]$ as well as on $[s, b]\times[c, d]$, and let g and h denote the restrictions of f to these two subrectangles respectively. Given any $\epsilon > 0$, let $Q_\epsilon := \{(x_i, y_j) : i = 0, 1, \ldots, n \text{ and } j = 0, 1, \ldots, k\}$ and $R_\epsilon := \{(u_i, v_j) : i = 0, 1, \ldots, m \text{ and } j = 0, 1, \ldots, \ell\}$ be partitions of $[a, s] \times [c, d]$ and of $[s, b] \times [c, d]$ respectively, such that

$$U(Q_\epsilon, g) - L(Q_\epsilon, g) < \epsilon/2 \quad \text{and} \quad U(R_\epsilon, h) - L(R_\epsilon, h) < \epsilon/2.$$

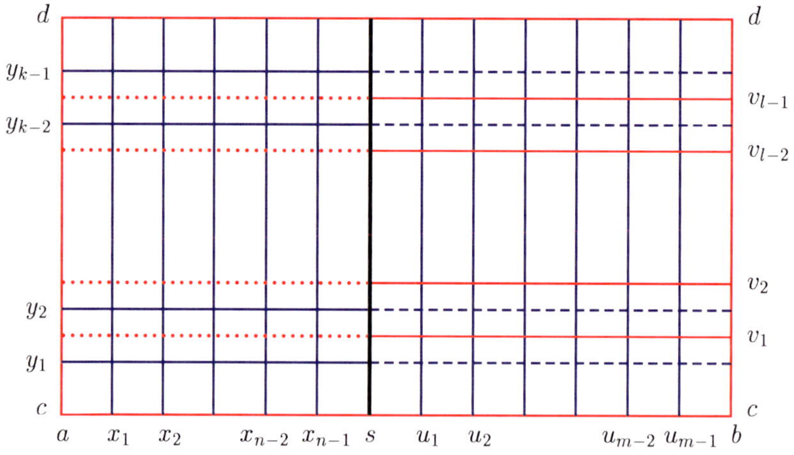

Fig. 5.6. Refinements of the partition $Q_\epsilon := \{(x_i, y_j)\}$ by adding the dotted lines and of $R_\epsilon := \{(u_i, v_j)\}$ by adding the dashed lines.

Let Q_ϵ^* denote the refinement of the partition Q_ϵ obtained by adding the points $\{(x_i, v_j) : i = 0, 1, \ldots, n \text{ and } j = 0, 1, \ldots, \ell\}$, and let R_ϵ^* denote the refinement of the partition R_ϵ obtained by adding the points $\{(u_i, y_j) : i = 0, 1, \ldots, m \text{ and } j = 0, 1, \ldots, k\}$. (See Figure 5.6.) Let P_ϵ^* denote the partition of $[a,b]\times[c,d]$ obtained by combining the points of the partitions Q^*_ϵ and R^*_ϵ. Then we have

$$U(P^*_\epsilon, f) = U(Q^*_\epsilon, g) + U(R^*_\epsilon, h) \quad \text{and} \quad L(P^*_\epsilon, f) = L(Q^*_\epsilon, g) + L(R^*_\epsilon, h).$$

Now by part (i) of Proposition 5.3, it follows that

$$U(P^*_\epsilon, f) \le U(Q_\epsilon, g) + U(R_\epsilon, h) \quad \text{and} \quad L(P^*_\epsilon, f) \ge L(Q_\epsilon, g) + L(R_\epsilon, h).$$

Hence $U(P^*_\epsilon, f) - L(P^*_\epsilon, f) < \epsilon/2 + \epsilon/2 = \epsilon$. Thus, by the Riemann Condition, f is integrable on $[a,b]\times[c,d]$.

To prove the last assertion in (i), suppose f is integrable on $[a,b]\times[c,d]$. Note that with $P^*_\epsilon, Q^*_\epsilon$, and R^*_ϵ as in the last paragraph, we have

$$L(P^*_\epsilon, f) \le I \le U(P^*_\epsilon, f), \quad \text{where } I := \iint_{[a,b]\times[c,d]} f(x,y)d(x,y).$$

Also, note that $U(P^*_\epsilon, f) = U(Q^*_\epsilon, g) + U(R^*_\epsilon, h)$ and $L(P^*_\epsilon, f) = L(Q^*_\epsilon, g) + L(R^*_\epsilon, h)$, and so we have

$$L(P^*_\epsilon, f) \le \iint_{[a,s]\times[c,d]} f(x,y)d(x,y) + \iint_{[s,b]\times[c,d]} f(x,y)d(x,y) \le U(P^*_\epsilon, f).$$

Since $U(P^*_\epsilon, f) - L(P^*_\epsilon, f) < \epsilon$, it follows that

$$\left| \iint_{[a,s]\times[c,d]} f(x,y)d(x,y) + \iint_{[s,b]\times[c,d]} f(x,y)d(x,y) - I \right| < \epsilon.$$

But $\epsilon > 0$ is arbitrary and therefore we must have

$$I = \iint_{[a,s]\times[c,d]} f(x,y)d(x,y) + \iint_{[s,b]\times[c,d]} f(x,y)d(x,y),$$

as desired.

(ii) This part can be proved using arguments similar to those in the proof of part (i) above. □

Corollary 5.10. *Let $f : [a,b]\times[c,d] \to \mathbb{R}$ be a bounded function. Consider $s \in (a,b)$ and $t \in (c,d)$. Then f is integrable on $[a,b]\times[c,d]$ if and only if f is integrable on each of the four rectangles $[a,s]\times[c,t]$, $[a,s]\times[t,d]$, $[s,b]\times[c,t]$, and $[s,b]\times[t,d]$. In this case,*

$$\iint_{[a,b]\times[c,d]} f = \iint_{[a,s]\times[c,t]} f + \iint_{[a,s]\times[t,d]} f + \iint_{[s,b]\times[c,t]} f + \iint_{[s,b]\times[t,d]} f.$$

Proof. The result follows from parts (i) and (ii) of Proposition 5.9. □

Remark 5.11. As mentioned at the beginning of this chapter, we have assumed $a < b$ and $c < d$ while defining the double integral of a function $f : [a, b] \times [c, d] \to \mathbb{R}$. In order to obtain uniformity of presentation, we adopt the following conventions. Suppose a, b, c, d are arbitrary real numbers and R is the rectangle in $\mathbb{R}^2$ with (a, c) and (b, d) as its diagonally opposite vertices. If $a = b$ or $c = d$, then every $f : R \to \mathbb{R}$ is integrable and

$$\iint_{[a,b]\times[c,d]} f(x, y)d(x, y) := 0.$$

If $a > b$ and $c < d$ or if $a < b$ and $c > d$ (so that $R = [b, a] \times [c, d]$ or $R = [a, b] \times [d, c]$), then an integrable function $f : R \to \mathbb{R}$ is also said to be integrable on $[a, b] \times [c, d]$ and we set

$$\iint_{[a,b]\times[c,d]} f(x, y)d(x, y) := -\iint_R f(x, y)d(x, y).$$

Finally, if $a > b$ and $c > d$ (so that $R = [b, a] \times [d, c]$), then an integrable function $f : R \to \mathbb{R}$ is also said to be integrable on $[a, b] \times [c, d]$ and we set

$$\iint_{[a,b]\times[c,d]} f(x, y)d(x, y) := \iint_R f(x, y)d(x, y).$$

Using these conventions together with Corollary 5.10, we obtain the following useful consequence of domain additivity. Suppose $a, b, c, d \in \mathbb{R}$ with $a < b$ and $c < d$, and suppose $f : [a, b] \times [c, d] \to \mathbb{R}$ is integrable. Given any $(u, v), (u_0, v_0) \in [a, b] \times [c, d]$, the double integral $I_{u,v}$ of f on $[a, u] \times [c, v]$ is given by

$$I_{u,v} = \iint_{[a,u_0]\times[c,v_0]} f + \iint_{[a,u_0]\times[v_0,v]} f + \iint_{[u_0,u]\times[c,v_0]} f + \iint_{[u_0,u]\times[v_0,v]} f.$$

Note that the above formula holds regardless of the relative positions of (u, v) and (u_0, v_0) in the rectangle $[a, b] \times [c, d]$. ◇

Integrability of Monotonic and Continuous Functions

Recall that in Chapter 1, we have defined the notion of monotonicity for functions of two variables. In effect, a function $f : [a, b] \times [c, d] \to \mathbb{R}$ is monotonically increasing (respectively monotonically decreasing) if it is monotonically increasing (respectively monotonically decreasing) in each of the two variables. We show below that such a function is always integrable. In fact, we prove a slightly more general result that permits the possibility that a function is monotonically increasing in one variable and monotonically decreasing in another. We also show that continuity implies integrability. As in one-variable calculus, proofs of both the results mainly use the Riemann Condition.

Proposition 5.12. *Given any $f : [a, b] \times [c, d] \to \mathbb{R}$, we have the following.*

(i) *If for every fixed $x \in [a, b]$, the function $\psi_x : [c, d] \to \mathbb{R}$ given by $\psi_x(y) := f(x, y)$ is monotonic, and for every fixed $y \in [c, d]$, the function $\phi_y : [a, b] \to \mathbb{R}$ given by $\phi_y(x) := f(x, y)$ is monotonic, then f is integrable. In particular, if f is monotonic, then f is integrable.*

(ii) *If f is continuous, then it is integrable.*

Proof. (i) First assume that for every fixed $x \in [a, b]$, the function ψ_x, and for every fixed $y \in [c, d]$, the function ϕ_y, are monotonically increasing. Then it is clear that $f(a, c) \leq f(x, y) \leq f(b, d)$ for all $(x, y) \in [a, b] \times [c, d]$. In particular, f is a bounded function. For $n \in \mathbb{N}$, consider the partition $\mathsf{P}_{n,n} := \{(x_i, y_j) : i, j = 0, 1, \ldots, n\}$ of $[a, b] \times [c, d]$ into $n \times n$ equal parts. Then for $i, j = 1, \ldots, n$, we have

$$M_{i,j}(f) = f(x_i, y_j) \quad \text{and} \quad m_{i,j}(f) = f(x_{i-1}, y_{j-1}).$$

Hence

$$\begin{aligned} U(\mathsf{P}_{n,n}, f) - L(\mathsf{P}_{n,n}, f) &= \sum_{i=1}^{n}\sum_{j=1}^{n}[M_{i,j}(f) - m_{i,j}(f)](x_i - x_{i-1})(y_j - y_{j-1}) \\ &= \frac{(b-a)(d-c)}{n^2}\sum_{i=1}^{n}\sum_{j=1}^{n}[f(x_i, y_j) - f(x_{i-1}, y_{j-1})]. \end{aligned}$$

The last double sum is telescopic and it is equal to

$$[f(x_n, y_n) - f(x_0, y_0)] + \sum_{i=1}^{n-1}[f(x_i, y_n) - f(x_i, y_0)] + \sum_{j=1}^{n-1}[f(x_n, y_j) - f(x_0, y_j)].$$

By our hypothesis, each of the differences in the square brackets in the above sum is at most $[f(b, d) - f(a, c)]$. Consequently,

$$\begin{aligned} U(\mathsf{P}_{n,n}, f) - L(\mathsf{P}_{n,n}, f) &\leq \frac{(b-a)(d-c)}{n^2}[f(b, d) - f(a, c)](1 + n - 1 + n - 1) \\ &= \frac{2n-1}{n^2}(b-a)(d-c)[f(b, d) - f(a, c)]. \end{aligned}$$

Thus, given any $\epsilon > 0$, there is $n \in \mathbb{N}$ such that $U(\mathsf{P}_{n,n}, f) - L(\mathsf{P}_{n,n}, f) < \epsilon$. By the Riemann Condition (Proposition 5.6), it follows that f is integrable.

A similar proof holds if the function f is monotonically decreasing in each of the two variables x and y, or if it is monotonically increasing in one variable and monotonically decreasing in the other.

(ii) Assume that $f : [a, b] \times [c, d] \to \mathbb{R}$ is continuous. Then f is bounded, thanks to Proposition 2.25. Also, by Proposition 2.37, f is uniformly continuous. Let $A := (b-a)(d-c)$ denote the area of $[a, b] \times [c, d]$ and let $\epsilon > 0$ be given. By Proposition 2.39, there is $\delta > 0$ such that

$$(x, y), (u, v) \in D \text{ and } |(x, y) - (u, v)| < \delta \Longrightarrow |f(x, y) - f(u, v)| < \frac{\epsilon}{A}.$$

Let $P = \{(x_i, y_j) : i = 0, 1, \ldots, n \text{ and } j = 0, 1, \ldots, k\}$ be a partition of $[a, b] \times [c, d]$ such that the area $A_{i,j} := (x_i - x_{i-1})(y_j - y_{j-1})$ of the (i, j)th subrectangle $R_{i,j} := [x_{i-1}, x_i] \times [y_{j-1}, y_j]$ is less than δ for $i = 1, \ldots, n$ and $j = 1, \ldots, k$. Now by Proposition 2.25, for each $i = 1, \ldots, n$ and $j = 1, \ldots, k$, there are (a_i, b_j) and (c_i, d_j) in $R_{i,j}$ such that $f(a_i, b_j) = M_{i,j}(f)$ and $f(c_i, d_j) = m_{i,j}(f)$. Hence we have

$$M_{i,j}(f) - m_{i,j}(f) = f(a_i, b_j) - f(c_i, d_j) < \frac{\epsilon}{A},$$

and so

$$U(P_\epsilon, f) - L(P_\epsilon, f) = \sum_{i=1}^{n} \sum_{j=1}^{k} [M_{i,j}(f) - m_{i,j}(f)]\, A_{i,j} < \frac{\epsilon}{A} \sum_{i=1}^{n} \sum_{j=1}^{k} A_{i,j} = \epsilon.$$

Thus, by the Riemann Condition (Proposition 5.6), f is integrable. □

In Proposition 5.43 we shall see that even if a function is discontinuous at a few points, it can be integrable.

Examples 5.13. (i) Let a, b, c, d, r, s be nonnegative real numbers such that $a < b$ and $c < d$. Define $f : [a, b] \times [c, d] \to \mathbb{R}$ by $f(x, y) := x^r y^s$. By either part (i) or part (ii) of Proposition 5.12, we see that f is integrable on $[a, b] \times [c, d]$.

(ii) Consider $f : [a, b] \times [c, d] \to \mathbb{R}$ defined by $f(x, y) := \sin(x + y)$. Then by part (ii) of Proposition 5.12, f is integrable on $[a, b] \times [c, d]$. ◇

Algebraic and Order Properties

First, we shall see that double integrals behave just like Riemann integrals with respect to algebraic operations on functions.

Proposition 5.14. *Let $f, g : [a, b] \times [c, d] \to \mathbb{R}$ be integrable functions. Then*

(i) *$f + g$ is integrable and $\iint_{[a,b]\times[c,d]} (f + g) = \iint_{[a,b]\times[c,d]} f + \iint_{[a,b]\times[c,d]} g$,*
(ii) *rf is integrable for any $r \in \mathbb{R}$ and $\iint_{[a,b]\times[c,d]} (rf) = r \iint_{[a,b]\times[c,d]} f$,*
(iii) *fg is integrable,*
(iv) *if there is $\delta > 0$ such that $|f(x, y)| \geq \delta$ for all $(x, y) \in [a, b] \times [c, d]$, then $1/f$ is integrable,*
(v) *if $f(x, y)) \geq 0$ for all $(x, y) \in [a, b] \times [c, d]$, then for any $k \in \mathbb{N}$, the function $f^{1/k}$ is integrable.*

Proof. Let $\epsilon > 0$ be given. By the Riemann Condition (Proposition 5.6), there are partitions Q and R of $[a, b] \times [c, d]$ such that $U(Q, f) - L(Q, f) < \epsilon$ and

$U(R, g) - L(R, g) < \epsilon$. Let P_ϵ denote the common refinement of Q and R. Then by part (i) of Proposition 5.3, we have

$$U(P_\epsilon, f) - L(P_\epsilon, f) < \epsilon \quad \text{and} \quad U(P_\epsilon, g) - L(P_\epsilon, g) < \epsilon.$$

(i) Let $P_\epsilon = \{(x_i, y_j) : i = 0, 1, \ldots, n \text{ and } j = 0, 1, \ldots, k\}$. For any $i = 1, \ldots, n$ and $j = 1, \ldots, k$, we have

$$M_{i,j}(f+g) \le M_{i,j}(f) + M_{i,j}(g) \quad \text{and} \quad m_{i,j}(f+g) \ge m_{i,j}(f) + m_{i,j}(g).$$

Multiplying both sides of these inequalities by $(x_i - x_{i-1})(y_j - y_{j-1})$ and summing from $i = 1$ to n and from $j = 1$ to k, we obtain

$$U(P_\epsilon, f+g) \le U(P_\epsilon, f) + U(P_\epsilon, g) \quad \text{and} \quad L(P_\epsilon, f+g) \ge L(P_\epsilon, f) + L(P_\epsilon, g).$$

Hence

$$U(P_\epsilon, f+g) - L(P_\epsilon, f+g) \le U(P_\epsilon, f) - L(P_\epsilon, f) + U(P_\epsilon, g) - L(P_\epsilon, g) < \epsilon + \epsilon = 2\epsilon.$$

Since $\epsilon > 0$ is arbitrary, the Riemann Condition shows that the function $f + g$ is integrable. Further, if we let $\alpha := U(P_\epsilon, f) + U(P_\epsilon, g)$ and $\beta := L(P_\epsilon, f) + L(P_\epsilon, g)$, then we have

$$\beta \le L(P_\epsilon, f+g) \le L(f+g) = \iint_{[a,b]\times[c,d]} (f+g) = U(f+g) \le U(P_\epsilon, f+g) \le \alpha.$$

Also, we have

$$\beta \le L(f) + L(g) = \iint_{[a,b]\times[c,d]} f + \iint_{[a,b]\times[c,d]} g = U(f) + U(g) \le \alpha.$$

Thus, we see that

$$\left| \iint_{[a,b]\times[c,d]} f + \iint_{[a,b]\times[c,d]} g - \iint_{[a,b]\times[c,d]} (f+g) \right| \le \alpha - \beta < 2\epsilon.$$

Since this is true for every $\epsilon > 0$, we obtain

$$\iint_{[a,b]\times[c,d]} (f+g) = \iint_{[a,b]\times[c,d]} f + \iint_{[a,b]\times[c,d]} g.$$

(ii) Let $r \in \mathbb{R}$. If $r = 0$, then $rf(x, y) = 0$ for all $(x, y) \in [a, b] \times [c, d]$ and (ii) follows easily. Now assume that $r > 0$. Then for any partition P of $[a, b] \times [c, d]$, we see that

$$L(P, rf) = rL(P, f) \quad \text{and} \quad U(P, rf) = rU(P, f).$$

Hence

$$L(rf) = rL(f) = rU(f) = U(rf).$$

On the other hand, if $r < 0$, then for any partition P of $[a, b] \times [c, d]$, we see that

$$L(P, rf) = rU(P, f) \quad \text{and} \quad U(P, rf) = rL(P, f),$$

and so

$$L(rf) = rU(f) = rL(f) = U(rf).$$

In both cases, we see that rf is integrable and

$$\iint_{[a,b]\times[c,d]} rf = r \iint_{[a,b]\times[c,d]} f.$$

(iii) For any $i = 1, \ldots, n$, and $(x, y), (u, v) \in [x_{i-1}, x_i] \times [y_{j-1}, y_j]$,

$$\begin{aligned}
&(fg)(x, y) - (fg)(u, v) \\
&\quad = f(x, y)[g(x, y) - g(u, v)] + [f(x, y) - f(u, v)]g(u, v) \\
&\quad \leq |f(x, y)|\, |g(x, y) - g(u, v)| + |g(u, v)|\, |f(x, y) - f(u, v)| \\
&\quad \leq M(|f|)[M_{i,j}(g) - m_{i,j}(g)] + M(|g|)[M_{i,j}(f) - m_{i,j}(f)].
\end{aligned}$$

Taking the supremum for (x, y) in $[x_{i-1}, x_i] \times [y_{j-1}, y_j]$ and the infimum for (u, v) in $[x_{i-1}, x_i] \times [y_{j-1}, y_j]$, we obtain

$$M_{i,j}(fg) - m_{i,j}(fg) \leq M(|f|)[M_{i,j}(g) - m_{i,j}(g)] + M(|g|)[M_{i,j}(f) - m_{i,j}(f)].$$

Multiplying both sides of this inequality by $(x_i - x_{i-1})(y_j - y_{j-1})$ and summing from $i = 1$ to n and from $j = 1$ to k, we obtain

$$\begin{aligned}
&U(P_\epsilon, fg) - L(P_\epsilon, fg) \\
&\quad \leq M(|f|)[U(P_\epsilon, g) - L(P_\epsilon, g)] + M(|g|)[U(P_\epsilon, f) - L(P_\epsilon, f)] \\
&\quad < [M(|f|) + M(|g|)]\epsilon.
\end{aligned}$$

Since $\epsilon > 0$ arbitrary, the Riemann Condition shows that fg is integrable.

(iv) Let $\delta > 0$ be such that $|f(x, y)| \geq \delta$ for all $(x, y) \in [a, b] \times [c, d]$. For $i = 1, \ldots, n$, $j = 1, \ldots, k$, and $(x, y), (u, v) \in [x_{i-1}, x_i] \times [y_{j-1}, y_j]$, we have

$$\begin{aligned}
\frac{1}{f(x, y)} - \frac{1}{f(u, v)} &= \frac{f(u, v) - f(x, y)}{f(x, y) f(u, v)} \\
&\leq \frac{|f(x, y) - f(u, v)|}{|f(x, y)|\, |f(u, v)|} \leq \frac{1}{\delta^2}[M_{i,j}(f) - m_{i,j}(f)].
\end{aligned}$$

Taking the supremum for (x, y) in $[x_{i-1}, x_i] \times [y_{j-1}, y_j]$ and the infimum for (u, v) in $[x_{i-1}, x_i] \times [y_{j-1}, y_j]$, we obtain

$$M_{i,j}(1/f) - m_{i,j}(1/f) \leq \frac{1}{\delta^2}[M_{i,j}(f) - m_{i,j}(f)],$$

and consequently

$$U(P_\epsilon, 1/f) - L(P_\epsilon, 1/f) \le \frac{1}{\delta^2}[U(P_\epsilon, f) - L(P_\epsilon, f)] < \frac{\epsilon}{\delta^2}.$$

Again, since $\epsilon > 0$ is arbitrary while $\delta > 0$ is fixed, the Riemann Condition shows that the function $1/f$ is integrable.

(v) Let $k \in \mathbb{N}$ and write $F = f^{1/k}$. First we assume that there is $\delta > 0$ such that $f(x, y) \ge \delta$ for all $(x, y) \in [a, b] \times [c, d]$. For any $(x, y), (u, v)$ in $[a, b] \times [c, d]$, we have $f(x, y) - f(u, v) = F(x, y)^k - F(u, v)^k$, and so

$$f(x, y) - f(u, v) = [F(x, y) - F(u, v)] \sum_{j=1}^{k} F(x, y)^{k-j} F(u, v)^{j-1}.$$

Now for $j = 1, \ldots, k$,

$$F(x, y)^{k-j} F(u, v)^{j-1} \ge \delta^{(k-j)/k} \delta^{(j-1)/k} = \delta^{(k-1)/k} > 0,$$

and so

$$F(x, y) - F(u, v) = \frac{f(x, y) - f(u, v)}{\sum_{j=1}^{k} F(x, y)^{k-j} F(u, v)^{j-1}} \le \frac{f(x, y) - f(u, v)}{k\delta^{(k-1)/k}}.$$

If $P = \{(x_i, y_j) : i = 0, 1, \ldots, n \text{ and } j = 0, 1, \ldots, k\}$ is any partition of $[a, b] \times [c, d]$ and $(x, y), (u, v) \in [x_{i-1}, x_i] \times [y_{j-1}, y_j]$ for some $i = 1, \ldots, n$ and $j = 1, \ldots, k$, then

$$F(x, y) - F(u, v) \le \frac{|f(x, y) - f(u, v)|}{k\delta^{(k-1)/k}} \le \frac{M_{i,j}(f) - m_{i,j}(f)}{k\delta^{(k-1)/k}}.$$

Taking the supremum for (x, y) in $[x_{i-1}, x_i] \times [y_{j-1}, y_j]$ and the infimum for (u, v) in $[x_{i-1}, x_i] \times [y_{j-1}, y_j]$, we obtain

$$M_{i,j}(F) - m_{i,j}(F) \le \frac{M_{i,j}(f) - m_{i,j}(f)}{k\delta^{(k-1)/k}} \quad \text{for } i = 1, \ldots, n \text{ and } j = 1, \ldots, k.$$

Multiplying both sides of this inequality by $(x_i - x_{i-1})(y_j - y_{j-1})$ and summing from $i = 1$ to n and from $j = 1$ to k, we obtain

$$U(P, F) - L(P, F) \le \frac{1}{k\delta^{(k-1)/k}}[U(P, f) - L(P, f)].$$

Since f is integrable, the Riemann Condition shows that F is also integrable.

Next, we consider the general case of any nonnegative integrable function f on $[a, b] \times [c, d]$. Let $\delta > 0$ and define $g : [a, b] \times [c, d] \to \mathbb{R}$ by $g(x, y) := f(x, y) + \delta$ and $G := g^{1/k}$. Then g is integrable by part (i) above, and $g(x, y) \ge \delta$ for all $x \in [a, b] \times [c, d]$. It follows from what we have proved above that G is integrable. Moreover, since f is nonnegative, we have

$$G - \delta^{1/k} = (f + \delta)^{1/k} - \delta^{1/k} \le f^{1/k} = F \le (f + \delta)^{1/k} = G,$$

and therefore,

$$L(G - \delta^{1/k}) \le L(F) \le U(F) \le U(G).$$

But $L(G - \delta^{1/k}) = L(G) - \delta^{1/k}(b-a)(d-c)$ and so

$$L(G - \delta^{1/k}) = \iint_{[a,b]\times[c,d]} G \; - \delta^{1/k}(b-a)(d-c) = U(G) - \delta^{1/k}(b-a)(d-c).$$

This shows that

$$0 \le U(F) - L(F) \le U(G) - L(G - \delta^{1/k}) = \delta^{1/k}(b-a)(d-c).$$

Since $\delta^{1/k} \to 0$ as $\delta \to 0$, we see that $F = f^{1/k}$ is integrable. □

With notation and hypotheses as in the above proposition, a combined application of its parts (i) and (ii) shows that the difference $f - g$ is integrable and

$$\iint_{[a,b]\times[c,d]} (f-g) = \iint_{[a,b]\times[c,d]} f - \iint_{[a,b]\times[c,d]} g.$$

Further, given any $n \in \mathbb{N}$, successive applications of part (iii) of the above proposition show that the nth power f^n is integrable. Likewise, a combined application of parts (iii) and (iv) shows that if there is $\delta > 0$ such that $|g(x,y)| \ge \delta$ for all $(x,y) \in [a,b] \times [c,d]$, then the quotient f/g is integrable. Also, a combined application of parts (iii) and (v) shows that if $f(x,y) \ge 0$ for all $(x,y) \in [a,b] \times [c,d]$, then given any positive $r \in \mathbb{Q}$, the rth power f^r is integrable since $r = n/k$, where $n, k \in \mathbb{N}$.

Example 5.15. Let $\phi : [a,b] \to \mathbb{R}$ and $\psi : [c,d] \to \mathbb{R}$ be Riemann integrable functions of one variable. Consider $f : [a,b] \times [c,d] \to \mathbb{R}$ defined by $f(x,y) := \phi(x) + \psi(y)$ for $(x,y) \in [a,b] \times [c,d]$. In view of Example 5.5 (ii) and part (i) of Proposition 5.14, we readily see that f is integrable on $[a,b] \times [c,d]$, and

$$\iint_{[a,b]\times[c,d]} f(x,y)d(x,y) = (d-c)\int_a^b \phi(x)dx + (b-a)\int_c^d \psi(y)dy.$$

In particular, given any $r, s \in \mathbb{R}$ with $r \ge 0$ and $s \ge 0$, we have

$$\iint_{[a,b]\times[c,d]} (x^r + y^s)\, d(x,y) = (d-c)\frac{b^{r+1} - a^{r+1}}{r+1} + (b-a)\frac{d^{s+1} - c^{s+1}}{s+1},$$

provided $0 \le a < b$ and $0 \le c < d$. ◇

Next, we consider how double integration behaves with respect to the order relation on functions.

Proposition 5.16. *Let $f, g : [a,b] \times [c,d] \to \mathbb{R}$ be integrable functions. Then*

(i) *If $f \le g$ on $[a,b] \times [c,d]$, then $\iint_{[a,b]\times[c,d]} f \le \iint_{[a,b]\times[c,d]} g$.*

(ii) *The function* $|f|$ *is integrable and* $\left|\iint_{[a,b]\times[c,d]} f\right| \le \iint_{[a,b]\times[c,d]} |f|$.

Proof. (i) Let $f(x,y) \le g(x,y)$ for all $(x,y) \in [a,b]\times[c,d]$. Then for any partition P of $[a,b]\times[c,d]$, we have $U(P,f) \le U(P,g)$, and so

$$\iint_{[a,b]\times[c,d]} f = U(f) \le U(g) = \iint_{[a,b]\times[c,d]} g.$$

(ii) Let $\epsilon > 0$ be given. By the Riemann Condition, there is a partition P_ϵ of $[a,b]\times[c,d]$ such that $U(P_\epsilon, f) - L(P_\epsilon, f) < \epsilon$. Let $P_\epsilon = \{(x_i, y_j) : i = 0, 1, \ldots, n \text{ and } j = 0, 1, \ldots, k\}$. For any $i = 1, \ldots, n$ and $j = 1, \ldots, k$, and any $(x,y), (u,v) \in [x_{i-1}, x_i] \times [y_{j-1}, y_j]$, we have

$$|f|(x,y) - |f|(u,v) \le |f(x,y) - f(u,v)| \le M_{i,j}(f) - m_{i,j}(f).$$

Taking the supremum for (x,y) in $[x_{i-1}, x_i] \times [y_{j-1}, y_j]$ and the infimum for (u,v) in $[x_{i-1}, x_i] \times [y_{j-1}, y_j]$, we obtain

$$M_{i,j}(|f|) - m_{i,j}(|f|) \le M_{i,j}(f) - m_{i,j}(f) \quad \text{for } i = 1, \ldots, n \text{ and } j = 1, \ldots, k.$$

Multiplying both sides of this inequality by $(x_i - x_{i-1})(y_j - y_{j-1})$ and summing from $i = 1$ to n and from $j = 1$ to k, we obtain

$$U(P_\epsilon, |f|) - L(P_\epsilon, |f|) \le U(P_\epsilon, f) - L(P_\epsilon, f) < \epsilon.$$

Now by the Riemann Condition, $|f|$ is integrable. Further, since $-|f|(x,y) \le f(x,y) \le |f|(x,y)$ for all $(x,y) \in [a,b]\times[c,d]$, by part (i) above we see that

$$\iint_{[a,b]\times[c,d]} -|f| \le \iint_{[a,b]\times[c,d]} f \le \iint_{[a,b]\times[c,d]} |f|.$$

But by part (ii) of Proposition 5.14, $\iint_{[a,b]\times[c,d]} -|f| = -\iint_{[a,b]\times[c,d]} |f|$. Hence

$$\left|\iint_{[a,b]\times[c,d]} f\right| \le \iint_{[a,b]\times[c,d]} |f|,$$

as desired. □

Remark 5.17. It may be noted that the converse of part (ii) of Proposition 5.16 is not true. In other words, if $f : [a,b]\times[c,d] \to \mathbb{R}$ is such that $|f|$ is integrable, then f need not be integrable. To see this, let $\tilde{f} : [a,b]\times[c,d] \to \mathbb{R}$ denote the bivariate Dirichlet function and consider $f := 2\tilde{f} - 1$. Then in view of Example 5.5 (iii), f is not integrable (lest $\tilde{f} = \frac{1}{2}(1+f)$ be integrable), but $|f(x,y)| = 1$ for all $(x,y) \in [a,b]\times[c,d]$, and thus $|f|$ is integrable. ◇

A Version of the Fundamental Theorem of Calculus

A central result of one-variable calculus is the Fundamental Theorem of Calculus, or in short, the FTC, which, roughly speaking, says that the processes of differentiation and integration are inverses of each other. A precise statement of the FTC is given below, and we remark that it is essentially the same as Proposition 6.21 of ACICARA, except in part (i) we require only that F be differentiable and satisfy $F' = f$ on the open interval (a, b) rather than on $[a, b]$, provided F is continuous on $[a, b]$. It may be observed that the proof given in ACICARA goes through verbatim even with this weaker hypothesis.

Fact 5.18 (FTC). *Let $f : [a, b] \to \mathbb{R}$ be a Riemann integrable function.*

(i) *If there is $F : [a, b] \to \mathbb{R}$ such that F is continuous on $[a, b]$, differentiable on (a, b), and $F' = f$ on (a, b), then $\int_a^b f(x)dx = F(b) - F(a)$.*
(ii) *If f is continuous at some $c \in [a, b]$, then the function $F : [a, b] \to \mathbb{R}$ defined by $F(x) := \int_a^x f(t)dt$ is differentiable at c and $F'(c) = f(c)$.*

There are analogues of the FTC to two dimensions which involve the notion of a "line integral." One of them, known as Green's Theorem, also involves the notion of "orientation." (See Theorems 10.3, 10.4, and 11.10 of [2, vol. II], and the Notes and Comments at the end of this chapter.) We describe below another analogue which does not involve either of these notions. It may be compared with Theorem 10.22 and Exercise 10-14 of [1, first ed.]. It says, roughly speaking, that the processes of mixed second-order partial differentiation and double integration are inverses of each other.

To begin with, we prove some basic properties of the function obtained from an integrable function defined on a rectangle R by integrating over varying subrectangles of R that share a vertex with R. It indicates already that (double) integration is a smoothing process in the sense that it converts an integrable (and possibly discontinuous) function into a continuous function, and a continuous function into a function whose partial derivatives exist.

Proposition 5.19. *Let $f : [a, b] \times [c, d] \to \mathbb{R}$ be an integrable function, and let $F : [a, b] \times [c, d] \to \mathbb{R}$ be defined by*

$$F(x, y) := \iint_{[a,x]\times[c,y]} f(s, t)d(s, t) \quad \text{for } (x, y) \in [a, b] \times [c, d].$$

Then we have the following.

(i) *F is continuous on $[a, b]\times[c, d]$. In fact, F satisfies a* **Lipschitz condition** *on $[a, b] \times [c, d]$, that is, there is $L \in \mathbb{R}$ such that*

$$|F(x, y) - F(u, v)| \le L\,|(x, y) - (u, v)| \quad \text{for all } (x, y), (u, v) \in [a, b]\times[c, d].$$

(ii) *Given any* $(x_0, y_0) \in [a, b] \times [c, d]$, *if we assume that for every* $\epsilon > 0$, *there is* $\delta > 0$ *such that*

$$x \in [a, b], \ |x - x_0| < \delta \Longrightarrow |f(x, y) - f(x_0, y)| < \epsilon \text{ for all } y \in [c, d],$$

and also assume that $\int_c^{y_0} f(x_0, t)dt$ *exists, then* $F_x(x_0, y_0)$ *exists and is equal to* $\int_c^{y_0} f(x_0, t)dt$.

Proof. (i) Since f is integrable on $[a, b] \times [c, d]$, it is bounded on $[a, b] \times [c, d]$, that is, there is $\alpha > 0$ such that $|f(s, t)| \leq \alpha$ for all $(s, t) \in [a, b] \times [c, d]$.

Let $(u, v) \in [a, b] \times [c, d]$. Given any $(x, y) \in [a, b] \times [c, d]$, in view of the version of domain additivity in Remark 5.11, we have

$$F(x, y) - F(u, v) = \iint_{[a,u]\times[v,y]} f + \iint_{[u,x]\times[c,v]} f + \iint_{[u,x]\times[v,y]} f.$$

Hence by the Basic Inequality (Proposition 5.4),

$$|F(x, y) - F(u, v)| \leq \alpha\big((b - a)|y - v| + (d - c)|x - u| + |x - u||y - v|\big).$$

Thus, if we let $K := \max\{b - a, d - c\}$ and observe that both $|x - u|$ and $|y - v|$ are $\leq |(x, y) - (u, v)|$, and also $\leq K$, then we can conclude that

$$|F(x, y) - F(u, v)| \leq L\,|(x, y) - (u, v)|, \quad \text{where } L := 3\alpha K.$$

Hence F satisfies a Lipschitz condition on $[a, b] \times [c, d]$, and consequently, F is continuous (in fact, uniformly continuous) on $[a, b] \times [c, d]$.

(ii) Fix $(x_0, y_0) \in [a, b] \times [c, d]$. Assume that the ϵ-δ condition in (ii) is satisfied and that $\int_c^{y_0} f(x_0, t)dt$ exists. Given any $x \in [a, b]$ with $x \neq x_0$, by Example 5.5 (ii), we see that

$$\int_c^{y_0} f(x_0, t)dt = \frac{1}{x - x_0}\iint_{[x_0,x]\times[c,y_0]} f(x_0, t)d(s, t).$$

Using this together with the conventions and a version of domain additivity stated in Remark 5.11, we obtain

$$\begin{aligned}&\frac{F(x, y_0) - F(x_0, y_0)}{x - x_0} - \int_c^{y_0} f(x_0, t)dt \\ &= \frac{1}{x - x_0}\left(\iint_{[x_0,x]\times[c,y_0]} \big(f(s, t) - f(x_0, t)\big)d(s, t)\right).\end{aligned}$$

Now let $\epsilon > 0$ be given. By our hypothesis, there is $\delta > 0$ such that

$$x \in [a, b], \ |x - x_0| < \delta \Longrightarrow |f(x, t) - f(x_0, t)| < \frac{\epsilon}{d - c} \text{ for all } t \in [c, d].$$

Thus, in view of the Basic Inequality (Proposition 5.4), for $x \in [a, b]$ with $0 < |x - x_0| < \delta$, we see that

$$\left| \frac{F(x,y_0) - F(x_0,y_0)}{x - x_0} - \int_c^{y_0} f(x_0,t)dt \right| < \frac{1}{|x-x_0|}\frac{\epsilon}{d-c}|x-x_0|(d-c) = \epsilon.$$

This proves that $F_x(x_0, y_0)$ exists and is equal to $\int_c^{y_0} f(x_0,t)dt$. □

We are now ready to prove an analogue of the FTC for double integrals.

Proposition 5.20. *Let $f : [a,b] \times [c,d] \to \mathbb{R}$ be an integrable function.*

(i) *Suppose there is $F : [a,b] \times [c,d] \to \mathbb{R}$ satisfying the following properties:*
- *For each fixed $y_0 \in [c,d]$, the function given by $x \longmapsto F(x,y_0)$ is continuous on $[a,b]$ and differentiable on (a,b).*
- *For each fixed $x_0 \in (a,b)$, the function given by $y \longmapsto F_x(x_0,y)$ is continuous on $[c,d]$ and differentiable on (c,d).*
- *F_{xy} exists and is equal to f on $(a,b) \times (c,d)$.*

Then

$$\iint_{[a,b]\times[c,d]} f(x,y)d(x,y) = \triangle_{(a,c)}^{(b,d)} F := F(b,d) - F(b,c) - F(a,d) + F(a,c).$$

(ii) *Let $F : [a,b] \times [c,d] \to \mathbb{R}$ be defined by*

$$F(x,y) := \iint_{[a,x]\times[c,y]} f(s,t)d(s,t) \quad \text{for } (x,y) \in [a,b] \times [c,d],$$

Suppose $(x_0,y_0) \in [a,b] \times [c,d]$ and f satisfies the following properties:
- *for every $\epsilon > 0$, there is $\delta > 0$ such that*

$$x \in [a,b],\ |x - x_0| < \delta \Longrightarrow |f(x,y) - f(x_0,y)| < \epsilon \text{ for all } y \in [c,d].$$

- *the function $\psi : [c,d] \to \mathbb{R}$ defined by $\psi(t) := f(x_0,t)$ is Riemann integrable on $[c,d]$ and continuous at y_0.*

Then $F_{xy}(x_0,y_0)$ exists and is equal to $f(x_0,y_0)$.

Proof. (i) Let $P = \{(x_i,y_j) : i = 0,1,\ldots,n \text{ and } j = 0,1,\ldots,k\}$ be a partition of $[a,b] \times [c,d]$. Given any $i = 1,\ldots,n$ and $j = 1,\ldots,k$, by applying the Rectangular Mean Value Theorem (Proposition 3.11) to the (restriction of) F on $[x_{i-1},x_i] \times [y_{j-1},y_j]$, we see that there is $(x_{i,j}^*, y_{i,j}^*) \in (x_{i-1},x_i) \times (y_{j-1},y_j)$ such that $F(x_i,y_j) - F(x_{i-1},y_j) - F(x_i,y_{j-1}) + F(x_{i-1},y_{j-1})$ is equal to

$$F_{xy}\left(x_{i,j}^*, y_{i,j}^*\right)(x_i - x_{i-1})(y_j - y_{j-1}) = f\left(x_{i,j}^*, y_{i,j}^*\right)(x_i - x_{i-1})(y_j - y_{j-1}).$$

Summing over $i = 1,\ldots,n$ and $j = 1,\ldots,k$, we see that

$$F(b,d) - F(b,c) - F(a,d) + F(a,c) = \sum_{i=1}^{n}\sum_{j=1}^{k} f\left(x_{i,j}^*, y_{i,j}^*\right)(x_i - x_{i-1})(y_j - y_{j-1}).$$

Consequently, $L(P,f) \le F(b,d) - F(b,c) - F(a,d) + F(a,c) \le U(P,f)$. Since this is true for every partition P of $[a,b] \times [c,d]$, it follows that

$$L(f) = \sup L(P, f) \le F(b,d) - F(b,c) - F(a,d) + F(a,c) \le \inf U(P, f) = U(f),$$

where the supremum and the infimum are taken over the set of all partitions of $[a,b] \times [c,d]$. Since f is integrable, we have $L(f) = U(f)$, and so

$$\iint_{[a,b]\times[c,d]} f(x,y)d(x,y) = F(b,d) - F(b,c) - F(a,d) + F(a,c).$$

(ii) Fix $(x_0, y_0) \in [a,b] \times [c,d]$. Since $\psi : [c,d] \to \mathbb{R}$ defined by $\psi(t) := f(x_0, t)$ is Riemann integrable on $[c,d]$, it is Riemann integrable on $[c,y]$, that is, $\int_c^y f(x_0,t)dt$ exists, for every $y \in [c,d]$. Hence by part (ii) of Proposition 5.19, we see that $F_x(x_0, y)$ exists and

$$F_x(x_0, y) = \int_c^y f(x_0,t)dt = \int_c^y \psi(t)dt \quad \text{for every } y \in [c,d].$$

Further, since ψ is continuous at y_0, by part (ii) of the FTC (Fact 5.18), we see that $F_{xy}(x_0, y_0)$ exists and is equal to $\psi(y_0) = f(x_0, y_0)$. □

Remark 5.21. Part (ii) of Proposition 5.19 admits a straightforward analogue with F_x replaced by F_y. More precisely, given any $(x_0, y_0) \in [a,b] \times [c,d]$, if we assume that for every $\epsilon > 0$, there is $\delta > 0$ such that

$$y \in [c,d], \ |y - y_0| < \delta \Longrightarrow |f(x,y) - f(x,y_0)| < \epsilon \text{ for all } x \in [a,b],$$

and also assume that $\int_a^{x_0} f(s,y_0)ds$ exists, then $F_y(x_0,y_0)$ exists and is equal to $\int_a^{x_0} f(s,y_0)ds$. The proof is similar.

Likewise, both the parts of Proposition 5.20 admit a straightforward analogue with F_x and F_{xy} replaced by F_y and F_{yx}, respectively, and the proofs are similar. ◇

Example 5.22. Suppose $\phi : [a,b] \to \mathbb{R}$ and $\psi : [c,d] \to \mathbb{R}$ are differentiable functions of one variable such that ϕ' and ψ' are Riemann integrable. Then

$$\iint_{[a,b]\times[c,d]} \phi'(x)\psi'(y)d(x,y) = \big(\phi(b) - \phi(a)\big)\big(\psi(d) - \psi(c)\big).$$

To see this, consider $F : [a,b]\times[c,d] \to \mathbb{R}$ defined by $F(x,y) := \phi(x)\psi(y)$. Note that F is continuous in the first variable, F_x exists and is continuous in the second variable, and F_{xy} exists. Indeed, $F_x(u,v) = \phi'(u)\psi(v)$ and $F_{xy}(u,v) = \phi'(v)\psi'(v)$ for $(u,v) \in [a,b] \times [c,d]$. Moreover, in view of Example 5.5 (ii) and part (iii) of Proposition 5.14, we see that F_{xy} is integrable on $[a,b] \times [c,d]$. Thus by part (i) of Proposition 5.20, we see that

$$\iint_{[a,b]\times[c,d]} F_{xy} = \iint_{[a,b]\times[c,d]} \phi'(x)\psi'(y)d(x,y) = \triangle_{(a,c)}^{(b,d)} F.$$

Finally, observe that $\triangle_{(a,c)}^{(b,d)} F = \big(\phi(b) - \phi(a)\big)\big(\psi(d) - \psi(c)\big)$. ◇

Corollary 5.23. *Let $f : [a,b] \times [c,d] \to \mathbb{R}$ be continuous and let $F : [a,b] \times [c,d] \to \mathbb{R}$ be defined by*

$$F(x,y) := \iint_{[a,x]\times[c,y]} f(s,t)d(s,t) \quad \text{for } (x,y) \in [a,b] \times [c,d].$$

Then F_x, F_y, F_{xy}, and F_{yx} exist on $[a,b] \times [c,d]$. Moreover,

$$F_x(x_0,y_0) = \int_c^{y_0} f(x_0,t)dt, \quad F_y(x_0,y_0) = \int_a^{x_0} f(s,y_0)ds,$$
$$F_{xy}(x_0,y_0) = f(x_0,y_0) = F_{yx}(x_0,y_0) \quad \text{for every } (x_0,y_0) \in [a,b] \times [c,d].$$

Proof. Follows from part (ii) of Proposition 5.19 and of Proposition 5.20, in light of Remark 5.21. □

Remark 5.24. For an analogue of Corollary 5.23 for the second-order partials F_{xx} and F_{yy}, see Exercise 34. An alternative and easier proof of Proposition 5.20, under an additional hypothesis on continuity, and in particular, of Corollary 5.23, is possible using a result known as Fubini's Theorem that we shall prove in the next subsection. (See Exercises 35 and 36.) ◇

Two of the most important applications of the FTC in one-variable calculus are results that are usually known as Integration by Parts and Integration by Substitution. (See, for example, Propositions 6.25 and 6.26 of ACICARA.) We prove below an analogous formula for Double Integration by Parts, and subsequently for Double Integration by Substitution.

Proposition 5.25 (Double Integration by Parts). *Let $R := [a,b] \times [c,d]$ and let $f, g, G : R \to \mathbb{R}$ be integrable functions satisfying the following:*

- *The functions f and G are continuous in the first variable.*
- *f_x, f_y, f_{xy}, G_x, and G_y exist and are integrable on R.*
- *The functions f_x and G_x are continuous in the second variable.*
- *G_{xy} exists and $G_{xy} = g$ on R.*

Then

$$\iint_R fg = \triangle_{(a,c)}^{(b,d)}(fG) - \iint_R \left(f_x G_y + f_y G_x + f_{xy} G\right),$$

where $\triangle_{(a,c)}^{(b,d)}(fG) := (fG)(b,d) - (fG)(b,c) - (fG)(a,d) + (fG)(a,c)$.

Proof. Let $H := fG$. By part (iii) of Proposition 5.14, H is integrable on R. Also, $H_x = fG_x + f_xG$ and $H_{xy} = fG_{xy} + f_yG_x + f_xG_y + f_{xy}G$. Define $h : R \to \mathbb{R}$ by $h := fg + f_xG_y + f_yG_x + f_{xy}G$. Then $H_{xy} = h$ on R. Hence by part (i) of Proposition 5.20, we have $\iint_R h = \triangle_{(a,c)}^{(b,d)} H = \triangle_{(a,c)}^{(b,d)}(fG)$, that is, $\iint_R fg = \triangle_{(a,c)}^{(b,d)}(fG) - \iint_R \left(f_xG_y + f_yG_x + f_{xy}G\right)$. □

It may be remarked that in contrast to one-variable calculus, the function G in the above result is not uniquely determined (up to addition by a constant) by the function g. Indeed, even if $G(x,y)$ is replaced by $G(x,y)+\phi(x)+\psi(y)$ for any differentiable functions ϕ and ψ of one variable, the condition $G_{xy}=g$ remains valid. Also, we remark that an analogous version of the above result is valid with f_x, f_{xy}, G_x, and G_{xy} replaced by f_y, f_{yx}, G_y, and G_{yx}, respectively.

Proposition 5.26 (Double Integration by Substitution). *Let $\alpha,\beta,\gamma,\delta$ be real numbers with $\alpha<\beta$ and $\gamma<\delta$, and let $E:=[\alpha,\beta]\times[\gamma,\delta]$ be a rectangle in $\mathbb{R}^2$. Suppose $\Phi: E\to\mathbb{R}^2$ is a transformation given by*

$$\Phi(u,v) := (\phi(u),\psi(v)) \text{ for all } (u,v)\in E,$$

where $\phi:[\alpha,\beta]\to\mathbb{R}$ and $\psi:[\gamma,\delta]\to R$ are differentiable functions such that ϕ' is integrable on $[\alpha,\beta]$ and ψ' is integrable on $[\gamma,\delta]$. Then $D:=\Phi(E)$ is a rectangle in $\mathbb{R}^2$ and the Jacobian $J(\Phi)$ of Φ is given by $J(\Phi)(u,v)=\phi'(u)\psi'(v)$ for all $(u,v)\in E$. Moreover, we have the following.

(i) *If $f: D\to\mathbb{R}$ is continuous, then $(f\circ\Phi)J(\Phi)$ is integrable on E and*

$$\iint_{[\phi(\alpha),\phi(\beta)]\times[\psi(\gamma),\psi(\delta)]} f(x,y)d(x,y) = \iint_E f(\Phi(u,v))\, J(\Phi)(u,v)d(u,v).$$

(ii) *If $f: D\to\mathbb{R}$ is integrable and if $J(\Phi)(u,v)\neq 0$ for all $(u,v)\in\mathbb{R}^2$ with $\alpha<u<\beta$ and $\gamma<v<\delta$, then $(f\circ\Phi)|J(\Phi)|$ is integrable on E and*

$$\iint_D f(x,y)d(x,y) = \iint_E f(\Phi(u,v))\,|J(\Phi)(u,v)|d(u,v).$$

Proof. Since ϕ and ψ are differentiable functions of one variable, they are continuous. Hence from one-variable calculus (for example, Propositions 3.8 and 3.13 of ACICARA), we see that $\phi([\alpha,\beta])$ and $\psi([\gamma,\delta])$ are closed and bounded intervals in $\mathbb{R}$. Let $[a,b]:=\phi([\alpha,\beta])$ and $[c,d]:=\psi([\gamma,\delta])$. Then it is clear that $D:=\Phi(E)$ is the rectangle $[a,b]\times[c,d]$ in $\mathbb{R}^2$. Also, it is clear that $J(\Phi)(u,v)=\phi'(u)\psi'(v)$ for all $(u,v)\in E$. We now prove (i) and (ii).

(i) Suppose $f: D\to\mathbb{R}$ is a continuous function. Define $F: D\to\mathbb{R}$ by

$$F(x,y) := \iint_{[a,x]\times[c,y]} f(s,t)d(s,t) \quad \text{for } (x,y)\in D=[a,b]\times[c,d].$$

Then by Corollary 5.19, F_x, F_y, and F_{xy} exist and $F_{xy}=f$ on D. Consider $H: E\to\mathbb{R}$ defined by $H:=F\circ\Phi$. By the Chain Rule of one-variable calculus (Proposition 4.9 of ACICARA), we see that for each fixed $v\in[\gamma,\delta]$, the function given by $u\longmapsto H(u,v)$ is differentiable on (α,β) and its derivative at $u_0\in(\alpha,\beta)$ is given by

$$H_u(u_0,v) = F_x(\phi(u_0),\psi(v))\,\phi'(u_0).$$

Consequently, again by the Chain Rule in one-variable calculus, we obtain

$$H_{uv}(u_0, v_0) = F_{xy}\left(\phi(u_0), \psi(v_0)\right)\psi'(v_0)\phi'(u_0) = (f \circ \Phi)(u_0, v_0)J(\Phi)(u_0, v_0)$$

for all $(u_0, v_0) \in \mathbb{R}^2$ with $\alpha < u_0 < \beta$ and $\gamma < v_0 < \delta$. Note also that since F_{xy} exists, F_x is continuous in the second variable, and therefore so is H_u. Hence by part (i) of Proposition 5.20, we see that

$$\iint_E f\left(\Phi(u, v)\right) J(\Phi)(u, v)d(u, v) = \triangle_{(\alpha,\gamma)}^{(\beta,\delta)} H,$$

and thus, in view of Corollary 5.10 and Remark 5.11, we obtain

$$\begin{aligned}\iint_E f\left(\Phi(u, v)\right) J(\Phi)(u, v)d(u, v) &= \triangle_{(\phi(\alpha),\psi(\gamma))}^{(\phi(\beta),\psi(\delta))} F \\ &= \iint_{[\phi(\alpha),\phi(\beta)]\times[\psi(\gamma),\psi(\delta)]} f(x, y)d(x, y).\end{aligned}$$

This proves (i).

(ii) Suppose $f : D \to \mathbb{R}$ is integrable and $J(\Phi)(u, v) \neq 0$ for all $(u, v) \in \mathbb{R}^2$ with $\alpha < u < \beta$ and $\gamma < v < \delta$. Let $g := (f \circ \Phi)\,|J(\Phi)|$. We will prove the integrability of g and the equality of $\iint_D f$ and $\iint_E g$ by showing that $L(f) \leq L(g)$ and $U(f) \geq U(g)$.

Since $\phi'(u)\psi'(v) = J(\Phi)(u, v) \neq 0$, we see that $\phi'(u) \neq 0$ for all $u \in (\alpha, \beta)$ and $\psi'(v) \neq 0$ for all $v \in (\gamma, \delta)$. Hence by the IVP of derivatives of functions of one variable (given, for example, in Proposition 4.14 of ACICARA), it follows that ϕ' does not change sign in the open interval (α, β) and ψ' does not change sign in the open interval (γ, δ). Hence we consider the following four cases.

Case 1. $\phi'(u) > 0$ for all $u \in (\alpha, \beta)$ and $\psi'(v) > 0$ for all $v \in (\gamma, \delta)$. In this case, ϕ is strictly increasing on $[\alpha, \beta]$, $\phi(\alpha) = a$, and $\phi(\beta) = b$. Also, ψ is strictly increasing on $[\gamma, \delta]$, $\psi(\gamma) = c$, and $\psi(\delta) = d$. Consider a partition $P := \{(x_i, y_j) : i = 0, 1, \ldots, n \text{ and } j = 0, 1, \ldots, k\}$ of $[a, b]\times[c, d]$. We shall now show that $L(P, f) \leq L(g)$. To begin with, let $u_i := \phi^{-1}(x_i)$ for $i = 0, 1, \ldots, n$ and $v_j = \psi^{-1}(y_j)$ for $j = 0, 1, \ldots, k$. Then $\{(u_i, v_j) : i = 0, 1, \ldots, n \text{ and } j = 0, 1, \ldots, k\}$ is a partition of $[\alpha, \beta]\times[\gamma, \delta]$. Moreover, $f\left([x_{i-1}, x_i] \times [y_{j-1}, y_j]\right) = (f \circ \Phi)\left([u_{i-1}, u_i] \times [v_{j-1}, v_j]\right)$ and so $m_{i,j}(f) = m_{i,j}\,(f \circ \Phi)$ for $i = 1, \ldots, n$ and $j = 1, \ldots, k$. Also, in view of Example 5.22, we see that

$$(x_i - x_{i-1})(y_j - y_{j-1}) = \iint_{[u_{i-1},u_i]\times[v_{j-1},v_j]} \phi'(u)\psi'(v)d(u, v).$$

Since $|J(\Phi)(u, v)| = \phi'(u)\psi'(v)$ for all $(u, v) \in [\alpha, \beta] \times [\gamma, \delta]$, we obtain

$$\begin{aligned}L(P, f) &= \sum_{i=1}^{n}\sum_{j=1}^{k} m_{i,j}(f)(x_i - x_{i-1})(y_j - y_{j-1}) \\ &= \sum_{i=1}^{n}\sum_{j=1}^{k} \iint_{[u_{i-1},u_i]\times[v_{j-1},v_j]} m_{i,j}\,(f \circ \Phi)\,|J(\Phi)(u, v)|d(u, v).\end{aligned}$$

For $i = 1, \ldots, n$ and $j = 1, \ldots, k$, let $g_{i,j}, h_{i,j} : [u_{i-1}, u_i] \times [v_{j-1}, v_j] \to \mathbb{R}$ be defined by $g_{i,j}(u,v) := g(u,v)$ and $h_{i,j}(u,v) := m_{i,j}(f \circ \Phi)\,|J(\Phi)(u,v)|$. Then $h_{i,j}$ is integrable on $[u_{i-1}, u_i] \times [v_{j-1}, v_j]$ and $h_{i,j} \le g_{i,j}$. Thus,

$$L(P,f) = \sum_{i=1}^{n}\sum_{j=1}^{k} \iint_{[u_{i-1},u_i]\times[v_{j-1},v_j]} h_{i,j} \le \sum_{i=1}^{n}\sum_{j=1}^{k} L(h_{i,j}) \le \sum_{i=1}^{n}\sum_{j=1}^{k} L(g_{i,j}).$$

Let $\epsilon > 0$ be given. Then for each $i = 1, \ldots, n$ and $j = 1, \ldots, k$, there is a partition $Q_{i,j}$ of $[u_{i-1}, u_i] \times [v_{j-1}, v_j]$ such that $L(g_{i,j}) - \frac{\epsilon}{nk} < L(Q_{i,j}, g_{i,j})$. Now let Q denote the partition of $[\alpha,\beta]\times[\gamma,\delta]$ obtained from putting together the partitions $Q_{i,j}$ of $[u_{i-1}, u_i] \times [v_{j-1}, v_j]$ for $i = 1, \ldots, n$ and $j = 1, \ldots, k$. In effect, Q is the union of suitable refinements $Q^*_{i,j}$ of $Q_{i,j}$ as i varies from 1 to n and j from 1 to k.[1] Thus, in view of Proposition 5.3, we see that

$$\sum_{i=1}^{n}\sum_{j=1}^{k}\left[L(g_{i,j}) - \frac{\epsilon}{nk}\right] < \sum_{i=1}^{n}\sum_{j=1}^{k} L(Q_{i,j}, g_{i,j}) \le \sum_{i=1}^{n}\sum_{j=1}^{k} L\left(Q^*_{i,j}, g_{i,j}\right) = L(Q,g).$$

It follows that

$$L(P,f) \le \sum_{i=1}^{n}\sum_{j=1}^{k} L(g_{i,j}) < L(Q,g) + \epsilon \le L(g) + \epsilon \quad \text{for every } \epsilon > 0,$$

and so $L(P,f) \le L(g)$. Taking the supremum over all partitions P of $[a,b] \times [c,d]$, we have $L(f) \le L(g)$. In a similar manner, we see that $U(f) \ge U(g)$.

Case 2. $\phi'(u) > 0$ for all $u \in (\alpha,\beta)$ and $\psi'(v) < 0$ for all $v \in (\gamma,\delta)$.

In this case, ϕ is strictly increasing on $[\alpha,\beta]$, $\phi(\alpha) = a$, and $\phi(\beta) = b$, whereas ψ is strictly decreasing on $[\gamma,\delta]$, $\psi(\gamma) = d$, and $\psi(\delta) = c$. Consider, as before, a partition $P := \{(x_i, y_j) : i = 0, 1, \ldots, n \text{ and } j = 0, 1, \ldots, k\}$ of $[a,b]\times[c,d]$. This time, let $u_i := \phi^{-1}(x_i)$ for $i = 0, 1, \ldots, n$ and $v_j = \psi^{-1}(y_{k-j})$ for $j = 0, 1, \ldots, k$. Then $\{(u_i, v_j) : i = 0, 1, \ldots, n \text{ and } j = 0, 1, \ldots, k\}$ is a partition of $[\alpha,\beta] \times [\gamma,\delta]$. Moreover, $f([x_{i-1}, x_i] \times [y_{k-j}, y_{k-j+1}]) = (f \circ \Phi)([u_{i-1}, u_i] \times [v_{j-1}, v_j])$ and so $m_{i,k-j+1}(f) = m_{i,j}(f \circ \Phi)$ for $i = 1, \ldots, n$ and $j = 1, \ldots, k$. Also, in view of Example 5.22, we see that

$$(x_i - x_{i-1})(y_{k-j} - y_{k-j+1}) = \iint_{[u_{i-1},u_i]\times[v_{j-1},v_j]} \phi'(u)\psi'(v)d(u,v).$$

Since $|J(\Phi)(u,v)| = -\phi'(u)\psi'(v)$ for all $(u,v) \in [\alpha,\beta] \times [\gamma,\delta]$, we obtain

[1] As illustrated in a special case in Figure 5.6 in the proof of Proposition 5.9, merely taking the union of the partitions $Q_{i,j}$ of subrectangles $[u_{i-1}, u_i] \times [v_{j-1}, v_j]$ may not yield a partition of the rectangle $[\alpha,\beta]\times[\gamma,\delta]$, and it is often necessary to add several points to each $Q_{i,j}$ so as to obtain a legitimate partition of $[\alpha,\beta] \times [\gamma,\delta]$.

$$L(P,f) = \sum_{i=1}^{n}\sum_{j=1}^{k} m_{i,k-j+1}(f)(x_i - x_{i-1})(y_{k-j+1} - y_{k-j})$$
$$= \sum_{i=1}^{n}\sum_{j=1}^{k} \iint_{[u_{i-1},u_i]\times[v_{j-1},v_j]} m_{i,j}\,(f\circ\Phi)\,|J(\Phi)(u,v)|d(u,v).$$

Now we proceed exactly as in Case 1 to conclude that $L(f) \le L(g)$. In a similar manner, we see that $U(f) \ge U(g)$.

Case 3. $\phi'(u) < 0$ for all $u \in (\alpha,\beta)$ and $\psi'(v) > 0$ for all $v \in (\gamma,\delta)$. This is similar to Case 2.

Case 4. $\phi'(u) < 0$ for all $u \in (\alpha,\beta)$ and $\psi'(v) < 0$ for all $v \in (\gamma,\delta)$. This is similar to Cases 2 and 3.

Thus, in each case we have $L(f) \le L(g) \le U(g) \le U(f)$. Since f is integrable, we have $L(f) = U(f)$, and hence $L(g) = U(g) = \iint_D f$. It follows that g is integrable and $\iint_D f = \iint_E g$, as desired. □

Remark 5.27. An analogue of Proposition 5.26 holds if Φ is instead a transformation given by

$$\Phi(u,v) := (\psi(v), \phi(u)) \text{ for all } (u,v) \in E,$$

where ϕ, ψ are as in Proposition 5.26. In this case, $J(\Phi)(u,v) = -\phi'(u)\psi'(v)$ for all $(u,v) \in E$, and the equality of the two double integrals in part (i) takes the form

$$\iint_{[\psi(\gamma),\psi(\delta)]\times[\phi(\alpha),\phi(\beta)]} f(x,y)d(x,y) = -\iint_E f\left(\Phi(u,v)\right) J(\Phi)(u,v)d(u,v),$$

whereas the statement in part (ii) remains the same. Proofs are similar. A result analogous to part (ii) of Proposition 5.26 for more general transformations Φ is known as the change of variables formula, and it will be discussed in greater detail in Section 5.3. ◇

Fubini's Theorem on Rectangles

The easiest and the most widely used method to evaluate double integrals is to reduce the problem to a repeated evaluation of Riemann integrals of functions of one variable. The following result shows when and how this can be done.

Proposition 5.28 (Fubini's Theorem on Rectangles). *Let $f : [a,b] \times [c,d] \to \mathbb{R}$ be an integrable function and let I denote the double integral of f on $[a,b]\times[c,d]$.*

(i) *If for each fixed $x \in [a,b]$, the Riemann integral $\int_c^d f(x,y)dy$ exists, then the* **iterated integral** $\int_a^b \left(\int_c^d f(x,y)dy\right) dx$ *exists and is equal to I.*

(ii) *If for each fixed* $y \in [c,d]$, *the Riemann integral* $\int_a^b f(x,y)dx$ *exists, then the* **iterated integral** $\int_c^d \left(\int_a^b f(x,y)dx\right) dy$ *exists and is equal to* I.

(iii) *If the hypotheses in both* (i) *and* (ii) *above hold, and in particular, if* f *is continuous on* $[a,b] \times [c,d]$, *then*

$$\int_a^b \left(\int_c^d f(x,y)dy\right) dx = \iint_{[a,b]\times[c,d]} f(x,y)d(x,y) = \int_c^d \left(\int_a^b f(x,y)dx\right) dy.$$

Proof. Let $\epsilon > 0$ be given. By the Riemann Condition (Proposition 5.6), there is a partition $P_\epsilon := \{(x_i, y_j) : i = 0, 1, \ldots, n \text{ and } j = 0, 1, \ldots, k\}$ of $[a,b]\times[c,d]$ such that

$$U(P_\epsilon, f) - L(P_\epsilon, f) < \epsilon.$$

(i) Assume that for each fixed $x \in [a,b]$, the Riemann integral $\int_c^d f(x,y)dy$ exists and consider the function $A : [a,b] \to \mathbb{R}$ defined by

$$A(x) = \int_c^d f(x,y)dy.$$

Since $m(f)(d-c) \le A(x) \le M(f)(d-c)$ for all $x \in [a,b]$, we see that A is a bounded function. Also, by Domain Additivity of Riemann integrals (Fact 5.8), we have

$$A(x) = \sum_{j=1}^{k} \int_{y_{j-1}}^{y_j} f(x,y)dy.$$

Hence for each fixed $i \in \{1, \ldots, n\}$, we obtain

$$\sum_{j=1}^{k} m_{i,j}(f)(y_j - y_{j-1}) \le A(x) \le \sum_{j=1}^{k} M_{i,j}(f)(y_j - y_{j-1}) \quad \text{for all } x \in [x_{i-1}, x_i].$$

Thus, upon letting $m_i(A) := \inf\{A(x) : x \in [x_{i-1}, x_i]\}$ and $M_i(A) := \sup\{A(x) : x \in [x_{i-1}, x_i]\}$, we see that

$$\sum_{j=1}^{k} m_{i,j}(f)(y_j - y_{j-1}) \le m_i(A) \le M_i(A) \le \sum_{j=1}^{k} M_{i,j}(f)(y_j - y_j).$$

Multiplying these inequalities by $x_i - x_{i-1}$ and summing over $i = 1, \ldots, n$, we obtain

$$
\begin{aligned}
L(P_\epsilon, f) &= \sum_{i=1}^{n} \Big[\sum_{j=1}^{k} m_{i,j}(f)(y_j - y_{j-1}) \Big] (x_i - x_{i-1}) \\
&\le \sum_{i=1}^{n} m_i(A)(x_i - x_{i-1}) \\
&\le \sum_{i=1}^{n} M_i(A)(x_i - x_{i-1}) \\
&\le \sum_{i=1}^{n} \Big[\sum_{j=1}^{k} M_{i,j}(f)(y_j - y_{j-1}) \Big] (x_i - x_{i-1}) = U(P_\epsilon, f).
\end{aligned}
$$

Thus the partition P_ϵ of $[a,b] \times [c,d]$ induces a partition $P := \{x_0, x_1, \ldots, x_n\}$ of $[a,b]$ such that the difference between the upper Riemann sum $U(P,A)$ and the lower Riemann sum $L(P,A)$ of A is given by

$$
U(P,A) - L(P,A) = \sum_{i=1}^{n} [M_i(A) - m_i(A)]\,(x_i - x_{i-1}) \le U(P_\epsilon, f) - L(P_\epsilon, f) < \epsilon.
$$

Hence by the Riemann Condition for functions of one variable (as given, for example, in Proposition 6.5 of ACICARA), the function A is integrable on $[a,b]$. Also, since

$$
L(P_\epsilon, f) \le L(P,A) \le \int_a^b A(x)dx \le U(P,A) \le U(P_\epsilon, f)
$$

and since

$$
L(P_\epsilon, f) \le \iint_{[a,b]\times[c,d]} f(x,y)d(x,y) \le U(P_\epsilon, f),
$$

we see that

$$
\left| \iint_{[a,b]\times[c,d]} f(x,y)d(x,y) - \int_a^b A(x)dx \right| < \epsilon.
$$

Since $\epsilon > 0$ is arbitrary, this proves (i).

(ii) Assume that for each fixed $y \in [c,d]$, the Riemann integral $\int_a^b f(x,y)dx$ exists and consider the function $B : [c,d] \to \mathbb{R}$ defined by

$$
B(y) = \int_a^b f(x,y)dx.
$$

A proof similar to the proof of (i) above can now be given.

(iii) If the hypotheses in both (i) and (ii) are satisfied, then the desired equalities are an immediate consequence of (i) and (ii). If f is continuous on $[a,b] \times [c,d]$, then for each fixed $x \in [a,b]$, the function given by $y \longmapsto f(x,y)$ is continuous on $[c,d]$, and the function given by $x \longmapsto f(x,y)$ is continuous on $[a,b]$, and consequently, both $\int_c^d f(x,y)dy$ and $\int_a^b f(x,y)dx$ exist, that is, the hypotheses in both (i) and (ii) are satisfied. □

Remark 5.29. Let $f : [a,b] \times [c,d] \to \mathbb{R}$ be a bounded function. Fubini's Theorem says that if f is integrable, then the double integral of f is equal to an iterated integral of f whenever all the Riemann integrals appearing in the latter exist. Geometrically this means that if f is a nonnegative bounded function defined on a rectangle and if the volume of the solid under the surface $z = f(x,y)$ and above the rectangle $[a,b] \times [c,d]$ is well defined, then it can be found either by calculating the areas

$$A(x) = \int_c^d f(x,y)dy, \quad x \in [a,b],$$

of cross sections of the solid perpendicular to the x-axis, or by calculating the areas

$$B(y) = \int_a^b f(x,y)dx, \quad y \in [c,d],$$

of cross sections of the solid perpendicular to the y-axis. ◇

In the examples below, we show that Fubini's Theorem can be used to quickly calculate some double integrals, and also that the conclusion of this theorem is not valid if any of its hypotheses is not satisfied.

Examples 5.30. (i) Let $\phi : [a,b] \to \mathbb{R}$ and $\psi : [c,d] \to \mathbb{R}$ be Riemann integrable functions of one variable. Consider $f : [a,b] \times [c,d] \to \mathbb{R}$ defined by $f(x,y) := \phi(x)\psi(y)$ for $(x,y) \in [a,b]\times[c,d]$. In view of Example 5.5 (ii), part (iii) of Proposition 5.14, and Fubini's Theorem (Proposition 5.28), we readily see that f is integrable on $[a,b] \times [c,d]$, and

$$\iint_{[a,b]\times[c,d]} f(x,y)d(x,y) = \int_a^b \left(\int_c^d \phi(x)\psi(y)dy \right) dx$$
$$= \left(\int_a^b \phi(x)dx \right) \left(\int_c^d \psi(y)dy \right).$$

In particular, given any $r, s \in \mathbb{R}$ with $r \geq 0$ and $s \geq 0$, we have

$$\iint_{[a,b]\times[c,d]} x^r y^s d(x,y) = \left(\frac{b^{r+1} - a^{r+1}}{r+1} \right) \left(\frac{d^{s+1} - c^{s+1}}{s+1} \right),$$

provided $0 \leq a < b$ and $0 \leq c < d$.

(ii) Let $R := [0,\pi]\times[0,\pi]$ and let $f : R \to \mathbb{R}$ be defined by $f(x,y) := \sin(x+y)$ for $(x,y) \in R$. We have seen in Example 5.13 (ii) that f is integrable. Applying Fubini's Theorem (Proposition 5.28), we see that

$$\iint_R \sin(x+y)d(x,y) = \int_0^\pi \left[\int_0^\pi \sin(x+y)dy \right] dx$$
$$= \int_0^\pi -\left[\cos(x+\pi) - \cos x\right] dx = 2\int_0^\pi \cos x \, dx = 0.$$

(iii) Consider the function $f : [0, 1] \times [0, 1] \to \mathbb{R}$ defined by

$$f(x, y) := \begin{cases} 1/x^2 & \text{if } 0 < y < x < 1, \\ -1/y^2 & \text{if } 0 < x < y < 1, \\ 0 & \text{otherwise.} \end{cases}$$

Note that if $x_n := 1/n$ and $y_n := 1/\sqrt{n}$ for $n \in \mathbb{N}$, then $f(x_n, y_n) = -n \to -\infty$, whereas if $x_n := 1/\sqrt{n}$ and $y_n := 1/n$ for $n \in \mathbb{N}$, then $f(x_n, y_n) = n \to \infty$. Thus f is neither bounded below nor bounded above. Hence the double integral of f is not defined. On the other hand, if $x = 0$ or if $x = 1$, then

$$A(x) := \int_0^1 f(x, y)dy = \int_0^1 0\, dy = 0,$$

whereas if $0 < x < 1$, then

$$\begin{aligned} A(x) := \int_0^1 f(x, y)dy &= \int_0^x f(x, y)dy + \int_x^1 f(x, y)dy \\ &= \int_0^x \frac{1}{x^2} dy + \int_x^1 \frac{-1}{y^2} dy = \frac{1}{x^2} \cdot x + \left[\frac{1}{y}\right]_x^1 = 1. \end{aligned}$$

Thus except at the two endpoints of $[0, 1]$, A is the constant function 1 on $[0, 1]$. So it follows (using, for example, Proposition 6.12 of ACICARA) that the function $A : [0, 1] \to \mathbb{R}$ is integrable. Moreover, by Proposition 5.28, we have

$$\int_0^1 \left[\int_0^1 f(x, y)dy\right] dx = \int_0^1 A(x)dx = \int_0^1 1\, dx = 1.$$

Similarly, if $y = 0$ or if $y = 1$, then

$$B(y) := \int_0^1 f(x, y)dx = \int_0^1 0\, dx = 0,$$

and if $0 < y < 1$, then

$$\begin{aligned} B(y) := \int_0^1 f(x, y)dx &= \int_0^y f(x, y)dx + \int_y^1 f(x, y)dx \\ &= \int_0^y \frac{-1}{y^2} dx + \int_y^1 \frac{1}{x^2} dx = -\frac{1}{y^2} \cdot y + \left[-\frac{1}{x}\right]_y^1 = -1. \end{aligned}$$

So, as before, the function $B : [0, 1] \to \mathbb{R}$ is integrable and

$$\int_0^1 \left[\int_0^1 f(x, y)dx\right] dy = \int_0^1 B(y)dy = \int_0^1 -1\, dy = -1.$$

This example shows that both the iterated integrals can exist without being equal. The reason Fubini's Theorem does not apply here is that f is not integrable on $[0, 1] \times [0, 1]$.

(iv) In one-variable calculus, we come across the **Thomae function**, namely, the function $\phi : [0,1] \to \mathbb{R}$ defined by

$$\phi(x) := \begin{cases} 1 & \text{if } x = 0, \\ 1/q & \text{if } x \in \mathbb{Q} \cap [0,1] \text{ and } x = p/q, \text{ where } p, q \in \mathbb{N} \\ & \text{are relatively prime,} \\ 0 & \text{otherwise.} \end{cases}$$

It is shown that ϕ is Riemann integrable on $[0,1]$ and $\int_0^1 \phi(x)\,dx = 0$. (See, for instance, Example 6.16 of ACICARA.) Let us consider a variant of this function, namely, the **bivariate Thomae function** $f : [0,1] \times [0,1] \to \mathbb{R}$ defined by

$$f(x,y) = \begin{cases} 1 & \text{if } x = 0 \text{ and } y \in \mathbb{Q} \cap [0,1], \\ 1/q & \text{if } x, y \in \mathbb{Q} \cap [0,1] \text{ and } x = p/q, \text{ where } p, q \in \mathbb{N} \\ & \text{are relatively prime,} \\ 0 & \text{otherwise.} \end{cases}$$

First we show that f is integrable. Let $\epsilon > 0$ be given. Let us assume without loss of generality that $\epsilon \le 2$. Then the set

$$\{x \in [0,1] : f(x,y) \ge \epsilon/2 \text{ for some } y \in [0,1]\}$$

is finite and it contains 0; thus we may write it as $\{c_1, \ldots, c_\ell\}$ for some $\ell \in \mathbb{N}$. Let $\{x_0, x_1, \ldots, x_n\}$ be a partition of $[0,1]$ such that $(x_i - x_{i-1}) < \epsilon/4\ell$ for $i = 1, \ldots, n$, and consider the partition

$$P_\epsilon := \{(x_0, 0), (x_0, 1), (x_1, 0), (x_1, 1), \ldots, (x_n, 0), (x_n, 1)\}$$

of $[0,1] \times [0,1]$. Since there is always an irrational number in $[x_{i-1}, x_i]$, we have $m_{i,1}(f) = 0$ for $i = 1, \ldots, n$, and so $L(P_\epsilon, f) = 0$. Also, noting that $f(x,y) \le 1$ for all $(x,y) \in [0,1] \times [0,1]$, and that the points $c_1, \ldots, c_\ell$ belong to at most 2ℓ subintervals among $[x_0, x_1], \ldots, [x_{n-1}, x_n]$ and also that $f(x,y) < \epsilon/2$ whenever (x,y) belongs to any of the remaining subrectangles, we obtain

$$\begin{aligned} U(P_\epsilon, f) &= \sum_{i=1}^{n} M_{i,1}(f)(x_i - x_{i-1})(1-0) \\ &< \frac{\epsilon}{4\ell} \cdot 2\ell + \frac{\epsilon}{2} \sum_{i=1}^{n} (x_i - x_{i-1})(1-0) = \epsilon. \end{aligned}$$

Thus $U(P_\epsilon, f) - L(P_\epsilon, f) < \epsilon - 0 = \epsilon$. The Riemann Condition implies that f is integrable. Moreover, since $f(x,y) \ge 0$ for all $x \in [0,1] \times [0,1]$, we see that

$$\iint_{[0,1]\times[0,1]} f(x,y) d(x,y) = \inf\{U(P,f) : P \text{ a partition of } [0,1]\times[0,1]\} = 0.$$

Next, let us consider one of the iterated integrals. To this end, fix $y \in [0,1]$. If $y \notin \mathbb{Q}$, then $f(x,y) = 0$ for all $x \in [0,1]$, and so

$$B(y) := \int_0^1 f(x,y)dx = 0,$$

and if $y \in \mathbb{Q}$, then the function $\phi_y : [0,1] \to \mathbb{R}$ defined by $\phi_y(x) = f(x,y)$ is Thomae's function on $[0,1]$. Hence ϕ_y is Riemann integrable and its integral is equal to zero. Thus,

$$B(y) := \int_0^1 f(x,y)dx = \int_0^1 \phi_y(x)dx = 0.$$

Consequently, the iterated integral

$$\int_0^1 \left(\int_0^1 f(x,y)dx \right) dy = \int_0^1 B(y)dy$$

exists and is equal to zero in conformity with Fubini's Theorem. On the other hand, consider a fixed $x \in [0,1]$. If $x \notin \mathbb{Q}$, then $f(x,y) = 0$ for all $y \in [0,1]$, and so

$$A(x) := \int_0^1 f(x,y)dy = 0.$$

Next, if $x := 0$, then the function $\psi_0 : [0,1] \to \mathbb{R}$ given by $\psi_0(y) := f(0,y)$ is the Dirichlet function on $[0,1]$, whereas if $x > 0$ and $x = p/q$, where $p, q \in \mathbb{N}$ have no common factor, then the function $\psi_x : [0,1] \to \mathbb{R}$ given by $\psi_x(y) := f(x,y)$ is the Dirichlet function on $[0,1]$ multiplied by $1/q$, and as remarked in Example 5.5 (iii), this function is not integrable on $[0,1]$. Hence for any $x \in \mathbb{Q} \cap [0,1]$, $\int_0^1 f(x,y)dy$ does not exist. Thus the iterated integral $\int_0^1 \left(\int_0^1 f(x,y)dy \right) dx$ is not defined. This example shows that a function $f : [a,b] \times [c,d] \to \mathbb{R}$ can be integrable and yet one of its iterated integrals may not exist. See also Exercise 32. ◇

Riemann Double Sums

Let $f : [a,b] \times [c,d] \to \mathbb{R}$ be a bounded function. We have seen that the integrability of f on $[a,b] \times [c,d]$ can be characterized by the Riemann Condition. Although we have made good use of the Riemann Condition to prove several interesting results earlier in this section, there are a number of difficulties in employing it to test the integrability of an arbitrary bounded function. To begin with, the calculation of $U(P,f)$ and $L(P,f)$, for a given partition P, involves finding suprema and infima of f over many subintervals of $[a,b] \times [c,d]$. This task is rarely easy. Next, it is not clear how one would go about finding a partition P for which $U(P,f) - L(P,f)$ is smaller than a prescribed positive value. Faced with these difficulties, we note that evaluating f at points of

$[a, b] \times [c, d]$ is much easier than finding suprema and infima of f over subintervals. With this in mind, we introduce the following variant of lower and upper double sums.

Let $P := \{(x_i, y_j) : i = 0, 1, \ldots, n \text{ and } j = 0, 1, \ldots, k\}$ be a partition of $[a, b] \times [c, d]$. Consider $s_i \in [x_{i-1}, x_i]$ for $i = 1, \ldots, n$ and $t_j \in [y_{j-1}, y_j]$ for $j = 1, \ldots, k$. Then

$$S(P, f) := \sum_{i=1}^{n} \sum_{j=1}^{k} f(s_i, t_j)(x_i - x_{i-1})(y_j - y_{j-1})$$

is called a **Riemann double sum** for f corresponding to P. It should be noted that $S(P, f)$ depends not only on P and f, but also on the choice of the points (s_i, t_j) in the (i, j)th subrectangle induced by P for $i = 1, \ldots, n$ and $j = 1, \ldots, k$. In any case, we always have $L(P, f) \le S(P, f) \le U(P, f)$.

It turns out that the integrability of f can be characterized in terms of Riemann double sums. To this end, we will use the notion of the mesh of a partition that was introduced at the beginning of this section and the upper bounds obtained in Lemma 5.2. It will follow that when a function f is integrable, the integral of f is, in some sense, a 'limit' of the Riemann double sums $S(P, f)$ as the mesh of P tends to zero. In particular, one can readily obtain a sequence (P_n) of partitions of $[a, b] \times [c, d]$ such that $S(P_n, f)$ converge to the integral of f.

Proposition 5.31 (Theorem of Darboux). *Let $f : [a, b] \times [c, d] \to \mathbb{R}$ be a bounded function. If f is integrable on $[a, b] \times [c, d]$, then given any $\epsilon > 0$, there is $\delta > 0$ such that for every partition P of $[a, b] \times [c, d]$ with $\mu(P) < \delta$, we have*

$$\left| S(P, f) - \iint_{[a,b]\times[c,d]} f(x, y)d(x, y) \right| < \epsilon,$$

where $S(P, f)$ is any Riemann double sum for f corresponding to P.

Conversely, assume that there is $r \in \mathbb{R}$ satisfying the following condition: Given $\epsilon > 0$, there is a partition P of $[a, b] \times [c, d]$ such that

$$|S(P, f) - r| < \epsilon,$$

where $S(P, f)$ is any Riemann double sum for f corresponding to P. Then f is double integrable and its double integral is equal to r.

Proof. First, assume that f is integrable on $[a, b] \times [c, d]$, and let $I(f)$ denote the double integral of f on $[a, b] \times [c, d]$. Let $\epsilon > 0$ be given. Since $I(f) = U(f)$, there is a partition P_1 of $[a, b] \times [c, d]$ such that $U(P_1, f) < I(f) + (\epsilon/2)$. Likewise, since $I(f) = L(f)$ as well, there is a partition P_2 of $[a, b] \times [c, d]$ such that $L(P_2, f) > I(f) - (\epsilon/2)$. Let P_0 be the common refinement of P_1 and P_2. Then in view of part (i) of Proposition 5.3, we have

$$U(P_0, f) \le U(P_1, f) < I(f) + \frac{\epsilon}{2} \quad \text{and} \quad L(P_0, f) \ge L(P_2, f) > I(f) - \frac{\epsilon}{2}.$$

Let $\alpha > 0$ be such that $|f(x,y)| \le \alpha$ for all $(x,y) \in [a,b] \times [c,d]$, and let $\ell := 2(b-a+d-c)$ be the perimeter of $[a,b] \times [c,d]$. Let m_0 denote the number of grid points of the partition P_0 in $[a,b) \times [c,d)$, and define $\delta := \epsilon/2\alpha\ell m_0$. Suppose P is any partition of $[a,b] \times [c,d]$ such that $\mu := \mu(P) < \delta$. Let P^* denote the common refinement of P and P_0. Then P^* is obtained from P by successive one-step refinements by points of P_0. By successively applying Lemma 5.2 to each of the m_0 points of P_0, we see that

$$U(P,f) \le U(P^*,f) + m_0\alpha\mu\ell \quad \text{and} \quad L(P,f) \ge L(P^*,f) - m_0\alpha\mu\ell.$$

Further, in view of part (i) of Proposition 5.3, we have

$$U(P^*,f) \le U(P_0,f) < I(f) + \frac{\epsilon}{2} \quad \text{and} \quad L(P^*,f) \ge L(P_0,f) > I(f) - \frac{\epsilon}{2}.$$

Combining the last two sets of inequalities displayed above and noting that $m_0\alpha\mu\ell < (\epsilon/2)$, thanks to our choice of δ, we see that

$$I(f) - \epsilon < L(P,f) \le S(P,f) \le U(P,f) < I(f) + \epsilon,$$

and hence $|S(P,f) - I(f)| < \epsilon$, as desired.

Conversely, suppose there is $r \in \mathbb{R}$ satisfying the condition in the second paragraph of the proposition. Let $\epsilon > 0$ be given, and let $P := \{(x_i, y_j) : i = 0,1,\dots,n \text{ and } j = 0,1,\dots,k\}$ be a partition of $[a,b] \times [c,d]$ such that $|S(P,f) - r| < (\epsilon/4)$ for any Riemann double sum $S(P,f)$ for f corresponding to P. Now, for each $i = 1,\dots,n$ and $j = 1,\dots,k$, we can find (s_i, t_j) and $(\tilde{s}_i, \tilde{t}_j)$ in $[x_{i-1}, x_i] \times [y_{j-1}, y_j]$ such that

$$M_{i,j}(f) < f(s_i,t_j) + \frac{\epsilon}{4A} \quad \text{and} \quad m_{i,j}(f) > f\left(\tilde{s}_i,\tilde{t}_j\right) - \frac{\epsilon}{4A},$$

where $A := (b-a)(d-c) = \sum_{i=1}^{n}\sum_{j=1}^{k}(x_i - x_{i-1})(y_j - y_{j-1})$. If we consider the specific Riemann double sums

$$S(P,f) := \sum_{i=1}^{n}\sum_{j=1}^{k} f(s_i,t_j)(x_i - x_{i-1})(y_j - y_{j-1})$$

and

$$\widetilde{S}(P,f) := \sum_{i=1}^{n}\sum_{j=1}^{k} f\left(\tilde{s}_i,\tilde{t}_j\right)(x_i - x_{i-1})(y_j - y_{j-1}),$$

then on the one hand,

$$U(P,f) < S(P,f) + \frac{\epsilon}{4} \quad \text{and} \quad L(P,f) > \widetilde{S}(P,f) - \frac{\epsilon}{4},$$

whereas on the other hand,

$$S(P,f) - \widetilde{S}(P,f) \le S(P,f) - r + r - \widetilde{S}(P,f) < \frac{\epsilon}{4} + \frac{\epsilon}{4} = \frac{\epsilon}{2}.$$

Consequently,

$$U(P,f) - L(P,f) < S(P,f) - \widetilde{S}(P,f) + \frac{\epsilon}{4} + \frac{\epsilon}{4} < \frac{\epsilon}{2} + \frac{\epsilon}{2} = \epsilon.$$

Since f satisfies the Riemann Condition, f is integrable. Furthermore, if $I(f)$ denotes the double integral of f on $[a,b] \times [c,d]$, then

$$r - \frac{\epsilon}{2} < \widetilde{S}(P,f) - \frac{\epsilon}{4} < L(P,f) \le I(f) \le U(P,f) < S(P,f) + \frac{\epsilon}{4} < r + \frac{\epsilon}{2}.$$

Thus $|r - I(f)| < (\epsilon/2)$. Since $\epsilon > 0$ is arbitrary, we obtain $r = I(f)$. □

Corollary 5.32. *If $f : [a,b] \times [c,d] \to \mathbb{R}$ is integrable and if (P_n) is a sequence of partitions of $[a,b] \times [c,d]$ such that $\mu(P_n) \to 0$, then*

$$S(P_n, f) \to \iint_{[a,b]\times[c,d]} f,$$

where $S(P_n, f)$ is any Riemann double sum for f corresponding to P_n.

Proof. Let $I(f)$ denote the double integral of f on $[a,b] \times [c,d]$, and let $\epsilon > 0$ be given. By Proposition 5.31, there is $\delta > 0$ such that $|S(P,f) - I(f)| < \epsilon$ for every partition P of $[a,b] \times [c,d]$ with $\mu(P) < \delta$. Since $\mu(P_n) \to 0$, there is $n_0 \in \mathbb{N}$ such that $\mu(P_n) < \delta$ for all $n \ge n_0$. Consequently, $|S(P_n, f) - I(f)| < \epsilon$ for all $n \ge n_0$. Thus $S(P_n, f) \to I(f)$. □

Remark 5.33. It may be tempting to define the mesh of a partition $P = \{(x_i, y_j) : i = 0, 1, \ldots, n \text{ and } j = 0, 1, \ldots, k\}$ of $[a,b] \times [c,d]$ to be the maximum of the areas of subrectangles induced by P, that is, to define $\mu(P)$ to be $\max\{(x_i - x_{i-1})(y_j - y_{j-1}) : i = 1, \ldots, n \text{ and } j = 1, \ldots, k\}$. However, with this definition, the Theorem of Darboux (Proposition 5.31) and Corollary 5.32 do not hold. To see this, consider the bivariate Thomae function defined in Example 5.30 (iv). This is an integrable function $f : [0,1] \times [0,1] \to \mathbb{R}$ with the property that $f(0,y) = 1 = f(1,y)$ for all $y \in \mathbb{Q} \cap [0,1]$ and $I(f) = 0$, where $I(f)$ denotes the double integral of f on $[0,1] \times [0,1]$. If for any $k \in \mathbb{N}$, we let $P_k := \{(i, j/k) : i = 0, 1 \text{ and } j = 0, 1, \ldots k\}$, then P_k is a partition of $[0,1] \times [0,1]$ such that the area of each subrectangle induced by P_k is $1/k$, which tends to 0 as $k \to \infty$. However, the Riemann double sum $S(P_k, f) = \sum_{i=1}^{1} \sum_{j=1}^{k} f(i, j/k)(1/k)$ is equal to 1 for every $k \in \mathbb{N}$. In particular, $S(P_k, f) \not\to I(f)$. This example shows why it is important to define the mesh of a partition as the maximum of the lengths of sides of the subrectangles induced by it. An alternative, and essentially equivalent, definition would be to take the mesh of a partition as the maximum of the diameters of the subrectangles induced by it. ◇

5.2 Double Integrals over Bounded Sets

In this section we extend the theory of double integrals on rectangles developed in Section 5.1 to double integrals over an arbitrary bounded subset D of $\mathbb{R}^2$. The approach followed here is to extend the function to a rectangle containing D by setting the extended function equal to zero outside D. While we have maintained a minor linguistic distinction, whereby we talk of integrals *over* D as opposed to integrals *on* a rectangle, it is quickly shown that the definition of integrals over D is independent of the choice of the rectangle containing D, and, in particular, consistent with the definition of integrals on rectangles. Algebraic and order properties are obtained as an immediate consequence of the corresponding results in Section 5.1. Further, Fubini's Theorem (Proposition 5.28) extends easily to integrals over regions that can be nicely sliced along one of the axes. Asserting the integrability of important classes of functions, such as continuous functions, does present some difficulties. The trouble is that even if a function is continuous on D, its extension to a rectangle containing D may well be discontinuous on the boundary of D, and can even fail to be integrable. To tackle this, we require a suitable notion to say that the boundary of D is "thin", in which case continuous functions on D become integrable. To this end, we introduce sets of content zero and prove a number of basic properties. This leads to a general definition of the "area" of a bounded subset of $\mathbb{R}^2$. We end this section with a general version of domain additivity (Proposition 5.9).

Let D be a bounded subset of $\mathbb{R}^2$ and let $f : D \to \mathbb{R}$ be a bounded function. Consider a rectangle $R := [a, b] \times [c, d]$ such that $D \subseteq R$ and the function $f^* : R \to \mathbb{R}$ defined by

$$f^*(x, y) := \begin{cases} f(x, y) & \text{if } (x, y) \in D, \\ 0 & \text{otherwise.} \end{cases}$$

We say that f is **integrable** over D if f^* is integrable on R, and in this case, the **double integral** of f (over D) is defined to be the double integral of f^* (on R), that is,

$$\iint_D f(x, y)d(x, y) := \iint_R f^*(x, y)d(x, y).$$

Let us first observe that the double integral of f is well defined. In other words, the integrability of f over D and the value of its double integral are independent of the choice of a rectangle R containing D and the corresponding extension f^* of f to R. This can be seen as follows. Let

$$\begin{aligned} a_1 &:= \inf\{x \in \mathbb{R} : (x, y) \in D \text{ for some } y \in \mathbb{R}\}, \\ b_1 &:= \sup\{x \in \mathbb{R} : (x, y) \in D \text{ for some } y \in \mathbb{R}\}, \\ c_1 &:= \inf\{y \in \mathbb{R} : (x, y) \in D \text{ for some } x \in \mathbb{R}\}, \\ d_1 &:= \sup\{y \in \mathbb{R} : (x, y) \in D \text{ for some } x \in \mathbb{R}\}. \end{aligned}$$

Consider $R_1 := [a_1, b_1] \times [c_1, d_1]$. Clearly, R_1 is uniquely determined by D. Since $D \subseteq R$, we see that $a \le a_1 \le b_1 \le b$ and $c \le c_1 \le d_1 \le d$, that is, $R_1 \subseteq R$. If $f_1^* := f^*_{|R_1}$, it is enough to show that f^* is integrable on R if and only if f_1^* is integrable on R_1, and in this case $\iint_R f^*(x,y)d(x,y) = \iint_{R_1} f_1^*(x,y)d(x,y)$.

Assume first that $a_1 = b_1$ or $c_1 = d_1$. Then $f^*(x,y) = 0$ for all $(x,y) \in \mathbb{R}^2$ except when either $x = a_1$ or $y = c_1$. The domain additivity (Proposition 5.9) and Example 5.7 show that f^* is integrable on R and $\iint_R f^*(x,y)d(x,y) = 0$. Also, by our definition, f_1^* is integrable on R_1 and $\iint_{R_1} f_1^*(x,y)d(x,y) = 0$.

Now assume that $a_1 < b_1$ and $c_1 < d_1$. If $R_1 = R$, then there is nothing to prove. Otherwise, the rectangle R gets divided into p subrectangles, where $p = 2, 3, 4, 6$ or 9, depending on whether $a = a_1$, $b = b_1$, $c = c_1$, $d = d_1$. One of these p subrectangles is the rectangle R_1. These cases are illustrated in Figure 5.7. By domain additivity (Proposition 5.9), f^* is integrable on R if and only if it is integrable on each of these subrectangles, and then the double integral of f^* on R is the sum of the double integrals of f^* on these subrectangles. If R_2 is any of these subrectangles of R, other than the subrectangle R_1, then $f^*(x,y) = 0$ for every $(x,y) \in R_2$ except possibly at some of the points on the sides of R_2. Example 5.7 shows that f^* is integrable on R_2 and the double integral of f^* on R_2 is equal to 0. This proves our assertion.

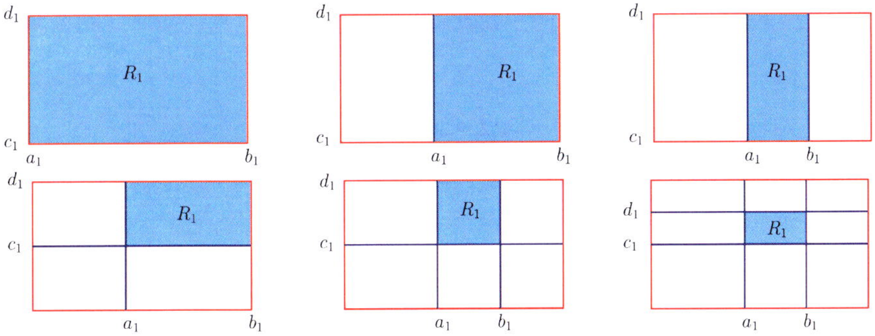

Fig. 5.7. Various possibilities for the subrectangle R_1.

If D is a bounded subset of $\mathbb{R}^2$ and $f : D \to \mathbb{R}$ is integrable, then we may denote the double integral

$$\iint_D f(x,y)d(x,y) \quad \text{simply by} \quad \iint_D f.$$

If, in addition, f is nonnegative, then the **volume** of the solid under the surface given by $z = f(x,y)$ and above the region D is defined to be the double integral of f over D. Thus

$$\mathrm{Vol}(E_f) := \iint_D f(x,y)d(x,y),$$

where

$$E_f := \{(x,y,z) \in \mathbb{R}^3 : (x,y) \in D \text{ and } 0 \le z \le f(x,y)\}.$$

We now consider the algebraic and the order properties of a double integral, and derive an analogue of Propositions 5.14 and 5.16.

Proposition 5.34. *Let D be a bounded subset of $\mathbb{R}^2$ and let $f, g : D \to \mathbb{R}$ be integrable functions. Also let $r \in \mathbb{R}$ and $k \in \mathbb{N}$. Then*

(i) *$f+g$ is integrable and $\iint_D (f+g) = \iint_D f + \iint_D g$,*
(ii) *rf is integrable and $\iint_D (rf) = r\iint_D f$,*
(iii) *fg is integrable,*
(iv) *if there is $\delta > 0$ such that $|f(x,y)| \ge \delta$ for all $(x,y) \in D$, then $1/f$ is integrable,*
(v) *if $f(x,y) \ge 0$ for all $(x,y) \in D$, then the function $f^{1/k}$ is integrable,*
(vi) *if $f \le g$ on D, then $\iint_D f \le \iint_D g$,*
(vii) *the function $|f|$ is integrable and $\left|\iint_D f\right| \le \iint_D |f|$.*

Proof. Let R be a rectangle such that $D \subseteq R$ and let $f^*, g^*, (f+g)^*, (rf)^*, (fg)^*$ denote, respectively, the extensions of $f, g, f+g, rf, fg$ to R by setting these equal to zero on points of $R \setminus D$. It is clear that

$$(f+g)^* = f^* + g^*, \quad (rf)^* = rf^*, \quad \text{and} \quad (fg)^* = f^* g^*.$$

Thus (i), (ii), and (iii) follow as an immediate consequence of parts (i), (ii), and (iii) of Proposition 5.14.

To prove (iv), assume that $\delta > 0$ such that $|f(x,y)| \ge \delta$ for all $(x,y) \in D$, and let $h := 1/f$. Define $h^* : R \to \mathbb{R}$ by extending the function $h : D \to \mathbb{R}$ as usual, that is,

$$h^*(x,y) = \begin{cases} h(x,y) & \text{if } (x,y) \in D, \\ 0 & \text{otherwise.} \end{cases}$$

Since $1/h^*$ is not even well defined unless $R = D$, we cannot give a proof as above. We therefore modify the proof of part (iv) of Proposition 5.14, and proceed as follows.

Let $P := \{(x_i, y_j) : i = 0, 1, \dots, n \text{ and } j = 0, 1, \dots, k\}$ be a partition of R. Fix $i, j \in \mathbb{N}$ with $i \le n$ and $j \le k$, and consider $(x,y), (u,v) \in [x_{i-1}, x_i] \times [y_{j-1}, y_j]$. We show that

$$h^*(x,y) - h^*(u,v) \le \frac{1}{\delta^2}\left[M_{i,j}(f^*) - m_{i,j}(f^*)\right]$$

by considering various cases separately.

Case 1. $(x,y) \in D$ and $(u,v) \in D$.

In this case we have

$$f^*(x,y) = f(x,y) \le M_{i,j}(f^*) \quad \text{and} \quad f^*(u,v) = f(u,v) \ge m_{i,j}(f^*).$$

Hence

$$\begin{aligned} h^*(x,y) - h^*(u,v) &= \frac{f(u,v) - f(x,y)}{f(x,y)f(u,v)} \le \frac{|f(u,v) - f(x,y)|}{|f(x,y)|\,|f(u,v)|} \\ &\le \frac{1}{\delta^2}\left[M_{i,j}(f^*) - m_{i,j}(f^*)\right]. \end{aligned}$$

Case 2. $(x,y) \notin D$ and $(u,v) \notin D$.
In this case $h^*(x,y) = 0 = h^*(u,v)$. Since $M_{i,j}(f^*) - m_{i,j}(f^*) \ge 0$, we are through.

Case 3. $(x,y) \in D$ and $(u,v) \notin D$.
In this case $h^*(x,y) - h^*(u,v) = 1/f(x,y)$. If $f(x,y) < 0$, then we are through. If $f(x,y) > 0$, then in fact $f(x,y) \ge \delta$, and so $M_{i,j}(f^*) \ge \delta$. On the other hand, since $(u,v) \notin D$, we have $m_{i,j}(f^*) \le 0$. Hence

$$h^*(x,y) - h^*(u,v) \le \frac{1}{\delta} \le \frac{M_{i,j}(f^*)}{\delta^2} \le \frac{1}{\delta^2}\left[M_{i,j}(f^*) - m_{i,j}(f^*)\right].$$

Case 4. Let $(x,y) \notin D$ and $(u,v) \in D$.
In this case $h^*(x,y) - h^*(u,v) = -1/f(u,v)$. If $f(u,v) > 0$, then we are through. If $f(u,v) < 0$, then in fact $f(u,v) \le -\delta$, and so $m_{i,j}(f^*) \le -\delta$. On the other hand, since $(x,y) \notin D$, we have $M_{i,j}(f^*) \ge 0$. Hence

$$h^*(x,y) - h^*(u,v) \le \frac{1}{\delta} \le -\frac{m_{i,j}(f^*)}{\delta^2} \le \frac{1}{\delta^2}\left[M_{i,j}(f^*) - m_{i,j}(f^*)\right].$$

Having established the desired inequality in all possible cases, we now proceed as in the proof of part (iv) of Proposition 5.14. Thus, taking the supremum for (x,y) in $[x_{i-1}, x_i] \times [y_{j-1}, y_j]$ and the infimum for (u,v) in $[x_{i-1}, x_i] \times [y_{j-1}, y_j]$, we obtain $M_{i,j}(h^*) - m_{i,j}(h^*) \le \left(1/\delta^2\right)\left[M_{i,j}(f^*) - m_{i,j}(f^*)\right]$, and consequently $U(P,h^*) - L(P,h^*) \le \left(1/\delta^2\right)\left[U(P,f^*) - L(P,f^*)\right]$. Now, since f^* satisfies the Riemann Condition, so does h^*. Thus h^* is integrable on R, that is, h is integrable over D. This proves (iv).

To prove (v), it suffices to observe that if $f \ge 0$ on D, then $f^* \ge 0$ on R and $(f^{1/k})^* = (f^*)^{1/k}$, and so the desired result is an immediate consequence of part (v) of Proposition 5.14.

Finally, the order properties (vi) and (vii) follow from Proposition 5.16 by noting that $f \le g$ implies $f^* \le g^*$ and also that $|f|^* = |f^*|$. □

With notation and hypotheses as in the above proposition, a combined application of its parts (i) and (ii) shows that the difference $f - g$ is integrable and $\iint_D (f-g) = \iint_D f - \iint_D g$. Further, given any $n \in \mathbb{N}$, successive applications of part (iii) of the above proposition show that the nth power f^n is integrable. Likewise, a combined application of parts (iii) and (iv) shows that if there is $\delta > 0$ such that $|g(x,y)| \geq \delta$ for all $(x,y) \in D$, then the quotient f/g is integrable. Also, a combined application of parts (iii) and (v) shows that if $f(x,y) \geq 0$ for all $(x,y) \in D$, then given any positive $r \in \mathbb{Q}$, the rth power f^r is integrable since $r = n/k$, where $n, k \in \mathbb{N}$.

Remark 5.35. Let D be a bounded subset of $\mathbb{R}^2$ and let $f : D \to \mathbb{R}$ be any function. Then

$$f = f^+ - f^-, \quad \text{where} \quad f^+ := \frac{|f| + f}{2} \quad \text{and} \quad f^- := \frac{|f| - f}{2}.$$

Note that both f^+ and f^- are nonnegative functions defined on D, and

$$f^+(x,y) = \max\{f(x,y), 0\} \text{ and } f^-(x,y) = -\min\{f(x,y), 0\} \text{ for } (x,y) \in D.$$

The functions f^+ and f^- are known as the **positive part** and the **negative part** of f, respectively. By parts (i), (ii), and (vii) of Proposition 5.34, we see that f is integrable if and only if f^+ and f^- are integrable, and then

$$\iint_D f = \iint_D f^+ - \iint_D f^- \quad \text{and} \quad \iint_D |f| = \iint_D f^+ + \iint_D f^-.$$

The integral of f over D may be interpreted as the "signed volume" of the solid in $\mathbb{R}^3$ delineated by the surface given by $z = f(x,y)$, $(x,y) \in D$. ◇

Fubini's Theorem over Elementary Regions

In Section 5.1, we have given a useful method of evaluating a double integral on a rectangle by converting it to an iterated integral. The relevant result of Fubini, when generalized to other subsets of $\mathbb{R}^2$, yields the most convenient way to calculate double integrals over a variety of regions. A precise definition of the kind of regions for which Fubini's Theorem is applicable is given below.

Let D be a bounded subset of $\mathbb{R}^2$. If there are $\phi_1, \phi_2 : [a,b] \to \mathbb{R}$ such that ϕ_1 and ϕ_2 are integrable, $\phi_1 \leq \phi_2$, and

$$D = \{(x,y) \in \mathbb{R}^2 : a \leq x \leq b \text{ and } \phi_1(x) \leq y \leq \phi_2(x)\},$$

or if there are $\psi_1, \psi_2 : [c,d] \to \mathbb{R}$ such that ψ_1 and ψ_2 are integrable, $\psi_1 \leq \psi_2$, and

$$D = \{(x,y) \in \mathbb{R}^2 : c \leq y \leq d \text{ and } \psi_1(y) \leq x \leq \psi_2(y)\},$$

then D is called an **elementary region**. (See Figure 5.8.)

Clearly, a rectangle is an elementary region in $\mathbb{R}^2$. Also, if $a > 0$, then the disk $D := \left\{(x,y) \in \mathbb{R}^2 : x^2 + y^2 \leq a^2\right\}$ is an elementary region in $\mathbb{R}^2$, since

$$D = \left\{(x,y) \in \mathbb{R}^2 : -a \leq x \leq a \text{ and } -\sqrt{a^2 - x^2} \leq y \leq \sqrt{a^2 - x^2}\right\},$$

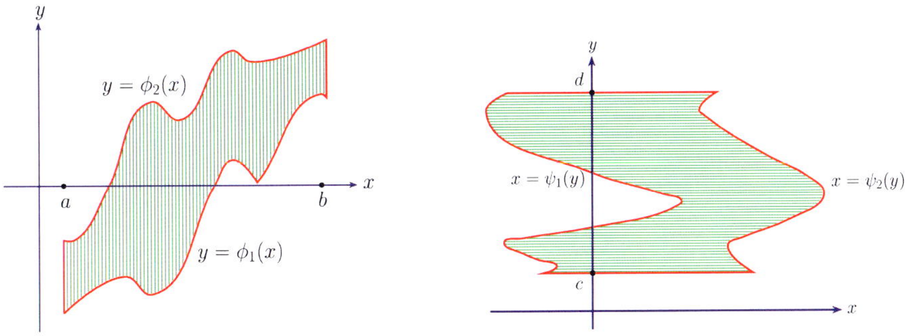

Fig. 5.8. Illustration of elementary regions

or alternatively,

$$D = \left\{(x, y) \in \mathbb{R}^2 : -a \leq y \leq a \text{ and } -\sqrt{a^2 - y^2} \leq x \leq \sqrt{a^2 - y^2}\right\}.$$

An essential feature of any elementary region D is the following: either there are $a, b \in \mathbb{R}$ such that for every $x \in [a, b]$, the vertical slice of D at x is a closed and bounded interval, or there are $c, d \in \mathbb{R}$ such that for every $y \in [c, d]$, the horizontal slice of D at y is a closed and bounded interval.

There do exist bounded subsets of $\mathbb{R}^2$ that are not elementary regions. For example, let D denote the star-shaped (closed and bounded) subset of $\mathbb{R}^2$ shown in Figure 5.9. Then D is not an elementary region, since for any $x \in \mathbb{R}$ with $1 < |x| < 2$, the vertical slice of D at x is not an interval, and for any $y \in \mathbb{R}$ with $1 < |y| < 2$, the horizontal slice of D at y is not an interval.

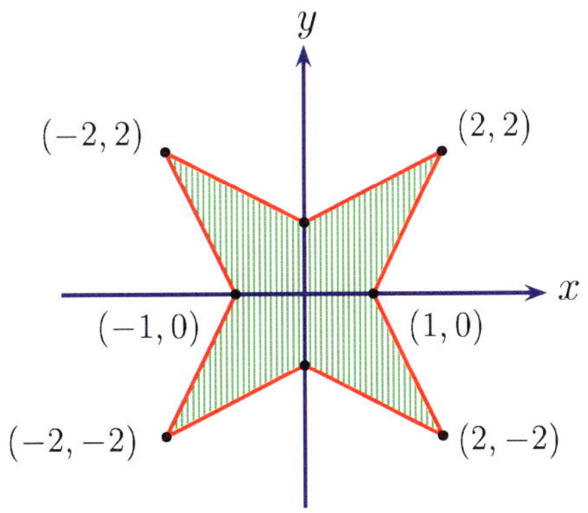

Fig. 5.9. A star-shaped subset of $\mathbb{R}^2$ that is not an elementary region

Proposition 5.36 (Fubini's Theorem over Elementary Regions). *Let D be an elementary region in $\mathbb{R}^2$ and let $f : D \to \mathbb{R}$ be an integrable function.*

(i) *If* $D := \{(x,y) \in \mathbb{R}^2 : a \le x \le b$ *and* $\phi_1(x) \le y \le \phi_2(x)\}$, *where* $\phi_1, \phi_2 : [a,b] \to \mathbb{R}$ *are Riemann integrable, and if for each fixed* $x \in [a,b]$, *the Riemann integral* $\int_{\phi_1(x)}^{\phi_2(x)} f(x,y)dy$ *exists, then*

$$\iint_D f(x,y)d(x,y) = \int_a^b \left(\int_{\phi_1(x)}^{\phi_2(x)} f(x,y)dy \right) dx.$$

(ii) *If* $D := \{(x,y) \in \mathbb{R}^2 : c \le y \le d$ *and* $\psi_1(y) \le x \le \psi_2(y)\}$, *where* $\psi_1, \psi_2 : [c,d] \to \mathbb{R}$ *are Riemann integrable, and if for each fixed* $y \in [c,d]$, *the Riemann integral* $\int_{\psi_1(y)}^{\psi_2(y)} f(x,y)dx$ *exists, then*

$$\iint_D f(x,y)d(x,y) = \int_c^d \left(\int_{\psi_1(y)}^{\psi_2(y)} f(x,y)dx \right) dy.$$

Proof. (i) Let D and ϕ_1, ϕ_2 be as stated in (i). In particular, assume that for each fixed $x \in [a,b]$, the Riemann integral $\int_{\phi_1(x)}^{\phi_2(x)} f(x,y)dy$ exists. Define

$$c := \inf\{\phi_1(x) : x \in [a,b]\} \quad \text{and} \quad d := \sup\{\phi_2(x) : x \in [a,b]\}.$$

Then $c \le d$. Let $R := [a,b] \times [c,d]$. Now for each fixed $x \in [a,b]$,

$$f^*(x,y) = \begin{cases} f(x,y) & \text{if } y \in [\phi_1(x), \phi_2(x)], \\ 0 & \text{if } y \in [c, \phi_1(x)) \text{ or } y \in (\phi_2(x), d]. \end{cases}$$

Hence by domain additivity of Riemann integrals (Fact 5.8), for each $x \in [a,b]$, the Riemann integral $\int_c^d f^*(x,y)dy$ exists, and we have

$$\begin{aligned} \int_c^d f^*(x,y)dy &= \int_c^{\phi_1(x)} f^*(x,y)dy + \int_{\phi_1(x)}^{\phi_2(x)} f^*(x,y)dy + \int_{\phi_2(x)}^d f^*(x,y)dy \\ &= \int_{\phi_1(x)}^{\phi_2(x)} f(x,y)dy. \end{aligned}$$

Thus by Fubini's Theorem for the function f^* (Proposition 5.28), we have

$$\iint_D f(x,y)d(x,y) = \iint_R f^*(x,y)d(x,y) = \int_a^b \left(\int_{\phi_1(x)}^{\phi_2(x)} f(x,y)dy \right) dx.$$

(ii) The proof is similar to the proof of part (i) above. □

Sets of Content Zero

If a function of one variable does not have too many discontinuities, then it is Riemann integrable. More precisely, if $f : [a,b] \to \mathbb{R}$ is bounded and if the set of discontinuities of f is of (one-dimensional) content zero, then f is Riemann integrable. (See Remark 5.42 below.) Recall that a bounded subset C of $\mathbb{R}$ is said to be of **one-dimensional content zero** if for every $\epsilon > 0$, there are finitely many closed intervals whose union contains C and the sum of whose lengths is less than ϵ. Examples of subsets of $\mathbb{R}$ of one-dimensional content zero include finite sets, the set $\{1/n : n \in \mathbb{N}\}$ and, in general, any $C \subseteq \mathbb{R}$ such that the interior of C is empty and ∂C is of content zero. (See, for example, Exercise 53 in Chapter 6 of ACICARA.) We shall now discuss an analogous notion for subsets of $\mathbb{R}^2$, which will turn out to be especially useful in the sequel. Let E be a bounded subset of $\mathbb{R}^2$. We say that E is of **(two-dimensional) content zero**, or that E has **(two-dimensional) content zero**, if the following condition holds: For every $\epsilon > 0$, there are finitely many rectangles whose union contains E and the sum of whose areas is less than ϵ.

We list below some basic properties of subsets of $\mathbb{R}^2$ of content zero. Here and hereinafter, we will simply speak of sets of content zero, that is, suppress the prefix "two-dimensional," while dealing with subsets of $\mathbb{R}^2$. On the other hand, when we consider subsets of $\mathbb{R}$ of content zero and, later in this chapter, subsets of $\mathbb{R}^3$ of content zero, we will explicitly mention the prefixes "one-dimensional" and "three-dimensional" as the case may be.

Proposition 5.37. *Let E be a subset of $\mathbb{R}^2$.*

(i) *If E is of content zero, then every subset of E is of content zero.*
(ii) *If E is a finite union of sets of content zero, then E has content zero.*
(iii) *If E is of content zero, then its closure $\overline{E}$ is of content zero.*
(iv) *E is of content zero if and only if E has no interior point and its boundary ∂E is of content zero.*
(v) *If $E = C \times D$, where $C \subseteq \mathbb{R}$ is of one-dimensional content zero and $D \subseteq \mathbb{R}$ is bounded, then E is of content zero.*

Proof. Both (i) and (ii) are obvious consequences of the definition of a set of content zero. To prove (iii), observe that rectangles are closed subsets of $\mathbb{R}^2$. It follows that if E is contained in the union of finitely many rectangles, then $\overline{E}$ is also contained in that union. Next, to prove (iv), first suppose E is of content zero. Then by (iii), $\overline{E}$ is of content zero. But by Proposition 2.7, $\overline{E} = E \cup \partial E$, and hence by (i), ∂E is of content zero. Further, if (x_0, y_0) is an interior point of E, then there is a rectangle R such that $(x_0, y_0) \in R \subseteq E$. Now if $\epsilon > 0$ is smaller than the area of R, then E cannot be covered by a finite union of rectangles having the sum of their areas less than ϵ. Thus E has no interior points. Conversely, suppose E has no interior points and its boundary ∂E is of content zero. Since every point of E is either an interior point of E or a boundary point of E, it follows that $E \subseteq \partial E$ and hence by (i), we conclude that E is of content zero. Finally, to prove (v) suppose $E = C \times D$, where $C \subseteq \mathbb{R}$ is of one-dimensional content zero and $D \subseteq \mathbb{R}$ is bounded. Then there are $\alpha, \beta \in \mathbb{R}$ with $\alpha < \beta$ such that $D \subseteq [\alpha, \beta]$. Now, given any $\epsilon > 0$, there are finitely many closed intervals $[a_1, b_1], \ldots, [a_n, b_n]$ whose union contains C

and the sum of whose lengths is less than $\epsilon/(\beta-\alpha)$. It follows that the sum of areas of the rectangles $[a_1,b_1]\times[\alpha,\beta],\ldots,[a_n,b_n]\times[\alpha,\beta]$ is less than ϵ and the union of these rectangles contains E. Thus E is of content zero. □

Corollary 5.38. *Let E_1 and E_2 be subsets of $\mathbb{R}^2$ such that ∂E_1 and ∂E_2 are of content zero. Then each of the sets $\partial(E_1\cup E_2)$, $\partial(E_1\cap E_2)$, and $\partial(E_1\setminus E_2)$ is of content zero.*

Proof. Observe that each of $\partial(E_1\cup E_2)$, $\partial(E_1\cap E_2)$, and $\partial(E_1\setminus E_2)$ is a subset of $\partial E_1\cup\partial E_2$, and use parts (i) and (ii) of Proposition 5.37. □

Examples 5.39. (i) Every finite subset of $\mathbb{R}^2$ is of content zero. But there also exist infinite subsets of $\mathbb{R}^2$ that are of content zero. For example, the infinite set

$$E := \left\{\left(\frac{1}{n},\frac{1}{k}\right)\in\mathbb{R}^2 : n,k\in\mathbb{N}\right\}$$

is of content zero. This follows by considering a square of arbitrarily small size with center $(0,0)$ and noting that only a finite number of points of the set E lie outside any such square. On the other hand, the infinite set $D := \{(x,y)\in[0,1]\times[0,1] : x,y\in\mathbb{Q}\}$ is not of content zero. This can be seen by noting that if the set D is contained in the union of finitely many rectangles, then this union also contains the square $[0,1]\times[0,1]$. Notice that both E and D are countable sets, that is, each of them is in one-to-one correspondence with $\mathbb{N}$.

(ii) The graph of a Riemann integrable function is of content zero. More precisely, if $\phi : [a,b]\to\mathbb{R}$ is Riemann integrable, then the set $E := \{(x,\phi(x))\in\mathbb{R}^2 : x\in[a,b]\}$ is of content zero. To prove this, let $\epsilon>0$ be given. By the Riemann Condition for functions of one variable (given, for example, in Proposition 6.5 of ACICARA), there is a partition $P=\{x_0,x_1,\ldots,x_n\}$ of $[a,b]$ such that $U(P,\phi)-L(P,\phi)<\epsilon$. Let $R_i := [x_{i-1},x_i]\times[m_i(\phi),M_i(\phi)]$ for $i=1,\ldots,n$. Clearly, E is contained in the union of $R_1,\ldots,R_n$, and

$$\sum_{i=1}^{n}\text{Area}(R_i)=\sum_{i=1}^{n}[M_i(\phi))-m_i(\phi)](x_i-x_{i-1})=U(P,\phi)-L(P,\phi)<\epsilon.$$

Thus E is of content zero. Similarly, it can be seen that if $\psi:[c,d]\to\mathbb{R}$ is an integrable function, then the set $\{(\psi(y),y)\in\mathbb{R}^2 : y\in[c,d]\}$ is of content zero. In particular, if L is any line segment (of finite length) in $\mathbb{R}^2$, then L is of content zero, and so is any subset of L. ◇

Remark 5.40. In contrast to Example 5.39 (ii) above, images of parametric curves in $\mathbb{R}^2$ need not be of content zero. More precisely, there can be continuous functions $x,y:[0,1]\to\mathbb{R}$ such that the set $C := \{(x(t),y(t)) : t\in[0,1]\}$ is not of content zero. In fact, C can be equal to the entire unit square $[0,1]\times[0,1]$; parametric curves with this property are called **space-filling curves**. For an explicit example, see Exercise 14 in Chapter 7 of Rudin's book [48]. For more on space-filling curves, see the book of Sagan [50]. ◇

Let us use the notion of sets of content zero to prove a neat generalization of part (ii) of Proposition 5.12.

Lemma 5.41. *Let R be a rectangle in $\mathbb{R}^2$ and $f : R \to \mathbb{R}$ a bounded function. If the set of discontinuities of f is of content zero, then f is integrable on R.*

Proof. Let E denote the set of discontinuities of f and let $\epsilon > 0$ be given. Since E is of content zero, there are finitely many rectangles $R_1, \dots, R_m$ such that

$$E \subseteq \bigcup_{i=1}^{m} R_i \quad \text{and} \quad \sum_{i=1}^{m} \text{Area}(R_i) < \frac{\epsilon}{2}.$$

We may assume without loss of generality that the rectangles $R_1, \dots, R_m$ are contained in the rectangle R. For each $i = 1, \dots, m$, we enlarge R_i slightly across each of its sides (except when a side of R_i lies on a side of R) and obtain a rectangle S_i such that

$$E \subseteq \bigcup_{i=1}^{m} R_i \subseteq \bigcup_{i=1}^{m} S_i \subseteq R \quad \text{and} \quad \sum_{i=1}^{m} \text{Area}(S_i) < \epsilon.$$

We then extend all four sides of each of the rectangles $S_1, \dots, S_m$ till they meet the boundary of R; this is indicated by the dashed lines in Figure 5.10. This gives a partition P_ϵ of R such that each subrectangle induced by P_ϵ is either contained in the union of $S_1, \dots, S_m$ or is disjoint from the union of $R_1, \dots, R_m$. Hence the sum of the areas of all the subrectangles induced by the partition P_ϵ that intersect E is less than or equal to the sum of the areas of the rectangles $S_1, \dots, S_m$, and hence less than ϵ.

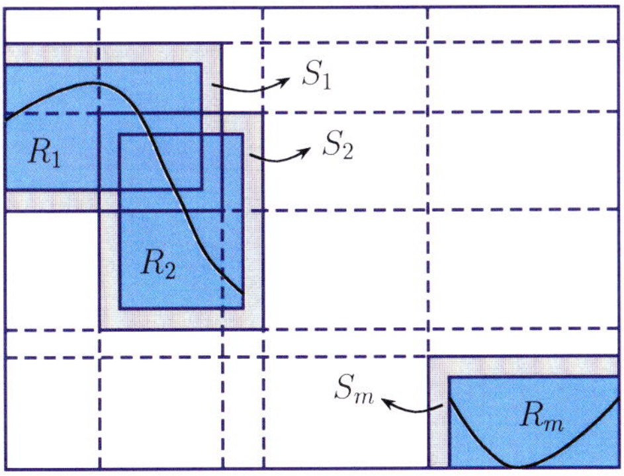

Fig. 5.10. The set E of content zero covered by small rectangles $R_1, \dots, R_m$ and by slightly larger rectangles $S_1, \dots, S_m$.

Let D_0 denote the union of all the subrectangles induced by P_ϵ that do not contain any point of E. Then f is continuous at every point of D_0, and since D_0 is a closed and bounded subset of $\mathbb{R}^2$, f is uniformly continuous on D_0 (Proposition 2.37), that is, there is $\delta > 0$ such that

$$(x, y),\ (u, v) \in D_0,\ |x - u| < \delta \text{ and } |y - v| < \delta \Longrightarrow |f(x, y) - f(u, v)| < \epsilon.$$

Let $P^*_\epsilon := \{(x_i, y_j) : i = 0, 1, \ldots, n \text{ and } j = 0, 1, \ldots, k\}$ be a refinement of P_ϵ such that $|x_i - x_{i-1}| < \delta$ for $i = 1, \ldots, n$ and $|y_j - y_{j-1}| < \delta$ for $j = 1, \ldots, k$. We will use the partition P^*_ϵ to show that f satisfies the Riemann Condition.

First, note that for $i = 1, \ldots, n$ and $j = 1, \ldots, k$, the subrectangles $R_{i,j} := [x_{i-1}, x_i] \times [y_{j-1}, y_j]$ induced by P^*_ϵ fall into two categories according as whether or not they intersect E. Thus the indexing set $\Lambda := \{(i, j) : i = 1, \ldots, n \text{ and } j = 1, \ldots, k\}$ is a disjoint union of Λ_1 and Λ_2, where

$$\Lambda_1 := \{(i, j) \in \Lambda : R_{i,j} \cap E = \emptyset\} \quad \text{and} \quad \Lambda_2 := \{(i, j) \in \Lambda : R_{i,j} \cap E \neq \emptyset\}.$$

Clearly, $\bigcup_{(i,j)\in\Lambda_1} R_{i,j} \subseteq D_0$. So the uniform continuity of f on D_0 implies that

$$f(x, y) - f(u, v) < \epsilon \quad \text{for all } (x, y), (u, v) \in R_{i,j}, \quad \text{provided } (i, j) \in \Lambda_1.$$

Consequently,

$$M_{i,j}(f) - m_{i,j}(f) \leq \epsilon \quad \text{for } (i, j) \in \Lambda_1.$$

On the other hand, since the function f is bounded on R, there is $\alpha > 0$ such that $-\alpha \leq f(x, y) \leq \alpha$ for all $(x, y) \in R$, and consequently, $M_{i,j}(f) \leq \alpha$ and $m_{i,j}(f) \geq -\alpha$ for $i = 1, \ldots, n$ and $j = 1, \ldots, k$. Thus,

$$M_{i,j}(f) - m_{i,j}(f) \leq 2\alpha \quad \text{for } (i, j) \in \Lambda_2 \quad \text{and} \quad \sum_{(i,j)\in\Lambda_2} \text{Area}(R_{i,j}) < \epsilon.$$

It follows that

$$\begin{aligned} U(P^*_\epsilon, f) - L(P^*_\epsilon, f) &= \sum_{i=1}^{n} \sum_{j=1}^{k} [M_{i,j}(f) - m_{i,j}(f)]\,(x_i - x_{i-1})(y_j - y_{j-1}) \\ &\leq \epsilon(b - a)(d - c) + 2\alpha\epsilon \\ &= [(b - a)(d - c) + 2\alpha]\epsilon. \end{aligned}$$

Since $\epsilon > 0$ is arbitrary, the Riemann Condition (Proposition 5.6) shows that f is integrable on R. □

Remark 5.42. An argument similar to (and in fact, simpler than) that in the proof of Proposition 5.41 proves the following one-variable analogue: If $\phi : [a, b] \to \mathbb{R}$ is a bounded function such that the set of discontinuities of ϕ is of one-dimensional content zero, then ϕ is Riemann integrable on $[a, b]$. (Compare Exercise 55 in Chapter 6 of ACICARA.) ◇

We are now ready to show that a continuous function on a bounded subset of $\mathbb{R}^2$ is integrable provided the boundary of its domain is thin, that is, it is of content zero. In fact, we shall prove a more general result, similar to Lemma 5.41, which gives a sufficient condition for a possibly discontinuous function defined on a bounded subset of $\mathbb{R}^2$ to be integrable. In practice, this condition is very useful for checking the integrability of a function and, in turn, ensuring that Fubini's Theorem is applicable.

Proposition 5.43. *Let D be a bounded subset of $\mathbb{R}^2$ and $f : D \to \mathbb{R}$ a bounded function. If the boundary ∂D of D is of content zero and if the set of discontinuities of f is also of content zero, then f is integrable over D. In particular, if $f : D \to \mathbb{R}$ is continuous and ∂D is of content zero, then f is integrable over D.*

In case the set D itself is of content zero, every bounded function is integrable over D and its double integral over D is equal to zero.

Proof. Let R be a rectangle containing the set D and let $f^* : R \to \mathbb{R}$ be the function obtained by extending the function $f : D \to \mathbb{R}$ as usual. Let E and E^* denote the sets of discontinuities of f and f^* respectively. It is easily seen that $E^* \subseteq E \cup \partial D$. Thus, if ∂D and E are of content zero, then by parts (ii) and (i) of Proposition 5.37, E^* is of content zero. Hence by Lemma 5.41, f^* is integrable on R, that is, f is integrable on D. In case f is continuous, then E is empty and so f is integrable over D.

Finally, assume that the set D itself is of content zero. In this case, with R and f^* as before, if $\epsilon > 0$ is given, then there is a partition Q_ϵ of R such that the sum of the areas of the subrectangles (induced by Q_ϵ) that intersect D is less than ϵ. Since the function f^* vanishes identically on the remaining subrectangles, we have

$$-\alpha\epsilon < L(Q_\epsilon, f^*) \leq U(Q_\epsilon, f^*) < \alpha\epsilon, \quad \text{where } \alpha := \sup\{|f(x, y)| : (x, y) \in D\}.$$

So, in view of the Riemann Condition, we see that f^* is integrable on R and its double integral is equal to zero, that is, the function f is integrable over D and its double integral is equal to zero. □

Remarks 5.44. (i) The above proposition gives two conditions that together imply the integrability of a bounded function f on a bounded subset D of $\mathbb{R}^2$. Neither of these two conditions is necessary. For example, if $f := 0$, then f is clearly integrable over D even if ∂D is not of content zero. Also, if $D := [0, 1] \times [0, 1]$ and $f : D \to \mathbb{R}$ is the bivariate Thomae function, then, as shown in Example 5.30 (iv), f is integrable on D, but the set of discontinuities of f is $D \cap \mathbb{Q}^2$, which is not of content zero. (See Exercise 23 of Chapter 2 and Example 5.39 (i).) On the other hand, neither of these two conditions can be dropped from the hypotheses of Proposition 5.43. For example, if $D := ([0, 1] \times [0, 1]) \cap \mathbb{Q}^2$ and $f : D \to \mathbb{R}$ is defined by $f(x, y) := 1$ for all $(x, y) \in D$, then f is continuous on D, so that the set of discontinuities of

f is of content zero (in fact, empty), but f is not integrable over D; indeed, if f^* denotes the extension of f to $[0,1] \times [0,1]$ defined as usual, then f^* is the bivariate Dirichlet function, which is not integrable on $[0,1] \times [0,1]$. (See Example 5.5 (iii).) Also, if $D := [0,1] \times [0,1]$, then ∂D is of content zero, but the bivariate Dirichlet function is not integrable over D.

(ii) Recall that if $f : [a,b] \times [c,d] \to \mathbb{R}$ is a bounded function, and if $a = b$ or $c = d$, then in Remark 5.11 we "declared" f to be integrable and its double integral to be equal to zero. This declaration is consistent with the last assertion of the preceding proposition because if $a = b$ or $c = d$, then $[a,b] \times [c,d]$ reduces to the line segment $\{(a,y) : y \in [c,d]\}$ or to the line segment $\{(x,c) : x \in [a,b]\}$, and any such line segment is of content zero (Example 5.39 (ii)). ◇

Corollary 5.45. *Let D be the elementary region given by*

$$D := \{(x,y) \in \mathbb{R}^2 : a \le x \le b \text{ and } \phi_1(x) \le y \le \phi_2(x)\},$$

where $\phi_1, \phi_2 : [a,b] \to \mathbb{R}$ are bounded functions such that $\phi_1 \le \phi_2$ and the set of discontinuities of ϕ_1 as well as of ϕ_2 is of one-dimensional content zero. Then ∂D is of content zero.

Further, if $f : D \to \mathbb{R}$ is a bounded function on D whose set of discontinuities is of content zero, then f is integrable over D, and if in addition, for each fixed $x \in [a,b]$, the iterated integral $\int_{\phi_1(x)}^{\phi_2(x)} f(x,y)dy$ exists, then

$$\iint_D f(x,y)d(x,y) = \int_a^b \left(\int_{\phi_1(x)}^{\phi_2(x)} f(x,y)dy \right) dx.$$

Proof. Let C_1 and C_2 denote, respectively, the subsets of $[a,b]$ consisting of the discontinuities of ϕ_1 and ϕ_2. Since C_1 and C_2 are of one-dimensional content zero, it follows (using, for example, Exercise 55 in Chapter 6 of ACICARA) that ϕ_1 and ϕ_2 are Riemann integrable. Hence, in view of Example 5.39 (ii), their graphs are of content zero. In other words, if we define

$$E_1 := \{(x, \phi_1(x)) \in \mathbb{R}^2 : a \le x \le b\} \text{ and } E_2 := \{(x, \phi_2(x)) \in \mathbb{R}^2 : a \le x \le b\},$$

then both E_1 and E_2 are of content zero. Next, since ϕ_1 and ϕ_2 are bounded with $\phi_1 \le \phi_2$, there are $\alpha, \beta \in \mathbb{R}$ with $\alpha < \beta$ such that $\alpha \le \phi_1(x) \le \phi_2(x) \le \beta$ for all $x \in [a,b]$. It follows that $D \subseteq R$, where $R := [a,b] \times [\alpha, \beta]$. Moreover, if we let $E_3 := \{(x,y) \in R : x = a \text{ or } x = b\}$ and $E_4 := \{(x,y) \in R : \phi_1 \text{ or } \phi_2 \text{ is discontinuous at } x\}$, then E_3 is a union of two line segments, whereas $E_4 \subseteq (C_1 \times [\alpha,\beta]) \cup (C_2 \times [\alpha,\beta])$. Thus, in view of parts (ii) and (v) of Proposition 5.37, we see that E_3 and E_4 are of content zero. Thus, to show that ∂D is of content zero, it suffices to prove that ∂D is contained in the union of the sets E_1, E_2, E_3, and E_4.

Let $(u,v) \in \partial D$. Then there is a sequence $((u_n, v_n))$ in D such that $(u_n, v_n) \to (u,v)$. Since $a \le u_n \le b$ and $\phi_1(u_n) \le v_n \le \phi_2(u_n)$ for all $n \in \mathbb{N}$, we see that $a \le u \le b$ and $\alpha \le v \le \beta$. If $u = a$ or $u = b$, then $(u, v) \in E_3$. Also, if either ϕ_1 or ϕ_2 is discontinuous at u, then $(u, v) \in E_4$. Now suppose that $a < u < b$ and that both ϕ_1 and ϕ_2 are continuous at u. Then $\phi_1(u_n) \to \phi_1(u)$ and $\phi_2(u_n) \to \phi_2(u)$, and so $\phi_1(u) \le v \le \phi_2(u)$. Moreover, if $v = \phi_1(u)$, then $(u, v) \in E_1$, whereas if $v = \phi_2(u)$, then $(u, v) \in E_2$. Finally, suppose $a < u < b$ and $\phi_1(u) < v < \phi_2(u)$ and also that both ϕ_1 and ϕ_2 are continuous at u. Let $\epsilon := \min\{(v - \phi_1(u))/2, (\phi_2(u) - v)/2\}$. Then there is $\delta > 0$ such that $(u - \delta, u + \delta) \subseteq [a, b]$ and moreover,

$$\begin{aligned} x \in (u - \delta, u + \delta) &\Longrightarrow |\phi_1(x) - \phi_1(u)| < \epsilon \ \text{ and } \ |\phi_2(x) - \phi_2(u)| < \epsilon \\ &\Longrightarrow \phi_1(x) < \phi_1(u) + \epsilon \le v - \epsilon < v + \epsilon \le \phi_2(u) - \epsilon < \phi_2(x). \end{aligned}$$

It follows that $(u - \delta, u + \delta) \times (v - \epsilon, v + \epsilon) \subseteq \{(x, y) \in \mathbb{R}^2 : a \le x \le b \text{ and } \phi_1(x) \le y \le \phi_2(x)\}$. In other words, (u, v) is an interior point of D. But this contradicts the assumption that $(u, v) \in \partial D$. Thus, we have proved that ∂D is contained in the union of the sets E_1, E_2, E_3, and E_4.

Now let $f : D \to \mathbb{R}$ be a bounded function such that the set of discontinuities of f is of content zero. By Proposition 5.43, f is integrable over D. The assertion about the equality of the double integral of f and the corresponding iterated integral follows from Proposition 5.36. □

Results similar to Corollary 5.45 hold for an elementary region D given by

$$D := \{(x, y) \in \mathbb{R}^2 : c \le y \le d \text{ and } \psi_1(y) \le x \le \psi_2(y)\},$$

where $\psi_1, \psi_2 : [c, d] \to \mathbb{R}$ are bounded functions such that $\psi_1 \le \psi_2$ and the sets of discontinuities of ψ_1 and ψ_2 are of one-dimensional content zero.

Examples 5.46. (i) Let $D := \{(x, y) \in \mathbb{R}^2 : y \ge 0 \text{ and } x^2 + 2y^2 \le 4\}$ be the semiellipsoidal region depicted on the left in Figure 5.11 and consider $f : D \to R$ defined by $f(x, y) := y$. Since

$$D = \left\{(x, y) \in \mathbb{R}^2 : -2 \le x \le 2 \text{ and } 0 \le y \le \sqrt{(4 - x^2)/2}\right\},$$

we see that D is an elementary region. Also, since f is continuous on D, it follows that f is integrable over D, and

$$\iint_D f = \int_{-2}^{2}\left(\int_0^{\sqrt{(4-x^2)/2}} y\,dy\right)dx = \frac{1}{2}\int_{-2}^{2}\frac{4 - x^2}{2}dx = \frac{8}{3}.$$

On the other hand, since

$$D = \left\{(x, y) \in \mathbb{R}^2 : 0 \le y \le \sqrt{2} \text{ and } -\sqrt{4 - 2y^2} \le x \le \sqrt{4 - 2y^2}\right\},$$

we have alternatively

$$\iint_D f = \int_0^{\sqrt{2}} \left(\int_{-\sqrt{4-2y^2}}^{\sqrt{4-2y^2}} y \, dx \right) dy = 2 \int_0^{\sqrt{2}} y\sqrt{4-2y^2} dy = \frac{8}{3}.$$

Note that the evaluation of the last integral is a bit more involved as compared to the earlier evaluation in this example.

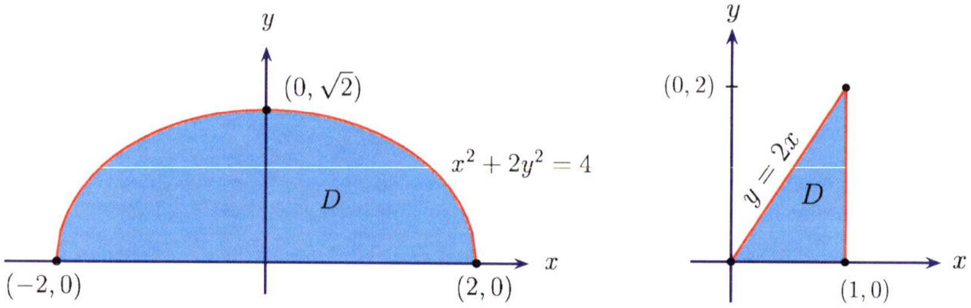

Fig. 5.11. The elementary regions in Example 5.46 (i) and (ii).

(ii) Let $D := \{(x, y) \in \mathbb{R}^2 : 0 \le x \le 1 \text{ and } 0 \le y \le 2x\}$ be the triangular region depicted on the right in Figure 5.11 and consider $f : D \to \mathbb{R}$ defined by $f(x, y) = \exp(x^2)$. Since D is clearly an elementary region and f is continuous on D, we see that the function f is integrable over D, and

$$\iint_D f = \int_0^1 \left(\int_0^{2x} \exp(x^2) dy \right) dx = \int_0^1 2x \exp(x^2) dx = e - 1.$$

Also, since $D = \{(x, y) \in \mathbb{R}^2 : 0 \le y \le 2 \text{ and } y/2 \le x \le 1\}$, we have alternatively

$$\iint_D f = \int_0^2 \left(\int_{y/2}^1 \exp(x^2) dx \right) dy.$$

However, the integral $\int_{y/2}^1 \exp(x^2) dx$ cannot be evaluated in terms of known functions. This example shows that an iterated integral may not always be useful in evaluating a double integral, and also that if one of the two ways of evaluating a double integral as an iterated integral does not work, then we should try the other. ◇

Concept of Area of a Bounded Subset of $\mathbb{R}^2$

We have seen earlier that the integrability of a function over a bounded subset D of $\mathbb{R}^2$ depends not only on the function, but also on the domain D. The simplest example of this is the constant function

$$1_D : D \to \mathbb{R} \quad \text{defined by} \quad 1_D(x, y) := 1 \text{ for all } (x, y) \in D.$$

In general, this is not an integrable function. For example, if $R := [0,1]\times[0,1]$ and $D := \{(x, y) \in R : x, y \in \mathbb{Q}\}$, then the function $1_D^* : R \to \mathbb{R}$, obtained by extending the function $1_D : D \to \mathbb{R}$ as usual, is the bivariate Dirichlet function. Thus, from Example 5.5 (iii), we see that 1_D^* is not integrable on R, that is, 1_D is not integrable over D. We shall presently see, however, that for a large class of bounded subsets D of $\mathbb{R}^2$, the function 1_D is integrable. In this case, it is natural to regard the double integral of 1_D over D to be the "area" of D. In light of this, we make the following general definition.

Let D be a bounded subset of $\mathbb{R}^2$. We say that D has an **area** if the function 1_D is integrable over D. In this case, the area of D is defined to be

$$\text{Area}(D) := \iint_D 1_D(x, y)d(x, y).$$

As an illustration, suppose D is a rectangle, say $D := [a, b] \times [c, d]$. From Example 5.5 (i), we see that D has an area, and $\text{Area}(D) = (b - a)(d - c)$. Thus, the general definition of area is consistent with the usual formula for the area of a rectangle given at the beginning of this chapter.

Proposition 5.47. *Let D be a bounded subset of $\mathbb{R}^2$. Then*

$$D \textit{ has an area} \iff \partial D \textit{ is of content zero.}$$

Furthermore,

$$D \textit{ has an area and } \text{Area}(D) = 0 \iff D \textit{ is of content zero.}$$

Proof. Suppose ∂D is of content zero. Since the function 1_D is continuous on D, by Proposition 5.43 we see that 1_D is integrable, that is, D has an area.

Conversely, assume that D has an area, that is, 1_D is integrable. Let R be a rectangle such that $D \subseteq R$ and let $1_D^* : R \to \mathbb{R}$ be the function obtained by extending the function $1_D : D \to \mathbb{R}$ as usual, that is,

$$1_D^*(x, y) := \begin{cases} 1 & \text{if } (x, y) \in D, \\ 0 & \text{if } (x, y) \notin D. \end{cases}$$

Let $\epsilon > 0$ be given. By the Riemann Condition (Proposition 5.6), there is a partition $P := \{(x_i, y_j) : i = 0, 1, \ldots, n \text{ and } j = 0, 1, \ldots, k\}$ of R such that $U(P, 1_D^*) - L(P, 1_D^*) < \epsilon/2$. For $i = 1, \ldots, n$ and $j = 1, \ldots, k$, let $R_{i,j}$ denote the (i, j)th subrectangle $[x_{i-1}, x_i] \times [y_{j-1}, y_j]$ induced by P. Then

$$\sum_{i=1}^{n}\sum_{j=1}^{k}[M_{i,j}(1_D^*) - m_{i,j}(1_D^*)] \cdot \text{Area}(R_{i,j}) < \frac{\epsilon}{2}.$$

We note that

$$M_{i,j}(1_D^*) = \begin{cases} 1 & \text{if } R_{i,j} \cap D \neq \emptyset, \\ 0 & \text{if } R_{i,j} \cap D = \emptyset, \end{cases} \quad \text{and} \quad m_{i,j}(1_D^*) = \begin{cases} 0 & \text{if } R_{i,j} \not\subseteq D, \\ 1 & \text{if } R_{i,j} \subseteq D. \end{cases}$$

Let $R_1, \ldots, R_p$ denote those subrectangles among $R_{i,j}$, $i = 1, \ldots, n$, $j = 1, \ldots, k$, for which $R_{i,j} \cap D \neq \emptyset$ and $R_{i,j} \not\subseteq D$. For such i, j, we have

$$M_{i,j}(1_D^*) - m_{i,j}(1_D^*) = 1 - 0 = 1 \quad \text{and thus} \quad \sum_{\ell=1}^{p} \text{Area}(R_\ell) < \frac{\epsilon}{2}.$$

Let E denote the union of the boundaries $\partial R_{i,j}$ of $R_{i,j}, i = 1, \ldots, n$ and $j = 1, \ldots, k$. Since any line segment is of content zero as shown in Example 5.39 (ii), and since a finite union of sets of content zero is a set of content zero, we see that E is of content zero. Hence there are rectangles $\widetilde{R}_1, \ldots, \widetilde{R}_q$ such that E is contained in their union and

$$\sum_{\ell=1}^{q} \text{Area}(\widetilde{R}_\ell) < \frac{\epsilon}{2}.$$

We shall show that ∂D is contained in the union of the rectangles $R_1, \ldots, R_p$, $\widetilde{R}_1, \ldots, \widetilde{R}_q$. Let $(x, y) \in \partial D$. Since the rectangle R containing D is closed, we see that $\partial D \subseteq R$, and so there are i and j such that $(x, y) \in R_{i,j}$. First suppose that (x, y) belongs to the interior of $R_{i,j}$. Then by the definition of a boundary point, the interior of $R_{i,j}$ contains a point belonging to D as well as a point not belonging to D, and so $R_{i,j} \cap D \neq \emptyset$ as well as $R_{i,j} \not\subseteq D$. Hence $R_{i,j}$ must be one of the subrectangles $R_1, \ldots, R_p$. On the other hand, if $(x, y) \in \partial R_{i,j}$, then $(x, y) \in E$ and hence (x, y) belongs to one of the rectangles $\widetilde{R}_1, \ldots, \widetilde{R}_q$. Thus $\partial D \subseteq R_1 \cup \cdots \cup R_p \cup \widetilde{R}_1 \cup \cdots \cup \widetilde{R}_q$ and

$$\sum_{\ell=1}^{p} \text{Area}(R_\ell) + \sum_{\ell=1}^{q} \text{Area}(\widetilde{R}_\ell) < \frac{\epsilon}{2} + \frac{\epsilon}{2} = \epsilon.$$

This shows that ∂D is of content zero.

To prove the second part, assume that D has an area and $\text{Area}(D) = 0$, that is, the double integral of 1_D over D is equal to zero. Then

$$\inf \{U(P, 1_D^*) : P \text{ is a partition of } R\} = \iint_R 1_D^*(x, y) d(x, y) = 0.$$

Thus for any given $\epsilon > 0$, there is a partition P_ϵ of R such that $U(P_\epsilon, 1_D^*) < \epsilon$. Let $R_1, \ldots, R_p$ denote the subrectangles induced by the partition P_ϵ that contain some point of D. Now D is contained in the union of $R_1, \ldots, R_p$, and

$$\sum_{\ell=1}^{p} \text{Area}(R_\ell) = U(P_\epsilon, 1_D^*) < \epsilon.$$

Hence the set D is of content zero. Conversely, if D is of content zero, then by Proposition 5.43, 1_D is integrable and its double integral over D is equal to zero, that is, D has an area and $\text{Area}(D) = 0$. □

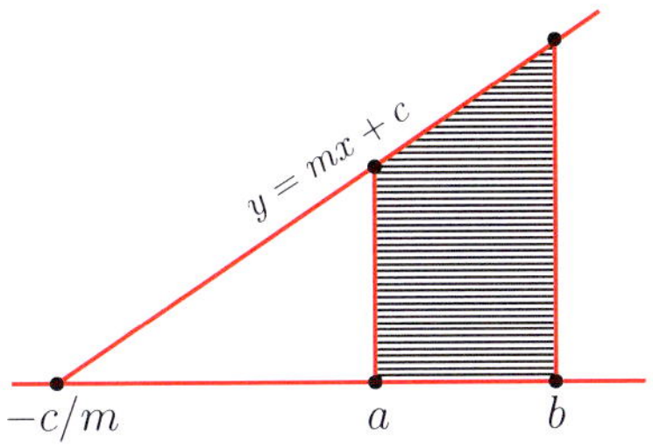

Fig. 5.12. Trapezoidal region in Example 5.48 (ii).

Examples 5.48. (i) We have seen already that if $R := [0,1] \times [0,1]$ and $D := \{(x,y) \in R : x, y \in \mathbb{Q}\}$, then 1_D is not integrable over D, that is, D does not have an area (Remark 5.44 (i)). Alternatively, we can reach this conclusion using Proposition 5.47 by observing that ∂D is R, which is not of content zero. For a closed and bounded subset of $\mathbb{R}^2$ that does not have an area, see Exercise 47.

(ii) Consider $D := \{(x,y) \in \mathbb{R}^2 : a \le x \le b \text{ and } 0 \le y \le mx + c\}$, where $a, b, c, m \in \mathbb{R}$ are such that $a < b$, $m > 0$, and $ma + c > 0$. (See Figure 5.12.) Then by Corollary 5.45, D has an area, which is given by

$$\iint_D 1_D = \int_a^b \left(\int_0^{mx+c} dy \right) dx = \int_a^b (mx + c)\, dx = m\frac{b^2 - a^2}{2} + c(b-a).$$

Consequently,

$$\text{Area}(D) = \frac{(b-a)}{2}\left[(ma + c) + (mb + c)\right].$$

It follows that the area of a trapezoid is half the height times the sum of the lengths of the two parallel sides. ◇

Corollary 5.49 (Basic Inequality). *Let D be a bounded subset of $\mathbb{R}^2$ such that its boundary ∂D is of content zero, and let $f : D \to \mathbb{R}$ be an integrable function. If there are $\alpha, \beta \in \mathbb{R}$ such that $\beta \le f \le \alpha$, then*

$$\beta \,\text{Area}(D) \le \iint_D f(x,y)d(x,y) \le \alpha \,\text{Area}(D).$$

In particular, if $|f| \le \alpha$, then we have

$$\left| \iint_D f(x,y)d(x,y) \right| \le \alpha \,\text{Area}(D).$$

Proof. By Proposition 5.47, D has an area, that is, the function 1_D is integrable. Hence the desired inequalities follow from parts (ii) and (vi) of Proposition 5.34. □

Corollary 5.50. *Let D be a bounded subset of $\mathbb{R}^2$ and let $f : D \to \mathbb{R}$ be an integrable function. If D_0 is a subset of D such that ∂D_0 is of content zero, then f is integrable over D_0.*

Proof. Let R be a rectangle such that $D \subseteq R$ and let $f^* : R \to \mathbb{R}$ be the function obtained by extending the function $f : D \to \mathbb{R}$ as usual. Let g denote the restriction of f to D_0. Since D_0 is contained in R, we may define g^*, $1^*_{D_0} : R \to \mathbb{R}$ by extending the functions g, 1_{D_0} as usual. It is easily seen that $g^* = f^* \cdot 1^*_{D_0}$. Now f^* is integrable on R by the definition of integrability of f, and since ∂D_0 is of content zero, Proposition 5.47 shows that $1^*_{D_0}$ is integrable on R. Hence part (iii) of Proposition 5.34 shows that g^* is integrable on R, that is, g is integrable over D_0, as desired. □

It may be noted that the requirement that ∂D_0 be of content zero cannot be omitted from the above corollary. For example, let $D = [0,1] \times [0,1]$ and $D_0 = \{(x, y) \in D : x, y \in \mathbb{Q}\}$. Then the function 1_D is integrable (Example 5.5 (i)), but its restriction to D_0 is not integrable (Example 5.5 (iii)).

Domain Additivity over Bounded Sets

Often a bounded subset D of $\mathbb{R}^2$ can be "decomposed" into several elementary regions. If a function is integrable over each of these elementary regions, then its double integral over D can be evaluated by splitting it over these regions. This is referred to as domain additivity, and we have already seen an instance of this in Proposition 5.9 in the context of double integrals on rectangles. The following two results give more general versions of domain additivity in the context of double integrals over bounded sets.

Proposition 5.51. *Let D be a bounded subset of $\mathbb{R}^2$ and let D_1, D_2 be subsets of D such that $D = D_1 \cup D_2$. Also, let $f : D \to \mathbb{R}$ be a bounded function. If f is integrable over D_1, over D_2, and over $D_1 \cap D_2$, then f is integrable over D and*

$$\iint_D f = \iint_{D_1} f + \iint_{D_2} f - \iint_{D_1 \cap D_2} f.$$

Conversely, if f is integrable over D and further, if ∂D_1 and ∂D_2 are both of content zero, then f is integrable over D_1, over D_2, and over $D_1 \cap D_2$.

Proof. Let R be a rectangle in $\mathbb{R}^2$ containing D and let $f^* : R \to \mathbb{R}$ be the function obtained by extending the function $f : D \to \mathbb{R}$ as usual. Let f_1, f_2, and g denote the restrictions of f to D_1, to D_2, and to $D_1 \cap D_2$ respectively. Since all these sets are contained in R, we may define the functions f_1^*, f_2^*, $g^* : R \to \mathbb{R}$ by extending the functions $f_1 : D_1 \to \mathbb{R}$, $f_2 : D_2 \to \mathbb{R}$, and $g : D_1 \cap D_2 \to \mathbb{R}$ as usual. It is easily seen that

$$f^* = f_1^* + f_2^* - g^*.$$

Now, if f is integrable over D_1, over D_2, and over $D_1 \cap D_2$, that is, if f_1^*, f_2^*, and g^* are integrable on R, then by parts (i) and (ii) of Proposition 5.34, the function f^* is integrable on R, and

$$\iint_R f^* = \iint_R f_1^* + \iint_R f_2^* - \iint_R g^*,$$

and this yields the desired formula. Conversely, suppose f is integrable over D and both ∂D_1 and ∂D_2 are of content zero. Then by Corollary 5.38, $\partial(D_1 \cap D_2)$ is of content zero. Hence from Corollary 5.50 it follows that f is integrable over D_1, over D_2, and over $D_1 \cap D_2$. □

In most applications, a bounded subset D of $\mathbb{R}^2$ is split as $D_1 \cup D_2$ and the overlap $D_1 \cap D_2$ is of content zero. In this case Proposition 5.51 takes the following simpler form.

Corollary 5.52 (Domain Additivity over Bounded Sets). *Let D be a bounded subset of $\mathbb{R}^2$ and let D_1, D_2 be subsets of D such that $D = D_1 \cup D_2$ and $D_1 \cap D_2$ are of content zero. Also, let $f : D \to \mathbb{R}$ be a bounded function such that f is integrable over D_1 and over D_2. Then f is integrable over D and*

$$\iint_D f(x,y)d(x,y) = \iint_{D_1} f(x,y)d(x,y) + \iint_{D_2} f(x,y)d(x,y).$$

Proof. By Proposition 5.43, we see that f is integrable over $D_1 \cap D_2$ and its double integral over $D_1 \cap D_2$ is equal to zero. Hence the desired result is an immediate consequence of Proposition 5.51. □

Examples 5.53. (i) Let $D := [a,b] \times [c,d]$ and $\phi : [a,b] \to \mathbb{R}$ be defined by

$$\phi(x) := \frac{d-c}{b-a}(b-x) + c.$$

Then $y = \phi(x)$ gives the line passing through (a,d) and (b,c), and it divides the rectangle D into nonoverlapping triangular regions D_1 and D_2 as shown in Figure 5.13. More precisely, $D_1 := \{(x,y) \in D : c \le y \le \phi(x)\}$ and $D_2 := \{(x,y) \in D : \phi(x) \le y \le d\}$. Thus D_1 and D_2 are elementary regions, and hence ∂D_1 and ∂D_2 are both of content zero. Further, $D_1 \cap D_2$ is the line segment joining the points (a,d) and (b,c), and so it is of content zero. Hence if a bounded function $f : D \to \mathbb{R}$ is integrable over D_1 as well as over D_2, then it is integrable over D, and its double integral over D is the sum of its double integrals over D_1 and over D_2.

(ii) Let D be the star-shaped (closed and bounded) region shown in Figure 5.9. Let $D_1 := \{(x,y) \in D : y \ge 0\}$ and $D_2 := \{(x,y) \in D : y \le 0\}$. Then D_1 and D_2 are elementary regions. Indeed, define $\phi_1, \phi_2 : [-2,2] \to \mathbb{R}$ by

$$\phi_1(x) := \begin{cases} -2(x+1) & \text{if } -2 \le x \le -1, \\ 0 & \text{if } -1 < x < 1, \\ 2(x-1) & \text{if } 1 \le x \le 2, \end{cases}$$

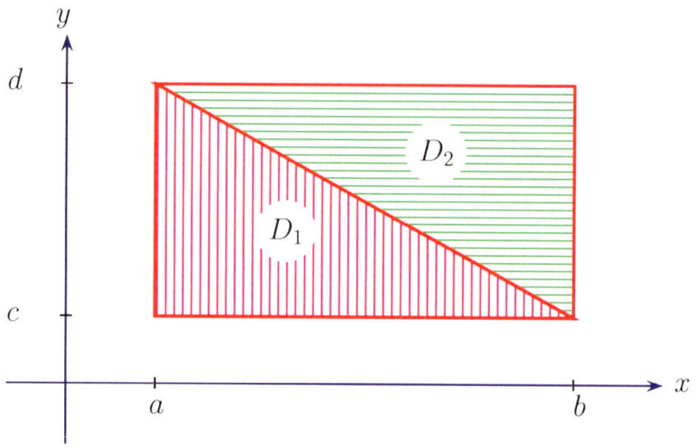

Fig. 5.13. Division of the rectangle $D := [a, b] \times [c, d]$ into nonoverlapping regions D_1 and D_2 by the line segment joining (a, d) and (b, c).

and

$$\phi_2(x) := \begin{cases} (2 - x)/2 & \text{if } -2 \le x \le 0, \\ (2 + x)/2 & \text{if } 0 < x \le 2. \end{cases}$$

Then ϕ_1 and ϕ_2 are continuous and $\phi_1 \le \phi_2$, and it is easily seen that $D_1 = \{(x, y) \in \mathbb{R}^2 : -2 \le x \le 2 \text{ and } \phi_1(x) \le y \le \phi_2(x)\}$ and $D_2 = \{(x, y) \in \mathbb{R}^2 : -2 \le x \le 2 \text{ and } -\phi_2(x) \le y \le -\phi_1(x)\}$. Also, the line segment $D_1 \cap D_2 = \{(x, y) \in \mathbb{R}^2 : -1 \le x \le 1,\ y = 0\}$ is of content zero. Hence if a bounded function $f : D \to \mathbb{R}$ is integrable over D_1 as well as over D_2, then it is integrable over D, and its double integral over D is the sum of its double integrals over D_1 and over D_2. ◇

As an application of domain additivity, we now prove an interesting property of double integrals that says, roughly speaking, that if the values of an integrable function are changed on a subset of content zero such that the modified function is bounded, then it is in fact integrable, and its double integral is equal to the double integral of the given function. This result may be compared with Proposition 6.12 and Exercise 57 in Chapter 6 of ACICARA.

Proposition 5.54. *Let D be a bounded subset of $\mathbb{R}^2$, let $f : D \to \mathbb{R}$ be integrable, and let $g : D \to \mathbb{R}$ be a bounded function such that the set $\{(x, y) \in D : g(x, y) \ne f(x, y)\}$ is of content zero. Then g is integrable over D and*

$$\iint_D g(x, y)d(x, y) = \iint_D f(x, y)d(x, y).$$

Proof. Let $D_1 := \{(x, y) \in D : g(x, y) \ne f(x, y)\}$ and $D_2 := D \setminus D_1$. Define $h : D \to \mathbb{R}$ by $h := g - f$, and let h_1 and h_2 denote the restrictions of h to D_1 and D_2 respectively. Let R be a rectangle such that $D \subseteq R$.

Since h_1 is a bounded function on D_1 and the set D_1 is of content zero, Proposition 5.43 shows that h_1 is integrable over D_1 and its double integral is equal to zero. Also, since the function h_2 vanishes identically on the set D_2, that is, the function h_2^* vanishes identically on R, we see that h_2 is integrable over D_2 and its double integral is also equal to zero. Further, the set $D_1 \cap D_2$ is of content zero, since it is the empty set. Hence by domain additivity (Corollary 5.52), the function h is integrable over $D = D_1 \cup D_2$ and

$$\iint_D h = \iint_{D_1} h + \iint_{D_2} h = 0 + 0 = 0.$$

Now it follows from part (i) of Proposition 5.34 that $g = h + f$ is integrable over D and

$$\iint_D g = \iint_D h + \iint_D f = \iint_D f,$$

as desired. □

5.3 Change of Variables

In this section, we shall examine the effect of a change of variables in a double integral. Thus, if f is a real-valued function on a bounded subset D of $\mathbb{R}^2$, and we change the variables x and y to new variables u and v by

$$x = \phi_1(u, v) \quad \text{and} \quad y = \phi_2(u, v) \qquad \text{or collectively} \qquad (x, y) = \Phi(u, v),$$

where the transformation $\Phi = (\phi_1, \phi_2)$ maps a bounded subset E of $\mathbb{R}^2$ onto D, then we would like to see how the double integral of $f(x, y)$ over D is related to the double integral of the function $g(u, v) := f(\Phi(u, v))$ over E. The precise relationship, which holds if f as well as Φ are sufficiently nice, is called the change of variables formula. It is an extremely useful result, but at the same time rather difficult to prove in the general case. In light of this, we shall proceed as follows. First, we consider the simplest of transformations, namely translations, and show that these do not alter the double integral. As an application, we shall determine a neat formula for the area of a parallelogram in $\mathbb{R}^2$. Next, we look at transformations that are a little more general than translations, but still of a simple kind, namely, affine transformations, and prove the corresponding change of variables formula. The case of affine transformations motivates the general result, which is stated precisely but not proved here. Finally, assuming the general result, we will prove a useful variant of it that will enable us to use polar coordinates.

Translation Invariance and Area of a Parallelogram

Translations are transformations of the form $\Phi(u, v) := (x^\circ + u, y^\circ + v)$, where (x°, y°) is a fixed point in $\mathbb{R}^2$ and (u, v) varies over a subset E of $\mathbb{R}^2$. These have the effect of interchanging the point (x°, y°) and the origin. It is quite natural to expect that translations do not alter the area of a bounded region. More generally, we have the following.

Lemma 5.55 (Translation Invariance). *Let D be a bounded subset of $\mathbb{R}^2$ and let $f : D \to \mathbb{R}$ be an integrable function. Fix $(x^\circ, y^\circ) \in \mathbb{R}^2$, and consider $E := \{(x - x^\circ, y - y^\circ) : (x, y) \in D\}$ and the function $g : E \to \mathbb{R}^2$ defined by $g(u, v) := f(x^\circ + u, y^\circ + v)$ for $(u, v) \in E$. Then E is a bounded subset of $\mathbb{R}^2$, g is integrable over E, and $\iint_D f(x, y)d(x, y) = \iint_E g(u, v)d(u, v)$.*

Proof. Let $R := [a, b] \times [c, d]$ be a rectangle such that $D \subseteq R$. Then $E \subseteq S$, where $S := [a-x^\circ, b-x^\circ] \times [c-y^\circ, d-y^\circ]$. In particular, E is a bounded subset of $\mathbb{R}^2$. Let $f^* : R \to \mathbb{R}$ and $g^* : S \to \mathbb{R}$ be obtained by extending the functions $f : D \to \mathbb{R}$ and $g : E :\to \mathbb{R}$ as usual. Partitions $P := \{(x_i, y_j) : i = 0, \ldots, n$ and $j = 0, \ldots, k\}$ of R and $Q := \{(u_i, v_j) : i = 0, \ldots, n$ and $j = 0, \ldots, k\}$ of S are in one-to-one correspondence given by the equations $u_i = x_i - x^\circ$ for $i = 0, \ldots, n$ and $v_j = y_j - y^\circ$ for $j = 0, \ldots, k$. Also, if for $i = 1, \ldots, n$ and $j = 1, \ldots, k$, we let $R_{i,j} := [x_{i-1}, x_i] \times [y_{j-1}, y_j]$ and $S_{i,j} := [u_{i-1}, u_i] \times [v_{j-1}, v_j]$ be the (i, j)th subrectangles induced by the corresponding partitions, then

$$\text{Area}(R_{i,j}) = (x_i - x_{i-1})(y_j - y_{j-1}) = (u_i - u_{i-1})(v_j - v_{j-1}) = \text{Area}(S_{i,j}).$$

Hence $L(P, f^*) = L(Q, g^*)$ as well as $U(P, f^*) = U(Q, g^*)$, and so

$$L(f^*) = L(g^*) \leq U(g^*) = U(f^*).$$

But since f is integrable on D, that is, $L(f^*) = U(f^*)$, we obtain $L(g^*) = U(g^*)$, that is, g is integrable on E, and moreover, $\iint_E g = \iint_D f$. □

Proposition 5.56. *Let $(x_i, y_i) \in \mathbb{R}^2$ for $i = 0, 1, 2$ be noncollinear and let D denote the parallelogram with one vertex at (x_0, y_0) and the vertices adjacent to (x_0, y_0) at (x_1, y_1) and (x_2, y_2). Then D has an area and*

$$\text{Area}(D) = \left| \det \begin{bmatrix} x_1 - x_0 & x_2 - x_0 \\ y_1 - y_0 & y_2 - y_0 \end{bmatrix} \right| = |(x_1-x_0)(y_2-y_0)-(x_2-x_0)(y_1-y_0)|.$$

Proof. Since ∂D is the union of four line segments of finite length, it is clear that ∂D is of content zero, and hence from Proposition 5.47, it follows that D has an area. To obtain the desired formula for Area(D), let us first assume that $(x_0, y_0) = (0, 0)$. Then our aim is to show that Area $(D) = |x_1 y_2 - x_2 y_1|$. We shall do this by describing the parallelogram D as an elementary region and using Fubini's Theorem to compute Area (D) as an iterated integral. However, such a description of D depends on the location of the vertices of D, and so we consider several cases as follows.

First, let us fix some notation that will be used in the rest of the proof. If $x_1 \neq 0$, then we let m_1 denote the slope y_1/x_1 of the line passing through $(0, 0)$ and (x_1, y_1), and if $x_2 \neq 0$, then we let m_2 denote the slope y_2/x_2 of the line passing through $(0, 0)$ and (x_2, y_2). Note that $m_1 \neq m_2$ whenever both m_1 and m_2 are defined.

Case 1. x_1 and x_2 are both nonzero and are of the same sign.

First assume that $x_1 > 0$ and $x_2 > 0$. Also, suppose $m_1 < m_2$. (See the parallelogram on the left in Figure 5.14.) Define $\phi_1, \phi_2 : [0, x_1 + x_2] \to \mathbb{R}$ by

$$\phi_1(x) := \begin{cases} m_1 x & \text{if } 0 \le x \le x_1, \\ m_2(x - x_1) + y_1 & \text{if } x_1 < x \le x_1 + x_2, \end{cases}$$

and

$$\phi_2(x) := \begin{cases} m_2 x & \text{if } 0 \le x \le x_2, \\ m_1(x - x_2) + y_2 & \text{if } x_2 < x \le x_1 + x_2. \end{cases}$$

Then it is easily seen that ϕ_1, ϕ_2 are continuous and $\phi_1 \le \phi_2$ and also that $D = \{(x, y) \in \mathbb{R}^2 : 0 \le x \le x_1 + x_2 \text{ and } \phi_1(x) \le y \le \phi_2(x)\}$. Since 1_D is continuous on D, in view of Corollary 5.45, we see that

$$\text{Area}(D) = \iint_D 1_D = \int_0^{x_1+x_2} \left(\int_{\phi_1(x)}^{\phi_2(x)} dy \right) dx = \int_0^{x_1+x_2} [\phi_2(x) - \phi_1(x)]\, dx.$$

Now $\int_0^{x_1+x_2} \phi_2(x) dx = \int_0^{x_2} m_2 x\, dx + \int_{x_2}^{x_1+x_2} [m_1(x - x_2) + y_2]\, dx$, whereas $\int_0^{x_1+x_2} \phi_1(x) dx = \int_0^{x_1} m_1 x\, dx + \int_{x_1}^{x_1+x_2} [m_2(x - x_1) + y_1]\, dx$, and thus

$$\text{Area}(D) = \left[m_2 \frac{x_2^2}{2} + m_1 \frac{x_1^2}{2} + y_2 x_1 \right] - \left[m_1 \frac{x_1^2}{2} + m_2 \frac{x_2^2}{2} + y_1 x_2 \right] = x_1 y_2 - x_2 y_1.$$

If instead we suppose $m_2 < m_1$, then we can easily see that $\phi_2 \le \phi_1$ and $D = \{(x, y) \in \mathbb{R}^2 : 0 \le x \le x_1 + x_2 \text{ and } \phi_2(x) \le y \le \phi_1(x)\}$, and proceeding as above, we obtain $\text{Area}(D) = x_2 y_1 - x_1 y_2$. Since $m_1 < m_2$ if and only if $x_1 y_2 - x_2 y_1 > 0$, we see that $\text{Area}(D) = |x_1 y_2 - x_2 y_1|$ when $x_1 > 0$ and $x_2 > 0$. If $x_1 < 0$ and $x_2 < 0$, then we obtain the same formula for $\text{Area}(D)$ by a similar argument.

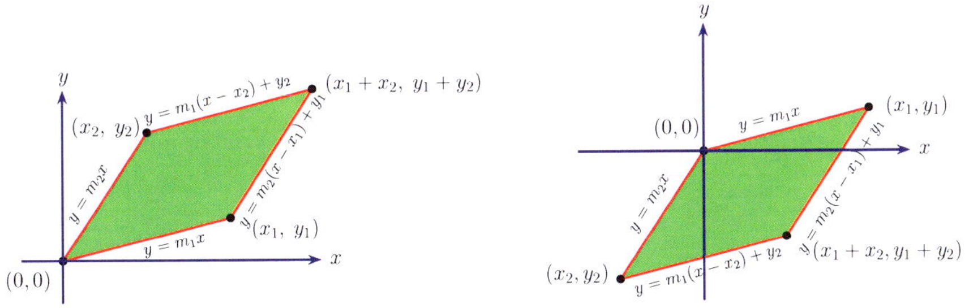

Fig. 5.14. Parallelograms in Proposition 5.56: (i) the case $x_1 > 0$ and $x_2 > 0$, and (ii) the case $x_1 > 0$ and $x_2 < 0$.

Case 2. x_1 and x_2 are both nonzero and are of opposite signs.

First assume that $x_1 > 0$ and $x_2 < 0$. Also, suppose $m_1 < m_2$. (See the parallelogram on the right in Figure 5.14.) Define $\phi_1, \phi_2 : [x_2, x_1] \to \mathbb{R}$ by

$$\phi_1(x) := \begin{cases} m_1(x - x_2) + y_2 & \text{if } x_2 \le x \le x_2 + x_1, \\ m_2(x - x_1) + y_1 & \text{if } x_2 + x_1 < x \le x_1, \end{cases}$$

and

$$\phi_2(x) := \begin{cases} m_2 x & \text{if } x_2 \le x \le 0, \\ m_1 x & \text{if } 0 < x \le x_1. \end{cases}$$

Then it is easily seen that ϕ_1, ϕ_2 are continuous and $\phi_1 \le \phi_2$, and also that $D = \{(x,y) \in \mathbb{R}^2 : x_2 \le x \le x_1 \text{ and } \phi_1(x) \le y \le \phi_2(x)\}$. Since 1_D is continuous on D, in view of Corollary 5.45, we see that

$$\text{Area}(D) := \iint_D 1_D = \int_{x_2}^{x_1} \left(\int_{\phi_1(x)}^{\phi_2(x)} dy \right) dx = \int_{x_2}^{x_1} [\phi_2(x) - \phi_1(x)]\, dx.$$

As before, a simple computation of Riemann integrals shows that

$$\text{Area}(D) = \left[-m_2 \frac{x_2^2}{2} + m_1 \frac{x_1^2}{2} \right] - \left[m_1 \frac{x_1^2}{2} + y_2 x_1 - m_2 \frac{x_2^2}{2} - y_1 x_2 \right] = x_2 y_1 - x_1 y_2.$$

If instead we suppose $m_2 < m_1$, then $\phi_2 \le \phi_1$ and $D = \{(x,y) \in \mathbb{R}^2 : x_2 \le x \le x_1 \text{ and } \phi_2(x) \le y \le \phi_1(x)\}$, and proceeding as above, we see that $\text{Area}(D) = x_1 y_2 - x_2 y_1$. Since $m_1 < m_2$ if and only if $x_2 y_1 - x_1 y_2 > 0$, we obtain $\text{Area}(D) = |x_1 y_2 - x_2 y_1|$. If $x_1 < 0$ and $x_2 > 0$, then we obtain the same formula for $\text{Area}(D)$ by a similar argument.

Case 3. $x_1 x_2 = 0$.

First assume that $x_2 = 0$. Then $x_1 \ne 0$ and $y_2 \ne 0$. Let us suppose that $x_1 > 0$. Define $\phi_1, \phi_2 : [0, x_1] \to \mathbb{R}$ by

$$\phi_1(x) := m_1 x \quad \text{and} \quad \phi_2(x) := m_1 x + y_2 \quad \text{for } 0 \le x \le x_1.$$

It is clear that ϕ_1, ϕ_2 are continuous. Moreover, if $y_2 > 0$, then $\phi_1 \le \phi_2$ and $D = \{(x,y) \in \mathbb{R}^2 : 0 \le x \le x_1 \text{ and } \phi_1(x) \le y \le \phi_2(x)\}$. Since 1_D is continuous on D, in view of Corollary 5.45, we see that

$$\text{Area}(D) := \iint_D 1_D = \int_0^{x_1} \left(\int_{\phi_1(x)}^{\phi_2(x)} dy \right) dx = \int_0^{x_1} y_2 dx = x_1 y_2 = |x_1 y_2|.$$

On the other hand, if $y_2 < 0$, then $\phi_2 \le \phi_1$ and we have $D = \{(x,y) \in \mathbb{R}^2 : 0 \le x \le x_1 \text{ and } \phi_2(x) \le y \le \phi_1(x)\}$, and a similar computation shows that $\text{Area}(D) = -x_1 y_2 = |x_1 y_2|$. If instead we suppose $x_1 < 0$, then proceeding as above, with the interval $[0, x_1]$ replaced by the interval $[x_1, 0]$, we obtain $\text{Area}(D) = |x_1 y_2|$. Hence when $x_2 = 0$, we have $\text{Area}(D) = |x_1 y_2|$. Finally, if $x_1 = 0$, then by a similar argument, we see that $\text{Area}(D) = |x_2 y_1|$. Thus, in any case, $\text{Area}(D) = |x_1 y_2 - x_2 y_1|$.

Finally, let us consider the general case in which (x_0, y_0) is not necessarily equal to $(0,0)$. Let $E := \{(x - x_0, y - y_0) : (x,y) \in D\}$. Then by Lemma 5.55, E has an area and

$$\text{Area}(D) = \iint_D 1_D(x,y)d(x,y) = \iint_E 1_E(u,v)d(u,v) = \text{Area}(E).$$

Now E is a parallelogram with one vertex at $(0,0)$ and the vertices adjacent to $(0,0)$ at $(x_1 - x_0, y_1 - y_0)$ and $(x_2 - x_0, y_2 - y_0)$. Hence using the result proved above, we see that

$$\text{Area}(D) = \text{Area}(E) = |(x_1 - x_0)(y_2 - y_0) - (x_2 - x_0)(y_1 - y_0)|,$$

as desired. □

Case of Affine Transformations

A function $\Phi : \mathbb{R}^2 \to \mathbb{R}^2$ is called an **affine transformation** if there are $(x^\circ, y^\circ) \in \mathbb{R}^2$ and $a_1, b_1, a_2, b_2 \in \mathbb{R}$ such that

$$\Phi(u,v) = (x^\circ + a_1u + b_1v,\ y^\circ + a_2u + b_2v) \quad \text{for all } (u,v) \in \mathbb{R}^2.$$

In *matrix notation*, this can be written as follows:

$$\Phi \begin{bmatrix} u \\ v \end{bmatrix} = \begin{bmatrix} x^\circ \\ y^\circ \end{bmatrix} + \begin{bmatrix} a_1 & b_1 \\ a_2 & b_2 \end{bmatrix} \begin{bmatrix} u \\ v \end{bmatrix} \quad \text{for all } (u,v) \in \mathbb{R}^2.$$

Let Φ be an affine transformation given as above. It can be readily seen that for any $t_1, \ldots, t_n \in \mathbb{R}$ and $(u_1, v_1), \ldots, (u_n, v_n) \in \mathbb{R}^2$,

$$\Phi\left(\sum_{i=1}^n t_i(u_i, v_i)\right) = \sum_{i=1}^n t_i\Phi(u_i, v_i) + \left(1 - \sum_{i=1}^n t_i\right)(x^\circ, y^\circ).$$

In particular, Φ preserves convex combinations. More precisely, Φ sends a convex combination $\sum_{i=1}^n t_i(u_i, v_i)$, where $t_i \geq 0$ for $i = 1, \ldots, n$ and $\sum_{i=1}^n t_i = 1$, of $(u_1, v_1), \ldots, (u_n, v_n)$ to the corresponding convex combination $\sum_{i=1}^n t_i\Phi(u_i, v_i)$ of $\Phi(u_1, v_1), \ldots, \Phi(u_n, v_n)$.

We say that Φ is an **invertible affine transformation** if

$$a_1b_2 - a_2b_1 = \det \begin{bmatrix} a_1 & b_1 \\ a_2 & b_2 \end{bmatrix} \neq 0.$$

This condition ensures that Φ is bijective, that is, for every $(x,y) \in \mathbb{R}^2$, there is a unique $(u,v) \in \mathbb{R}^2$ such that $\Phi(u,v) = (x,y)$. In other words, for every $(x,y) \in \mathbb{R}^2$, the equations

$$x^\circ + a_1u + b_1v = x \quad \text{and} \quad y^\circ + a_2u + b_2v = y$$

have a unique solution $(u,v) \in \mathbb{R}^2$. In fact, if we let $d := a_1b_2 - a_2b_1$, then it is easy to see that the unique solution is given by

$$u = \frac{1}{d}\left[(b_1y^\circ - b_2x^\circ) + (b_2x - b_1y)\right] \text{ and } v = \frac{1}{d}\left[(a_2x^\circ - a_1y^\circ) + (a_1y - a_2x)\right].$$

This shows that if Φ is an invertible affine transformation, then $\Phi^{-1} : \mathbb{R}^2 \to \mathbb{R}^2$ is also an affine transformation. The important quantity $a_1b_2 - a_2b_1$ associated to Φ can be easily recognized as the Jacobian $J(\Phi)$ of Φ. Indeed, if we write $\Phi = (\phi_1, \phi_2)$, where $\phi_1, \phi_2 : \mathbb{R}^2 \to \mathbb{R}$ are defined by

$$\phi_1(u,v) := x^\circ + a_1u + b_1v \quad \text{and} \quad \phi_2(u,v) := y^\circ + a_2u + b_2v \quad \text{for all } (u,v) \in \mathbb{R}^2,$$

then both ϕ_1 and ϕ_2 have continuous partial derivatives and

$$J(\Phi)(u,v) = \det \begin{bmatrix} \dfrac{\partial \phi_1}{\partial u} & \dfrac{\partial \phi_1}{\partial v} \\[2ex] \dfrac{\partial \phi_2}{\partial u} & \dfrac{\partial \phi_2}{\partial v} \end{bmatrix} = \det \begin{bmatrix} a_1 & b_1 \\ a_2 & b_2 \end{bmatrix} = a_1b_2 - a_2b_1$$

for all $(u,v) \in \mathbb{R}^2$. In particular, the Jacobian of the affine transformation Φ is a constant, which may be denoted simply by $J(\Phi)$. Moreover,

$$\Phi \text{ is an invertible affine transformation} \iff J(\Phi) \neq 0.$$

Now assume that Φ is an invertible affine transformation. Then Φ maps distinct points in $\mathbb{R}^2$ to distinct points in $\mathbb{R}^2$. Moreover, since Φ preserves convex combinations, it maps the line segment joining two points (u_1, v_1) and (u_2, v_2) of $\mathbb{R}^2$ to the line segment joining the two points $\Phi(u_1, v_1)$ and $\Phi(u_2, v_2)$. Consequently, Φ maps a straight line onto a straight line. Further, Φ maps two parallel straight lines onto two parallel straight lines. To see this, note that a line L in $\mathbb{R}^2$ is given parametrically by

$$\begin{cases} x = rt + s, \\ y = pt + q, \end{cases} \quad \text{where } r, s, p, q \in \mathbb{R} \text{ with } (r, p) \neq (0, 0),$$

and the parameter t varies over $\mathbb{R}$. The slope of L is determined, up to proportionality, by the pair (r, p). One checks easily that the image L' of L under Φ is the line given parametrically by

$$\begin{cases} x = r't + s', \\ y = p't + q', \end{cases} \text{where} \begin{bmatrix} r' \\ p' \end{bmatrix} = \begin{bmatrix} a_1 & b_1 \\ a_2 & b_2 \end{bmatrix} \begin{bmatrix} r \\ p \end{bmatrix} \text{and} \begin{bmatrix} s' \\ q' \end{bmatrix} = \begin{bmatrix} x^\circ \\ y^\circ \end{bmatrix} + \begin{bmatrix} a_1 & b_1 \\ a_2 & b_2 \end{bmatrix} \begin{bmatrix} s \\ q \end{bmatrix},$$

and the parameter t varies over $\mathbb{R}$. Now suppose L_1 is another line in $\mathbb{R}^2$ and L_1' is its image under Φ. Then their parametric equations

$$\begin{cases} x = r_1t + s_1, \\ y = p_1t + q_1, \end{cases} \quad \text{and} \quad \begin{cases} x = r_1't + s_1', \\ y = p_1't + q_1', \end{cases}$$

are related to each other in the same manner as those of L and L'. It follows that if (r, p) is proportional to (r_1, p_1), then (r', p') is proportional to (r_1', p_1'). In other words, Φ maps parallel lines onto parallel lines. As a consequence, Φ maps a parallelogram onto a parallelogram.

Remark 5.57. In case $J(\Phi) = 0$, then it can be seen that Φ maps $\mathbb{R}^2$ onto a straight line when not all a_1, b_1, a_2, b_2 are zero, whereas Φ maps $\mathbb{R}^2$ to the single point (x°, y°) when $a_1 = b_1 = a_2 = b_2 = 0$. ◇

It is natural to ask what is the effect of an invertible affine transformation on the area of a parallelogram, or more generally, on a bounded region that has an area. This question has an elegant answer given by the following.

Proposition 5.58. *Let $\Phi : \mathbb{R}^2 \to \mathbb{R}^2$ be an affine transformation such that $J(\Phi) \neq 0$. If E is a bounded subset of $\mathbb{R}^2$ that has an area, then $D := \Phi(E)$ is a bounded subset of $\mathbb{R}^2$ such that D has an area and*

$$\text{Area}(D) = |J(\Phi)|\text{Area}(E).$$

Proof. Let the affine transformation $\Phi : \mathbb{R}^2 \to \mathbb{R}^2$ be given by

$$\Phi \begin{bmatrix} u \\ v \end{bmatrix} = \begin{bmatrix} x^\circ \\ y^\circ \end{bmatrix} + \begin{bmatrix} a_1 & b_1 \\ a_2 & b_2 \end{bmatrix} \begin{bmatrix} u \\ v \end{bmatrix} \quad \text{for all } (u, v) \in \mathbb{R}^2,$$

where $(x^\circ, y^\circ) \in \mathbb{R}^2$ and $a_1, b_1, a_2, b_2 \in \mathbb{R}$ with $J(\Phi) = a_1 b_2 - a_2 b_1 \neq 0$.

We shall first consider the case in which E is a parallelogram. Then as noted above, $D := \Phi(E)$ is also a parallelogram, and so D is bounded. Moreover, in view of Proposition 5.56, both E and D have an area. Let one vertex of the parallelogram E be at (u_0, v_0) and let the vertices adjacent to (u_0, v_0) be at (u_1, v_1) and (u_2, v_2). Let $\Phi(u_i, v_i) := (x_i, y_i)$ for $i = 0, 1, 2$. Then (x_0, y_0) is a vertex of the parallelogram $D := \Phi(E)$ and the vertices adjacent to (x_0, y_0) are at (x_1, y_1) and (x_2, y_2). For $i = 1, 2$,

$$\begin{bmatrix} x_i - x_0 \\ y_i - y_0 \end{bmatrix} = \begin{bmatrix} x_i \\ y_i \end{bmatrix} - \begin{bmatrix} x_0 \\ y_0 \end{bmatrix} = \begin{bmatrix} a_1 & b_1 \\ a_2 & b_2 \end{bmatrix} \left(\begin{bmatrix} u_i \\ v_i \end{bmatrix} - \begin{bmatrix} u_0 \\ v_0 \end{bmatrix} \right) = \begin{bmatrix} a_1 & b_1 \\ a_2 & b_2 \end{bmatrix} \begin{bmatrix} u_i - u_0 \\ v_i - v_0 \end{bmatrix}$$

and thus

$$\begin{bmatrix} x_1 - x_0 & x_2 - x_0 \\ y_1 - y_0 & y_2 - y_0 \end{bmatrix} = \begin{bmatrix} a_1 & b_1 \\ a_2 & b_2 \end{bmatrix} \begin{bmatrix} u_1 - u_0 & u_2 - u_0 \\ v_1 - v_0 & v_2 - v_0 \end{bmatrix}.$$

Since the determinant of the product of two matrices is equal to the product of the determinants of those matrices, we obtain

$$\left| \det \begin{bmatrix} x_1 - x_0 & x_2 - x_0 \\ y_1 - y_0 & y_2 - y_0 \end{bmatrix} \right| = \left| \det \begin{bmatrix} a_1 & b_1 \\ a_2 & b_2 \end{bmatrix} \right| \cdot \left| \det \begin{bmatrix} u_1 - u_0 & u_2 - u_0 \\ v_1 - v_0 & v_2 - v_0 \end{bmatrix} \right|.$$

Thus, using Proposition 5.56, we see that $\text{Area}(D) = |J(\Phi)|\text{Area}(E)$.

Now let us consider the general case in which E is a bounded subset of $\mathbb{R}^2$ that has an area. Let S be a rectangle such that $E \subseteq S$. Then $D = \Phi(E) \subseteq \Phi(S)$ and $\Phi(S)$ is a parallelogram. In particular, $\Phi(S)$ is bounded and hence so is D. To prove that D has an area that equals $|J(\Phi)|\text{Area}(E)$, we proceed as follows. Let $1_E^* : S \to \mathbb{R}$ be obtained by extending the function $1_E : E \to \mathbb{R}$

as usual. Since E has an area, 1_E^* is integrable. Let $\epsilon > 0$ be given and let $\delta := |J(\Phi)| > 0$. By the Riemann Condition (Proposition 5.6), we can find a partition $Q := \{(u_i, v_j) : i = 0, 1, \ldots, n \text{ and } j = 0, 1, \ldots, k\}$ of S such that

$$\text{Area}(E) - L(Q, 1_E^*) < \frac{\epsilon}{2\delta} \quad \text{and} \quad U(Q, 1_E^*) - \text{Area}(E) < \frac{\epsilon}{2\delta}.$$

Let E_0 denote the union of all those subrectangles induced by Q that are contained in E, and let E_1 denote the union of all those subrectangles induced by Q that intersect E. Then $E_0 \subseteq E \subseteq E_1$. (See Figure 5.15.)

If we let $D_0 := \Phi(E_0)$ and $D_1 := \Phi(E_1)$, then clearly $D_0 \subseteq D \subseteq D_1$. Let R be a rectangle such that $D_1 \subseteq R$, and let $1_{D_0}^*, 1_D^*, 1_{D_1}^* : R \to \mathbb{R}$ be obtained by extending the functions $1_{D_0} : D_0 \to \mathbb{R}$, $1_D : D \to \mathbb{R}$ and $1_{D_1} : D_1 \to \mathbb{R}$ as usual. For $i = 1, \ldots, n$ and $j = 1, \ldots, k$, let $S_{i,j}$ denote the (i, j)th subrectangle induced by Q. Now, from the case of parallelograms considered above, Φ maps the parallelogram (in fact, the rectangle) $S_{i,j}$ onto a parallelogram with the effect that $\text{Area}(S_{i,j})$ is multiplied by $\delta := J(\Phi)$. Thus, using domain additivity (Corollary 5.52), we see, as in the proof of Proposition 5.47, that

$$\text{Area}(D_0) = \delta \sum_{S_{i,j} \subseteq E} \text{Area}(S_{i,j}) = \delta L(Q, 1_E^*) > \delta \text{Area}(E) - \frac{\epsilon}{2}$$

and

$$\text{Area}(D_1) = \delta \sum_{S_{i,j} \cap E \neq \phi} \text{Area}(S_{i,j}) = \delta U(Q, 1_E^*) < \delta \text{Area}(E) + \frac{\epsilon}{2}.$$

Since $1_{D_0}^* \leq 1_D^* \leq 1_{D_1}^*$, we obtain

$$\delta \text{Area}(E) - \frac{\epsilon}{2} < \text{Area}(D_0) = L(1_{D_0}^*) \leq L(1_D^*)$$

and

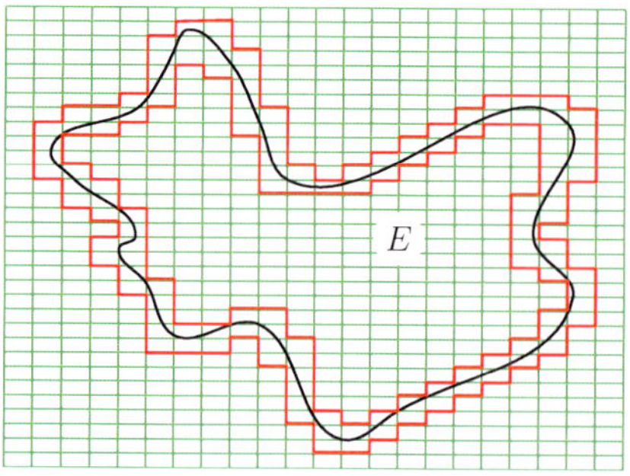

Fig. 5.15. Surrounding ∂E by the union of subrectangles contained in E and by the union of subrectangles that intersect E.

$$U(1_D^*) \leq U(1_{D_1}^*) = \text{Area}(D_1) < \delta\text{Area}(E) + \frac{\epsilon}{2},$$

so that

$$0 \leq U(1_D^*) - L(1_D^*) < \left(\delta\text{Area}(E) + \frac{\epsilon}{2}\right) - \left(\delta\text{Area}(E) - \frac{\epsilon}{2}\right) = \epsilon.$$

Since this is true for every $\epsilon > 0$, we see that $L(1_D^*) = \delta\text{Area}(E) = U(1_D^*)$. Hence the function 1_D^* is integrable, that is, D has an area and

$$\text{Area}(D) = \iint_R 1_D^*(x, y)d(x, y) = \delta\text{Area}(E) = |J(\Phi)|\text{Area}(E),$$

as desired. □

Proposition 5.59 (Change of Variables by Affine Transformations). *Let D be a bounded subset of $\mathbb{R}^2$ such that ∂D is of content zero, and let $f : D \to \mathbb{R}$ be a bounded function whose set of discontinuities is of content zero. Suppose $\Phi : \mathbb{R}^2 \to \mathbb{R}^2$ is an affine transformation with $J(\Phi) \neq 0$ and $E \subseteq \mathbb{R}^2$ is such that $\Phi(E) = D$. Then E is bounded and ∂E is of content zero. Moreover, $f \circ \Phi : E \to \mathbb{R}$ is a bounded function whose set of discontinuities is of content zero, and*

$$\iint_D f(x, y)d(x, y) = \iint_E f(\Phi(u, v))|J(\Phi)|d(u, v).$$

Proof. Since $E = \Phi^{-1}(D)$, by Proposition 5.47 together with Proposition 5.58 applied to the affine transformation Φ^{-1}, we see that E is bounded and ∂E is of content zero. Also, it is clear that $f \circ \Phi$ is a bounded function. Moreover, if we write $\Phi := (\phi_1, \phi_2)$, then clearly the component functions $\phi_1, \phi_2 : \mathbb{R}^2 \to \mathbb{R}$ are continuous, and hence from part (iii) of Proposition 2.17, we see that if C is the set of discontinuities of f, then $\Phi^{-1}(C)$ is the set of discontinuities of $f \circ \Phi$. So again, by Proposition 5.47 together with Proposition 5.58 applied to the affine transformation Φ^{-1}, it follows that $\Phi^{-1}(C)$ is of content zero.

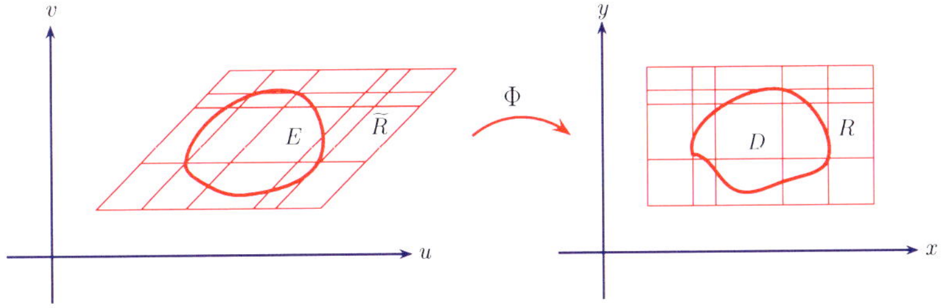

Fig. 5.16. Typical view of the sets D, R, E, and $\widetilde{R}$ in the proof of Proposition 5.59.

Let R be a rectangle containing D, let $f^* : R \to \mathbb{R}$ be obtained by extending f as usual, and let $\widetilde{R} := \Phi^{-1}(R)$. (See Figure 5.16.) Define $g : E \to \mathbb{R}$ and

$\tilde{g} : \widetilde{R} \to \mathbb{R}$ by $g(u,v) := f \circ \Phi((u,v))$ for $(u,v) \in E$ and $\tilde{g}(u,v) := f^*(\Phi(u,v))$ for $(u,v) \in \widetilde{R}$. Since $\tilde{g}_{|E} = g$ and moreover, $\tilde{g}$ is zero on $\widetilde{R} \setminus E$, we see that the set of discontinuities of $\tilde{g}$ is contained in the union of ∂E and the set of discontinuities of g. Also, $\partial \widetilde{R}$ is of content zero, being a union of four line segments. Hence by Proposition 5.43, $\tilde{g}$ is integrable on $\widetilde{R}$. Now let $P := \{(x_i, y_j) : i = 0, 1, \ldots, n \text{ and } j = 0, 1, \ldots, k\}$ be any partition of R. Fix integers i, j such that $1 \le i \le n$ and $1 \le j \le k$, and let $R_{i,j} := [x_{i-1}, x_i] \times [y_{j-1}, y_j]$ denote the (i,j)th subrectangle induced by P, and let $\widetilde{R}_{i,j} = \Phi^{-1}(R_{i,j})$ denote the corresponding parallelogram in $\widetilde{R}$. Since $\partial \widetilde{R}_{i,j}$ is of content zero, it follows from Corollary 5.50 that $\tilde{g}$ is integrable on $\widetilde{R}_{i,j}$ as well. Given any $(u,v) \in \widetilde{R}_{i,j}$, we have $\Phi(u,v) \in R_{i,j}$ and $f^*(\Phi(u,v)) = \tilde{g}(u,v)$, and hence

$$m_{i,j}(f^*) \le \tilde{g}(u,v) \le M_{i,j}(f^*) \quad \text{for all } (u,v) \in \widetilde{R}_{i,j}.$$

Integrating over $\widetilde{R}_{i,j}$ and using the order property (part (vi) of Proposition 5.34), we obtain

$$\iint_{\widetilde{R}_{i,j}} m_{i,j}(f^*) d(u,v) \le \iint_{\widetilde{R}_{i,j}} \tilde{g}(u,v) d(u,v) \le \iint_{\widetilde{R}_{i,j}} M_{i,j}(f^*) d(u,v).$$

But $\iint_{\widetilde{R}_{i,j}} d(u,v) = \text{Area}(\widetilde{R}_{i,j})$, and since $R_{i,j} = \Phi(\widetilde{R}_{i,j})$, by Proposition 5.58 we see that $\text{Area}(R_{i,j}) = |J(\Phi)| \text{Area}(\widetilde{R}_{i,j})$. Hence multiplying throughout by $|J(\Phi)|$, we obtain

$$m_{i,j}(f^*) \text{Area}(R_{i,j}) \le |J(\Phi)| \iint_{\widetilde{R}_{i,j}} \tilde{g}(u,v) d(u,v) \le M_{i,j}(f^*) \text{Area}(R_{i,j}).$$

Summing from $i = 1$ to n and from $j = 1$ to k, and using domain additivity (Corollary 5.52), we see that

$$L(P, f^*) \le |J(\Phi)| \iint_{\widetilde{R}} \tilde{g}(u,v) d(u,v) \le U(P, f^*).$$

Since P is an arbitrary partition of R, it follows that

$$L(f^*) \le |J(\Phi)| \iint_{\widetilde{R}} \tilde{g}(u,v) d(u,v) \le U(f^*).$$

But f^* is integrable on R, and so $L(f^*) = U(f^*) = \iint_R f^* = \iint_D f$. Thus

$$\iint_D f(x,y) d(x,y) = |J(\Phi)| \iint_{\widetilde{R}} \tilde{g}(u,v) d(u,v).$$

Now, ∂E and $\partial \widetilde{R}$ are both of content zero, and so by Corollary 5.38, $\partial(\widetilde{R} \setminus E)$ is also of content zero. Hence by Corollary 5.50, the integrability of $\tilde{g}$ on $\widetilde{R}$

implies the integrability of $\tilde{g}$ on $\widetilde{R} \setminus E$ as well as on E. Consequently, by domain additivity (Corollary 5.52), we have

$$\iint_{\widetilde{R}} \tilde{g} = \iint_{\widetilde{R}\setminus E} \tilde{g} + \iint_{E} \tilde{g} = 0 + \iint_{E} g = \iint_{E} g.$$

Thus we obtain

$$\iint_D f(x,y)d(x,y) = |J(\Phi)| \iint_E g(u,v)d(u,v) = \iint_E g(u,v)|J(\Phi)|d(u,v),$$

as desired. □

Example 5.60. Let $D := \{(x,y) \in \mathbb{R}^2 : \pi \le x+y \le 3\pi \text{ and } -\pi \le x-y \le \pi\}$ and let $f : D \to \mathbb{R}$ be defined by $f(x,y) = (x-y)^2 \sin^2(x+y)$. Since D is an elementary region, one can evaluate the double integral of f over D using Fubini's Theorem. However, changing the variables x and y to u and v, so as to have $u = x+y$ and $v = x-y$, that is, letting $x := (u+v)/2$ and $y := (u-v)/2$, simplifies the region of integration as well as the integrand. Consider $\Phi : \mathbb{R}^2 \to \mathbb{R}^2$ given by $\Phi(u,v) := \big((u+v)/2, (u-v)/2\big)$. The conditions $\pi \le x+y \le 3\pi$ and $-\pi \le x-y \le \pi$ yield $\pi \le u \le 3\pi$ and $-\pi \le v \le \pi$. If we let $E := [\pi, 3\pi] \times [-\pi, \pi]$, then $\Phi(E) = D$. (See Figure 5.17.)

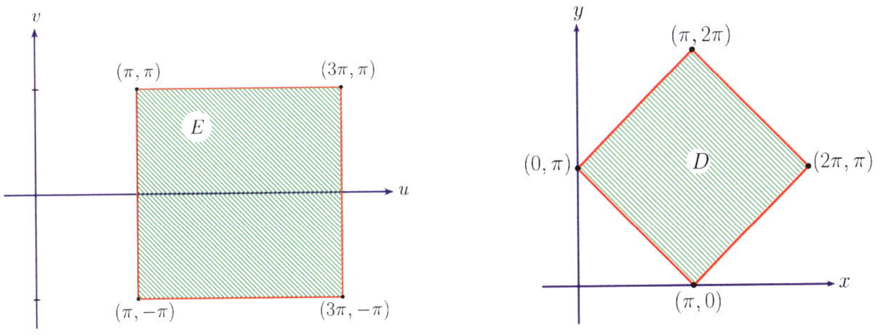

Fig. 5.17. Illustration of the sets D and E in Example 5.60.

Since

$$J(\Phi) = \det \begin{bmatrix} \frac{1}{2} & \frac{1}{2} \\ \frac{1}{2} & -\frac{1}{2} \end{bmatrix} = -\frac{1}{2} \ne 0,$$

Proposition 5.59 shows that

$$\begin{aligned}\iint_D f(x,y)d(x,y) &= \frac{1}{2}\iint_E v^2 \sin^2 u \, d(u,v) = \frac{1}{2}\int_{\pi}^{3\pi} \left(\int_{-\pi}^{\pi} v^2 \sin^2 u \, dv \right) du \\ &= \frac{1}{2} \cdot \frac{2}{3}\pi^3 \int_{\pi}^{3\pi} \sin^2 u \, du = \frac{\pi^3}{3} \int_{\pi}^{3\pi} \sin^2 u \, du = \frac{\pi^4}{3}.\end{aligned}$$

The simplification in the calculation of the above double integral due to an appropriate change of variables is noteworthy. ◇

General Case

While we have proved a satisfactory result for a change of variables by an affine transformation, its use is limited. We therefore look for a change of variables result involving a more general transformation. The basic idea is to utilize the fact that any "nice" transformation from a subset of $\mathbb{R}^2$ to $\mathbb{R}^2$ can be approximated, at least locally, by an affine transformation from $\mathbb{R}^2$ to $\mathbb{R}^2$.

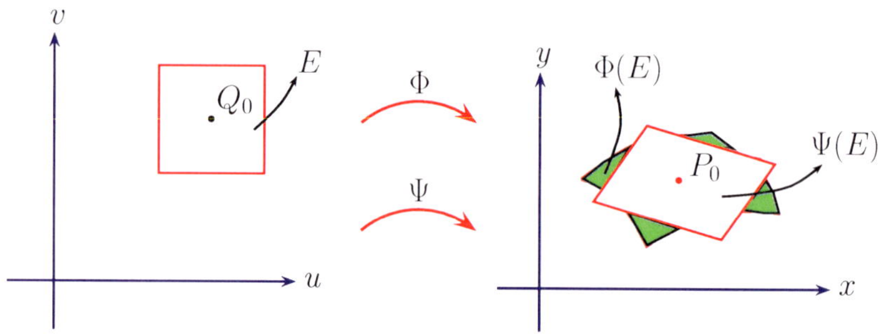

Fig. 5.18. The idea behind the change of variables formula for double integrals.

To be more precise, consider $Q_0 := (u_0, v_0) \in \mathbb{R}^2$, a square neighborhood E of Q_0, and a transformation $\Phi : E \to \mathbb{R}^2$. Let $\Phi := (\phi_1, \phi_2)$. Assume that ϕ_1 and ϕ_2 have partial derivatives at all points in E and they are continuous at Q_0, and moreover, $J(\Phi)(Q_0) \neq 0$. Let $P_0 := \Phi(u_0, v_0)$. For $(u, v) \in \mathbb{R}^2$, let

$$\psi_1(u,v) := \phi_1(u_0,v_0) + \frac{\partial \phi_1}{\partial u}\Big|_{(u_0,v_0)}(u-u_0) + \frac{\partial \phi_1}{\partial v}\Big|_{(u_0,v_0)}(v-v_0)$$

be the linear approximation to ϕ_1 around (u_0, v_0), and

$$\psi_2(u,v) = \phi_2(u_0,v_0) + \frac{\partial \phi_2}{\partial u}\Big|_{(u_0,v_0)}(u-u_0) + \frac{\partial \phi_2}{\partial v}\Big|_{(u_0,v_0)}(v-v_0)$$

the linear approximation to ϕ_2 around (u_0, v_0). Then by Proposition 4.18, we have for all $(u, v) \in E$,

$$\phi_1(u,v) = \psi_1(u,v) + \epsilon_1(u,v) \quad \text{and} \quad \phi_2(u,v) = \psi_2(u,v) + \epsilon_2(u,v),$$

where $\epsilon_1(u, v) \to 0$ and $\epsilon_2(u, v) \to 0$ as $(u, v) \to (u_0, v_0)$. Thus the transformation $\Psi : \mathbb{R}^2 \to \mathbb{R}^2$ given by $\Psi := (\psi_1, \psi_2)$ is affine, maps Q_0 to P_0, and approximates the transformation Φ around Q_0. Also,

$$J(\Psi) = \left(\frac{\partial \psi_1}{\partial u}\frac{\partial \psi_2}{\partial v} - \frac{\partial \psi_1}{\partial v}\frac{\partial \psi_2}{\partial u}\right) = \left(\frac{\partial \phi_1}{\partial u}\frac{\partial \phi_2}{\partial v} - \frac{\partial \phi_1}{\partial v}\frac{\partial \phi_2}{\partial u}\right)(u_0,v_0) = J(\Phi)(u_0,v_0).$$

Since $J(\Psi) \neq 0$, Proposition 5.58 shows that $\Psi(E)$ has an area and

$$\text{Area}(\Psi(E)) = |J(\Psi)|\text{Area}(E) = |J(\Phi)(u_0, v_0)|\text{Area}(E).$$

(See Figure 5.18.) It is therefore reasonable to expect that if E is a small square neighborhood of the point (u_0, v_0), then $\text{Area}(\Phi(E))$ would be approximately equal to $\text{Area}(\Psi(E))$, which equals $|J(\Phi)(u_0, v_0))|\text{Area}(E)$ as we have seen above, that is, the scaling factor around (u_0, v_0) would be $|J(\Phi)(u_0, v_0)|$. Keeping the above motivation in mind, we now state a version of the change of variables result. Typically, the sets appearing in the statement of this result may be viewed as in Figure 5.19.

Proposition 5.61 (Change of Variables Formula). *Let D be a closed and bounded subset of $\mathbb{R}^2$ such that ∂D is of content zero, and let $f : D \to \mathbb{R}$ be a bounded function whose set of discontinuities is of content zero. Suppose Ω is an open subset of $\mathbb{R}^2$ and $\Phi : \Omega \to \mathbb{R}^2$ is a one-one transformation such that $D \subseteq \Phi(\Omega)$. Also, suppose $\Phi := (\phi_1, \phi_2)$, where both ϕ_1 and ϕ_2 have continuous partial derivatives in Ω and $J(\Phi)(u, v) \neq 0$ for all $(u, v) \in \Omega$. Let $E \subseteq \Omega$ be such that $\Phi(E) = D$. Then E is a closed and bounded subset of Ω such that ∂E is of content zero. Moreover, $f \circ \Phi : E \to \mathbb{R}$ is a bounded function whose set of discontinuities is of content zero, and*

$$\iint_D f(x, y)d(x, y) = \iint_E (f \circ \Phi)(u, v)|J(\Phi)(u, v)|d(u, v).$$

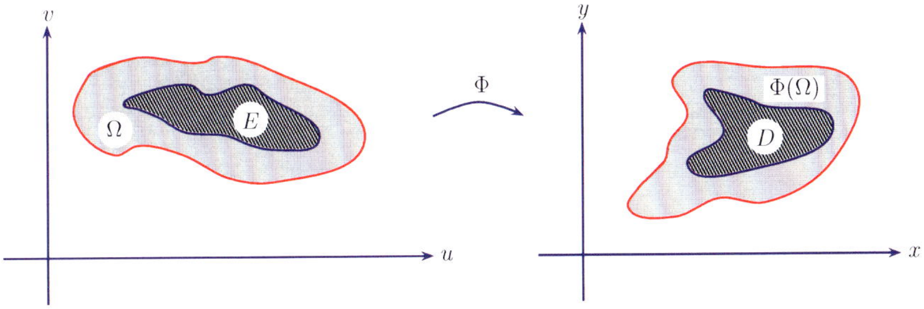

Fig. 5.19. Typical view of the sets D, E, Ω, and $\Phi(\Omega)$ in Proposition 5.61.

Proof of Proposition 5.61 is omitted. Among the various proofs available in the literature, we suggest the proof given in the book of Pugh [45] for interested readers. In fact, the statement and proof of the change of variables formula in [45] assumes that the set E in Proposition 5.61 is a rectangle. However, using domain additivity and the fact that E is closed and bounded, it is not difficult to see that Proposition 5.61 can be deduced from Proposition 33 in Section 8 of Chapter 5 in the book of Pugh [45]. Further remarks and references for the change of variables formula and its proof can be found in the Notes and Comments at the end of this chapter.

Remark 5.62. The change of variables formula is analogous to the principle of Integration by Substitution in one-variable calculus (given, for example, in Proposition 6.26 of ACICARA). To understand this better, let us recall that the latter states that if $\phi : [\alpha, \beta] \to \mathbb{R}$ is a differentiable function such that ϕ' is integrable on $[\alpha, \beta]$, $\phi([\alpha, \beta]) = [a, b]$ and $\phi'(t) \neq 0$ for every $t \in (\alpha, \beta)$, then for any integrable function $f : [a, b] \to \mathbb{R}$, the function $(f \circ \phi)|\phi'| : [\alpha, \beta] \to \mathbb{R}$ is integrable and

$$\int_a^b f(x)dx = \int_\alpha^\beta f(\phi(t))|\phi'(t)|dt, \quad \text{or} \quad \int_a^b f(x)dx = \int_\alpha^\beta f(\phi(t)) \left| \frac{dx}{dt} \right| dt,$$

where in the second formula we have written $x = \phi(t)$. If ϕ is strictly increasing, then $\phi(\alpha) = a$, $\phi(\beta) = b$, and $\phi'(t) > 0$ for $t \in (\alpha, \beta)$, and so the above formula becomes

$$\int_{\phi(\alpha)}^{\phi(\beta)} f(x)dx = \int_\alpha^\beta f(\phi(t))\phi'(t)dt.$$

Notice that the scaling factor is the absolute value of the derivative in the general case. The analogy with the change of variables formula becomes clearer if we write the Jacobian in a more suggestive notation and rewrite the conclusion of Proposition 5.61 as follows:

$$\iint_D f(x, y)d(x, y) = \iint_E (f \circ \Phi)(u, v) \left| \frac{\partial(x, y)}{\partial(u, v)} \right| d(u, v).$$

Proving the one-variable Integration by Substitution formula is relatively easy. The difficulty in the case of functions of two variables may be attributed to two factors. First, there is no two-dimensional analogue of the one-dimensional result that says that if the derivative of a function is nonzero, then the function is strictly monotonic, and second, we treat double integrals over a variety of regions in $\mathbb{R}^2$. On the other hand, we can use the change of variables formula not only to simplify the integrand, but more importantly to simplify the region of integration. This is illustrated in Examples 5.63 (i) and (ii).

The change of variables formula stated in Proposition 5.61 above may also be compared with the result in part (ii) of Proposition 5.26, where the function f was assumed only to be integrable, albeit on a rectangle, but the transformation Φ was of a restrictive kind. ◇

Examples 5.63. (i) Consider $D := \{(x, y) \in \mathbb{R}^2 : x \geq 0,\ y \geq 0 \text{ and } 1 \leq 2(x + y) \leq 2\}$ and $f : D \to \mathbb{R}$ defined by

$$f(x, y) := \frac{y}{x + y} \quad \text{for } (x, y) \in D.$$

To find $\iint_D f$, one can use Fubini's Theorem. But it is much more efficient to consider the following change of variables:

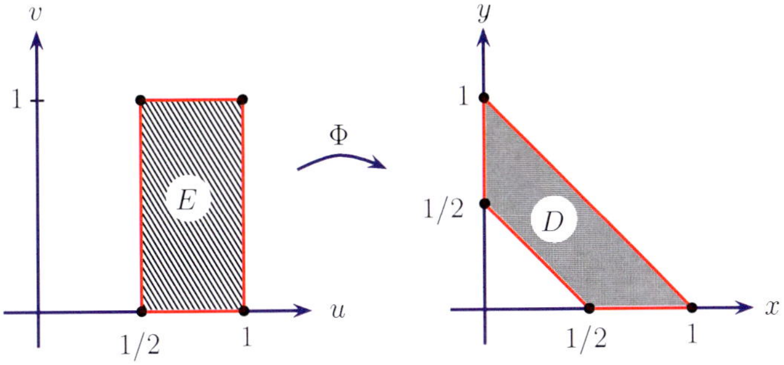

Fig. 5.20. Illustration of the sets $D := \{(x, y) \in \mathbb{R}^2 : x \geq 0,\ y \geq 0 \text{ and } 1 \leq 2(x + y) \leq 2\}$ and $E := [1/2, 1] \times [0, 1]$ in Example 5.63 (i).

$$u := x + y \text{ and } v := \frac{y}{x + y} \quad \text{or equivalently,} \quad x := u(1 - v) \text{ and } y := uv.$$

More precisely, we let $\Omega := \{(u, v) \in \mathbb{R}^2 : u > 0\}$ and let $\Phi : \Omega \to \mathbb{R}^2$ be defined by $\Phi(u, v) = \big(u(1 - v), uv\big)$. Then Φ gives a one-to-one correspondence from Ω to $\Phi(\Omega) = \{(x, y) \in \mathbb{R}^2 : x + y > 0\}$. Also, if $\Phi = (\phi_1, \phi_2)$, then clearly, the partial derivatives of ϕ_1 and ϕ_2 exist and are continuous, and

$$J(\Phi)(u, v) = \det \begin{bmatrix} 1 - v & -u \\ v & u \end{bmatrix} = u \neq 0 \quad \text{for all } (u, v) \in \Omega.$$

Further, if we let E denote the rectangle $[1/2, 1] \times [0, 1]$, then it can be seen that $\Phi(E) = D$. (See Figure 5.20.) Since f is continuous on D, we obtain

$$\begin{aligned} \iint_D f(x, y) d(x, y) &= \iint_E f(u(1 - v), uv)|u| d(u, v) = \iint_E uv\, d(u, v) \\ &= \left(\int_{1/2}^{1} u\, du \right) \left(\int_0^1 v\, dv \right) = \frac{3}{16}. \end{aligned}$$

(ii) Let D denote the subset of $\mathbb{R}^2$ bounded by the curves given by $y = -x^2$, $y = x$, and $x = 1$, and let $f : D \to \mathbb{R}$ be defined by $f(x, y) := x - y$. Consider $\Omega := \{(u, v) \in \mathbb{R}^2 : v > -1/2\}$ and $\Phi : \Omega \to \mathbb{R}^2$ defined by

$$\Phi(u, v) = (u + v, u - v^2).$$

If $x := u + v$ and $y := u - v^2$ for $(u, v) \in \Omega$, then $x - y = v + v^2$ and $x - v = u$, that is,

$$u = x + \frac{1}{2} - \frac{\sqrt{1 + 4(x - y)}}{2} \quad \text{and} \quad v = -\frac{1}{2} + \frac{\sqrt{1 + 4(x - y)}}{2},$$

provided $1 + 4(x - y) > 0$. This shows that the function Φ gives a one-to-one correspondence from Ω to $\Phi(\Omega) = \{(x, y) \in \mathbb{R}^2 : 1 + 4(x - y) > 0\}$.

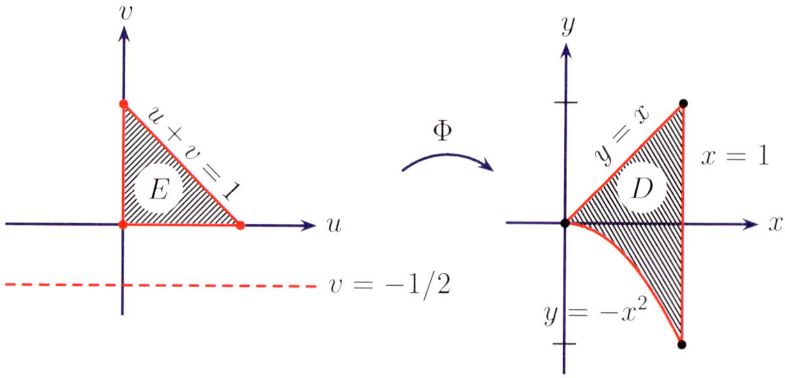

Fig. 5.21. Illustration of the sets D and E in Example 5.63 (ii).

Also, if $\Phi = (\phi_1, \phi_2)$, then the partial derivatives of ϕ_1 and ϕ_2 are clearly continuous and

$$J(\Phi)(u,v) = \det \begin{bmatrix} 1 & 1 \\ 1 & -2v \end{bmatrix} = -(2v+1) \neq 0 \quad \text{for all } (u,v) \in \Omega.$$

Further, if we let E denote the triangular region bounded by the lines given by $u = 0$, $v = 0$, and $u + v = 1$, then it can be seen that $\Phi(E) = D$. (See Figure 5.21.) Since f is continuous on D, we obtain

$$\begin{aligned}\iint_D f(x,y)d(x,y) &= \iint_E f(u+v, u-v^2)|-(2v+1)|d(u,v)\\ &= \iint_E (v+v^2)(2v+1)d(u,v)\\ &= \int_0^1 \left(\int_0^{1-u} (v+v^2)(2v+1)dv\right) du\\ &= \int_0^1 \Big[\frac{(1-u)^4}{2} + (1-u)^3 + \frac{(1-u)^2}{2}\Big] du\\ &= \frac{1}{10} + \frac{1}{4} + \frac{1}{6} = \frac{31}{60}.\end{aligned}$$

As a check on our calculations, we obtain

$$\iint_D f(x,y)d(x,y) = \int_0^1 \left[\int_{-x^2}^{x} (x-y)dy\right] dx = \int_0^1 \left(\frac{x^2}{2} + x^3 + \frac{x^4}{2}\right) dx = \frac{31}{60},$$

as before. ◇

Remark 5.64. The hypothesis $J(\Phi)(u,v) \neq 0$ for all $(u,v) \in \Omega$ of Proposition 5.61 may be weakened by assuming only that $J(\Phi)(u,v) \geq 0$ for all $(u,v) \in \Omega$ or $J(\Phi)(u,v) \leq 0$ for all $(u,v) \in \Omega$, and $J(\Phi)(u,v) = 0$ only for (u,v) in a subset of Ω having content zero. (See the hint for Exercise 10-30 in the first edition of [1].) This weakening is reasonable, since the values of a function on a set of content zero do not affect either the integrability of the function or the value of its double integral (Proposition 5.54). ◇

We now discuss an important change of variables carried out by switching to polar coordinates. Let $\Phi : \mathbb{R}^2 \to \mathbb{R}^2$ be defined by

$$\Phi(r, \theta) = (r\cos\theta,\ r\sin\theta).$$

If $\Phi = (\phi_1, \phi_2)$, then ϕ_1 and ϕ_2 have continuous partial derivatives in $\mathbb{R}^2$ and

$$J(\Phi)(r, \theta) = \det\begin{bmatrix} \cos\theta & -r\sin\theta \\ \sin\theta & r\cos\theta \end{bmatrix} = r \quad \text{for all } (r, \theta) \in \mathbb{R}^2.$$

Thus the Jacobian of Φ is nonzero on $\{(r, \theta) \in \mathbb{R}^2 : r \neq 0\}$. Also, it follows from Proposition 1.26 that given any $\theta_0 \in \mathbb{R}$, the function Φ gives a one-to-one correspondence from the set $\{(r, \theta) \in \mathbb{R}^2 : r > 0 \text{ and } \theta_0 - \pi < \theta \leq \theta_0 + \pi\}$ to the set $\{(x, y) \in \mathbb{R}^2 : (x, y) \neq (0, 0)\}$. In this case, Proposition 5.61 will therefore be directly applicable to a closed and bounded subset D of $\mathbb{R}^2$ if D does not intersect a "cone" with vertex at $(0, 0)$, as illustrated in Figure 5.22.

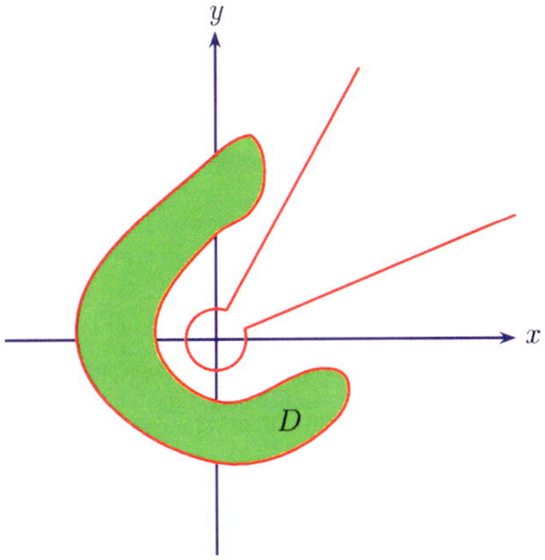

Fig. 5.22. A closed and bounded subset D of $\mathbb{R}^2$ disjoint from a "cone" with vertex at the origin.

We shall now prove a result that shows that switching to polar coordinates is possible even when the above conditions on D are not satisfied. Its proof uses Proposition 5.61 for suitable subsets of D.

Proposition 5.65. *Let D be a closed and bounded subset of $\mathbb{R}^2$ such that ∂D is of content zero, and let $f : D \to \mathbb{R}$ be a continuous function. Suppose*

$E := \{(r,\theta) \in \mathbb{R}^2 : r \geq 0, -\pi \leq \theta \leq \pi$ and $(r\cos\theta, r\sin\theta) \in D\}$, and also suppose ∂E is of content zero. Then the function from E to $\mathbb{R}$ given by $(r,\theta) \longmapsto f(r\cos\theta, r\sin\theta)$ is continuous and

$$\iint_D f(x,y)d(x,y) = \iint_E f(r\cos\theta, r\sin\theta)\, r\, d(r,\theta).$$

Proof. Let $g : E \to \mathbb{R}$ be defined by $g(r,\theta) := f(r\cos\theta, r\sin\theta)$. From the continuity of f and part (iii) of Proposition 2.17, we see that g is continuous.

Define $D^+ := \{(x,y) \in D : y \geq 0\}$ and $D^- := \{(x,y) \in D : y \leq 0\}$. Then D^+ and D^- are closed and bounded subsets of $\mathbb{R}^2$ such that $D = D^+ \cup D^-$ and the set $D^+ \cap D^-$ is of content zero. Also, since the set ∂D is of content zero, the sets ∂D^+ and ∂D^- are of content zero.

Let $E^+ := \{(r,\theta) \in E : 0 \leq \theta \leq \pi\}$ and $E^- := \{(r,\theta) \in E : -\pi \leq \theta \leq 0\}$. Then E^+ and E^- are bounded subsets of $\mathbb{R}^2$ such that $E = E^+ \cup E^-$, and the set $E^+ \cap E^-$ is of content zero. Also, since the set ∂E is of content zero, the sets ∂E^+ and ∂E^- are of content zero.

First we show that

$$\iint_{D^+} f(x,y)d(x,y) = \iint_{E^+} f(r\cos\theta, r\sin\theta)\, r\, d(r,\theta).$$

Note that since f and g are continuous and since ∂D^+ and ∂E^+ are of content zero, both the integrals above exist. Let $\epsilon > 0$ be given and define

$$D^+_\epsilon := \left\{(x,y) \in D^+ : \epsilon \leq \sqrt{x^2+y^2}\right\} \quad \text{and} \quad E^+_\epsilon := \left\{(r,\theta) \in E^+ : \epsilon \leq r\right\}.$$

(See Figure 5.23.) By Domain Additivity (Corollary 5.52),

$$\iint_{D^+} f(x,y)d(x,y) = \iint_{D^+_\epsilon} f(x,y)d(x,y) + \iint_{D^+ \setminus D^+_\epsilon} f(x,y)d(x,y).$$

The continuous function f is bounded on the closed and bounded set D by part (ii) of Proposition 2.25. Thus there is $\alpha \in \mathbb{R}$ such that $|f| \leq \alpha$. Also, since $D^+ \setminus D^+_\epsilon \subseteq [-\epsilon, \epsilon] \times [0, \epsilon]$, we see that $\text{Area}(D^+ \setminus D^+_\epsilon) \leq 2\epsilon^2$. Hence

$$\left| \iint_{D^+ \setminus D^+_\epsilon} f(x,y)d(x,y) \right| \leq 2\alpha\epsilon^2.$$

It follows that

$$\iint_{D^+_\epsilon} f(x,y)d(x,y) \to \iint_{D^+} f(x,y)d(x,y) \quad \text{as } \epsilon \to 0.$$

Similarly, we see that

$$\iint_{E^+_\epsilon} f(r\cos\theta, r\sin\theta)\, r\, d(r,\theta) \to \iint_{E^+} f(r\cos\theta, r\sin\theta) r\, d(r,\theta) \quad \text{as } \epsilon \to 0.$$

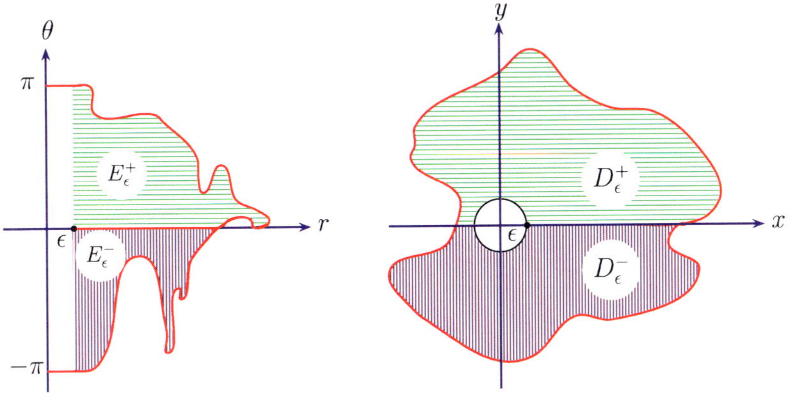

Fig. 5.23. Illustration of the sets $E_\epsilon^+, E_\epsilon^-$ and their polar transforms $D_\epsilon^+, D_\epsilon^-$.

Thus it is enough to prove that

$$\iint_{D_\epsilon^+} f(x,y)d(x,y) = \iint_{E_\epsilon^+} f(r\cos\theta, r\sin\theta)\, r\, d(r,\theta).$$

But this follows by appealing to Proposition 5.61 if we note the following: Define $\Phi : \mathbb{R}^2 \to \mathbb{R}^2$ by

$$\Phi(r,\theta) := (r\cos\theta,\ r\sin\theta) \quad \text{for } (r,\theta) \in \mathbb{R}^2$$

and let $\Omega^+ := \{(r,\theta) \in \mathbb{R}^2 : r > 0, -\pi/2 < \theta < 3\pi/2\}$. Then Ω^+ is an open set in $\mathbb{R}^2$, Φ is one-one on Ω^+, $J(\Phi)(r,\theta) = r \neq 0$ for all $(r,\theta) \in \Omega^+$, D_ϵ^+ is a closed and bounded subset of $\mathbb{R}^2$ such that ∂D_ϵ^+ is of content zero, E_ϵ^+ is a bounded subset of Ω^+ such that ∂E_ϵ^+ is of content zero, and $\Phi(E_\epsilon^+) = D_\epsilon^+$. Thus we obtain

$$\iint_{D^+} f(x,y)d(x,y) = \iint_{E^+} f(r\cos\theta, r\sin\theta)\, r\, d(r,\theta).$$

Similarly, we obtain

$$\iint_{D^-} f(x,y)d(x,y) = \iint_{E^-} f(r\cos\theta, r\sin\theta)\, r\, d(r,\theta)$$

by defining the sets D_ϵ^- and E_ϵ^- analogously and letting $\Omega^- := \{(r,\theta) \in \mathbb{R}^2 : r > 0, -3\pi/2 < \theta < \pi/2\}$.

Domain Additivity (Corollary 5.52) now implies that

$$\iint_D f(x,y)d(x,y) = \iint_E f(r\cos\theta, r\sin\theta)\, r\ d(r,\theta),$$

as desired. □

Examples 5.66. (i) Let $D := \{(x, y) \in \mathbb{R}^2 : x^2 + y^2 \le 1\}$ and let $f : D \to \mathbb{R}$ be defined by $f(x, y) := \sqrt{1 - x^2 - y^2}$. As in Proposition 5.65, let

$$E := \{(r, \theta) \in \mathbb{R}^2 : r \ge 0, -\pi \le \theta \le \pi \text{ and } (r\cos\theta, r\sin\theta) \in D\}.$$

Then $E := [0, 1] \times [-\pi, \pi]$ and we have

$$\begin{aligned}\iint_D f(x, y) d(x, y) &= \iint_E f(r\cos\theta, r\sin\theta)\, r\, d(r, \theta) \\ &= \int_0^1 \left(\int_{-\pi}^{\pi} \left(\sqrt{1 - r^2}\right) r\, d\theta \right) dr \\ &= 2\pi \int_0^1 r\sqrt{1 - r^2} dr = \pi \int_0^1 \sqrt{s}\, ds = \frac{2\pi}{3}.\end{aligned}$$

(ii) Let a, b be positive real numbers, let D denote the ellipsoidal region $\{(x, y) \in \mathbb{R}^2 : (x^2/a^2) + (y^2/b^2) \le 1\}$, and let $f : D \to \mathbb{R}$ be defined by $f(x, y) = y^2$. We can first change D to the unit disk $D_1 := \{(u, v) \in \mathbb{R}^2 : u^2 + v^2 \le 1\}$ by the simple affine transformation $(x, y) \mapsto (x/a, y/b)$, that is, consider the change of variables given by $u = x/a$ and $v = y/b$. This shows that

$$\iint_D f = \iint_{D_1} f(au, bv)\, ab\, d(u, v), \quad \text{since } \frac{\partial(x, y)}{\partial(u, v)} = ab.$$

Now transform D_1 to $E := [0, 1] \times [-\pi, \pi]$ by switching to polar coordinates as in (i) above. This yields

$$\iint_D f = \iint_{D_1} f(au, bv)\, ab\, d(u, v) = \iint_E f(ar\cos\theta, br\sin\theta)\, abr\, d(r, \theta).$$

Hence we have

$$\iint_D f(x, y) d(x, y) = \int_0^1 \left(\int_{-\pi}^{\pi} (br\sin\theta)^2\, abr\, d\theta \right) dr = \frac{ab^3\pi}{4}.$$

(iii) Let us evaluate the improper integral $\int_0^\infty e^{-t^2} dt$ using double integrals. For $b \ge 0$, let $D_b := \{(s, t) \in \mathbb{R}^2 : s \ge 0, t \ge 0 \text{ and } s^2 + t^2 \le b^2\}$ and $I_b := \int_0^b e^{-t^2} dt$. Note that $D_b \subseteq R_b$, where $R_b := [0, b] \times [0, b]$. Now switching to polar coordinates, we obtain

$$\iint_{D_b} e^{-(s^2+t^2)} d(s, t) = \int_0^{\pi/2} \left(\int_0^b re^{-r^2} dr \right) d\theta = \frac{\pi}{4}\left(1 - \frac{e^{-b^2}}{2}\right).$$

On the other hand, by Fubini's Theorem, we have

$$\iint_{R_b} e^{-(s^2+t^2)} d(s, t) = \left(\int_0^b e^{-s^2} ds \right) \left(\int_0^b e^{-t^2} dt \right) = I_b^2.$$

Since $D_b \subseteq R_b \subseteq D_{b\sqrt{2}}$ and $e^{-(s^2+t^2)} \geq 0$ for all $(s,t) \in \mathbb{R}^2$, we see that

$$\frac{\pi}{4}\left(1 - \frac{e^{-b^2}}{2}\right) \leq I_b^2 \leq \frac{\pi}{4}\left(1 - \frac{e^{-2b^2}}{2}\right).$$

Letting $b \to \infty$, we obtain

$$\int_0^\infty e^{-t^2} dt = \lim_{b\to\infty} \int_0^b e^{-t^2} dt = \lim_{b\to\infty} I_b = \sqrt{\frac{\pi}{4}} = \frac{\sqrt{\pi}}{2}.$$

This result can be used to show that $\Gamma(1/2) = \sqrt{\pi}$. (See, for example, Exercise 49 in Chapter 9 of ACICARA.) ◇

5.4 Triple Integrals

In this section we shall extend the considerations of the last three sections to functions defined on subsets of $\mathbb{R}^3$. All statements made in Sections 5.1 and 5.2 about bounded subsets of $\mathbb{R}^2$ and functions defined on them can be carried over to bounded subsets of $\mathbb{R}^3$ (or more generally, to bounded subsets of $\mathbb{R}^n$, where $n \geq 3$) and functions defined on them in a straightforward manner. There is no need to introduce any new ideas. We shall therefore merely mention some important points without giving detailed proofs.

Let us recall that by a **cuboid** we mean a subset of $\mathbb{R}^3$ of the form

$$[a,b] \times [c,d] \times [p,q] := \{(x,y,z) \in \mathbb{R}^3 : a \leq x \leq b,\ c \leq y \leq d, \text{ and } p \leq z \leq q\},$$

where $a, b, c, d, p, q \in \mathbb{R}$ with $a < b$, $c < d$, and $p < q$, and by the volume of this cuboid we mean the number $(b-a)(d-c)(q-p)$. Let $K := [a,b] \times [c,d] \times [p,q]$ and let $f : K \to \mathbb{R}$ be a bounded function. Let us consider a **partition** $P := \{(x_i, y_j, z_\ell) : i = 0, 1, \ldots, n,\ j = 0, 1, \ldots, k, \text{ and } \ell = 0, 1, \ldots, r\}$ of K, where $a = x_0 < x_1 < \cdots < x_{n-1} < x_n = b$, $c = y_0 < y_1 < \cdots < y_{k-1} < y_k = d$, and $p = z_0 < z_1 < \cdots < z_{r-1} < z_r = q$. For $i = 1, \ldots, n$, $j = 1, \ldots, k$, and $\ell = 1, \ldots, r$, the cuboid $[x_{i-1}, x_i] \times [y_{j-1}, y_j] \times [z_{\ell-1}, z_\ell]$, is called the (i,j,ℓ)th **subcuboid induced by** the partition P; let $m_{i,j,\ell}(f)$ and $M_{i,j,\ell}(f)$ denote respectively the infimum and the supremum of the values of f on this subcuboid. We define the **lower triple sum** $L(P,f)$, the **upper triple sum** $U(P,f)$, the **lower triple integral** $L(f)$, and the **upper triple integral** $U(f)$ exactly as we did in Section 5.1 for a function defined on a rectangle $[a,b] \times [c,d]$. A refinement of a partition of K is defined similarly, and it easy to see that analogues of Lemma 5.2 and Proposition 5.3 hold. The function f is said to be **integrable** on K if $L(f) = U(f)$, and then the common value is called the **triple integral**, or simply the **integral**, of f (on K), and it is denoted by

$$\iiint_K f(x,y,z)d(x,y,z) \quad \text{or simply by} \quad \iiint_K f.$$

The simplest example of an integrable function on K is given by the constant function $f(x, y, z) := 1$ for all $(x, y, z) \in K$. For any partition P of K, we see that $L(P, f) = U(P, f) = (b-a)(d-c)(q-p)$, and so the triple integral of f is equal to $(b-a)(d-c)(q-p)$. On the other hand, consider the **trivariate Dirichlet function** given by $f : K \to \mathbb{R}$, where

$$f(x, y, z) = \begin{cases} 1 & \text{if } (x, y, z) \in K \text{ and all } x, y, \text{ and } z \text{ are rational numbers,} \\ 0 & \text{if } (x, y, z) \in K \text{ and } x, y, \text{ or } z \text{is an irrational number.} \end{cases}$$

For any partition P of K, we see that $U(P, f) = (b-a)(d-c)(q-p)$, while $L(P, f) = 0$. Hence f is not integrable on K.

We remark that analogues of refinement results (Lemma 5.2 and Proposition 5.3), Basic Inequality (Proposition 5.4), the Riemann Condition (Proposition 5.6), Domain Additivity on Rectangles (Proposition 5.9), the integrability of a monotonic and of a continuous function (Proposition 5.12), the algebraic and order properties (Propositions 5.14 and 5.16), the Fundamental Theorem of Calculus (Proposition 5.20), and the Theorem of Darboux together with its consequence (Proposition 5.31 and Corollary 5.32) can be obtained for triple integrals on a cuboid. It may be noted that the Cuboidal Mean Value Theorem (Exercise 43) is useful for formulating an analogue of the first part of the Fundamental Theorem of Calculus for functions of three variables. A version of Fubini's Theorem for triple integrals on a cuboid goes as follows.

Proposition 5.67 (Fubini's Theorem on Cuboids). *Let $K := [a, b] \times [c, d] \times [p, q]$ be a cuboid in $\mathbb{R}^3$ and $f : K \to \mathbb{R}$ be an integrable function. Let I denote the triple integral of f on K.*

(i) *If for each fixed $x \in [a, b]$, the double integral $\iint_{[c,d]\times[p,q]} f(x, y, z)d(y, z)$ exists, then the* **iterated integral** $\int_a^b \left(\iint_{[c,d]\times[p,q]} f(x, y, z)d(y, z)\right)dx$ *exists and is equal to I.*

(ii) *If for each fixed $(x, y) \in [a, b] \times [c, d]$, the Riemann integral $\int_p^q f(x, y, z)dz$ exists, then the* **iterated integral** $\iint_{[a,b]\times[c,d]} \left(\int_p^q f(x, y, z)dz\right)d(x, y)$ *exists and is equal to I.*

(iii) *If the hypotheses in both* (i) *and* (ii) *above hold, then the* **iterated integral** $\int_a^b \left[\int_c^d \left(\int_p^q f(x, y, z)dz\right)dy\right]dx$ *exists and is equal to I.*

Proof. (i) Assume that for each $x \in [a, b]$, the double integral

$$A(x) := \iint_{[c,d]\times[p,q]} f(x, y, z)d(y, z)$$

exists. Since $m(f)(d-c)(q-p) \le A(x) \le M(f)(d-c)(q-p)$ for all $x \in [a, b]$, it follows that A is a bounded function on $[a, b]$. Proceeding exactly as in the proof of part (i) of Proposition 5.28, using Domain Additivity on the rectangle $[c,d] \times [p,q]$ (Proposition 5.9) as well as the Riemann Condition for the function A on $[a,b]$ (Proposition 6.5 of ACICARA), we see that the function A is integrable on $[a,b]$ and

$$I = \int_a^b A(x)dx,$$

as desired.

(ii) Assume that for each $(x,y) \in [a,b] \times [c,d]$, the Riemann integral

$$\mathcal{A}(x,y) := \int_p^q f(x,y,z)dz$$

exists. Since $m(f)(q-p) \le \mathcal{A}(x,y) \le M(f)(q-p)$ for all $(x,y) \in [a,b] \times [c,d]$, it follows that $\mathcal{A}$ is a bounded function on $[a,b]\times[c,d]$. Proceeding exactly as in the proof of part (i) of Proposition 5.28, using domain additivity of Riemann integrals on the interval $[p,q]$ (Fact 5.8) as well as the Riemann Condition for the function $\mathcal{A}$ on $[a,b] \times [c,d]$ (Proposition 5.6), we see that the function $\mathcal{A}$ is integrable on $[a,b] \times [c,d]$ and

$$I = \iint_{[a,b]\times[c,d]} \mathcal{A}(x,y)d(x,y),$$

as desired.

(iii) Under the assumptions in both (i) and (ii) above, we proceed as follows. Fix $x \in [a,b]$ and let $g_x : [c,d] \times [p,q] \to \mathbb{R}$ be given by $g_x(y,z) := f(x,y,z)$. By the assumption in (i) above, the function g_x is integrable on $[c,d] \times [p,q]$, and by the assumption in (ii) above, $\int_p^q g_x(y,z)dz$ exists for each fixed $y \in [c,d]$. Hence by applying Fubini's Theorem for double integrals on the rectangle $[c,d] \times [p,q]$ (Proposition 5.28) to the function g_x, we obtain

$$A(x) := \iint_{[c,d]\times[p,q]} g_x(y,z)d(y,z) = \int_c^d \left(\int_p^q g_x(y,z)dz \right) dy.$$

But $I = \int_a^b A(x)dx$ by (i) above. Thus the desired result follows. □

There are other versions of Fubini's Theorem obtained by interchanging the roles of the three variables x, y, and z. In each case, an iterated integral is shown to be equal to the triple integral. This implies that the order of integration can be reversed under suitable conditions.

Triple Integrals over Bounded Sets

Let D be a bounded subset of $\mathbb{R}^3$ and $f : D \to \mathbb{R}$ a bounded function. Let K be a cuboid in $\mathbb{R}^3$ such that $D \subseteq K$ and define $f^* : K \to \mathbb{R}$ by

$$f^*(x,y,z) = \begin{cases} f(x,y,z) & \text{if } (x,y,z) \in D, \\ 0 & \text{otherwise.} \end{cases}$$

We say that f is **integrable** (over D) if f^* is integrable (on K), and in this case, the triple integral of f (over D) is defined to be the **triple integral** of f^* on K, that is, $\iiint_D f = \iiint_K f^*$. The integrability of the function f^* and the value of the triple integral of f^* are independent of the choice of the cuboid K containing D.

An analogue of Proposition 5.34 regarding algebraic and order properties holds for triple integrals as well. Let us now consider Fubini's Theorem for a function defined on a bounded subset D of $\mathbb{R}^3$ that is of one of the following two special kinds: (i) Each slice of D by a plane perpendicular to one of the three coordinate axes has a boundary of (two-dimensional) content zero, (ii) D is the region in $\mathbb{R}^3$ lying between the graphs of two functions defined on a subset of one of the three coordinate planes whose boundary has (two-dimensional) content zero. These cases are illustrated in Figures 5.24 and 5.25.

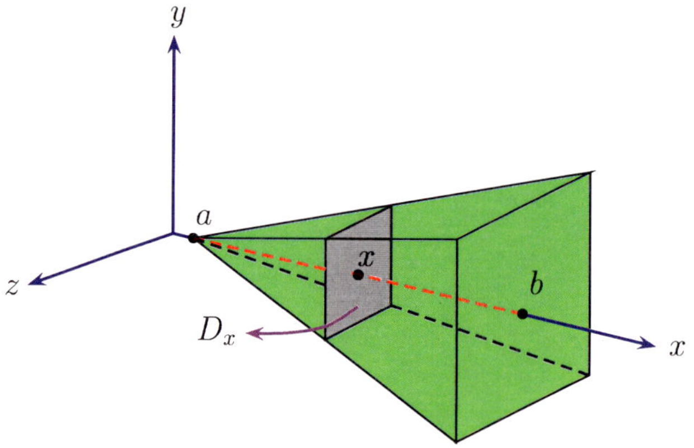

Fig. 5.24. Illustration of Cavalieri's Principle (i): the slice D_x of a solid D.

Proposition 5.68 (Cavalieri's Principle). *Let D be a bounded subset of $\mathbb{R}^3$ and $f : D \to \mathbb{R}$ an integrable function. Let I denote the triple integral of f over D.*

(i) *Suppose $D := \{(x, y, z) \in \mathbb{R}^3 : a \le x \le b$ and $(y, z) \in D_x\}$, where for each fixed $x \in [a, b]$, D_x is a subset of $\mathbb{R}^2$ whose boundary is of (two-dimensional) content zero, and for each fixed $x \in [a, b]$, the double integral $\iint_{D_x} f(x, y, z)d(y, z)$ exists. Then the* **iterated integral**

$$\int_a^b \left(\iint_{D_x} f(x, y, z)d(y, z) \right) dx$$

exists and is equal to I.

(ii) *Suppose* $D := \{(x, y, z) \in \mathbb{R}^3 : (x, y) \in D_0$ *and* $\phi_1(x, y) \leq z \leq \phi_2(x, y)\}$, *where* D_0 *is a subset of* $\mathbb{R}^2$ *whose boundary is of (two-dimensional) content zero,* $\phi_1, \phi_2 : D_0 \to \mathbb{R}$ *are integrable functions such that* $\phi_1 \leq \phi_2$, *and for each fixed* $(x, y) \in D_0$, *the Riemann integral* $\int_{\phi_1(x,y)}^{\phi_1(x,y)} f(x, y, z)dz$ *exists. Then the* **iterated integral**

$$\iint_{D_0} \left(\int_{\phi_1(x,y)}^{\phi_2(x,y)} f(x, y, z)dz \right) d(x, y)$$

exists and is equal to I.

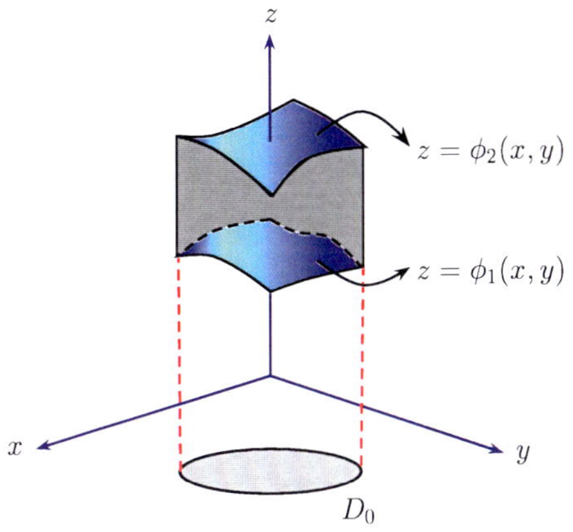

Fig. 5.25. Illustration of Cavalieri's Principle (ii): a solid between two surfaces defined over D_0.

Proof. Let $K := [a, b] \times [c, d] \times [p, q]$ be a cuboid containing D, and let $f^* : K \to \mathbb{R}$ be obtained by extending $f : D \to \mathbb{R}$ as usual.

(i) The iterated integral $\int_a^b \left(\iint_{[c,d]\times[p,q]} f^*(x, y, z)d(y, z) \right) dx$ exists and is equal to the triple integral of f^* on K by part (i) of Proposition 5.67. Hence

$$I = \int_a^b \left(\iint_{[c,d]\times[p,q]} f^*(x, y, z)d(y, z) \right) dx.$$

Fix $x \in [a, b]$. Then $D_x \subseteq [c, d] \times [p, q]$, and since the boundaries of $[c, d] \times [p, q]$ and D_x are of content zero, the boundary of $[c, d] \times [p, q] \setminus D_x$ is also of content zero. Since $f^*(x, y, z) = 0$ for all $(y, z) \in [c, d] \times [p, q] \setminus D_x$, Domain Additivity (Corollary 5.52) shows that

$$\iint_{[c,d]\times[p,q]} f^*(x, y, z)d(y, z) = \iint_{D_x} f(x, y, z)d(y, z).$$

This proves (i).

(ii) The iterated integral $\iint_{[a,b]\times[c,d]}\left(\int_p^q f^*(x,y,z)dz\right)d(x,y)$ exists and is equal to the triple integral of f^* on K by part (ii) of Proposition 5.67. Hence

$$I = \iint_{[a,b]\times[c,d]}\left(\int_p^q f^*(x,y,z)dz\right)d(x,y).$$

Now $D_0 \subseteq [a,b]\times[c,d]$ is such that ∂D_0 is of content zero, and $f^*(x,y,z)=0$ if $(x,y)\in[a,b]\times[c,d]\setminus D_0$ and $z\in[p,q]$. Thus, as in (i) above,

$$\iint_{[a,b]\times[c,d]}\left(\int_p^q f^*(x,y,z)dz\right)d(x,y) = \iint_{D_0}\left(\int_p^q f^*(x,y,z)dz\right)d(x,y).$$

But $f^*(x,y,z)=0$ if $(x,y)\in D_0$ and $z\in[p,q]\setminus[\phi_1(x,y),\phi_2(x,y)]$. Hence for each $(x,y)\in D_0$, we have

$$\int_p^q f^*(x,y,z)dz = \int_{\phi_1(x,y)}^{\phi_2(x,y)} f(x,y,z)dz.$$

This proves (ii). □

Other versions of Cavalieri's Principle can be obtained by interchanging the roles of the three variables x, y, and z. In each case, an iterated integral is shown to be equal to the triple integral. This implies that the order of integration can be changed under suitable conditions.

Remark 5.69. In part (i) of Proposition 5.68, if D_x is an elementary region in the yz-plane for every fixed $x\in[a,b]$, then Fubini's Theorem (Proposition 5.36) can be used to evaluate the double integral $\iint_{D_x} f(x,y,z)d(y,z)$, provided the function $(y,z)\longmapsto f(x,y,z)$ from D_x to $\mathbb{R}$ satisfies the required additional hypotheses. Similarly, in part (ii) of Proposition 5.68, if D_0 is an elementary region in the xy-plane, then Fubini's Theorem (Proposition 5.36) can be used to evaluate the double integral $\iint_{D_0}\left(\int_{\phi_1(x,y)}^{\phi_2(x,y)} f(x,y,z)dz\right)d(y,z)$, provided the function $(x,y)\longmapsto f(x,y,z)$ from D_0 to $\mathbb{R}$ satisfies the required additional hypotheses. In this manner, the evaluation of a triple integral can be reduced to the evaluation of several Riemann integrals. (See Exercise 21.)

For example, if $D := \{(x,y,z)\in\mathbb{R}^3 : x^2+y^2+z^2\le 1\}$, and $f: D\to\mathbb{R}$ is a continuous function, then we have

$$\iiint_D f(x,y)d(x,y,z) = \int_{-1}^{1}\left[\int_{-\sqrt{1-x^2}}^{\sqrt{1-x^2}}\left(\int_{-\sqrt{1-x^2-y^2}}^{\sqrt{1-x^2-y^2}} f(x,y,z)dz\right)dy\right]dx.$$

Similar expressions for the triple integral of the function f over the set D can be obtained by interchanging the orders of integrations with respect to x, y, and z. ◇

Sets of Three-Dimensional Content Zero

Let E be a bounded subset of $\mathbb{R}^3$. We say that E is of **three-dimensional content zero** if the following condition holds: For every $\epsilon > 0$, there are finitely many cuboids whose union contains E and the sum of whose volumes is less than ϵ.

It is easily seen that each of the properties in Proposition 5.37 admits a straightforward analogue for subsets E of $\mathbb{R}^3$. This will also have a consequence similar to Corollary 5.38, and moreover, examples similar to those listed in Examples 5.39 can also readily be given. In particular, if D_0 is a bounded subset of $\mathbb{R}^2$ and $\phi : D_0 \to \mathbb{R}$ is an integrable function of two variables, then the graph of ϕ is of three-dimensional content zero.

A proof similar to the proof of Proposition 5.43 can be given to show that if D is a bounded subset of $\mathbb{R}^3$ such that ∂D is of three-dimensional content zero and $f : D \to \mathbb{R}$ is a bounded function such that the set of discontinuities of f is also of three-dimensional content zero, then f is integrable over D.

Using Proposition 5.43 and part (ii) of Cavalieri's Principle (Proposition 5.68), we can obtain the following analogue of Corollary 5.45 for triple integrals: Let D_0 be a bounded subset of $\mathbb{R}^2$ such that ∂D_0 is of (two-dimensional) content zero, and let $\phi_1, \phi_2 : D_0 \to \mathbb{R}$ be bounded functions such that $\phi_1 \le \phi_2$ and the sets of discontinuities of ϕ_1 and ϕ_2 are of (two-dimensional) content zero. If $D := \{(x, y, z) \in \mathbb{R}^3 : (x, y) \in D_0 \text{ and } \phi_1(x, y) \le z \le \phi_2(x, y)\}$, then ∂D is of three-dimensional content zero. Moreover, if $f : D \to \mathbb{R}$ is a bounded function such that the set of discontinuities of f is of three-dimensional content zero, then f is integrable on D and

$$\iiint_D f(x, y, z)d(x, y, z) = \iint_{D_0} \left(\int_{\phi_1(x,y)}^{\phi_2(x,y)} f(x, y, z)dz \right) d(x, y).$$

Concept of Volume of a Bounded Subset of $\mathbb{R}^3$

The integrability of a function over a bounded subset D of $\mathbb{R}^3$ depends not only on the function, but also on its domain D. For example, consider the constant function

$$1_D : D \to \mathbb{R} \quad \text{defined by} \quad 1_D(x, y, z) := 1 \text{ for all } (x, y, z) \in D.$$

In general, this is not an integrable function. For example, if $K := [a, b] \times [c, d] \times [p, q]$ and $D := \{(x, y, z) \in K : x, y, z \in \mathbb{Q}\}$, then the function $1_D^* : K \to \mathbb{R}$, obtained by extending the function $1_D : D \to \mathbb{R}$ as usual, is in fact the trivariate Dirichlet function. We have seen that it is not integrable on K, that is, 1_D is not integrable over D. However, for a large class of bounded subsets D of $\mathbb{R}^3$, the function 1_D is integrable. It is natural to regard the triple integral of 1_D over D to be the "volume" of D. With this in mind, we make the following general definition.

Let D be a bounded subset of $\mathbb{R}^3$. We say that D has a **volume** if the function 1_D is integrable over D, and then the volume of D is defined to be

$$\mathrm{Vol}(D) := \iiint_D 1_D(x, y, z)d(x, y, z).$$

If D is a cuboid, say $D := [a, b] \times [c, d] \times [p, q]$, then we have seen that 1_D is integrable on D and its triple integral is equal to $(b-a)(d-c)(q-p)$, that is, D has a volume and $\mathrm{Vol}(D) = (b-a)(d-c)(q-p)$. Thus, the general definition of volume is consistent with the usual formula for the volume of a cuboid given at the beginning of this chapter.

The following analogue of Proposition 5.47 holds. If D is a bounded subset of $\mathbb{R}^3$, then D has a volume if and only if ∂D is of three-dimensional content zero. Moreover, if D has a volume, then $\mathrm{Vol}(D) = 0$ if and only if D is of three-dimensional content zero. In particular, it follows that if D is a bounded subset of $\mathbb{R}^3$ and $f : D \to \mathbb{R}$ is an integrable function, then f is integrable over every subset D_0 of D for which ∂D_0 is of content zero. This result allows us to deduce domain additivity of a triple integral over a bounded set in $\mathbb{R}^3$ as in Proposition 5.51 and Corollary 5.52.

Change of Variables in Triple Integrals

Translations are among the simplest transformations of $\mathbb{R}^3$ onto itself. Along the lines of Lemma 5.55 and its proof, it can be easily shown that triple integrals remain invariant under translations. As an application, we can obtain a formula for the volume of a parallelepiped similar to the formula for the area of a parallelogram given in Proposition 5.56. More precisely, consider noncoplanar points $(x_0, y_0, z_0), (x_1, y_1, z_1), (x_2, y_2, z_2)$, and (x_3, y_3, z_3) in $\mathbb{R}^3$ and let D denote the parallelepiped with one vertex at (x_0, y_0, z_0) and the vertices adjacent to (x_0, y_0, z_0) at (x_1, y_1, z_1), (x_2, y_2, z_2), and (x_3, y_3, z_3). Then

$$\mathrm{Vol}(D) = \left| \det \begin{bmatrix} x_1 - x_0 & y_1 - y_0 & z_1 - z_0 \\ x_2 - x_0 & y_2 - y_0 & z_2 - z_0 \\ x_3 - x_0 & y_3 - y_0 & z_3 - z_0 \end{bmatrix} \right|.$$

A function $\Phi : \mathbb{R}^3 \to \mathbb{R}^3$ is called an **affine transformation** if there are $(x^\circ, y^\circ, z^\circ) \in \mathbb{R}^3$ and $a_i, b_i, c_i \in \mathbb{R}$ for $i = 1, 2, 3$ such that

$$\Phi(u, v, w) = (x^\circ + a_1u + b_1v + c_1w, y^\circ + a_2u + b_2v + c_2w, z^\circ + a_3u + b_3v + c_3w)$$

for all $(u, v, w) \in \mathbb{R}^3$. The Jacobian of this function is given by

$$J(\Phi) = \det \begin{bmatrix} a_1 & b_1 & c_1 \\ a_2 & b_2 & c_2 \\ a_3 & b_3 & c_3 \end{bmatrix}.$$

Let $\Phi : \mathbb{R}^3 \to \mathbb{R}^3$ be an affine transformation. As in the case of an affine transformation from $\mathbb{R}^2$ to $\mathbb{R}^2$, it can be seen that if $J(\Phi) \neq 0$, then Φ is bijective and $\Phi^{-1} : \mathbb{R}^3 \to \mathbb{R}^3$ is also an affine transformation. In this case, Φ maps a parallelepiped E onto a parallelepiped, and the volume changes by a factor of $|J(\Phi)|$. More generally, we can show, as in Proposition 5.58, that if E is a bounded subset of $\mathbb{R}^3$ that has a volume and if $D := \Phi(E)$, then D has a volume and

$$\mathrm{Vol}(D) = |J(\Phi)|\mathrm{Vol}(E).$$

With this preparation we obtain the following analogue of Proposition 5.59 for triple integrals by a straightforward modification of its proof.

Proposition 5.70. *Let D be a bounded subset of $\mathbb{R}^3$ such that ∂D is of three-dimensional content zero, and let $f : D \to \mathbb{R}$ be a bounded function whose set of discontinuities is of three-dimensional content zero. Suppose $\Phi : \mathbb{R}^3 \to \mathbb{R}^3$ is an affine transformation with $J(\Phi) \neq 0$, and $E \subseteq \mathbb{R}^3$ is such that $\Phi(E) = D$. Then E is bounded and ∂E is of three-dimensional content zero. Moreover, $f \circ \Phi : E \to \mathbb{R}$ is a bounded function, the set of discontinuities of f is of three-dimensional content zero, and*

$$\iiint_D f(x,y,z)d(x,y,z) = \iiint_E f(\Phi(u,v,w))|J(\Phi)|d(u,v,w).$$

Finally, we have the following analogue of Proposition 5.61.

Proposition 5.71 (Change of Variables Formula for Triple Integrals). *Let D be a closed and bounded subset of $\mathbb{R}^3$ such that ∂D is of three-dimensional content zero, and let $f : D \to \mathbb{R}$ be a bounded function such that the set of discontinuities of f is of three-dimensional content zero. Suppose Ω is an open subset of $\mathbb{R}^3$ and $\Phi : \Omega \to \mathbb{R}^3$ is a one-one transformation such that $D \subseteq \Phi(\Omega)$. Also, suppose $\Phi := (\phi_1, \phi_2, \phi_3)$, where ϕ_1, ϕ_2, and ϕ_3 have continuous partial derivatives in Ω and $J(\Phi)(u,v,w) \neq 0$ for all $(u,v,w) \in \Omega$. Let $E \subseteq \Omega$ be such that $\Phi(E) = D$. Then E is a closed and bounded subset of Ω such that ∂E is of three-dimensional content zero. Moreover, $f \circ \Phi : E \to \mathbb{R}$ is a bounded function such that the set of discontinuities of $f \circ \Phi$ is of three-dimensional content zero, and*

$$\iiint_D f(x,y,z)d(x,y,z) = \iiint_E (f \circ \Phi)(u,v,w)|J(\Phi)(u,v,w)|d(u,v,w).$$

Two important cases involving a change of variables in triple integrals are given by switching to cylindrical coordinates or to spherical coordinates. First, we consider cylindrical coordinates. Let $\Phi : \mathbb{R}^3 \to \mathbb{R}^3$ be defined by

$$\Phi(r, \theta, z) := (r\cos\theta,\ r\sin\theta,\ z) \quad \text{for } (r, \theta, z) \in \mathbb{R}^3.$$

Then for all $(r, \theta, z) \in \mathbb{R}^3$, we have

$$J(\Phi)(r,\theta,z) = \det \begin{bmatrix} \cos\theta & \sin\theta & 0 \\ -r\sin\theta & r\cos\theta & 0 \\ 0 & 0 & 1 \end{bmatrix} = r.$$

The Jacobian of Φ is nonzero on the set $\{(r,\theta,z) \in \mathbb{R}^3 : r \neq 0\}$. Also, it follows from what we have seen in Section 1.3 that given any $\theta_0 \in \mathbb{R}$, the function Φ gives a one-to-one correspondence from the set

$$E_c := \{(r,\theta,z) \in \mathbb{R}^3 : r > 0 \text{ and } \theta_0 - \pi < \theta \leq \theta_0 + \pi\}$$

to the set $\{(x,y,z) \in \mathbb{R}^3 : (x,y) \neq (0,0)\}$.

Next, we consider spherical coordinates. Let $\Phi : \mathbb{R}^3 \to \mathbb{R}^3$ be defined by

$$\Phi(\rho,\varphi,\theta) := (\rho\sin\varphi\cos\theta,\ \rho\sin\varphi\sin\theta,\ \rho\cos\varphi) \quad \text{for } (\rho,\varphi,\theta) \in \mathbb{R}^3.$$

Then for all $(\rho,\varphi,\theta) \in \mathbb{R}^3$, we have

$$J(\Phi)(\rho,\varphi,\theta) = \det \begin{bmatrix} \sin\varphi\cos\theta & \sin\varphi\sin\theta & \cos\varphi \\ \rho\cos\varphi\cos\theta & \rho\cos\varphi\sin\theta & -\rho\sin\varphi \\ -\rho\sin\varphi\sin\theta & \rho\sin\varphi\cos\theta & 0 \end{bmatrix} = \rho^2\sin\varphi.$$

The Jacobian of Φ is nonzero on the set $\{(\rho,\varphi,\theta) \in \mathbb{R}^3 : \rho \neq 0 \text{ and } \varphi \neq m\pi \text{ for any } m \in \mathbb{Z}\}$. Also, it follows from Proposition 1.27 that given any $\theta_0 \in \mathbb{R}$, the function Φ gives a one-to-one correspondence from the set

$$E_s := \{(\rho,\varphi,\theta) \in \mathbb{R}^3 : \rho > 0,\ 0 < \varphi < \pi \text{ and } \theta_0 - \pi < \theta \leq \theta_0 + \pi\}$$

to the set $\{(x,y,z) \in \mathbb{R}^3 : (x,y) \neq (0,0)\}$.

The above observations show that in switching to cylindrical or spherical coordinates, the three-dimensional change of variables result will be directly applicable for a closed and bounded subset D of $\mathbb{R}^3$ if D does not intersect a triangular wedge based on the z-axis. However, the following proposition shows that switching to cylindrical or spherical coordinates is possible even when the above condition on D is not satisfied.

Proposition 5.72. *Let D be a closed and bounded subset of $\mathbb{R}^3$ such that ∂D is of three-dimensional content zero and let $f : D \to \mathbb{R}$ be a continuous function.*

(i) *If $E := \{(r,\theta,z) \in \mathbb{R}^3 : r \geq 0, -\pi \leq \theta \leq \pi$ and $(r\cos\theta,\ r\sin\theta,\ z) \in D\}$ and if ∂E is of three-dimensional content zero, then the triple integral of f over D is equal to*

$$\iiint_E f(r\cos\theta, r\sin\theta, z)\, r\, d(r,\theta,z).$$

(ii) *If $E := \{(\rho,\varphi,\theta) \in \mathbb{R}^3 : \rho \geq 0, 0 \leq \varphi \leq \pi, -\pi \leq \theta \leq \pi$ and $(\rho\sin\varphi\cos\theta, \rho\sin\varphi\sin\theta, \rho\cos\varphi) \in D\}$ and if ∂E is of three-dimensional content zero, then the triple integral of f over D is equal to*

$$\iiint_E f(\rho\sin\varphi\cos\theta, \rho\sin\varphi\sin\theta, \rho\cos\varphi)\, \rho^2\sin\varphi\, d(\rho,\varphi,\theta).$$

Proof. As in the proof of Proposition 5.65, let $D^+ := \{(x,y,z) \in D : y \geq 0\}$ and $D^- := \{(x,y,z) \in D : y \leq 0\}$.

(i) Let $E^+ := \{(r,\theta,z) \in E : 0 \leq \theta \leq \pi\}$. Given $\epsilon > 0$, let $D_\epsilon^+ := \left\{(x,y,z) \in D^+ : \epsilon \leq \sqrt{x^2+y^2}\right\}$ and $E_\epsilon^+ := \{(r,\theta,z) \in E^+ : \epsilon \leq r\}$. If we let $\Omega^+ := \{(r,\theta,z) \in \mathbb{R}^3 : r > 0 \text{ and } -\pi/2 < \theta < 3\pi/2\}$, then the desired result follows using arguments similar to those given in the proof of Proposition 5.65.

(ii) Let $E^+ := \{(\rho,\varphi,\theta) \in E : 0 \leq \theta \leq \pi\}$. Given any $\epsilon \in \mathbb{R}$ with $0 < \epsilon < \pi$, let D_ϵ^+ denote the set of all elements (x,y,z) in D^+ that satisfy

$$\epsilon \leq \sqrt{x^2+y^2+z^2} \quad \text{and} \quad \sqrt{x^2+y^2+z^2}\cos(\pi-\epsilon) \leq z \leq \sqrt{x^2+y^2+z^2}\cos\epsilon,$$

and let $E_\epsilon^+ := \{(\rho,\varphi,\theta) \in E^+ : \epsilon \leq \rho \text{ and } \epsilon \leq \varphi \leq \pi-\epsilon\}$. If we let $\Omega^+ := \{(\rho,\varphi,\theta) \in \mathbb{R}^3 : \rho > 0,\ 0 < \varphi < \pi \text{ and } -\pi/2 < \theta < 3\pi/2\}$, then the desired result follows using arguments similar to those given in the proof of Proposition 5.65. □

As a consequence of the above proposition, we can determine the volume of a solid cylinder and of a solid ball. The formulas for the volumes of these solids are well known in high school geometry.

Corollary 5.73. *Let a and h be positive real numbers.*

(i) *The volume of* $\left\{(x,y,z) \in \mathbb{R}^3 : x^2+y^2 \leq a^2 \text{ and } 0 \leq z \leq h\right\}$ *is* $\pi a^2 h$.
(ii) *The volume of* $\left\{(x,y,z) \in \mathbb{R}^3 : x^2+y^2+z^2 \leq a^2\right\}$ *is* $4\pi a^3/3$.

Proof. (i) Let $D := \left\{(x,y,z) \in \mathbb{R}^3 : x^2+y^2 \leq a^2 \text{ and } 0 \leq z \leq h\right\}$, $f := 1_D$, let $E := [0,a] \times [-\pi,\pi] \times [0,h]$, and apply part (i) of Proposition 5.72.

(ii) Let $D := \left\{(x,y,z) \in \mathbb{R}^3 : x^2+y^2+z^2 \leq a^2\right\}$, $f := 1_D$, and $E := [0,a] \times [0,\pi] \times [-\pi,\pi]$, and apply part (ii) of Proposition 5.72. □

Examples 5.74. (i) Let $D := \{(x,y,z) \in \mathbb{R}^3 : x \geq 0,\ y \geq 0,\ z \geq 0,\ x \leq y+z \text{ and } 1 \leq 2(x+y+z) \leq 2\}$, and consider $f : D \to \mathbb{R}$ defined by $f(x,y,z) := z/(y+z)$. As in Example 5.63 (i), we can find the triple integral of f over D by making the following change of variables:

$$u := x+y+z, \quad v := \frac{y+z}{x+y+z}, \quad w := \frac{z}{y+z},$$

or equivalently,

$$x := u(1-v), \quad y := uv(1-w), \quad z := uvw.$$

More precisely, consider $\Omega := \{(u,v,w) \in \mathbb{R}^3 : u > 0 \text{ and } v > 0\}$, and $\Phi : \Omega \to \mathbb{R}^3$ defined by $\Phi(u,v,w) = \big(u(1-v), uv(1-w), uvw\big)$. Then Φ gives a one-to-one correspondence from Ω to $\Phi(\Omega) = \{(x,y,z) \in \mathbb{R}^2 : x > -y-z \text{ and } y+z > 0\}$. Also, if $\Phi = (\phi_1, \phi_2, \phi_3)$, then the partial derivatives of ϕ_1, ϕ_2, and ϕ_3 are clearly continuous and

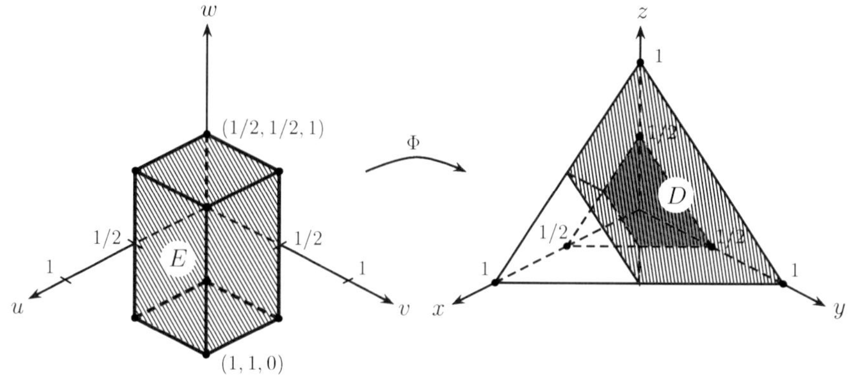

Fig. 5.26. The sets $D := \{(x, y, z) \in \mathbb{R}^3 : x \geq 0,\ y \geq 0,\ z \geq 0,\ x \leq y + z$, and $1 \leq 2(x + y + z) \leq 2\}$ and $E := [\frac{1}{2}, 1] \times [\frac{1}{2}, 1] \times [0, 1]$ in Example 5.74 (i).

$$J(\Phi)(u, v, w) = \det \begin{bmatrix} 1-v & -u & 0 \\ v(1-w) & u(1-w) & -uv \\ vw & uw & uv \end{bmatrix} = u^2 v \neq 0$$

for all $(u, v, w) \in \Omega$. Further, if we let E denote the cuboid $[1/2, 1] \times [1/2, 1] \times [0, 1]$, then it can be seen that $\Phi(E) = D$. (See Figure 5.26.) Since f is continuous on D, we obtain

$$\begin{aligned} \iiint_D f(x, y, z) d(x, y, z) &= \iiint_E f(u(1-v), uv(1-w), uvw)|u^2 v| d(u, v, w) \\ &= \iiint_E u^2 vw\, d(u, v, w) \\ &= \left(\int_{1/2}^1 u^2\, du \right) \left(\int_{1/2}^1 v\, dv \right) \left(\int_0^1 w\, dw \right) = \frac{7}{128}. \end{aligned}$$

(ii) Let a, h be positive real numbers. Consider the solid cylinder $D := \{(x, y, z) \in \mathbb{R}^3 : x^2 + y^2 \leq a^2 \text{ and } 0 \leq z \leq h\}$ and the function $f : D \to \mathbb{R}$ defined by $f(x, y, z) := z\sqrt{a^2 - x^2 - y^2}$. As in part (i) of Proposition 5.72, let $E := \{(r, \theta, z) \in \mathbb{R}^3 : r \geq 0,\ -\pi \leq \theta \leq \pi \text{ and } (r\cos\theta, r\sin\theta, z) \in D\}$. Then $E = [0, a] \times [-\pi, \pi] \times [0, h]$ and we obtain

$$\begin{aligned} \iiint_D f &= \iiint_E f(r\cos\theta, r\sin\theta, z)\, r\, d(r, \theta, z) \\ &= \int_0^a \left[\int_{-\pi}^{\pi} \left(\int_0^h zr\sqrt{a^2 - r^2}\, dz \right) d\theta \right] dr \\ &= \frac{2\pi h^2}{2} \int_0^a r\sqrt{a^2 - r^2}\, dr = \frac{\pi h^2}{2} \int_0^{a^2} \sqrt{s}\, ds = \frac{\pi a^3 h^2}{3}. \end{aligned}$$

(iii) Let $a \in \mathbb{R}$ with $a > 0$ and $D := \{(x, y, z) \in \mathbb{R}^3 : x^2 + y^2 + z^2 \le a^2\}$. Consider $f : D \to \mathbb{R}$ defined by $f(x, y, z) = z^2$. As in part (ii) of Proposition 5.72, let $E := \{(\rho, \varphi, \theta) \in \mathbb{R}^3 : \rho \ge 0,\ 0 \le \varphi \le \pi,\ -\pi \le \theta \le \pi$ and $(\rho \sin\varphi \cos\theta, \rho \sin\varphi \sin\theta, \rho\cos\varphi) \in D\}$. Then $E = [0, a] \times [0, \pi] \times [-\pi, \pi]$ and the triple integral of f over D is equal to

$$\begin{aligned}\iiint_E & f(\rho\sin\varphi\cos\theta, \rho\sin\varphi\sin\theta, \rho\cos\varphi)\, \rho^2 \sin\varphi\, d(\rho, \varphi, \theta) \\ &= \int_0^a \left[\int_0^\pi \left(\int_{-\pi}^{\pi} (\rho^2\cos^2\varphi)\, \rho^2 \sin\varphi\, d\theta\right) d\varphi\right] d\rho \\ &= 2\pi \int_0^a \rho^4 \left(\int_0^\pi \cos^2\varphi \sin\varphi\, d\varphi\right) d\rho = \frac{2\pi a^5}{5} \cdot \frac{2}{3} = \frac{4\pi a^5}{15}.\end{aligned}$$

Alternatively, we may use part (i) of Proposition 5.72, and observe that if $E := \{(r, \theta, z) \in \mathbb{R}^3 : r \ge 0,\ -\pi \le \theta \le \pi$ and $(r\cos\theta, r\sin\theta, z) \in D\}$, then $E = \{(r, \theta, z) \in \mathbb{R}^3 : -a \le z \le a,\ 0 \le r \le \sqrt{a^2 - z^2}$ and $-\pi \le \theta \le \pi\}$. This shows that the triple integral of f over D is equal to

$$\int_{-a}^{a} z^2 \left[\int_0^{\sqrt{a^2-z^2}} \left(\int_{-\pi}^{\pi} d\theta\right) r dr\right] dz = \pi \int_{-a}^{a} z^2(a^2 - z^2) dz = \frac{4\pi a^5}{15}.$$

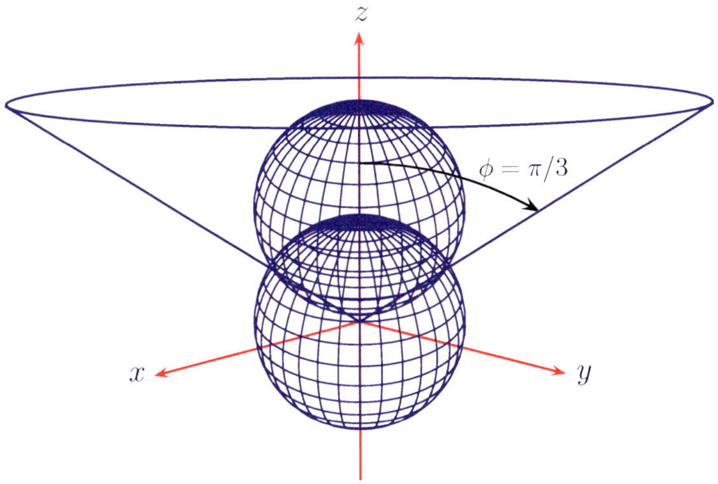

Fig. 5.27. Illustration of the solid in Example 5.74 (iv).

(iv) Let $a \in \mathbb{R}$ with $a > 0$ and $D := \{(x, y, z) \in \mathbb{R}^3 : a^2 \le x^2 + y^2 + z^2 \le 2az\}$. Consider $f : D \to \mathbb{R}$ defined by $f(x, y, z) := z$. We note that the set D consists of points in $\mathbb{R}^3$ that are outside the sphere given by $x^2 + y^2 + z^2 = a^2$ and are inside the sphere given by $x^2 + y^2 + (z - a)^2 = a^2$. (See

Figure 5.27.) As in part (ii) of Proposition 5.65, let $E := \{(\rho, \varphi, \theta) \in \mathbb{R}^3 : \rho \geq 0, 0 \leq \varphi \leq \pi, -\pi \leq \theta \leq \pi$ and $(\rho \sin\varphi \cos\theta, \rho \sin\varphi \sin\theta, \rho\cos\varphi) \in D\}$. Now if $a^2 = x^2 + y^2 + z^2 = 2az$ for $(x, y, z) \in \mathbb{R}^3$, then $a^2 = \rho^2 = 2a\rho\cos\varphi$, that is, $\cos\varphi = 1/2$, and if $0 \leq \varphi \leq \pi$, then it follows that $\varphi = \pi/3$. Thus $E = \{(\rho, \varphi, \theta) \in \mathbb{R}^3 : 0 \leq \varphi \leq \pi/3,\ a \leq \rho \leq 2a\cos\varphi$ and $-\pi \leq \theta \leq \pi\}$. Hence the triple integral of f over D is equal to

$$\begin{aligned}
&\iiint_E f(\rho\sin\varphi\cos\theta, \rho\sin\varphi\sin\theta, \rho\cos\varphi)\,\rho^2\sin\varphi\, d(\rho,\varphi,\theta)\\
&= \int_0^{\pi/3}\left[\int_a^{2a\cos\varphi}\left(\int_{-\pi}^{\pi}(\rho\cos\varphi)\,\rho^2\sin\varphi\, d\theta\right)d\rho\right]d\varphi\\
&= 2\pi\int_0^{\pi/3}\cos\varphi\sin\varphi\frac{(16a^4\cos^4\varphi - a^4)}{4}\,d\varphi\\
&= \frac{\pi a^4}{2}\int_0^{\pi/3}(16\cos^5\varphi - \cos\varphi)\sin\varphi\, d\varphi = \frac{9\pi a^4}{8}. \qquad \diamond
\end{aligned}$$

Notes and Comments

It is customary in textbooks on multivariable calculus to define a double integral as some kind of "limit" of Riemann double sums. The fact that such a "limit" is of a different genre and needs to be handled carefully is often ignored. We have chosen instead to define double integrals over rectangles by considering suprema and infima of lower and upper Riemann double sums. This mimics a standard approach in one-variable calculus and the one that was followed in ACICARA. *The Riemann condition and domain additivity remain the key tools to derive most of the basic properties of double integrals over rectangles. Alternative approaches to defining the double integral of a real-valued function on a rectangle are possible. One such would be to consider more general types of partitions of a rectangle rather than the kind of "product partitions" considered in this book. For example, a partition of a rectangle could be defined as a finite collection of nonoverlapping subrectangles or, more generally, a finite collection of nonoverlapping subsets whose boundaries are of content zero. While this might make the initial definition somewhat difficult to assimilate, it can lead to simpler proofs of such basic results as Lemma 5.2, where the effect of a one-step refinement on upper and lower sums is studied. For a treatment using a general notion of a partition along the above lines, we refer the reader to Section 4.2 of Courant and John [12, vol. II].*

As an application of the Rectangular Mean Value Theorem proved in Chapter 3, we obtain here a version of the Fundamental Theorem of Calculus for double integrals on rectangles. There are other analogues of the Fundamental Theorem of Calculus that involve the notion of "line integral" and more generally, the notion of an integral of a "differential form," and are known as Green's Theorem and Stokes's Theorem, respectively. To learn more about these, one can begin by consulting the books of Apostol [2, vol. II] and Courant and John [12, vol. II], moving on to somewhat more advanced books such as Fleming [19], Munkres [39], and Spivak [54]. The concept of "orientability" plays an important role in these results. They are not discussed in this book.

On the real line, the intervals are the only connected subsets, and hence there isn't an acute need to extend the theory of Riemann integration to functions defined on arbitrary bounded subsets of $\mathbb{R}$ *other than intervals. However, on* $\mathbb{R}^2$ *there are far too many bounded connected subsets other than rectangles. Thus it is necessary to extend the theory of double integrals on rectangles to doubles integrals over arbitrary bounded subsets* D *of* $\mathbb{R}^2$*. We have done this by extending a real-valued function on* D *to a rectangle containing* D *by setting it equal to zero outside* D*. In particular, the integrability of the constant function* 1_D *leads to a general definition of "area" of* D*. It may be noted that our treatment avoids the use of relatively sophisticated notions such as Jordan measurability and Lebesgue measurability. Instead we use a relatively simpler notion of sets of (two-dimensional) content zero. Bounded subsets of* $\mathbb{R}^2$ *having an area can be characterized as those for which the boundary is of content zero. More generally, we show that if* D *is a bounded subset of* $\mathbb{R}^2$ *such that the boundary of* D *is of content zero, then a bounded real-valued function on* D *is integrable over* D *if its set of discontinuities is of content zero. It may be remarked that a subset* D *of* $\mathbb{R}^2$ *is said to be of* **(Lebesgue) measure zero** *if for every* $\epsilon > 0$*, there are countably many rectangles whose union contains* D *and the sum of whose areas is less than* ϵ*. It can be shown that a bounded function defined on a rectangle in* $\mathbb{R}^2$ *is integrable if and only if the set of points at which the function is discontinuous is of measure zero. (See, for example, Theorem 14.5 in the second edition of Apostol [1].) A set of content zero is clearly of measure zero, but the converse need not be true. In fact, the closure of a set of content zero is of content zero, but this is not so for a set of measure zero. However, a closed and bounded subset of* $\mathbb{R}^2$ *is of measure zero if and only if it is of content zero.*

In our view, a proof of the general result regarding a change of variables is too involved to be included in this book. This is perhaps the only result in the book that we have stated and used in the sequel without giving a proof. Nonetheless, we have given a proof for the special case of the result involving an affine transformation. Even in this special case, the proof is by no means simple-minded and gives an indication of the level of difficulty of a proof of the general result. The case of affine transformations is used to motivate the change of variables formula in the general case. For a proof of the latter that seems closest to the spirit of our book, we have referred the reader to Section 8 of Chapter 5 in Pugh [45]. The general change of variables result does not directly apply when we change Cartesian coordinates to polar coordinates in two dimensions, and to cylindrical or spherical coordinates in three dimensions. In these cases, we have shown why such a switch of coordinates is justified.

For alternative approaches to the change of variables formula, each involving significantly different techniques, one may consult Theorem 6 of Chapter 8 in Buck [8], Theorem 10-30 in the first edition of Apostol [1], Theorem 10.9 of Rudin [48], and the articles [34] and [35] of Lax. Proofs of some versions of the change of variables formula, based on the article of Schwartz [51], are given in Section 12.7 of Corwin and Szczarba [11], Theorem 6.42 of Webb [57], and Theorem 9.3.1 of Marsden and Hoffman [38].

Exercises

Part A

1. Let $a, b, c, d \in \mathbb{R}$ with $0 \le a < b$, $0 \le c < d$, and let $f, g : [a, b] \times [c, d] \to \mathbb{R}$ be defined by $f(x, y) := x^2y^2$ and $g(x, y) := x^2 + y^2$. From first principles, show that f and g are integrable and find their double integrals.
2. Show that $\iint_{[0,\pi]\times[0,\pi]} |\cos(x + y)|d(x, y) = 2\pi$.
3. Let $a, b \in \mathbb{R}$ with $a > 0$ and $b > 0$, and let $f : [0, \infty] \to \mathbb{R}$ be defined by
$$f(t) := \begin{cases} 0 & \text{if } t = 0, \\ (e^{-at} - e^{-bt})/t & \text{if } t > 0. \end{cases}$$
Show that the improper integral $\int_0^\infty f(t)dt$ converges to $\ln(b/a)$. (Hint: Observe that $f(t) = \int_a^b e^{-tu}du$ for $t > 0$. Use Proposition 5.28.)
4. Let $f : [a, b] \times [c, d] \to \mathbb{R}$ be a function. Show that f is integrable if and only if there is $r \in \mathbb{R}$ satisfying the following condition: For every $\epsilon > 0$, there is a partition P_ϵ of $[a, b] \times [c, d]$ such that $|S(P, f) - r| < \epsilon$, where P is any partition of $[a, b] \times [c, d]$ finer than P_ϵ and $S(P, f)$ is any Riemann double sum corresponding to P and f.
5. Let D and $\widetilde{D}$ be bounded subsets of $\mathbb{R}^2$ and let $f : D \to \mathbb{R}$ be a function. Suppose $D \subseteq \widetilde{D}$ and $\tilde{f} : \widetilde{D} \to \mathbb{R}$ is defined by
$$\tilde{f}(x, y) = \begin{cases} f(x, y) & \text{if } (x, y) \in D, \\ 0 & \text{otherwise.} \end{cases}$$
Show that f is integrable over D if and only if $\tilde{f}$ is integrable over $\widetilde{D}$.
6. Let D denote the triangular region bounded by the line segments joining $(0, 0)$, $(0, 1)$, and $(2, 2)$. If $f(x, y) := (x + y)^2$ for all $(x, y) \in D$, find the double integral of f over D.
7. Let D denote the region bounded by the lines given by $y = 0$, $x = 1$, $y = 2x$, and let $f(x, y) := e^{x^2}$ for all $(x, y) \in D$. Find the double integral of f over D.
8. Let $D := \{(x, y) \in \mathbb{R}^2 : 0 \le x \le y \le 1, \}$, and let $f : D \to \mathbb{R}$ be defined by
$$f(x, y) = \begin{cases} (\sin x)/x & \text{if } (x, y) \in D \text{ with } (x, y) \neq (0, 0), \\ 1 & \text{if } (x, y) = (0, 0). \end{cases}$$

Show that f is integrable over D and find the double integral of f over D.

9. Let $D := \{(x, y) \in \mathbb{R}^2 : 0 \le x \le 1 \text{ and } x^2 \le y \le 2x^2\}$ and let $f(x, y) := x + y$ for all $(x, y) \in D$. Find the double integral of f over D.
10. In each of the following cases, write the iterated integral with the order of integration reversed.

$$\text{(i)} \int_0^1 \left(\int_1^{e^x} dy \right) dx, \quad \text{(ii)} \int_0^1 \left(\int_{x^2}^{x} dy \right) dx,$$

$$\text{(iii)} \int_0^2 \left(\int_0^{x^3} dy \right) dx, \quad \text{(iv)} \int_0^1 \left(\int_{-\sqrt{y}}^{\sqrt{y}} dx \right) dy.$$

11. Evaluate the following integrals.

$$\text{(i)} \int_0^1 \left(\int_y^1 x^2 e^{xy} dx \right) dy, \quad \text{(ii)} \int_0^8 \left(\int_{\sqrt[3]{x}}^2 \frac{dy}{y^4+1} \right) dx,$$

$$\text{(iii)} \int_0^2 (\tan^{-1} \pi x - \tan^{-1} x) dx.$$

12. Let D be a bounded subset of $\mathbb{R}^2$ such that ∂D is of content zero, and let $\epsilon > 0$ be given. Prove the following.
    (i) There are finitely many rectangles $R_1, \ldots, R_\ell$ such that each of them is contained in D and the sum of their areas is greater than $\text{Area}(D) - \epsilon$.
    (ii) There are finitely many rectangles $S_1, \ldots, S_m$ such that D is contained in their union and the sum of their areas is less than $\text{Area}(D) + \epsilon$.

    Further, show that each rectangle R_i in (i) above can be assumed to be one of the rectangles among $S_1, \ldots, S_m$ in (ii) above.
13. Let D be a bounded subset of $\mathbb{R}^2$, $f : D \to \mathbb{R}$ an integrable function, and define $f^+, f^- : D \to \mathbb{R}$ as in Remark 5.35. Also, let $D^+ := \{(x, y) \in D : f(x, y) \ge 0\}$ and $D^- := \{(x, y) \in D : f(x, y) \le 0\}$. Assume that ∂D^+ and ∂D^- are both of content zero. Show that f is integrable over D^+ as well as over D^-, and moreover, $\iint_{D^+} f = \iint_D f^+$ and $\iint_{D^-} f = -\iint_D f^-$. Deduce that

$$\iint_D f = \iint_{D^+} f + \iint_{D^-} f \quad \text{and} \quad \iint_D |f| = \iint_{D^+} f - \iint_{D^-} f.$$

    Further, show that either $\iint_D |f| \le 2 \left| \iint_{D^+} f \right|$ or $\iint_D |f| \le 2 \left| \iint_{D^-} f \right|$.
14. Let $0 < a < b$ and $D := \left\{ (x, y) \in \mathbb{R}^2 : a \le \sqrt{x^2 + y^2} \le b \right\}$. Show that D is not an elementary region, but there are elementary regions D_1 and D_2 such that $D = D_1 \cup D_2$, while $D_1 \cap D_2$ is of content zero. If $f : D \to \mathbb{R}$ is defined by $f(x, y) := x + y$, then find the double integral of f over D.
15. Let D denote the parallelogram with $(\pi, 0)$, $(2\pi, \pi)$, $(\pi, 2\pi)$, and $(0, \pi)$ as its vertices, and let $f : D \to \mathbb{R}$ be defined by $f(x, y) := \sin^2(x + y)$. Find the double integral of f over D. (Hint: Let $u := x + y$.)
16. Let D denote the subset of $\mathbb{R}^2$ bounded by the lines given by $y = x$, $y = -x$, $y = -x+4$, and $y = x+2$. If $f : D \to \mathbb{R}$ is defined by $f(x, y) := xy$, then find the double integral of f over D. (Hint: Let $u := y - x$ and $v := x + y$.)

17. Let $D := \{(x, y) \in \mathbb{R}^2 : 0 \le y \le 2 \text{ and } y \le 2x \le y + 4\}$ and let $f : D \to \mathbb{R}$ be defined by $f(x, y) := y^3(2x - y)e^{(2x-y)^2}$. Show that the double integral of f over D is equal to $e^{16} - 1$. (Hint: Let $u := 2x - y$.)
18. If $D \subseteq \mathbb{R}^3$ and

$$\iiint_D d(x, y, z) = \int_{-1}^{1} \left[\int_{x^2}^{1} \left(\int_{0}^{1-y} dz \right) dy \right] dx,$$

then describe D. Rewrite the triple integral as an iterated integral in which dx, dy, and dz appear in each of the following orders (i) dz, dx, dy, (ii) dx, dy, dz, (iii) dx, dz, dy, (iv) dy, dz, dx, and (v) dy, dx, dz.
19. Let D denote the subset of $\mathbb{R}^3$ bounded by the plane given by $z = 0$, the circular cylinder given by the polar equation $r = \cos\theta$, and the paraboloid given by the polar equation $z = 3r^2$. Write

$$\iiint_D r \, d(r, \theta, z)$$

as an iterated integral.
20. Let D denote the subset of $\mathbb{R}^3$ bounded by the planes given by $x = 0$, $y = 0$, $z = 2$, and the paraboloid given by $z = x^2 + y^2$. Find

$$\iiint_D x \, d(x, y, z).$$

21. **(Fubini's Theorem for Triple Integrals)** Let D be a bounded subset of $\mathbb{R}^3$. Also, let $f : D \to \mathbb{R}$ be an integrable function and let I denote the triple integral of f over D.
(i) With notation and hypotheses as in part (i) of Proposition 5.68, suppose further that for every $x \in [a, b]$, there are $c_x, d_x \in \mathbb{R}$ with $c_x \le d_x$ and integrable functions $\phi_x, \psi_x : [c_x, d_x] \to \mathbb{R}$ such that $\phi_x \le \psi_x$ and

$$D_x = \{(y, z) \in \mathbb{R}^2 : c_x \le y \le d_x \text{ and } \phi_x(y) \le z \le \psi_x(y)\}.$$

If the Riemann integral $\int_{\phi_x(y)}^{\psi_x(y)} f(x, y, z)dz$ exists for each fixed $x \in [a, b]$ and each fixed $y \in [c_x, d_x]$, then show that

$$I = \int_a^b \left[\int_{c_x}^{d_x} \left(\int_{\phi_x(y)}^{\psi_x(y)} f(x, y, z)dz \right) dy \right] dx.$$

(ii) With notation and hypotheses as in part (ii) of Proposition 5.68, suppose further that there are $a, b \in \mathbb{R}$ with $a \le b$ and integrable functions $\phi, \psi : [a, b] \to \mathbb{R}$ such that $\phi \le \psi$ and

$$D_0 = \{(x, y) \in \mathbb{R}^2 : a \le x \le b \text{ and } \phi(x) \le y \le \psi(x)\}.$$

If the iterated integral $\int_{\phi(x)}^{\psi(x)} \left(\int_{\phi_1(x,y)}^{\phi_2(x,y)} f(x, y, z)dz \right) dy$ exists for each fixed $x \in [a, b]$, then show that

$$I = \int_a^b \left[\int_{\phi(x)}^{\psi(x)} \left(\int_{\phi_1(x,y)}^{\phi_2(x,y)} f(x,y,z)dz \right) dy \right] dx.$$

22. Let D_0 be a bounded subset of $\mathbb{R}^2$ and $\varphi : D_0 \to \mathbb{R}$ an integrable function. Show that the set $\{(x, y, \varphi(x, y)) : (x, y) \in D_0\}$ is of three-dimensional content zero.
23. Let D be a bounded subset of $\mathbb{R}^3$, and D_0 its projection on the xy-plane, that is, $D_0 = \{(x, y) \in \mathbb{R}^2 : (x, y, z) \in D \text{ for some } z \in \mathbb{R}\}$. If D_0 is of (two-dimensional) content zero, then show that D is of three-dimensional content zero.
24. Let $E := \{(x, y, z) \in \mathbb{R}^3 : x, y, z \geq 0 \text{ and } x + y + z \leq 1\}$ denote the tetrahedron with $(0, 0, 0), (1, 0, 0), (0, 1, 0)$, and $(0, 0, 1)$ as its vertices. If i, j, k are nonnegative integers, then show that
$$\iiint_E x^i y^j z^k d(x, y, z) = \frac{i!j!k!}{(3 + i + j + k)!}.$$
In particular, conclude that $\text{Vol}(E) = \frac{1}{6}$.
25. Let $D := \{(x, y, z) \in \mathbb{R}^3 : x, y, z \geq 0 \text{ and } x+y+z \leq 1\}$ and let $f : D \to \mathbb{R}$ be given by $f(x, y, z) := e^{x+2y+3z}$. Show that the triple integral of f over D is equal to $(e - 1)^3/6$.
26. Let D be a subset of $\mathbb{R}^3$ as in Example 5.74 (i) and let $f : D \to \mathbb{R}$ be defined by $f(x, y, z) := (y + z)/(x + y + z)$. Show that the triple integral of f over D is equal to $49/192$.
27. Let $D := \left\{(x, y, z) \in \mathbb{R}^3 : (x^2/a^2) + (y^2/b^2) + (z^2/c^2) \leq 1\right\}$, where a, b, c are positive real numbers, and let $f : D \to \mathbb{R}$ be defined by $f(x, y, z) := |xyz|$. Show that the triple integral of f over D is equal to $a^2b^2c^2/6$. (Hint: Let $x = a\rho \sin\varphi \cos\theta$, $y = b\rho \sin\varphi \sin\theta$, and $z = c\rho \cos\varphi$.)
28. Let $D := \left\{(x, y, z) \in \mathbb{R}^3 : x^2 + y^2 \leq z^2 \text{ and } 0 \leq z \leq 1\right\}$ denote the cone with vertex at $(0, 0, 0)$ and height 1, having its axis on the positive z-axis. If $f : D \to \mathbb{R}$ is defined by $f(x, y, z) := x^2 + y^2 + z^2$, show that the triple integral of f over D is equal to $3\pi/10$, using (i) cylindrical coordinates and (ii) spherical coordinates.
29. Let $a \in \mathbb{R}$ with $a > 0$ and $D := \left\{(x, y, z) \in \mathbb{R}^3 : x^2 + y^2 + z^2 \leq a^2\right\}$. Consider $f : D \to \mathbb{R}$ defined by $f(x, y, z) := z^4$. Show that the triple integral of f over D is equal to $4\pi a^7/35$.
30. Let $D := \left\{(x, y, z) \in \mathbb{R}^3 : a^2 \leq x^2 + y^2 + z^2 \leq b^2\right\}$, where $a, b \in \mathbb{R}$ with $0 < a < b$. Consider $f : D \to \mathbb{R}$ defined by $f(x, y, z) := (x^2 + y^2 + z^2)^{3/2}$. Show that the triple integral of f over D is equal to $4\pi \ln(b/a)$.
31. Let a, b, c be positive real numbers and let
$$D := \left\{(x, y, z) \in \mathbb{R}^3 : x \geq 0,\ y \geq 0,\ z \geq 0, \text{ and } \frac{x^2}{a^2} + \frac{y^2}{b^2} + \frac{z^2}{c^2} \leq 1\right\}.$$
Consider $f : D \to \mathbb{R}$ defined by $f(x, y, z) := xyz$. Show that the triple integral of f over D is equal to $a^2b^2c^2/48$.

Part B

32. Let $f : [a,b] \times [c,d] \to \mathbb{R}$ be a bounded function. For each fixed $x \in [a,b]$, define $\psi_x : [c,d] \to \mathbb{R}$ by $\psi_x(y) := f(x,y)$ and let $A_U(x) := U(\psi_x)$, $A_L(x) := L(\psi_x)$. Also, for each fixed $y \in [c,d]$, define $\phi_y : [a,b] \to \mathbb{R}$ by $\phi_y(x) := f(x,y)$ and let $B_U(y) := U(\phi_y)$, $B_L(y) := L(\phi_y)$. Prove the following:

$$L(f) \le L(A_U) \le U(A_U) \le U(f), \quad L(f) \le L(A_L) \le U(A_L) \le U(f),$$

$$L(f) \le L(B_U) \le U(B_U) \le U(f), \quad L(f) \le L(B_L) \le U(B_L) \le U(f).$$

Further, if f is integrable, then show that each of the following integrals exists and is equal to the double integral of f:

$$\int_a^b A_U(x)dx, \quad \int_a^b A_L(x)dx, \quad \int_c^d B_U(y)dy, \quad \int_c^d B_L(y)dy.$$

(Hint: Proof of Proposition 5.28 and Exercise 41 in Chapter 6 of ACICARA)

33. **(Domain Additivity of Lower Double Integral and Upper Double Integral)** Let $f : [a,b] \times [c,d] \to \mathbb{R}$ be a bounded function, $s \in (a,b)$, and $t \in (c,d)$. Let $f_{1,1} = f_{|[a,s]\times[c,t]}$, $f_{1,2} = f_{|[a,s]\times[t,d]}$, $f_{2,1} = f_{|[s,b]\times[c,t]}$ and $f_{2,2} = f_{|[s,b]\times[t,d]}$. Show that

$$U(f) = \sum_{i=1}^{2}\sum_{j=1}^{2} U(f_{i,j}) \text{ and } L(f) = \sum_{i=1}^{2}\sum_{j=1}^{2} L(f_{i,j}).$$

34. Let $f : [a,b] \times [c,d] \to \mathbb{R}$ be continuous. Define $F : [a,b] \times [c,d] \to \mathbb{R}$ by

$$F(x,y) := \iint_{[a,x]\times[c,y]} f(s,t)d(s,t) \quad \text{for } (x,y) \in [a,b] \times [c,d].$$

Show that the first partials as well as the mixed second-order partials of F exist and are continuous on $[a,b] \times [c,d]$. Further, if, in addition, f_x exists and is continuous on $[a,b] \times [c,d]$, then show that F_{xx} exists and is continuous on $[a,b] \times [c,d]$. Likewise, if, in addition, f_y exists and is continuous on $[a,b] \times [c,d]$, then show that F_{yy} exists and is continuous on $[a,b] \times [c,d]$. Also, show that if the additional hypotheses on f_x and f_y are satisfied, then for each $(x_0, y_0) \in [a,b] \times [c,d]$, we have

$$F_{xx}(x_0,y_0) = \int_c^{y_0} f_x(x_0,t)dt \quad \text{and} \quad F_{yy}(x_0,y_0) = \int_a^{x_0} f_y(s,y_0)ds.$$

(Hint: Exercise 30 of Chapter 3)

35. Let $F : [a,b] \times [c,d] \to \mathbb{R}$ be a continuous function such that both F_x and F_{xy} exist and are continuous on $[a,b] \times [c,d]$. Use Fubini's Theorem (Proposition 5.28) and the FTC (Fact 5.18) to show that

$$\iint_{[a,b]\times[c,d]} F_{xy} = F(b,d) - F(b,c) - F(a,d) + F(a,c).$$

(Compare part (i) of Proposition 5.20.)

36. Let $f : [a,b] \times [c,d] \to \mathbb{R}$ be continuous. Define $F : [a,b] \times [c,d] \to \mathbb{R}$ by

$$F(x,y) := \iint_{[a,x]\times[c,y]} f(s,t)d(s,t) \quad \text{for } (x,y) \in [a,b] \times [c,d].$$

Use Fubini's Theorem (Proposition 5.28) and the FTC (Fact 5.18) to show that the first-order as well as the mixed second-order partial derivatives of F exist and moreover, $F_{xy}(x_0,y_0) = f(x_0,y_0) = F_{yx}(x_0,y_0)$ for every $(x_0,y_0) \in [a,b] \times [c,d]$. (Compare part (ii) of Proposition 5.20.)

37. Let $R := [a,b] \times [c,d]$ and let $f, g, G : R \to \mathbb{R}$ satisfy the following four properties: (i) f_x, f_y, and f_{xy} exist and are continuous on R, (ii) g is integrable on R, (iii) G_x and G_y exist and are continuous on R, and (iv) G_{xy} exists and $G_{xy} = g$ on R. Show that

$$\iint_R fg = \triangle_{(a,c)}^{(b,d)}(fG) - \int_a^b [(f_xG)(s,d) - (f_xG)(s,c)]\, ds$$
$$- \int_c^d [(f_yG)(b,t) - (f_yG)(a,t)]\, dt + \iint_R f_{xy}G,$$

where $\triangle_{(a,c)}^{(b,d)}(fG) := (fG)(b,d) - (fG)(b,c) - (fG)(a,d) + (fG)(a,c)$.

38. **(Cauchy Condition)** Let $f : [a,b] \times [c,d] \to \mathbb{R}$ be a bounded function. Show that f is double integrable on $[a,b]\times[c,d]$ if and only if the following "Cauchy Condition" is satisfied: For every $\epsilon > 0$, there is a partition P_ϵ of $[a,b] \times [c,d]$ such that $|S(P_\epsilon, f) - T(P_\epsilon, f)| < \epsilon$ for any Riemann double sums $S(P_\epsilon, f)$ and $T(P_\epsilon, f)$ for f corresponding to P_ϵ.

39. **(Theorem of Darboux for Lower and Upper Integrals)** Let $f : [a,b] \times [c,d] \to \mathbb{R}$ be a bounded function. Show that given any $\epsilon > 0$, there is $\delta > 0$ such that for every partition P of $[a,b] \times [c,d]$ with $\mu(P) < \delta$, we have $0 \le U(P,f) - U(f) < \epsilon$ and $0 \le L(f) - L(P,f) < \epsilon$.

40. Let $f : [a,b] \times [c,d] \to \mathbb{R}$ be such that f, f_x, and f_{xy} exist and are continuous on $[a,b] \times [c,d]$. Show that f is of bounded bivariation and

$$W(f) = \iint_{[a,b]\times[c,d]} |f_{xy}(s,t)|\, d(s,t).$$

(Hint: Part (i) of Proposition 5.20 and Proposition 5.9.)

41. Let $f : [a,b] \times [c,d] \to \mathbb{R}$ be a bounded function, and let (P_n) be any sequence of partitions of $[a,b] \times [c,d]$ such that $\mu(P_n) \to 0$. Show that $U(P_n, f) \to U(f)$ and $L(P_n, f) \to L(f)$. Deduce that if f is integrable and $I(f)$ denotes the double integral of f on $[a,b]\times[c,d]$, then $U(P_n, f) \to I(f)$ and $L(P_n, f) \to I(f)$. (Compare Corollary 5.32.)

42. Let D be a bounded subset of $\mathbb{R}^2$ and let $f : D \to \mathbb{R}$ be a bounded function. Suppose there is $r \in \mathbb{R}$ satisfying the following condition: For every $\epsilon > 0$, there are integrable functions $g_\epsilon, h_\epsilon : D \to \mathbb{R}$ such that $g_\epsilon \leq f \leq h_\epsilon$ and

$$\iint_D g_\epsilon(x,y)d(x,y) > r - \epsilon, \quad \text{whereas} \quad \iint_D h_\epsilon(x,y)d(x,y) < r + \epsilon.$$

Show that f is integrable over D and the double integral of f equals r.

43. Let D be a bounded subset of $\mathbb{R}^2$ such that ∂D is of content zero and let $f : D \to \mathbb{R}$ be integrable over D.
(i) If $D_0 \subseteq D$ is such that ∂D_0 is of content zero, show that f is integrable over D_0 as well as over $D \setminus D_0$, and the double integral of f over D is equal to the sum of the double integrals of f over D_0 and over $D \setminus D_0$.
(ii) If $g : D \to \mathbb{R}$ is integrable over D and there is $\beta > 0$ such that $|f - g| \leq \beta$, then show that

$$\left| \iint_D f(x,y)d(x,y) - \iint_D g(x,y)d(x,y) \right| \leq \beta \text{Area}(D).$$

(iii) If D_1 and D_2 are bounded subsets of D such that ∂D_1 and ∂D_2 are of content zero, then show that

$$\left| \iint_{D_1} f(x,y)d(x,y) - \iint_{D_2} f(x,y)d(x,y) \right| \leq \alpha \text{Area}(S),$$

where $S := (D_1 \setminus D_2) \cup (D_2 \setminus D_1)$ and $\alpha := \sup\{|f(x,y)| : (x,y) \in S\}$.

44. Let D be a bounded subset of $\mathbb{R}^2$ and let $f : D \to \mathbb{R}$ be a function. Let R be a rectangle such that D is contained in the interior of R. Define $f^* : R \to \mathbb{R}$ by extending the function f as usual. Let $(x_0, y_0) \in R$. Show that f^* is discontinuous at (x_0, y_0) if and only if either $(x_0, y_0) \in D$ and f is discontinuous at (x_0, y_0), or $(x_0, y_0) \in \partial D$ and there is a sequence (x_n, y_n) in D such that $f(x_n, y_n) \not\to 0$.

45. Let D be a bounded subset of $\mathbb{R}^2$ and let $f : D \to \mathbb{R}$ be a bounded function. Define $\hat{f} : \mathbb{R}^2 \to \mathbb{R}$ by

$$\hat{f}(x,y) := \begin{cases} f(x,y) & \text{if } (x,y) \in D, \\ 0 & \text{otherwise.} \end{cases}$$

If $\widehat{E}$ denotes the set of discontinuities of $\hat{f}$, then show that $\widehat{E}$ is a bounded set, and if $\widehat{E}$ is of content zero, then f is integrable over D. In particular, if D is a closed set, $f(x,y) := 0$ for every $(x,y) \in \partial D$, f is continuous at every $(x,y) \in \partial D$, and the set of discontinuities of f has content zero, then show that f is integrable over D.

46. Let D be a bounded subset of $\mathbb{R}^2$ and let $1_D : D \to \mathbb{R}$ be defined by $1_D(x,y) = 1$ for all $(x,y) \in D$. If R is a rectangle containing D and 1_D^* is the function obtained by extending the function 1_D as usual, then for any partition $P := \{(x_i, y_j) : i = 0, \ldots, n \text{ and } j = 0, 1, \ldots, k\}$, show that

$$U(P, 1_D^*) - L(P, 1_D^*) = \sum\sum_{(i,j)\in S} (x_i - x_{i-1})(y_j - y_{j-1}),$$

where S is the set of all pairs (i, j) of nonnegative integers with $i \le n$ and $j \le k$ such that the (i, j)th subrectangle $[x_{i-1}, x_i] \times [y_{j-1}, y_j]$ of P has a nonempty intersection with D as well as with $R \setminus D$.

47. Let $R := [0, 1] \times [0, 1]$ and write $R \cap \mathbb{Q}^2 = \{(p_n, q_n) : n \in \mathbb{N}\}$. Let $\delta \in \mathbb{R}$ with $0 < \delta < 1$, and for $n \in \mathbb{N}$, let S_n denote the open square of area $\delta/2^n$ centered at (p_n, q_n). If $D := R\setminus\bigcup_{n=1}^{\infty} S_n$, then show that D is a closed and bounded subset of $\mathbb{R}^2$ but D does not have an area. Further, if $f : R \to \mathbb{R}$ is defined by $f(x, y) := \inf\{|(x, y) - (u, v)| : (u, v) \in D\}$, then show that f is continuous and the boundary of the set $\{(x, y) \in R : f(x, y) = 0\}$ is not of content zero. (Hint: $\partial D = D$)
48. Show that if $\Phi : \mathbb{R}^2 \to \mathbb{R}^2$ is an affine transformation, then Φ satisfies a Lipschitz condition, that is, there is a constant $L > 0$ such that

$$|\Phi(u_1, v_1) - \Phi(u_2, v_2)| \le L\, |(u_1, v_1) - (u_2, v_2)| \text{ for all } (u_1, v_1), (u_2, v_2) \in \mathbb{R}^2.$$

49. Let Ω be an open subset of $\mathbb{R}^2$ and let $\Phi = (\phi_1, \phi_2) : \Omega \to \mathbb{R}^2$ be such that both ϕ_1 and ϕ_2 have continuous partial derivatives on Ω. Let D be a bounded subset of $\Phi(\Omega)$ such that ∂D is of content zero and let $f : D \to \mathbb{R}$ be an integrable function. Suppose there is a sequence (D_n) of closed subsets of D such that for each $n \in \mathbb{N}$, the boundary ∂D_n of D_n is of content zero, f is continuous on D_n, and $\text{Area}(D \setminus D_n) \to 0$ as $n \to \infty$. Further, suppose (E_n) is a sequence of subsets of Ω such that $\Phi(E_n) = D_n$ and ∂E_n is of content zero for each $n \in \mathbb{N}$. Show that

$$\iint_D f(x, y) d(x, y) = \lim_{n\to\infty} \iint_{E_n} f(\Phi(u, v))|J(\Phi)| d(u, v).$$

Formulate and prove a similar result for subsets of $\mathbb{R}^3$.
50. Let $D := \{(x, y) \in \mathbb{R}^2 : x \ge 0,\ y \ge 0 \text{ and } x + y \le 1\}$ and let $f : D \to \mathbb{R}$ be defined by $f(0, 0) := 0$ and $f(x, y) := y/(x + y)$ for $(x, y) \in D$ with $(x, y) \ne (0, 0)$. Use the change of variables given by

$$u := x + y \quad \text{and} \quad v := \frac{y}{x + y} \quad \text{for } (x, y) \in D \text{ with } (x, y) \ne (0, 0)$$

to show that $\iint_D f = 1/4$. (Hint: Consider $D_n := \{(x, y) \in D : x + y \ge \frac{1}{n}\}$ for $n \in \mathbb{N}$ and use Exercise 49. Compare 5.63 (i).)
51. Let $D := \{(x, y, z) \in \mathbb{R}^3 : x \ge 0,\ y \ge 0,\ z \ge 0, \text{ and } x + y + z \le 1\}$ and let $f : D \to \mathbb{R}$ be defined by $f(x, y, z) := z/(y + z)$ for $(x, y, z) \in D$ with $y + z \ne 0$ and $f(x, y, z) := 0$ for $(x, y, z) \in D$ with $y + z = 0$. Use the change of variables given by

$$u := x + y + z, \quad v := \frac{y + z}{x + y + z}, \quad \text{and} \quad w := \frac{z}{y + z}$$

for $(x, y, z) \in D$ with $y+z \neq 0$ to show that $\iint_D f = 1/12$. (Hint: Consider $D_n := \{(x, y, z) \in D : n(x + y + z) \geq 1$ and $n(y + z) \geq x + y + z\}$ for $n \in \mathbb{N}$ and use Exercise 49 for subsets of $\mathbb{R}^3$. Compare 5.74 (i).)

52. Let Ω, Φ, D, f, and E be as in Proposition 5.61. Let Λ be an open subset of $\mathbb{R}^2$ and let $\Psi : \Lambda \to \mathbb{R}^2$ be one-one and such that $\Psi(\Lambda) \subseteq \Omega$. Let $\Psi := (\psi_1, \psi_2)$, and assume that ψ_1 and ψ_2 have continuous partial derivatives in Λ and $J(\Psi)(s, t) \neq 0$ for all $(s, t) \in \Lambda$. If G is a bounded subset of Λ such that ∂G is of content zero and $\Psi(G) = E$, then show that the double integral of f over D is equal to

$$\iint_G (f \circ \Phi)(\Psi(s, t))|J(\Phi)(\Psi_1(s, t), \Psi_2(s, t))|\,|J(\Psi)(s, t)|d(s, t).$$

53. **(Double Polar Coordinates)** Let $\Phi : \mathbb{R}^4 \to \mathbb{R}^4$ be the function given by $\Phi(r, \theta, \rho, \varphi) := (r\cos\theta,\ r\sin\theta,\ \rho\cos\varphi,\ \rho\sin\varphi)$. Show that the Jacobian of Φ is given by $J(\Phi)(r, \theta, \rho, \varphi) = r\rho$ for all $(r, \theta, \rho, \varphi) \in \mathbb{R}^4$. Prove a result analogous to Proposition 5.65 and deduce that if $a > 0$ and $D := \left\{(x_1, x_2, x_3, x_4) \in \mathbb{R}^4 : x_1^2 + x_2^2 + x_3^2 + x_4^2 \leq a^2\right\}$, then the quadruple integral of the function 1_D is equal to $\pi^2 a^4/2$.

54. **(Spherical Polar Coordinates)** Let $\Phi : \mathbb{R}^5 \to \mathbb{R}^5$ be the function given by $\Phi(\rho, \varphi, \theta, r, \psi) := (\rho\sin\varphi\cos\theta,\ \rho\sin\varphi\sin\theta,\ \rho\cos\varphi,\ r\cos\psi,\ r\sin\psi)$. Show that the Jacobian of Φ is given by $J(\Phi)(\rho, \varphi, \theta, r, \psi) = r\rho^2\sin\varphi$. Prove a result analogous to Proposition 5.72 and deduce that if $a > 0$ and $D := \left\{(x_1, x_2, x_3, x_4, x_5) \in \mathbb{R}^5 : x_1^2 + x_2^2 + x_3^2 + x_4^2 + x_5^2 \leq a^2\right\}$, then the quintuple integral of the function 1_D is equal to $8\pi^2 a^5/15$.

55. For $n \in \mathbb{N}$, let $D := \{(x_1, \ldots, x_n) \in \mathbb{R}^n : x_1, \ldots, x_n \geq 0,\ x_1 + \cdots + x_n \leq 1\}$ denote the **standard n-simplex** in $\mathbb{R}^n$. If $i_1, \ldots, i_n$ are any nonnegative integers, then show that

$$\int \cdots \int_D x_1^{i_1} \cdots x_n^{i_n} d(x_1, \ldots, x_n) = \frac{i_1! \cdots i_n!}{(n + i_1 + \cdots + i_n)!}.$$

In particular, conclude that

$$\int \cdots \int_D 1_D(x_1, \ldots, x_n)d(x_1, \ldots, x_n) = \frac{1}{n!}.$$

56. Let $n \in \mathbb{N}$. For $i = 0, 1, \ldots, n$, let $(x_1^{(i)}, \ldots, x_n^{(i)})$ be points in $\mathbb{R}^n$ that do not lie in any translate of an $(n-1)$-dimensional subspace of $\mathbb{R}^n$, and let D denote the n-simplex having these points as its vertices. Show that

$$\int \cdots \int_D 1_D(x_1, \ldots, x_n)d(x_1, \ldots, x_n) = \frac{|d|}{n!},$$

where d is the determinant of the $n \times n$ matrix whose (i, j)th entry is $x_j^{(i)} - x_j^{(0)}$ for $i, j = 1, \ldots, n$.

6

Applications and Approximations of Multiple Integrals

It is customary in one-variable calculus to include geometric applications of Riemann integration so as to give definitions and methods for the evaluation of area of a planar region, arc length of a curve, volume of a solid of revolution, and area of a surface of revolution. (See, for example, Chapter 8 of ACICARA.) While the definition of arc length thus obtained is quite general, the definitions of area, volume, and surface area are applicable only to a restricted class of planar regions, solids, and surfaces. In fact, general definitions are obtained using the notions of double integrals and triple integrals developed in Chapter 5. In Section 6.1 below, we discuss the general notions of area and volume, and show that these are consistent with the definitions given in one-variable calculus for certain regions in $\mathbb{R}^2$ and solids in $\mathbb{R}^3$. Areas of surfaces in $\mathbb{R}^3$ are discussed in Section 6.2, and it is shown that areas of surfaces of revolution are a special case. Subsequently, a general treatment of centroids of planar regions, solids, and surfaces is given in Section 6.3, and this includes a theorem of Pappus relating the volume of a solid of revolution with the area of the corresponding planar region and its centroid. In the last section of this chapter, we consider cubature rules, which are higher-dimensional analogues of quadrature rules given in Section 8.6 of ACICARA. These are useful in finding approximations of double and triple integrals.

6.1 Area and Volume

We begin by considering areas of subsets of $\mathbb{R}^2$. This discussion will be followed by a discussion of volumes of subsets of $\mathbb{R}^3$.

Area of a Bounded Subset of $\mathbb{R}^2$

Let D be a bounded subset of $\mathbb{R}^2$ such that ∂D is of content zero, and let $1_D : D \to \mathbb{R}$ denote the function given by $1_D(x, y) := 1$ for $(x, y) \in D$. By Proposition 5.47, we see that D has an area and moreover,

S.R. Ghorpade and B.V. Limaye, *A Course in Multivariable Calculus and Analysis*, Undergraduate Texts in Mathematics, DOI 10.1007/978-1-4419-1621-1_6,

$$\text{Area}(D) := \iint_D 1_D(x,y)d(x,y).$$

Let D_1 and D_2 be bounded subsets of $\mathbb{R}^2$ such that ∂D_1 and ∂D_2 are of content zero. Then $\partial(D_1 \cup D_2)$ and $\partial(D_1 \cap D_2)$ are also of content zero, since each of these sets is contained in the union of ∂D_1 and ∂D_2. Hence letting f be the function $1_{D_1 \cup D_2}$ in Proposition 5.51, we obtain

$$\text{Area}(D_1 \cup D_2) = \text{Area}(D_1) + \text{Area}(D_2) - \text{Area}(D_1 \cap D_2).$$

If, in particular, $D_1 \cap D_2$ itself is of content zero, then we have

$$\text{Area}(D_1 \cup D_2) = \text{Area}(D_1) + \text{Area}(D_2).$$

In one-variable calculus, it is customary to define the area between two curves, say $y = f_1(x)$ and $y = f_2(x)$, where $f_1, f_2 : [a,b] \to \mathbb{R}$ are Riemann integrable functions satisfying $f_1 \le f_2$, to be

$$\int_a^b [f_2(x) - f_1(x)]dx.$$

To show that this is consistent with the general definition of area in terms of double integrals, we proceed as follows. Let f_1, f_2 be as above and consider the elementary region

$$D := \{(x,y) \in \mathbb{R}^2 : a \le x \le b \text{ and } f_1(x) \le y \le f_2(x)\}.$$

Then D is the region between the two curves $y = f_1(x)$ and $y = f_2(x)$, where $x \in [a,b]$. Assume that the sets of discontinuities of the functions f_1 and f_2 are of one-dimensional content zero. Then from Corollary 5.45, we see that ∂D is of content zero and moreover,

$$\text{Area}(D) = \int_a^b \left(\int_{f_1(x)}^{f_2(x)} dy \right) dx = \int_a^b [f_2(x) - f_1(x)]dx.$$

For example, if $a = 0$ and if b, h are any positive real numbers and we take $f_1(x) := 0$ and $f_2(x) := hx/b$ for $x \in [0,b]$, then D is the triangular region with $(0,0)$, $(b,0)$, and (b,h) as its vertices. Further, $\text{Area}(D) = \int_0^b f_2(x)dx = \frac{1}{2}bh$. In other words, the area of a triangle is half the base times the height.

In a similar manner, if an elementary region D is given by

$$D := \{(x,y) \in \mathbb{R}^2 : c \le y \le d \text{ and } g_1(y) \le x \le g_2(y)\},$$

where $g_1, g_2 : [c,d] \to \mathbb{R}$ are such that $g_1 \le g_2$, and the sets of discontinuities of g_1 and g_2 are of one-dimensional content zero, then by Corollary 5.45, D has an area and moreover,

$$\text{Area}(D) = \int_c^d \left(\int_{g_1(y)}^{g_2(y)} dx \right) dy = \int_c^d [g_2(y) - g_1(y)]dy.$$

Thus the definitions usually given in one-variable calculus (for example, in Section 8.1 of ACICARA) of the area of a region between two curves given by Cartesian equations are consistent with our general definition of an area.

Remark 6.1. Let $f_1, f_2 : [a, b] \to \mathbb{R}$ be bounded functions with $f_1 \le f_2$ and let $D := \{(x, y) \in \mathbb{R}^2 : a \le x \le b \text{ and } f_1(x) \le y \le f_2(x)\}$. We have shown that $\int_a^b [f_2(x) - f_1(x)]dx$ equals $\text{Area}(D) := \iint_D 1_D(x, y)d(x, y)$ under the assumption that the set of discontinuities of f_1 and f_2 are of one-dimensional content zero. Of course, this assumption implies Riemann integrability of f_1 and f_2, and it is satisfied by most functions that one comes across. However, equality also holds if we assume only that f_1 and f_2 are Riemann integrable. This can be shown using a result (given, for example, in Theorem 11.53 of Rudin [48]) that a bounded function of one variable is Riemann integrable if and only if its set of discontinuities is of (Lebesgue) measure zero. Indeed, using this and proceeding as in the proof of Corollary 5.45, we see that ∂D is of (Lebesgue) measure zero, and hence, as observed in the Notes and Comments on Chapter 5, ∂D is of content zero. So the desired equality follows from Fubini's Theorem over elementary regions (Proposition 5.36). A similar remark applies to elementary regions bounded by curves of the type $x = g_1(y)$ and $x = g_2(y)$, where $y \in [c, d]$. $\diamond$

Regions between Polar Curves

Our aim is to show that the general definition of area using double integrals is consistent with the formula given in one-variable calculus for the area of the region between two polar curves, that is, curves in $\mathbb{R}^2$ defined by an equation given in polar coordinates. For simplicity we shall restrict to continuous polar curves. A basic fact needed is the formula worked out in the example below.

Example 6.2. Let $a \in \mathbb{R}$ with $a > 0$, $\theta_0 \in [0, \pi]$, and let D_0 denote the sector of a disk of radius a that subtends an angle θ_0 at the center; for example, let D_0 denote the set of $(x, y) \in \mathbb{R}^2$ such that $x^2 + y^2 \le a^2$ and $0 \le \cos^{-1}\left(x/\sqrt{x^2 + y^2}\right) \le \theta_0$ for $(x, y) \ne (0, 0)$. If $\theta_0 = 0$, then D_0 reduces to a line segment, and its area is clearly equal to 0. Now let $\theta_0 \in (0, \pi/2]$. Then $D_0 = T \cup E$, where T is the triangular region with $(0, 0)$, $(a \cos\theta_0, 0)$, and $(a \cos\theta_0, a \sin\theta_0)$ as its vertices, whereas E is the region below the curve $y = \sqrt{a^2 - x^2}$, $x \in [a \cos\theta_0, a]$. (See Figure 6.1.) Also, $T \cap E$ is clearly of content zero. It follows that D_0 has an area and $\text{Area}(D_0)$ is equal to

$$\text{Area}(T) + \text{Area}(E) = \frac{1}{2}(a \cos\theta_0)(a \sin\theta_0) + \int_{a\cos\theta_0}^{a} \sqrt{a^2 - x^2}\, dx = \frac{a^2\theta_0}{2}.$$

Next, if $\theta_0 \in (\pi/2, \pi]$, then we can consider $\theta_1 := \theta_0 - (\pi/2)$ and observe that $\theta_1 \in (0, \pi/2]$ and also that

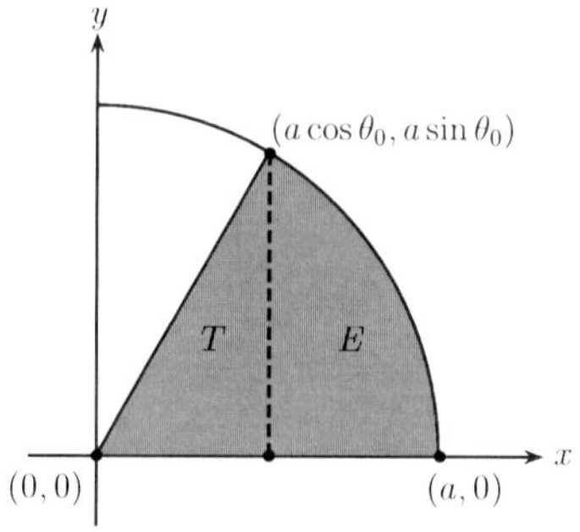

Fig. 6.1. Illustration of the regions T and E in Example 6.2.

$$\frac{a^2(\pi/2)}{2} + \frac{a^2\theta_1}{2} = \frac{a^2\pi}{4} + \frac{a^2}{2}\left(\theta_0 - \frac{\pi}{2}\right) = \frac{a^2\theta_0}{2},$$

which leads to the conclusion that $\text{Area}(D_0) = a^2\theta_0/2$ for any $\theta_0 \in [0, \pi]$. $\diamond$

It may be noted that it would have been much easier to obtain the formula $\text{Area}(D_0) = a^2\theta_0/2$ in the above example by directly using double integrals, switching to polar coordinates, and using Fubini's Theorem. However, to ensure that this switching is valid, that is, that Proposition 5.65 is applicable, one has to check that the relevant sets have boundaries of content zero. We shall show below that such a thing holds in a greater generality.

Proposition 6.3. *Let $p : [\alpha, \beta] \to \mathbb{R}$ be a nonnegative Riemann integrable function. Then the set $C := \{(p(\theta)\cos\theta,\ p(\theta)\sin\theta) \in \mathbb{R}^2 : \alpha \le \theta \le \beta\}$ is of content zero.*

Proof. Since the function p is nonnegative and bounded, there is a positive real number M such that $0 \le p(\theta) \le M$ for all $\theta \in [\alpha, \beta]$. Let $\epsilon > 0$ be given. By the Riemann Condition for functions of one variable (given, for example, in Proposition 6.5 of ACICARA), there exists a partition $Q := \{\theta_0, \theta_1, \ldots, \theta_n\}$ of $[\alpha, \beta]$ such that $U(Q,p) - L(Q,p) < \epsilon/2M$, that is,

$$\sum_{i=1}^{n} [M_i(p) - m_i(p)](\theta_i - \theta_{i-1}) < \frac{\epsilon}{2M}.$$

For $i = 1, \ldots, n$, let $D_i := \{(r\cos\theta, r\sin\theta) \in \mathbb{R}^2 : \theta_{i-1} \le \theta \le \theta_i \text{ and } m_i(p) \le r \le M_i(p)\}$. (See Figure 6.2.) Fix $i \in \{1, \ldots, n\}$. Since ∂D_i consists of two line segments and two circular arcs, it is of content zero. Also, by the formula in Example 6.2 for the area of a sector of a circle and by Domain Additivity (Proposition 5.51),

$$\begin{aligned}\text{Area}(D_i) &= \frac{1}{2}M_i(p)^2(\theta_i - \theta_{i-1}) - \frac{1}{2}m_i(p)^2(\theta_i - \theta_{i-1}) \\ &= \frac{1}{2}\,[M_i(p) + m_i(p)]\,[M_i(p) - m_i(p)](\theta_i - \theta_{i-1}) \\ &\le M[M_i(p) - m_i(p)](\theta_i - \theta_{i-1}).\end{aligned}$$

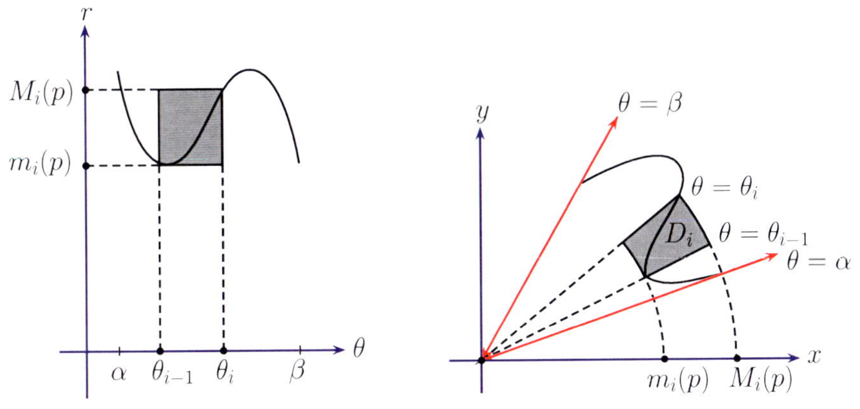

Fig. 6.2. Transformation of a polar rectangle.

Now let $D := D_1 \cup \cdots \cup D_n$. Since for every $\theta \in [\alpha, \beta]$, there is $i \in \{1, \ldots, n\}$ such that $\theta \in [\theta_{i-1}, \theta_i]$ and then $p(\theta) \in [m_i(p), M_i(p)]$, we see that

$$C = \{(p(\theta)\cos\theta, p(\theta)\sin\theta) \in \mathbb{R}^2 : \alpha \le \theta \le \beta\} \subseteq D.$$

In view of Corollary 5.38, ∂D is of content zero. Thus D has an area and

$$\text{Area}(D) \le \sum_{i=1}^{n} \text{Area}(D_i) \le \sum_{i=1}^{n} M[M_i(p) - m_i(p)](\theta_i - \theta_{i-1}) < M \cdot \frac{\epsilon}{2M} = \frac{\epsilon}{2}.$$

Let R be a rectangle containing D and consider the function $1_D^* : R \to \mathbb{R}$ obtained by extending the function $1_D : D \to \mathbb{R}$ as usual. Then

$$\text{Area}(D) = \iint_D 1_D d(x, y) = \iint_R 1_D^* d(x, y).$$

Since 1_D^* is integrable, there is a partition P of the rectangle R such that $U(P, 1_D^*) < \text{Area}(D) + \frac{\epsilon}{2}$. Among the subrectangles induced by the partition P, let $R_1, \ldots, R_k$ be the ones that intersect with D. Then D is contained in the union of $R_1, \ldots, R_k$, and the sum of the areas of these subrectangles is equal to $U(P, 1_D^*)$. Thus we have $C \subseteq D \subseteq R_1 \cup \cdots \cup R_n$, and the sum of the areas of the rectangles $R_1, \ldots, R_n$ is less than $\text{Area}(D) + \frac{\epsilon}{2} < \frac{\epsilon}{2} + \frac{\epsilon}{2} = \epsilon$. Hence the set C is of content zero. □

Consider a region D between two polar curves. (See Figure 6.3.) More precisely, let $\alpha, \beta \in \mathbb{R}$ be such that either $-\pi < \alpha < \beta \le \pi$ or $\alpha = -\pi$, $\beta = \pi$. Suppose $p_1, p_2 : [\alpha, \beta] \to \mathbb{R}$ are nonnegative continuous functions such that $p_1 \le p_2$. In case $\alpha = -\pi$, $\beta = \pi$, assume that $p_i(\pi) = p_i(-\pi)$ for $i = 1, 2$. Let

$$D := \{(r\cos\theta,\ r\sin\theta) \in \mathbb{R}^2 : \alpha \le \theta \le \beta \text{ and } p_1(\theta) \le r \le p_2(\theta)\}.$$

Note that $\partial D = L_\alpha \cup L_\beta \cup C_1 \cup C_2$, where

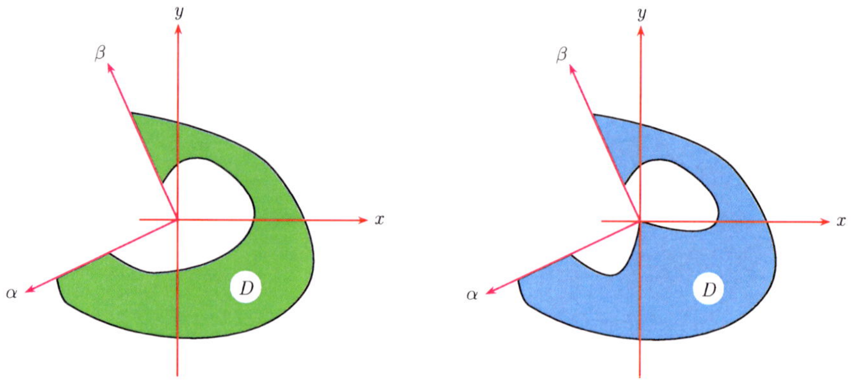

Fig. 6.3. Illustrations of regions between two polar curves.

$$L_\gamma := \{(r\cos\gamma,\ r\sin\gamma) \in \mathbb{R}^2 : p_1(\gamma) \le r \le p_2(\gamma)\} \quad \text{for } \gamma \in \{\alpha, \beta\},$$
$$C_i := \{(p_i(\theta)\cos\theta,\ p_i(\theta)\sin\theta) \in \mathbb{R}^2 : \alpha \le \theta \le \beta\} \quad \text{for } i \in \{1, 2\}.$$

Moreover, since L_α and L_β are line segments, they are of content zero, whereas by Proposition 6.3, both C_1 and C_2 are of content zero. Hence ∂D is of content zero. To find the area of D, we shall use the change of variables result for polar coordinates. Thus, as in Proposition 5.65, let

$$E := \{(r, \theta) \in \mathbb{R}^2 : r \ge 0,\ -\pi \le \theta \le \pi \text{ and } (r\cos\theta, r\sin\theta) \in D\}.$$

Consider the set $E_0 := \{(r, \theta) \in \mathbb{R}^2 : \alpha \le \theta \le \beta \text{ and } p_1(\theta) \le r \le p_2(\theta)\}$. Then $E = E_0$ if $(0, 0) \notin D$, while $E = E_0 \cup \{(0, \theta) : \theta \in [-\pi, \pi]\}$ if $(0, 0) \in D$. (See Figure 6.3.) Also, as in Corollary 5.45, we see that ∂E_0 is of content zero. It follows that ∂E is of content zero. Now by the change of variables formula for polar coordinates (Proposition 5.65) and by Fubini's Theorem for elementary regions (Proposition 5.36), we have

$$\text{Area}(D) = \iint_E r\, d(r, \theta) = \int_\alpha^\beta \int_{p_1(\theta)}^{p_2(\theta)} r\, dr d\theta = \frac{1}{2} \int_\alpha^\beta [p_2^2(\theta) - p_1^2(\theta)] d\theta.$$

This shows that the definition sometimes given in one-variable calculus (for example, in Section 8.1 of ACICARA) of the area of a region between two continuous curves given by polar equations of the form $r = p(\theta)$ is consistent with our general definition of an area. The area between curves given by polar equations of the form $\theta = \alpha(r)$ is treated in Exercise 33.

Finally, we observe that the area of a bounded set in $\mathbb{R}^2$ is invariant under a translation and a rotation of the set. In fact, invariance under a translation has already been proved in Lemma 5.55. The following result proves invariance under a rotation.

Proposition 6.4. *Let E be a bounded subset of $\mathbb{R}^2$ that has an area. Also, let $\alpha \in (-\pi, \pi]$ and $D := \{(u\cos\alpha - v\sin\alpha,\ u\sin\alpha + v\cos\alpha) : (u,v) \in E\}$. Then D has an area and* $\text{Area}(D) = \text{Area}(E)$.

Proof. Consider the affine transformation $\Phi : \mathbb{R}^2 \to \mathbb{R}^2$ defined by $\Phi(u,v) := (u\cos\alpha - v\sin\alpha,\ u\sin\alpha + v\cos\alpha)$. It is clear that $\Phi(E) = D$. Moreover,

$$J(\Phi)(u,v) = \det\begin{bmatrix} \cos\alpha & -\sin\alpha \\ \sin\alpha & \cos\alpha \end{bmatrix} = \cos^2\alpha + \sin^2\alpha = 1 \text{ for all } (u,v) \in \mathbb{R}^2.$$

Hence by Proposition 5.58, D has an area and $\text{Area}(D) = \text{Area}(E)$. □

Volume of a Bounded Subset of $\mathbb{R}^3$

Let D be a bounded subset of $\mathbb{R}^3$ such that ∂D is of three-dimensional content zero, and let $1_D : D \to \mathbb{R}$ denote the function given by $1_D(x,y) := 1$ for $(x,y) \in D$. In Section 5.4, we have defined the volume of D by

$$\text{Vol}(D) := \iiint_D 1_D(x,y,z)d(x,y,z).$$

Let D_1 and D_2 be bounded subsets of $\mathbb{R}^3$ such that ∂D_1 and ∂D_2 are of three-dimensional content zero. Then $\partial(D_1 \cup D_2)$ and $\partial(D_1 \cap D_2)$ are also of three-dimensional content zero, since each of these sets is contained in the union of ∂D_1 and ∂D_2. Hence by the domain additivity of the triple integral of the function $1_{D_1 \cup D_2}$, we obtain

$$\text{Vol}(D_1 \cup D_2) = \text{Vol}(D_1) + \text{Vol}(D_2) - \text{Vol}(D_1 \cap D_2).$$

If, in particular, $D_1 \cap D_2$ itself is of three-dimensional content zero, then

$$\text{Vol}(D_1 \cup D_2) = \text{Vol}(D_1) + \text{Vol}(D_2).$$

In analogy with "area between two curves," we now proceed to discuss the notion of "volume between two surfaces." Let D_0 be a bounded subset of $\mathbb{R}^2$ such that ∂D_0 is of (two-dimensional) content zero. Let $f_1, f_2 : D_0 \to \mathbb{R}$ be bounded functions such that $f_1 \le f_2$ and the sets of discontinuities of f_1 and f_2 are of (two-dimensional) content zero. Then by Proposition 5.43, the functions f_1 and f_2 are integrable over D_0. If D is the solid between the surfaces given by $z = f_1(x,y)$ and $z = f_2(x,y)$, that is, if

$$D := \{(x,y,z) \in \mathbb{R}^3 : (x,y) \in D_0 \text{ and } f_1(x,y) \le z \le f_2(x,y)\},$$

then ∂D is of three-dimensional content zero, as seen in Section 5.4. By part (ii) of Cavalieri's Principle (Proposition 5.68), the volume of D is equal to

$$\text{Vol}(D) = \iint_{D_0} [f_2(x,y) - f_1(x,y)]d(x,y).$$

Similar results hold for solids between surfaces given by equations of the type $y = g(x, z)$ or of the type $x = h(y, z)$. Moreover, comments analogous to those in Remark 6.1 are applicable. Thus, in particular, the above formulas for volume hold in greater generality when the functions $f_1, f_2 : D_0 \to \mathbb{R}$ are assumed only to be integrable.

Example 6.5. Let $a \in \mathbb{R}$ with $a > 0$ and let D denote the subset of $\mathbb{R}^3$ enclosed by the cylinders given by $x^2 + y^2 = a^2$ and $x^2 + z^2 = a^2$. Then $D = \left\{(x, y, z) : (x, y) \in D_0 \text{ and } -\sqrt{a^2 - x^2} \le z \le \sqrt{a^2 - x^2}\right\}$, where $D_0 := \{(x, y) \in \mathbb{R}^2 : x^2 + y^2 \le a^2\}$. Hence

$$\text{Vol}(D) = \iint_{D_0} \left(\int_{-\sqrt{a^2-x^2}}^{\sqrt{a^2-x^2}} dz \right) d(x, y) = \iint_{D_0} 2\sqrt{a^2 - x^2}\, d(x, y).$$

Also, $D_0 = \left\{(x, y) \in \mathbb{R}^2 : -a \le x \le a \text{ and } -\sqrt{a^2 - x^2} \le y \le \sqrt{a^2 - x^2}\right\}$. So

$$\text{Vol}(D) = 2 \int_{-a}^{a} \left(\int_{-\sqrt{a^2-x^2}}^{\sqrt{a^2-x^2}} \sqrt{a^2 - x^2}\, dy \right) dx = 4 \int_{-a}^{a} (a^2 - x^2) dx = \frac{16a^3}{3}.$$

This example may be compared with Example 8.4 (ii) of ACICARA. ◇

Solids between Cylindrical or Spherical Surfaces

Analogous to the formula for the area of a region between two polar curves, we shall obtain formulas for the volume of a solid between two cylindrical surfaces, that is, surfaces given by equations in cylindrical coordinates (r, θ, z), and a solid between two spherical surfaces, that is, surfaces given by equations in spherical coordinates (ρ, θ, φ). Basic facts needed for these formulas are worked out in the examples below. These examples are analogous to Example 6.2.

Examples 6.6. (i) Let $a \in \mathbb{R}$ with $a > 0$ and $\theta_0 \in [0, \pi]$. Let D_0 denote the sector of a disk of radius a that subtends an angle θ_0 at the center; for

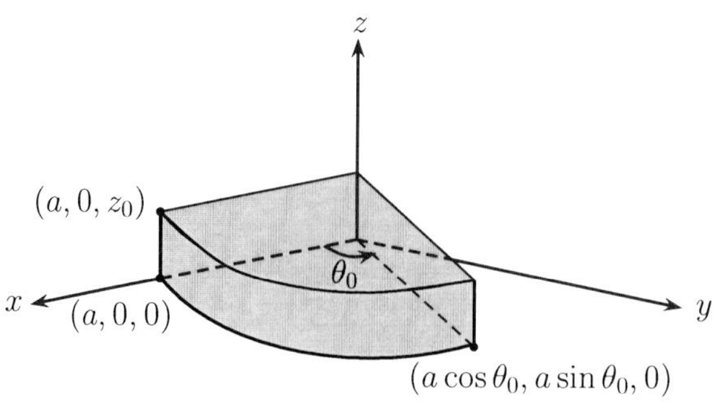

Fig. 6.4. Sector of a cylindrical solid considered in Example 6.6 (i).

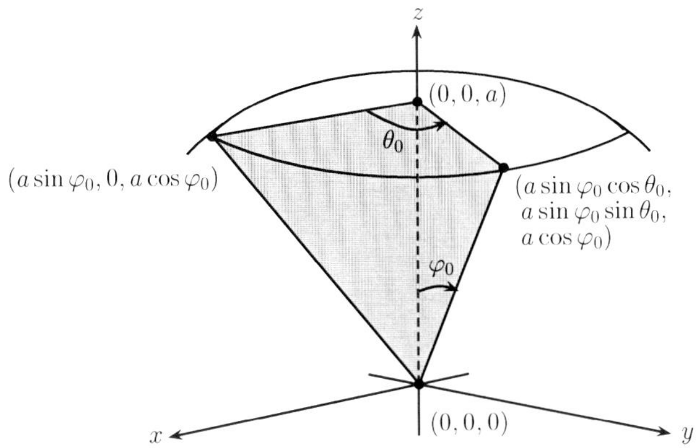

Fig. 6.5. Sector of a spherical solid considered in Example 6.6 (ii).

example, let $D_0 \subseteq \mathbb{R}^2$ be exactly as in Example 6.2. Given $z_0 > 0$, we shall refer to the set $D := \{(x, y, z) \in \mathbb{R}^3 : (x, y) \in D_0 \text{ and } 0 \le z \le z_0\}$ as a **sector of a cylindrical solid**. (See Figure 6.4.) It is clear that D has a volume and

$$\text{Vol}(D) = \iint_{D_0} z_0 d(x, y) = z_0 \text{Area}(D_0) = \frac{a^2\theta_0 z_0}{2}.$$

(ii) Let $a \in \mathbb{R}$ with $a > 0$, $\theta_0 \in [0, \pi]$, and $\varphi_0 \in [0, \pi]$. Let D denote the set of all $(x, y, z) \in \mathbb{R}^3$ such that $x^2+y^2+z^2 \le a^2$, $0 \le \cos^{-1}\left(x/\sqrt{x^2+y^2}\right) \le \theta_0$ whenever $(x, y) \ne (0, 0)$, and $0 \le \cos^{-1}\left(z/\sqrt{x^2+y^2+z^2}\right) \le \varphi_0$ whenever $(x, y, z) \ne (0, 0, 0)$. We shall refer to D as a **sector of a spherical solid**. (See Figure 6.5.) Observe that if $\theta_0 = 0$, then D reduces to a surface in the xz-plane, whereas if $\varphi_0 = 0$, then D reduces to a line segment on the z-axis, and thus in either of these two cases, D is of three-dimensional content zero, and so $\text{Vol}(D) = 0$. Assume that $\theta_0 \in (0, \pi/2]$ and $\varphi_0 \in (0, \pi/2]$. Now, $(x, y, z) \in D$ if and only if $z \le \sqrt{a^2 - x^2 - y^2}$, $0 \le \cos^{-1}\left(x/\sqrt{x^2+y^2}\right) \le \theta_0$ whenever $(x, y) \ne (0, 0)$, and $\cos\varphi_0 \le z/\sqrt{x^2+y^2+z^2}$ whenever $(x, y, z) \ne (0, 0, 0)$. Let $(x, y, z) \in \mathbb{R}^3$ with $(x, y, z) \ne (0, 0, 0)$. Then it is easy to see that $\cos\varphi_0 \le z/\sqrt{x^2+y^2+z^2}$ if and only if $x^2+y^2 \le (x^2+y^2+z^2)\sin^2\varphi_0$, that is, $\cot\varphi_0\sqrt{x^2+y^2} \le z$. Thus, if we let $E_0 := \left\{(x, y) \in \mathbb{R}^2 : x^2 + y^2 \le a^2\sin^2\varphi_0 \text{ and } 0 \le \cos^{-1}\left(x/\sqrt{x^2+y^2}\right) \le \theta_0 \text{ for } (x, y) \ne (0, 0)\right\}$, then $D = \{(x, y, z) \in \mathbb{R}^3 : (x, y) \in E_0 \text{ and } \cot\varphi_0\sqrt{x^2+y^2} \le z \le \sqrt{a^2 - x^2 - y^2}\}$, that is, D is the solid between the surfaces $z = \cot\varphi_0\sqrt{x^2+y^2}$ and $z = \sqrt{a^2-x^2-y^2}$ as (x,y) ranges over E_0. Note also that $\cot\varphi_0\sqrt{x^2+y^2} \le \sqrt{a^2-x^2-y^2}$ for all $(x,y) \in E_0$. It follows that D has a volume and

$$\mathrm{Vol}(D) = \iint_{E_0} \left[\sqrt{a^2-x^2-y^2} - \cot\varphi_0\sqrt{x^2+y^2}\right] d(x,y).$$

To compute the above double integral, we use polar coordinates and obtain

$$\begin{aligned}\mathrm{Vol}(D) &= \int_0^{\theta_0}\int_0^{a\sin\varphi_0} \left(\sqrt{a^2-r^2}\right) r\,dr\,d\theta - \int_0^{\theta_0}\int_0^{a\sin\varphi_0} (r\cot\varphi_0)\,r\,dr\,d\theta \\ &= -\frac{\theta_0}{2}\left[\frac{(a^2-r^2)^{3/2}}{3/2}\right]_0^{a\sin\varphi_0} - \theta_0\frac{a^3\sin^3\varphi_0\cot\varphi_0}{3} \\ &= \frac{a^3\theta_0}{3}\left(1-\cos\varphi_0\right).\end{aligned}$$

The case in which $\theta_0 \in (\pi/2, \pi]$ or $\varphi_0 \in (\pi/2,\pi]$ can be handled by symmetry considerations or by arguing as in Example 6.2. At any rate, the sector D of a spherical solid has a volume and $\mathrm{Vol}(D) = a^3\theta_0\left(1-\cos\varphi_0\right)/3$ for any $\theta_0 \in [0,\pi]$ and $\varphi_0 \in [0,\pi]$. $\diamond$

It may be noted that the above examples are in consonance with the formulas given in Corollary 5.73 for a solid cylinder and a solid ball. Using these examples, we can derive analogues of Proposition 6.3 for surfaces given by equations in cylindrical coordinates (r,θ,z) or in spherical coordinates (ρ,θ,φ). For example, if $h : [a,b]\times[\alpha,\beta] \to \mathbb{R}$ and $p : [\alpha,\beta]\times[\gamma,\delta] \to \mathbb{R}$ are integrable, then the subsets $\{(r\cos\theta,\ r\sin\theta,\ h(r,\theta)) : (r,\theta) \in [a,b]\times[\alpha,\beta]\}$ and $\{(p(\theta,\varphi)\sin\varphi\cos\theta,\ p(\theta,\varphi)\sin\varphi\sin\theta,\ p(\theta,\varphi)\cos\varphi) : (\theta,\varphi) \in [\alpha,\beta]\times[\gamma,\delta]\}$ of $\mathbb{R}^3$ have three-dimensional content zero. In other words, the surface given in cylindrical coordinates by $z = h(r,\theta)$, $(r,\theta) \in [a,b]\times[\alpha,\beta]$ and the surface given in spherical coordinates by $\rho = p(\theta,\varphi)$, $(\theta,\varphi) \in [\alpha,\beta]\times[\gamma,\delta]$ are of three-dimensional content zero. As a consequence, we can derive formulas analogous to those for the area of the region between two curves given by polar equations. Let $a, b, \alpha, \beta, \gamma, \delta \in \mathbb{R}$ with $0 < a < b$, $-\pi < \alpha < \beta \le \pi$ or $\alpha = -\pi$ and $\beta = \pi$, and $0 \le \gamma < \delta \le \pi$, and let $h_1, h_2 : [a,b]\times[\alpha,\beta] \to \mathbb{R}$ and $p_1, p_2 : [\alpha,\beta]\times[\gamma,\delta] \to \mathbb{R}$ be continuous functions satisfying $h_1 \le h_2$ and $p_1 \le p_2$. In case $\alpha = -\pi$ and $\beta = \pi$, assume that for $i = 1, 2$, we have $h_i(r,-\pi) = h_i(r,\pi)$ for all $r \in [a,b]$ and $p_i(-\pi,\varphi) = p_i(\pi,\varphi)$ for all $\varphi \in [\gamma,\delta]$. If D is the solid in $\mathbb{R}^3$ bounded by the surfaces given in cylindrical coordinates by $z = h_1(r,\theta)$ and $z = h_2(r,\theta)$, where $(r,\theta) \in [a,b]\times[\alpha,\beta]$, then

$$\mathrm{Vol}(D) = \int_\alpha^\beta\int_a^b \left[h_2(r,\theta) - h_1(r,\theta)\right] r\,dr\,d\theta.$$

Likewise, if D is the solid bounded by the surfaces given in spherical coordinates by $\rho = p_1(\theta,\varphi)$ and $\rho = p_2(\theta,\varphi)$, where $(\theta,\varphi) \in [\alpha,\beta]\times[\gamma,\delta]$, then

$$\mathrm{Vol}(D) = \frac{1}{3}\int_{\gamma}^{\delta}\int_{\alpha}^{\beta} \left(p_2^3(\theta,\varphi) - p_1^3(\theta,\varphi)\right) \sin\varphi \, d\theta \, d\varphi.$$

To see that these formulas are correct, we use the change of variables formula for cylindrical or spherical coordinates (Proposition 5.72) together with Cavalieri's Principle and Fubini's Theorem (Proposition 5.68 and Remark 5.69). Note that the change of variables is justified, since the boundaries of the relevant sets are of three-dimensional content zero, thanks to the above-mentioned analogues of Proposition 6.3. In a similar manner, one can obtain formulas for the volume of solids bounded by surfaces given by equations in cylindrical coordinates of the form $\theta = \alpha(r, z)$ or $r = g(\theta, z)$, or by equations in spherical coordinates of the form $\theta = \alpha(\rho, \varphi)$ or $\varphi = \gamma(\rho, \theta)$.

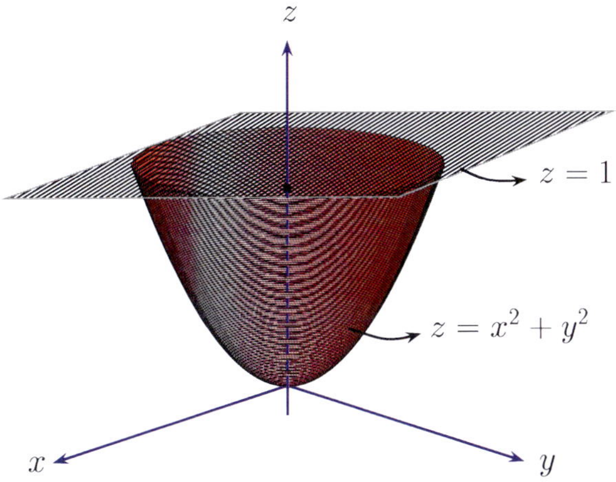

Fig. 6.6. The solid between the plane $z = 1$ and the paraboloid $z = x^2 + y^2$.

Examples 6.7. (i) Let D denote the subset of $\mathbb{R}^3$ between the plane given by $z = 1$ and the paraboloid given by $z = x^2 + y^2$. (See Figure 6.6.) Using cylindrical coordinates, we see that D is the solid bounded by the surfaces given by $z = r^2$ and $z = 1$, where $(r, \theta) \in [0, 1] \times [-\pi, \pi]$. Thus the formula obtained above yields

$$\mathrm{Vol}(D) = \int_{-\pi}^{\pi}\int_{0}^{1} (1 - r^2) r \, dr \, d\theta = 2\pi\left(\frac{1}{2} - \frac{1}{4}\right) = \frac{\pi}{2}.$$

(ii) Let $a \in \mathbb{R}$ with $a > 0$ and let D denote the subset of $\mathbb{R}^3$ consisting of points that are outside the sphere given by $x^2 + y^2 + z^2 = a^2$ and inside the sphere given by $x^2 + y^2 + (z - a)^2 = a^2$. (See Figure 5.27.) Using spherical coordinates, we see that D is the solid bounded by the surfaces given by $\rho = a$ and $\rho = 2a\cos\varphi$, where $(\varphi, \theta) \in [0, \pi/3] \times [-\pi, \pi]$. [Observe that if $a = \rho = 2a\cos\varphi$, then $\cos\varphi = 1/2$, that is, $\varphi = \pi/3$.] Thus the formula obtained above yields

$$\mathrm{Vol}(D) = \int_{-\pi}^{\pi}\int_{0}^{\pi/3} \frac{1}{3}\left[8a^3\cos^3\varphi - a^3\right]\sin\varphi\, d\varphi\, d\theta = \frac{2\pi a^3}{3}\int_{1/2}^{1}(8t^3-1)dt.$$

It follows that $\mathrm{Vol}(D) = 11\pi a^3/12$. ◇

Slicing by Planes and the Washer Method

The Washer Method is typically used in one-variable calculus for finding the volume of a solid obtained by revolving about the x-axis a region in $\mathbb{R}^2$ bounded by two curves, say $y = f_1(x)$ and $y = f_2(x)$, where $x \in [a,b]$. According to this method, the volume of such a solid is given by the Riemann integral

$$\int_a^b \pi[f_2(x)^2 - f_1(x)^2]dx,$$

where it is assumed that $f_1, f_2 : [a,b] \to \mathbb{R}$ are Riemann integrable and $0 \le f_1 \le f_2$. Here, the integrand $A(x) := \pi[f_2(x)^2 - f_1(x)^2]$ represents the area of a slice of the solid by a plane perpendicular to the x-axis. We shall now relate the formulas used in the Washer Method and the more general Slice Method to the general definition of volume as a triple integral.

Let D be a bounded subset of $\mathbb{R}^3$ such that ∂D is of three-dimensional content zero, and let $a, b \in \mathbb{R}$ with $a \le b$ be such that $a \le x \le b$ for all $(x,y,z) \in D$. For each $x \in [a,b]$, let D_x denote the corresponding cross section of D, yz-plane, that is, let $D_x := \{(y,z) \in \mathbb{R}^2 : (x,y,z) \in D\}$. Clearly, D_x is a bounded subset of $\mathbb{R}^2$. Assume that ∂D_x is of (two-dimensional) content zero. Then by part (i) of Proposition 5.68, we obtain

$$\mathrm{Vol}(D) = \int_a^b \left(\iint_{D_x} d(y,z)\right) dx = \int_a^b A(x)dx.$$

This shows that the formula for calculating the volume of a solid by slicing it by planes perpendicular to the x-axis (as given, for example, in Section 8.2 of ACICARA) is consistent with our general definition of volume.

Now suppose $f_1, f_2 : [a,b] \to \mathbb{R}$ are bounded functions whose sets of discontinuities are of one-dimensional content zero and that satisfy $0 \le f_1 \le f_2$. Let D be the subset of $\mathbb{R}^3$ obtained by revolving the planar region between the graphs of f_1 and f_2, namely $\left\{(x,y) \in \mathbb{R}^2 : a \le x \le b \text{ and } f_1(x) \le y \le f_2(x)\right\}$, about the x-axis. It is clear that

$$D = \left\{(x,y,z) \in \mathbb{R}^3 : a \le x \le b \text{ and } f_1(x)^2 \le y^2 + z^2 \le f_2(x)^2\right\}.$$

Let $M \in \mathbb{R}$ be such that $f_2(x) \le M$ for all $x \in [a,b]$. Then D is clearly a subset of $\left\{(x,y,z) \in \mathbb{R}^3 : a \le x \le b \text{ and } y^2 + z^2 \le M^2\right\}$, and in particular, D is bounded. Moreover, ∂D is contained in the union of the sets

$$\{(x,y,z) \in \mathbb{R}^3 : a \le x \le b \text{ and } y^2+z^2 = f_i(x)^2\} \quad \text{for } i \in \{1,2\},$$
$$\{(c,y,z) \in \mathbb{R}^3 : f_1(c)^2 \le y^2+z^2 \le f_2(c)^2\} \quad \text{for } c \in \{a,b\},$$
$$\{(x,y,z) \in \mathbb{R}^3 : a \le x \le b,\ f_1 \text{ or } f_2 \text{ is discontinuous at } x \text{ and } y^2+z^2 \le M^2\}.$$

It can be easily seen that all these sets are of three-dimensional content zero. Hence ∂D is of three-dimensional content zero. Now, for any $x \in [a,b]$, the corresponding cross section is $D_x = \{(y,z) \in \mathbb{R}^2 : f_1(x)^2 \le y^2+z^2 \le f_2(x)^2\}$, and $\partial D_x = \{(y,z) \in \mathbb{R}^2 : y^2+z^2 = f_1(x)^2\} \cup \{(y,z) \in \mathbb{R}^2 : y^2+z^2 = f_2(x)^2\}$. So, in view of Proposition 5.37 and Example 5.39 (ii), we see that ∂D_x is of (two-dimensional) content zero for every $x \in [a,b]$. Also, since the set

$$E_x := \{(r,\theta) \in \mathbb{R}^2 : r \ge 0, -\pi \le \theta \le \pi \text{ and } (r\cos\theta, r\sin\theta) \in D_x\}$$

is the rectangle $[-\pi,\pi] \times [f_1(x), f_2(x)]$, we see that ∂E_x is of (two-dimensional) content zero for each $x \in [a,b]$. Hence by Proposition 5.65, we have

$$A(x) := \iint_{D_x} d(y,z) = \int_{-\pi}^{\pi} \left(\int_{f_1(x)}^{f_2(x)} r\, dr \right) d\theta = \pi[f_2(x)^2 - f_1(x)^2].$$

Thus we conclude that

$$\text{Vol}(D) = \int_a^b A(x)dx = \pi \int_a^b [f_2(x)^2 - f_1(x)^2]dx.$$

In other words, the formula for calculating the volume of a solid of revolution by the Washer Method (as given, for example, in Section 8.2 of ACICARA) is consistent with our general definition of volume.

Example 6.8. Let $a,b,c,d \in \mathbb{R}$ with $a < b$ and $0 \le c \le d$, and consider $f_1, f_2 : [a,b] \to \mathbb{R}$ defined by $f_1(x) := c$ and $f_2(x) := d$ for all $x \in [a,b]$. Let $D := \{(x,y,z) \in \mathbb{R}^3 : a \le x \le b \text{ and } c^2 \le y^2+z^2 \le d^2\}$. Then D consists of a portion of a solid circular cylinder of radius d and height $b-a$ with a solid circular cylinder of radius c of the same height removed from the center. As a special case of the discussion and the formula above, we see that ∂D is of three-dimensional content zero, that is, D has a volume, and the volume of this solid is equal to $\pi(b-a)(d^2-c^2)$. ◇

Slivering by Cylinders and the Shell Method

In one-variable calculus, the Washer Method is usually studied alongside the Shell Method. Just as the Washer Method is a special case of the Slice Method, the Shell Method is a special case of the method of slivering by coaxial right circular cylinders (as explained in Section 8.2 of ACICARA), in which one considers the slivers[1] of a solid lying between two right circular cylinders, and

[1] A **sliver** of a solid D is a cross section of D by a cylinder.

then the volume is given as the Riemann integral of the surface area of such slivers. More precisely, let D be a bounded subset of $\mathbb{R}^3$ and suppose there are $p, q \in \mathbb{R}$ with $0 \le p < q$ such that $p^2 \le x^2 + y^2 \le q^2$ for all $(x, y, z) \in D$. Consider the counterpart of D in cylindrical coordinates, namely,

$$E := \{(r, \theta, z) \in \mathbb{R}^3 : r \ge 0, -\pi \le \theta \le \pi \text{ and } (r\cos\theta, r\sin\theta, z) \in D\}.$$

For each $r \in [p, q]$, let E_r be the corresponding cross section of E given by

$$E_r := \{(\theta, z) \in [-\pi, \pi] \times \mathbb{R} : (r\cos\theta, r\sin\theta, z) \in D\}.$$

Note that E_r is a bounded subset of $\mathbb{R}^2$ for each $r \in [p, q]$, and E is a bounded subset of $\mathbb{R}^3$. Now suppose ∂D and ∂E are of three-dimensional content zero, while ∂E_r is of (two-dimensional) content zero. Then by part (i) of Proposition 5.72 and part (i) of Proposition 5.68, we obtain

$$\text{Vol}(D) = \iiint_E r\, d(r, \theta, z) = \int_p^q \left(r \iint_{E_r} d(\theta, z) \right) dr.$$

Thus, if we let $B(r) := \text{Area}(E_r) = \iint_{E_r} d(\theta, z)$ for $r \in [p, q]$, then we have

$$\text{Vol}(D) = \int_p^q rB(r)dr.$$

This shows that the formula for calculating the volume of a solid of revolution by the method of slivering by coaxial cylinders whose axes lie on the z-axis is consistent with our general definition of volume.

Now let us consider solids of revolution in $\mathbb{R}^3$ to which the Shell Method is usually applied. More precisely, let $a, b \in \mathbb{R}$ with $0 \le a < b$ and let $f_1, f_2 : [a, b] \to \mathbb{R}$ be continuous functions such that $f_1 \le f_2$. Let D be the solid obtained by revolving the region $\{(x, y) \in \mathbb{R}^2 : a \le x \le b \text{ and } f_1(x) \le y \le f_2(x)\}$ about the y-axis. Then

$$D = \left\{(x, y, z) \in \mathbb{R}^3 : a^2 \le x^2 + z^2 \le b^2 \text{ and } f_1(x) \le y \le f_2(x)\right\}.$$

Since the functions f_1 and f_2 are continuous, the set D is a closed and bounded subset of $\mathbb{R}^3$. Also, ∂D is the union of the sets

$$\left\{(x, y, z) \in \mathbb{R}^2 : x^2 + z^2 = c^2 \text{ and } f_1(x) \le y \le f_2(x)\right\} \quad \text{for } c \in \{a, b\},$$
$$\left\{(x, y, z) \in \mathbb{R}^2 : a^2 \le x^2 + z^2 \le b^2 \text{ and } y = f_i(x)\right\} \quad \text{for } i \in \{1, 2\}.$$

Since each of these sets is of three-dimensional content zero, we see that ∂D is of three-dimensional content zero. Further, since the set

$$E := \{(x, \theta, y) \in \mathbb{R}^3 : x \ge 0, \; -\pi \le \theta \le \pi \text{ and } (x\cos\theta, \; y, \; x\sin\theta) \in D\}$$

is the same as the set

$$\{(x,\theta,y) \in \mathbb{R}^3 : a \le x \le b,\ -\pi \le \theta \le \pi \text{ and } f_1(x) \le y \le f_2(x)\},$$

we see that E is a closed and bounded subset of $\mathbb{R}^3$, and ∂E is of three-dimensional content zero. Lastly, for each $x \in [a,b]$, the corresponding cross section of E is given by $E_x = [-\pi,\pi] \times [f_1(x), f_2(x)]$, and so it is clear that ∂E_x is of (two-dimensional) content zero. Thus D and E satisfy the conditions in our discussion of the method of slivering by coaxial cylinders. Also, since

$$B(x) := \text{Area}(E_x) = \iint_{E_x} d(\theta,z) = 2\pi[f_2(x) - f_1(x)],$$

we conclude that

$$\text{Vol}(D) = \int_a^b xB(x)dx = 2\pi \int_a^b x[f_2(x) - f_1(x)]dx.$$

This shows that the formula for calculating the volume of a solid of revolution by the Shell Method (as given in Section 8.2 of ACICARA) is consistent with our general definition of volume. It may be noted that the equivalence of the general definition of volume (as a triple integral) and of the definition in terms of a Riemann integral as in the Washer Method or the Shell Method continues to hold if the roles of the x-axis, the y-axis, and the z-axis are interchanged. In particular, if for a solid of revolution, both the Washer Method and the Shell Method are applicable, then the volume can be calculated by either of these methods.

We shall now see how to find the volume of a solid generated by revolving a bounded closed subset in $\mathbb{R}^2$ about an arbitrary line in its plane. Its proof will involve the use of cylindrical coordinates as well as a change of variables under an affine map. First we prove a preliminary result that says that if a subset of $\mathbb{R}^2$ having (two-dimensional) content zero is revolved about the x-axis, then it generates a subset of $\mathbb{R}^3$ having three-dimensional content zero. The method of proof is similar to that used in proving Proposition 6.3.

Lemma 6.9. *Let B_0 be a bounded subset of $\mathbb{R}^2$ having (two-dimensional) content zero. Suppose B_0 lies in the upper half-plane, that is, $y \ge 0$ for every $(x,y) \in B_0$, and let B denote the set generated by revolving B_0 about the x-axis. Then B is of three-dimensional content zero.*

Proof. Since B_0 is bounded and lies in the upper half-plane, there is $M > 0$ such that $0 \le y \le M$ for all $(x,y) \in B_0$. Let $\epsilon > 0$ be given. Since B_0 is of (two-dimensional) content zero, there are finitely many rectangles $R_i := [a_i, b_i] \times [c_i, d_i], i = 1, \ldots, n$, such that B_0 is contained in their union and the sum of their areas, namely, $\sum_{i=1}^n (b_i - a_i)(d_i - c_i)$, is less than $\epsilon/4\pi M$. We may assume without loss of generality that $c_i \ge 0$ and $d_i \le M$ for each $i = 1, \ldots, n$. If the rectangle R_i is revolved about the x-axis, then we obtain the solid $D_i := \{(x,y,z) \in \mathbb{R}^3 : a_i \le x \le b_i \text{ and } c_i^2 \le y^2 + z^2 \le d_i^2\}$ for $i = 1, \ldots, n$. In view of Example 6.8, ∂D_i is of three-dimensional content zero and the volume $\text{Vol}(D_i)$ of D_i is equal to $\pi(b_i - a_i)(d_i^2 - c_i^2)$ for $i = 1, \ldots, n$.

Let W denote the union of $D_1, \ldots, D_n$. Since B_0 is contained in the union of $R_1, \ldots, R_n$, the set B generated by revolving the set B_0 about the x-axis is contained in W. Also, in view of Corollary 5.38, ∂W is of three-dimensional content zero, and

$$\mathrm{Vol}(W) \le \sum_{i=1}^{n} \mathrm{Vol}(D_i) = \sum_{i=1}^{n} \pi(b_i - a_i)(d_i - c_i)(d_i + c_i) \le \pi \cdot \frac{\epsilon}{4\pi M} \cdot 2M = \frac{\epsilon}{2}.$$

Let K denote the cuboid $[a,b] \times [-M, M] \times [-M, M]$. Clearly, K contains W. Consider the function $1_W^* : K \to \mathbb{R}$ obtained by extending the function $1_W : W \to \mathbb{R}$ as usual. Then

$$\mathrm{Vol}(W) = \iiint_W 1_W d(x,y,z) = \iiint_K 1_W^* d(x,y,z).$$

Now there is a partition P of the cuboid K such that $U(P, 1_W^*) < \mathrm{Vol}(W) + \epsilon/2$. Among the subcuboids induced by the partition P, let $K_1, \ldots, K_m$ be the ones that intersect with W. Then W is contained in the union of $K_1, \ldots, K_m$, and the sum of the volumes of these subcuboids is equal to $U(P, 1_W^*)$. (Compare Exercise 12 of Chapter 5.) Thus $B \subseteq W \subseteq K_1 \cup \cdots \cup K_m$, and the sum of the areas of the cuboids $K_1, \ldots, K_m$ is less than $\mathrm{Vol}(W) + (\epsilon/2) \le (\epsilon/2) + (\epsilon/2) = \epsilon$. It follows that B is of three-dimensional content zero. □

Proposition 6.10. *Let D_0 be a closed and bounded subset of $\mathbb{R}^2$ such that ∂D_0 is of content zero, and let L be a line given by $ax + by + c = 0$, where $a, b, c \in \mathbb{R}$ and $a^2 + b^2 \neq 0$. If L does not cross D_0 and if D_0 is revolved about L, then the volume of the solid D so generated is given by*

$$\mathrm{Vol}(D) = 2\pi \iint_{D_0} \frac{|ax + by + c|}{\sqrt{a^2 + b^2}} d(x,y).$$

Proof. Since the line L does not cross D_0, we obtain $ax + by + c \ge 0$ for all $(x,y) \in D_0$ or $ax + by + c \le 0$ for all $(x,y) \in D_0$. Let us assume the former.

First we prove the proposition in the special case in which the line L is the x-axis, that is, when $a = 0$, $b = 1$, and $c = 0$. In this case, $y \ge 0$ for all $(x,y) \in D_0$. We shall consider the cylindrical coordinates (x, r, θ) in (x,y,z)-space by letting $y := r\cos\theta$ and $z := r\sin\theta$. First we note that $D = \{(x,y,z) \in \mathbb{R}^3 : (x, \sqrt{y^2 + z^2}) \in D_0\}$. Since D_0 is a closed and bounded subset of $\mathbb{R}^2$, we see that D is a closed and bounded subset of $\mathbb{R}^3$. Also, it is easy to see that $\partial D \subseteq \{(x,y,z) \in \mathbb{R}^2 : (x, \sqrt{y^2 + z^2}) \in \partial D_0\}$. Since ∂D_0 is of (two-dimensional) content zero, it follows from Lemma 6.9 that ∂D is of three-dimensional content zero. Further, the set $E := \{(x, r, \theta) \in \mathbb{R}^3 : r \ge 0,\ -\pi \le \theta \le \pi \text{ and } (x, r\cos\theta, r\sin\theta) \in D\}$ is the same as the set $\{(x, y, \theta) \in \mathbb{R}^3 : (x,y) \in D_0 \text{ and } -\pi \le \theta \le \pi\}$. Now, since ∂D_0 is of (two-dimensional) content zero, it follows that ∂E is of three-dimensional content zero. Hence by part (i) of Proposition 5.72 and part (ii) of Proposition 5.68, we obtain

$$\begin{aligned}\operatorname{Vol}(D) &:= \iiint_D d(x,y,z) = \iiint_E r\, d(x,r,\theta)\\ &= \iint_{D_0}\left(\int_{-\pi}^{\pi} d\theta\right) y\, d(x,y) = 2\pi \iint_{D_0} y\, d(x,y).\end{aligned}$$

This proves the desired formula for $\operatorname{Vol}(D)$ when the line L is the x-axis.

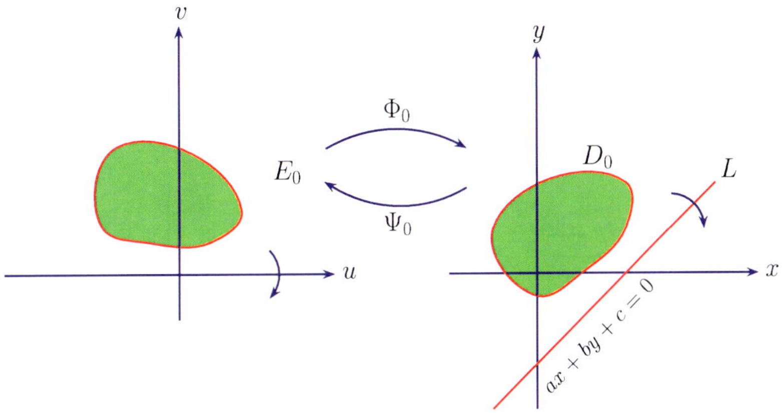

Fig. 6.7. Adjusting the axis of revolution of a planar region.

Let us now consider the case in which $a, b, c \in \mathbb{R}$ satisfy $a^2 + b^2 = 1$. We show that by a suitable change of variables, the line L can be assumed to be the x-axis. Consider an affine function $\Phi_0 : \mathbb{R}^2 \to \mathbb{R}^2$ defined by

$$\Phi_0(u,v) := \big(bu + av - c(a-b),\ -au + bv - c(a+b)\big).$$

Then

$$J(\Phi_0)(u,v) = \det\begin{bmatrix} b & a\\ -a & b\end{bmatrix} = a^2 + b^2 = 1 \quad \text{for all } (u,v) \in \mathbb{R}^2.$$

Also, it is easy to see that if we define

$$x := bu + av - c(a-b) \quad \text{and} \quad y := -au + bv - c(a+b),$$

then

$$ax + by + c = v \quad \text{and} \quad bx - ay - c = u.$$

Thus, letting $\Psi_0 := \Phi_0^{-1}$, it follows that

$$\Psi_0(x,y) = (bx - ay - c,\ ax + by + c) \quad \text{for all } (x,y) \in \mathbb{R}^2.$$

It can be easily checked that $J(\Psi_0)(x,y) = 1$ for all $(x,y) \in \mathbb{R}^2$. Let $E_0 := \Psi_0(D_0)$. (See Figure 6.7.) Then Proposition 5.59 shows that ∂E_0 is of content zero, and since $\Phi_0(E_0) = D_0$,

$$\iint_{D_0} (ax + by + c)d(x, y) = |J(\Phi_0)| \iint_{E_0} v\, d(u, v) = \iint_{E_0} v\, d(u, v).$$

Note that $v \geq 0$ whenever $(u, v) \in E_0$. Let E denote the solid in $\mathbb{R}^3$ generated by revolving the region E_0 about the u-axis, that is,

$$E := \left\{(u, v, w) \in \mathbb{R}^3 : \left(u, \sqrt{v^2 + w^2}\right) \in E_0\right\}.$$

Since D_0 is a bounded, closed subset of $\mathbb{R}^2$ and $E_0 := \Psi_0(D_0)$, the set E_0 is also bounded and closed. As we have seen in the special case at the beginning of this proof, it follows that E is a closed subset of $\mathbb{R}^3$ and ∂E is of three-dimensional content zero, and moreover,

$$\text{Vol}(E) := \iiint_E d(u, v, w) = 2\pi \iint_{E_0} v\, d(u, v).$$

It remains to show that D has a volume and $\text{Vol}(D) = \text{Vol}(E)$. To this end, we use a suitable change of variables in $\mathbb{R}^3$. Consider $\Phi : \mathbb{R}^3 \to \mathbb{R}^3$ defined by

$$\Phi(u, v, w) := \big(\Phi_0(u, v), w\big) = \big(bu + av - c(a - b), -au + bv - c(a + b), w\big).$$

Then Φ is an affine function, $\Phi(E) = D$, and $J(\Phi)(u, v, w) = 1$ for all (u, v, w) in $\mathbb{R}^3$. It follows by the three-dimensional analogue of Proposition 5.59 that ∂D is of three-dimensional content zero and $\text{Vol}(D) = |J(\Phi)|\text{Vol}(E) = \text{Vol}(E)$. Now, since $ax + by + c \geq 0$ for all $(x, y) \in D_0$, we conclude that

$$\text{Vol}(D) = 2\pi \iint_{D_0} |ax + by + c| d(x, y).$$

If $a^2 + b^2 \neq 1$, we replace a, b, and c by $a/\sqrt{a^2 + b^2}$, $b/\sqrt{a^2 + b^2}$, and $c/\sqrt{a^2 + b^2}$, respectively, and obtain the desired formula for $\text{Vol}(D)$. The case in which $ax + by + c \leq 0$ for all $(x, y) \in D_0$ is proved similarly. □

Remark 6.11. The Washer Method and the Shell Method are particular cases of Proposition 6.10. To see this, let $D_0 := \{(x, y) \in \mathbb{R}^2 : a \leq x \leq b \text{ and } f_1(x) \leq y \leq f_2(x)\}$, where $f_1, f_2 : [a, b] \to \mathbb{R}$ are continuous functions. Then D_0 is a closed and bounded subset of $\mathbb{R}^2$, and ∂D_0 is of (two-dimensional) content zero. If $0 \leq f_1 \leq f_2$ and L is the x-axis given by $y = 0$, then by Proposition 6.10, the volume of the solid generated by revolving the region D_0 about L is equal to

$$2\pi \iint_{D_0} |y| d(x, y) = 2\pi \int_a^b \left(\int_{f_1(x)}^{f_2(x)} y\, dy\right) dx = \pi \int_a^b [f_2(x)^2 - f_1(x)^2] dx,$$

as in the Washer Method. Also, if $a \geq 0$ and L is the y-axis given by $x = 0$, then by Proposition 6.10, the volume of the solid generated by revolving the region D_0 about L is equal to

$$2\pi \iint_{D_0} |x| d(x,y) = 2\pi \int_a^b \left(\int_{f_1(x)}^{f_2(x)} x\, dy \right) dx = 2\pi \int_a^b x[f_2(x) - f_1(x)]dx,$$

as in the Shell Method. $\diamond$

Before concluding this section, we mention that the volume of a solid in $\mathbb{R}^3$ is invariant under a translation and a rotation just as the area of a region in $\mathbb{R}^2$ is invariant under a translation and a rotation. A translation in $\mathbb{R}^3$ is carried out by an affine function $\Phi : \mathbb{R}^3 \to \mathbb{R}^3$ given by

$$\Phi(u, v, w) = (x^\circ + u,\ y^\circ + v,\ z^\circ + w),$$

where $(x^\circ, y^\circ, z^\circ) \in \mathbb{R}^3$ is fixed. The absolute value of the Jacobian of this affine function is equal to 1. A rotation in $\mathbb{R}^3$ by an arbitrary angle about an arbitrary line passing through the origin can be carried out by a composition of affine functions $\Phi, \Psi, \Xi : \mathbb{R}^3 \to \mathbb{R}^3$ given by

$$\begin{aligned}
\Phi(u, v, w) &= (u,\ v\cos\alpha - w\sin\alpha,\ v\sin\alpha + w\cos\alpha),\\
\Psi(u, v, w) &= (u\cos\beta - w\sin\beta,\ v,\ u\sin\beta + w\cos\beta),\\
\Xi(u, v, w) &= (u\cos\gamma - v\sin\gamma,\ u\sin\gamma + v\cos\gamma,\ w),
\end{aligned}$$

where $\alpha, \beta, \gamma \in (-\pi, \pi]$. These α, β, γ are known as the **Euler angles** in the xyz-convention. (See pages 31–33 of [41], or pages 143–148 and 608 of [25].) Now for all $(u, v, w) \in \mathbb{R}^3$, the Jacobians $J(\Phi)(u, v, w)$, $J(\Psi)(u, v, w)$, and $J(\Xi)(u, v, w)$ are given by the 3×3 determinants

$$\begin{vmatrix} 1 & 0 & 0 \\ 0 & \cos\alpha & -\sin\alpha \\ 0 & \sin\alpha & \cos\alpha \end{vmatrix}, \quad \begin{vmatrix} \cos\beta & 0 & -\sin\beta \\ 0 & 1 & 0 \\ \sin\beta & 0 & \cos\beta \end{vmatrix}, \quad \text{and} \quad \begin{vmatrix} \cos\gamma & -\sin\gamma & 0 \\ \sin\gamma & \cos\gamma & 0 \\ 0 & 0 & 1 \end{vmatrix},$$

respectively. Since the absolute value of these Jacobians is equal to 1, the invariance result follows as in Proposition 6.4.

6.2 Surface Area

In this section, we discuss how to measure the extent of a surface in 3-space, that is, how to calculate the "area of a surface." In the special case of a planar surface, that is, when the surface lies entirely in a plane, say the xy-plane, the question is identical to that of measuring the area of a region in $\mathbb{R}^2$. We have discussed this already in Chapter 5 and noted that the area is given by a double integral. We shall now proceed to motivate and formulate the general definition of surface area, which will be a natural extension of the definition of the area of a planar region. It will be convenient and useful to restrict to what are known as parametrically defined surfaces.

A **parametrically defined surface** S in $\mathbb{R}^3$ is given by $\big(x(u,v),\, y(u,v),\, z(u,v)\big)$, $(u,v) \in E$, where E is a bounded subset of $\mathbb{R}^2$ such that ∂E is of content zero, and $x, y, z : E \to \mathbb{R}$ are real-valued functions defined on E. [2] We shall refer to E as the **parameter domain** of the surface S and the triple (x, y, z) of real-valued functions as the **parametrization** of S. Note that the surface S is determined by its parametrization, that is, by the three functions $x, y, z : E \to \mathbb{R}$, and not by the subset $\{\big(x(u,v),\, y(u,v),\, z(u,v)\big) : (u,v) \in E\}$ of $\mathbb{R}^3$ traced by S. For example, the surface S_1 in $\mathbb{R}^3$ given by $(\cos u, \sin u, v)$, $(u,v) \in [-\pi, \pi] \times [0,1]$, and the surface S_2 in $\mathbb{R}^3$ given by $(\cos 2u, \sin 2u, v)$, $(u,v) \in [-\pi, \pi] \times [0,1]$, have the same parameter domain and they trace the same subset of $\mathbb{R}^3$, but they are obviously different surfaces, since S_1 goes around the z-axis once, while S_2 goes around the z-axis twice.

In order to motivate the general definition of the area of a surface, let us first consider a special case in which the parameter domain is a parallelogram in $\mathbb{R}^2$ and the parametrization is given by affine functions, that is, the parametrization is obtained by restricting an affine transformation from $\mathbb{R}^2$ to $\mathbb{R}^3$ to the parameter domain. Recall that as in Section 5.3, $\Phi : \mathbb{R}^2 \to \mathbb{R}^3$ is an **affine transformation** if there are $(x^\circ, y^\circ, z^\circ) \in \mathbb{R}^3$ and $a_i, b_i \in \mathbb{R}$ for $i = 1, 2, 3$ such that

$$\Phi(u,v) := (x^\circ + a_1 u + b_1 v,\; y^\circ + a_2 u + b_2 v,\; z^\circ + a_3 u + b_3 v) \quad \text{for } (u,v) \in \mathbb{R}^2.$$

In *matrix notation*, this can be written as follows:

$$\Phi \begin{bmatrix} u \\ v \end{bmatrix} := \begin{bmatrix} x^\circ \\ y^\circ \\ z^\circ \end{bmatrix} + \begin{bmatrix} a_1 & b_1 \\ a_2 & b_2 \\ a_3 & b_3 \end{bmatrix} \begin{bmatrix} u \\ v \end{bmatrix} \qquad \text{for } (u,v) \in \mathbb{R}^2.$$

It is easy to see that Φ is injective if and only if the "rank" of the 3×2 matrix above is 2, that is, at least one of the three determinants

$$d_1 := \det \begin{bmatrix} a_1 & b_1 \\ a_2 & b_2 \end{bmatrix}, \quad d_2 := \det \begin{bmatrix} a_2 & b_2 \\ a_3 & b_3 \end{bmatrix}, \quad \text{and} \quad d_3 := \det \begin{bmatrix} a_3 & b_3 \\ a_1 & b_1 \end{bmatrix}$$

is not equal to zero. Moreover, if $(u_i, v_i) \in \mathbb{R}^2$ and $t_i \in \mathbb{R}$ for $i = 1, \ldots, n$ (where $n \in \mathbb{N}$), then as in Section 5.3, we have

$$\Phi\left(\sum_{i=1}^n t_i (u_i, v_i)\right) = \sum_{i=1}^n t_i \Phi(u_i, v_i) + \left(1 - \sum_{i=1}^n t_i\right)(x^\circ, y^\circ, z^\circ).$$

This implies that Φ maps a line segment in $\mathbb{R}^2$ onto a line segment in $\mathbb{R}^3$. In fact, Φ maps parallel lines in $\mathbb{R}^2$ onto parallel lines in $\mathbb{R}^3$. Also, Φ maps a

[2] To be pedantic, a **parametrically defined surface** is a (vector-valued) map from E to $\mathbb{R}^3$ that sends (u,v) in E to $(x(u,v), y(u,v), z(u,v))$ in $\mathbb{R}^3$, where $E \subseteq \mathbb{R}^2$ and $x, y, z : E \to \mathbb{R}$ are as above. Generally, one requires that E be a rectangle and that the three functions $x, y, z : E \to \mathbb{R}$ be continuous. In most applications this is so, but we do not make it a part of the definition.

parallelogram in $\mathbb{R}^2$ onto a parallelogram lying in a plane in $\mathbb{R}^3$. Thus, if the parameter domain E is a parallelogram in $\mathbb{R}^2$, then $\Phi(E)$ is a parallelogram in $\mathbb{R}^3$. With this in view, to define the surface area in this special case, we undertake an analysis of the area of a parallelogram in $\mathbb{R}^2$ or in $\mathbb{R}^3$.

Parallelograms in $\mathbb{R}^2$ and in $\mathbb{R}^3$

Let $\mathbf{p}_i := (u_i, v_i)$, $i = 1, 2, 3$, be noncollinear points in $\mathbb{R}^2$, and let E denote the parallelogram with $\mathbf{p}_1, \mathbf{p}_2, \mathbf{p}_3$ as vertices such that $\mathbf{p}_2$ and $\mathbf{p}_3$ are adjacent to $\mathbf{p}_1$. (See Figure 6.8.) By Proposition 5.56,

$$\text{Area}(E) := |(u_2 - u_1)(v_3 - v_1) - (v_2 - v_1)(u_3 - u_1)|.$$

Using the identity $(ad - bc)^2 = (a^2 + b^2)(c^2 + d^2) - (ac + bd)^2$, we obtain

$$\begin{aligned}[\text{Area}(E)]^2 &= \left[(u_2 - u_1)^2 + (v_2 - v_1)^2\right]\left[(u_3 - u_1)^2 + (v_3 - v_1)^2\right] \\ &\quad - \left[(u_2 - u_1)(u_3 - u_1) + (v_2 - v_1)(v_3 - v_1)\right]^2.\end{aligned}$$

The last term on the right-hand side can be expressed in terms of the cosine of the angle between the line segments $\mathbf{p}_1\mathbf{p}_2$ and $\mathbf{p}_1\mathbf{p}_3$. Indeed, if $\alpha \in [0, \pi]$ is this angle, then as in Section 1.1, $\cos\alpha$ is equal to

$$\frac{(\mathbf{p}_2 - \mathbf{p}_1) \cdot (\mathbf{p}_3 - \mathbf{p}_1)}{|\mathbf{p}_2 - \mathbf{p}_1||\mathbf{p}_3 - \mathbf{p}_1|} = \frac{(u_2 - u_1)(u_3 - u_1) + (v_2 - v_1)(v_3 - v_1)}{\sqrt{(u_2 - u_1)^2 + (v_2 - v_1)^2}\sqrt{(u_3 - u_1)^2 + (v_3 - v_1)^2}}.$$

Consequently, $[\text{Area}(E)]^2 = |\mathbf{p}_2 - \mathbf{p}_1|^2|\mathbf{p}_3 - \mathbf{p}_1|^2(1 - \cos^2\alpha)$, that is,

$$\text{Area}(E) = |\mathbf{p}_2 - \mathbf{p}_1||\mathbf{p}_3 - \mathbf{p}_1| \sin\alpha.$$

Thus the area of a parallelogram in $\mathbb{R}^2$ is the length of its base times its height.

Now consider noncollinear points $\mathbf{q}_i := (x_i, y_i, z_i)$ for $i = 1, 2, 3$ in $\mathbb{R}^3$ and let D denote the parallelogram in $\mathbb{R}^3$ with $\mathbf{q}_1, \mathbf{q}_2, \mathbf{q}_3$ as vertices such that $\mathbf{q}_2$ and $\mathbf{q}_3$ are adjacent to $\mathbf{q}_1$. (See Figure 6.8.) In analogy with the area of a parallelogram in $\mathbb{R}^2$, let us tentatively define the "area" of the parallelogram D in $\mathbb{R}^3$ to be equal to the length of its base times its height. Thus, if $\beta \in [0, \pi]$ is the angle between the line segments $\mathbf{q}_1\mathbf{q}_2$ and $\mathbf{q}_1\mathbf{q}_3$, then

$$\text{Area}(D) := |\mathbf{q}_2 - \mathbf{q}_1|\,|\mathbf{q}_3 - \mathbf{q}_1| \sin\beta.$$

Since $\cos\beta = (\mathbf{q}_2 - \mathbf{q}_1) \cdot (\mathbf{q}_3 - \mathbf{q}_1)/|\mathbf{q}_2 - \mathbf{q}_1||\mathbf{q}_3 - \mathbf{q}_1|$ and $\sin^2\beta = 1 - \cos^2\beta$, squaring both sides of the above formula for $\text{Area}(D)$, we obtain

$$[\text{Area}(D)]^2 = |\mathbf{q}_2 - \mathbf{q}_1|^2|\mathbf{q}_3 - \mathbf{q}_1|^2 - [(\mathbf{q}_2 - \mathbf{q}_1) \cdot (\mathbf{q}_3 - \mathbf{q}_1)]^2.$$

Further, using the algebraic identity

$$(a^2 + b^2 + c^2)(p^2 + q^2 + r^2) - (ap + bq + cr)^2 = (aq - bp)^2 + (br - cq)^2 + (cp - ar)^2,$$

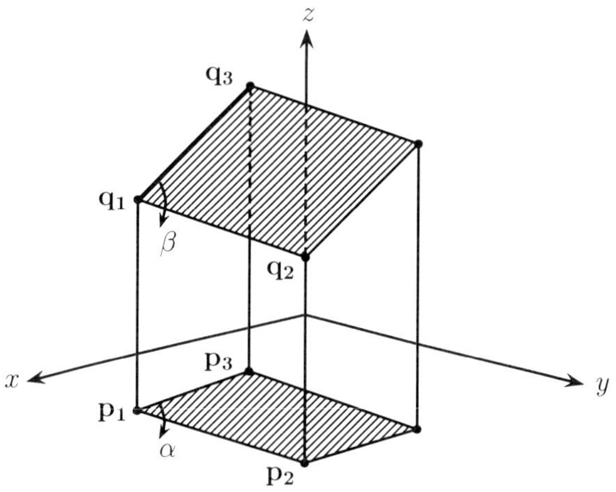

Fig. 6.8. Finding the "area" of a parallelogram in $\mathbb{R}^2$ and in $\mathbb{R}^3$.

we can conclude that

$$\begin{aligned}[\text{Area}(D)]^2 = & \left[(y_2 - y_1)(z_3 - z_1) - (z_2 - z_1)(y_3 - y_1)\right]^2 \\ & + \left[(z_2 - z_1)(x_3 - x_1) - (x_2 - x_1)(z_3 - z_1)\right]^2 \\ & + \left[(x_2 - x_1)(y_3 - y_1) - (y_2 - y_1)(x_3 - x_1)\right]^2.\end{aligned}$$

It follows that if D_1, D_2, and D_3 denote projections of the parallelogram D onto the yz-plane, the zx-plane, and the xy-plane respectively, then

$$\text{Area}(D) = \sqrt{\text{Area}(D_1)^2 + \text{Area}(D_2)^2 + \text{Area}(D_3)^2}.$$

We are now ready to prove the following analogue of Proposition 5.58 for affine transformations of the plane into 3-space.

Proposition 6.12. *Let $\Phi : \mathbb{R}^2 \to \mathbb{R}^3$ be an affine transformation and let ϕ_1, ϕ_2, ϕ_3 be its components, that is, $\Phi := (\phi_1, \phi_2, \phi_3)$. Define the transformations $\Phi_1, \Phi_2, \Phi_3 : \mathbb{R}^2 \to \mathbb{R}^2$ by $\Phi_1(u, v) := (\phi_2(u, v), \phi_3(u, v))$, $\Phi_2(u, v) := (\phi_3(u, v), \phi_1(u, v))$, and $\Phi_3(u, v) := (\phi_1(u, v), \phi_2(u, v))$. Assume that at least one of the Jacobians $J(\Phi_1), J(\Phi_2)$, and $J(\Phi_3)$ is not equal to zero. Let E be a parallelogram in $\mathbb{R}^2$ and let D denote the parallelogram $\Phi(E)$ in $\mathbb{R}^3$. Then*

$$\text{Area}(D) = \left(\sqrt{J(\Phi_1)^2 + J(\Phi_2)^2 + J(\Phi_3)^2}\right) \text{Area}(E).$$

Proof. Let the affine transformation $\Phi : \mathbb{R}^2 \to \mathbb{R}^3$ be given by

$$\Phi(u, v) = (x^\circ + a_1 u + b_1 v,\ y^\circ + a_2 u + b_2 v,\ z^\circ + a_3 u + b_3 v) \quad \text{for } (u, v) \in \mathbb{R}^2,$$

where $(x^\circ, y^\circ, z^\circ) \in \mathbb{R}^3$ and $a_i, b_i, c_i \in \mathbb{R}$ for $i = 1, 2, 3$ are fixed. Then it is easily seen that

$$J(\Phi_1) = a_2b_3 - b_2a_3, \quad J(\Phi_2) = a_3b_1 - b_3a_1, \quad \text{and} \quad J(\Phi_3) = a_1b_2 - b_1a_2.$$

Let (u_1, v_1) denote one of the vertices of the parallelogram E in $\mathbb{R}^2$, and let $(u_2, v_2), (u_3, v_3)$ denote the vertices of E adjacent to (u_1, v_1). Then by Proposition 5.56,

$$\text{Area}(E) = |(u_2 - u_1)(v_3 - v_1) - (u_3 - u_1)(v_2 - v_1)|.$$

Also, if $\Phi(u_i, v_i) = (x_i, y_i, z_i)$ for $i = 1, 2, 3$, then (x_1, y_1, z_1) is a vertex of the parallelogram $D := \Phi(E)$ in $\mathbb{R}^3$ and (x_2, y_2, z_2), (x_3, y_3, z_3) are the vertices of D adjacent to (x_1, y_1, z_1). As we have noted just before stating this proposition, $\text{Area}(D) = \sqrt{\text{Area}(D_1)^2 + \text{Area}(D_2)^2 + \text{Area}(D_3)^2}$, where

$$\begin{aligned}
\text{Area}(D_1) &= |(y_2 - y_1)(z_3 - z_1) - (y_3 - y_1)(z_2 - z_1)|, \\
\text{Area}(D_2) &= |(z_2 - z_1)(x_3 - x_1) - (z_3 - z_1)(x_2 - x_1)|, \\
\text{Area}(D_3) &= |(x_2 - x_1)(y_3 - y_1) - (x_3 - x_1)(y_2 - y_1)|.
\end{aligned}$$

For $i = 2, 3$, we have $(x_i, y_i, z_i) - (x_1, y_1, z_1) = \Phi(u_i, v_i) - \Phi(u_1, v_1)$, and hence $x_i - x_1 = a_1(u_i - u_1) + b_1(v_i - v_1)$, $y_i - y_1 = a_2(u_i - u_1) + b_2(v_i - v_1)$, and $z_i - z_1 = a_3(u_i - u_1) + b_3(v_i - v_1)$. Thus, in *matrix notation*, we have

$$\begin{bmatrix} x_2 - x_1 & x_3 - x_1 \\ y_2 - y_1 & y_3 - y_1 \end{bmatrix} = \begin{bmatrix} a_1 & b_1 \\ a_2 & b_2 \end{bmatrix} \begin{bmatrix} u_2 - u_1 & u_3 - u_1 \\ v_2 - v_1 & v_3 - v_1 \end{bmatrix}.$$

Since the determinant of the product of two square matrices is the product of the determinants of those matrices, we see that $\text{Area}(D_1) = |J(\Phi_1)|\text{Area}(E)$. Similarly, $\text{Area}(D_2) = |J(\Phi_2)|\text{Area}(E)$ and $\text{Area}(D_3) = |J(\Phi_3)|\text{Area}(E)$. Thus,

$$\text{Area}(D) = \left(\sqrt{J(\Phi_1)^2 + J(\Phi_2)^2 + J(\Phi_3)^2}\right) \text{Area}(E),$$

as desired. □

Area of a Smooth Surface

In Proposition 6.12, we have shown that if a parallelogram (and, in particular, a rectangle) in $\mathbb{R}^3$ is transformed to a parallelogram in $\mathbb{R}^3$ by an affine transformation Φ, then the area will have to be scaled by the "Jacobian factor" $\sqrt{J(\Phi_1)^2 + J(\Phi_2)^2 + J(\Phi_3)^2}$. We shall presently see that this result is quite crucial in developing the notion of the area of any "smooth" surface. The key idea is that any such surface can be approximated locally by a plane. To explain this, let $E \subseteq \mathbb{R}^2$ and consider a surface S in $\mathbb{R}^3$ parametrically given by $(x(u, v), y(u, v), z(u, v))$, $(u, v) \in E$. Let $Q_0 := (u_0, v_0)$ be an interior point of E and let $P_0 = (x_0, y_0, z_0) := \big(x(Q_0),\ y(Q_0),\ z(Q_0)\big)$. Assume that the functions x, y, and z are differentiable at Q_0. For $(u, v) \in \mathbb{R}^2$, let

$$\phi_1(u,v) := x_0 + x_u(u_0,v_0)(u-u_0) + x_v(u_0,v_0)(v-v_0),$$
$$\phi_2(u,v) := y_0 + y_u(u_0,v_0)(u-u_0) + y_v(u_0,v_0)(v-v_0),$$
$$\phi_3(u,v) := z_0 + z_u(u_0,v_0)(u-u_0) + z_v(u_v,v_0)(v-v_0).$$

Then by Proposition 4.18, we see that

$$x(u,v) - \phi_1(u,v) \to 0, \quad y(u,v) - \phi_2(u,v) \to 0, \quad z(u,v) - \phi_3(u,v) \to 0,$$

as $(u,v) \to (u_0,v_0)$. Thus the tangent plane to S at P_0, parametrically given by $(\phi_1(u,v), \phi_2(u,v), \phi_3(u,v))$, $(u,v) \in \mathbb{R}^2$, approximates the surface S around (u_0,v_0). It is therefore reasonable to expect that if R is a small rectangle centered at Q_0, then the "area" of the small surface in $\mathbb{R}^3$ given by $(x(u,v), y(u,v), z(u,v))$, $(u,v) \in R$, will be approximated by the area of the parallelogram $\Phi(R)$, namely $\sqrt{J(\Phi_1)^2 + J(\Phi_2)^2 + J(\Phi_3)^2}\,\text{Area}(R)$, where $\Phi := (\phi_1, \phi_2, \phi_3)$, $\Phi_1 := (\phi_2, \phi_3)$, $\Phi_2 := (\phi_3, \phi_1)$, and $\Phi_3 := (\phi_1, \phi_2)$, as before. Here, $J(\Phi_1)$ is equal to the Jacobian of the function given by $(u,v) \longmapsto (y(u,v), z(u,v))$ at Q_0, and similarly for $J(\Phi_2)$ and $J(\Phi_3)$. We observe that this parallelogram is in a plane that is tangent to the surface at the point $Q_0 := \Phi(P_0)$. (See Figure 6.9.)

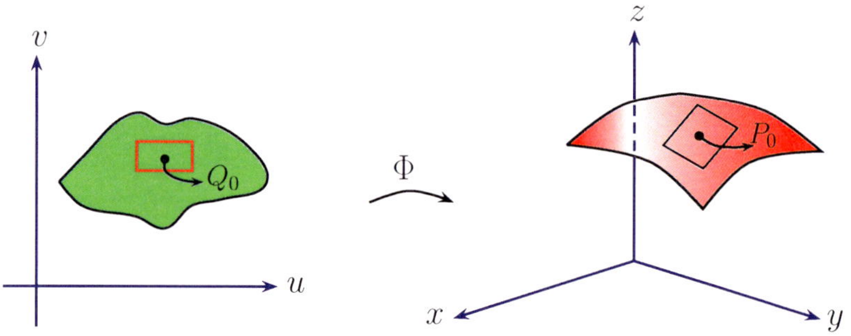

Fig. 6.9. Motivating the definition of surface area.

Keeping the above motivation in mind, we proceed to define the area of a surface. Let a parametrically defined surface S in $\mathbb{R}^3$ be given by $(x(u,v), y(u,v), z(u,v))$, $(u,v) \in E$, where E is a bounded subset of $\mathbb{R}^2$ such that ∂E is of content zero. We say that S is a **smooth surface** if the functions x, y, z are defined on an open subset Ω of $\mathbb{R}^2$ containing E and have continuous first-order partial derivatives on Ω. In this case, the **surface area** of S is defined to be

$$\text{Area}(S) := \iint_E \sqrt{J(\Phi_1)^2 + J(\Phi_2)^2 + J(\Phi_3)^2}\, d(u,v),$$

where $\Phi_1, \Phi_2, \Phi_3 : E \to \mathbb{R}^2$ are given by $\Phi_1(u,v) := (y(u,v),\, z(u,v))$, $\Phi_2(u,v) := (z(u,v),\, x(u,v))$, and $\Phi_3(u,v) := (x(u,v),\, y(u,v))$. Note that

$$J(\Phi_1)^2 + J(\Phi_2)^2 + J(\Phi_3)^2 = (y_u z_v - y_v z_u)^2 + (z_u x_v - z_v x_u)^2 + (x_u y_v - x_v y_u)^2.$$

Thus, if we define

$$U := x_u^2 + y_u^2 + z_u^2, \quad V := x_v^2 + y_v^2 + z_v^2, \quad \text{and} \quad W := x_u x_v + y_u y_v + z_u z_v,$$

then using the algebraic identity quoted earlier, it can be easily seen that

$$\text{Area}(S) = \iint_E \sqrt{UV - W^2}\, d(u,v).$$

This formulation is useful in calculating the area of a surface.

We now show that the area of a surface does not change under certain "reparametrizations."

Proposition 6.13. *Let E be a bounded subset of $\mathbb{R}^2$ such that ∂E is of content zero, and let S be a smooth surface in $\mathbb{R}^3$ given by $(x(u,v), y(u,v), z(u,v))$, $(u,v) \in E$. Let Ω be an open set containing E on which the functions x, y, and z have continuous first-order partial derivatives. Suppose $\widetilde{\Omega}$ is an open subset of $\mathbb{R}^2$ and $\mathbf{\Psi} := (\psi_1, \psi_2) : \widetilde{\Omega} \to \mathbb{R}^2$ is a one-one function such that $\mathbf{\Psi}(\widetilde{\Omega}) \subseteq \Omega$, ψ_1 and ψ_2 have continuous first-order partial derivatives, and $J(\mathbf{\Psi})(\tilde{u}, \tilde{v}) \neq 0$ for all $(\tilde{u}, \tilde{v}) \in \widetilde{\Omega}$. Let $\widetilde{E} := \Phi^{-1}(E)$, and let $\widetilde{S}$ denote the surface given by $(\tilde{x}(\tilde{u},\tilde{v}),\ \tilde{y}(\tilde{u},\tilde{v}),\ \tilde{z}(\tilde{u},\tilde{v}))$, $(\tilde{u},\tilde{v}) \in \widetilde{E}$, where $\tilde{x} := x \circ \Phi$, $\tilde{y} := y \circ \Phi$, and $\tilde{z} := z \circ \Phi$. Then $\widetilde{S}$ is a smooth surface and* $\text{Area}(\widetilde{S}) = \text{Area}(S)$.

Proof. Let $\Phi_1, \Phi_2, \Phi_3 : \Omega \to \mathbb{R}^2$ be defined by $\Phi_1(u,v) := (y(u,v),\ z(u,v))$, $\Phi_2(u,v) := (z(u,v),\ x(u,v))$, and $\Phi_3(u,v) := (x(u,v),\ y(u,v))$ for $(u,v) \in \Omega$. Define $\widetilde{\Phi}_1, \widetilde{\Phi}_2, \widetilde{\Phi}_3 : \widetilde{\Omega} \to \mathbb{R}$ by $\widetilde{\Phi}_i := \Phi_i \circ \mathbf{\Psi}$ for $i = 1, 2, 3$. By the Chain Rule (Proposition 3.51) together with Remark 3.52, we obtain

$$J(\widetilde{\Phi}_i)(\tilde{u},\tilde{v}) = J(\Phi_i)(\mathbf{\Psi}(\tilde{u},\tilde{v}))J(\mathbf{\Psi})(\tilde{u},\tilde{v}), \qquad \text{for } (\tilde{u},\tilde{v}) \in \widetilde{\Omega} \text{ and } i = 1, 2, 3.$$

Hence Proposition 5.59 (Change of Variables by Affine Transformations) gives

$$\begin{aligned}
\text{Area}(S) &= \iint_E \sqrt{J(\Phi_1)^2 + J(\Phi_2)^2 + J(\Phi_3)^2}\, d(u,v) \\
&= \iint_{\widetilde{E}} \sqrt{J(\Phi_1 \circ \mathbf{\Psi})^2 + J(\Phi_2 \circ \mathbf{\Psi})^2 + J(\Phi_3 \circ \mathbf{\Psi})^2}\, |J(\mathbf{\Psi})|\, d(\tilde{u},\tilde{v}) \\
&= \iint_{\widetilde{E}} \sqrt{J(\widetilde{\Phi}_1)^2 + J(\widetilde{\Phi}_2)^2 + J(\widetilde{\Phi}_3)^2}\, d(\tilde{u},\tilde{v}) \\
&= \text{Area}(\widetilde{S}),
\end{aligned}$$

as desired. □

Let us consider three special cases of parametrically defined surfaces involving Cartesian, cylindrical, or spherical coordinates.

Cartesian Coordinates: Let D be a bounded subset of $\mathbb{R}^2$ such that ∂D is of content zero, and let Ω be an open subset of $\mathbb{R}^2$ containing D, and $f : \Omega \to \mathbb{R}$ a function having continuous first-order partial derivatives. Then for the smooth surface S given by $z = f(x, y)$, $(x, y) \in D$,

$$\text{Area}(S) = \iint_D \sqrt{1 + f_x^2 + f_y^2}\, d(x, y).$$

This follows by considering the parametrization of S given by $x(u, v) := u$, $y(u, v) := v$, $z(u, v) := f(u, v)$ for $(u, v) \in D$.

Similar expressions can be written down for smooth surfaces given by equations of the form $x = f(y, z)$ and $y = f(z, x)$.

Cylindrical Coordinates: Let E be a bounded subset of $[-\pi, \pi] \times \mathbb{R}$ such that ∂E is of content zero, and let Ω be an open subset of $\mathbb{R}^2$ containing E, and $g : \Omega \to \mathbb{R}$ a nonnegative function having continuous first-order partial derivatives. Assume that $g(-\pi, z) = g(\pi, z)$ whenever $z \in \mathbb{R}$, and both $(-\pi, z)$ and (π, z) are in E. Then for the smooth surface S given, in cylindrical coordinates, by $r = g(\theta, z)$, $(\theta, z) \in E$,

$$\text{Area}(S) = \iint_E \sqrt{g_\theta^2 + g^2(g_z^2 + 1)}\, d(\theta, z).$$

This follows by considering the parametrization of S given by $x(u, v) := g(u, v) \cos u$, $y(u, v) := g(u, v) \sin u$, $z(u, v) := v$ for $(u, v) \in E$ and noting that $x_u y_v - x_v y_u = -g g_v$, $y_u z_v - y_v z_u = g_u \sin v + g \cos v$, and $z_u x_v - z_v x_u = g_u \cos v - g \sin v$.

Similar expressions can be written down for smooth surfaces given, in cylindrical coordinates, by equations of the form $\theta = \alpha(r, z)$ and $z = h(r, \theta)$. (See Exercises 34 and 35.)

Spherical Coordinates: Let E be a bounded subset of $[0, \pi] \times [-\pi, \pi]$ such that ∂E is of content zero, and let Ω be an open subset of $\mathbb{R}^2$ containing E, and $p : \Omega \to \mathbb{R}$ a nonnegative function having continuous first-order partial derivatives. Then for the smooth surface S given, in spherical coordinates, by $\rho = p(\varphi, \theta)$, $(\varphi, \theta) \in E$,

$$\text{Area}(S) = \iint_E p\sqrt{\sin^2 \varphi (p^2 + p_\varphi^2) + p_\theta^2}\, d(\varphi, \theta).$$

This follows by considering the parametrization of S given by $x(u, v) = p(u, v) \sin u \cos v$, $y(u, v) = p(u, v) \sin u \sin v$, and $z(u, v) = p(u, v) \cos u$ for $(u, v) \in E$ and noting that $x_u^2 + y_u^2 + z_u^2 = p^2 + p_u^2$, $x_v^2 + y_v^2 + z_v^2 = p^2 \sin^2 u + p_v^2$, and $x_u x_v + y_u y_v + z_u z_v = p_u p_v$.

Similar expressions can be written down for smooth surfaces given, in spherical coordinates, by equations of the form $\theta = \alpha(\rho, \varphi)$ and $\varphi = \gamma(\rho, \theta)$. (See Exercises 36 and 37.)

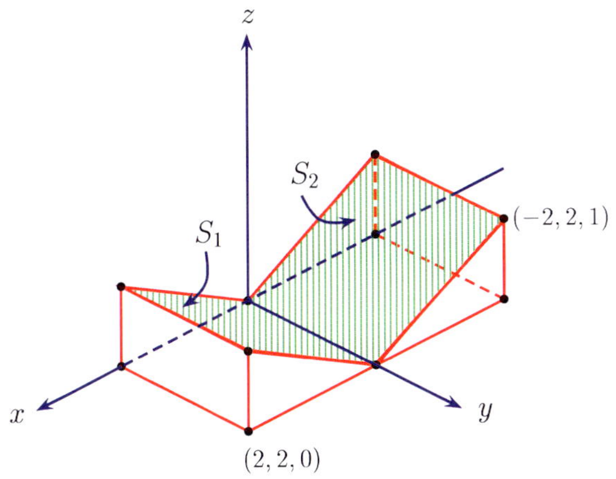

Fig. 6.10. A piecewise smooth surface that is not smooth.

Remark 6.14. The notion of the area of a smooth surface can be extended to more general surfaces as follows. Let a parametrically defined surface S in $\mathbb{R}^3$ be given by $(x(u,v), y(u,v), z(u,v))$, $(u,v) \in E$, where E is a bounded subset of $\mathbb{R}^2$ such that ∂E is of content zero. Then S is said to be **piecewise smooth** if the functions $x, y, z : E \to \mathbb{R}$ are continuous and there are finitely many smooth surfaces that together constitute the surface S. More precisely, there is $n \in \mathbb{N}$ and for $i = 1, \ldots, n$, there is a surface S_i given by $(x_i(u,v), y_i(u,v), z_i(u,v))$, $(u,v) \in E_i$, such that $E = \bigcup_{i=1}^n E_i$, ∂E_i is of content zero for $i = 1, \ldots, n$, $E_i \cap E_j$ is of content zero whenever $i \neq j$, and we have $x_i(u,v) = x(u,v)$, $y_i(u,v) = y(u,v)$, $z_i(u,v) = z(u,v)$ for all $(u,v) \in E_i$ and $i = 1, \ldots, n$. In this case, the area of S is defined to be

$$\text{Area}(S) := \sum_{i=1}^{n} \text{Area}(S_i).$$

In view of Propositions 5.51 and 5.54, we may write

$$\text{Area}(S) := \iint_E \sqrt{J(\Phi_1)^2 + J(\Phi_2)^2 + J(\Phi_3)^2}\, d(u,v),$$

where $\Phi_1, \Phi_2, \Phi_3 : E \to \mathbb{R}^2$ are given by $\Phi_1(u,v) := (y(u,v), z(u,v))$, $\Phi_2(u,v) := (z(u,v), x(u,v))$, and $\Phi_3(u,v) := (x(u,v), y(u,v))$. A simple example of a piecewise smooth surface that is not smooth is depicted in Figure 6.10. More precisely, let $E := [-2,2] \times [0,2]$ and let $x, y, z : E \to \mathbb{R}$ be given by $x(u,v) := u$, $y(u,v) := v$, and $z(u,v) := u/2$ if $u \geq 0$, while $z(u,v) := -u/2$ if $u < 0$. Let $E_1 := [0,2] \times [0,2]$ and consider the surface S_1 given by $x(u,v) := u$, $y(u,v) := v$, $z(u,v) := u/2$ for $(u,v) \in E_1$. Also, let $E_2 := [-2,0] \times [0,2]$ and consider the surface S_2 given by $x(u,v) := u$, $y(u,v) := v$, $z(u,v) := -u/2$ for $(u,v) \in E_2$. Then it is clear that S_1 and S_2 are smooth surfaces, and together they define the piecewise smooth surface S. $\diamond$

Examples 6.15. (i) Let $a \in \mathbb{R}$ with $a > 0$ and let S denote the part of the paraboloid given by $z = x^2 + y^2$ that is cut out by the cylinder given by $x^2 + y^2 = a^2$. If we let $D := \{(x, y) \in \mathbb{R}^2 : x^2 + y^2 \leq a^2\}$ and $f(x, y) := x^2 + y^2$ for $(x, y) \in D$, then

$$\text{Area}(S) = \iint_D \sqrt{1 + f_x^2 + f_y^2}\, d(x, y) = \iint_D \sqrt{1 + 4x^2 + 4y^2}\, d(x, y),$$

and using polar coordinates, we obtain

$$\text{Area}(S) = \int_{-\pi}^{\pi} \left(\int_0^a \left(\sqrt{1 + 4r^2} \right) r\, dr \right) d\theta = \frac{\pi}{6} \left[(1 + 4a^2)^{3/2} - 1 \right].$$

(ii) Let $a \in \mathbb{R}$ with $a > 0$ and let E_a denote a bounded subset of $[-\pi, \pi] \times \mathbb{R}$ such that ∂E_a is of content zero. Let S denote the surface given, in cylindrical coordinates, by $r = a$ and $(\theta, z) \in E_a$. Note that S is a part of a right circular cylinder of radius a. If we let $p(\theta, z) := a$ for $(\theta, z) \in E_a$, then $p_\theta = p_z = 0$ and hence

$$\text{Area}(S) = \iint_{E_a} \sqrt{p_\theta^2 + p^2(p_z^2 + 1)}\, d(\theta, z) = \iint_{E_a} a\, d(\theta, z) = a\, \text{Area}(E_a).$$

In particular, if $h \in \mathbb{R}$ with $h > 0$ and $E_a := [-\pi, \pi] \times [0, h]$, then $\text{Area}(E_a) = 2\pi h$ and thus we see that the surface area of a right circular cylinder of radius a and height h is $2\pi h a$.

(iii) Let $a \in \mathbb{R}$ with $a > 0$ and let E_a denote a bounded subset of $[0, \pi] \times [-\pi, \pi]$ such that ∂E_a is of content zero. Let S denote the surface given, in spherical coordinates, by $\rho = a$ for $(\varphi, \theta) \in E_a$. Note that S is a part of a sphere of radius a. If we let $p(\varphi, \theta) := a$ for $(\varphi, \theta) \in E_a$, then $p_\varphi = p_\theta = 0$ and hence

$$\text{Area}(S) = \iint_E p \sqrt{\sin^2 \varphi (p^2 + p_\varphi^2) + p_\theta^2}\, d(\varphi, \theta) = a^2 \iint_{E_a} \sin \varphi\, d(\varphi, \theta).$$

In particular, if we recall (from Remark 8.13 of ACICARA, for example) that the **solid angle** Θ subtended by the surface S at the center of the sphere of radius a is, by definition, the ratio $\text{Area}(S)/a^2$, then we have the following integral formula for the solid angle:

$$\Theta = \iint_{E_a} \sin \varphi\, d(\varphi, \theta).$$

In particular, if $\varphi_0 \in [0, \pi]$ and $E_a := [0, \varphi_0] \times [-\pi, \pi]$, then the area of the spherical cap S is given by

$$\text{Area}(S) = a^2 \int_{-\pi}^{\pi} \left(\int_0^{\varphi_0} \sin \varphi\, d\varphi \right) d\theta = 2\pi a^2 (1 - \cos \varphi_0).$$

As a special case, by taking $\varphi_0 = \pi$, we obtain that the surface area of the sphere of radius a is $4\pi a^2$. ◇

Surfaces of Revolution

In one-variable calculus, the definition of the area of a surface of revolution obtained by revolving a piecewise smooth curve[3] about a line is usually given in terms of a Riemann integral. (See, for example, Section 8.4 of ACICARA.) The following result shows that this definition is consistent with the general definition given in Remark 6.14 of the area of a piecewise smooth surface.

Proposition 6.16. *Let C be a piecewise smooth curve C in $\mathbb{R}^2$ given by $(x(t), y(t))$, $t \in [\alpha, \beta]$, and let L be a line given by $ax + by + c = 0$, where $a, b, c \in \mathbb{R}$, and not both a and b are zero. If L does not cross C and if C is revolved about L, then the area of the surface S so generated is given by*

$$\text{Area}(S) := 2\pi \int_\alpha^\beta \frac{|ax(t) + by(t) + c|}{\sqrt{a^2 + b^2}} \sqrt{x'(t)^2 + y'(t)^2}\, dt.$$

Proof. Because of the domain additivity of Riemann integrals (Proposition 6.7 of ACICARA) and the domain additivity of double integrals (Proposition 5.51), there is no loss of generality in assuming that C is a smooth curve.

Since the line L does not cross the curve C, we have $ax(t) + by(t) + c \geq 0$ for all $t \in [\alpha, \beta]$ or $ax(t) + by(t) + c \leq 0$ for all $t \in [\alpha, \beta]$. Let us assume the former. Let R denote the rectangle $[\alpha, \beta] \times [-\pi, \pi]$.

First we prove the proposition in the special case in which the line L is the x-axis, that is, when $a = 0$, $b = 1$, and $c = 0$. In this case, $y(t) \geq 0$ for all $t \in [\alpha, \beta]$ and S is given by $(\xi(t, \theta),\ \eta(t, \theta),\ \zeta(t, \theta))$, $(t, \theta) \in R$, where

$$\xi(t, \theta) := x(t), \quad \eta(t, \theta) := y(t)\cos\theta, \quad \text{and} \quad \zeta(t, \theta) := y(t)\sin\theta.$$

Note that S is a smooth surface. Moreover, for all $(t, \theta) \in R$, we have

$$\begin{aligned}
U &:= \xi_t^2 + \eta_t^2 + \zeta_t^2 = x'(t)^2 + \big(y'(t)\cos\theta\big)^2 + \big(y'(t)\sin\theta\big)^2 = x'(t)^2 + y'(t)^2,\\
V &:= \xi_\theta^2 + \eta_\theta^2 + \zeta_\theta^2 = 0^2 + \big(-y(t)\sin\theta\big)^2 + \big(y(t)\cos\theta\big)^2 = y(t)^2,\\
W &:= \xi_t\xi_\theta + \eta_t\eta_\theta + \zeta_t\zeta_\theta\\
&= x'(t)\cdot 0 + \big(y'(t)\cos\theta\big)\big(-y(t)\sin\theta\big) + \big(y'(t)\sin\theta\big)\big(y(t)\cos\theta\big) = 0.
\end{aligned}$$

Hence we obtain

---

[3] Recall that a parametrically defined curve C given by $(x(t), y(t))$, $t \in [\alpha, \beta]$, is said to be **smooth** if the functions $x, y : [\alpha, \beta] \to \mathbb{R}$ are differentiable and their derivatives are continuous. It is is said to be **piecewise smooth** if the functions x and y are continuous on $[\alpha, \beta]$ and if there are finitely many points $\gamma_0 < \gamma_1 < \cdots < \gamma_n$ in $[\alpha, \beta]$, where $\gamma_0 = \alpha$ and $\gamma_n = \beta$, such that for each $i = 1, \ldots, n$, the curve given by $(x(t), y(t))$, $t \in [\gamma_{i-1}, \gamma_i]$, is smooth. If the curve C is piecewise smooth, then the **length** of C is defined to be $\ell(C) = \int_\alpha^\beta \sqrt{x'(t)^2 + y'(t)^2}\, dt := \sum_{i=1}^n \int_{\gamma_{i-1}}^{\gamma_i} \sqrt{x'(t)^2 + y'(t)^2}\, dt$.

$$\text{Area}(S) = \iint_R \sqrt{UV - W^2}\, d(t, \theta) = \int_{-\pi}^{\pi} \left(\int_{\alpha}^{\beta} |y(t)| \sqrt{x'(t)^2 + y'(t)^2}\, dt \right) d\theta,$$

and thus $\text{Area}(S) = 2\pi \int_\alpha^\beta y(t)\sqrt{x'(t)^2 + y'(t)^2}\, dt$, as desired.

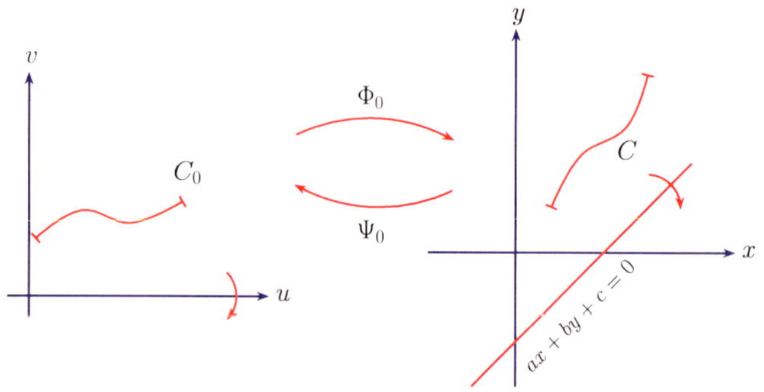

Fig. 6.11. Adjusting the axis of revolution of a smooth curve.

Let us now consider the case in which $a, b, c \in \mathbb{R}$ satisfy $a^2 + b^2 = 1$. We show that by a suitable change of variables, the line L can be assumed to be the x-axis. Let us use the affine functions $\Phi_0, \Psi_0 : \mathbb{R}^2 \to \mathbb{R}^2$ and $\Phi : \mathbb{R}^3 \to \mathbb{R}^3$ introduced in the proof of Proposition 6.10. If $\Psi_0 := (\psi_1, \psi_2)$, then we have

$$\psi_1(x, y) = bx - ay - c \quad \text{and} \quad \psi_2(x, y) = ax + by + c \quad \text{for } (x, y) \in \mathbb{R}^2.$$

Consider the curve C_0 given by $(u(t), v(t))$, $t \in [\alpha, \beta]$, where

$$u(t) := \psi_1(x(t), y(t)) \quad \text{and} \quad v(t) := \psi_2(x(t), y(t)), \quad t \in [\alpha, \beta].$$

(See Figure 6.11.) Note that $v(t) \geq 0$ for all $t \in [\alpha, \beta]$ and let S_0 denote the surface in $\mathbb{R}^3$ obtained by revolving the curve C_0 about the u-axis. Thus by the special case considered earlier,

$$\text{Area}(S_0) = \int_\alpha^\beta |v(t)| \sqrt{u'(t)^2 + v'(t)^2} dt.$$

Now, for $t \in [\alpha, \beta]$, we have $v(t) = ax(t) + by(t) + c$ and

$$u'(t)^2 + v'(t)^2 = \big(bx'(t) - ay'(t)\big)^2 + \big(ax'(t) + by'(t)\big)^2 = x'(t)^2 + y'(t)^2,$$

where the last equality follows since $a^2 + b^2 = 1$. Hence

$$\text{Area}(S_0) = \int_\alpha^\beta |ax(t) + by(t) + c| \sqrt{x'(t)^2 + y'(t)^2} dt.$$

On the other hand, since S_0 is obtained by revolving C_0 around the u-axis, it is given by $(\xi_0(t, \theta), \eta_0(t, \theta), \zeta_0(t, \theta))$, $(t, \theta) \in R$, where

$$\xi_0(t,\theta) := u(t), \quad \eta_0(t,\theta) := v(t)\cos\theta, \quad \text{and} \quad \zeta_0(t,\theta) := v(t)\sin\theta.$$

Further, the surface S obtained by revolving the curve C about the line L is given by

$$(\xi(t,\theta), \eta(t,\theta), \zeta(t,\theta)) := \Phi(\xi_0(t,\theta), \eta_0(t,\theta), \zeta_0(t,\theta)), \quad (t,\theta) \in R.$$

Since $\Phi(u,v,w) = \big(bu+av-c(a-b), -au+bv-c(a+b), w\big)$ for $(u,v,w) \in \mathbb{R}^3$, we see that

$$\begin{aligned}
\xi(t,\theta) &= b\xi_0(t,\theta) + a\eta_0(t,\theta) - c(a-b),\\
\eta(t,\theta) &= -a\xi_0(t,\theta) + b\eta_0(t,\theta) - c(a+b),\\
\zeta(t,\theta) &= \zeta_0(t,\theta) \quad \text{for all } (t,\theta) \in R.
\end{aligned}$$

Thus, if we let

$$U_0 := \left(\frac{\partial \xi_0}{\partial t}\right)^2 + \left(\frac{\partial \eta_0}{\partial t}\right)^2 + \left(\frac{\partial \zeta_0}{\partial t}\right)^2, \quad V_0 := \left(\frac{\partial \xi_0}{\partial \theta}\right)^2 + \left(\frac{\partial \eta_0}{\partial \theta}\right)^2 + \left(\frac{\partial \zeta_0}{\partial \theta}\right)^2$$

and

$$W_0 := \left(\frac{\partial \xi_0}{\partial t}\right)\left(\frac{\partial \xi_0}{\partial \theta}\right) + \left(\frac{\partial \eta_0}{\partial t}\right)\left(\frac{\partial \eta_0}{\partial \theta}\right) + \left(\frac{\partial \zeta_0}{\partial t}\right)\left(\frac{\partial \zeta_0}{\partial \theta}\right),$$

then we see that

$$U := \xi_t^2 + \eta_t^2 + \zeta_t^2 = \left(b\frac{\partial \xi_0}{\partial t} + a\frac{\partial \eta_0}{\partial t}\right)^2 + \left(-a\frac{\partial \xi_0}{\partial t} + b\frac{\partial \eta_0}{\partial t}\right)^2 + \left(\frac{\partial \zeta_0}{\partial t}\right)^2 = U_0,$$

where the last equality follows since $a^2 + b^2 = 1$. In a similar manner we see that $V := \xi_\theta^2 + \eta_\theta^2 + \zeta_\theta^2 = V_0$ and $W := \xi_t\xi_\theta + \eta_t\eta_\theta + \zeta_t\zeta_\theta = W_0$. Consequently, $UV - W^2 = U_0V_0 - W_0^2$, and therefore $\text{Area}(S) = \text{Area}(S_0)$. This yields the desired formula for $\text{Area}(S)$ in the case $a^2 + b^2 = 1$. If $a^2 + b^2 \neq 1$, then we replace a, b, and c by by $a/\sqrt{a^2+b^2}$, $b/\sqrt{a^2+b^2}$, and $c/\sqrt{a^2+b^2}$, respectively, and obtain the desired result. □

Examples in which areas of surfaces of revolution are computed can be found in books on one-variable calculus; in particular, see Examples 8.14 (i), (ii), and (iii) of ACICARA.

Remark 6.17. Let a, b, c be positive real numbers and consider the ellipsoid $\big\{(x,y,z) \in \mathbb{R}^3 : (x^2/a^2) + (y^2/c^2) + (z^2/c^2) = 1\big\}$. If $R := [0,\pi] \times [-\pi,\pi]$, then this ellipsoid is the image of the surface S parametrically given by $(x(u,v),\ y(u,v),\ z(u,v))$, $(u,v) \in \mathbb{R}$, where

$$x(u,v) := a\sin u\cos v, \quad y(u,v) := b\sin u\sin v, \quad z(u,v) := c\cos u.$$

In general, S is not a surface of revolution. We can easily check that

$$U := x_u^2 + y_u^2 + z_u^2 = \cos^2 u\big(a^2\cos^2 v + b^2\sin^2 v\big) + c^2\sin^2 u,$$
$$V := x_v^2 + y_v^2 + z_v^2 = \sin^2 u\big(a^2\sin^2 v + b^2\cos^2 v\big),$$
$$W := x_u x_v + y_u y_v + z_u z_v = \big(b^2 - a^2\big)\cos u\sin u\cos v\sin v$$

for all $(u, v) \in R$, and so

$$UV - W^2 = \sin^2 u\big[a^2b^2\cos^2 u + c^2\sin^2 u\big(a^2\sin^2 v + b^2\cos^2 v\big)\big].$$

To calculate the area of the ellipsoid S, we need to evaluate the double integral

$$\text{Area}(S) = \iint_R \sin u\sqrt{a^2b^2\cos^2 u + c^2\sin^2 u\big(a^2\sin^2 v + b^2\cos^2 v\big)}\, d(u, v).$$

If $a, b,$ and c are distinct, then this integral cannot be evaluated in terms of elementary functions. In fact, we are led to consider the so-called elliptic functions. If $c = b = a$, then S is a sphere of radius a, and its area is equal to

$$\text{Area}(S) = \iint_R \sin u\sqrt{a^4}\, d(u, v) = a^2\int_{-\pi}^{\pi}\left(\int_0^{\pi}\sin u\, du\right)dv = 4\pi a^2.$$

Now let $c = b$. Then S is the spheroid obtained by revolving the ellipse given by $(x^2/a^2) + (y^2/b^2) = 1$ about the x-axis. It can be shown (as in Example 8.14 (ii) of ACICARA) that if $b < a$, then

$$\text{Area}(S) = 2\pi b^2 + \frac{2\pi a^2 b}{\sqrt{a^2 - b^2}}\sin^{-1}\left(\frac{\sqrt{a^2 - b^2}}{a}\right),$$

whereas if $b > a$, then

$$\text{Area}(S) = 2\pi b^2 + \frac{2\pi a^2 b}{\sqrt{b^2 - a^2}}\ln\left(\frac{b + \sqrt{b^2 - a^2}}{a}\right).$$

Note that $\text{Area}(S) \to 2\pi a^2 + 2\pi a^2 = 4\pi a^2$ as $b \to a$. ◇

6.3 Centroids of Surfaces and Solids

The centroid, also known as the center of gravity or the barycenter, of a body is the center of its mass or the point at which the body will balance itself when placed on a needle. For example, the centroid of a triangular region is the point of intersection of its three medians. For planar regions or more generally, for surfaces and for solids, the centroid can be precisely defined and effectively calculated using integrals. In effect, the coordinates of a centroid are the weighted averages of the corresponding coordinate functions. We begin this section with a brief discussion of averages and weighted averages, and follow it up by the definitions and examples of centroids of bodies of the following types: (i) planar regions, (ii) surfaces in 3-space, and (iii) solids. It may be noted that the treatment here extends the notion of centroid to more general situations than those given in Section 8.5 of ACICARA.

Averages and Weighted Averages

The notion of average of finitely many values is elementary and well known. In turn, this leads to the notion of average value of a function $f : D \to \mathbb{R}$ when the domain D is finite, say the finite subset $\{(a_1, b_1), \dots, (a_n, b_n)\}$ of $\mathbb{R}^2$, where $n \in \mathbb{N}$. The average of f is then given by

$$\mathrm{Av}(f) := \frac{f(a_1, b_1) + \cdots + f(a_n, b_n)}{n}.$$

Alternatively, if $z_1, \dots, z_k$ are the distinct values of f and if $w_1, \dots, w_k$ are the corresponding weights, that is, w_i is the number of elements in the set $\{(a, b) \in D : f(a, b) = z_i\}$ for $i = 1, \dots, k$, then $w_1 + \cdots + w_k = n$ and

$$\mathrm{Av}(f) := \frac{w_1 z_1 + \cdots + w_k z_k}{w_1 + \cdots + w_k}.$$

Thus, $\mathrm{Av}(f)$ could also be viewed as the "weighted average" of the k values $z_1, \dots, z_k$, where the "weight function" is the map $w : \{1, \dots, k\} \to \mathbb{R}$ given by $i \longmapsto w_i$. Simple examples show that $\mathrm{Av}(f)$ need not be a value of f.

To pass from the discrete to the continuous case, first suppose $D := [a, b] \times [c, d]$ is a rectangle in $\mathbb{R}^2$ and $f : D \to \mathbb{R}$ is any function. We can subdivide D into small subrectangles and assume, for simplicity, that on each of these subrectangles, f is a constant function. For example, let $P := \{(x_i, y_j) : i = 0, 1, \dots, n \text{ and } j = 0, 1, \dots, k\}$ be a partition of $[a, b] \times [c, d]$ and let (s_i, t_j) for $i = 0, 1, \dots, n$ and $j = 0, 1, \dots, k$ be points in the (i, j)th subrectangle $[x_{i-1}, x_i] \times [y_{j-1}, y_j]$ induced by P. Then the quotient

$$\frac{\sum_{i=1}^{n} \sum_{j=1}^{k} f(s_i, t_j)\,(x_i - x_{i-1})\,(y_j - y_{j-1})}{\sum_{i=1}^{n} \sum_{j=1}^{k} (x_i - x_{i-1})\,(y_j - y_{j-1})}$$

can be construed as the "average value" of f over D. The denominator in the quotient above is simply $(b - a)(d - c)$, that is, the area of D, whereas the numerator approaches the (double) integral of f over D as the partition P becomes finer and finer, provided f is integrable. Moreover, for such a partition P, the assumption that f is a constant function on each of the subrectangles induced by P appears reasonable even for an arbitrary integrable function $f : D \to \mathbb{R}$. With this in view, we make the following definition.

Let D be a bounded subset of $\mathbb{R}^2$ such that D has an area, that is, ∂D is of content zero. Assume that $\mathrm{Area}(D) \neq 0$. The **average** of an integrable function $f : D \to \mathbb{R}$ is defined to be the real number

$$\mathrm{Av}(f) := \frac{1}{\mathrm{Area}(D)} \iint_D f(x, y)d(x, y).$$

More generally, if D is a bounded subset of $\mathbb{R}^2$ and $w : D \to \mathbb{R}$ is an integrable function such that $w \geq 0$ and $\iint_D w(x, y)d(x, y) \neq 0$, then for an integrable function $f : D \to \mathbb{R}$, the **weighted average** of f with respect to w is defined to be the real number

$$\mathrm{Av}(f; w) := \frac{1}{W} \iint_D f(x,y)w(x,y)d(x,y), \quad \text{where } W := \iint_D w(x,y)d(x,y).$$

Note that if ∂D is of content zero and $w := 1$, then $\mathrm{Av}(f; w) = \mathrm{Av}(f)$. If D is a "nice" region (for example, a rectangle) and $f : D \to \mathbb{R}$ is continuous, then it can be easily seen that in contrast to the discrete case, $\mathrm{Av}(f)$ is a value of f (at some point of D). However, in general, simple examples show that a (weighted) average of f need not be a value of f. (Exercises 16 and 17.)

The average and the weighted average of a function defined on a subset of $\mathbb{R}^3$ are defined analogously using triple integrals instead of double integrals.

Centroids of Planar Regions

Consider a bounded subset D of $\mathbb{R}^2$ such that D has an area, that is, ∂D is of content zero. Assume that its area is not equal to zero, that is,

$$\mathrm{Area}(D) := \iint_D d(x,y) \neq 0.$$

Let $f, g : D \to \mathbb{R}$ denote the coordinate functions on D given by $f(x,y) := x$ and $g(x,y) := y$. The **centroid** of D is defined to be $(\overline{x}, \overline{y}) \in \mathbb{R}^2$, where

$$\overline{x} := \mathrm{Av}(f) \quad \text{and} \quad \overline{y} := \mathrm{Av}(g).$$

Thus

$$\overline{x} := \frac{1}{\mathrm{Area}(D)} \iint_D x\, d(x,y) \quad \text{and} \quad \overline{y} := \frac{1}{\mathrm{Area}(D)} \iint_D y\, d(x,y).$$

It may be worthwhile to note a special case of an elementary region given by $D := \{(x,y) \in \mathbb{R}^2 : a \le x \le b \text{ and } f_1(x) \le y \le f_2(x)\}$, where $f_1, f_2 : [a,b] \to \mathbb{R}$ are bounded functions whose sets of discontinuities are of one-dimensional content zero and that satisfy $f_1 \le f_2$. In this case, by Corollary 5.45, ∂D is of (two-dimensional) content zero. Moreover, by Fubini's Theorem for elementary regions (Proposition 5.36), we have

$$\mathrm{Area}(D) = \int_a^b [f_2(x) - f_1(x)]dx,$$

and in case $\mathrm{Area}(D) \neq 0$, the coordinates of the centroid of D are given by

$$\overline{x} = \frac{1}{\mathrm{Area}(D)} \int_a^b \left(\int_{f_1(x)}^{f_2(x)} x\, dy \right) dx = \frac{1}{\mathrm{Area}(D)} \int_a^b x[f_2(x) - f_1(x)]\, dx,$$

$$\overline{y} = \frac{1}{\mathrm{Area}(D)} \int_a^b \left(\int_{f_1(x)}^{f_2(x)} y\, dy \right) dx = \frac{1}{2\mathrm{Area}(D)} \int_a^b [f_2(x)^2 - f_1(x)^2]\, dx.$$

Similar results hold if $D := \{(x, y) \in \mathbb{R}^2 : c \le y \le d \text{ and } g_1(y) \le x \le g_2(y)\}$, where $g_1, g_2 : [c, d] \to \mathbb{R}$ are bounded functions whose sets of discontinuities are of one-dimensional content zero and that satisfy $g_1 \le g_2$. This shows that the definition of centroid we have given is consistent with the definition and/or formulas usually given in texts on one-variable calculus. (See, for example, Section 8.5 of ACICARA.)

Example 6.18. Let D denote a parallelogram in $\mathbb{R}^2$ with vertices (x_i, y_i), $i = 1, 2, 3, 4$. Assume that no three of these four points are collinear and that (x_2, y_2) and (x_3, y_3) are the vertices adjacent to (x_1, y_1). Then it is clear that

$$x_4 = x_2 + x_3 - x_1 \quad \text{and} \quad y_4 = y_2 + y_3 - y_1.$$

As noted in Proposition 5.56, D has an area and

$$\text{Area}(D) = |(x_2 - x_1)(y_3 - y_1) - (x_3 - x_1)(y_2 - y_1)|.$$

To determine the centroid $(\overline{x}, \overline{y})$ of D, we need to compute $\iint_D x\, d(x, y)$ and $\iint_D y\, d(x, y)$. To this end, we transform D to the unit square $E := [0, 1] \times [0, 1]$. This can be done using the affine transformation $\Phi : \mathbb{R}^2 \to \mathbb{R}^2$ given by

$$\Phi(u, v) = \big(x_1 + (x_2 - x_1)u + (x_3 - x_1)v,\ y_1 + (y_2 - y_1)u + (y_3 - y_1)v\big).$$

Observe that Φ maps $(0, 0), (1, 0), (0, 1)$, and $(1, 1)$ to $(x_1, y_1), (x_2, y_2), (x_3, y_3)$, and (x_4, y_4), respectively, and since $\Phi(E)$ must be a parallelogram, it follows that $\Phi(E) = D$. Also, it is easily seen that

$$J(\Phi) = (x_2 - x_1)(y_3 - y_1) - (x_3 - x_1)(y_2 - y_1), \quad \text{and so} \quad |J(\Phi)| = \text{Area}(D).$$

Let $f : D \to \mathbb{R}$ and $g : E \to \mathbb{R}$ be defined by $f(x, y) := x$ and $g := f \circ \Phi$. Using the Change of Variables formula (Proposition 5.59), we obtain

$$\iint_D f(x, y) d(x, y) = \iint_E g(u, v)|J(\Phi)| d(u, v) = \text{Area}(D) \iint_E g(u, v) d(u, v).$$

Now $g(u, v) = x_1 + (x_2 - x_1)u + (x_3 - x_1)v$ for $(u, v) \in E$, and so

$$\begin{aligned}\overline{x} &= \int_0^1 \left(\int_0^1 \big(x_1 + (x_2 - x_1)u + (x_3 - x_1)v\big) du \right) dv \\ &= x_1 + \frac{1}{2}(x_2 - x_1) + \frac{1}{2}(x_3 - x_1) = \frac{x_1 + x_2 + x_3 + x_4}{4}.\end{aligned}$$

In a similar manner, we see that $\overline{y} = (y_1 + y_2 + y_3 + y_4)/4$. ◇

As indicated at the beginning of this section, if D is any triangular region in $\mathbb{R}^2$ with vertices (x_i, y_i), $i = 1, 2, 3$, then the centroid of D is the point $\big((x_1 + x_2 + x_3)/3,\ (y_1 + y_2 + y_3)/3\big)$, that is, the point of intersection of the medians of D. This will be proved in the next section (Corollary 6.30), and subsequently, we shall indicate a method to determine the centroid of a large class of planar regions known as polygonal regions. Using this method, we shall show that the centroid $(\overline{x}, \overline{y})$ of a quadrilateral with vertices (x_i, y_i), $i = 1, \ldots, 4$, need not be given by $\overline{x} = (x_i + x_2 + x_3 + x_4)/4$ and $\overline{y} = (y_1 + y_2 + y_3 + y_4)/4$ (Remark 6.31). A method of finding the centroid of a planar region based on a result of Pappus is given in Exercise 20.

Remark 6.19. Symmetry considerations are often useful in the calculation of centroids. For example, suppose $D \subseteq \mathbb{R}^2$ is a bounded subset of $\mathbb{R}^2$ that has an area and $\text{Area}(D) \neq 0$. If D is invariant under reflection with respect to the y-axis, that is, if $(-x, y) \in D$ whenever $(x, y) \in D$, then the centroid $(\overline{x}, \overline{y})$ of D will necessarily be on the y-axis, that is, $\overline{x} = 0$. To see this, consider the affine transformation $\Phi : \mathbb{R}^2 \to \mathbb{R}^2$ defined by $\Phi(u, v) = (-u, v)$. Then $J(\Phi) = -1$ and $\Phi(D) = D$. Hence by the Change of Variables formula (Proposition 5.59) applied to $f : D \to \mathbb{R}$ defined by $f(x, y) := x$, we see that

$$\iint_D x \, d(x, y) = \iint_D f \circ \Phi(u, v) |J(\Phi)| d(u, v) = \iint_D -u \, d(u, v).$$

Consequently, $\overline{x} = -\overline{x}$, that is, $\overline{x} = 0$. In a similar manner, we see that if $(x, -y) \in D$ whenever $(x, y) \in D$, then we have $\overline{y} = 0$.

For example, let $a \in \mathbb{R}$ with $a > 0$, and let $(\overline{x}, \overline{y})$ denote the centroid of the semidisk $D := \{(x, y) \in \mathbb{R}^2 : y \geq 0 \text{ and } x^2 + y^2 \leq a^2\}$. By symmetry, we can immediately conclude that $\overline{x} = 0$. On the other hand, $\overline{y} \neq 0$, and in fact, an easy calculation shows that $\overline{y} = 4a/3\pi$. ◇

Centroids of Surfaces

Let a smooth surface S in $\mathbb{R}^3$ be given by $(x(u, v), y(u, v), z(u, v))$, $(u, v) \in E$, where E is a bounded subset of $\mathbb{R}^2$ such that ∂E is of content zero. Assume that the surface area of S is not equal to zero, that is,

$$\text{Area}(S) := \iint_E \sqrt{UV - W^2} \, d(u, v) \neq 0,$$

where $U, V, W : E \to \mathbb{R}$ are defined, as usual, by

$$U := x_u^2 + y_u^2 + z_u^2, \quad V := x_v^2 + y_v^2 + z_v^2, \quad \text{and} \quad W := x_u x_v + y_u y_v + z_u z_v.$$

Let $w : E \to \mathbb{R}$ be defined by $w := \sqrt{UV - W^2}$. Then the **centroid** of S is defined to be $(\overline{x}, \overline{y}, \overline{z}) \in \mathbb{R}^3$, where

$$\overline{x} := \text{Av}(x; w), \quad \overline{y} := \text{Av}(y; w), \quad \text{and} \quad \overline{z} := \text{Av}(z; w).$$

Thus, for example,

$$\overline{x} = \frac{1}{\text{Area}(S)} \iint_E x(u, v) \, w(u, v) \, d(u, v) = \frac{1}{\text{Area}(S)} \iint_E x\sqrt{UV - W^2}.$$

As with the area of a surface (Remark 6.14), the above definition of the centroid readily extends to the more general case in which S is a piecewise smooth surface. This may be tacitly assumed in some of the examples below.

As in Remark 6.19, symmetry considerations can be used in the calculation of centroids of surfaces, and these can be justified using the Change of Variables formula (Proposition 5.59). Roughly speaking, if the surface is invariant with respect to reflection along the yz-plane, that is, if $(-x, y, z)$ is on the surface whenever (x, y, z) is on it, then $\overline{x} = 0$. More precisely, if S, E, and $x, y, z, U, V, W, w : E \to \mathbb{R}$ are as above and if $\Phi : \mathbb{R}^2 \to \mathbb{R}^2$ is an affine transformation such that $|J(\Phi)| = 1$ and $\Phi(E) = E$ and moreover, $(x(\Phi(u,v)),\ y(\Phi(u,v)),\ z(\Phi(u,v))) = (-x(u,v),\ y(u,v),\ z(u,v))$, and $w(\Phi(u,v)) = w(u,v)$ for all $(u,v) \in E$, then the x-coordinate $\overline{x}$ of the centroid of S is zero. Similar results hold for $\overline{y}$ and $\overline{z}$.

It may be worthwhile to note a special case of the general formula for the centroid of a surface. Suppose the surface S is given by $z = f(x,y)$, $(x,y) \in D$, where D is a bounded subset of $\mathbb{R}^2$ that has an area and f is a real-valued function of two variables defined on an open subset of $\mathbb{R}^2$ containing D such that f has continuous first-order partial derivatives. Then as in Section 6.2,

$$\text{Area}(S) = \iint_D \sqrt{1 + f_x^2 + f_y^2}\, d(x,y).$$

Moreover, if we let $A := \text{Area}(S)$ and assume that $A \neq 0$, then it is readily seen that

$$\overline{x} = \frac{1}{A} \iint_D x\sqrt{1 + f_x^2 + f_y^2}\, d(x,y) \quad \text{and} \quad \overline{y} = \frac{1}{A} \iint_D y\sqrt{1 + f_x^2 + f_y^2}\, d(x,y),$$

whereas

$$\overline{z} = \frac{1}{A} \iint_D f(x,y)\sqrt{1 + f_x^2 + f_y^2}\, d(x,y).$$

Examples 6.20. (i) Let S be the surface in Example 6.15 (i), that is, let S be the surface given by $z = f(x,y)$, $(x,y) \in D$, where $D := \{(x,y) \in \mathbb{R}^2 : x^2 + y^2 \le a^2\}$ and $f : D \to \mathbb{R}$ is defined by $f(x,y) := x^2 + y^2$. Denote the surface area of S by A. We have seen that $A = \pi\big[(1 + 4a^2)^{3/2} - 1\big]/6$. Moreover, if $(\overline{x}, \overline{y}, \overline{z})$ denotes the centroid of S, then as noted above,

$$\overline{x} = \frac{1}{A} \iint_D x\sqrt{1 + f_x^2 + f_y^2}\, d(x,y) = \frac{1}{A} \iint_D x\sqrt{1 + 4x^2 + 4y^2}\, d(x,y) = 0,$$

where the last equality follows from switching to polar coordinates and noting that $\int_{-\pi}^{\pi} \cos\theta\, d\theta = 0$. In a similar manner, we see that $\overline{y} = 0$. Equivalently, we could have deduced from symmetry that $\overline{x} = \overline{y} = 0$. On the other hand, as noted above,

$$\overline{z} = \frac{1}{A} \iint_D f(x,y)\sqrt{1 + f_x^2 + f_y^2}\, d(x,y) = \frac{1}{A} \int_{-\pi}^{\pi} \left(\int_0^a r^2 \sqrt{1 + 4r^2}\; r\, dr \right) d\theta.$$

Since the integral inside the parentheses is equal to

$$\frac{1}{32}\int_1^{1+4a^2}\sqrt{s}(s-1)ds = \frac{1}{80}\left((1+4a^2)^{5/2}-1\right) - \frac{1}{48}\left((1+4a^2)^{3/2}-1\right),$$

we conclude that

$$\overline{z} = \frac{3}{20}\frac{\left((1+4a^2)^{5/2}-1\right)}{\left((1+4a^2)^{3/2}-1\right)} - \frac{1}{4}.$$

(ii) Let $a \in \mathbb{R}$ with $a > 0$ and let S denote the right circular cylinder given by $x^2 + y^2 = a^2$ and $0 \le z \le h$. If we let $E := [-\pi, \pi] \times [0, h]$, then S is parametrically given by $(a\cos\theta,\ a\sin\theta,\ z)$, $(\theta, z) \in E$. We have seen in Example 6.15 (ii) that $\sqrt{UV - W^2} = a$ and Area$(S) = 2\pi ah$. Now, using symmetry (or alternatively, a direct calculation), we have $\overline{x} = \overline{y} = 0$, while

$$\overline{z} = \frac{1}{\text{Area}(S)}\iint_E az\, d(\theta, z) = \frac{a}{2\pi ah}\left(\int_{-\pi}^{\pi} d\theta\right)\left(\int_0^h z\, dz\right) = \frac{h}{2}.$$

Thus $(0, 0, h/2)$ is the centroid of S.

(iii) Let $a \in \mathbb{R}$ with $a > 0$ and let S denote the sphere given by $x^2 + y^2 + z^2 = a^2$. If we let $E := [0, \pi] \times [-\pi, \pi]$, then S is parametrically given by $(a\sin\varphi\cos\theta,\ a\sin\varphi\sin\theta,\ a\cos\varphi)$, $(\varphi, \theta) \in E$. We have seen in Example 6.15 (iii) that $\sqrt{UV - W^2} = a^2\sin\varphi$ for $(\varphi, \theta) \in E$ and Area$(S) = 4\pi a^2$. Using symmetry (or alternatively, a direct calculation), we see in this case that $\overline{x} = \overline{y} = \overline{z} = 0$, that is, the origin is the centroid of S. $\diamond$

We remark that in all three examples given above, the centroid $(\overline{x}, \overline{y}, \overline{z})$ of the surface S does not lie on S, that is, if the surface S is given by $(x(u, v), y(u, v), z(u, v))$, $(u, v) \in E$, then there is no $(u_0, v_0) \in E$ such that $(\overline{x}, \overline{y}, \overline{z}) = (x(u_0, v_0), y(u_0, v_0), z(u_0, v_0))$.

Let us consider the case in which the surface S is a surface of revolution. Let C be a piecewise smooth curve in $\mathbb{R}^2$ given by $(x(t), y(t))$, $t \in [\alpha, \beta]$, and let L be a line in $\mathbb{R}^2$ that does not cross C. Let the line L be given by the equation $ax + by + c = 0$, where $a, b, c \in \mathbb{R}$ and we assume, for simplicity, that $a^2 + b^2 = 1$. Let S denote the surface obtained by revolving the curve C about the line L. If $R := [\alpha, \beta] \times [-\pi, \pi]$, then as in the proof of Proposition 6.16, the surface S is given by $(\xi(t, \theta),\ \eta(t, \theta),\ \zeta(t, \theta))$, $(t, \theta) \in R$, where

$$\begin{aligned}
\xi(t, \theta) &:= b\big(bx(t) - ay(t) - c\big) + a\big(ax(t) + by(t) + c\big)\cos\theta - c(a - b),\\
\eta(t, \theta) &:= -a\big(bx(t) - ay(t) - c\big) + b\big(ax(t) + by(t) + c\big)\cos\theta - c(a + b),\\
\zeta(t, \theta) &:= \big(ax(t) + by(t) + c\big)\sin\theta
\end{aligned}$$

for $(t, \theta) \in R$. By Proposition 6.16,

$$\begin{aligned}\text{Area}(S) &= \iint_R \big|ax(t)+by(t)+c\big|\sqrt{x'(t)^2+y'(t)^2}\,d(t,\theta)\\ &= 2\pi\int_\alpha^\beta \big|ax(t)+by(t)+c\big|\sqrt{x'(t)^2+y'(t)^2}\,dt.\end{aligned}$$

Assume that $\text{Area}(S)\neq 0$. Since $\int_{-\pi}^{\pi}\cos\theta\,d\theta = 0 = \int_{-\pi}^{\pi}\sin\theta\,d\theta$, we obtain

$$\begin{aligned}\overline{x} &= \frac{1}{\text{Area}(S)}\iint_R \xi(t,\theta)\big|ax(t)+by(t)+c\big|\sqrt{x'(t)^2+y'(t)^2}\,d(t,\theta)\\ &= \frac{2\pi}{\text{Area}(S)}\int_\alpha^\beta \big[b\big(bx(t)-ay(t)\big)-ac\big]\big|ax(t)+by(t)+c\big|\sqrt{x'(t)^2+y'(t)^2}\,dt,\\ \overline{y} &= \frac{1}{\text{Area}(S)}\iint_R \eta(t,\theta)\big|ax(t)+by(t)+c\big|\sqrt{x'(t)^2+y'(t)^2}\,d(t,\theta)\\ &= \frac{2\pi}{\text{Area}(S)}\int_\alpha^\beta \big[a\big(ay(t)-bx(t)\big)-bc\big]\big|ax(t)+by(t)+c\big|\sqrt{x'(t)^2+y'(t)^2}\,dt,\\ \overline{z} &= \frac{1}{\text{Area}(S)}\iint_R \zeta(t,\theta)\big|ax(t)+by(t)+c\big|\sqrt{x'(t)^2+y'(t)^2}\,d(t,\theta) = 0.\end{aligned}$$

Of course, we could as well have concluded that $\overline{z}=0$ by symmetry. It can be checked easily that $(\overline{x},\overline{y})$ lies on the line given by $ax+by+c=0$.

In case a^2+b^2 is nonzero, but not necessarily equal to 1, we may replace a, b, and c by $a/\sqrt{a^2+b^2}$, $b/\sqrt{a^2+b^2}$, and $c/\sqrt{a^2+b^2}$, respectively, and obtain

$$\begin{aligned}\overline{x} &= \frac{\int_\alpha^\beta \big[b\big(bx(t)-ay(t)\big)-ac\big]\big|ax(t)+by(t)+c\big|\sqrt{x'(t)^2+y'(t)^2}\,dt}{(a^2+b^2)\int_\alpha^\beta \big|ax(t)+by(t)+c\big|\sqrt{x'(t)^2+y'(t)^2}\,dt},\\ \overline{y} &= \frac{\int_\alpha^\beta \big[a\big(ay(t)-bx(t)\big)-bc\big]\big|ax(t)+by(t)+c\big|\sqrt{x'(t)^2+y'(t)^2}\,dt}{(a^2+b^2)\int_\alpha^\beta \big|ax(t)+by(t)+c\big|\sqrt{x'(t)^2+y'(t)^2}\,dt},\\ \overline{z} &= 0.\end{aligned}$$

Special cases of the above formulas for $\overline{x}$ and $\overline{y}$ and corresponding examples are given in Section 8.5 of ACICARA, notably Examples 8.15 (iii) and (iv).

Centroids of Solids

Let D be a bounded subset of $\mathbb{R}^3$ such that D has a volume, that is, ∂D is of three-dimensional content zero. Suppose the volume of D is nonzero, that is,

$$\text{Vol}(D) := \iiint_D d(x,y,z)\neq 0.$$

Let $f,g,h: D\to\mathbb{R}$ be the coordinate functions on D given by $f(x,y,z):=x$, $g(x,y,z):=y$, and $h(x,y,z):=z$. Then the centroid of D is defined to be $(\overline{x},\overline{y},\overline{z})\in\mathbb{R}^3$, where $\overline{x}:=\text{Av}(f)$, $\overline{y}:=\text{Av}(g)$, and $\overline{z}:=\text{Av}(h)$. Thus, for example,

$$\overline{x} := \frac{1}{\text{Vol}(D)} \iiint_D x \, d(x, y, z).$$

As in Remark 6.19, symmetry considerations can be used in the calculation of centroids of solids and these can be justified using the Change of Variables formula for triple integrals (Proposition 5.70). In effect, if a solid $D \subseteq \mathbb{R}^3$ is invariant with respect to reflection along the yz-plane, that is, if $(-x, y, z) \in D$ whenever $(x, y, z) \in D$, then $\overline{x} = 0$. Similar results hold for $\overline{y}$ and $\overline{z}$.

It may be worthwhile to consider reductions of the general formulas for $\overline{x}$, $\overline{y}$, and $\overline{z}$ in three special cases.

Case 1. Suppose D is a solid lying between two surfaces given by $z = f_1(x, y)$ and $z = f_2(x, y)$, that is, $D := \{(x, y, z) : (x, y) \in D_0, f_1(x, y) \le z \le f_2(x, y)\}$, where D_0 is a subset of $\mathbb{R}^2$. Let us assume that the set D_0 is bounded, ∂D_0 is of content zero, $f_1, f_2 : D_0 \to \mathbb{R}$ are bounded functions whose sets of discontinuities are of (two-dimensional) content zero, and $f_1 \le f_2$. In this case, by Corollary 5.45, ∂D_0 is of (two-dimensional) content zero, and by Cavalieri's Principle (part (ii) of Proposition 5.68), we have

$$\text{Vol}(D) = \iint_{D_0} [f_2(x, y) - f_1(x, y)] d(x, y).$$

Now let $V := \text{Vol}(D)$ and suppose $V \ne 0$. Then by Cavalieri's Principle (part (ii) of Proposition 5.68), we also see that $\overline{x}$ and $\overline{y}$, that is, the x-coordinate and the y-coordinate of the centroid of D, are given, respectively, by

$$\frac{1}{V} \iint_{D_0} x[f_2(x, y) - f_1(x, y)] d(x, y) \quad \text{and} \quad \frac{1}{V} \iint_{D_0} y[f_2(x, y) - f_1(x, y)] d(x, y),$$

whereas $\overline{z}$, that is, the z-coordinate of the centroid of D, is given by

$$\frac{1}{V} \iint_{D_0} \left(\int_{f_1(x,y)}^{f_2(x,y)} z \, dz \right) d(x, y) = \frac{1}{2V} \iint_{D_0} \left[f_2(x, y)^2 - f_1(x, y)^2 \right] d(x, y).$$

Similar results hold for solids lying between surfaces given by $y = g_1(x, z)$ and $y = g_2(x, z)$, or given by $x = h_1(y, z)$ and $x = h_2(y, z)$.

Case 2. Suppose D is a solid lying between two vertical planes given by $x = a$ and $x = b$, that is, $D \subseteq \mathbb{R}^3$ and $a, b \in \mathbb{R}$ with $a \le b$ are such that $a \le x \le b$ for all $(x, y, z) \in D$. For each $x \in [a, b]$, let D_x be the corresponding cross section of D given by $D_x = \{(y, z) \in \mathbb{R}^2 : (x, y, z) \in D\}$. Let us assume that the set D is bounded, ∂D is of three-dimensional content zero, and ∂D_x is of (two-dimensional) content zero for each $x \in [a, b]$. Now let $V := \text{Vol}(D)$ and assume that $V \ne 0$. Also let $(\overline{x}, \overline{y}, \overline{z})$ denote the centroid of D. Then by Cavalieri's Principle (part (i) of Proposition 5.68),

$$V = \int_a^b \left(\iint_{D_x} d(y, z) \right) dx, \quad \overline{x} = \frac{1}{V} \int_a^b \left(\iint_{D_x} x \, d(y, z) \right) dx,$$

$$\overline{y} = \frac{1}{V} \int_a^b \left(\iint_{D_x} y \, d(y, z) \right) dx, \quad \overline{z} = \frac{1}{V} \int_a^b \left(\iint_{D_x} z \, d(y, z) \right) dx.$$

Let us interpret the above formulas in terms of the areas and the centroids of the cross sections D_x, $x \in [a,b]$, of the solid D by vertical planes. Assume that for each $x \in [a,b]$, the area

$$A(x) := \iint_{D_x} d(y,z)$$

of D_x is not equal to zero. Then the centroid $(\tilde{y}(x), \tilde{z}(x))$ of D_x is given by

$$\tilde{y}(x) = \frac{1}{A(x)} \iint_{D_x} y\, d(y,z) \quad \text{and} \quad \tilde{z}(x) = \frac{1}{A(x)} \iint_{D_x} z\, d(y,z),$$

and hence

$$V = \int_a^b A(x)dx, \quad \overline{x} = \frac{1}{V}\int_a^b xA(x)dx,$$
$$\overline{y} = \frac{1}{V}\int_a^b \tilde{y}(x)A(x)dx, \quad \overline{z} = \frac{1}{V}\int_a^b \tilde{z}(x)A(x)dx.$$

In particular, suppose $f_1, f_2 : [a,b] \to \mathbb{R}$ are bounded functions whose sets of discontinuities are of one-dimensional content zero and $0 \le f_1 \le f_2$. Let D denote the solid generated by revolving the region

$$\left\{(x,y) \in \mathbb{R}^2 : a \le x \le b \text{ and } f_1(x) \le y \le f_2(x)\right\}$$

about the x-axis. Then for each $x \in [a,b]$, the corresponding cross section is given by $D_x := \left\{(y,z) \in \mathbb{R}^2 : f_1(x)^2 \le y^2 + z^2 \le f_2(x)^2\right\}$. We have seen in Section 6.1 that $A(x) = \pi[f_2(x)^2 - f_1(x)^2]$ for all $x \in [a,b]$, and consequently,

$$V = \int_a^b A(x)dx = \pi \int_a^b [f_2(x)^2 - f_1(x)^2]dx.$$

Now it can be easily seen that $\tilde{y}(x) = \tilde{z}(x) = 0$ for each $x \in [a,b]$, and so $\overline{y} = 0 = \overline{z}$. On the other hand,

$$\overline{x} = \frac{1}{V}\int_a^b xA(x)dx = \frac{\pi}{V}\int_a^b x[f_2(x)^2 - f_1(x)^2]dx.$$

This shows that our definition of the centroid of a solid is consistent with the formulas given in Section 8.5 of ACICARA. Similar results hold for solids lying between two vertical planes given by $y = c$ and $y = d$, or between two horizontal planes given by $z = p$ and $z = q$.

Case 3. Suppose D is a solid lying between two cylinders whose common axis is the z-axis, that is, there are $p, q \in \mathbb{R}$ with $0 \le p < q$ such that $p^2 \le x^2 + y^2 \le q^2$ for all $(x,y,z) \in D$. Let

$$E := \{(r, \theta, z) \in \mathbb{R}^3 : r \ge 0, -\pi \le \theta \le \pi \text{ and } (r\cos\theta, r\sin\theta, z) \in D\},$$

and for each $r \in [p, q]$, let $E_r := \{(\theta, z) \in [-\pi, \pi] \times \mathbb{R} : (r, \theta, z) \in E\}$. Let us assume that D and E are closed and bounded subsets of $\mathbb{R}^3$, ∂D and ∂E are of three dimensional content zero, ∂E_r is of (two-dimensional) content zero for each $r \in [p, q]$, and $V \neq 0$, where $V := \text{Vol}(D)$. By part (i) of Proposition 5.72 and Cavalieri's Principle (part (i) of Proposition 5.68), we obtain

$$V = \int_p^q r \left(\iint_{E_r} d(\theta, z) \right) dr.$$

Moreover, Cavalieri's Principle (part (i) of Proposition 5.68) also shows that

$$\overline{x} = \frac{1}{V} \int_p^q r^2 \left(\iint_{E_r} \cos\theta \, d(\theta, z) \right) dr, \quad \overline{y} = \frac{1}{V} \int_p^q r^2 \left(\iint_{E_r} \sin\theta \, d(\theta, z) \right) dr,$$

and

$$\overline{z} = \frac{1}{V} \int_p^q r \left(\iint_{E_r} z \, d(\theta, z) \right) dr.$$

Let us interpret the above formulas in terms of areas and centroids of the slivers $S_r := \{(x, y, z) \in D : x^2 + y^2 = r^2\}$, $r \in [p, q]$, of D by coaxial cylinders whose common axis is the z-axis. For $r \in [p, q]$, the surface S_r is given by $(r\cos\theta, r\sin\theta, z)$, $(\theta, z) \in E_r$. Assume that the area

$$B(r) := \iint_{E_r} d(\theta, z)$$

of the parameter domain E_r is nonzero for each $r \in [p, q]$. Define $x, y : E_r \to \mathbb{R}$ by $x(\theta, z) := r\cos\theta$ and $y(\theta, z) := r\sin\theta$. Then S_r is parametrically given by $(x(\theta, z),\ y(\theta, z), z)$, $(\theta, z) \in E_r$. Correspondingly, we have $U = (-r\sin\theta)^2 + (r\cos\theta)^2 + 0^2 = r^2$, $V = 0^2 + 0^2 + 1^2 = 1$, $W = (-r\sin\theta)(0) + (r\cos\theta)(0) + (0)(1) = 0$, and so $\sqrt{UV - W^2} = r$. Hence the surface area of S_r is equal to

$$A(r) = \iint_{E_r} r \, d(\theta, z) = rB(r).$$

The centroid $(\tilde{x}(r), \tilde{y}(r), \tilde{z}(r))$ of S_r is determined by the equations

$$A(r)\tilde{x}(r) = r^2 \iint_{E_r} \cos\theta \, d(\theta, z), \quad A(r)\tilde{y}(r) = r^2 \iint_{E_r} \sin\theta \, d(\theta, z),$$

and

$$A(r)\tilde{z}(r) = r \iint_{E_r} z \, d(\theta, z)$$

for each $r \in (p, q]$. Thus

$$V = \int_p^q A(r)dr, \quad \overline{x} = \frac{1}{V} \int_p^q \tilde{x}(r)A(r)dr,$$

$$\overline{y} = \frac{1}{V} \int_p^q \tilde{y}(r)A(r)dr, \quad \overline{z} = \frac{1}{V} \int_p^q \tilde{z}(r)A(r)dr.$$

Similar results hold if D is a solid lying between two cylinders whose common axis is the x-axis or the y-axis. We consider a special case of the latter.

Let $0 \le a < b$ and consider continuous functions $f_1, f_2 : [a,b] \to \mathbb{R}$ such that $f_1 \le f_2$ and let D be the solid generated by revolving the region

$$\{(x,y) \in \mathbb{R}^2 : a \le x \le b \text{ and } f_1(x) \le y \le f_2(x)\}$$

about the y-axis. Now, with notation as above, $E_x := [-\pi,\pi] \times [f_1(x), f_2(x)]$ and $B(x) = 2\pi[f_2(x) - f_1(x)]$, and therefore, $A(x) = 2\pi x[f_2(x) - f_1(x)]$ for all $x \in [a,b]$. Hence

$$V = \int_a^b A(x)dx = 2\pi \int_a^b x[f_2(x) - f_1(x)]dx.$$

Further, it can be easily seen that $\overline{x} = 0 = \overline{z}$ and

$$\overline{y} = \frac{1}{V}\int_a^b x\left(\iint_{E_x} y\, d(\theta, y)\right) dx = \frac{\pi}{V}\int_a^b x[f_2(x)^2 - f_1(x)^2]dx.$$

Note that

$$\overline{y} = \frac{2\pi}{V}\int_a^b \frac{[f_1(x) + f_2(x)]}{2} \cdot x[f_2(x) - f_1(x)]dx,$$

where $[f_1(x) + f_2(x)]/2$ is the y-coordinate of the centroid of the vertical cut of the region $\{(x,y) \in \mathbb{R}^2 : a \le x \le b \text{ and } f_1(x) \le y \le f_2(x)\}$ at $x \in [a,b]$. It may be noted that this fact was used in Section 8.5 of ACICARA as a motivation for defining the centroid of D.

Examples 6.21. (i) Let $a \in \mathbb{R}$ with $a > 0$ and let D denote the subset of $\mathbb{R}^3$ enclosed by the cylinders given by $x^2 + y^2 = a^2$ and $x^2 + z^2 = a^2$. Let $(\overline{x}, \overline{y}, \overline{z})$ be the centroid of D. Proceeding as in Example 6.5, if we let $D_0 := \{(x,y) \in \mathbb{R}^2 : x^2 + y^2 \le a^2\}$, then $D = \{(x,y,z) \in \mathbb{R}^3 : (x,y) \in D_0$ and $-\sqrt{a^2 - x^2} \le z \le \sqrt{a^2 - x^2}\}$ and $\text{Vol}(D) = 16a^2/3 \ne 0$. Hence, as in Case 1 above, $\text{Vol}(D)\,\overline{x}$ is equal to

$$\iint_{D_0} x\left[2\sqrt{a^2 - x^2}\right] d(x,y) = \int_{-a}^{a}\left(\int_{-\sqrt{a^2-y^2}}^{\sqrt{a^2-y^2}} 2x\sqrt{a^2 - x^2}\, dx\right) dy = 0,$$

and $\text{Vol}(D)\,\overline{y}$ is equal to

$$\iint_{D_0} y\left[2\sqrt{a^2 - x^2}\right] d(x,y) = \int_{-a}^{a} 2\sqrt{a^2 - x^2}\left(\int_{-\sqrt{a^2-x^2}}^{\sqrt{a^2-x^2}} y\, dy\right) dx = 0,$$

and moreover,

$$2\text{Vol}(D)\,\overline{z} = \iint_{D_0} \left[(a^2 - x^2) - (a^2 - x^2)\right] d(x,y) = 0.$$

Thus $(0,0,0)$ is the centroid of D.

(ii) Let D denote the subset of $\mathbb{R}^3$ between the plane given by $z = 1$ and the paraboloid given by $z = x^2 + y^2$. Let $(\overline{x}, \overline{y}, \overline{z})$ be the centroid of D. It is clear from the symmetry of D that $\overline{x} = 0 = \overline{y}$. In Example 6.7 (i), we have found that $\mathrm{Vol}(D) = \pi/2$. If for $r \in [0, 1]$, we let $E_r := \{(\theta, z) \in \mathbb{R}^2 : -\pi \le \theta \le \pi \text{ and } r^2 \le z \le 1\}$, then we see that $D = \{(r\cos\theta,\ r\sin\theta, z) : 0 \le r \le 1 \text{ and } (\theta, z) \in E_r\}$. Hence, as in Case 3 above, $\mathrm{Vol}(D)\,\overline{z}$ equals

$$\int_0^1 r\left(\iint_{E_r} z\, d(\theta, z)\right) dr = \int_0^1 r\pi(1 - r^4)dr = \pi\left(\frac{1}{2} - \frac{1}{6}\right) = \frac{\pi}{3},$$

and consequently, $\overline{z} = 2/3$. Thus $(0, 0, 2/3)$ is the centroid of D.

(iii) Let $a \in \mathbb{R}$ with $a > 0$ and let D denote the subset of $\mathbb{R}^3$ consisting of points that are outside the sphere given, in spherical coordinates, by $\rho = a$ and are inside the sphere given, in spherical coordinates, by $\rho = 2a\cos\varphi$. Thus, as in Example 5.74 (iii), $D = \{(\rho\sin\varphi\cos\theta, \rho\sin\varphi\sin\theta, \rho\cos\varphi) : (\rho, \varphi, \theta) \in E\}$, where $E := \{(\rho, \varphi, \theta) \in \mathbb{R}^3 : 0 \le \varphi \le \pi/3,\ a \le \rho \le 2a\cos\varphi \text{ and } -\pi \le \theta \le \pi\}$. Let $(\overline{x}, \overline{y}, \overline{z})$ be the centroid of D. It is clear from the symmetry of D that $\overline{x} = 0 = \overline{y}$. In view of Example 5.74 (iv),

$$\mathrm{Vol}(D)\,\overline{z} := \iiint_D z\, d(x, y, z) = \iiint_E (\rho\cos\varphi)\rho^2 \sin\varphi\, d(\rho, \varphi, \theta) = \frac{9\pi a^4}{8}.$$

Now, as noted in Example 6.7 (ii), $\mathrm{Vol}(D) = 11\pi a^3/12$. Thus it follows that $(0, 0, 27a/22)$ is the centroid of D.

(iv) Consider the solid $D := \{(r\cos\theta, r\sin\theta, z) \in \mathbb{R}^3 : (r, \theta, z) \in E\}$, where $E := \{(r, \theta, z) \in \mathbb{R}^3 : 1 \le r \le 2,\ 0 \le \theta \le \pi/2 \text{ and } 0 \le z \le r\}$. It is clear that D and E are closed and bounded subsets of $\mathbb{R}^3$, and that ∂D and ∂E are of three-dimensional content zero. Also, if $E_r := [0, \pi/2] \times [0, r]$ for $1 \le r \le 2$, then ∂E_r is of (two-dimensional) content zero for each $r \in [1, 2]$. Let $(\overline{x}, \overline{y}, \overline{z})$ be the centroid of D. Then, as in Case 3 above,

$$\mathrm{Vol}(D) = \int_1^2 r\left(\iint_{E_r} d(\theta, z)\right) dr = \int_1^2 r\left(\frac{\pi}{2}\right)(r)dr = \frac{7\pi}{6},$$

and moreover,

$$\mathrm{Vol}(D)\,\overline{x} = \int_1^2 r\left(\iint_{E_r} r\cos\theta\, d(\theta, z)\right) dr = \int_1^2 r^2\,(1)\,(r)dr = \frac{15}{4},$$

$$\mathrm{Vol}(D)\,\overline{y} = \int_1^2 r\left(\iint_{E_r} r\sin\theta\, d(\theta, z)\right) dr = \int_1^2 r^2\,(1)\,(r)dr = \frac{15}{4},$$

$$\mathrm{Vol}(D)\,\overline{z} = \int_1^2 r\left(\iint_{E_r} z\, d(\theta, z)\right) dr = \int_1^2 r\left(\frac{\pi}{2}\right)\left(\frac{r^2}{2}\right) dr = \frac{15\pi}{16}.$$

Thus $45(4, 4, \pi)/56\pi$ is the centroid of D. ◇

Centroids of Solids of Revolution

When a solid in $\mathbb{R}^3$ is obtained by revolving a planar region about a line in its plane, we can obtain simpler formulas for its centroid in terms of double integrals rather than triple integrals. We shall now proceed to derive these formulas and deduce a theorem of Pappus that relates the volume of a solid of revolution with the area of the corresponding planar region and the centroid.

Let D_0 be a closed and bounded subset of $\mathbb{R}^2$ such that D_0 has an area, that is, ∂D_0 is of (two-dimensional) content zero, and let L be a line in $\mathbb{R}^2$ that does not cross D_0. Assume that L is given by $ax+by+c=0$, where $a,b,c \in \mathbb{R}$ with $a^2+b^2=1$ and $ax+by+c \geq 0$ for all $(x,y) \in D_0$. If D denotes the solid generated by revolving D_0 about the line L, then by Proposition 6.10,

$$\mathrm{Vol}(D) = 2\pi \iint_{D_0} (ax+by+c)d(x,y).$$

Now let $(\overline{x}, \overline{y}, \overline{z})$ denote the centroid of D. It is clear from the symmetry of D that $\overline{z}=0$. To obtain simpler formulas for $\overline{x}$ and $\overline{y}$, we will transform the line L to the x-axis. To this end, let us use the affine functions Φ_0, Ψ_0, and Φ as well as the sets $E_0 := \Psi_0(D_0)$ and $E := \left\{(u,v,w) \in \mathbb{R}^3 : \left(u, \sqrt{v^2+w^2}\right) \in E_0\right\}$ introduced in the proof of Proposition 6.10. The assumption $ax+by+c \geq 0$ for all $(x,y) \in D_0$ corresponds to the condition $v \geq 0$ for all $(u,v) \in E_0$. Also, $\Phi(E) = D$ and $J(\Phi)(u,v,w) = 1$ for all $(u,v,w) \in \mathbb{R}^3$. Hence by the Change of Variables Formula (Proposition 5.70), we obtain

$$\iiint_D x\, d(x,y,z) = \iiint_E \big(bu + av - c(a-b)\big)d(u,v,w).$$

Switching to cylindrical coordinates (u,r,θ) in (u,v,w)-space by letting $v := r\cos\theta$ and $w := r\sin\theta$ and using part (i) of Proposition 5.72, we obtain

$$\begin{aligned}\iiint_D x\, d(x,y,z) &= \iint_{E_0} \left(\int_{-\pi}^{\pi} \big(bu + ar\cos\theta - c(a-b)\big)d\theta\right) r\, d(u,r)\\ &= 2\pi \iint_{E_0} \big(bu - c(a-b)\big) r\, d(u,r),\end{aligned}$$

where the last equality follows since $\int_{-\pi}^{\pi} \cos\theta\, d\theta = 0$. Thus

$$\begin{aligned}\iiint_D x\, d(x,y,z) &= 2\pi \iint_{E_0} \big(bu - c(a-b)\big) v\, d(u,v)\\ &= 2\pi \iint_{D_0} \big(b(bx - ay - c) - c(a-b)\big)(ax+by+c)d(x,y)\\ &= 2\pi \iint_{D_0} \big(b(bx-ay) - ac\big)(ax+by+c)d(x,y).\end{aligned}$$

Similarly,

$$\iiint_D y\, d(x,y,z) = 2\pi \iint_{D_0} \big(a(ay-bx)-bc\big)\big)(ax+by+c)d(x,y).$$

Hence we obtain

$$\overline{x} = \frac{2\pi}{\text{Vol}(D)} \iint_{D_0} \big(b(bx-ay)-ac\big)(ax+by+c)d(x,y),$$
$$\overline{y} = \frac{2\pi}{\text{Vol}(D)} \iint_{D_0} \big(a(ay-bx)-bc\big)(ax+by+c)d(x,y).$$

In case a^2+b^2 is not necessarily equal to 1, we may replace a, b, and c by $a/\sqrt{a^2+b^2}$, $b/\sqrt{a^2+b^2}$, and $c/\sqrt{a^2+b^2}$, respectively, in the above formulas and obtain

$$\overline{x} = \frac{\iint_{D_0} \big(b(bx-ay)-ac\big)\big|ax+by+c\big|d(x,y)}{(a^2+b^2)\iint_{D_0} |ax+by+c|d(x,y)},$$
$$\overline{y} = \frac{\iint_{D_0} \big(a(ay-bx)-bc\big)\big|ax+by+c\big|d(x,y)}{(a^2+b^2)\iint_{D_0} |ax+by+c|d(x,y)},$$
$$\overline{z} = 0.$$

We remark that the following special cases of the above formulas are often considered in one-variable calculus (for example, in Section 8.5 of ACICARA). Let $D_0 = \{(x,y) \in \mathbb{R}^2 : a \le x \le b \text{ and } f_1(x) \le y \le f_2(x)\}$, where $f_1, f_2 : [a,b] \to \mathbb{R}$ are continuous functions. First, if $0 \le f_1 \le f_2$ and L denotes the x-axis, that is, $a = 0 = c$ and $b = 1$, then we have

$$\overline{x} = \frac{2\pi}{\text{Vol}(D)} \iint_{D_0} xy\, d(x,y) = \frac{2\pi}{\text{Vol}(D)} \int_a^b x \left(\int_{f_1(x)}^{f_2(x)} y\, dy \right) dx$$
$$= \frac{\pi}{\text{Vol}(D)} \int_a^b x\big[f_2(x)^2 - f_1(x)^2\big]dx$$

and $\overline{y} = 0 = \overline{z}$. Next, if $a \ge 0$ and L denotes the y-axis, that is, $b = 0 = c$ and $a = 1$, then we have $\overline{x} = 0 = \overline{z}$ and

$$\overline{y} = \frac{2\pi}{\text{Vol}(D)} \iint_{D_0} yx\, d(x,y) = \frac{\pi}{\text{Vol}(D)} \int_a^b x\big[f_2(x)^2 - f_1(x)^2\big]dx.$$

Before concluding this section, we prove a theorem of Pappus for solids of revolution. Two special cases of this result were treated in Proposition 8.18 of ACICARA.

Proposition 6.22 (Theorem of Pappus). *Let D_0 be a bounded subset of $\mathbb{R}^2$ such that ∂D_0 is of (two-dimensional) content zero and let L be a line in $\mathbb{R}^2$ that does not cross D_0. If D_0 is revolved about L, then the volume of the solid so generated is equal to the product of the area of D_0 and the distance traveled by the centroid of D_0. Symbolically, we have*

Volume of Solid of Revolution = Area × Distance Traveled by Centroid.

Proof. Let $(\overline{x}, \overline{y})$ denote the centroid of D_0, and let D denote the solid generated by revolving D_0 about the line L. Then by Proposition 6.10, we have

$$\text{Vol}(D) = 2\pi \iint_{D_0} \frac{|ax + by + c|}{\sqrt{a^2 + b^2}} d(x, y).$$

On the other hand, by the definition of a centroid, we have

$$\overline{x} = \frac{1}{\text{Area}(D_0)} \iint_{D_0} x \, d(x, y)) \quad \text{and} \quad \overline{y} = \frac{1}{\text{Area}(D_0)} \iint_{D_0} y \, d(x, y).$$

Further, the distance d traveled by $(\overline{x}, \overline{y})$ about the line L is equal to 2π times the distance of the point $(\overline{x}, \overline{y})$ from the line L. Thus if the line L is given by $ax + by + c = 0$, where $a, b, c \in \mathbb{R}$ with $a^2 + b^2 \neq 0$, then

$$\begin{aligned} d = \frac{2\pi |a\overline{x} + b\overline{y} + c|}{\sqrt{a^2 + b^2}} &= \frac{2\pi}{\text{Area}(D_0)\sqrt{a^2 + b^2}} \left| \iint_{D_0} (ax + by + c) d(x, y) \right| \\ &= \frac{2\pi}{\text{Area}(D_0)} \iint_{D_0} \frac{|ax + by + c|}{\sqrt{a^2 + b^2}} d(x, y), \end{aligned}$$

where the last equality follows since either $ax + by + c \geq 0$ for all $(x, y) \in D$ or $ax + by + c \leq 0$ for all $(x, y) \in D$. Thus $\text{Vol}(D) = \text{Area}(D_0) \times d$. This proves the proposition. □

Example 6.23. Let $p, q \in \mathbb{R}$ with $p > 0$ and $0 < q \leq p/\sqrt{2}$. Consider the disk $D_0 := \{(x, y) \in \mathbb{R}^2 : (x - p)^2 + y^2 \leq q^2\}$. Let L be the line given by $x + y = 0$. Then L does not cross D_0. Let us find the volume of the solid D generated by revolving D_0 about L. (See Figure 6.12.) The area of D_0 is πq^2. By symmetry, the centroid of D_0 is at $(p, 0)$, and its distance from L is equal to $p/\sqrt{2}$. By the Theorem of Pappus (Proposition 6.22),

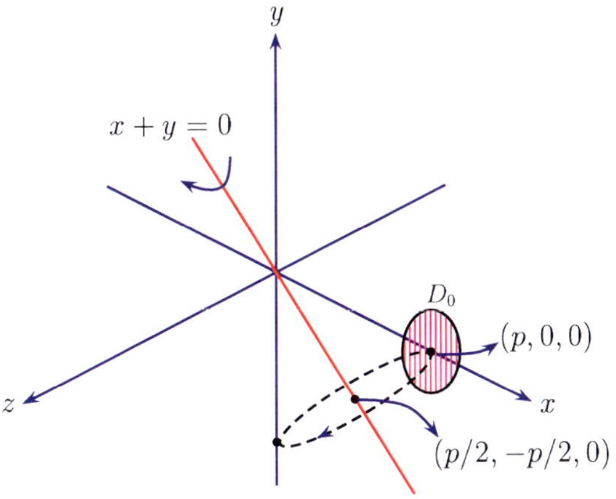

Fig. 6.12. Region D_0 in Example 6.23 revolved about the line $x + y = 0$.

$$\mathrm{Vol}(D) = \pi q^2 \cdot 2\pi \frac{p}{\sqrt{2}} = \sqrt{2}\pi^2 p q^2.$$

Let us also determine the centroid $(\overline{x}, \overline{y}, \overline{z})$ of the solid D. Letting $a = b = 1/\sqrt{2}$ and $V := \mathrm{Vol}(D)$ in the formulas for the centroid of a solid of revolution, we obtain

$$\overline{x} = \frac{2\pi}{V} \iint_{D_0} \frac{1}{\sqrt{2}} \Big(\frac{x}{\sqrt{2}} - \frac{y}{\sqrt{2}}\Big)\Big(\frac{x}{\sqrt{2}} + \frac{y}{\sqrt{2}}\Big) d(x,y) = \frac{\pi}{V\sqrt{2}} \iint_{D_0} (x^2 - y^2) d(x,y).$$

Switching to polar coordinates, we obtain

$$\begin{aligned}
\iint_{D_0} (x^2 - y^2) d(x,y) &= \int_0^q \left(\int_{-\pi}^{\pi} \left[(p + r\cos\theta)^2 - r^2 \sin^2\theta\right] d\theta \right) r\, dr \\
&= \int_0^q \left(\int_{-\pi}^{\pi} \left[p^2 - 2pr\cos\theta + r^2 \cos 2\theta\right] d\theta \right) r\, dr \\
&= 2\pi \int_0^q p^2 r\, dr = \pi p^2 q^2.
\end{aligned}$$

Thus it follows that

$$\overline{x} = \frac{\pi\left(\pi p^2 q^2\right)}{\sqrt{2}\left(\sqrt{2}\pi^2 p q^2\right)} = \frac{p}{2}.$$

Since the point $(\overline{x}, \overline{y})$ lies on the line L, we have $\overline{y} = -\overline{x} = -p/2$. Since $\overline{z} = 0$ by symmetry, we see that $(p/2, -p/2, 0)$ is the centroid of the solid D. ◇

6.4 Cubature Rules

The actual evaluation of a double integral by analytical methods is in general a formidable task. In the case of a double integral over an elementary region (or more generally, over a finite union of nonoverlapping elementary regions), one may reduce it to an iterated integral using Fubini's Theorem, but even then it is necessary to evaluate several Riemann integrals, a task that is by no means easy. It is therefore useful to develop methods that will yield at least an approximation of a given double integral. Thus we seek analogues of the quadrature rules for approximate evaluation of Riemann integrals that are usually studied in one-variable calculus (for example, Section 8.6 of ACICARA). It may be remarked that the case of double integrals is more difficult than the corresponding one-variable situation, for a Riemann integral, because one would like to evaluate double integrals over a variety of regions in $\mathbb{R}^2$, while one is usually content with evaluating Riemann integrals on closed and bounded intervals in $\mathbb{R}$.

Thanks to the Theorem of Darboux (Proposition 5.31) and Corollary 5.32, Riemann double sums can be employed to find approximate values of a double integral. This leads to a procedure, known as a cubature rule, for approximation of a double integral. Given a bounded subset D of $\mathbb{R}^2$, a **cubature rule** over D associates to an integrable function $f : D \to \mathbb{R}$ the real number

$$\sum_{i=1}^{n} \sum_{j=1}^{k} w_{i,j} f(s_{i,j}, t_{i,j}),$$

where $n, k \in \mathbb{N}$, $w_{i,j} \in \mathbb{R}$ and $(s_{i,j}, t_{i,j}) \in D$ for $i = 1, \ldots, n$ and $j = 1, \ldots, k$. The real numbers $w_{i,j}$ are known as the **weights** and the points $(s_{i,j}, t_{i,j})$ are known as the **nodes** of this cubature rule.

In this section, we shall discuss two kinds of cubature rules: (i) product rules and (ii) rules based on a triangulation of the region of integration.

Product Rules on Rectangles

A product cubature rule is obtained by constructing a "product" of two quadrature rules as described below. Let us first assume that $D := [a, b] \times [c, d]$ and consider quadrature rules Q and R on $[a, b]$ and on $[c, d]$, respectively, that associate to Riemann integrable functions $\phi : [a, b] \to \mathbb{R}$ and $\psi : [c, d] \to \mathbb{R}$ the real numbers

$$Q(\phi) := \sum_{i=1}^{n} u_i \phi(x_i) \quad \text{and} \quad R(\psi) := \sum_{j=1}^{k} v_j \psi(y_j),$$

where $u_1, \ldots, u_n \in \mathbb{R}$, $v_1, \ldots, v_k \in \mathbb{R}$, $x_1, \ldots, x_n \in [a, b]$, and $y_1, \ldots, y_k \in [c, d]$. We define $Q \times R$ to be the cubature rule that associates to an integrable function $f : [a, b] \times [c, d] \to \mathbb{R}$ the real number

$$(Q \times R)(f) := \sum_{i=1}^{n} \sum_{j=1}^{k} u_i v_j f(x_i, y_j).$$

We may refer to $Q \times R$ as the **product cubature rule on a rectangle** corresponding to the quadrature rules Q and R. The rule $Q \times R$ is obtained by successive applications of the rules Q and R to appropriate functions. To see this, suppose $f : [a, b] \times [c, d] \to \mathbb{R}$ is integrable and for each fixed $x \in [a, b]$, the function $\psi_x : [c, d] \to \mathbb{R}$ defined by $\psi_x(y) := f(x, y)$ is Riemann integrable. If $\phi : [a, b] \to \mathbb{R}$ defined by

$$\phi(x) := R(\psi_x) = \sum_{j=1}^{k} v_j \psi_x(y_j) = \sum_{j=1}^{k} v_j f(x, y_j) \quad \text{for } x \in [a, b]$$

is Riemann integrable on $[a, b]$, then

$$(Q \times R)(f) = Q(\phi).$$

Likewise, if $f : [a,b] \times [c,d] \to \mathbb{R}$ is integrable and for each fixed $y \in [c,d]$, the function $\phi_y : [a,b] \to \mathbb{R}$ defined by $\phi_y(x) := f(x,y)$ is Riemann integrable, and further, $\psi : [c,d] \to \mathbb{R}$ defined by $\psi(y) := Q(\phi_y)$ is Riemann integrable on $[c,d]$, then $(Q \times R)(f) = R(\psi)$.

Examples 6.24. (i) For $n, k \in \mathbb{N}$, let $P_{n,k} := \{(x_i, y_j) : i = 0, 1, \ldots, n$ and $j = 0, 1, \ldots, k\}$ be a partition of $[a,b] \times [c,d]$. Consider any $s_i \in [x_{i-1}, x_i]$ for $i = 1, \ldots, n$ and $t_j \in [y_{j-1}, y_j]$ for $j = 1, \ldots, k$. Then the Riemann double sum

$$S(P_{n,k}, f) := \sum_{i=1}^{n} \sum_{j=1}^{k} f(s_i, t_j)(x_i - x_{i-1})(y_j - y_{j-1})$$

gives a product cubature rule $Q_n \times R_k$, where

$$Q_n(\phi) := \sum_{i=1}^{n} (x_i - x_{i-1})\phi(s_i) \quad \text{and} \quad R_k(\psi) := \sum_{j=1}^{k} (y_j - y_{j-1})\psi(t_j)$$

for integrable functions $\phi : [a,b] \to \mathbb{R}$ and $\psi : [c,d] \to \mathbb{R}$. Let $k = n$ and assume that the mesh $\mu(P_{n,n})$ tends to 0 as $n \to \infty$. Then by Corollary 5.32, we see that

$$(Q_n \times R_n)(f) \to \iint_{[a,b]\times[c,d]} f(x,y)d(x,y)$$

for every integrable function $f : [a,b] \times [c,d] \to \mathbb{R}$.

(ii) Let $n, k \in \mathbb{N}$. Consider the partition $\{x_0, x_1, \ldots, x_n\}$ of $[a,b]$ into n equal parts and the partition $\{y_0, y_1, \ldots, y_k\}$ of $[c,d]$ into k equal parts given by

$$x_i := a + h_n i, \quad i = 0, 1, \ldots, n, \quad \text{and} \quad y_j := c + h'_k j, \quad j = 0, 1, \ldots, k,$$

where $h_n := (b-a)/n$ and $h'_k := (d-c)/k$. Let T_n and T'_k denote the Compound Trapezoidal Rules on $[a,b]$ and on $[c,d]$ respectively, that is,

$$T_n(\phi) = \frac{h_n}{2} \sum_{i=1}^{n} [\phi(x_{i-1}) + \phi(x_i)] \quad \text{and} \quad T'_k(\psi) = \frac{h'_k}{2} \sum_{j=1}^{k} [\psi(y_{j-1}) + \psi(y_j)]$$

for integrable functions $\phi : [a,b] \to \mathbb{R}$ and $\psi : [c,d] \to \mathbb{R}$. If f is an integrable function on $[a,b] \times [c,d]$, then $(T_n \times T'_k)(f)$ is equal to

$$\frac{h_n h'_k}{4} \sum_{i=1}^{n} \sum_{j=1}^{k} \Big[f(x_{i-1}, y_{j-1}) + f(x_{i-1}, y_j) + f(x_i, y_{j-1}) + f(x_i, y_j) \Big].$$

For computational purposes, it is convenient to rearrange the terms appearing in the above expression and obtain

$$
\begin{aligned}
(T_n \times T'_k)(f) = \frac{h_n h'_k}{4} \Big[& f(a,c) + f(a,d) + f(b,c) + f(b,d) \\
& + 2\sum_{i=1}^{n-1} \big(f(x_i,c) + f(x_i,d)\big) + 2\sum_{j=1}^{k-1} \big(f(a,y_j) + f(b,y_j)\big) \\
& + 4\sum_{i=1}^{n-1}\sum_{j=1}^{k-1} f(x_i,y_j)\Big].
\end{aligned}
$$

As a check on our calculations, consider $f := 1$ on $[a,b] \times [c,d]$. Then

$$
\begin{aligned}
(T_n \times T'_k)(f) &= \frac{h_n h'_k}{4}\Big[4 + 4(n-1) + 4(k-1) + 4(n-1)(k-1)\Big] \\
&= \frac{(b-a)(d-c)}{nk}\Big[1 + (n-1) + (k-1) + (n-1)(k-1)\Big] \\
&= (b-a)(d-c),
\end{aligned}
$$

as expected. Reverting to the general case, let us note that each of the double sums

$$
h_n h'_k \sum_{i=1}^{n}\sum_{j=1}^{k} f(x_{i-1}, y_{j-1}), \quad h_n h'_k \sum_{i=1}^{n}\sum_{j=1}^{k} f(x_{i-1}, y_j),
$$

$$
h_n h'_k \sum_{i=1}^{n}\sum_{j=1}^{k} f(x_i, y_{j-1}), \quad h_n h'_k \sum_{i=1}^{n}\sum_{j=1}^{k} f(x_i, y_j),
$$

is a Riemann double sum for f. Letting $k = n$, it follows from Corollary 5.32 that for any integrable function $f : [a,b] \times [c,d] \to \mathbb{R}$, we have

$$
(T_n \times T'_n)(f) \to \frac{1}{4}\left(4\iint_{[a,b]\times[c,d]} f(x,y)d(x,y)\right) = \iint_{[a,b]\times[c,d]} f(x,y)d(x,y).
$$

In a similar manner, we can obtain the product cubature rules $M_n \times M'_k$ and $S_n \times S'_k$, where M_n and S_n are the compound Midpoint Rule and the compound Simpson Rule on $[a,b]$, respectively, and where M'_k and S'_k are the compound Midpoint Rule and the compound Simpson Rule on $[c,d]$, respectively. (See Exercises 26 and 27.) ◇

We now show that error estimates for quadrature rules can be used to obtain error estimates for product cubature rules.

Proposition 6.25. *Let $Q \times R$ be a product cubature rule on $[a,b] \times [c,d]$ obtained from quadrature rules Q and R given by*

$$
Q(\phi) := \sum_{i=1}^{n} u_i \phi(x_i) \quad \textit{and} \quad R(\psi) := \sum_{j=1}^{k} v_j \psi(y_j),
$$

where $\phi : [a,b] \to \mathbb{R}$ *and* $\psi : [c,d] \to \mathbb{R}$ *are Riemann integrable functions,* $x_1, \ldots, x_n \in [a,b]$, $y_1, \ldots, y_k \in [c,d]$, $u_1, \ldots, u_n \in \mathbb{R}$, *and* $v_1, \ldots, v_k \in \mathbb{R}$ *with* $\sum_{j=1}^{k} |v_j| \le d-c$. *Let* $f : [a,b] \times [c,d] \to \mathbb{R}$ *be an integrable function such that for every fixed* $y \in [c,d]$, *the function* $\phi_y : [a,b] \to \mathbb{R}$ *defined by* $\phi_y(x) := f(x,y)$ *is Riemann integrable on* $[a,b]$, *and for every fixed* $x \in [a,b]$, *the function* $\psi_x : [c,d] \to \mathbb{R}$ *defined by* $\psi_x(y) := f(x,y)$ *is Riemann integrable on* $[c,d]$. *Assume that there are* $p \in \mathbb{N}$, $r \in \mathbb{R}$ *with* $r \ge 0$, *and for each* $y \in [c,d]$, *a constant* $\alpha(\phi_y)$ *depending on* ϕ_y *such that*

$$\left| \int_a^b \phi_y(x)dx - Q(\phi_y) \right| \le \frac{r\,\alpha(\phi_y)}{n^p} \quad \textit{for all } y \in [c,d],$$

and that there are $q \in \mathbb{N}$, $s \in \mathbb{R}$ *with* $s \ge 0$, *and for each* $x \in [a,b]$, *a constant* $\beta(\psi_x)$ *depending on* ψ_x *such that*

$$\left| \int_a^b \psi_x(y)dy - R(\psi_x) \right| \le \frac{s\,\beta(\psi_x)}{k^q} \quad \textit{for all } x \in [a,b].$$

Finally, assume that there are $\alpha_0, \beta_0 \in \mathbb{R}$ *such that* $\alpha(\phi_y) \le \alpha_0$ *for all* $y \in [c,d]$ *and* $\beta(\psi_x) \le \beta_0$ *for all* $x \in [a,b]$. *Then*

$$\left| \iint_{[a,b]\times[c,d]} f(x,y)d(x,y) - (Q \times R)(f) \right| \le \frac{(d-c)r\,\alpha_0}{n^p} + \frac{(b-a)s\,\beta_0}{k^q}.$$

Proof. Consider the functions $g, \phi : [a,b] \to \mathbb{R}$ defined by

$$g(x) := \int_c^d f(x,y)dy \quad \text{and} \quad \phi(x) := \sum_{j=1}^{k} v_j f(x, y_j).$$

Note that g is well defined, since ψ_x is Riemann integrable $[c,d]$. for each $x \in [a,b]$. Also, by part (i) of Proposition 5.28, g is Riemann integrable on $[a,b]$ and

$$\iint_{[a,b]\times[c,d]} f(x,y)d(x,y) = \int_a^b g(x)dx.$$

Moreover, since $\phi = \sum_{j=1}^{k} v_j \phi_{y_j}$, the function ϕ is Riemann integrable on $[a,b]$. Further, for each $x \in [a,b]$, we have

$$|g(x) - \phi(x)| = \left| \int_c^d \psi_x(y)dy - \sum_{j=1}^{k} v_j \psi_x(y_j) \right| = \left| \int_c^d \psi_x(y)dy - R(\psi_x) \right| \le \frac{s}{k^q}\beta(\psi_x).$$

Consequently,

$$\left| \iint_{[a,b]\times[c,d]} f(x,y)d(x,y) - \int_a^b \phi(x)dx \right| \le \int_a^b |g(x) - \phi(x)|\, dx \le \frac{(b-a)s\,\beta_0}{k^q}.$$

On the other hand, we have

$$\begin{aligned}\left|\int_a^b \phi(x)dx - (Q\times R)(f)\right| &= \left|\sum_{j=1}^k v_j\left(\int_a^b f(x,y_j)dx - \sum_{i=1}^n u_i f(x_i,y_j)\right)\right| \\ &\le \sum_{j=1}^k |v_j|\left|\int_a^b \phi_{y_j}(x)dx - \sum_{i=1}^n u_i\phi_{y_j}(x_i)\right| \\ &= \sum_{j=1}^k |v_j|\left|\int_a^b \phi_{y_j}(x)dx - Q\left(\phi_{y_j}\right)\right|.\end{aligned}$$

Hence using the hypothesis on ϕ_y, we obtain

$$\left|\int_a^b \phi(x)dx - (Q\times R)(f)\right| \le \sum_{j=1}^k |v_j|\frac{r}{n^p}\alpha(\phi_{y_j}) \le \frac{(d-c)r\,\alpha_0}{n^p}.$$

By the Triangle Inequality, it follows that

$$\left|\iint_{[a,b]\times[c,d]} f(x,y)d(x,y) - (Q\times R)(f)\right| \le \frac{(d-c)r\,\alpha_0}{n^p} + \frac{(b-a)s\,\beta_0}{k^q},$$

as desired. □

Example 6.26. Let T_n and T'_k denote the Compound Trapezoidal Rules on $[a,b]$ and on $[c,d]$ respectively. We know from one-variable calculus (for example, part (i) of Proposition 8.23 of ACICARA) that if $\phi : [a,b] \to \mathbb{R}$ and $\psi : [c,d] \to \mathbb{R}$ are twice differentiable and if their second derivatives are bounded, then

$$\left|\int_a^b \phi(x)dx - T_n(\phi)\right| \le \frac{(b-a)^3}{12n^2}\alpha(\phi), \text{ where } \alpha(\phi) := \sup\{|\phi''(x)| : x\in(a,b)\}$$

and

$$\left|\int_c^d \psi(y)dy - T'_k(\psi)\right| \le \frac{(d-c)^3}{12k^2}\beta(\psi), \text{ where } \beta(\psi) := \sup\{|\psi''(y)| : y\in(c,d)\}.$$

Now let $f : [a,b]\times[c,d] \to \mathbb{R}$ be an integrable function such that the first- and second-order partial derivatives of f exist and are bounded. Then letting $r = (b-a)^3/12$, $s := (d-c)^3/12$, and $p = q = 2$ in Proposition 6.25, we have

$$\left|\iint_{[a,b]\times[c,d]} f \;-\; (T_n\times T'_k)(f)\right| \le \frac{(b-a)(d-c)}{12}\left[\frac{(b-a)^2\alpha_0}{n^2} + \frac{(d-c)^2\beta_0}{k^2}\right],$$

where $\alpha_0 := \sup\{|f_{xx}(s,t)| : s\in(a,b), t\in[c,d]\}$ and $\beta_0 := \sup\{|f_{yy}(s,t)| : s\in[a,b], t\in(c,d)\}$.

Error estimates for the product Midpoint Rule and the product Simpson Rule are given in Exercises 28 and 29. ◇

Remark 6.27. If quadrature rules Q and R involve n and k nodes respectively, then the product cubature rule $Q \times R$ involves nk nodes. Thus from the point of view of numerical computation, a product cubature rule is far more expensive as compared to the two individual quadrature rules. All the same, Proposition 6.25 shows that if (Q_n) and (R_n) are sequences of quadrature rules each involving n nodes and if $Q_n(\phi)$ is an approximation of $\int_a^b \phi(x)dx$ of order $O(1/n^p)$ and $R_n(\psi)$ is an approximation of $\int_a^b \psi(x)dx$ of order $O(1/n^q)$, then $(Q_n \times R_n)(f)$ is an approximation of $\iint_{[a,b]\times[c,d]} f(x,y)d(x,y)$ only of order $O(1/n^{\min\{p,q\}})$. Thus a product cubature rule is much less efficient in approximating a double integral as compared to the approximation of a Riemann integral by one of its component quadrature rules. This is an example of what is often called the "curse of dimensionality."

Product Rules over Elementary Regions

We shall now proceed to show how a product cubature rule on a rectangle can be adapted to a rule over an elementary region. Let D be an elementary region given by

$$D := \left\{(x,y) \in \mathbb{R}^2 : a \le x \le b \text{ and } \phi_1(x) \le y \le \phi_2(x)\right\},$$

where $\phi_1, \phi_2 : [a,b] \to \mathbb{R}$ are Riemann integrable functions. Let $f : D \to \mathbb{R}$ be an integrable function such that for each fixed $x \in [a,b]$, the function from $[\phi_1(x), \phi_2(x)]$ to $\mathbb{R}$ given by $y \longmapsto f(x,y)$ is continuous. Then by Fubini's Theorem for elementary regions (Proposition 5.36), we have

$$\iint_D f(x,y)d(x,y) = \int_a^b \left(\int_{\phi_1(x)}^{\phi_2(x)} f(x,y)dy\right) dx.$$

Let Q and R be quadrature rules on $[a,b]$ and on $[c,d]$ respectively given by

$$Q(\phi) = \sum_{i=1}^{n} u_i \phi(x_i) \quad \text{and} \quad R(\psi) = \sum_{j=1}^{k} v_j \psi(y_j).$$

To obtain a cubature rule over D, we transform, for each $x \in [a,b]$, the Riemann integral $\int_{\phi_1(x)}^{\phi_2(x)} f(x,y)dy$ to a Riemann integral on $[c,d]$ as follows. For $x \in [a,b]$ with $\phi_1(x) < \phi_2(x)$, consider the function $\gamma_x : [c,d] \to \mathbb{R}$ defined by

$$\gamma_x(t) := \frac{\phi_2(x) - \phi_1(x)}{d-c}(t-c) + \phi_1(x) \quad \text{for } t \in [c,d].$$

Then $\gamma_x([c,d]) = [\phi_1(x), \phi_2(x)]$ and $\gamma_x'(t) = [\phi_2(x) - \phi_1(x)]/(d-c) \neq 0$ for all $t \in [c,d]$. Hence using a suitable substitution in Riemann integrals (justified, for example, by part (ii) of Proposition 6.26 of ACICARA), we obtain

$$\int_{\phi_1(x)}^{\phi_2(x)} f(x,y)dy = \frac{\phi_2(x)-\phi_1(x)}{d-c}\int_c^d f(x,\gamma_x(t))dt.$$

This formula holds even for $x \in [a,b]$ with $\phi_1(x) = \phi_2(x)$. Thus we have

$$\iint_D f(x,y)d(x,y) = \iint_{[a,b]\times[c,d]} \tilde{f}(x,y)d(x,y),$$

where $\tilde{f} : [a,b] \times [c,d] \to \mathbb{R}$ is defined by

$$\tilde{f}(x,y) := \frac{\phi_2(x)-\phi_1(x)}{d-c} f(x,\gamma_x(y)) \quad \text{for } (x,y) \in [a,b]\times[c,d].$$

Hence $(Q \times R)(\tilde{f})$ can be considered an approximation of the double integral of f over D. But

$$\begin{aligned}(Q \times R)(\tilde{f}) &= \sum_{i=1}^{n}\sum_{j=1}^{k} u_i v_j \tilde{f}(x_i, y_j)\\ &= \sum_{i=1}^{n}\sum_{j=1}^{k} u_i v_j \frac{\phi_2(x_i)-\phi_1(x_i)}{d-c} f(x_i, \gamma_{x_i}(y_j))\\ &= \sum_{i=1}^{n}\sum_{j=1}^{k} u_i v_{i,j} f(x_i, y_{i,j}),\end{aligned}$$

where for $i = 1, \ldots, n$ and $j = 1, \ldots, k$, we have put

$$v_{i,j} = \frac{\phi_2(x_i)-\phi_1(x_i)}{d-c} v_j \quad \text{and} \quad y_{i,j} = \frac{\phi_2(x_i)-\phi_1(x_i)}{d-c}(y_j - c) + \phi_1(x_i).$$

Notice that if $\sum_{j=1}^{k} v_j = d - c$, then $\sum_{j=1}^{k} v_{i,j} = \phi_2(x_i) - \phi_1(x_i)$ for each $i = 1, \ldots, n$. Now for any integrable function f over D, let us define

$$\widetilde{C}(f) := (Q \times R)(\tilde{f}).$$

This yields a cubature rule $\widetilde{C}$ for the elementary region D, which is referred to as a **product cubature rule** over D and denoted by $Q \times \widetilde{R}$. Thus

$$(Q \times \widetilde{R})(f) := (Q \times R)(\tilde{f}) \quad \text{for any integrable function } f : D \to \mathbb{R}.$$

If D is an elementary domain given by

$$D := \{(x,y) \in \mathbb{R}^2 : c \le y \le d \text{ and } \psi_1(y) \le x \le \psi_2(y)\},$$

where $\psi_1, \psi_2 : [c,d] \to \mathbb{R}$ are integrable functions, then we can similarly construct a product cubature rule $\widetilde{Q} \times R$ over D from the quadrature rules Q and R on $[a,b]$ and on $[c,d]$ respectively. (See Exercise 31.) ◇

Example 6.28. Let $D := \{(x, y) \in \mathbb{R}^2 : 0 \le x \le 1 \text{ and } 0 \le y \le \sqrt{1-x^2}\}$ and for $n \in \mathbb{N}$, let T_n denote the Compound Trapezoidal Rule on $[0,1]$ with n nodes. In this case, we have $a = 0 = c$ and $b = 1 = d$, while $\phi_1(x) = 0$ and $\phi_2(x) = \sqrt{1-x^2}$ for all $x \in [0,1]$, and so $\gamma_x(t) = t\sqrt{1-x^2}$ for $t \in [0,1]$. Given an integrable function $f : D \to \mathbb{R}$, define $\tilde{f} : [0,1] \times [0,1] \to \mathbb{R}$ by

$$\tilde{f}(x, y) := \frac{\phi_2(x) - \phi_1(x)}{d - c} f\left(x, \gamma_x(y)\right) = \sqrt{1-x^2}\, f\left(x, y\sqrt{1-x^2}\right).$$

Then as in Example 6.24 (ii),

$$\begin{aligned}
(T_n \times \widetilde{T}_k)(f) &= (T_n \times T_k)(\tilde{f}) \\
&= \frac{1}{4nk}\Bigg[f(0,0) + f(0,1) + 2\sum_{i=1}^{n-1}\sqrt{1-x_i^2}\left(f(x_i, 0) + f\left(x_i, \sqrt{1-x_i^2}\right)\right) \\
&\qquad + 2\sum_{j=1}^{k-1} f(0, y_j) + 4\sum_{i=1}^{n-1}\sum_{j=1}^{k-1}\sqrt{1-x_i^2}\, f\left(x_i, y_j\sqrt{1-x_i^2}\right)\Bigg],
\end{aligned}$$

where $x_i = i/n$ for $i = 1, \ldots, n$ and $y_j = j/k$ for $j = 1, \ldots, k$. ◇

Triangular Prism Rules

A **polygonal region** is a subset of $\mathbb{R}^2$ given by a connected union of finitely many nonoverlapping triangular regions. Here, by nonoverlapping we mean that whenever any two of these triangular regions intersect, the intersection is either a common vertex or a common edge. (See Figure 6.13.) A polygonal region D can be so partitioned in many different ways, and any such partition of D is called a **triangulation** of D.

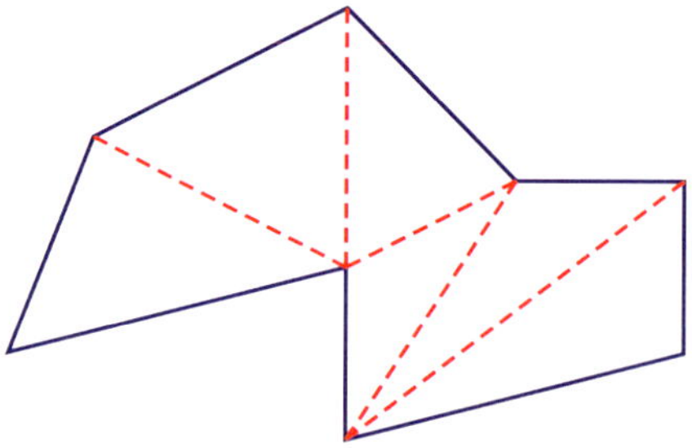

Fig. 6.13. Illustration of a polygonal region.

If $D \subseteq \mathbb{R}^2$ is a polygonal region, then evaluation of double integrals over D can be reduced to evaluation of finitely many double integrals over triangular regions. Indeed, if we have a triangulation of D into p triangular regions

$D_1, \ldots, D_p$, and if $f : D \to \mathbb{R}$ is an integrable function, then by Domain Additivity (Proposition 5.51), we see that f is integrable on each D_i, $i = 1, \ldots, p$, and we have

$$\iint_D f = \sum_{i=1}^{p} \iint_{D_i} f.$$

Of course, evaluating the double integral of an arbitrary integrable function over a triangular region may not be easy. However, we can always obtain an approximate evaluation using the following simple idea. Subdivide the triangular region further into several small triangular regions. On each of these small pieces, approximate the given function by a simpler function, such as, for example, a linear or a quadratic function, whose integral can be readily evaluated. This leads to cubature rules over a triangular region and, in turn, to cubature rules over a polygonal region.

To pursue the above-mentioned simple idea, we first prove a basic result about integrals of linear and quadratic functions over triangular regions.

Proposition 6.29. *For $i = 1, 2, 3$, let (x_i, y_i) be noncollinear points in $\mathbb{R}^2$ and D the triangular region in $\mathbb{R}^2$ having these points as its vertices. Then*

$$\text{Area}(D) = \frac{1}{2}\left|(x_2 - x_1)(y_3 - y_1) - (x_3 - x_1)(y_2 - y_1)\right|.$$

Further, let $A := \text{Area}(D)$ and let $f : D \to \mathbb{R}$ be a polynomial function in two variables of total degree m.

(i) *If $m \le 1$, then the double integral of f over D is equal to*

$$\frac{A}{3}\big[f(x_1, y_1) + f(x_2, y_2) + f(x_3, y_3)\big].$$

(ii) *If $m \le 2$, then the double integral of f over D is equal to*

$$\frac{A}{3}\left[f\Big(\frac{x_1 + x_2}{2}, \frac{y_1 + y_2}{2}\Big) + f\Big(\frac{x_2 + x_3}{2}, \frac{y_2 + y_3}{2}\Big) + f\Big(\frac{x_3 + x_1}{2}, \frac{y_3 + y_1}{2}\Big)\right].$$

Proof. Consider the triangular region E in $\mathbb{R}^2$ having $(0,0), (1,0)$, and $(0,1)$ as its vertices, and the affine transformation $\Phi : \mathbb{R}^2 \to \mathbb{R}^2$ given by

$$\Phi(u, v) = \big(x_1 + (x_2 - x_1)u + (x_3 - x_1)v,\ y_1 + (y_2 - y_1)u + (y_3 - y_1)v\big).$$

Then $\Phi(0,0) = (x_1, y_1), \Phi(1,0) = (x_2, y_2)$, and $\Phi(0,1) = (x_3, y_3)$. Since $\Phi(E)$ must be a triangular region, it follows that $\Phi(E) = D$. Also, it is easily seen that $J(\Phi) = (x_2 - x_1)(y_3 - y_1) - (x_3 - x_1)(y_2 - y_1)$. Further, since $\text{Area}(E) = 1/2$, Proposition 5.58 shows that

$$\text{Area}(D) = |J(\Phi)|\text{Area}(E) = \frac{1}{2}|(x_2 - x_1)(y_3 - y_1) - (x_3 - x_1)(y_2 - y_1)|.$$

Next, consider the function $g := f \circ \Phi$. It is clear that g is also a polynomial function of total degree m. By Proposition 5.59, we have

$$\iint_D f(x,y)d(x,y) = |J(\Phi)| \iint_E g(u,v)d(u,v) = 2A \iint_E g(u,v)d(u,v).$$

(i) Let $m \le 1$. Then there are $c_{0,0}, c_{1,0}, c_{0,1} \in \mathbb{R}$ such that

$$g(u,v) := c_{0,0} + c_{1,0}u + c_{0,1}v \quad \text{for all } (u,v) \in E.$$

By Fubini's Theorem (Proposition 5.36), we have

$$\begin{aligned}
\iint_E 1_E \, d(u,v) &= \int_0^1 \left(\int_0^{1-u} dv \right) du = \frac{1}{2}, \\
\iint_E u \, d(u,v) &= \int_0^1 \left(\int_0^{1-u} u \, dv \right) du = \int_0^1 u(1-u)du = \frac{1}{6}, \\
\iint_E v \, d(u,v) &= \int_0^1 \left(\int_0^{1-v} v \, du \right) dv = \int_0^1 v(1-v)dv = \frac{1}{6}.
\end{aligned}$$

Hence we see that

$$\iint_E g(u,v)d(u,v) = \frac{c_{0,0}}{2} + \frac{c_{1,0} + c_{0,1}}{6} = \frac{1}{6}\left[c_{0,0} + (c_{0,0} + c_{1,0}) + (c_{0,0} + c_{0,1})\right].$$

Since $g(0,0) = c_{0,0}$, $g(1,0) = c_{0,0} + c_{1,0}$, and $g(0,1) = c_{0,0} + c_{0,1}$, we obtain

$$\iint_D f(x,y)d(x,y) = 2A \iint_E g(u,v)d(u,v) = \frac{A}{3}\big[g(0,0) + g(1,0) + g(0,1)\big].$$

Since $g = f \circ \Phi$, we conclude that

$$\iint_D f(x,y)d(x,y) = \frac{A}{3}\big[f(x_1,y_1) + f(x_2,y_2) + f(x_3,y_3)\big].$$

(ii) Let $m \le 2$. Then there are $c_{0,0}, c_{1,0}, c_{0,1}, c_{2,0}, c_{2,0}, c_{1,1} \in \mathbb{R}$ such that

$$g(u,v) := c_{0,0} + c_{1,0}u + c_{0,1}v + c_{1,1}uv + c_{2,0}u^2 + c_{0,2}v^2 \quad \text{for all } (u,v) \in E.$$

By Fubini's Theorem, we have

$$\begin{aligned}
\iint_E uv \, d(u,v) &= \int_0^1 \left(\int_0^{1-u} uv \, dv \right) du = \frac{1}{2}\int_0^1 u(1-u)^2 du = \frac{1}{24}, \\
\iint_E u^2 d(u,v) &= \int_0^1 \left(\int_0^{1-u} u^2 dv \right) du = \int_0^1 u^2(1-u)du = \frac{1}{12}, \\
\iint_E v^2 d(u,v) &= \int_0^1 \left(\int_0^{1-v} v^2 du \right) dv = \int_0^1 v^2(1-v)dv = \frac{1}{12}.
\end{aligned}$$

Hence we see that

$$\iint_E g(u,v)d(u,v) = \frac{1}{2}c_{0,0} + \frac{1}{6}(c_{1,0} + c_{0,1}) + \frac{1}{24}c_{1,1} + \frac{1}{12}(c_{2,0} + c_{0,2}).$$

Since

$$g\Big(\frac{1}{2},0\Big) = c_{0,0} + \frac{1}{2}c_{1,0} + \frac{1}{4}c_{2,0}, \quad g\Big(0,\frac{1}{2}\Big) = c_{0,0} + \frac{1}{2}c_{0,1} + \frac{1}{4}c_{0,2},$$
$$g\Big(\frac{1}{2},\frac{1}{2}\Big) = c_{0,0} + \frac{1}{2}\left(c_{1,0} + c_{0,1}\right) + \frac{1}{4}c_{1,1} + \frac{1}{4}\left(c_{2,0} + c_{0,2}\right),$$

we obtain

$$\iint_D f = 2A\iint_E g = \frac{A}{3}\left[g\Big(\frac{1}{2},0\Big) + g\Big(\frac{1}{2},\frac{1}{2}\Big) + g\Big(0,\frac{1}{2}\Big)\right].$$

Since $g = f\circ\Phi$, we conclude that the double integral of f over D is equal to

$$\frac{A}{3}\left[f\Big(\frac{x_1+x_2}{2},\frac{y_1+y_2}{2}\Big) + f\Big(\frac{x_2+x_3}{2},\frac{y_2+y_3}{2}\Big) + f\Big(\frac{x_3+x_1}{2},\frac{y_3+y_1}{2}\Big)\right],$$

as desired. □

Corollary 6.30. *For $i = 1,2,3$, let (x_i, y_i) be noncollinear points in $\mathbb{R}^2$ and let D be the triangular region in $\mathbb{R}^2$ having these points as its vertices. Then the centroid $(\overline{x}, \overline{y})$ of D is given by*

$$\overline{x} = \frac{x_1+x_2+x_3}{3} \quad \textit{and} \quad \overline{y} = \frac{y_1+y_2+y_3}{3}.$$

Proof. Use part (i) of Proposition 6.29 with $f(x,y) := x$ for $(x,y)\in D$ and also with $f(x,y) := y$ for $(x,y)\in D$. □

Remark 6.31. As a consequence of Corollary 6.30, we can obtain the centroid of any polygonal region as follows. Suppose D is a polygonal region and $D = D_1\cup\cdots\cup D_n$ is a partition of D into nonoverlapping triangular regions $D_1,\ldots,D_n$. Let $(\overline{x}_i,\overline{y}_i)$ denote the centroid of D_i for $i = 1,\ldots,n$. Then by domain additivity (Proposition 5.51), we have $\text{Area}(D) = \sum_{i=1}^n \text{Area}(D_i)$ and

$$\iint_D x\,d(x,y) = \sum_{i=1}^n \iint_{D_i} x\,d(x,y) = \sum_{i=1}^n \text{Area}(D_i)\overline{x}_i,$$
$$\iint_D y\,d(x,y) = \sum_{i=1}^n \iint_{D_i} y\,d(x,y) = \sum_{i=1}^n \text{Area}(D_i)\overline{y}_i.$$

Hence the the coordinates of the centroid of D are given by

$$\overline{x} = \frac{\sum_{i=1}^n \text{Area}(D_i)\overline{x}_i}{\sum_{i=1}^n \text{Area}(D_i)} \quad \text{and} \quad \overline{y} = \frac{\sum_{i=1}^n \text{Area}(D_i)\overline{y}_i}{\sum_{i=1}^n \text{Area}(D_i)}.$$

For example, consider the quadrilateral D with $(0,0), (1,0)$, (0,1), and $(1,2)$ as its vertices. We have $D = D_1\cup D_2$, where D_1 is the triangular region with $(0,0), (1,0)$, and $(0,1)$ as its vertices, while D_2 is the triangular region with $(1,0), (0,1)$, and $(1,2)$ as its vertices. Evidently, $\text{Area}(D_1) = 1/2$ and $\text{Area}(D_2) = 1$. By Corollary 6.30, the centroid of D_1 is $(1/3, 1/3)$ and that of D_2 is $(2/3, 1)$. Hence the centroid of D is $(5/9, 7/9)$. This shows that in contrast to the case of a parallelogram, the centroid $(\overline{x}, \overline{y})$ of a quadrilateral D with vertices (x_i, y_i), $i = 1, \ldots, 4$, may not be given by $\overline{x} = (x_i+x_2+x_3+x_4)/4$ and $\overline{y} = (y_1 + y_2 + y_3 + y_4)/4$. ◇

We shall now obtain some simple cubature rules over a triangular region D by replacing a given integrable function by an appropriate polynomial function (of two variables) of total degree 0, 1, or 2, and calculating the "signed volume" of the corresponding surface. For reasons that shall soon become apparent, these cubature rules are called **Triangular Prism Rules**, and they can be viewed as analogues of quadrature rules (given, for example, in Section 8.6 of ACICARA) for Riemann integrals.

Let D be a triangular region in $\mathbb{R}^2$ with vertices (x_1, y_1), (x_2, y_2), and (x_3, y_3). As we have seen in Proposition 6.29, the area of D is given by

$$\text{Area}(D) := \frac{1}{2}|(x_2 - x_1)(y_3 - y_1) - (x_3 - x_1)(y_2 - y_1)|.$$

Consider an integrable function $f : D \to \mathbb{R}$.

1. Let us fix $(s,t) \in D$, and replace the function f by the constant function p_0, where $p_0 = f(s,t)$. The "signed volume" under the surface given by $z = p_0$, $(x,y) \in D$, is the "volume" of the triangular prism with base D and "height" $f(s,t)$. This gives a cubature rule that associates to f the real number

$$\text{Area}(D) f(s,t).$$

This is analogous to the Rectangular Rule for Riemann integrals.

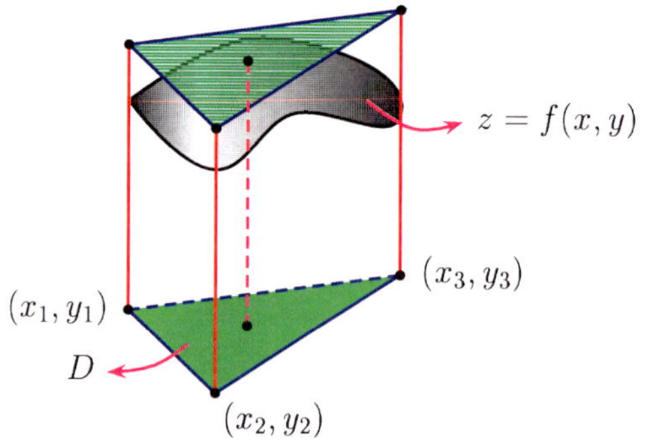

Fig. 6.14. Illustration of a Triangular Prism Rule analogous to the Midpoint Rule.

In particular, if (s,t) is the centroid $\big((x_1+x_2+x_3)/3, (y_1+y_2+y_3)/3\big)$ of D (as given in Corollary 6.30 and illustrated in Figure 6.14), then we obtain the cubature rule that associates to f the real number

$$\mathsf{C}(f) := \text{Area}(D) f\left(\frac{x_1 + x_2 + x_3}{3}, \frac{y_1 + y_2 + y_3}{3}\right).$$

This is analogous to the Midpoint Rule for Riemann integrals.

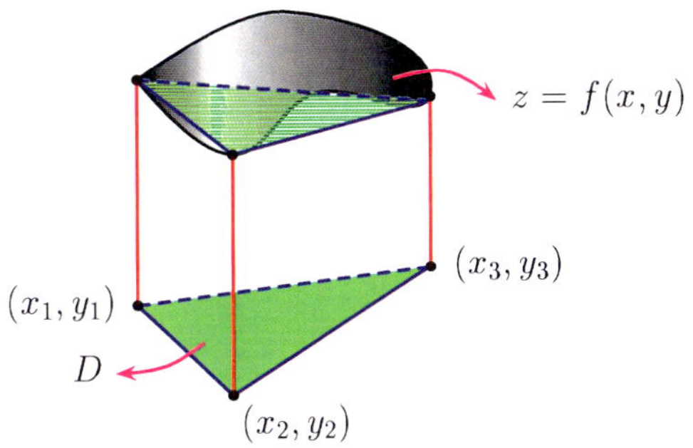

Fig. 6.15. Illustration of a Triangular Prism Rule analogous to the Trapezoidal Rule.

2. Let us replace the function f by a polynomial function p_1 (of two variables) of total degree 1 whose value at (x_i, y_i) is equal to $f(x_i, y_i)$ for $i = 1, 2, 3$. The "signed volume" under the surface given by $z = p_1(x, y)$, $(x, y) \in D$, is the "volume" of the obliquely cut triangular prism with base D, and the "lengths" of the three parallel edges are equal to $f(x_1, y_1)$, $f(x_2, y_2)$, and $f(x_3, y_3)$. (See Figure 6.15.) In view of part (i) of Proposition 6.29, this gives a cubature rule that associates to f the real number

$$\mathsf{T}(f) := \frac{\text{Area}(D)}{3}[f(x_1, y_1) + f(x_2, y_2) + f(x_3, y_3)].$$

This is analogous to the Trapezoidal Rule for Riemann integrals.

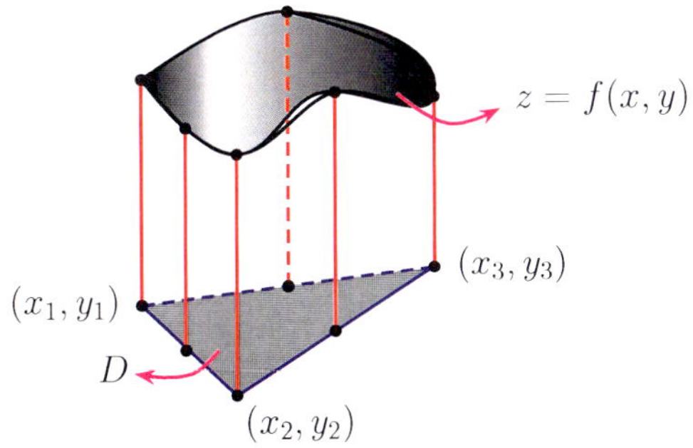

Fig. 6.16. Illustration of a Triangular Prism Rule analogous to Simpson's Rule.

3. Let us replace the function f by a polynomial function p_2 (of two variables) of total degree 2 whose values at (x_1, y_1), (x_2, y_2), (x_3, y_3), and at

$$\left(\frac{x_1+x_2}{2}, \frac{y_1+y_2}{2}\right), \left(\frac{x_2+x_3}{2}, \frac{y_2+y_3}{2}\right), \text{ and } \left(\frac{x_3+x_1}{2}, \frac{y_3+y_1}{2}\right)$$

are equal to the values of f at the corresponding points. The "signed volume" under the surface given by $z = p_2(x,y)$, $(x,y) \in D$, is the "volume" of the paraboloidal triangular prism with base D, the "lengths" of the three parallel edges are equal to $f(x_1,y_1), f(x_2,y_2), f(x_3,y_3)$, and the "heights" at the midpoints of the sides of D are equal to the values of f at those midpoints. (See Figure 6.16.) In view of part (ii) of Proposition 6.29, this gives a cubature rule that associates to f the real number $\mathsf{S}(f)$ given by

$$\frac{\text{Area}(D)}{3}\left[f\left(\frac{x_1+x_2}{2}, \frac{y_1+y_2}{2}\right)+f\left(\frac{x_2+x_3}{2}, \frac{y_2+y_3}{2}\right)+f\left(\frac{x_3+x_1}{2}, \frac{y_3+y_1}{2}\right)\right].$$

This is analogous to Simpson's Rule for Riemann integrals.

As in the case of simple quadrature rules, the simple cubature rules given above can be expected to yield only rough approximations of a double integral of a function over a triangular region D. To obtain more precise approximations, we may partition the triangular region D into smaller triangular regions and apply the above cubature rules to the function f restricted to each smaller region and then sum up the "signed volumes" so obtained. It is often convenient, and also efficient, to partition the triangular region D symmetrically in the following manner. Joining the midpoints of the three sides of D, we may construct four congruent triangular regions, the area of each being equal to Area$(D)/4$. This procedure can be repeated for each of the four smaller triangular regions. Continuing in this manner, after the nth step, we shall have a **symmetric triangulation** of D consisting of 4^n triangular regions, each of which has area equal to Area$(D)/4^n$. (See Figure 6.17.)

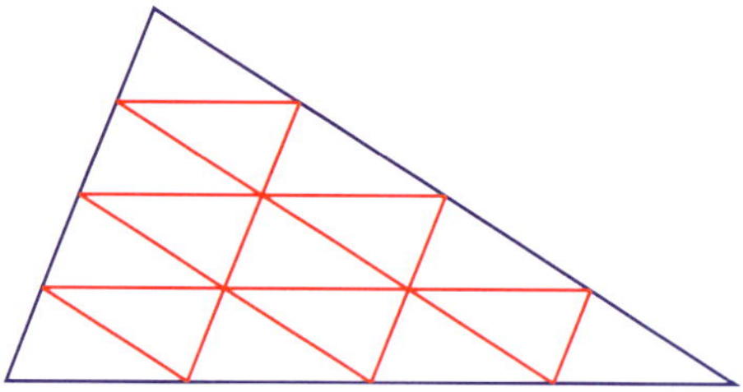

Fig. 6.17. A symmetric triangulation of D into $4^2 = 16$ triangular regions.

To obtain various **compound cubature rules**, fix $n \in \mathbb{N}$ and let $D_1, \ldots, D_{4^n}$ be the congruent triangular subregions of D described above.

1. Let (c_i, d_i) denote the centroid of the triangular region D_i for $i = 1, \ldots, 4^n$. (See Figure 6.18 for the case $n = 2$.) Then we obtain a compound cubature rule given by

$$\mathsf{C}_n(f) = \frac{\text{Area}(D)}{4^n} \sum_{i=1}^{4^n} f(c_i, d_i).$$

It is analogous to the Compound Midpoint Rule for Riemann integrals. The number of evaluations needed for calculating C_n is 4^n.

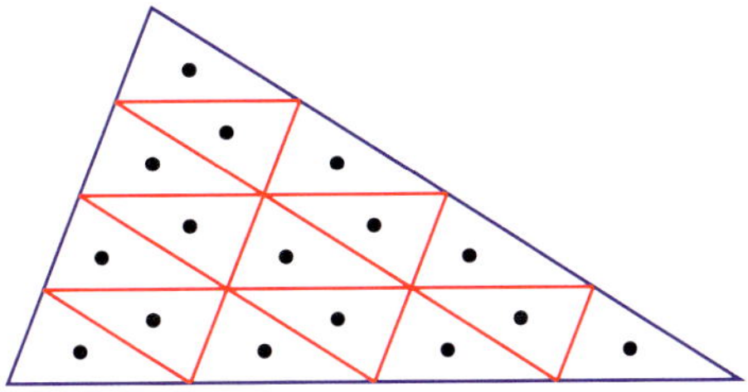

Fig. 6.18. A triangulation of D with centroids (c_i, d_i) marked by $\bullet$.

2. The total number of vertices in the triangulation of D given by $D_1, \ldots, D_{4^n}$ is $\left(2^{n-1}+1\right)(2^n+1)$. Let (q_i, r_i) denote a vertex that is on one of the three sides of D, but is not a vertex of D itself, $i = 1, \ldots, m$, where $m := 3(2^n - 1)$. Also, let (s_i, t_i) denote a vertex that is in the interior of D, $i = 1, \ldots, k$, where $k := (2^{n-1}+1)(2^n+1) - 3 \cdot 2^n = (2^{n-1}-1)(2^n-1)$. (See Figure 6.19 for the case $n = 2$.) Observe that (x_i, y_i) is a vertex of only one of the triangular regions $D_1, \ldots, D_{4^n}$ for $i = 1, 2, 3$. On the other hand, (q_i, r_i) is a vertex of exactly three of the triangular regions $D_1, \ldots, D_{4^n}$ for $i = 1, \ldots, m$, whereas (s_i, t_i) is a vertex of exactly six of the triangular regions $D_1, \ldots, D_{4^n}$ for $i = 1, \ldots, k$. Thus we obtain a compound cubature rule given by

$$\mathsf{T}_n(f) = \frac{\text{Area}(D)}{3 \cdot 4^n} \left[\sum_{i=1}^{3} f(x_i, y_i) + 3 \sum_{i=1}^{m} f(q_i, r_i) + 6 \sum_{i=1}^{k} f(s_i, t_i) \right].$$

It is analogous to the Compound Trapezoid Rule for Riemann integrals. The number of evaluations needed for calculating T_n is $(2^{n-1}+1)(2^n+1)$.

3. The total number of sides in the triangulation of D given by $D_1, \ldots, D_{4^n}$ is $3 \cdot 2^{n-1}(2^n+1)$. Let (q_i, r_i) denote the midpoint of a side that is on one of the three sides of D, $i = 1, \ldots, m$, where $m := 3 \cdot 2^n$. Also, let (s_i, t_i) denote the midpoint of a side that lies in the interior of D, $i = 1, \ldots, k$, where $k := 3 \cdot 2^{n-1}(2^n+1) - 3 \cdot 2^n = 3 \cdot 2^{n-1}(2^n-1)$. (See Figure 6.20 for the case $n = 2$.) Observe that (q_i, r_i) is the midpoint of exactly one side of the triangular regions $D_1, \ldots, D_{4^n}$ for $i = 1, \ldots, m$, whereas (s_i, t_i) is the midpoint of exactly two sides of the triangular regions $D_1, \ldots, D_{4^n}$ for $i = 1, \ldots, k$. Thus we obtain a compound cubature rule given by

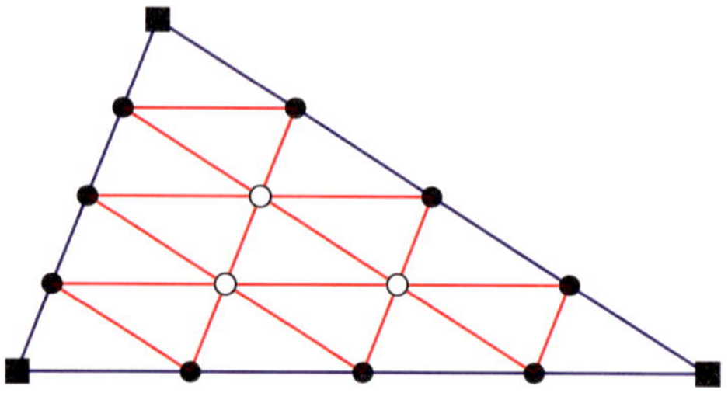

Fig. 6.19. A triangulation of D with vertices (x_i, y_i) of D marked by ■, interior vertices (s_i, t_i) marked by ○, and the remaining vertices (q_i, r_i) marked by ●.

$$\mathsf{S}_n(f) = \frac{\text{Area}(D)}{3 \cdot 4^n} \left[\sum_{i=1}^{m} f(q_i, r_i) + 2 \sum_{i=1}^{k} f(s_i, t_i) \right].$$

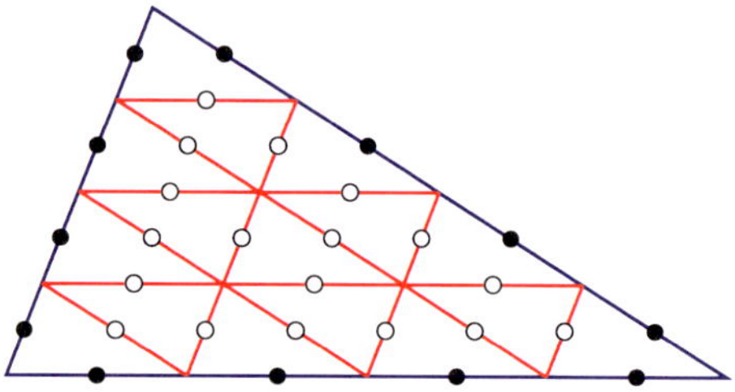

Fig. 6.20. A triangulation of D with midpoints (q_i, r_i) on the sides of D marked by ● and the other midpoints (s_i, t_i) marked by ○.

It is analogous to the Compound Simpson Rule (given, for example, in Section 8.6 of ACICARA) for Riemann integrable functions defined on an interval. The number of evaluations needed for calculating S_n is $3 \cdot 2^{n-1}(2^n + 1)$.

We shall now prove that the compound cubature rules given above converge to the double integral of an integrable function f defined on a triangular region. For simplicity, we first restrict to a triangular region with vertices $(0,0)$, $(1,0)$, and $(0,1)$, and comment later on how the general case can be deduced from this.

Proposition 6.32. *Let* $E := \{(x, y) \in \mathbb{R} : 0 \le x \le 1$ *and* $0 \le y \le 1 - x\}$ *and* $g : E \to \mathbb{R}$ *an integrable function. Then the sequences* $(\mathsf{C}_n(g)), (\mathsf{T}_n(g))$, *and* $(\mathsf{S}_n(g))$ *of compound cubature rules converge to* $\iint_E g(x, y)d(x, y)$ *as* $n \to \infty$.

Proof. Let $R := [0, 1] \times [0, 1]$ and let $g^* : R \to \mathbb{R}$ be defined by

$$g^*(x, y) = \begin{cases} g(x, y) & \text{if } (x, y) \in E \\ 0 & \text{if } (x, y) \notin E. \end{cases}$$

Then g^* is integrable on R, and by Corollary 5.32, any sequence of Riemann double sums converges to the double integral of g^* on R, that is, to the double integral of g over E, provided the mesh of the corresponding partition tends to zero. Since g is bounded on E, there is $\alpha > 0$ such that $|g^*(x,y)| \leq \alpha$ for all $(x,y) \in R$. For $n \in \mathbb{N}$, let $x_i = y_i = i/2^n$ for $i = 0, 1, \ldots, 2^n$, and consider the partition $P_n := \{(x_i, y_j) : i, j = 0, \ldots, 2^n\}$ of R into $2^n \times 2^n$ equal parts. Note that P_n induces a symmetric triangulation of E consisting of triangular regions $E_1, \ldots, E_{4^n}$. Also, in view of Proposition 6.29, we have $\text{Area}(E) = 1/2$. We shall now show, one by one, that each of the sequences $(\mathsf{C}_n(g)), (\mathsf{T}_n(g))$, and $(\mathsf{S}_n(g))$ of compound cubature rules corresponding to the symmetric triangulation above converges to $\iint_E g(x,y)d(x,y)$.

1. Let (c_i, d_i) be the centroid of E_i for $i = 1, \ldots, 4^n$. Then we have

$$\mathsf{C}_n(g) = \frac{1}{2 \cdot 4^n} \sum_{i=1}^{4^n} g(c_i, d_i).$$

Let $\widetilde{E}$ denote the complementary triangular region

$$\widetilde{E} := \left\{(x,y) \in \mathbb{R}^2 : 0 \leq x \leq 1 \text{ and } 1 - x \leq y \leq 1\right\}.$$

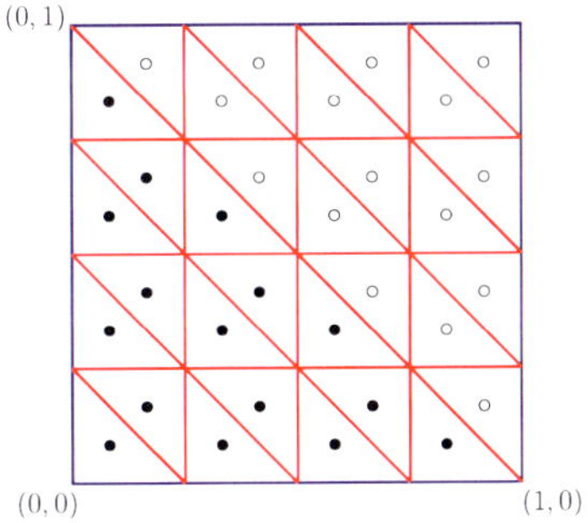

Fig. 6.21. A symmetric triangulation of E with centroids (c_i, d_i) marked by $\bullet$ and of the complementary triangular region $\widetilde{E}$ with centroids $(\tilde{c}_i, \tilde{d}_i)$ marked by $\circ$.

Consider the corresponding symmetric triangulation of $\widetilde{E}$ consisting of triangular regions $\widetilde{E}_1, \ldots, \widetilde{E}_{4^n}$, and let $(\tilde{c}_i, \tilde{d}_i)$ denote the centroid of $\widetilde{E}_i$ for $i = 1, \ldots, 4^n$. (See Figure 6.21.) Since $g^*(c_i, d_i) = g(c_i, d_i)$ and $g^*(\tilde{c}_i, \tilde{d}_i) = 0$ for $i = 1, \ldots, 4^n$, we see that

$$\mathsf{C}_n(g) = \frac{1}{2 \cdot 4^n} \sum_{i=1}^{4^n} [g^*(c_i, d_i) + g^*(\tilde{c}_i, \tilde{d}_i)].$$

Now for each $i, j = 1, \ldots, 2^n$, exactly two of the points $(c_1, d_1), \ldots, (c_{4^n}, d_{4^n})$, $(\tilde{c}_1, \tilde{d}_1), \ldots, (\tilde{c}_{4^n}, \tilde{d}_{4^n})$ lie in the subrectangle $[x_{i-1}, x_i] \times [y_{j-1}, y_j]$ induced by the partition P_n of R. Also, $(x_i - x_{i-1})(y_j - y_{j-1}) = 1/4^n$ for each $i, j = 1, \ldots, 2^n$. Hence we have

$$\mathsf{C}_n(g) = \frac{1}{2}\left[S_1(P_n, g^*) + S_2(P_n, g^*)\right],$$

where $S_1(P_n, g^*)$ and $S_2(P_n, g^*)$ are both Riemann double sums for the function g^* defined on R. Since the mesh $\mu(P_n) = 1/2^n$ tends to 0 as $n \to \infty$, we see that $S_j(P_n, g^*) \to \iint_R g^*(x,y)d(x,y)$ for $j = 1, 2$. Thus it follows that

$$\mathsf{C}_n(g) \to \frac{1}{2}\left[\iint_R g^*(x,y)d(x,y) + \iint_R g^*(x,y)d(x,y)\right] = \iint_E g(x,y)d(x,y).$$

2. Note that the vertices of E are $(0,0)$, $(1,0)$, and $(0,1)$, whereas the other vertices of any of the triangular regions $E_1, \ldots, E_{4^n}$ that are on one of the three sides of E are $(0, y_i)$, $(x_i, 0)$, and $(x_i, y_{2^n - i})$ for $i = 1, \ldots, 2^n - 1$. Thus

$$\mathsf{T}_n(g) = \frac{1}{6 \cdot 4^n}\Bigg[g(0,0) + g(1,0) + g(0,1) + 3\sum_{i=1}^{2^n-1}\Big(g(0,y_i) + g(x_i,0) + g(x_i, y_{2^n-i})\Big) + 6\sum_{i=1}^{2^n-2}\sum_{j=1}^{2^n-i-1} g(x_i, y_j)\Bigg].$$

Define $\mathsf{T}_n^*(g^*)$ to be equal to

$$\frac{1}{6 \cdot 4^n}\sum_{i=1}^{2^n}\sum_{j=1}^{2^n}\Big(g^*(x_{i-1}, y_{j-1}) + 2g^*(x_{i-1}, y_j) + 2g^*(x_i, y_{j-1}) + g^*(x_i, y_j)\Big).$$

(See Figure 6.22.) Since $g^*(x_i, y_j) = g(x_i, y_j)$ if $(x_i, y_j) \in E$ and $g^*(x_i, y_j) = 0$ if $(x_i, y_j) \notin E$, it can be seen that

$$\mathsf{T}_n^*(g^*) - \mathsf{T}_n(g) = \frac{1}{6 \cdot 4^n}\left[g(1,0) + g(0,1) + 3\sum_{i=1}^{2^n-1} g(x_i, y_{2^n-i})\right].$$

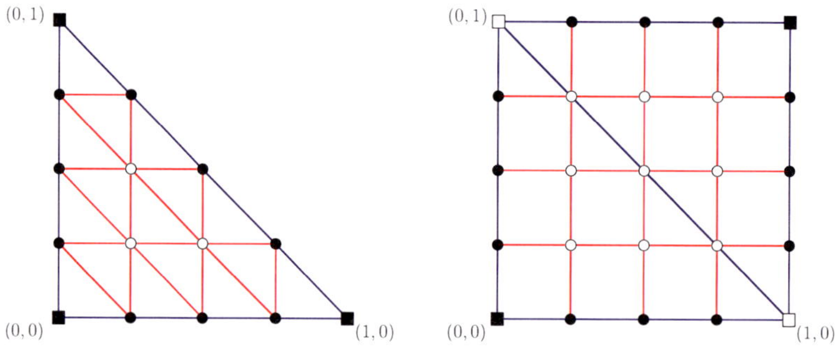

Fig. 6.22. Number of times a node appears in $\mathsf{T}_n(g)$ and in $\mathsf{T}_n^*(g^*)$, where ■, □, ●, and ○ indicate that the node appears once, twice, thrice, and six times respectively.

Hence $|\mathsf{T}_n^*(g^*) - \mathsf{T}_n(g)| \le (3(2^n) - 1)\,\alpha/6 \cdot 4^n$, which tends to 0 as $n \to \infty$. Since the mesh $\mu(P_n) = 1/2^n$ tends to 0 as $n \to \infty$, we see that

$$\mathsf{T}_n^*(g^*) \to \frac{1}{6}\left[\iint_R g^* + 2\iint_R g^* + 2\iint_R g^* + \iint_R g^*\right] = \iint_R g^*.$$

Thus it follows that

$$\mathsf{T}_n(g) \to \iint_R g^* = \iint_E g.$$

3. We have already identified the vertices of the triangular regions $E_1, \ldots, E_{4^n}$ that lie on one of the three sides of E and also those that do not lie on any of the sides of E. Considering their midpoints, we see that $\mathsf{S}_n(g)$ is equal to

$$\frac{1}{6 \cdot 4^n}\left[\sum_{i=1}^{2^n}\left(g\Big(\frac{x_{i-1}+x_i}{2}, c\Big) + g\Big(a, \frac{y_{i-1}+y_i}{2}\Big) + g\Big(\frac{x_{i-1}+x_i}{2}, \frac{y_{2^n-i}+y_{2^n-i+1}}{2}\Big)\right)\right.$$
$$\left. + 2\sum_{i=1}^{2^n-1}\sum_{j=1}^{2^n-i}\left(g\Big(x_i, \frac{y_{j-1}+y_j}{2}\Big) + g\Big(\frac{x_{i-1}+x_i}{2}, y_j\Big) + g\Big(\frac{x_{i-1}+x_i}{2}, \frac{y_{j-1}+y_j}{2}\Big)\right)\right].$$

Define $\mathsf{S}_n^*(g^*)$ to be equal to

$$\frac{1}{6 \cdot 4^n}\sum_{i=1}^{2^n}\sum_{j=1}^{2^n}\left(g^*\Big(x_{i-1}, \frac{y_{j-1}+y_j}{2}\Big) + g^*\Big(x_i, \frac{y_{j-1}+y_j}{2}\Big)\right.$$
$$\left. + g^*\Big(\frac{x_{i-1}+x_i}{2}, y_{j-1}\Big) + g^*\Big(\frac{x_{i-1}+x_i}{2}, y_j\Big) + 2g^*\Big(\frac{x_{i-1}+x_i}{2}, \frac{y_{j-1}+y_j}{2}\Big)\right).$$

(See Figure 6.23.) Since $g^*(x,y) = g(x,y)$ if $(x,y) \in E$ and $g^*(x,y) = 0$ if $(x,y) \notin E$, it can be seen that

$$\mathsf{S}_n^*(g^*) - \mathsf{S}_n(g) = \frac{1}{6 \cdot 4^n}\sum_{i=1}^{2^n} g\Big(\frac{x_{i-1}+x_i}{2}, \frac{y_{2^n-i}+y_{2^n-i+1}}{2}\Big).$$

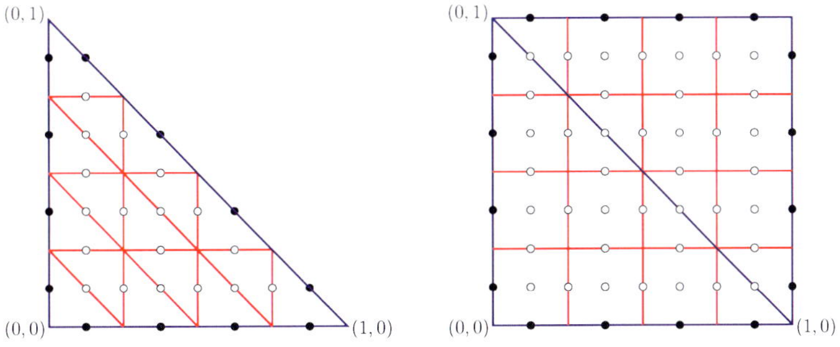

Fig. 6.23. Number of times a node appears in $\mathsf{S}_n(g)$ and in $\mathsf{S}_n^*(g^*)$, where • and ∘ indicate that the node appears once and twice respectively.

Hence $|\mathsf{S}_n^*(g^*) - \mathsf{S}_n(g)| \le 2^n\alpha/6\cdot 4^n = \alpha/6\cdot 2^n$, which tends to 0 as $n\to\infty$. Since the mesh $\mu(P_n) = 1/2^n$ tends to 0 as $n\to\infty$, we see that

$$\mathsf{S}_n^*(g^*) \to \frac{1}{6}\left[\iint_R g^* + \iint_R g^* + \iint_R g^* + \iint_R g^* + 2\iint_R g^*\right] = \iint_R g^*.$$

Thus it follows that

$$\mathsf{S}_n(g) \to \iint_R g^* = \iint_E g,$$

as desired. □

Remark 6.33. Proposition 6.32 and its proof readily extend to the case in which E is a triangular region with vertices (a,c), (b,c), and (a,d), where $a,b,c,d\in\mathbb{R}$ with $a<b$ and $c<d$. More generally, if D is any triangular region whose vertices are noncollinear points (x_i,y_i), $i=1,2,3$, and $f: D\to\mathbb{R}$ is continuous, then the sequences $(\mathsf{C}_n(f))$, $(\mathsf{T}_n(f))$, and $(\mathsf{S}_n(f))$ of compound cubature rules converge to $\iint_D f(x,y)d(x,y)$ as $n\to\infty$. To see this, consider an affine transformation $\Phi:\mathbb{R}^2\to\mathbb{R}^2$ as in the proof of Proposition 6.29 and let E be as in Proposition 6.32. Then $\Phi(E)=D$ and $\text{Area}(D) = |J(\Phi)|\text{Area}(E) = |J(\Phi)|/2$. Further, let $g: E\to\mathbb{R}$ be defined by $g(u,v) := f(\Phi(u,v))$ for $(u,v)\in E$. Now observe that the invertible affine transformation Φ preserves symmetric triangulations as well as the midpoints of sides and the centroids of triangular subregions, and hence

$$\mathsf{C}_n(f) = \frac{\text{Area}(D)}{\text{Area}(E)}\mathsf{C}_n(g),\quad \mathsf{T}_n(f) = \frac{\text{Area}(D)}{\text{Area}(E)}\mathsf{T}_n(g),\quad \mathsf{S}_n(f) = \frac{\text{Area}(D)}{\text{Area}(E)}\mathsf{S}_n(g).$$

On the other hand, g is continuous on E, and so by Proposition 5.59, we have

$$\iint_D f(x,y)d(x,y) = |J(\Phi)|\iint_E g(u,v)d(u,v) = \frac{\text{Area}(D)}{\text{Area}(E)}\iint_E g(u,v)d(u,v).$$

Thus, the desired result follows from Proposition 6.32. ◇

Remark 6.34. Error estimates for the compound cubature rules considered above can be obtained in analogy with those for compound quadrature rules. We briefly indicate the key facts.

First, by the very method of our construction, the cubature rules C, T, and S on a triangular region give exact values of the double integrals of polynomial functions (in two variables) of total degree 0, 1, and 2 respectively. In fact, the centroid rule C gives the exact value of the double integral of any polynomial function (in two variables) of total degree 1. This follows by noting that if D is a triangular region in $\mathbb{R}^2$ and $(\overline{x},\overline{y})$ denotes its centroid, then by definition

$$\overline{x} = \frac{1}{\text{Area}(D)}\iint_D x\,d(x,y) \quad\text{and}\quad \overline{y} = \frac{1}{\text{Area}(D)}\iint_D y\,d(x,y).$$

Using these facts and the Classical Version of the Bivariate Taylor Theorem (Proposition 3.47), and noting that the diameter of each triangular region in the symmetric triangulation of D consisting of 4^n congruent triangular regions is $d/2^n$, where $d = \text{diam}(D)$, the following results can be proved.

1. Let $f : D \to \mathbb{R}$ be a function whose partial derivatives of the second order exist and are continuous on an open subset of $\mathbb{R}^2$ containing D. Then for all $n \in \mathbb{N}$, we have

$$\left| \iint_D f(x,y)d(x,y) - \mathsf{C}_n(f) \right| \le c_1 \left(\frac{d}{2^n} \right)^2 \alpha$$

and

$$\left| \iint_D f(x,y)d(x,y) - \mathsf{T}_n(f) \right| \le c_2 \left(\frac{d}{2^n} \right)^2 \alpha,$$

where $\alpha := \max\{|f_{xx}(s,t)|, |f_{xy}(s,t)|, |f_{yy}(s,t)| : (s,t) \in D\}$, and c_1, c_2 are constants independent of the function f and $n \in \mathbb{N}$.

2. Let $f : D \to \mathbb{R}$ be a function whose partial derivatives of the fourth order exist and are continuous on an open set containing D. Then for all $n \in \mathbb{N}$, we have

$$\left| \iint_D f(x,y)d(x,y) - \mathsf{S}_n(f) \right| \le c_3 \left(\frac{d}{2^n} \right)^4 \beta,$$

where

$$\beta := \max \left\{ \left| \frac{\partial^4 f}{\partial x^i \partial y^j}(s,t) \right| : (s,t) \in D \text{ and } i,j \ge 0,\ i+j = 4 \right\},$$

and c_3 is a constant independent of the function f and $n \in \mathbb{N}$.

The interested reader is referred to Theorem 5.1.3 and the discussion on symmetric triangulations on pages 173–175 of [3]. ◇

Approximations of triple integrals can be constructed on the same lines as the approximations we have constructed for double integrals. For example, if f is an integrable function on a cuboid $[a,b] \times [c,d] \times [p,q]$, then triple Riemann sums for f can be used for approximating the triple integral of f on $[a,b] \times [c,d] \times [p,q]$. Further, a product rule can be defined using a quadrature rule on each of the three intervals $[a,b]$, $[c,d]$, $[p,q]$, or using a quadrature rule on one of them and a cubature rule on the product of the other two. Also, if f is an integrable function on a tetrahedron D, then analogues of the Midpoint Rule, the Trapezoidal Rule, and Simpson's Rule can be obtained for approximating the triple integral of f over D. (See Exercise 42.) If D is a polyhedron in $\mathbb{R}^3$, then it can be partitioned into finitely many nonoverlapping tetrahedrons and we can use domain additivity to develop approximate methods for calculating a triple integral over D.

Notes and Comments

A recurrent theme in the first three sections of this chapter is reconciliation. It is standard in courses on one-variable calculus to define and determine the areas of planar regions between two curves by means of Riemann integrals. Now that we have a more general definition of area by means of double integrals, it seems imperative to indicate that the two seemingly different definitions are equivalent. We have shown this equivalence, in most cases, using Fubini's Theorem. Also, it is common in one-variable calculus to determine volumes of solids of revolution using two distinct methods, known as the washer method and the shell method. It is something of a mystery and a matter of faith why the volume calculated by either of these methods turns out to be the same. We have used Cavalieri's Principle to relate the formulas given by these methods to the general definition of volume in terms of triple integrals, thereby proving the equivalence of the two methods. We have also carried out a similar exercise for the more general variants of the washer method and the shell method, namely the slice method and the method of slivering by coaxial cylinders, that are sometimes discussed in one-variable calculus (for example, Section 8.2 of ACICARA*). It may be noted that even when these general variants are discussed, or when only the washer method and the shell method are discussed, the treatment in one-variable calculus is restricted to the case in which the axes of revolution are the coordinates axes, or at best, lines parallel to the coordinate axes. The reason, not normally revealed, is that the volume of a solid obtained by revolving a region in* $\mathbb{R}^2$ *about an arbitrary line not crossing it can be reduced to a double integral as given in Proposition 6.10, and not a Riemann integral. Moreover, we have also proved in Section 6.1 the invariance of area of a planar region under rotations and translations (Proposition 6.4) and outlined how a similar result holds for the volume of a solid in* $\mathbb{R}^3$*. The latter involves the so-called Euler angles, which appear to have been forgotten in books on calculus and analysis, but still seem to survive in books on mechanics and robotics, such as [25] and [41].*

While defining the area of a surface, we have restricted to parametrically defined piecewise smooth surfaces. Our discussion here runs parallel to our discussion of the length of a parametrically defined piecewise smooth curve given in Section 8.3 of ACICARA*. The motivation for the definition of the area of such a surface comes from considering the area of a parallelogram on the tangent plane of the surface, the area being equal to the square root of the sum of the squares of the areas of its projections on the three coordinate planes. An alternative approach could have been based on considering the surface areas of inscribed polyhedra formed of triangles. One expects that the sum of the areas of these triangles would tend to a limit, which could then be defined as the area of the surface. However, such a limit may not exist even for a simple-looking surface such as a cylinder. The interested readers may see Appendix A.4 of Chapter 4 in Courant and John [12, vol. II].*

The discussion of areas, volumes, and surface areas in the first two sections of this chapter paves the way, in Section 6.3, for a definition of the centroid of a far more general variety of planar regions, solids, and surfaces than what is usually done in one-variable calculus. Further, this enables us to prove a theorem of Pappus for solids of revolution, thus extending a version of this theorem proved in Section 8.5 of ACICARA *in the special cases in which the axis of revolution is the x-axis or the y-axis.*

In the last section of this chapter, we have discussed a topic that is often not covered in books on multivariable calculus, namely, methods for computing double and triple integrals approximately. We restrict mainly to double integrals and discuss two distinct ways for their approximate evaluations. One way is to use "products" of the quadrature rules for Riemann integrals. The other way is to develop approximations on the same principle that was followed in developing quadrature rules in Section 8.6 of ACICARA*, which is to approximate the given function by a piecewise constant function, or a piecewise linear function, or a piecewise quadratic function. This leads us to three types of so-called Triangular Prism Rules for approximating the double integral of a function defined on a triangular region or, more generally, on a polygonal region in $\mathbb{R}^2$. The three types are analogous to the Midpoint Rule, the Trapezoidal Rule, and Simpson's Rule for Riemann integrals. Error bounds for all the cubature rules are also discussed, albeit briefly. We have also pointed out the so-called curse of dimensionality, which makes approximating a double integral by product quadrature rules much less efficient compared to the approximation of a Riemann integral by the constituent quadrature rules. In any case, numerical methods for approximate evaluations of double and triple integrals are still useful, since one often comes across integrals that cannot be evaluated exactly. With this in view, the methods developed in the last section complement the formulas for areas, volumes, surface areas, and centroids given in the first three sections of this chapter. For more on the subject of approximations of multiple integrals, we refer to the books of Engles [17], Sobolev [52], and Stroud [55] as well as the article of Lyness and Jesperson [37] and the more advanced text of Sobolev and Vaskevich [53].*

Exercises

Part A

1. Find the area of the region bounded by the curves in $\mathbb{R}^2$ given by
   (i) $2y + x = 0$, $y + 2x = 0$, and $x + y = 1$,
   (ii) $x = y^2$ and $x = 2y - y^2$,
   (iii) $x + y = 0$ and $x = y - y^2$.
2. Let D denote the region in the first quadrant of the xy-plane bounded by the hyperbolas given by $xy = 1$ and $xy = 9$ as well as the lines given by $y = x$ and $y = 4x$. Find the area of D by effecting a change of variables given by $x := u/v$, $y := uv$, where $v > 0$.

3. Find the area of the region enclosed by one petal of the rose given by the polar equation $r = \cos 3\theta$.
4. Find the volume of the solid under the surface in $\mathbb{R}^3$ given by $z := x + 4$ and above the region in the xy-plane bounded by the parabola given by $y = 4 - x^2$ and the line given by $y = 3x$.
5. Let $a \in \mathbb{R}$ with $a > 0$. Find the volume of the solid
   (i) under the surface in $\mathbb{R}^3$ given by $z := (x^2+y^2)/a$ and above the region in the xy-plane bounded by the circle given by $x^2 + y^2 = a^2$,
   (ii) bounded by the sphere given by $x^2+y^2+z^2 = 2a^2$ and the paraboloid given by $az = x^2 + y^2$.
6. Find the volume of the solid in the first octant bounded by the coordinate planes, the cylinder given by $x^2+y^2 = 4$, and the plane given by $z+y = 3$.
7. A hemispherical bowl of radius 5 cm is filled with water to within 3 cm of the top. Find the volume of the water in the bowl.
8. Let $a \in \mathbb{R}$ with $a > 0$. Find the volume of the solid bounded by the surfaces given, in spherical coordinates, by
   (i) the sphere given by $\rho = a$ and the planes given by $\theta = 0$, $\theta = \pi/3$,
   (ii) the sphere given by $\rho = a$ and the cone given by $\varphi = \pi/3$,
   (iii) the sphere given by $\rho = a$ and the cones given by $\varphi = \pi/3$, $\varphi = 2\pi/3$,
   (iv) the sphere given by $\rho = 2$ and the surface given by $\rho = 1 + \cos\varphi$.
9. Let D be the disk in $\mathbb{R}^3$ given, in spherical coordinates, by the inequality $\rho \le 2\sin\varphi$ and the equation $\theta = \pi/2$. Find the volume of the solid generated by revolving D about the z-axis. Verify your answer using the Theorem of Pappus (Proposition 6.22).
10. Let $P_i := (x_i, y_i)$, $i = 1, 2, 3$, be three noncollinear points in $\mathbb{R}^2$, and let E denote the triangular region in $\mathbb{R}^2$ with P_1, P_2, P_3 as vertices. Fix $a, b, c \in \mathbb{R}$ with $(a, b) \ne (0, 0)$. Consider points $Q_i := (x_i, y_i, z_i)$, $i = 1, 2, 3$, in $\mathbb{R}^3$ lying in the plane given by $z = ax + by + c$. If D denotes the triangular region in $\mathbb{R}^3$ with Q_1, Q_2, Q_3 as vertices, then show that $\text{Area}(D) = \sqrt{1 + a^2 + b^2}\,\text{Area}(E)$.
11. Let $a \in \mathbb{R}$ with $a > 0$. Find the surface area of the part of a paraboloid given by $z^2 + x^2 = 2ay$ that is cut out by a plane given by $y = a$.
12. Let $a \in \mathbb{R}$ with $a > 0$. Find the area of the surface S_1 given, in cylindrical coordinates, by $r = a$ for $0 \le \theta \le \pi/2$, $0 \le z \le \theta$. Also, find the area of the surface S_2 given, in spherical coordinates, by $\rho = a$ for $0 \le \varphi \le \pi/2$, $0 \le \theta \le (\pi/2) - \varphi$.
13. Let $a \in \mathbb{R}$ with $a > 0$. Find the area of the surface in $\mathbb{R}^3$ given by $(a \sin\varphi\cos\theta,\ a\sin\varphi\sin\theta,\ a\cos\varphi)$, $(\varphi, \theta) \in [0, \pi] \times [0, \pi]$.
14. Let $a, b, h, \varphi \in \mathbb{R}$ with $0 < b < a$, $h > 0$ and $\varphi > 0$. Find the area of the surfaces given by the following.
   (i) $(a\cos t,\ a\sin t\cos\theta,\ a\sin t\sin\theta)$, $(t, \theta) \in [0, \varphi] \times [-\pi, \pi]$.
   (ii) $(t,\ at\cos\theta,\ at\sin\theta)$, $(t, \theta) \in [0, h] \times [-\pi, \pi]$.
   (iii) $((a + b\cos t)\cos\theta,\ b\sin t,\ (a + b\cos t)\sin\theta)$, $(t, \theta) \in [-\pi, \pi] \times [-\pi, \pi]$.

15. Let $a, \alpha, \beta, \gamma, \delta \in \mathbb{R}$ with $a > 0$, $0 \le \alpha < \beta \le \pi$, and $-\pi \le \gamma < \delta \le \pi$. Show that the area of the surface given, in spherical coordinates, by $\rho = a$, $(\varphi, \theta) \in [\alpha, \beta] \times [\gamma, \delta]$, is equal to $a^2(\cos\alpha - \cos\beta)(\delta - \gamma)$.
16. Let D be a bounded subset of $\mathbb{R}^2$, $f : D \to \mathbb{R}$ an integrable function, and $w : D \to \mathbb{R}$ a nonnegative integrable function such that $\iint_D w(x, y) d(x, y) \neq 0$. Prove the following statements.
    (i) If $m := \inf\{f(x, y) : (x, y) \in D\}$ and $M := \sup\{f(x, y) : (x, y) \in D\}$, then $m \le \text{Av}(f; w) \le M$.
    (ii) If D is closed as well as path-connected and f is continuous on D, then there is $(x_0, y_0) \in D$ such that $\text{Av}(f; w) = f(x_0, y_0)$.
17. Let D be a bounded subset of $\mathbb{R}^2$ and $f : D \to \mathbb{R}$ an integrable function. Give examples to show that $\text{Av}(f)$ need not be a value of f, where (i) f is continuous but D is not path-connected, and (ii) D is path-connected, but f is not continuous.
18. Find the centroid of each of the surfaces given in Exercises 11, 12, and 13.
19. Let $a \in \mathbb{R}$ with $a > 0$. Find the centroid of the triangular region bounded by the lines given by $x = 0$, $y = 0$, and $x + y = a$.
20. **(Formula of Pappus)** Let D be a bounded subset of $\mathbb{R}^2$ and let D_1, D_2 be subsets of D such that $D = D_1 \cup D_2$ and $D_1 \cap D_2$, ∂D_1, ∂D_2 are of content zero, but neither D_1 nor D_2 is of content zero. If $(\overline{x}, \overline{y})$ is the centroid of D and $(\overline{x}_i, \overline{y}_i)$ is the centroid of D_i for $i = 1, 2$, then show that

$$\overline{x} = \frac{\text{Area}(D_1)\overline{x}_1 + \text{Area}(D_2)\overline{x}_2}{\text{Area}(D)} \quad \text{and} \quad \overline{y} = \frac{\text{Area}(D_1)\overline{y}_1 + \text{Area}(D_2)\overline{y}_2}{\text{Area}(D)}.$$

Deduce that $(\overline{x}, \overline{y})$ lies on the line joining $(\overline{x}_1, \overline{y}_1)$ and $(\overline{x}_2, \overline{y}_2)$. Use this result to find the centroid of a quadrilateral whose vertices are (x_1, y_1), (x_2, y_2), (x_3, y_3), and (x_4, y_4). (Hint: Proposition 5.51)
21. For $i = 1, \ldots, 4$, let (x_i, y_i, z_i) be noncoplanar points in $\mathbb{R}^3$ and let D denote the tetrahedron having these points as its vertices. Show that the volume $\text{Vol}(D)$ of D is equal to $|d|/6$, where

$$d := \det \begin{bmatrix} x_2 - x_1 & y_2 - y_1 & z_2 - z_1 \\ x_3 - x_1 & y_3 - y_1 & z_3 - z_1 \\ x_4 - x_1 & y_4 - y_1 & z_4 - z_1 \end{bmatrix}.$$

Further, let $f : D \to \mathbb{R}$ be a polynomial function in three variables of total degree m, where m is a nonnegative integer, and let I denote the triple integral of f over D. Prove the following results.
(i) If $m \le 1$, then

$$I = \frac{\text{Vol}(D)}{4} \sum_{i=1}^{4} f(x_i, y_i, z_i).$$

(ii) If $m \le 2$, then

$$I = \frac{\text{Vol}(D)}{20} \left[4 \sum_{i=1}^{3} \sum_{j=i+1}^{4} f\left(\frac{x_i + x_j}{2}, \frac{y_i + y_j}{2}, \frac{z_i + z_j}{2} \right) - \sum_{i=1}^{4} f(x_i, y_i, z_i) \right].$$

(Hint: Consider the tetrahedron with $(0,0,0)$, $(1,0,0)$, $(0,1,0)$, and $(0,0,1)$ as its vertices, and the affine transformation $\Phi : \mathbb{R}^3 \to \mathbb{R}^3$ given by $\Phi := (\phi_1, \phi_2, \phi_3)$, where $\phi_1(u,v,w) := x_1 + (x_2 - x_1)u + (x_3 - x_1)v + (x_4 - x_1)w$, and ϕ_2, ϕ_3 are defined similarly with x replaced by y, z respectively. Use Exercise 24 of Chapter 5. Compare Proposition 6.29.)

22. Let D denote a tetrahedron in $\mathbb{R}^3$ with vertices (x_i, y_i, z_i), $i = 1, \ldots, 4$. Show that the centroid $(\overline{x}, \overline{y}, \overline{z})$ of D is given by

$$\overline{x} = \frac{1}{4}\sum_{i=1}^{4} x_i, \quad \overline{y} = \frac{1}{4}\sum_{i=1}^{4} y_i, \quad \overline{z} = \frac{1}{4}\sum_{i=1}^{4} z_i.$$

23. Find the x-coordinate of the centroid of the solid bounded by the planes given by $z = 0$, $z = x + 2$, and the elliptic cylinder given by $x^2 + 4y^2 = 4$.
24. Let $a \in \mathbb{R}$ with $a > 0$. Find the centroid of the solid bounded by the sphere given by $x^2 + y^2 + z^2 = 2a^2$ and the paraboloid given by $az = x^2 + y^2$. (Hint: Use cylindrical coordinates.)
25. Let $f : [a,b] \times [c,d] \to \mathbb{R}$ be an integrable function. For $n, k \in \mathbb{N}$, let $P_{n,k} := \{(x_{n,k,i}, y_{n,k,j}) : i = 0, 1, \ldots, n \text{ and } j = 0, 1, \ldots, k\}$ be a partition of $[a,b] \times [c,d]$ and let $(s_{n,k,i}, t_{n,k,j}) \in [x_{n,k,i-1}, x_{n,k,i}] \times [y_{n,k,j-1}, y_{n,k,j}]$ for $i = 1, \ldots, n$ and $j = 1, \ldots, k$. Define

$$S_{n,k} := \sum_{i=1}^{n}\sum_{j=1}^{k} f(s_{n,k,i}, t_{n,k,j})(x_{n,k,i} - x_{n,k,i-1})(y_{n,k,j} - y_{n,k,j-1}).$$

Suppose that for every $\epsilon > 0$, there is $(n_0, k_0) \in \mathbb{N}^2$ such that the mesh $\mu(P_{n,k})$ is less than ϵ for all $(n,k) \geq (n_0, k_0)$. Show that for every $\epsilon > 0$, there is $(n_1, k_1) \in \mathbb{N}^2$ such that

$$\left| \iint_{[a,b]\times[c,d]} f(x,y)d(x,y) - S_{n,k} \right| < \epsilon \quad \text{for all } (n,k) \geq (n_1, k_1).$$

26. For $n, k \in \mathbb{N}$, let M_n and M'_k denote the Compound Midpoint Rules on $[a,b]$ and $[c,d]$ respectively. Find the formula for the product cubature rule $M_n \times M'_k$ on $[a,b] \times [c,d]$.
27. For $n, k \in \mathbb{N}$, let S_n and S'_k denote the Compound Simpson Rules on $[a,b]$ and $[c,d]$ respectively. Find the formula for the product cubature rule $S_n \times S'_k$ on $[a,b] \times [c,d]$.
28. Let $f : [a,b] \times [c,d] \to \mathbb{R}$ be a function satisfying the conditions given in Example 6.26. Show that for $n, k \in \mathbb{N}$,

$$\left| \iint_{[a,b]\times[c,d]} f - (M_n \times M'_k)(f) \right| \leq \frac{(b-a)(d-c)}{24}\left[\frac{(b-a)^2\alpha}{n^2} + \frac{(d-c)^2\beta}{k^2}\right],$$

where α and β are defined as in Example 6.26.

29. State suitable conditions on a function $f : [a, b] \times [c, d] \to \mathbb{R}$ under which we have for $n, k \in \mathbb{N}$,

$$\left| \iint_{[a,b]\times[c,d]} f - (S_n \times S'_k)(f) \right| \le \frac{(b-a)(d-c)}{180} \left[\frac{(b-a)^3\alpha}{n^4} + \frac{(d-c)^3\beta}{k^4} \right],$$

where α and β denote, respectively, the suprema of the sets

$$\left\{ \left| \frac{\partial^4 f}{\partial x^4}(s, t) \right| : s \in (a, b),\ t \in [c, d] \right\} \text{ and } \left\{ \left| \frac{\partial^4 f}{\partial y^4}(s, t) \right| : s \in [a, b], t \in (c, d) \right\}.$$

30. Let $D := \{(x, y) \in \mathbb{R}^2 : 0 \le x \le 1 \text{ and } 0 \le y \le 1 - x\}$. If T_n denotes the Compound Trapezoidal Rule on $[0, 1]$ with n nodes, show that the product cubature rule $T_n \times \widetilde{T}_k$ over D associates to an integrable function $f : D \to \mathbb{R}$ the real number $(T_n \times \widetilde{T}_k)(f)$ given by

$$\frac{1}{4nk} \Bigg[f(0,0) + f(0,1) + 2\sum_{i=1}^{n-1}(1 - x_i)\big(f(x_i, 0) + f(x_i, 1 - x_i)\big)$$
$$+ 2\sum_{j=1}^{k-1} f(0, y_j) + 4\sum_{i=1}^{n-1}\sum_{j=1}^{k-1}(1 - x_i) f(x_i, (1 - x_i)y_j) \Bigg],$$

where $x_i = i/n$ for $i = 1, \ldots, n$ and $y_j = j/k$ for $j = 1, \ldots, k$. If $f := 1$ on D, check that $(T_n \times \widetilde{T}_k)(f)$ gives the area of the triangle D.

31. Let $D := \{(x, y) \in \mathbb{R}^2 : c \le y \le d \text{ and } \psi_1(y) \le x \le \psi_2(y)\}$ be an elementary domain, where $c, d \in \mathbb{R}$ with $c < d$ and $\psi_1, \psi_2 : [c, d] \to \mathbb{R}$ are integrable functions. For $\phi : [a, b] \to \mathbb{R}$ and $\psi : [c, d] \to \mathbb{R}$, let quadrature rules Q and R on $[a, b]$ and on $[c, d]$ be given by $Q(\phi) = \sum_{i=1}^{n} u_i \phi(x_i)$ and $R(\psi) = \sum_{j=1}^{k} v_j \psi(y_j)$ respectively. Construct a product cubature rule $\widetilde{Q} \times R$ over D by adapting the quadrature rule Q on $[a, b]$ to functions on the interval $[\psi_1(y_j), \psi_2(y_j)]$ for each $j = 1, \ldots, k$. Show that $(\widetilde{Q} \times R)(f) := \sum_{i=1}^{n}\sum_{j=1}^{k} u_{i,j} v_j f(x_{i,j}, y_j)$ for $f : [a, b] \times [c, d] \to \mathbb{R}$, where for $i = 1, \ldots, n$ and $j = 1, \ldots, k$,

$$u_{i,j} := \frac{\psi_2(y_j) - \psi_1(y_j)}{b - a} u_i \quad \text{and} \quad x_{i,j} := \frac{\psi_2(y_j) - \psi_1(y_j)}{b - a}(x_i - a)\psi_1(y_j).$$

(Compare the formula for the rule $Q \times \widetilde{R}$ given in the text.)

32. Let $D := \{(x, y) \in \mathbb{R}^2 : 0 \le y \le 1 \text{ and } 0 \le x \le 1 - y\}$. If T_n denotes the Trapezoidal Rule on $[0, 1]$ with n nodes, construct a product cubature rule $\widetilde{T}_n \times T_k$ over D similar to the cubature rule $T_n \times \widetilde{T}_k$ given in Exercise 30.

Part B

33. Let $a, b \in \mathbb{R}$ with $0 \le a < b$ and let $\alpha_1, \alpha_2 : [a, b] \to [-\pi, \pi]$ be continuous functions such that $\alpha_1 \le \alpha_2$. Let $D := \{(r\cos\theta,\ r\sin\theta) \in \mathbb{R}^2 : a \le r \le$

b and $\alpha_1(r) \le \theta \le \alpha_2(r)\}$ denote the region between the polar curves given by $\theta = \alpha_1(r)$, $\theta = \alpha_2(r)$ and between the circles given, in polar coordinates, by $r = a$, $r = b$. Show that $\text{Area}(D) := \int_a^b r[\alpha_2(r) - \alpha_1(r)]dr$. (Hint: Use the argument given in the proof of Proposition 6.3. Let $E := \{(r,\theta) \in \mathbb{R}^2 : r \ge 0, -\pi \le \theta \le \pi \text{ and } (r\cos\theta, r\sin\theta) \in D\}$. Then E is closed and ∂E is of content zero: If $E_0 := \{(r,\theta) \in \mathbb{R}^2 : a \le r \le b \text{ and } \alpha_1(r) \le \theta \le \alpha_2(r)\}$, $E_1 := \{(r,\pi) \in \mathbb{R}^2 : a \le r \le b \text{ and } \alpha_1(r) = -\pi\}$, and $E_2 := \{(r,-\pi) \in \mathbb{R}^2 : a \le r \le b \text{ and } \alpha_2(r) = \pi\}$, then $E = E_0 \cup E_1 \cup E_2$, provided $(0,0) \notin D$, and $E = E_0 \cup E_1 \cup E_2 \cup (\{0\} \times [-\pi,\pi])$, provided $(0,0) \in D$.)

34. Let $a, b \in \mathbb{R}$ with $0 \le a \le b$. Also, let $E \subseteq [a,b] \times \mathbb{R}$ and let Ω be an open subset of $\mathbb{R}^2$ containing E, and $\alpha : \Omega \to \mathbb{R}$ a function having continuous first-order partial derivatives with that $\alpha(E) \subseteq [-\pi,\pi]$. Show that if S is the surface given, in cylindrical coordinates, by $\theta = \alpha(r,z)$, $(r,z) \in E$, then
$$\text{Area}(S) := \iint_E \sqrt{1 + r^2(\alpha_r^2 + \alpha_z^2)}\, d(r,z).$$
Deduce that the surface area of the part of the spiral ramp in $\mathbb{R}^3$ given, in cylindrical coordinates, by $\theta = \alpha(r,z)$, where $\alpha(r,z) := z$ for $(r,z) \in [0,1] \times [0,\pi]$, is equal to $[\sqrt{2} + \ln(1+\sqrt{2})]\pi/2$.

35. Let $a, b, \alpha, \beta \in \mathbb{R}$ with $0 \le a < b$ and $-\pi \le \alpha < \beta \le \pi$. Also, let $E \subseteq [a,b] \times [\alpha,\beta]$ and let Ω be an open subset of $\mathbb{R}^2$ containing E, and $h : \Omega \to \mathbb{R}$ a function having continuous first-order partial derivatives. Show that if S is the surface given, in cylindrical coordinates, by $z = h(r,\theta)$, $(r,\theta) \in E$, then
$$\text{Area}(S) := \iint_E \sqrt{r^2(1 + h_r^2) + h_\theta^2}\, d(r,\theta).$$
Deduce that the surface area of the part of the paraboloid in $\mathbb{R}^3$ given, in cylindrical coordinates, by $z = h(r,\theta)$, where $h(r,\theta) := r^2$ for $(r,\theta) \in [0,1] \times [-\pi,\pi]$, is equal to $\pi(5\sqrt{5} - 1)/6$.

36. Let $a, b, \alpha, \beta \in \mathbb{R}$ with $0 \le a \le b$ and $-\pi \le \alpha < \beta \le \pi$. Also, let $E \subseteq [a,b] \times [\alpha,\beta]$ and let Ω be an open subset of $\mathbb{R}^2$ containing E, and $\gamma : \Omega \to \mathbb{R}$ a function having continuous first-order partial derivatives such that $\gamma(E) \subseteq [0,\pi]$. Show that if S is the surface given, in spherical coordinates, by $\varphi = \gamma(\rho,\theta)$, $(\rho,\theta) \in E$, then
$$\text{Area}(S) := \iint_E \rho\sqrt{\gamma_\theta^2 + \sin^2\gamma(1 + \rho^2\gamma_\rho^2)}\, d(\rho,\theta).$$
If $\varphi_0, \ell \in \mathbb{R}$ with $0 < \varphi_0 < \pi/2$ and $\ell > 0$, and if $E = [0,\ell] \times [-\pi,\pi]$, then deduce that the surface area of a part of the cone given, in spherical coordinates, by $\varphi = \gamma(\rho,\theta)$, where $\gamma(\rho,\theta) := \varphi_0$, $(\rho,\theta) \in E$, is equal to $\pi\ell^2 \sin\varphi_0$.

37. Let $a, b, \gamma, \delta \in \mathbb{R}$ with $0 \le a < b$ and $-\pi \le \gamma < \delta \le \pi$. Also, let $E \subseteq [a,b] \times [\gamma,\delta]$ and let Ω be an open subset of $\mathbb{R}^2$ containing E, and $\alpha : \Omega \to \mathbb{R}$ a function having continuous first-order partial derivatives such that $\alpha(E) \subseteq [-\pi,\pi]$. Show that if S is the surface given, in spherical

coordinates, by $\theta = \alpha(\rho, \varphi)$, $(\rho, \varphi) \in E$, then

$$\text{Area}(S) := \iint_E \rho\sqrt{1 + \sin^2\varphi(\alpha_\varphi^2 + \rho^2\alpha_\rho^2)}\, d(\rho, \varphi).$$

If $a \in \mathbb{R}$ with $a > 0$, $\varphi_0 \in [0, \pi]$, and $\theta_0 \in [-\pi, \pi]$, and if $E := [0, a] \times [0, \varphi_0]$, then deduce that the surface area of the sector of the disk given, in spherical coordinates, by $\theta = \alpha(\rho, \varphi)$, where $\alpha(\rho, \varphi) := \theta_0$, $(\rho, \varphi) \in E$, is equal to $a^2\varphi_0/2$.

38. Let C be a smooth curve in $\mathbb{R}^2$ given by $(x(t), y(t))$, $t \in [\alpha, \beta]$, and let L be a line given by $ax + by + c = 0$, where $a, b \in \mathbb{R}$ with $a^2 + b^2 = 1$ and $ax(t) + by(t) + c \geq 0$ for all $t \in [\alpha, \beta]$. Show that the surface S obtained by revolving C about L is given by $(\xi(t, \theta), \eta(t, \theta), \zeta(t, \theta))$, $(t, \theta) \in [\alpha, \beta] \times [-\pi, \pi]$, where

$$\begin{aligned}
\xi(t, \theta) &:= b\big(bx(t) - ay(t)\big) - ac + a\big(ax(t) + by(t) + c\big)\cos\theta,\\
\eta(t, \theta) &:= a\big(ay(t) - bx(t)\big) - bc + b\big(ax(t) + by(t) + c\big)\cos\theta,\\
\zeta(t, \theta) &:= (ax(t) + by(t) + c)\sin\theta.
\end{aligned}$$

Further, show that $(\xi_t^2 + \eta_t^2 + \zeta_t^2)(t, \theta) = x'(t)^2 + y'(t)^2$, $(\xi_\theta^2 + \eta_\theta^2 + \zeta_\theta^2)(t, \theta) = (ax(t) + by(t) + c)^2$ and $(\xi_t\xi_\theta + \eta_t\eta_\theta + \zeta_t\zeta_\theta)(t, \theta) = 0$ for all $(t, \theta) \in [\alpha, \beta] \times [-\pi, \pi]$. Deduce that

$$\text{Area}(S) = 2\pi \int_\alpha^\beta (ax(t) + by(t) + c)\sqrt{x'(t)^2 + y'(t)^2}\, dt.$$

39. Let D be a bounded, path-connected subset of $\mathbb{R}^2$, let $f : D \to \mathbb{R}$ be a continuous function, and $w : D \to \mathbb{R}$ a nonnegative integrable function such that $\iint_D w(x, y)d(x, y) \neq 0$. If D has an interior point at which w is continuous and positive, then show that there is $(x_0, y_0) \in D$ such that $\text{Av}(f; w) = f(x_0, y_0)$. (Hint: Consider the cases $m < \text{Av}(f, w) < M$, $\text{Av}(f, w) = m$, and $\text{Av}(f, w) = M$ separately.)
40. Let D be a bounded subset of $\mathbb{R}^2$ such that ∂D is of content zero, but D itself is not of content zero. If D is path-connected and $f : D \to \mathbb{R}$ is a continuous function, then show that there is $(x_0, y_0) \in D$ such that $\text{Av}(f) = f(x_0, y_0)$. (Hint: Exercise 39)
41. Let $a, b \in \mathbb{R}$ with $0 \leq a < b$ and let $\alpha_1, \alpha_2, f_1, f_2 : [a, b] \to \mathbb{R}$ be continuous functions such that $-\pi \leq \alpha_1 \leq \alpha_2 \leq \pi$ and $f_1 \leq f_2$. Let $D := \big\{(x\cos\theta, y, x\sin\theta) \in \mathbb{R}^3 : a \leq x \leq b,\ \alpha_1(x) \leq \theta \leq \alpha_2(x) \text{ and } f_1(x) \leq y \leq f_2(x)\big\}$. Show that the volume of D is given by

$$\text{Vol}(D) = \int_a^b x[\alpha_2(x) - \alpha_1(x)][f_2(x) - f_1(x)]dx$$

and the centroid $(\overline{x}, \overline{y}, \overline{z})$ of D is given by

$$\overline{x} = \frac{1}{\text{Vol}(D)} \int_a^b x^2[\sin \alpha_2(x) - \sin \alpha_1(x)][f_2(x) - f_1(x)]dx,$$
$$\overline{y} = \frac{1}{2\text{Vol}(D)} \int_a^b x[\alpha_2(x) - \alpha_1(x)][f_2(x)^2 - f_1(x)^2]dx,$$
$$\overline{z} = \frac{1}{\text{Vol}(D)} \int_a^b x^2[\cos \alpha_1(x) - \cos \alpha_2(x)][f_2(x) - f_1(x)]dx.$$

(Hint: For $x \in [a,b]$, consider $E_x := [\alpha_1(x), \alpha_2(x)] \times [f_1(x), f_2(x)]$.)

42. Let D denote a tetrahedron in $\mathbb{R}^3$ with four noncoplanar points (x_i, y_i, z_i), $i = 1, \ldots, 4$, as its vertices, and let $f : D \to \mathbb{R}$ be an integrable function. Show that the analogues of the Midpoint Rule, the Trapezoidal Rule, and Simpson's Rule for approximating triple integrals over D are given by the following rules.

(i) $\mathsf{C}(f) := \text{Vol}(D) f(\overline{x}, \overline{y}, \overline{z})$, where $\overline{x} = \frac{1}{4}\sum_{i=1}^{4} x_i,\ \overline{y} = \frac{1}{4}\sum_{i=1}^{4} y_i,\ \overline{z} = \frac{1}{4}\sum_{i=1}^{4} z_i.$

(ii) $\mathsf{T}(f) := \dfrac{\text{Vol}(D)}{4} \displaystyle\sum_{i=1}^{4} f(x_i, y_i, z_i).$

(iii) $\mathsf{S}(f) := \dfrac{\text{Vol}(D)}{20}\left[4\displaystyle\sum_{i=1}^{3}\sum_{j=i+1}^{4} f(x_{i,j}, y_{i,j}, z_{i,j}) - \sum_{i=1}^{4} f(x_i, y_i, z_i)\right]$, where

$x_{i,j} := \dfrac{x_i + x_j}{2},\ y_{i,j} := \dfrac{y_i + y_j}{2},\ z_{i,j} := \dfrac{z_i + z_j}{2}$ for $1 \le i < j \le 4$.

(Hint: Exercises 21 and 22.)

43. Let $n \in \mathbb{N}$ and $D \subseteq \mathbb{R}^n$ be as in Exercise 56 of Chapter 5. Further, let $f : D \to \mathbb{R}$ be a polynomial function in n variables of total degree m, where m is a nonnegative integer, and let I denote the n-fold integral of f over D. Prove the following results.

(i) If $m \le 1$, then
$$I = \frac{\text{Vol}(D)}{n+1} \sum_{i=1}^{n+1} f(x_1^{(i)}, \ldots, x_n^{(i)}).$$

(ii) If $m \le 2$, then
$$I = \frac{\text{Vol}(D)}{(n+1)(n+2)} \left[4\sum_{i=1}^{n}\sum_{j=i+1}^{n+1} f\left(\frac{x_1^{(i)} + x_1^{(j)}}{2}, \ldots, \frac{x_n^{(i)} + x_n^{(j)}}{2}\right) - (n-2)\sum_{i=1}^{n+1} f(x_1^{(i)}, \ldots, x_n^{(i)})\right].$$

(Compare Exercise 21.)

7

Double Series and Improper Double Integrals

In this chapter, we shall develop the theory of double sequences, double series, and improper double integrals. Our treatment will be analogous to the treatment of sequences, series, and improper integrals of functions of one variable given in Chapter 9 of ACICARA. Much of this chapter can be read independently of the previous chapters of this book.

In the preamble to Chapter 2, we mentioned that the notion of sequences in $\mathbb{R}$, that is, functions from $\mathbb{N}$ to $\mathbb{R}$, admits two generalizations in the setting of two variables: pairs of sequences and double sequences, that is, functions from $\mathbb{N}$ to $\mathbb{R}^2$ and functions from $\mathbb{N}^2$ to $\mathbb{R}$. The former were discussed in Section 2.1 and we shall now take up a study of the latter. Thus, in Section 7.1 below, we outline the theory of double sequences and the associated notions of convergence, boundedness, monotonicity, etc. Double series and their convergence is discussed in Section 7.2. Various tests for determining the convergence or divergence of a double series are given in Section 7.3. In Section 7.4, double power series are treated as a special case of double series, and Taylor double series of infinitely differentiable functions are treated as a special case of double power series. We then turn, in Section 7.5, to a "continuous" analogue of double series, namely improper double integrals of functions defined on a set of the form $[a, \infty) \times [c, \infty)$, where $a,\ c \in \mathbb{R}$. Tests for the convergence of an improper double integral are given in Section 7.6. Finally, in Section 7.7, the process of double integration is extended to functions defined on an unbounded subset of $\mathbb{R}^2$, and to unbounded functions defined on a bounded subset of $\mathbb{R}^2$.

7.1 Double Sequences

This section on double sequences is meant as a preparation for the subsequent sections on double series.

A **double sequence** (in $\mathbb{R}$) is a real-valued function whose domain is the set $\mathbb{N}^2 := \{(m, n) : m, n \in \mathbb{N}\}$ of all pairs of positive integers. We shall denote

S.R. Ghorpade and B.V. Limaye, *A Course in Multivariable Calculus and Analysis*, Undergraduate Texts in Mathematics, DOI 10.1007/978-1-4419-1621-1_7,

double sequences by $(a_{m,n})$, $(b_{m,n})$, and so on, or by $(A_{m,n})$, $(B_{m,n})$, and so on. The value of a double sequence $(a_{m,n})$ at $(m,n) \in \mathbb{N}^2$ is $a_{m,n}$, and this is called the (m,n)th **term** of that double sequence.

We shall use the attributes "bounded above," or "bounded below," and "bounded" for a double sequence just as we use them for a function of two variables. We shall use the componentwise partial order on $\mathbb{N}^2$ given by

$$(m_1, n_1) \le (m_2, n_2) \iff m_1 \le m_2 \text{ and } n_1 \le n_2$$

for (m_1, n_1) and (m_2, n_2) in $\mathbb{N}^2$.

We say that a double sequence $(a_{m,n})$ is **convergent** if there is $a \in \mathbb{R}$ satisfying the following condition: For every $\epsilon > 0$, there is $(m_0, n_0) \in \mathbb{N}^2$ such that

$$|a_{m,n} - a| < \epsilon \quad \text{for all } (m,n) \ge (m_0, n_0).$$

In this case, we say that $(a_{m,n})$ **converges** to a and write $a_{m,n} \to a$ (as $(m,n) \to (\infty, \infty)$). It is easy to see that the real number a is then unique; it is called the **limit** or the **double limit** of $(a_{m,n})$, and is denoted by

$$\lim_{(m,n)\to(\infty,\infty)} a_{m,n}.$$

A double sequence that is not convergent is said to be **divergent**. In particular, if for every $\alpha \in \mathbb{R}$, there is $(m_0, n_0) \in \mathbb{N}^2$ such that $a_{m,n} > \alpha$ for all $(m,n) \ge (m_0, n_0)$, then we say that $(a_{m,n})$ **diverges** to ∞ and we write $a_{m,n} \to \infty$. Similarly, $(a_{m,n})$ **diverges** to $-\infty$ if for every $\beta \in \mathbb{R}$, there is $(m_0, n_0) \in \mathbb{N}^2$ such that $a_{m,n} < \beta$ for all $(m,n) \ge (m_0, n_0)$. For example, if $a_{m,n} := 1/(m+n)$, $b_{m,n} := m+n$, and $c_{m,n} = (-1)^{m+n}$ for $(m,n) \in \mathbb{N}^2$, then $a_{m,n} \to 0$ and $(b_{m,n})$ diverges to ∞, while the double sequence $(c_{m,n})$ is bounded, but divergent.

We recall that a convergent sequence is bounded. However, a convergent double sequence may not be bounded. For example, let $a_{m,n} := n$ if $m = 1$, $a_{m,n} := m$ if $n = 1$, and $a_{m,n} := 0$ if $m \ne 1$ and $n \ne 1$. Then $a_{m,n} \to 0$, since $a_{m,n} = 0$ for all $(m,n) \ge (2,2)$, but clearly $(a_{m,n})$ is not bounded.

As indicated by the above example, the convergence of a double sequence $(a_{m,n})$ is not altered if some of the $a_{m,n}$'s are changed, provided there is $(m_0, n_0) \in \mathbb{N}^2$ such that either $m \le m_0$ or $n \le n_0$ whenever a term $a_{m,n}$ is changed. Let us write a double sequence $(a_{m,n})$ schematically as follows:

$$\begin{array}{cccc} a_{1,1} & a_{1,2} & a_{1,3} & \cdots \\ a_{2,1} & a_{2,2} & a_{2,3} & \cdots \\ a_{3,1} & a_{3,2} & a_{3,3} & \cdots \\ \vdots & \vdots & \vdots & \end{array}$$

Then some or all $a_{m,n}$'s written in a finite number of rows and/or in a finite number of columns can be changed without altering the convergence of $(a_{m,n})$. (Note that each row and each column contains an infinite number of $a_{m,n}$'s.)

The **Limit Theorem** for double sequences says that if $a_{m,n} \to a$ and $b_{m,n} \to b$, then $a_{m,n} + b_{m,n} \to a + b$, $ra_{m,n} \to ra$ for any $r \in \mathbb{R}$, $a_{m,n}b_{m,n} \to ab$, and if $a \neq 0$, then there is $(m_0, n_0) \in \mathbb{N}^2$ such that $a_{m,n} \neq 0$ for all $(m, n) \geq (m_0, n_0)$ and $1/a_{m,n} \to 1/a$; further, if there is $(m_1, n_1) \in \mathbb{N}^2$ such that $a_{m,n} \leq b_{m,n}$ for all $(m, n) \geq (m_1, n_1)$, then $a \leq b$, and if $a_{m,n} \geq 0$ for all $(m, n) \in \mathbb{N}^2$, then $a_{m,n}^{1/k} \to a^{1/k}$ for any $k \in \mathbb{N}$. Also, if $a_{m,n} \to a$, then $|a_{m,n}| \to |a|$, but the converse does not hold unless $a = 0$. Proofs of these results are routine.

Another useful result is the **Sandwich Theorem** for double sequences: If $(a_{m,n})$, $(b_{m,n})$, and $(c_{m,n})$ are double sequences such that $a_{m,n} \leq c_{m,n} \leq b_{m,n}$, and if $c \in \mathbb{R}$ is such that $a_{m,n} \to c$ and $b_{m,n} \to c$, then $c_{m,n} \to c$ as well. Again, the proof is routine.

A double sequence $(a_{m,n})$ is called a **Cauchy double sequence** if for every $\epsilon > 0$, there is $(m_0, n_0) \in \mathbb{N}^2$ such that

$$|a_{m,n} - a_{p,q}| < \epsilon \quad \text{for all } (m, n), (p, q) \geq (m_0, n_0).$$

The following result allows us to prove the convergence of a double sequence without having to guess its limit beforehand.

Proposition 7.1 (Cauchy Criterion for Double Sequences). *A double sequence is convergent if and only if it is a Cauchy double sequence.*

Proof. It is easy to see that a convergent double sequence is a Cauchy double sequence. Conversely, let $(a_{m,n})$ be a Cauchy double sequence and consider the diagonal sequences (b_n) defined by $b_n := a_{n,n}$ for $n \in \mathbb{N}$. Then clearly (b_n) is a Cauchy sequence, and by the Cauchy criterion for sequences given in one-variable calculus (for example, Proposition 2.19 of ACICARA), (b_n) is convergent. Let $b_n \to b$ and let $\epsilon > 0$ be given. Then there is $n_0 \in \mathbb{N}$ such that

$$|b_n - b| < \frac{\epsilon}{2} \quad \text{for all } n \geq n_0.$$

Since $(a_{m,n})$ is Cauchy, there is $n_1 \in \mathbb{N}$ such that $n_1 \geq n_0$ and

$$|a_{m,n} - a_{p,q}| < \frac{\epsilon}{2} \quad \text{for all } (m, n), (p, q) \geq (n_1, n_1),$$

and consequently,

$$|a_{m,n} - b| \leq |a_{m,n} - a_{n_1,n_1}| + |b_{n_1} - b| < \frac{\epsilon}{2} + \frac{\epsilon}{2} = \epsilon \quad \text{for all } (m, n) \geq (n_1, n_1).$$

Thus $(a_{m,n})$ converges to b. □

We shall now consider iterated limits of a double sequence. The following result is similar to Exercise 28 of Chapter 2 on iterated limits of a function of two real variables.

Proposition 7.2 (Iterated Limits of Double Sequences). *Suppose* $(a_{m,n})$ *is a convergent double sequence and let* $a_{m,n} \to a$.

(i) *If* $\lim_{n\to\infty} a_{m,n}$ *exists for each* $m \in \mathbb{N}$, *then the* **iterated limit**

$$\lim_{m\to\infty}\left(\lim_{n\to\infty} a_{m,n}\right)$$

exists and it is equal to a.

(ii) *If* $\lim_{m\to\infty} a_{m,n}$ *exists for each* $n \in \mathbb{N}$, *then the* **iterated limit**

$$\lim_{n\to\infty}\left(\lim_{m\to\infty} a_{m,n}\right)$$

exists and it is equal to a.

(iii) *If the hypotheses in* (i) *and* (ii) *above hold, then the double sequence* $(a_{m,n})$ *is bounded and*

$$\lim_{m\to\infty}\left(\lim_{n\to\infty} a_{m,n}\right) = a = \lim_{n\to\infty}\left(\lim_{m\to\infty} a_{m,n}\right).$$

Proof. Let $\epsilon > 0$ be given. Since $a_{m,n} \to a$, there is $(m_0, n_0) \in \mathbb{N}^2$ such that

$$|a_{m,n} - a| < \frac{\epsilon}{2} \quad \text{for all } (m,n) \geq (m_0, n_0).$$

Assume that $\lim_{n\to\infty} a_{m,n}$ exists for each $m \in \mathbb{N}$ and let us denote it by b_m. Then for each fixed $m \in \mathbb{N}$, there is $k_m \in \mathbb{N}$ such that

$$|a_{m,n} - b_m| < \frac{\epsilon}{2} \quad \text{for all } n \geq k_m.$$

For $m \geq m_0$, if we let $n_1 := \max\{n_0, k_m\}$, then

$$|b_m - a| \leq |b_m - a_{m,n_1}| + |a_{m,n_1} - a| < \frac{\epsilon}{2} + \frac{\epsilon}{2} = \epsilon.$$

Thus $\lim_{m\to\infty} b_m$ exists and it is equal to a. This proves (i). The proof of (ii) is similar.

Suppose now that the hypotheses in (i) and (ii) hold. Since $|a_{m,n}| \to |a|$, there is $(m_1, n_1) \in \mathbb{N}^2$ such that

$$|a_{m,n}| < 1 + |a| \quad \text{for all } (m,n) \geq (m_1, n_1).$$

Also, for each fixed m with $1 \leq m < m_1$, the sequence $(a_{m,n})$ is bounded since $\lim_{n\to\infty} a_{m,n}$ exists. Thus there is $\alpha \geq 0$ such that $|a_{m,n}| \leq \alpha$ if $1 \leq m < m_1$ and $n \in \mathbb{N}$. Similarly, there is $\beta \geq 0$ such that $|a_{m,n}| \leq \beta$ if $m \in \mathbb{N}$ and $1 \leq n < n_1$. Hence $|a_{m,n}| \leq \max\{1 + |a|, \alpha, \beta\}$ for all $(m,n) \in \mathbb{N}^2$. Thus $(a_{m,n})$ is bounded. The last part of (iii) follows from (i) and (ii). □

We give examples to show that if any of the hypotheses in the above proposition is not satisfied, then the conclusion(s) may not hold.

Examples 7.3. (i) Let $a_{m,n} := (-1)^{m+n}(m+n)/mn$ for $(m,n) \in \mathbb{N}^2$. Since $|a_{m,n}| \le (1/n) + (1/m)$ for $(m,n) \in \mathbb{N}^2$, we see that $a_{m,n} \to 0$. However, $\lim_{n\to\infty} a_{m,n}$ does not exist for any fixed $m \in \mathbb{N}$. Indeed,

$$a_{m,n} = (-1)^m \left[\frac{(-1)^n}{n} + \frac{(-1)^n}{m}\right] \quad \text{for all } (m,n) \in \mathbb{N}^2$$

and $(-1)^n/n \to 0$ as $n \to \infty$, while $\lim_{n\to\infty}(-1)^n/m$ does not exist.

(ii) Let $a_{m,n} := mn/(m^2+n^2)$ for $(m,n) \in \mathbb{N}^2$. Then for each fixed $m \in \mathbb{N}$, $\lim_{n\to\infty} a_{m,n}$ exists and it is equal to 0, since $|a_{m,n}| \le m/n$ for all $n \in \mathbb{N}$. Similarly, for each fixed $n \in \mathbb{N}$, $\lim_{m\to\infty} a_{m,n}$ exists and it is equal to 0. However, $(a_{m,n})$ is not convergent, since $a_{m,n} = 1/2$ if $m = n$ and $a_{m,n} = 2/5$ if $m = 2n$.

(iii) Let $a_{m,n} := m/(m+n)$ for $(m,n) \in \mathbb{N}^2$. Then for each fixed $m \in \mathbb{N}$, $\lim_{n\to\infty} a_{m,n} = 0$ and for each fixed $n \in \mathbb{N}$, $\lim_{m\to\infty} a_{m,n} = 1$. Hence $\lim_{m\to\infty}(\lim_{n\to\infty} a_{m,n}) = 0$, whereas $\lim_{n\to\infty}(\lim_{m\to\infty} a_{m,n}) = 1$. Notice that $(a_{m,n})$ is not convergent, since $a_{m,n} = 1/2$ if $m = n$ and $a_{m,n} = 2/3$ if $m = 2n$. ◇

Monotonicity and Bimonotonicity

We now consider monotonicity of a double sequence in analogy with the monotonicity of real-valued functions defined on $I \times J$, where I and J are intervals in $\mathbb{R}$. (See Section 1.2.) We say that a double sequence $(a_{m,n})$ is **monotonically increasing** if $a_{m,n} \le a_{m+1,n}$ and $a_{m,n} \le a_{m,n+1}$ for all $(m,n) \in \mathbb{N}^2$. Likewise we say that it is **monotonically decreasing** if $a_{m,n} \ge a_{m+1,n}$ and $a_{m,n} \ge a_{m,n+1}$ for all $(m,n) \in \mathbb{N}^2$. Observe that a double sequence $(a_{m,n})$ is monotonically increasing if and only if

$$a_{m,n} \le a_{p,q} \quad \text{for all } (m,n), (p,q) \in \mathbb{N}^2 \text{ with } (m,n) \le (p,q).$$

Also, a double sequence $(a_{m,n})$ is monotonically increasing if and only if for each fixed $m \in \mathbb{N}$, the sequence given by $n \longmapsto a_{m,n}$ is (monotonically) increasing and for each fixed $n \in \mathbb{N}$, the sequence given by $m \longmapsto a_{m,n}$ is (monotonically) increasing. Likewise for monotonically decreasing double sequences. A double sequence is said to be **monotonic** if it is monotonically increasing or monotonically decreasing.

We have noted earlier that a convergent double sequence may not be bounded and a bounded double sequence may not be convergent. With this in view, the following result is noteworthy.

Proposition 7.4. (i) *A monotonically increasing double sequence* $(a_{m,n})$ *is convergent if and only if it is bounded above. In this case,*

$$a_{m,n} \to \sup\{a_{m,n} : (m,n) \in \mathbb{N}^2\},$$

If $(a_{m,n})$ *is monotonically increasing, but not bounded above, then* $a_{m,n} \to \infty$.

(ii) *A monotonically decreasing double sequence* $(a_{m,n})$ *is convergent if and only if it is bounded below. In this case,*

$$a_{m,n} \to \inf\{a_{m,n} : (m,n) \in \mathbb{N}^2\}.$$

If $(a_{m,n})$ *is monotonically decreasing, but not bounded below, then* $a_{m,n} \to -\infty$.

Proof. Let $(a_{m,n})$ be a monotonically increasing double sequence. Suppose it is bounded above, and let $a := \sup\{a_{m,n} : (m,n) \in \mathbb{N}^2\}$. Given $\epsilon > 0$, there is $(m_0, n_0) \in \mathbb{N}^2$ such that $a - \epsilon < a_{m_0,n_0}$. Hence

$$a - \epsilon < a_{m_0,n_0} \le a_{m,n} \le a < a + \epsilon \quad \text{for all } (m,n) \ge (m_0,n_0).$$

Thus $a_{m,n} \to a$.

Conversely, suppose $(a_{m,n})$ is convergent and $a_{m,n} \to a$. Then there is $(m_0, n_0) \in \mathbb{N}^2$ such that

$$a_{m,n} < a + 1 \quad \text{for all } (m,n) \ge (m_0,n_0).$$

Now given any $(m,n) \in \mathbb{N}^2$, we have $(m + m_0, n + n_0) \ge (m,n)$ as well as $(m + m_0, n + n_0) \ge (m_0, n_0)$, and so

$$a_{m,n} \le a_{m+m_0,n+n_0} < a + 1.$$

Thus $(a_{m,n})$ is bounded above by $a + 1$.

If $(a_{m,n})$ is not bounded above, then given $\alpha \in \mathbb{R}$, there is $(m_0, n_0) \in \mathbb{N}^2$ such that $a_{m_0,n_0} > \alpha$. But then $a_{m,n} \ge a_{m_0,n_0} > \alpha$ for all $(m,n) \ge (m_0,n_0)$. Thus $a_{m,n} \to \infty$. This completes the proof of (i).

A similar proof can be given for (ii). □

Corollary 7.5. *A monotonic double sequence* $(a_{m,n})$ *is convergent if and only if the sequence* $(a_{p,p})$ *of its diagonal terms is convergent. In this case,*

$$\lim_{(m,n)\to(\infty,\infty)} a_{m,n} = \lim_{p\to\infty} a_{p,p}.$$

Proof. Suppose $(a_{m,n})$ is a monotonically increasing sequence. If for any $(m,n) \in \mathbb{N}^2$, we let $p := \max\{m,n\}$, then $a_{m,n} \le a_{p,p}$. Consequently, $\{a_{m,n} : (m,n) \in \mathbb{N}^2\}$ is bounded above if and only if $\{a_{p,p} : p \in \mathbb{N}\}$ is bounded above, and in this case, $\sup\{a_{m,n} : (m,n) \in \mathbb{N}^2\} = \sup\{a_{p,p} : p \in \mathbb{N}\}$. Hence Proposition 7.4 and its analogue for sequences (given, for example, in Proposition 2.8 of ACICARA) yield the desired result. The case of when $(a_{m,n})$ is a monotonically decreasing double sequence is proved similarly. □

Finally, we consider a bivariate version of monotonicity in analogy with the bimonotonicity of real-valued functions defined on $I \times J$, where I and J are intervals in $\mathbb{R}$. (See Section 1.2.) This notion will be useful in treating conditional convergence of a double series in Section 7.3.

We say that a double sequence $(a_{m,n})$ is **bimonotonically increasing** if $a_{m,n+1} + a_{m+1,n} \le a_{m,n} + a_{m+1,n+1}$ for all $(m,n) \in \mathbb{N}^2$. Likewise we say that it is **bimonotonically decreasing** if $a_{m,n+1} + a_{m+1,n} \ge a_{m,n} + a_{m+1,n+1}$ for all $(m,n) \in \mathbb{N}^2$. Observe that for any $(m,n), (p,q) \in \mathbb{N}^2$ with $(m,n) \le (p,q)$,

$$a_{m,n} + a_{p,q} - a_{m,q} - a_{p,n} = \sum_{i=m}^{p-1} \sum_{j=n}^{q-1} \left(a_{i,j} + a_{i+1,j+1} - a_{i,j+1} - a_{i+1,j}\right),$$

and so a double sequence $(a_{m,n})$ is bimonotonically increasing if and only if

$$a_{m,q} + a_{p,n} \le a_{m,n} + a_{p,q} \quad \text{for all } (m,n), (p,q) \in \mathbb{N}^2 \text{ with } (m,n) \le (p,q).$$

It is readily seen that a similar characterization holds for bimonotonically decreasing sequences. A double sequence is said to be **bimonotonic** if it is bimonotonically increasing or bimonotonically decreasing.

The following proposition is often useful for constructing several examples of monotonic and bimonotonic double sequences.

Proposition 7.6. *Given any sequences (α_n) and (β_n) in $\mathbb{R}$, consider the double sequences $(a_{m,n})$ and $(b_{m,n})$ defined by*

$$a_{m,n} := \alpha_m + \beta_n \quad \text{and} \quad b_{m,n} := \alpha_m \beta_n \qquad \text{for } (m,n) \in \mathbb{N}^2.$$

The following results hold.

(i) *$(a_{m,n})$ is monotonically increasing if and only if both (α_n) and (β_n) are increasing.*
(ii) *Assume that $\alpha_n \ge 0$ and $\beta_n \ge 0$ for all $n \in \mathbb{N}$, and also that $\alpha_{m_0} > 0$ and $\beta_{n_0} > 0$ for some $m_0, n_0 \in \mathbb{N}$. Then $(b_{m,n})$ is monotonically increasing if and only if both (α_n) and (β_n) are increasing.*
(iii) *$(a_{m,n})$ is always bimonotonically increasing as well as bimonotonically decreasing.*
(iv) *If (α_n) and (β_n) are monotonic, then $(b_{m,n})$ is bimonotonic. More specifically, if (α_n) and (β_n) are both increasing or both decreasing, then $(b_{m,n})$ is bimonotonically increasing, whereas if (α_n) is increasing and (β_n) is decreasing, or vice versa, then $(b_{m,n})$ is bimonotonically decreasing.*

Proof. Both (i) and (ii) are straightforward consequences of the definitions, whereas (iii) and (iv) follow from noting that $a_{p,q} + a_{m,n} = a_{m,q} + a_{p,n}$ and $b_{p,q} + b_{m,n} - b_{m,q} - b_{p,n} = (\alpha_p - \alpha_m)(\beta_q - \beta_n)$ for all $(m,n), (p,q) \in \mathbb{N}^2$. □

Results similar to parts (i) and (ii) of Proposition 7.6 hold for monotonically decreasing double sequences.

Examples 7.7. (i) Let $a_{m,n} := m+n$ and $b_{m,n} := mn$ for $(m,n) \in \mathbb{N}^2$. Then both the double sequences $(a_{m,n})$ and $(b_{m,n})$ are monotonically increasing as well as bimonotonically increasing. On the other hand, if we let $c_{m,n} :=$

$m - n$ for $(m, n) \in \mathbb{N}^2$, then the double sequence $(c_{m,n})$ is bimonotonic, but not monotonic, whereas if for $(m, n) \in \mathbb{N}^2$, we let

$$d_{m,n} := \begin{cases} -1 & \text{if } m = 1 = n, \\ mn & \text{if } m > 1 \text{ or } n > 1, \end{cases}$$

then the double sequence $(d_{m,n})$ is monotonic, but not bimonotonic. Indeed, $(d_{m,n})$ is clearly monotonically increasing, but it is not bimonotonic since $d_{1,2}+d_{2,1} = 4 > 3 = d_{1,1}+d_{2,2}$ and $d_{2,3}+d_{3,2} = 12 < 13 = d_{2,2}+d_{3,3}$.

(ii) Let $p \in \mathbb{R}$ and $a_{m,n} := (m + n)^p$ for $(m, n) \in \mathbb{N}^2$. Using the results in Example 1.8 (iii), we can easily see that the double sequence $(a_{m,n})$ is monotonically decreasing and bimonotonically increasing if $p \le 0$, monotonically increasing and bimonotonically decreasing if $0 \le p \le 1$, and both monotonically and bimonotonically increasing if $p \ge 1$. For another example of this kind, see Exercise 9. ◇

In Exercises 52 and 53, we introduce the concepts of **bounded variation** and **bounded bivariation** for a double sequence and explore their relationships with monotonic and bimonotonic double sequences.

7.2 Convergence of Double Series

A **double series** of real numbers is an ordered pair $\big((a_{k,\ell}), (A_{m,n})\big)$ of double sequences of real numbers such that

$$A_{m,n} = \sum_{k=1}^{n} \sum_{\ell=1}^{m} a_{k,\ell} \quad \text{for all } (m, n) \in \mathbb{N}^2.$$

(We note that for each $(m, n) \in \mathbb{N}^2$, the finite double sum given above is independent of the order in which it is taken.) Equivalently, a double series is an ordered pair $\big((a_{k,\ell}), (A_{m,n})\big)$ of double sequences of real numbers such that

$$a_{k,\ell} = A_{k,\ell} - A_{k,\ell-1} - A_{k-1,\ell} + A_{k-1,\ell-1} \quad \text{for all } (k, \ell) \in \mathbb{N}^2,$$

where $A_{k,0} := 0$ for all $k = 0, 1, 2, \ldots$ and $A_{0,\ell} := 0$ for all $\ell = 0, 1, 2, \ldots$ with the standard convention that an empty sum is equal to zero. The first double sequence $(a_{k,\ell})$ is called the **double sequence of terms**, and the second double sequence $(A_{m,n})$ is called the **double sequence of partial double sums** of the double series $\big((a_{k,\ell}), (A_{m,n})\big)$. The two double sequences $(a_{k,\ell})$ and $(A_{m,n})$ determine each other uniquely. We shall use an informal but suggestive notation $\sum\sum_{(k,\ell)} a_{k,\ell}$ for the double series $\big((a_{k,\ell}), (A_{m,n})\big)$. Sometimes it is convenient to allow the indices k and ℓ to take the values $k = k_0, k_0 + 1, \ldots$ and $\ell = \ell_0, \ell_0 + 1, \ldots$ for some fixed pair (k_0, ℓ_0) of integers.

We say that a double series $\sum\sum_{(k,\ell)} a_{k,\ell}$ is **convergent** if the double sequence $(A_{m,n})$ of its partial double sums is convergent. If $(A_{m,n})$ converges to A, then the (unique) real number A is called the **double sum** of the double series $\sum\sum_{(k,\ell)} a_{k,\ell}$ and it is denoted by the same symbol used to denote the double series. Thus, when we write

$$\sum\sum_{(k,\ell)} a_{k,\ell} = A,$$

we mean that the double series on the left is convergent and its double sum is the real number A. In this case, we may also say that $\sum\sum_{(k,\ell)} a_{k,\ell}$ **converges** to A. A double series that is not convergent is said to be **divergent**. In particular, we say that the double series **diverges** to ∞ or to $-\infty$ according as the double sequence of its partial double sums tends to ∞ or to $-\infty$. The convergence of a double series is not affected if we change a finite number of its terms, although its double sum may be altered by doing so. On the other hand, if we change an infinite number of terms (even if they belong to a single row or a single column) we may affect the convergence of the double series. For example, if $a_{k,\ell} := 0$ for all $(k,\ell) \in \mathbb{N}^2$, then clearly $\sum\sum_{(k,\ell)} a_{k,\ell}$ is convergent. But if we let $b_{k,1} := 1$ for $k \in \mathbb{N}$ and $b_{k,\ell} := 0$ for all $(k,\ell) \geq (1,2)$, then $\sum\sum_{(k,\ell)} b_{k,\ell}$ diverges to ∞. (This is in contrast to the effect on the convergence of a double sequence when some of its terms are changed, as discussed in Section 7.1.)

It may be noted that a (single) series $\sum_k a_k$ can be thought of as a double series $\sum\sum_{(k,\ell)} a_{k,\ell}$ if we define $a_{k,1} := a_k$ for $k \in \mathbb{N}$ and $a_{k,\ell} := 0$ for $k \in \mathbb{N}$ and $\ell \geq 2$. In this case, $A_{m,n} = \sum_{k=1}^{m} a_k$ for all $(m,n) \in \mathbb{N}^2$. Consequently, examples considered in the theory of (single) series work for double series as well. As in the case of series, a quick and useful way to show that a double series is divergent is to use the following result, which gives a necessary condition for the convergence of a double series.

Proposition 7.8 ((k,ℓ)th Term Test). *If $\sum\sum_{(k,\ell)} a_{k,\ell}$ is convergent, then $a_{k,\ell} \to 0$ as $k,\ell \to \infty$. In other words, if $a_{k,\ell} \not\to 0$ as $k,\ell \to \infty$, then $\sum\sum_{(k,\ell)} a_{k,\ell}$ is divergent.*

Proof. Let $\sum\sum_{(k,\ell)} a_{k,\ell}$ be a convergent double series. If $(A_{m,n})$ is the double sequence of its partial double sums and A is its double sum, then we have $a_{k,\ell} = A_{k,\ell} - A_{k-1,\ell} - A_{k,\ell-1} + A_{k-1,\ell-1} \to A - A - A + A = 0$. □

It will be seen in Example 7.10 (iii) below that the two series $\sum\sum_{(k,\ell)} 1/k\ell$ and $\sum\sum_{(k,\ell)} 1/(k+\ell)$ diverge to ∞ even though their (k,ℓ)th terms tend to 0 as $(k,\ell) \to (\infty,\infty)$. Thus the converse of the (k,ℓ)th Term Test (Proposition 7.8) is not true. A variant of the (k,ℓ)th Term Test, which may be called **Abel's (k,ℓ)th Term Test**, is given in Exercise 10.

The following result gives a sufficient condition for the convergence of certain "product series," and is often helpful. A refined version of this result is given in Exercise 8.

Proposition 7.9. *Let* $\sum_k b_k$ *and* $\sum_\ell c_\ell$ *be series of real numbers and let* $a_{k,\ell} := b_k c_\ell$ *for* $(k,\ell) \in \mathbb{N}^2$. *Then the following results hold.*

(i) *If* $\sum_k b_k$ *and* $\sum_\ell c_\ell$ *are both convergent, then the double series* $\sum\sum_{(k,\ell)} a_{k,\ell}$ *is convergent and moreover,* $\sum\sum_{(k,\ell)} a_{k,\ell} = \left(\sum_k b_k\right)\left(\sum_\ell c_\ell\right)$.

(ii) *If both* $\sum_k b_k$ *and* $\sum_\ell c_\ell$ *diverge to* ∞, *then the double series* $\sum\sum_{(k,\ell)} a_{k,\ell}$ *diverges to* ∞.

(iii) *If* $\sum_k b_k$ *converges to* B *and* $B \neq 0$, *while* $\sum_\ell c_\ell$ *is divergent, then the double series* $\sum\sum_{(k,\ell)} b_k c_\ell$ *is divergent.*

Proof. Let (B_m) and (C_n) denote the sequences of partial sums of the series $\sum_k b_k$ and $\sum_\ell c_\ell$, respectively. Also, let $(A_{m,n})$ denote the double sequence of partial double sums of $\sum\sum_{(k,\ell)} a_{k,\ell}$. Then

$$A_{m,n} = \left(\sum_{k=1}^{m} b_k\right)\left(\sum_{\ell=1}^{n} c_\ell\right) = B_m C_n \quad \text{for all } (m,n) \in \mathbb{N}^2.$$

Consequently, if $B_m \to B$ and $C_n \to C$ for some $B, C \in \mathbb{R}$, then $A_{m,n} \to BC$. Also, if $B_m \to \infty$ and $C_n \to \infty$, then $A_{m,n} \to \infty$. This proves (i) and (ii). Moreover, if $B_m \to B$ with $B \neq 0$ and if the double sequence $(A_{m,n})$ converges to A, then (C_n) converges to A/B. This proves (iii). □

In the following examples as well as the rest of this chapter, we shall adopt the convention that $x^0 = 1$ for any $x \in \mathbb{R}$ (including $x = 0$).

Examples 7.10. (i) **(Geometric Double Series)** Let $x, y \in \mathbb{R}$. Define $a_{k,\ell} := x^k y^\ell$ for nonnegative integers k, ℓ. Note that according to the convention mentioned above, $a_{0,0} := 1$, $a_{k,0} := x^k$, and $a_{0,\ell} := y^\ell$ for $k, \ell \in \mathbb{N}$. The double series $\sum\sum_{(k,\ell)} a_{k,\ell}$, where the index (k,ℓ) varies over pairs of nonnegative integers, is called the **geometric double series**. Now recall that the geometric series $\sum_{k=0}^{\infty} x^k$ is convergent if and only if $|x| < 1$. (See, for instance, Example 9.1 (i) of ACICARA.) Hence from part (i) of Proposition 7.9, we see that the geometric double series is convergent if $|x| < 1$ and $|y| < 1$; moreover,

$$\sum\sum_{(k,\ell)} a_{k,\ell} = \sum\sum_{(k,\ell)\geq(0,0)} x^k y^\ell = \frac{1}{(1-x)(1-y)} \qquad \text{for } |x| < 1 \text{ and } |y| < 1.$$

Further, if $|x| \geq 1$ and $|y| \geq 1$, then $|x^k y^\ell| = |x|^k |y|^\ell \geq 1$ for all nonnegative integers k, ℓ. Hence from the (k,ℓ)th Term Test (Proposition 7.8), we see that the geometric double series is divergent. Finally, since $1/(1-z)$ is nonzero whenever $z \in \mathbb{R}$ with $|z| < 1$, it follows from part (iii) of Proposition 7.9 that if only one of $|x|$ and $|y|$ is less than 1, then the geometric double series is divergent. Thus we see that the geometric double series $\sum\sum_{(k,\ell)} x^k y^\ell$ is convergent if and only if $|x| < 1$ and $|y| < 1$.

(ii) **(Exponential Double Series)** Let $x, y \in \mathbb{R}$. Define $a_{k,\ell} = x^k y^\ell / k!\ell!$ for nonnegative integers k, ℓ. The double series $\sum\sum_{(k,\ell)} a_{k,\ell}$, where the index

(k, ℓ) varies over pairs of nonnegative integers, is called the **exponential double series**. From part (i) of Proposition 7.9, we readily see that the exponential double series is always convergent and

$$\sum\sum_{(k,\ell)} a_{k,\ell} = \sum\sum_{(k,\ell)\geq(0,0)} \frac{x^k}{k!}\frac{y^\ell}{\ell!} = (\exp x)(\exp y) = \exp(x+y) \quad \text{for } x, y \in \mathbb{R}.$$

(iii) **(Harmonic Double Series and Their Variants)** The double series $\sum\sum_{(k,\ell)} 1/k\ell$ and $\sum\sum_{(k,\ell)} 1/(k+\ell)$ can be considered as analogues of the harmonic series $\sum_k 1/k$, and either of the two double series may be referred to as a **harmonic double series**. We know from the theory of (single) series that the harmonic series diverges to ∞. (See, for instance, Example 2.10 (iii) of ACICARA). Hence by part (ii) of Proposition 7.9, we see that the double series $\sum\sum_{(k,\ell)} 1/k\ell$ diverges to ∞. More generally, for any $p \in \mathbb{R}$, we know that the series $\sum_k 1/k^p$ is convergent for $p > 1$ (in which case, the sum will obviously be nonzero) and it diverges to ∞ for $p \leq 1$. (See, for instance, Example 2.10 (v) of ACICARA). Thus, using parts (i), (ii), and (iii) of Proposition 7.9, we see that for any $p, q \in \mathbb{R}$,

$$\sum\sum_{(k,\ell)} \frac{1}{k^p \ell^q} \text{ is convergent } \iff p > 1 \text{ and } q > 1.$$

As for the other variant of the harmonic series, namely the double series $\sum\sum_{(k,\ell)} 1/(k+\ell)$, we also find that it diverges to ∞, since

$$\sum_{k=1}^{m}\sum_{\ell=1}^{n} \frac{1}{k+\ell} \geq \frac{1}{2}\sum_{k=1}^{m}\sum_{\ell=1}^{n} \frac{1}{k\ell} \quad \text{for all } (m, n) \in \mathbb{N}^2.$$

Let us now consider the double series $\sum\sum_{(k,\ell)} 1/(k+\ell)^2$. For $n \in \mathbb{N}$, let $A_{n,n} := \sum_{k=1}^{n}\sum_{\ell=1}^{n} 1/(k+\ell)^2$. For $n \geq 3$ and $i = 2, \ldots, n-1$, each of the $i-1$ summands $1/[1+(i-1)]^2, 1/[2+(i-2)]^2, \ldots, 1/[(i-2)+2]^2, 1/[(i-1)+1]^2$ of $A_{n,n}$ is equal to $1/i^2$, and so

$$A_{n,n} \geq \sum_{i=2}^{n-1} \frac{i-1}{i^2} \geq \frac{1}{2}\sum_{i=2}^{n-1} \frac{1}{i}.$$

Since $\sum_{i=2}^{n-1} 1/i \to \infty$ as $n \to \infty$, we see that $A_{n,n} \to \infty$ as $n \to \infty$. Thus the double series $\sum\sum_{(k,\ell)} 1/(k+\ell)^2$ diverges to ∞. This indicates that the threshold for the convergence of $\sum\sum_{(k,\ell)} 1/(k+\ell)^p$ is not $p = 1$. Indeed, it will be shown later in Examples 7.17 (i) and 7.58 that $\sum\sum_{(k,\ell)} 1/(k+\ell)^p$ is convergent if and only if $p > 2$.

(iv) **(Alternating Double Series)** We know from the theory of (single) series that if we alternate the signs of the terms of the harmonic series, then the alternating series thus obtained, namely $\sum_k (-1)^{k-1}/k$, is convergent. (See, for instance, Example 2.10 (iii) of ACICARA). Hence from part (i) of Proposition 7.9, we see that the corresponding alternating double series $\sum\sum_{(k,\ell)}(-1)^{k+\ell}/k\ell$ is convergent. More generally, let $p \in \mathbb{R}$. We know that the series $\sum_k(-1)^{k-1}/k^p$ is convergent if and only if $p > 0$. (See, for instance, Examples 9.7 (i) and 9.23 (i) of ACICARA). Furthermore, if for $n \in \mathbb{N}$, we denote by A_n the nth partial sum of the series $\sum_k(-1)^{k-1}/k^p$, then

$$A_{2n} = \left(1 - \frac{1}{2^p}\right) + B_n, \quad \text{where } B_n := \sum_{k=2}^{n}\left(\frac{1}{(2k-1)^p} - \frac{1}{(2k)^p}\right),$$

and since $B_n \geq 0$, we see that $A_{2n} \geq (1 - 2^{-p})$. Consequently, if $p > 1$, then the sum of the series $\sum_k(-1)^{k-1}/k^p$ is at least $(1 - 2^{-p})$, and hence nonzero. With this in view, it follows from parts (i) and (iii) of Proposition 7.9 together with the (k,ℓ)th Term Test (Proposition 7.8) that for any $p, q \in \mathbb{R}$,

$$\sum\sum_{(k,\ell)} \frac{(-1)^{k+\ell}}{k^p \ell^q} \text{ is convergent } \iff p > 0 \text{ and } q > 0.$$

As for the other variant, it will be seen in Example 7.42 that the alternating double series $\sum\sum_{(k,\ell)}(-1)^{k+\ell}/(k+\ell)^p$ is convergent if and only if $p > 0$. $\diamond$

The following statements about the convergence of a double series follow from the corresponding statements for the convergence of a double sequence given in Section 7.1.

1. **(Limit Theorem)** Let $\sum\sum_{(k,\ell)} a_{k,\ell} = A$ and $\sum\sum_{(k,\ell)} b_{k,\ell} = B$. Then

$$\sum\sum_{(k,\ell)}(a_{k,\ell} + b_{k,\ell}) = A + B \quad \text{and} \quad \sum\sum_{(k,\ell)}(r a_{k,\ell}) = rA \quad \text{for any } r \in \mathbb{R}.$$

   Further, if $a_{k,\ell} \leq b_{k,\ell}$ for all $(k,\ell) \in \mathbb{N}^2$, then $A \leq B$.
2. **(Sandwich Theorem)** If $(a_{k,\ell})$, $(b_{k,\ell})$, and $(c_{k,\ell})$ are double sequences of real numbers such that $a_{k,\ell} \leq c_{k,\ell} \leq b_{k,\ell}$ for each $(k,\ell) \in \mathbb{N}^2$, and further $\sum\sum_{(k,\ell)} a_{k,\ell} = A$ as well as $\sum\sum_{(k,\ell)} b_{k,\ell} = A$, then $\sum\sum_{(k,\ell)} c_{k,\ell} = A$.
3. **(Cauchy Criterion)** A double series $\sum\sum_{(k,\ell)} a_{k,\ell}$ is convergent if and only if for every $\epsilon > 0$, there is $(m_0, n_0) \in \mathbb{N}^2$ such that

$$\left|\sum_{k=p+1}^{m}\sum_{\ell=q+1}^{n} a_{k,\ell} + \sum_{k=1}^{p}\sum_{\ell=q+1}^{n} a_{k,\ell} + \sum_{k=p+1}^{m}\sum_{\ell=1}^{q} a_{k,\ell}\right| < \epsilon$$

   for all $(m,n) \geq (p,q) \geq (m_0,n_0)$. (Compare Proposition 7.1 and note that the three sums above together are equal to $A_{m,n} - A_{p,q}$ for $(m,n) \geq (p,q)$.)

We shall now relate the convergence of a double series $\sum\sum_{(k,\ell)} a_{k,\ell}$ to the convergence of the two series $\sum_{k=1}^{\infty}\left(\sum_{\ell=1}^{\infty} a_{k,\ell}\right)$ and $\sum_{\ell=1}^{\infty}\left(\sum_{k=1}^{\infty} a_{k,\ell}\right)$. For this purpose and for later use, it is convenient to use the following terminology: For each fixed $k \in \mathbb{N}$, the (single) series $\sum_\ell a_{k,\ell}$ is called a **row-series**, and for each fixed $\ell \in \mathbb{N}$, the (single) series $\sum_k a_{k,\ell}$ is called a **column-series** (corresponding to the double series $\sum\sum_{(k,\ell)} a_{k,\ell}$). The following result may be compared with Proposition 5.28.

Proposition 7.11 (Fubini's Theorem for Double Series). *Assume that $\sum\sum_{(k,\ell)} a_{k,\ell}$ is a convergent double series and let A denote its double sum.*

(i) *If each row-series is convergent, then the corresponding* **iterated series** $\sum_{k=1}^{\infty}\left(\sum_{\ell=1}^{\infty} a_{k,\ell}\right)$ *is convergent and its sum is equal to A.*

(ii) *If each column-series is convergent, then the corresponding* **iterated series** $\sum_{\ell=1}^{\infty}\left(\sum_{k=1}^{\infty} a_{k,\ell}\right)$ *is convergent and its sum is equal to A.*

(iii) *If each row-series as well as each column-series is convergent, then the double sequence of partial double sums of $\sum\sum_{(k,\ell)} a_{k,\ell}$ is bounded, and*

$$\sum_{k=1}^{\infty}\left(\sum_{\ell=1}^{\infty} a_{k,\ell}\right) = \sum\sum_{(k,\ell)} a_{k,\ell} = \sum_{\ell=1}^{\infty}\left(\sum_{k=1}^{\infty} a_{k,\ell}\right).$$

Proof. Let $(A_{m,n})$ denote, as usual, the double sequence of partial double sums of $\sum\sum_{(k,\ell)} a_{k,\ell}$. By our assumption, $(A_{m,n})$ converges to A.

Suppose each row-series is convergent. Then for each fixed $m \in \mathbb{N}$,

$$\lim_{n\to\infty} A_{m,n} = \lim_{n\to\infty} \sum_{k=1}^{m}\sum_{\ell=1}^{n} a_{k,\ell} = \sum_{k=1}^{m}\left(\lim_{n\to\infty}\sum_{\ell=1}^{n} a_{k,\ell}\right) = \sum_{k=1}^{m}\left(\sum_{\ell=1}^{\infty} a_{k,\ell}\right).$$

Hence by Proposition 7.2, the iterated limit $\lim_{m\to\infty}(\lim_{n\to\infty} A_{m,n})$ exists and is equal to A, that is,

$$\lim_{m\to\infty} \sum_{k=1}^{m}\left(\sum_{\ell=1}^{\infty} a_{k,\ell}\right) = A.$$

Thus the iterated series $\sum_{k=1}^{\infty}\left(\sum_{\ell=1}^{\infty} a_{k,\ell}\right)$ converges and its sum equals A. This proves (i). The proof of (ii) is similar.

Finally, suppose each row-series as well as each column-series is convergent. Then for each fixed $m \in \mathbb{N}$, the limit $\lim_{n\to\infty} A_{m,n}$ exists, and for each fixed $n \in \mathbb{N}$, the limit $\lim_{m\to\infty} A_{m,n}$ exists. Hence by part (iii) of Proposition 7.2, $(A_{m,n})$ is bounded. The last part of (iii) follows from (i) and (ii). □

Examples 7.12. (i) Even if a double series $\sum\sum_{(k,\ell)} a_{k,\ell}$ converges, both the iterated series may diverge. For instance, consider a double sequence $(a_{k,\ell})$ given schematically as follows:

$$\begin{array}{ccccc} 1 & 1 & 1 & 1 & \cdots \\ 1 & -3 & -1 & -1 & \cdots \\ 1 & -1 & 0 & 0 & \cdots \\ 1 & -1 & 0 & 0 & \cdots \\ \vdots & \vdots & \vdots & \vdots & \end{array}$$

Then $A_{1,n} = n$ for all $n \in \mathbb{N}$ and $A_{m,1} = m$ for all $m \in \mathbb{N}$, while $A_{m,n} = 0$ for all $(m,n) \geq (2,2)$. Hence $\sum\sum_{(k,\ell)} a_{k,\ell} = \lim_{(m,n)\to(\infty,\infty)} A_{m,n} = 0$. But $\sum_{\ell=1}^{n} a_{1,\ell} = n$ for all $n \in \mathbb{N}$ and $\sum_{\ell=1}^{n} a_{2,\ell} = -n$ for all $n \geq 2$, while $\sum_{k=1}^{m} a_{k,1} = m$ for all $m \in \mathbb{N}$ and $\sum_{k=1}^{m} a_{k,2} = -m$ for all $m \geq 2$. Hence $\sum_{\ell=1}^{\infty} a_{1,\ell}$ and $\sum_{k=1}^{\infty} a_{k,1}$ diverge to ∞, whereas $\sum_{\ell=1}^{\infty} a_{2,\ell}$ and $\sum_{k=1}^{\infty} a_{k,2}$ diverge to $-\infty$. Clearly, none of the iterated series is even well defined.

(ii) Even if both iterated series $\sum_{k=1}^{\infty}(\sum_{\ell=1}^{\infty} a_{k,\ell})$ and $\sum_{\ell=1}^{\infty}(\sum_{k=1}^{\infty} a_{k,\ell})$ converge and have the same sum, the double series may diverge. For instance, consider a double sequence $(a_{k,\ell})$ given schematically as follows:

$$\begin{array}{rrrrrl} 2 & 0 & -1 & 0 & 0 & \cdots \\ 0 & 2 & 0 & -1 & 0 & \cdots \\ -1 & 0 & 2 & 0 & -1 & \cdots \\ 0 & -1 & 0 & 2 & 0 & \cdots \\ \vdots & \vdots & \vdots & \vdots & \vdots & \end{array}$$

Then $\sum_{\ell=1}^{\infty} a_{k,\ell}$ is equal to 1 if $k = 1$ or 2, and it is equal to 0 if $k \geq 3$. Similarly, $\sum_{k=1}^{\infty} a_{k,\ell}$ is equal to 1 if $\ell = 1$ or 2, and it is equal to 0 if $\ell \geq 3$. Thus $\sum_{k=1}^{\infty}(\sum_{\ell=1}^{\infty} a_{k,\ell}) = 2 = \sum_{\ell=1}^{\infty}(\sum_{k=1}^{\infty} a_{k,\ell})$. But $A_{m,m} = 4$ for all $m \geq 2$ and $A_{m,m-1} = 3$ for all $m \geq 3$, so that the double sequence $(A_{m,n})$ is divergent, that is, the double series $\sum\sum_{(k,\ell)} a_{k,\ell}$ is divergent.

(iii) Even if both iterated series $\sum_{k=1}^{\infty}(\sum_{\ell=1}^{\infty} a_{k,\ell})$ and $\sum_{\ell=1}^{\infty}(\sum_{k=1}^{\infty} a_{k,\ell})$ converge, their sums may be unequal. For instance, consider a double sequence $(a_{k,\ell})$ given schematically as follows:

$$\begin{array}{rrrrrl} 0 & 1 & 0 & 0 & 0 & \cdots \\ -1 & 0 & 1 & 0 & 0 & \cdots \\ 0 & -1 & 0 & 1 & 0 & \cdots \\ 0 & 0 & -1 & 0 & 1 & \cdots \\ \vdots & \vdots & \vdots & \vdots & \vdots & \end{array}$$

Then $\sum_{\ell=1}^{\infty} a_{k,\ell}$ is equal to 1 if $k = 1$ and it is equal to 0 if $k \geq 2$, whereas $\sum_{k=1}^{\infty} a_{k,\ell}$ is equal to -1 if $\ell = 1$ and it is equal to 0 if $\ell \geq 2$. Hence $\sum_{k=1}^{\infty}(\sum_{\ell=1}^{\infty} a_{k,\ell}) = 1$, while $\sum_{\ell=1}^{\infty}(\sum_{k=1}^{\infty} a_{k,\ell}) = -1$. Of course, $\sum\sum_{(k,\ell)} a_{k,\ell}$ is divergent, since $A_{m,m} = 0$ for all $m \in \mathbb{N}$ and $A_{m,m-1} = -1$ for all $m \geq 2$. ◇

Telescoping Double Series

If $(b_{k,\ell})$ is a double sequence of real numbers, then the double series

$$\sum_{(k,\ell)}\sum (b_{k,\ell} - b_{k+1,\ell} - b_{k,\ell+1} + b_{k+1,\ell+1})$$

is known as a **telescoping double series**. We have the following result regarding its convergence.

Proposition 7.13. *A telescoping double series* $\sum\sum_{(k,\ell)}(b_{k,\ell}-b_{k+1,\ell}-b_{k,\ell+1}+b_{k+1,\ell+1})$ *is convergent if and only if the double sequence* $(b_{k,1}+b_{1,\ell}-b_{k,\ell})$ *is convergent, and in this case*

$$\sum\sum_{(k,\ell)}(b_{k,\ell}-b_{k+1,\ell}-b_{k,\ell+1}+b_{k+1,\ell+1}) = b_{1,1} - \lim_{(k,\ell)\to(\infty,\infty)}(b_{k,1}+b_{1,\ell}-b_{k,\ell}).$$

Proof. Let $(m,n)\in\mathbb{N}^2$. Then

$$\sum_{k=1}^{m}\sum_{\ell=1}^{n}(b_{k,\ell}-b_{k+1,\ell}-b_{k,\ell+1}+b_{k+1,\ell+1}) = b_{1,1}-b_{m+1,1}-b_{1,n+1}+b_{m+1,n+1}.$$

This yields the desired result. □

It may be noted that every double series can be written as a telescoping double series. In fact, if $A_{m,n}$ is the (m,n)th partial double sum of a double series $\sum\sum_{(k,\ell)} a_{k,\ell}$, then letting $b_{k,\ell} := A_{k-1,\ell-1}$ for all $(k,\ell)\geq(1,1)$ with the usual convention that $A_{0,0}$, $A_{k,0}$, and $A_{0,\ell}$ are all equal to zero, we obtain $a_{k,\ell} = A_{k,\ell}-A_{k-1,\ell}-A_{k,\ell-1}+A_{k-1,\ell-1} = b_{k+1,\ell+1}-b_{k,\ell+1}-b_{k+1,\ell}+b_{k,\ell}$ for all $(k,\ell)\in\mathbb{N}^2$. However, Proposition 7.13 is particularly useful when it is possible to write $\sum\sum_{(k,\ell)} a_{k,\ell}$ as a telescoping double series without involving its partial double sums. For example, consider the series $\sum\sum_{(k,\ell)} a_{k,\ell}$, where

$$a_{k,\ell} := \frac{1}{k\ell(k+1)(\ell+1)} = \left(\frac{1}{k}-\frac{1}{k+1}\right)\left(\frac{1}{\ell}-\frac{1}{\ell+1}\right).$$

If we let $b_{k,\ell} := 1/k\ell$ for all $(k,\ell)\in\mathbb{N}^2$, then $a_{k,\ell} = b_{k,\ell}-b_{k+1,\ell}-b_{k,\ell+1}+b_{k+1,\ell+1}$ for all $(k,\ell)\in\mathbb{N}^2$. Since $b_{1,1}=1$ and $b_{k,1}+b_{1,\ell}-b_{k,\ell}\to 0$, it follows from Proposition 7.13 that the double series $\sum\sum_{(k,\ell)} 1/k\ell(k+1)(\ell+1)$ is convergent and its double sum is equal to $1-0=1$.

Double Series with Nonnegative Terms

The following necessary and sufficient condition for the convergence of a double series with nonnegative terms is very useful.

Proposition 7.14. *Let* $(a_{k,\ell})$ *be a double sequence such that* $a_{k,\ell}\geq 0$ *for all* $(k,\ell)\in\mathbb{N}^2$. *Then* $\sum\sum_{(k,\ell)} a_{k,\ell}$ *is convergent if and only if the double sequence* $(A_{m,n})$ *of its partial double sums is bounded above, and in this case*

$$\sum\sum_{(k,\ell)} a_{k,\ell} = \sup\{A_{m,n} : (m,n)\in\mathbb{N}^2\}.$$

If $(A_{m,n})$ *is not bounded above, then* $\sum\sum_{(k,\ell)} a_{k,\ell}$ *diverges to* ∞.

Proof. Since $a_{k,\ell} \geq 0$ for all $(k,\ell) \in \mathbb{N}^2$, we have $A_{m+1,n} = A_{m,n} + a_{m+1,1} + \cdots + a_{m+1,n} \geq A_{m,n}$ and $A_{m,n+1} = A_{m,n} + a_{1,n+1} + \cdots + a_{1,n+1} \geq A_{m,n}$ for all $(m,n) \in \mathbb{N}^2$. Hence the double sequence $(A_{m,n})$ is monotonically increasing. By part (i) of Proposition 7.4, we see that $(A_{m,n})$ is convergent if and only if it is bounded above, and in this case

$$\sum\sum_{(k,\ell)} a_{k,\ell} = \lim_{(m,n)\to(\infty,\infty)} A_{m,n} = \sup\{A_{m,n} : (m,n) \in \mathbb{N}^2\}.$$

Further, if $(A_{m,n})$ is not bounded above, then $A_{m,n} \to \infty$, that is, the double series $\sum\sum_{(k,\ell)} a_{k,\ell}$ diverges to ∞. □

In view of the above result, when $a_{k,\ell} \geq 0$ for all $(k,\ell) \in \mathbb{N}^2$, we write $\sum\sum_{(k,\ell)} a_{k,\ell} < \infty$ if the double series $\sum\sum_{(k,\ell)} a_{k,\ell}$ is convergent and $\sum\sum_{(k,\ell)} a_{k,\ell} = \infty$ if the double series $\sum\sum_{(k,\ell)} a_{k,\ell}$ is divergent.

Proposition 7.14 shows that if a double series $\sum\sum_{(k,\ell)} a_{k,\ell}$ with nonnegative terms is convergent and A is its double sum, then the double sequence $(a_{k,\ell})$ of its terms as well as the double sequence $(A_{m,n})$ of its partial double sums is bounded. This follows by observing that in this case, $0 \leq a_{k,\ell} \leq A_{k,\ell} \leq A$ for all $(k,\ell) \in \mathbb{N}^2$.

An interesting application of Proposition 7.14, known as **Cauchy's Condensation Test**, is given in Exercise 17.

A result similar to Proposition 7.14 holds for double series with nonpositive terms. More generally, when the terms $a_{k,\ell}$ have the same sign except possibly for a finite number of them, then $\sum\sum_{(k,\ell)} a_{k,\ell}$ is convergent if and only if $(A_{m,n})$ is bounded. However, if infinitely many $a_{k,\ell}$'s are positive and infinitely many $a_{k,\ell}$'s are negative, then $\sum\sum_{(k,\ell)} a_{k,\ell}$ may diverge even though $(A_{m,n})$ is bounded, and $(A_{m,n})$ may be unbounded even though $\sum\sum_{(k,\ell)} a_{k,\ell}$ is convergent. These two statements are illustrated respectively by the double series given schematically as follows:

$$\begin{array}{rccc} 1 & 0 & 0 & \cdots \\ -1 & 0 & 0 & \cdots \\ 1 & 0 & 0 & \cdots \\ -1 & 0 & 0 & \cdots \\ \vdots & \vdots & \vdots & \end{array} \quad \text{and} \quad \begin{array}{crccc} 1 & -1 & 0 & 0 & \cdots \\ 1 & -1 & 0 & 0 & \cdots \\ 1 & -1 & 0 & 0 & \cdots \\ 1 & -1 & 0 & 0 & \cdots \\ \vdots & \vdots & \vdots & \vdots & \end{array}$$

For double series with nonnegative terms, the following result is an improvement over Fubini's Theorem for double series (Proposition 7.11).

Proposition 7.15 (Tonelli's Theorem for Double Series). *Let $(a_{k,\ell})$ be a double sequence such that $a_{k,\ell} \geq 0$ for all $(k,\ell) \in \mathbb{N}^2$. Then the following statements are equivalent.*

(i) *The double series $\sum\sum_{(k,\ell)} a_{k,\ell}$ is convergent.*

(ii) *Each row-series is convergent and the iterated series* $\sum_{k=1}^{\infty}\left(\sum_{\ell=1}^{\infty} a_{k,\ell}\right)$ *is convergent.*

(iii) *Each column-series is convergent and the iterated series* $\sum_{\ell=1}^{\infty}\left(\sum_{k=1}^{\infty} a_{k,\ell}\right)$ *is convergent.*

In this case,

$$\sum_{k=1}^{\infty}\left(\sum_{\ell=1}^{\infty} a_{k,\ell}\right) = \sum\sum_{(k,\ell)} a_{k,\ell} = \sum_{\ell=1}^{\infty}\left(\sum_{k=1}^{\infty} a_{k,\ell}\right).$$

Proof. Suppose (i) holds. If A denotes the double sum of $\sum\sum_{(k,\ell)} a_{k,\ell}$, then in view of Proposition 7.14, $\sum_{\ell=1}^{n} a_{k,\ell} \le A_{k,n} \le A$ for each fixed $k \in \mathbb{N}$ and all $n \in \mathbb{N}$. Thus each row-series is a (single) series with nonnegative terms whose partial sums are bounded, and hence it convergent. By Fubini's Theorem (Proposition 7.11), it follows that the iterated series $\sum_{k=1}^{\infty}\left(\sum_{\ell=1}^{\infty} a_{k,\ell}\right)$ is convergent and its sum is equal to A.

Suppose (ii) holds. Then for $(m,n) \in \mathbb{N}^2$,

$$A_{m,n} = \sum_{k=1}^{m}\sum_{\ell=1}^{n} a_{k,\ell} \le \sum_{k=1}^{m}\left(\sum_{\ell=1}^{\infty} a_{k,\ell}\right) \le \sum_{k=1}^{\infty}\left(\sum_{\ell=1}^{\infty} a_{k,\ell}\right) = \alpha, \quad \text{say.}$$

Hence by Proposition 7.14, the double series $\sum\sum_{(k,\ell)} a_{k,\ell}$ is convergent.

This establishes the equivalence of the statements (i) and (ii). The proof of the equivalence of the statements (i) and (iii) is similar. The equality of the double sum and the sum of either of the two iterated series is also established in this process. □

Examples 7.12 show that the nonnegativity of the terms of the double series in Tonelli's Theorem cannot be omitted.

The question of the convergence of a double series $\sum\sum_{(k,\ell)} a_{k,\ell}$ with nonnegative terms can be reduced to the question of the convergence of each of the following two (single) series, which correspond to summing the double series $\sum\sum_{(k,\ell)} a_{k,\ell}$ "by squares" or "by diagonals" as illustrated in Figure 7.1.

1. The (single) series $\sum_{j=1}^{\infty} b_j$, where for each $j \in \mathbb{N}$, b_j is the sum of all those terms $a_{k,\ell}$ such that one of k and ℓ is equal to j and the other is at most j, that is, $b_j := \sum_{i=1}^{j} a_{i,j} + \sum_{i=1}^{j-1} a_{j,i}$. Thus $b_1 = a_{1,1}$, $b_2 = a_{1,2} + a_{2,2} + a_{2,1}$, $b_3 = a_{1,3} + a_{2,3} + a_{3,3} + a_{3,1} + a_{3,2}$, and so on.
2. The (single) series $\sum_{j=1}^{\infty} c_j$, where for each $j \in \mathbb{N}$, c_j is the sum of all those terms $a_{k,\ell}$ such that $k + \ell = j + 1$, that is, $c_j := \sum_{i=1}^{j} a_{j-i+1,i}$. Thus $c_1 = a_{1,1}$, $c_2 = a_{2,1} + a_{1,2}$, $c_3 = a_{3,1} + a_{2,2} + a_{1,3}$, and so on.

The series $\sum_{j=1}^{\infty} c_j$ is sometimes referred to as the **diagonal series** corresponding to the double series $\sum\sum_{(k,\ell)} a_{k,\ell}$.

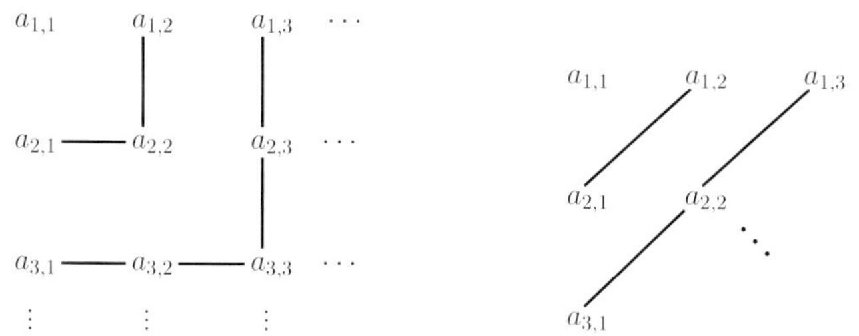

Fig. 7.1. Summing a double series "by squares" and "by diagonals."

Proposition 7.16. *Let $\sum\sum_{(k,\ell)} a_{k,\ell}$ be a double series with nonnegative terms, and b_j, c_j be as above. Then the following statements are equivalent.*

(i) **(Summing by Rectangles)** $\sum\sum_{(k,\ell)} a_{k,\ell}$ *is convergent.*
(ii) **(Summing by Squares)** $\sum_{j=1}^{\infty} b_j$ *is convergent.*
(iii) **(Summing by Diagonals)** $\sum_{j=1}^{\infty} c_j$ *is convergent.*

In this case,

$$\sum\sum_{(k,\ell)} a_{k,\ell} = \sum_{j=1}^{\infty} b_j = \sum_{j=1}^{\infty} c_j.$$

Proof. For $(m,n) \in \mathbb{N}^2$, let $A_{m,n} := \sum_{k=1}^{m}\sum_{\ell=1}^{n} a_{k,\ell}$ as usual, and also let $B_n := \sum_{j=1}^{n} b_j$. Then it is easy to see that $B_n = A_{n,n}$ for all $n \in \mathbb{N}$. Thus, in view of Corollary 7.5, (i) and (ii) are equivalent, and in this case

$$\sum\sum_{(k,\ell)} a_{k,\ell} = \sup\{A_{m,n} : (m,n) \in \mathbb{N}^2\} = \sup\{B_n : n \in \mathbb{N}\} = \sum_{j=1}^{\infty} b_j.$$

Next, for $n \in \mathbb{N}$, let $C_n := \sum_{j=1}^{n} c_j$. Note that if $(k,\ell) \in \mathbb{N}^2$ is such that $k+\ell \le n+1$, then $k \le n$ and $\ell \le n$. This implies that $C_n \le A_{n,n}$ for all $n \in \mathbb{N}$. Also, if $(k,\ell) \in \mathbb{N}^2$ and $(k,\ell) \le (m,n)$, then $k+\ell \le m+n = (m+n-1)+1$. This implies that $A_{m,n} \le C_{m+n-1}$ for all $(m,n) \in \mathbb{N}^2$. In view of these relations, we see that (i) and (iii) are equivalent, and in this case

$$\sum\sum_{(k,\ell)} a_{k,\ell} = \sup\{A_{m,n} : (m,n) \in \mathbb{N}^2\} = \sup\{C_n : n \in \mathbb{N}\} = \sum_{j=1}^{\infty} c_j,$$

as desired. □

Examples 7.17. (i) Let $p > 0$ and for $(k,\ell) \in \mathbb{N}^2$, let $a_{k,\ell} := 1/(k+\ell)^p$. Then $c_j = \sum_{i=1}^{j} 1/(j+1)^p = j/(j+1)^p$ for $j \in \mathbb{N}$. Since

$$\frac{1}{2(j+1)^{p-1}} \le \frac{j}{(j+1)^p} < \frac{1}{(j+1)^{p-1}} \quad \text{for all } j \in \mathbb{N},$$

the series $\sum_{j=1}^{\infty} c_j$ is convergent if and only if $p > 2$. So by Proposition 7.16, we see that $\sum\sum_{(k,\ell)} 1/(k+\ell)^p$ is convergent if and only if $p > 2$.

(ii) Since $A_{m,n} \to A$ as $(m,n) \to (\infty,\infty)$ implies that $A_{n,n} \to A$ as $n \to \infty$, we see that in Proposition 7.16 the statement (i) implies the statement (ii) irrespective of the sign of the terms of the double series. However, the converse does not hold in general, as Example 7.12 (ii) shows. In this example, $A_{n,n} = 4$ for all $n \geq 2$, so that $\sum_{j=1}^{\infty} b_j = 4$. However, the double series does not converge. This follows by noting that $A_{n,n+1} = 3$ for $n \geq 2$.

(iii) Let a double sequence $(a_{k,\ell})$ be schematically given as follows:

$$\begin{array}{cccccc} 0 & 1 & 1 & 1 & 1 & \cdots \\ 1 & -2 & -1 & -1 & -1 & \cdots \\ 0 & 0 & 0 & 0 & 0 & \cdots \\ 1 & -1 & 0 & 0 & 0 & \cdots \\ 0 & 0 & 0 & 0 & 0 & \cdots \\ 1 & -1 & 0 & 0 & 0 & \cdots \\ \vdots & \vdots & \vdots & \vdots & \vdots & \end{array}$$

Here $A_{m,n} = 0$ for all $(m,n) \geq (2,2)$ and so $\sum\sum_{(k,\ell)} a_{k,\ell}$ is convergent and its double sum is equal to 0. However, since $c_1 = 0$, $c_2 = 2$, and $c_j = (-1)^j$ for $j \geq 3$, we see that $\sum_{j=1}^{\infty} c_j$ is divergent. On the other hand, Example 7.12 (ii) shows that $\sum\sum_{(k,\ell)} a_{k,\ell}$ may be divergent, while $\sum_{j=1}^{\infty} c_j$ is convergent. In this example, we note that $c_1 = 2$ and $c_j = 0$ for all $j \geq 2$, so that $\sum_{j=1}^{\infty} c_j = 2$. It is also possible that both $\sum\sum_{(k,\ell)} a_{k,\ell}$ and $\sum_j c_j$ are convergent but the double sum is not equal to the "sum by diagonals." To illustrate this, consider the double sequence $(a_{n,k})$ schematically given as follows:

$$\begin{array}{cccccc} 1 & 1 & 1 & 1 & 1 & \cdots \\ 1 & -1 & -1 & -1 & -1 & \cdots \\ 1 & -1 & 0 & 0 & 0 & \cdots \\ 1 & -1 & 0 & 0 & 0 & \cdots \\ 1 & -1 & 0 & 0 & 0 & \cdots \\ \vdots & \vdots & \vdots & \vdots & & \end{array}$$

Here $A_{m,n} = 2$ for $(m,n) \neq (1,1)$, and so the double sum is equal to 2. But since $c_1 = 1, c_2 = 2\ c_3 = 1$, $c_j = 0$ for all $j \geq 4$, we have $\sum_j c_j = 4$. $\diamond$

Absolute Convergence and Conditional Convergence

In this subsection we shall discuss the convergence and divergence of the double series $\sum\sum_{(k,\ell)} |a_{k,\ell}|$ formed by considering the absolute values of the terms of a double series $\sum\sum_{(k,\ell)} a_{k,\ell}$. A double series $\sum\sum_{(k,\ell)} a_{k,\ell}$ is said to be **absolutely convergent** if the double series $\sum\sum_{(k,\ell)} |a_{k,\ell}|$ is convergent.

Proposition 7.18. *An absolutely convergent double series is convergent.*

Proof. Let $\sum\sum_{(k,\ell)} a_{k,\ell}$ be an absolutely convergent double series. For each $(k,\ell) \in \mathbb{N}^2$, define

$$a_{k,\ell}^+ := \frac{|a_{k,\ell}| + a_{k,\ell}}{2} \quad \text{and} \quad a_{k,\ell}^- := \frac{|a_{k,\ell}| - a_{k,\ell}}{2}.$$

Let $(A_{m,n})$, $(A_{m,n}^+)$, $(A_{m,n}^-)$, and $(\widetilde{A}_{m,n})$ denote the double sequences of the partial double sums of $\sum\sum_{(k,\ell)} a_{k,\ell}$, $\sum\sum_{(k,\ell)} a_{k,\ell}^+$, $\sum\sum_{(k,\ell)} a_{k,\ell}^-$, and $\sum\sum_{(k,\ell)} |a_{k,\ell}|$, respectively. By Proposition 7.14, $(\widetilde{A}_{m,n})$ is bounded. Also, $0 \le a_{k,\ell}^+ \le |a_{k,\ell}|$ and $0 \le a_{k,\ell}^- \le |a_{k,\ell}|$ for all $(k,\ell) \in \mathbb{N}^2$, and so

$$0 \le A_{m,n}^+ \le \widetilde{A}_{m,n} \quad \text{and} \quad 0 \le A_{m,n}^- \le \widetilde{A}_{m,n} \quad \text{for all } (m,n) \in \mathbb{N}^2,$$

and therefore, the double sequences $(A_{m,n}^+)$ and $(A_{m,n}^-)$ are bounded. Using Proposition 7.14 once again, we see that the double series $\sum\sum_{(k,\ell)} a_{k,\ell}^+$ and $\sum\sum_{(k,\ell)} a_{k,\ell}^-$ are convergent. But $a_{k,\ell} = a_{k,\ell}^+ - a_{k,\ell}^-$ for all $(k,\ell) \in \mathbb{N}^2$. Hence the double series $\sum\sum_{(k,\ell)} a_{k,\ell}$ is convergent. □

The converse of the above result does not hold, as can be seen by considering the double series $\sum\sum_{(k,\ell)}(-1)^{k\ell}/(k\ell)$, which is convergent, but not absolutely convergent. (See Example 7.10 (iv).) A convergent double series that is not absolutely convergent is said to be **conditionally convergent**.

The notions of row-series and column-series introduced earlier can be used to obtain the following useful characterization of absolute convergence.

Proposition 7.19. *A double series $\sum\sum_{(k,\ell)} a_{k,\ell}$ is absolutely convergent if and only if the following conditions hold:*

(i) *There are $(k_0,\ell_0) \in \mathbb{N}^2$ and $\alpha_0 > 0$ such that*

$$\sum_{k=k_0}^{m} \sum_{\ell=\ell_0}^{n} |a_{k,\ell}| \le \alpha_0 \quad \textit{for all } (m,n) \ge (k_0,\ell_0).$$

(ii) *Each row-series as well as each column-series is absolutely convergent.*

Proof. Suppose $\sum\sum_{(k,\ell)} a_{k,\ell}$ is absolutely convergent. Since $|a_{k,\ell}| \ge 0$ for all $(k,\ell) \in \mathbb{N}^2$, Proposition 7.14 shows that condition (i) holds with $(k_0,\ell_0) := (1,1)$, and Tonelli's Theorem for Double Series (Proposition 7.15) shows that condition (ii) also holds.

Conversely, suppose conditions (i) and (ii) hold. Let (k_0,ℓ_0) and α_0 be as in (i). By (ii), we see that for each fixed $k \in \mathbb{N}$, there is $\beta_k > 0$ such that $\sum_\ell |a_{k,\ell}| \le \beta_k$ and for each fixed $\ell \in \mathbb{N}$, there is $\gamma_\ell > 0$ such that $\sum_k |a_{k,\ell}| \le \gamma_\ell$. Let $(\widetilde{A}_{m,n})$ be the double sequence of partial double sums of the double series $\sum\sum_{(k,\ell)} |a_{k,\ell}|$, and let $p_0 := \max\{k_0,\ell_0\}$. Then

$$\widetilde{A}_{p,p} = \sum_{k=1}^{p}\sum_{\ell=1}^{p}|a_{k,\ell}| = \sum_{k=p_0}^{p}\sum_{\ell=p_0}^{p}|a_{k,\ell}| + \sum_{k=1}^{p_0-1}\sum_{\ell=1}^{p}|a_{k,\ell}| + \sum_{\ell=1}^{p_0-1}\sum_{k=p_0}^{p}|a_{k,\ell}|$$
$$\leq \alpha_0 + \sum_{k=1}^{p_0-1}\beta_k + \sum_{\ell=1}^{p_0-1}\gamma_\ell \qquad \text{for } p \in \mathbb{N} \text{ with } p \geq p_0.$$

This implies that the diagonal sequence $(\widetilde{A}_{p,p})$ is bounded, and therefore by Corollary 7.5, the monotonically increasing double sequence $(\widetilde{A}_{m,n})$ is bounded. Hence by Proposition 7.14, $(\widetilde{A}_{m,n})$ is convergent, that is, the double series $\sum\sum_{(k,\ell)} a_{k,\ell}$ is absolutely convergent. □

The above result will be used in Section 7.3 for obtaining several tests for the absolute convergence of a double series.

Remark 7.20. Conditions (i) and (ii) in Proposition 7.19 are *both* needed to characterize absolute convergence. For example, if we let $a_{k,1} := 1$ for all $k \in \mathbb{N}$ and $a_{k,\ell} := 0$ for all $(k,\ell) \geq (1,2)$, then condition (i) is satisfied with $k_0 := 1$ and $\ell_0 := 2$, but $\sum\sum_{(k,\ell)} a_{k,\ell}$ is not (absolutely) convergent, since $A_{m,n} = m$ for all $(m,n) \in \mathbb{N}^2$. On the other hand, if we let $a_{k,\ell} := 1/(k+\ell)^2$ for $(k,\ell) \in \mathbb{N}^2$, then condition (ii) is satisfied, but $\sum\sum_{(k,\ell)} a_{k,\ell}$ is not (absolutely) convergent, as is seen in Example 7.10 (iii). This shows also that none of the conditions (i) and (ii) in Proposition 7.19 implies the other. ◇

We now show that several results for convergent double series with nonnegative terms remain valid for absolutely convergent double series.

Proposition 7.21. *Let $\sum\sum_{(k,\ell)} a_{k,\ell}$ be an absolutely convergent double series. Then the following hold.*

(i) *The double sequence $(A_{m,n})$ of partial double sums is bounded.*
(ii) *Each row-series as well as each column-series is absolutely convergent, and*
$$\sum_{k=1}^{\infty}\left(\sum_{\ell=1}^{\infty} a_{k,\ell}\right) = \sum\sum_{(k,\ell)} a_{k,\ell} = \sum_{\ell=1}^{\infty}\left(\sum_{k=1}^{\infty} a_{k,\ell}\right).$$
(iii) *The corresponding diagonal series $\sum_{j=1}^{\infty} c_j$ is absolutely convergent, and*
$$\sum_{j=1}^{\infty} c_j = \sum\sum_{(k,\ell)} a_{k,\ell}.$$

Proof. For $(m,n) \in \mathbb{N}^2$, let $A_{m,n}$ and $\widetilde{A}_{m,n}$ denote the (m,n)th partial double sums of $\sum\sum_{(k,\ell)} a_{k,\ell}$ and $\sum\sum_{(k,\ell)} |a_{k,\ell}|$, and let A and $\widetilde{A}$ denote their double sums, respectively.

Now (i) follows from Proposition 7.14, since $|A_{m,n}| \leq \widetilde{A}_{m,n}$ for all $(m,n) \in \mathbb{N}^2$, while (ii) follows from Proposition 7.19 and Fubini's Theorem (Proposition 7.11).

To prove (iii), let $\sum_{j=1}^{\infty} c_j$ and $\sum_{j=1}^{\infty} d_j$ denote the diagonal series corresponding to the double series $\sum\sum_{(k,\ell)} a_{k,\ell}$ and $\sum\sum_{(k,\ell)} |a_{k,\ell}|$ respectively, and for $n \in \mathbb{N}$, let C_n and D_n denote the corresponding nth partial sums. By Proposition 7.16, it follows that $D_n \to \widetilde{A}$. But since $|c_j| \le d_j$ for all $j \in \mathbb{N}$ and the sequence (D_n) is bounded, we see that the sequence $\left(\sum_{j=1}^{n} |c_j|\right)$ is bounded, and so the series $\sum_{j=1}^{\infty} c_j$ converges absolutely. Now it can be easily seen that $|A_{n,n} - C_n| \le |\widetilde{A}_{n,n} - D_n|$ for all $n \in \mathbb{N}$. Since $\widetilde{A}_{n,n} \to \widetilde{A}$ and also $D_n \to \widetilde{A}$, we see that the sequences $(A_{n,n})$ and (C_n) have the same limit, that is, $\sum_{j=1}^{\infty} c_j = \sum\sum_{(k,\ell)} a_{k,\ell}$, as desired. □

Example 7.12 (ii) shows that a double series may diverge even if the corresponding diagonal series is absolutely convergent. (Also, see Example 7.50 (ii) and Exercises 32, 63, 64.)

Unconditional Convergence

The notion of convergence of a double series developed in this chapter is dependent on the order in which the terms are summed, or more precisely, on the manner in which the partial double sums are formed. Roughly speaking, we have been "summing by rectangles." We have shown in Proposition 7.21 that for an absolutely convergent series, "summing by diagonals" also leads to the same double sum. The notion of unconditional convergence defined below is an extension of the idea that the existence of a double sum ought to be independent of the manner in which the partial double sums are formed. Some authors adopt this as *the* definition of convergence of a double series. It will be seen, however, that this seemingly different notion is, in fact, equivalent to absolute convergence.

For this purpose, we shall say that a sequence (S_n) of subsets of $\mathbb{N}^2$ is **exhausting** if S_n is finite and $S_n \subseteq S_{n+1}$ for each $n \in \mathbb{N}$, and $\bigcup_{n=1}^{\infty} S_n = \mathbb{N}^2$. For example, if we let $S_n := \{(k,\ell) \in \mathbb{N}^2 : k \le n \text{ and } \ell \le n\}$ for $n \in \mathbb{N}$, then clearly (S_n) is an exhausting sequence of subsets of $\mathbb{N}^2$. We say that a double series $\sum\sum_{(k,\ell)} a_{k,\ell}$ is **unconditionally convergent** if there is $A \in \mathbb{R}$ such that for every exhausting sequence (S_n) of subsets of $\mathbb{N}^2$, the limit

$$\lim_{n\to\infty} \sum\sum_{(k,\ell)\in S_n} a_{k,\ell}$$

exists and is equal to A. In this case A is called the **unconditional double sum** of $\sum\sum_{(k,\ell)} a_{k,\ell}$.

It is easily seen that if $\sum\sum_{(k,\ell)} a_{k,\ell}$ and $\sum\sum_{(k,\ell)} b_{k,\ell}$ are unconditionally convergent double series, with A and B as their unconditional double sums, then so are $\sum\sum_{(k,\ell)} (a_{k,\ell} + b_{k,\ell})$ and $\sum\sum_{(k,\ell)} (ra_{k,\ell})$ for any $r \in \mathbb{R}$, and moreover, their unconditional double sums are $A + B$ and rA, respectively.

We show below that for a double series with nonnegative terms, the notions of convergence and unconditional convergence are equivalent. This result may be viewed as a generalization of Proposition 7.16.

Proposition 7.22. *A double series with nonnegative terms is unconditionally convergent if and only if it is convergent. In this case, its double sum coincides with its unconditional double sum.*

Proof. Let $\sum\sum_{(k,\ell)} a_{k,\ell}$ be a double series with $a_{k,\ell} \geq 0$ for all $(k,\ell) \in \mathbb{N}^2$, and for $(m,n) \in \mathbb{N}^2$, let $A_{m,n}$ denote its (m,n)th partial double sum.

Suppose $\sum\sum_{(k,\ell)} a_{k,\ell}$ is unconditionally convergent and let A be its unconditional double sum. Consider $S_n := \{(k,\ell) \in \mathbb{N}^2 : k \leq n \text{ and } \ell \leq n\}$ and $A_n := \sum\sum_{(k,\ell)\in S_n} a_{k,\ell}$ for $n \in \mathbb{N}$. Since (S_n) is an exhausting sequence of subsets of $\mathbb{N}^2$, we obtain $A_n \to A$ as $n \to \infty$. Hence by Proposition 7.16, the double series $\sum\sum_{(k,\ell)} a_{k,\ell}$ is convergent and its double sum is A.

Conversely, suppose $\sum\sum_{(k,\ell)} a_{k,\ell}$ is convergent and A is its double sum. By Proposition 7.14, the double sequence $(A_{m,n})$ is bounded and moreover, $A = \sup\{A_{m,n} : (m,n) \in \mathbb{N}^2\}$. Let (S_n) be any exhausting sequence of subsets of $\mathbb{N}^2$, and let $A_n := \sum\sum_{(k,\ell)\in S_n} a_{k,\ell}$ for $n \in \mathbb{N}$. Then the sequence (A_n) is monotonically increasing. Moreover, since each S_n is finite, $A_n \leq A_{r,s}$ for some $(r,s) \in \mathbb{N}^2$, and therefore $A_n \leq A$ for all $n \in \mathbb{N}$. Consequently, (A_n) is convergent and $\lim_{n\to\infty} A_n = \sup\{A_p : p \in \mathbb{N}\} \leq A$. On the other hand, since $\bigcup_{p=1}^{\infty} S_p = \mathbb{N}^2$, for any $(m,n) \in \mathbb{N}^2$, there is $p \in \mathbb{N}$ such that $(k,\ell) \in S_p$ for all $(k,\ell) \in \mathbb{N}^2$ with $(k,\ell) \leq (m,n)$, and therefore $A_{m,n} \leq A_p$. It follows that $\sup\{A_p : p \in \mathbb{N}\} = \sup\{A_{m,n} : (m,n) \in \mathbb{N}^2\} = A$. Thus $\sum\sum_{(k,\ell)} a_{k,\ell}$ is unconditionally convergent and its unconditional double sum is A. □

To obtain an analogue of the characterization in Proposition 7.22 for double series having terms of mixed signs, we require the following auxiliary result.

Lemma 7.23. *Let $\sum\sum_{(k,\ell)} a_{k,\ell}$ be an unconditionally convergent double series. Then there is $\alpha \in \mathbb{R}$ such that*

$$\sum\sum_{(k,\ell)\in S} |a_{k,\ell}| \leq \alpha \quad \textit{for every finite subset } S \textit{ of } \mathbb{N}^2.$$

Proof. First we show that there is $\beta \in \mathbb{R}$ satisfying $\left|\sum\sum_{(k,\ell)\in S} a_{k,\ell}\right| \leq \beta$ for every finite subset S of $\mathbb{N}^2$. Assume for a moment that this is not the case. Since the set $\mathbb{N}^2$ is countable, we can find $(k_1,\ell_1),(k_2,\ell_2),\ldots$ in $\mathbb{N}^2$ such that $\mathbb{N}^2 = \{(k_j,\ell_j) : j \in \mathbb{N}\}$. Let $D_n := \{(k_j,\ell_j) : j = 1,\ldots,n\}$ for $n \in \mathbb{N}$. Set $U_1 := D_1$. Then there is a finite subset T_1 of $\mathbb{N}^2$ such that $\left|\sum\sum_{(k,\ell)\in T_1} a_{k,\ell}\right| \geq 1 + |a_{k_1,\ell_1}|$, and for each $n \geq 2$, there is a finite subset T_n of $\mathbb{N}^2$ such that

$$\left|\sum\sum_{(k,\ell)\in T_n} a_{k,\ell}\right| \geq n + \sum\sum_{(k,\ell)\in U_n} |a_{k,\ell}|, \quad \text{where } U_n := D_n \cup T_1 \cup \cdots \cup T_{n-1}.$$

Define $S_n := T_n \cup U_n$ for $n \in \mathbb{N}$. Then it is easily seen that (S_n) is an exhausting sequence of subsets of $\mathbb{N}^2$. If for $n \in \mathbb{N}$, we let $V_n := S_n \setminus T_n$, then $V_n \subseteq U_n$ and

$$\Big|\sum\sum_{(k,\ell)\in S_n} a_{k,\ell}\Big| = \Big|\sum\sum_{(k,\ell)\in T_n} a_{k,\ell}+\sum\sum_{(k,\ell)\in V_n} a_{k,\ell}\Big| \geq \Big|\sum\sum_{(k,\ell)\in T_n} a_{k,\ell}\Big|-\sum\sum_{(k,\ell)\in U_n} |a_{k,\ell}| \geq n.$$

Hence $\lim_{n\to\infty}\sum\sum_{(k,\ell)\in S_n} a_{k,\ell}$ cannot exist, which is a contradiction. This proves that there is $\beta \in \mathbb{R}$ satisfying the inequality stated at the beginning of the proof.

Now, given any finite subset S of $\mathbb{N}^2$, if we let $S^+ := \{(k,\ell) \in S : a_{k,\ell} \geq 0\}$ and $S^- := \{(k,\ell) \in S : a_{k,\ell} \leq 0\}$, then

$$\sum\sum_{(k,\ell)\in S} |a_{k,\ell}| = \Big|\sum\sum_{(k,\ell)\in S^+} a_{k,\ell}\Big| + \Big|\sum\sum_{(k,\ell)\in S^-} a_{k,\ell}\Big| \leq 2\beta.$$

We obtain the desired result upon letting $\alpha := 2\beta$. □

Proposition 7.24. *A double series is unconditionally convergent if and only if it is absolutely convergent.*

Proof. Suppose $\sum\sum_{(k,\ell)} a_{k,\ell}$ is an unconditionally convergent double series. By Lemma 7.23, the partial double sums of $\sum\sum_{(k,\ell)} |a_{k,\ell}|$ are bounded, and hence by Proposition 7.14, we see that $\sum\sum_{(k,\ell)} |a_{k,\ell}|$ is convergent, that is, $\sum\sum_{(k,\ell)} a_{k,\ell}$ is absolutely convergent.

Conversely, suppose $\sum\sum_{(k,\ell)} a_{k,\ell}$ is an absolutely convergent double series. For each $(k,\ell) \in \mathbb{N}^2$, let $a^+_{k,\ell}$ and $a^-_{k,\ell}$ be as in the proof of Proposition 7.18. Then $\sum\sum_{(k,\ell)} a^+_{k,\ell}$ and $\sum\sum_{(k,\ell)} a^-_{k,\ell}$ are convergent double series with nonnegative terms, and therefore by Proposition 7.22, both of them are unconditionally convergent. Since $a_K l = a^+_{k,\ell} - a^-_{k,\ell}$ for all $(k,\ell) \in \mathbb{N}^2$, it follows that $\sum\sum_{(k,\ell)} a_{k,\ell}$ is unconditionally convergent. □

In view of Propositions 7.18, 7.22, and 7.24, we see that an unconditionally convergent double series is always convergent, and its unconditional double sum is equal to its double sum. However, since there do exist conditionally convergent series (Example 7.10 (iv)), we also see that a convergent double series need not be unconditionally convergent.

7.3 Convergence Tests for Double Series

In this section, we discuss several practical tests for deciding the convergence or divergence of a double series. We have already encountered the simplest among these, namely the (k,ℓ)th Term Test. In what follows we first consider tests for absolute convergence and later, tests for conditional convergence.

Tests for Absolute Convergence

The following simple test is widely used for determining the absolute convergence of a double series.

Proposition 7.25 (Comparison Test for Double Series). *Let $a_{k,\ell}$ and $b_{k,\ell}$ be real numbers such that $|a_{k,\ell}| \le b_{k,\ell}$ for all $(k,\ell) \in \mathbb{N}^2$. If $\sum\sum_{(k,\ell)} b_{k,\ell}$ is convergent, then $\sum\sum_{(k,\ell)} a_{k,\ell}$ is absolutely convergent and*

$$\Big|\sum\sum_{(k,\ell)} a_{k,\ell}\Big| \le \sum\sum_{(k,\ell)} b_{k,\ell}.$$

Proof. Suppose $\sum\sum_{(k,\ell)} b_{k,\ell}$ is convergent. For $(m,n) \in \mathbb{N}^2$, we have

$$\left|\sum_{k=1}^{m}\sum_{\ell=1}^{n} a_{k,\ell}\right| \le \sum_{k=1}^{m}\sum_{\ell=1}^{n} |a_{k,\ell}| \le \sum_{k=1}^{m}\sum_{\ell=1}^{n} b_{k,\ell}.$$

Since $b_{k,\ell} \ge 0$ for all $(k,\ell) \in \mathbb{N}^2$, the double sequence of the partial double sums of $\sum\sum_{(k,\ell)} b_{k,\ell}$ is bounded above (Proposition 7.14). By the second of the above inequalities, the same holds for $\sum\sum_{(k,\ell)} |a_{k,\ell}|$. Also, since $|a_{k,\ell}| \ge 0$ for all $(k,\ell) \in \mathbb{N}^2$, it follows from Proposition 7.14 that $\sum\sum_{(k,\ell)} |a_{k,\ell}|$ is convergent, that is, $\sum\sum_{(k,\ell)} a_{k,\ell}$ is absolutely convergent. The above inequalities also imply that $\left|\sum\sum_{(k,\ell)} a_{k,\ell}\right| \le \sum\sum_{(k,\ell)} |a_{k,\ell}| \le \sum\sum_{(k,\ell)} b_{k,\ell}$. □

The above result can also be stated as follows: If $|a_{k,\ell}| \le b_{k,\ell}$ for all $(k,\ell) \in \mathbb{N}^2$ and if $\sum\sum_{(k,\ell)} |a_{k,\ell}|$ diverges to ∞, then so does $\sum\sum_{(k,\ell)} b_{k,\ell}$. The geometric double series and the double series $\sum\sum_{(k,\ell)} 1/k^p\ell^q$, where $p,q \in \mathbb{R}$, are often useful in employing the Comparison Test for Double Series.

Examples 7.26. (i) For $p \in \mathbb{R}$ and $(k,\ell) \in \mathbb{N}^2$, let $a_{k,\ell} := 1/(k^p + \ell^p)$. Assume that $p > 2$ and let $b_{k,\ell} := 1/2(k\ell)^{p/2}$ for $(k,\ell) \in \mathbb{N}^2$. Then $|a_{k,\ell}| \le b_{k,\ell}$ for $(k,\ell) \in \mathbb{N}^2$ by the A.M.-G.M. inequality. Hence by the Comparison Test, $\sum\sum_{(k,\ell)} a_{k,\ell}$ is convergent. Next, let $p \le 2$ and define $b_{k,\ell} := 1/(k+\ell)^2$ for $(k,\ell) \in \mathbb{N}^2$. Then $|a_{k,\ell}| \ge 1/(k^2+\ell^2) \ge b_{k,\ell}$ for $(k,\ell) \in \mathbb{N}^2$. Hence by the Comparison Test, $\sum\sum_{(k,\ell)} a_{k,\ell}$ is divergent. For a more general result along these lines, see Exercise 70.

(ii) For $(k,\ell) \in \mathbb{N}^2$, let $a_{k,\ell} := (2^k 5^\ell + k\ell^2)/(3^k 7^\ell + k^3 + \ell^4)$. Define $b_{k,\ell} := (2/3)^k (5/7)^\ell$ for $(k,\ell) \in \mathbb{N}^2$. Since $k\ell^2 < 2^k 5^\ell$ and $k^3 + \ell^4 > 0$, we see that

$$|a_{k,\ell}| < \frac{2^k 5^\ell + 2^k 5^\ell}{3^k 7^\ell} = 2b_{k,\ell} \text{ for all } (k,\ell) \in \mathbb{N}^2.$$

Hence by the Comparison Test, $\sum\sum_{(k,\ell)} a_{k,\ell}$ is convergent.

(iii) Let $a_{k,\ell} := 1/(1 + k + \ell + k\ell + k^3\ell^4)^{1/2}$ for $(k,\ell) \in \mathbb{N}^2$. Define $b_{k,\ell} := 1/k^{3/2}\ell^2$ for $(k,\ell) \in \mathbb{N}^2$. Then $|a_{k,\ell}| \le b_{k,\ell}$ for all $(k,\ell) \in \mathbb{N}^2$. Hence by the Comparison Test, $\sum\sum_{(k,\ell)} a_{k,\ell}$ is (absolutely) convergent. ◇

We shall now consider analogues of the limit comparison test, the root test, and the ratio test for double series. We first state some basic results in the case of a (single) series for ease of reference. For proofs of these results, see, for example, Corollary 9.12, and Propositions 9.15 and 9.16 of ACICARA.

Fact 7.27. *Let (a_k) be a sequence of real numbers.*

(i) *Assume that $a_k > 0$ for all $k \in \mathbb{N}$. Let (b_k) be a sequence of positive real numbers such that $a_k/b_k \to r$ as $k \to \infty$, where $r \in \mathbb{R}$ with $r \neq 0$. Then the series $\sum_k a_k$ is convergent if and only if the series $\sum_k b_k$ is convergent.*
(ii) *If there is $\alpha \in \mathbb{R}$ with $\alpha < 1$ such that $|a_k|^{1/k} \leq \alpha$ for all large k, then the series $\sum_k a_k$ is absolutely convergent. If $|a_k|^{1/k} \geq 1$ for infinitely many $k \in \mathbb{N}$, then the series $\sum_k a_k$ is divergent.*
(iii) *If there is $\alpha \in \mathbb{R}$ with $\alpha < 1$ such that $|a_{k+1}| \leq \alpha |a_k|$ for all large k, then the series $\sum_k a_k$ is absolutely convergent. If $|a_{k+1}| \geq |a_k| > 0$ for all large $k \in \mathbb{N}$, then the series $\sum_k a_k$ is divergent.*

The following result will lead us to the limit comparison test for double series, which is often easier to use than the comparison test.

Proposition 7.28. *Let $(a_{k,\ell})$ and $(b_{k,\ell})$ be double sequences such that $b_{k,\ell} \neq 0$ for all $(k,\ell) \in \mathbb{N}^2$. Suppose each row-series as well as each column-series corresponding to both $\sum\sum_{(k,\ell)} a_{k,\ell}$ and $\sum\sum_{(k,\ell)} b_{k,\ell}$ is absolutely convergent, and $a_{k,\ell}/b_{k,\ell} \to r$ as $(k,\ell) \to (\infty,\infty)$, where $r \in \mathbb{R} \cup \{\pm\infty\}$.*

(i) *If $b_{k,\ell} > 0$ for all $(k,\ell) \in \mathbb{N}^2$, $\sum\sum_{(k,\ell)} b_{k,\ell}$ is convergent, and $r \in \mathbb{R}$, then $\sum\sum_{(k,\ell)} a_{k,\ell}$ is absolutely convergent.*
(ii) *If $a_{k,\ell} > 0$ for all $(k,\ell) \in \mathbb{N}^2$, $\sum\sum_{(k,\ell)} a_{k,\ell}$ is convergent, and $r \neq 0$, then $\sum\sum_{(k,\ell)} b_{k,\ell}$ is absolutely convergent.*

Proof. (i) Suppose $b_{k,\ell} > 0$ for all $(k,\ell) \in \mathbb{N}^2$ and the double series $\sum\sum_{(k,\ell)} b_{k,\ell}$ is convergent. Let $r \in \mathbb{R}$ be such that Since $a_{k,\ell}/b_{k,\ell} \to r$ as $(k,\ell) \to (\infty,\infty)$. Then there is $(k_0,\ell_0) \in \mathbb{N}^2$ such that for all $(k,\ell) \geq (k_0,\ell_0)$,

$$(r-1)b_{k,\ell} < a_{k,\ell} < (r+1)b_{k,\ell} \quad \text{and so} \quad |a_{k,\ell}| < \max\{|r-1|,|r+1|\}b_{k,\ell}.$$

Also, by Proposition 7.14, there is $\beta > 0$ such that $\sum_{k=1}^{m}\sum_{\ell=1}^{n} b_{k,\ell} \leq \beta$ for $(m,n) \in \mathbb{N}^2$. Hence for all $(m,n) \geq (k_0,\ell_0)$, we have

$$\sum_{k=k_0}^{m}\sum_{\ell=\ell_0}^{n} |a_{k,\ell}| < \max\{|r-1|,|r+1|\}\sum_{k=k_0}^{m}\sum_{\ell=\ell_0}^{n} b_{k,\ell} \leq \max\{|r-1|,|r+1|\}\beta.$$

By Proposition 7.19, the double series $\sum\sum_{(k,\ell)} a_{k,\ell}$ is absolutely convergent.

(ii) Suppose $a_{k,\ell} > 0$ for all $(k,\ell) \in \mathbb{N}^2$ and $r \neq 0$. Then the limit of $b_{k,\ell}/a_{k,\ell}$ as $(k,\ell) \to (\infty,\infty)$ is $1/r$ or 0 according as $r \in \mathbb{R}$ or $r = \pm\infty$. By interchanging $a_{k,\ell}$ and $b_{k,\ell}$ in (i) above, the desired result follows. □

Corollary 7.29 (Limit Comparison Test for Double Series). *Let $(a_{k,\ell})$ and $(b_{k,\ell})$ be double sequences of positive real numbers. Suppose each row-series as well as each column-series corresponding to both $\sum\sum_{(k,\ell)} a_{k,\ell}$ and $\sum\sum_{(k,\ell)} b_{k,\ell}$ is convergent, and*

$$\lim_{(k,\ell)\to(\infty,\infty)} \frac{a_{k,\ell}}{b_{k,\ell}} = r, \quad \textit{where } r \in \mathbb{R} \textit{ and } r \neq 0.$$

Then

$$\sum_{(k,\ell)} a_{k,\ell} \textit{ is convergent} \iff \sum_{(k,\ell)} b_{k,\ell} \textit{ is convergent.}$$

Proof. The implication $\Longrightarrow$ follows from part (ii) of Proposition 7.28, while the reverse implication $\Longleftarrow$ follows from part (i) of Proposition 7.28. □

Remark 7.30. In the Limit Comparison Test for Double Series, the condition $r \in \mathbb{R}$ and $r \neq 0$ cannot be dropped. To see that $r = 0$ will not work, let

$$a_{k,\ell} := \frac{1}{k^2\ell^2} \quad \text{and} \quad b_{k,\ell} := \frac{1}{(k+\ell)^2} \quad \text{for } (k,\ell) \in \mathbb{N}^2.$$

Then $\lim_{\ell\to\infty}(a_{k,\ell}/b_{k,\ell}) = 1/k^2$ for each $k \in \mathbb{N}$ and $\lim_{k\to\infty}(a_{k,\ell}/b_{k,\ell}) = 1/\ell$ for each $\ell \in \mathbb{N}$. Hence by Fact 7.27 (i), we see that each row-series as well as each column-series corresponding to both $\sum\sum_{(k,\ell)} a_{k,\ell}$ and $\sum\sum_{(k,\ell)} b_{k,\ell}$ is convergent. However, $\lim_{(k,\ell)\to(\infty,\infty)}(a_{k,\ell}/b_{k,\ell}) = 0$, and as shown in Examples 7.10 (iii), the double series $\sum\sum_{(k,\ell)} a_{k,\ell}$ converges, while the double series $\sum\sum_{(k,\ell)} b_{k,\ell}$ diverges. By interchanging the definitions of $a_{k,\ell}$ and $b_{k,\ell}$, we see that $r = \infty$ will also not work. ◇

Examples 7.31. (i) Let $a_{k,\ell} := \sin(1/k^2\ell^2)$ for all $(k,\ell) \in \mathbb{N}^2$. Consider $b_{k,\ell} := 1/k^2\ell^2$ for all $(k,\ell) \in \mathbb{N}^2$, and observe that $(a_{k,\ell}/b_{k,\ell}) \to 1$ as $(k,\ell) \to (\infty,\infty)$. Since $\sum\sum_{(k,\ell)} b_{k,\ell}$ is convergent, Corollary 7.29 shows that $\sum\sum_{(k,\ell)} a_{k,\ell}$ is convergent.

(ii) Let $a_{k,\ell} := \sin\left(1/(k+\ell)^2\right)$ for all $(k,\ell) \in \mathbb{N}^2$. Consider $b_{k,\ell} := 1/(k+\ell)^2$ for all $(k,\ell) \in \mathbb{N}^2$, and observe that $(a_{k,\ell}/b_{k,\ell}) \to 1$ as $(k,\ell) \to (\infty,\infty)$. Since $\sum\sum_{(k,\ell)} b_{k,\ell}$ is divergent, Corollary 7.29 now shows $\sum\sum_{(k,\ell)} a_{k,\ell}$ is divergent. ◇

The following result will lead us to Cauchy's root test, or simply the root test, which is one of the most basic tests to determine the absolute convergence of a double series.

In what follows, we shall say that a statement holds whenever "**both k and ℓ are large**" to mean that there is $(k_0, \ell_0) \in \mathbb{N}^2$ such that the statement holds for all $(k,\ell) \in \mathbb{N}^2$ with $(k,\ell) \geq (k_0,\ell_0)$.

Proposition 7.32. *Let $(a_{k,\ell})$ be a double sequence of real numbers.*

(i) *Suppose each row-series as well as each column-series corresponding to $\sum\sum_{(k,\ell)} a_{k,\ell}$ is absolutely convergent. If there is $\alpha \in \mathbb{R}$ with $\alpha < 1$ such that $|a_{k,\ell}|^{1/(k+\ell)} \leq \alpha$ whenever both k and ℓ are large, then $\sum\sum_{(k,\ell)} a_{k,\ell}$ is absolutely convergent.*

(ii) *If for each $(k_0,\ell_0) \in \mathbb{N}^2$, there is $(k,\ell) \in \mathbb{N}^2$ such that $(k,\ell) \geq (k_0,\ell_0)$ and $|a_{k,\ell}|^{1/(k+\ell)} \geq 1$, then $\sum\sum_{(k,\ell)} a_{k,\ell}$ is divergent.*

Proof. (i) Suppose there are $\alpha \in \mathbb{R}$ with $\alpha < 1$ and $(k_0, \ell_0) \in \mathbb{N}^2$ such that $|a_{k,\ell}|^{1/(k+\ell)} \le \alpha$ for all $(k, \ell) \ge (k_0, \ell_0)$. Then $\alpha \ge 0$ and

$$\sum_{k=k_0}^{m} \sum_{\ell=\ell_0}^{n} |a_{k,\ell}| < \Big(\sum_{k=k_0}^{m} \alpha^k \Big) \Big(\sum_{\ell=\ell_0}^{n} \alpha^\ell \Big) \le \frac{1}{(1-\alpha)^2} \quad \text{for } (m,n) \ge (k_0, \ell_0).$$

Hence (i) follows from Proposition 7.19.

(ii) Suppose for each $(k_0, \ell_0) \in \mathbb{N}^2$, there is $(k, \ell) \in \mathbb{N}^2$ such that $(k, \ell) \ge (k_0, \ell_0)$ and $|a_{k,\ell}|^{1/(k+\ell)} \ge 1$, that is, $|a_{k,\ell}| \ge 1$. Hence $a_{k,\ell} \not\to 0$ as $(k, \ell) \to (\infty, \infty)$. By the (k, ℓ)th Term Test (Proposition 7.8), it follows that $\sum\sum_{(k,\ell)} a_{k,\ell}$ is divergent.

□

Corollary 7.33 (Root Test for Double Series). *Let $(a_{k,\ell})$ be a double sequence of real numbers such that $|a_{k,\ell}|^{1/(k+\ell)} \to a$ as $(k, \ell) \to (\infty, \infty)$, where $a \in \mathbb{R} \cup \{\infty\}$. If each row-series as well as each column-series corresponding to $\sum\sum_{(k,\ell)} a_{k,\ell}$ is absolutely convergent and $a < 1$, then $\sum\sum_{(k,\ell)} a_{k,\ell}$ is absolutely convergent. On the other hand, if $a > 1$, then $\sum\sum_{(k,\ell)} a_{k,\ell}$ is divergent, and all but finitely many row-series and column-series are also divergent.*

Proof. The first assertion follows from part (i) of Proposition 7.32 with $\alpha := (1 + a)/2$. Now suppose $a > 1$. Then there is $(k_0, \ell_0) \in \mathbb{N}^2$ such that $|a_{k,\ell}|^{1/(k+\ell)} \ge 1$ for all $(k, \ell) \ge (k_0, \ell_0)$. Part (ii) of Proposition 7.32 shows that $\sum\sum_{(k,\ell)} a_{k,\ell}$ is divergent. Also, for each fixed $k \ge k_0$, we see that $a_{k,\ell} \not\to 0$ as $\ell \to \infty$, and hence the row-series $\sum_\ell a_{k,\ell}$ diverges. Similarly, for each fixed $\ell \ge \ell_0$, the column-series $\sum_k a_{k,\ell}$ diverges. □

The following result will lead us to D'Alembert's ratio test, or simply the ratio test, which is another basic test to determine the absolute convergence of a double series.

Proposition 7.34. *Let $(a_{k,\ell})$ be a double sequence of real numbers.*

(i) *Suppose each row-series as well as each column-series corresponding to the double series $\sum\sum_{(k,\ell)} a_{k,\ell}$ is absolutely convergent. If there is $\alpha \in \mathbb{R}$ with $\alpha < 1$ such that either $|a_{k,\ell+1}| \le \alpha |a_{k,\ell}|$ whenever both k and ℓ are large, or $|a_{k+1,\ell}| \le \alpha |a_{k,\ell}|$ whenever both k and ℓ are large, then $\sum\sum_{(k,\ell)} a_{k,\ell}$ is absolutely convergent.*
(ii) *If $\min\{|a_{k,\ell+1}|, |a_{k+1,\ell}|\} \ge |a_{k,\ell}| > 0$ whenever both k and ℓ are large, then $\sum\sum_{(k,\ell)} a_{k,\ell}$ is divergent, and all but finitely many row-series and column-series are also divergent.*

Proof. (i) Suppose there are $\alpha \in \mathbb{R}$ with $\alpha < 1$ and $(k_0, \ell_0) \in \mathbb{N}^2$ such that $|a_{k,\ell+1}| \le \alpha |a_{k,\ell}|$ for all $(k, \ell) \ge (k_0, \ell_0)$. We may assume that $\alpha > 0$. Now

$$|a_{k,\ell}| \le \alpha |a_{k,\ell-1}| \le \cdots \le \alpha^{\ell-\ell_0} |a_{k,\ell_0}| \quad \text{for all } (k, \ell) \ge (k_0, \ell_0).$$

Since $0 < \alpha < 1$, we see that $\sum_{\ell=1}^{n} \alpha^{\ell} \le 1/(1-\alpha)$ for all $n \in \mathbb{N}$. Also, since the series $\sum_k a_{k,\ell_0}$ is absolutely convergent, there is $\beta > 0$ such that $\sum_{k=1}^{m} |a_{k,\ell_0}| \le \beta$ for all $m \in \mathbb{N}$. Consequently,

$$\sum_{k=k_0}^{m} \sum_{\ell=\ell_0}^{n} |a_{k,\ell}| \le \frac{\alpha^{-\ell_0}\beta}{1-\alpha} \quad \text{for all } (m,n) \ge (k_0,\ell_0).$$

Hence by Proposition 7.19, $\sum\sum_{(k,\ell)} a_{k,\ell}$ is absolutely convergent. A similar argument holds if there are $\alpha \in \mathbb{R}$ with $\alpha < 1$ and $(k_0,\ell_0) \in \mathbb{N}^2$ such that $|a_{k+1,\ell}| \le \alpha |a_{k,\ell}|$ for all $(k,\ell) \ge (k_0,\ell_0)$.

(ii) Suppose there is $(k_0,\ell_0) \in \mathbb{N}^2$ such that $\min\{|a_{k,\ell+1}, a_{k+1,\ell}|\} \ge |a_{k,\ell}| > 0$ for all $(k,\ell) \ge (k_0,\ell_0)$. Then

$$|a_{k,\ell}| \ge |a_{k,\ell-1}| \ge \cdots \ge |a_{k,\ell_0}| \ge |a_{k-1,\ell_0}| \ge \cdots \ge |a_{k_0,\ell_0}| > 0$$

for all $(k,\ell) \ge (k_0,\ell_0)$. Since $a_{k_0,\ell_0} \ne 0$, we see that $a_{k,\ell} \not\to 0$ as $(k,\ell) \to (\infty,\infty)$, and further, for each fixed $k \ge k_0$, $a_{k,\ell} \not\to 0$ as $\ell \to \infty$ and for each fixed $\ell \ge \ell_0$, $a_{k,\ell} \not\to 0$ as $k \to \infty$. The desired results now follow from the (k,ℓ)th Term Test for double series (Proposition 7.8) and the kth Term Test for (single) series (given, for example, in Proposition 9.6 of ACICARA). □

Corollary 7.35 (Ratio Test for Double Series). *Let $(a_{k,\ell})$ be a double sequence of nonzero real numbers such that either $|a_{k,\ell+1}|/|a_{k,\ell}| \to a$ or $|a_{k+1,\ell}|/|a_{k,\ell}| \to \tilde{a}$ as $(k,\ell) \to (\infty,\infty)$, where $a, \tilde{a} \in \mathbb{R} \cup \{\infty\}$. If each row-series as well as each column-series corresponding to $\sum\sum_{(k,\ell)} a_{k,\ell}$ is absolutely convergent and $a < 1$ or $\tilde{a} < 1$, then $\sum\sum_{(k,\ell)} a_{k,\ell}$ is absolutely convergent. On the other hand, if $a > 1$, then $\sum\sum_{(k,\ell)} a_{k,\ell}$ is divergent and all but finitely many row-series are also divergent, while if $\tilde{a} > 1$, then $\sum\sum_{(k,\ell)} a_{k,\ell}$ is divergent and all but finitely many column-series are also divergent.*

Proof. The first result is a consequence of part (i) of Proposition 7.34 with $\alpha := (1+a)/2$ or $\alpha := (1+\tilde{a})/2$ according as $a < 1$ or $\tilde{a} < 1$.

Now suppose $a > 1$. Then there are $\alpha \in \mathbb{R}$ with $\alpha > 1$ and $(k_0,\ell_0) \in \mathbb{N}^2$ such that $|a_{k,\ell+1}|/|a_{k,\ell}| \ge \alpha$ for all $(k,\ell) \ge (k_0,\ell_0)$. Then

$$|a_{k,\ell}| \ge \alpha |a_{k,\ell-1}| \ge \cdots \ge \alpha^{\ell-\ell_0} |a_{k,\ell_0}| \quad \text{for all } (k,\ell) \ge (k_0,\ell_0).$$

Given any $(k_1,\ell_1) \in \mathbb{N}^2$, let $k := \max\{k_0,k_1\}$. Since $\alpha > 1$ and $a_{k,\ell_0} \ne 0$, we can find $\ell \ge \max\{\ell_0,\ell_1\}$ such that $\alpha^{\ell-\ell_0}|a_{k,\ell_0}| \ge 1$. Then $k \ge k_1$, $\ell \ge \ell_1$, and $|a_{k,\ell}| \ge 1$. This shows that $a_{k,\ell} \not\to 0$ as $(k,\ell) \to (\infty,\infty)$, and so $\sum\sum_{(k,\ell)} a_{k,\ell}$ is divergent by the (k,ℓ)th Term Test. Also, for each fixed $k \ge k_0$, we have $|a_{k,\ell}| \ge \alpha^{\ell-\ell_0}|a_{k,\ell_0}| \ge |a_{k,\ell_0}| > 0$ for all $\ell \ge \ell_0$, and so $a_{k,\ell} \not\to 0$ as $\ell \to \infty$, which implies that $\sum_\ell a_{k,\ell}$ is divergent. Similar arguments hold if $\tilde{a} > 1$. □

A variant of the comparison test involving ratios of successive terms of two double series, called the ratio comparison test, for is given in Exercise 18.

Applying this test to the double series $\sum\sum_{(k,\ell)} 1/(k\ell)^p$, where $p > 0$, one can obtain an analogue of Raabe's test (stated, for example, in Exercise 13 of Chapter 9 of ACICARA) for double series. It is particularly useful when $|a_{k,\ell+1}|/|a_{k,\ell}| \to 1$ and $|a_{k+1,\ell}|/|a_{k,\ell}| \to 1$ as $(k,\ell) \to (\infty,\infty)$. See Exercises 19 and 21. Finally, we remark that there is a very useful test for the convergence for a double series of nonnegative terms, known as the integral test. It is based on "improper double integrals" and will be given in Proposition 7.57.

Examples 7.36. (i) If the limit a in the Root Test (Corollary 7.33) is equal to 1, then the double series $\sum\sum_{(k,\ell)} a_{k,\ell}$ may converge absolutely or it may diverge. The same holds if the limits a and $\tilde{a}$ in the Ratio Test (Corollary 7.35) are equal to 1. For example, let $a_{k,\ell} := 1/k^2\ell^2$ and $b_{k,\ell} := 1/(k+\ell)^2$ for $(k,\ell) \in \mathbb{N}^2$. Then it is easy to see that each row-series as well as each column-series corresponding to both $\sum\sum_{(k,\ell)} a_{k,\ell}$ and $\sum\sum_{(k,\ell)} b_{k,\ell}$ is (absolutely) convergent and all the above-mentioned limits are equal to 1 for both cases. However, as we have seen in Example 7.10 (iii), $\sum\sum_{(k,\ell)} a_{k,\ell}$ is (absolutely) convergent, but $\sum\sum_{(k,\ell)} b_{k,\ell}$ is divergent.

(ii) Let $p > 0$ and for $(k,\ell) \in \mathbb{N}^2$, let $a_{k,\ell} := (k+\ell)^p/2^k 3^\ell$. It is easy to see (using Fact 7.27 (iii), for example) that each row-series as well as each column-series corresponding to $\sum\sum_{(k,\ell)} a_{k,\ell}$ is (absolutely) convergent. Since $|a_{k,\ell+1}|/|a_{k,\ell}| \to 1/3$ as $(k,\ell) \to (\infty,\infty)$, Corollary 7.35 shows that $\sum\sum_{(k,\ell)} a_{k,\ell}$ is (absolutely) convergent. Alternatively, the same conclusion follows by noting that $|a_{k+1,\ell}|/|a_{k,\ell}| \to 1/2$ as $(k,\ell) \to (\infty,\infty)$.

(iii) For $(k,\ell) \in \mathbb{N}^2$, let $a_{k,\ell} := (k+\ell)!/2^k 3^\ell$. Since $a_{k,\ell+1}/a_{k,\ell} \to \infty$ as $(k,\ell) \to (\infty,\infty)$, Corollary 7.35 shows that $\sum\sum_{(k,\ell)} a_{k,\ell}$ is divergent. Alternatively, observe that $a_{k,\ell} \geq (k!/2^k)(\ell!/3^\ell) \geq 1$ for $(k,\ell) \geq (4,7)$, and so the (k,ℓ)th Term Test shows that $\sum\sum_{(k,\ell)} a_{k,\ell}$ is divergent.

(iv) For $(k,\ell) \in \mathbb{N}^2$, let $a_{k,\ell} := (k+\ell)!/(k+\ell)^{k+\ell}$. Since $\left(1+(1/n)\right)^n \to e$ as $n \to \infty$, where e is the base of the natural logarithm, we see that $a_{k,\ell+1}/a_{k,\ell} \to 1/e$ as $(k,\ell) \to (\infty,\infty)$. Also, for each fixed $k \in \mathbb{N}$, we have $\lim_{\ell\to\infty} a_{k,\ell+1}/a_{k,\ell} = 1/e$, and for each fixed $\ell \in \mathbb{N}$, we have $\lim_{k\to\infty} a_{k+1,\ell}/a_{k,\ell} = 1/e$. Since $e > 1$, Corollary 7.35 and Fact 7.27 (iii) show that $\sum\sum_{(k,\ell)} a_{k,\ell}$ is (absolutely) convergent.

(v) For $(k,\ell) \in \mathbb{N}^2$, let

$$a_{k,\ell} := \begin{cases} \dfrac{1}{2^{k+\ell}} & \text{if } k+\ell \text{ is even,} \\ \dfrac{1}{3^{k+\ell}} & \text{if } k+\ell \text{ is odd.} \end{cases}$$

Since $|a_{k,\ell+1}|/|a_{k,\ell}| = |a_{k+1,\ell}|/|a_{k,\ell}| = 2^{k+\ell}/3^{k+\ell+1} \leq 4/27$ if $k+\ell$ is even, and $|a_{k,\ell+1}|/|a_{k,\ell}| = |a_{k+1,\ell}|/|a_{k,\ell}| = 3^{k+\ell}/2^{k+\ell+1} \geq 27/16$ if $k+\ell$ is odd, the Ratio test for Double Series (Corollary 7.35) is not applicable to this example. For the same reason, Proposition 7.34 is also not applicable. Further, since the double sequence $(|a_{k,\ell}|^{1/(k+\ell)})$ does not converge, the Root Test for Double Series (Corollary 7.33) is not applicable. However, since $|a_{k,\ell}|^{1/\ell}$ and $|a_{k,\ell}|^{1/k}$ are less than or equal to 1/2 for all $(k,\ell) \in \mathbb{N}^2$, we see that each row-series as well as each column-series corresponding to $\sum\sum_{(k,\ell)} a_{k,\ell}$ is (absolutely) convergent. Also, $|a_{k,\ell}|^{1/(k+\ell)} \le \frac{1}{2} < 1$ for all $(k,\ell) \in \mathbb{N}^2$, and hence Proposition 7.32 is applicable. Thus $\sum\sum_{(k,\ell)} a_{k,\ell}$ is (absolutely) convergent. ◇

Tests for Conditional Convergence

Now we turn to tests for conditional (that is, nonabsolute) convergence. They are based on the following result, which may be compared with the well-known partial summation formula, which states that if $n \in \mathbb{N}$ with $n \ge 2$, a_k and b_k are real numbers for $k = 1, \ldots, n$, and we let $B_n := \sum_{k=1}^{n} b_k$, then

$$\sum_{k=1}^{n} a_k b_k = a_n B_n + \sum_{k=1}^{n-1} (a_k - a_{k+1}) B_k.$$

(See, for example, Proposition 9.19 of ACICARA.)

Proposition 7.37 (Partial Double Summation Formula). *Consider $(m,n) \in \mathbb{N}^2$ with $(m,n) \ge (2,2)$. For $k = 1, \ldots, m$ and $\ell = 1, \ldots, n$, let $a_{k,\ell}$ and $b_{k,\ell}$ be real numbers, and let $B_{m,n} := \sum_{k=1}^{m}\sum_{\ell=1}^{n} b_{k,\ell}$. Then*

$$\sum_{k=1}^{m}\sum_{\ell=1}^{n} a_{k,\ell} b_{k,\ell} = a_{m,n} B_{m,n} + \sum_{k=1}^{m-1}\sum_{\ell=1}^{n-1} (a_{k,\ell} - a_{k+1,\ell} - a_{k,\ell+1} + a_{k+1,\ell+1}) B_{k,\ell}$$
$$+ \sum_{k=1}^{m-1} (a_{k,n} - a_{k+1,n}) B_{k,n} + \sum_{\ell=1}^{n-1} (a_{m,\ell} - a_{m,\ell+1}) B_{m,\ell}.$$

Proof. Since $b_{k,\ell} = B_{k,\ell} - B_{k-1,\ell} - B_{k,\ell-1} + B_{k-1,\ell-1}$ (with the usual convention that $B_{0,0} = 0$, $B_{0,\ell} = 0$, and $B_{k,0} = 0$ for $(k,\ell) \in \mathbb{N}^2$), we obtain

$$\sum_{\ell=1}^{n} a_{k,\ell} b_{k,\ell} = \sum_{\ell=1}^{n} a_{k,\ell} B_{k,\ell} - \sum_{\ell=1}^{n} a_{k,\ell} B_{k-1,\ell} - \sum_{\ell=1}^{n} a_{k,\ell} B_{k,\ell-1} + \sum_{\ell=1}^{n} a_{k,\ell} B_{k-1,\ell-1}.$$

Hence

$$\sum_{k=1}^{m}\sum_{\ell=1}^{n} a_{k,\ell} b_{k,\ell} = \sum_{k=1}^{m}\sum_{\ell=1}^{n} a_{k,\ell} B_{k,\ell} - \sum_{k=1}^{m-1}\sum_{\ell=1}^{n} a_{k+1,\ell} B_{k,\ell}$$
$$- \sum_{k=1}^{m}\sum_{\ell=1}^{n-1} a_{k,\ell+1} B_{k,\ell} + \sum_{k=1}^{m-1}\sum_{\ell=1}^{n-1} a_{k+1,\ell+1} B_{k,\ell}.$$

Upon rewriting the double sum $\sum_{k=1}^{m}\sum_{\ell=1}^{n} a_{k,\ell}B_{k,\ell}$ in the above equality as

$$a_{m,n}B_{m,n} + \sum_{k=1}^{m-1}\sum_{\ell=1}^{n-1} a_{k,\ell}B_{k,\ell} + \sum_{k=1}^{m-1} a_{k,n}B_{k,n} + \sum_{\ell=1}^{n-1} a_{m,\ell}B_{m,\ell}\,,$$

and collating appropriate terms, we obtain the desired identity. □

We shall now consider an important test for conditional convergence of a double series that is analogous to Dirichlet's test for series, given, for example, in Proposition 9.20 of ACICARA.

Proposition 7.38 (Dirichlet's Test for Double Series). *Let $(a_{k,\ell})$ and $(b_{k,\ell})$ be double sequences of real numbers such that*

(i) *$(a_{k,\ell})$ is bimonotonic,*
(ii) *for each fixed $\ell \in \mathbb{N}$, the sequence given by $k \longmapsto a_{k,\ell}$ is monotonic, and for each fixed $k \in \mathbb{N}$, the sequence given by $\ell \longmapsto a_{k,\ell}$ is monotonic,*
(iii) *$\lim_{k\to\infty} a_{k,k}$, $\lim_{k\to\infty} a_{k,1}$, and $\lim_{\ell\to\infty} a_{1,\ell}$ exist, and each is equal to 0,*
(iv) *the double sequence of partial double sums of $\sum\sum_{(k,\ell)} b_{k,\ell}$ is bounded.*

Then the double series $\sum\sum_{(k,\ell)} a_{k,\ell}b_{k,\ell}$ is convergent and the double sequence of its partial double sums is bounded.

Proof. First we show that $a_{k,\ell} \to 0$ as $(k,\ell) \to (\infty,\infty)$ and that $(a_{k,\ell})$ is bounded. Let $\epsilon > 0$ be given. By hypothesis (iii), there is $k_0 \in \mathbb{N}$ such that

$$(k,\ell) \in \mathbb{N}^2 \text{ with } (k,\ell) \geq (k_0,k_0) \Longrightarrow |a_{k,k}| < \epsilon, \quad |a_{k,1}| < \epsilon, \quad \text{and} \quad |a_{1,\ell}| < \epsilon.$$

Let us consider $(k,\ell) \in \mathbb{N}^2$ with $k \geq k_0$ and $\ell \leq k$. By hypothesis (ii), either $a_{k,1} \leq a_{k,\ell} \leq a_{k,k}$ or $a_{k,1} \geq a_{k,\ell} \geq a_{k,k}$. Since both $a_{k,1}$ and $a_{k,k}$ are in the open interval $(-\epsilon,\epsilon)$, we see that $a_{k,\ell} \in (-\epsilon,\epsilon)$. Similarly, if $(k,\ell) \in \mathbb{N}^2$ with $\ell \geq k_0$ and $k \leq \ell$, then $a_{k,\ell} \in (-\epsilon,\epsilon)$. Thus for all $(k,\ell) \in \mathbb{N}^2$ with either $k \geq k_0$ or $\ell \geq k_0$, we have $|a_{k,\ell}| < \epsilon$. Since $\epsilon > 0$ is arbitrary, this implies that $a_{k,\ell} \to 0$ as $(k,\ell) \to (\infty,\infty)$. Also, letting $\epsilon := 1$ and $\alpha := \max\{|a_{k,\ell}| : 1 \leq k,\ell \leq k_0\}$, we obtain $|a_{k,\ell}| \leq \max\{1,\alpha\}$, showing that $(a_{k,\ell})$ is bounded.

We now examine each term on the right side of the Partial Double Summation Formula (Proposition 7.37). Recall that the first term is $a_{m,n}B_{m,n}$, where $B_{m,n} := \sum_{k=1}^{m}\sum_{\ell=1}^{n} b_{k,\ell}$ for $(m,n) \in \mathbb{N}^2$. Since $(B_{m,n})$ is bounded, thanks to hypothesis (iv), there is $\beta > 0$ such that $|B_{m,n}| \leq \beta$ for all $(m,n) \in \mathbb{N}^2$. Since $a_{m,n} \to 0$ as $(m,n) \to (\infty,\infty)$, it follows that $a_{m,n}B_{m,n} \to 0$ as $(m,n) \to (\infty,\infty)$.

As for the second term, note that by hypothesis (i), the double sequence $(a_{k,\ell})$ is bimonotonic, and hence for all $(m,n) \geq (2,2)$,

$$\begin{aligned}\sum_{k=1}^{m-1}\sum_{\ell=1}^{n-1} & |(a_{k,\ell} - a_{k+1,\ell} - a_{k,\ell+1} + a_{k+1,\ell+1})B_{k,\ell}| \\ & \le \beta\left|\sum_{k=1}^{m-1}\sum_{\ell=1}^{n-1}(a_{k,\ell} - a_{k+1,\ell} - a_{k,\ell+1} + a_{k+1,\ell+1})\right| \\ & = \beta|a_{1,1} - a_{m,1} - a_{1,n} + a_{m,n}|,\end{aligned}$$

as we have seen in the proof of Proposition 7.13. Since the double sequence $(a_{m,n})$ is bounded, it follows from Proposition 7.14 that the double series $\sum\sum_{(k,\ell)}(a_{k,\ell} - a_{k+1,\ell} - a_{k,\ell+1} + a_{k+1,\ell+1})B_{k,\ell}$ is absolutely convergent. By part (i) of Proposition 7.21, its partial double sums are bounded, and by Proposition 7.18, it is convergent. Let C denote its double sum.

As for the third term, note that since for each fixed $n \in \mathbb{N}$, the sequence $k \longmapsto a_{k,n}$ is monotonic, it follows that for all $(m,n) \ge (2,2)$,

$$\left|\sum_{k=1}^{m-1}(a_{k,n} - a_{k+1,n})B_{k,n}\right| \le \beta\left|\sum_{k=1}^{m-1}(a_{k,n} - a_{k+1,n})\right| = \beta|a_{1,n} - a_{m,n}|.$$

Since $a_{1,n} \to 0$ as $n \to \infty$, and $a_{m,n} \to 0$ as $(m,n) \to (\infty,\infty)$, we see that $|a_{1,n} - a_{m,n}| \to 0$, and so $\sum_{k=1}^{m-1}(a_{k,n} - a_{k+1,n})B_{k,n} \to 0$ as $(m,n) \to (\infty,\infty)$. Similarly, it follows that $\sum_{\ell=1}^{n-1}(a_{m,\ell} - a_{m,\ell+1})B_{m,\ell} \to 0$ as $(m,n) \to (\infty,\infty)$.

By the Partial Double Summation Formula, we obtain

$$\sum_{k=1}^{m}\sum_{\ell=1}^{n} a_{k,\ell}b_{k,\ell} \to 0 + C + 0 + 0 = C \quad \text{as } (m,n) \to (\infty,\infty).$$

Thus the double series $\sum\sum_{(k,\ell)} a_{k,\ell}b_{k,\ell}$ is convergent. Also, since each of the four terms on the right side of the Partial Double Summation Formula is bounded, we see that the double sequence of the partial double sums of $\sum\sum_{(k,\ell)} a_{k,\ell}b_{k,\ell}$ is bounded. □

For a similar result, which is analogous to Abel's test for series (given, for example, in Exercise 17 of Chapter 9 of ACICARA), see Exercise 48. For generalizations of both these results, which are analogous to Dedekind's tests for (single) series (given, for example, in Exercise 19 of Chapter 9 of ACICARA), see Exercise 49.

Corollary 7.39 (Leibniz's Test for Double Series). *Let $(a_{k,\ell})$ be a double sequence of real numbers satisfying conditions* (i), (ii), *and* (iii) *given in Proposition 7.38. Then the double series $\sum\sum_{(k,\ell)}(-1)^{k+\ell}a_{k,\ell}$ is convergent.*

Proof. Define $b_{k,\ell} := (-1)^{k+\ell}$ for $(k,\ell) \in \mathbb{N}^2$ and $B_{m,n} := \sum_{k=1}^{m}\sum_{\ell=1}^{n} b_{k,\ell}$ for $(m,n) \in \mathbb{N}^2$. Then

$$B_{m,n} = \left(\sum_{k=1}^{m}(-1)^k\right)\left(\sum_{\ell=1}^{n}(-1)^\ell\right) = \begin{cases} 0 & \text{if either } m \text{ or } n \text{ is even,} \\ 1 & \text{if both } m \text{ and } n \text{ are odd.}\end{cases}$$

Hence the double sequence $(B_{m,n})$ is bounded. Now Proposition 7.38 shows that the series $\sum\sum_{(k,\ell)}(-1)^{k+\ell}a_{k,\ell}$ is convergent. □

In the next corollary, we shall use the trigonometric identities

$$\sin\frac{B}{2}\sum_{j=1}^{p}\sin(A+jB)=\sin\left(A+\frac{p+1}{2}B\right)\sin\frac{pB}{2}$$

and

$$\sin\frac{B}{2}\sum_{j=1}^{p}\cos(A+jB)=\cos\left(A+\frac{p+1}{2}B\right)\sin\frac{pB}{2},$$

where $A,B\in\mathbb{R}$ and $p\in\mathbb{N}$. These can be easily derived by expressing the left hand sides as telescoping sums.

Corollary 7.40 (Convergence Test for Trigonometric Double Series). *Let* $(a_{k,\ell})$ *be a double sequence of real numbers satisfying conditions* (i), (ii), *and* (iii) *of Proposition 7.38. Let* θ *and* φ *be real numbers, neither of which is an integral multiple of* 2π. *Then the double series*

$$\sum\sum_{(k,\ell)} a_{k,\ell}\sin(k\theta+\ell\varphi) \quad \textit{and} \quad \sum\sum_{(k,\ell)} a_{k,\ell}\cos(k\theta+\ell\varphi)$$

are convergent.

Proof. For $(m,n)\in\mathbb{N}^2$, define

$$B_{m,n}:=\sum_{k=1}^{m}\sum_{\ell=1}^{n}\sin(k\theta+\ell\varphi) \quad\text{and}\quad C_{m,n}:=\sum_{k=1}^{m}\sum_{\ell=1}^{n}\cos(k\theta+\ell\varphi).$$

Since neither of θ and φ is an integral multiple of 2π, we have $\sin(\theta/2)\neq 0$ and $\sin(\varphi/2)\neq 0$. Using the above-mentioned trigonometric identities, we obtain

$$B_{m,n}=\sum_{k=1}^{m}\left(\frac{\sin\left(k\theta+\frac{n+1}{2}\varphi\right)\sin\frac{n\varphi}{2}}{\sin\frac{\varphi}{2}}\right)=\frac{\sin\frac{n\varphi}{2}}{\sin\frac{\varphi}{2}}\cdot\frac{\sin\left(\frac{n+1}{2}\varphi+\frac{m+1}{2}\varphi\right)\sin\frac{m\theta}{2}}{\sin\frac{\theta}{2}}$$

and

$$C_{m,n}=\sum_{k=1}^{m}\left(\frac{\cos\left(k\theta+\frac{n+1}{2}\varphi\right)\sin\frac{n\varphi}{2}}{\sin\frac{\varphi}{2}}\right)=\frac{\sin\frac{n\varphi}{2}}{\sin\frac{\varphi}{2}}\cdot\frac{\cos\left(\frac{n+1}{2}\varphi+\frac{m+1}{2}\varphi\right)\sin\frac{m\theta}{2}}{\sin\frac{\theta}{2}}.$$

It follows that

$$|B_{m,n}|\leq\frac{1}{|\sin\frac{\theta}{2}\sin\frac{\varphi}{2}|} \quad\text{and}\quad |C_{m,n}|\leq\frac{1}{|\sin\frac{\theta}{2}\sin\frac{\varphi}{2}|} \quad\text{for all } (m,n)\in\mathbb{N}^2.$$

Letting $b_{k,\ell}:=\sin(k\theta+\ell\varphi)$ for $(k,\ell)\in\mathbb{N}^2$ in Proposition 7.38, we see that $\sum\sum_{(k,\ell)}a_{k,\ell}\sin(k\theta+\ell\varphi)$ is convergent. Similarly, $\sum\sum_{(k,\ell)}a_{k,\ell}\cos(k\theta+\ell\varphi)$ is convergent. □

We observe that Corollary 7.39 is a special case of Corollary 7.40 with $\theta = \pi = \varphi$.

Remark 7.41. If both θ and φ are integral multiples of 2π, then $\sin(k\theta + \ell\varphi) = 0$ and $\cos(k\theta + \ell\varphi) = 1$ for all $(k, \ell) \in \mathbb{N}^2$, and so each term of the double series $\sum\sum_{(k,\ell)} a_{k,\ell} \sin(k\theta+\ell\varphi)$ is equal to zero, while the double series $\sum\sum_{(k,\ell)} a_{k,\ell} \cos(k\theta + \ell\varphi)$ is just the double series $\sum\sum_{(k,\ell)} a_{k,\ell}$, which may converge or diverge. Next, assume that one of θ and φ is an integral multiple of 2π but the other is not. Suppose $\theta = 2p\pi$ for some $p \in \mathbb{Z}$ and $\varphi \neq 2q\pi$ for any $q \in \mathbb{Z}$. Then $\sin(k\theta + \ell\varphi) = \sin(\ell\varphi)$ and $\cos(k\theta + \ell\varphi) = \cos(\ell\varphi)$ for all $(k, \ell) \in \mathbb{N}^2$. Depending on the choice of the double sequence $(a_{k,\ell})$ (satisfying conditions (i), (ii), and (iii) given in Proposition 7.38) and on the choice of φ, the double series $\sum\sum_{(k,\ell)} a_{k,\ell} \sin(\ell\varphi)$ and $\sum\sum_{(k,\ell)} a_{k,\ell} \cos(\ell\varphi)$ may converge absolutely, may converge conditionally, or may diverge. Examples of these cases are given in Exercise 23. ◇

Example 7.42. Let $p > 0$ and $a_{k,\ell} := 1/(k + \ell)^p$ for $(k, \ell) \in \mathbb{N}^2$. Then $(a_{k,\ell})$ satisfies conditions (i), (ii), and (iii) of Proposition 7.38, thanks to Example 7.7 (ii). Hence by Leibniz's Test, the double series

$$\sum\sum_{(k,\ell)} \frac{(-1)^{k+\ell}}{(k+\ell)^p}$$

is convergent. In fact, by Example 7.17 (i), this double series is absolutely convergent if $p > 2$ and it is conditionally convergent if $0 < p \leq 2$. On the other hand, the (k, ℓ)th Term Test shows that it is divergent if $p \leq 0$.

Further, if θ and φ are real numbers neither of which is an integral multiple of 2π, then Corollary 7.40 shows that the double series

$$\sum\sum_{(k,\ell)} \frac{\sin(k\theta + \ell\varphi)}{(k+\ell)^p} \quad \text{and} \quad \sum\sum_{(k,\ell)} \frac{\cos(k\theta + \ell\varphi)}{(k+\ell)^p}$$

are convergent. ◇

7.4 Double Power Series

For nonnegative integers k and ℓ, let $c_{k,\ell} \in \mathbb{R}$. The double series

$$\sum\sum_{(k,\ell)\geq(0,0)} c_{k,\ell} x^k y^\ell, \qquad \text{where } (x, y) \in \mathbb{R}^2,$$

is called a **double power series** (around $(0, 0)$), and for $(k, \ell) \geq (0, 0)$, the real number $c_{k,\ell}$ is called its (k, ℓ)th **coefficient**. Henceforth when we consider a double power series $\sum\sum_{(k,\ell)} c_{k,\ell} x^k y^\ell$, it will be tacitly assumed that the index (k, ℓ) varies over the set of all pairs of nonnegative integers (and not over $\mathbb{N}^2$) and that the coefficients $c_{k,\ell}$ are in $\mathbb{R}$. For $(m, n) \geq (0, 0)$, the (m, n)th partial double sum of the double power series $\sum\sum_{(k,\ell)} c_{k,\ell} x^k y^\ell$ is

$$A_{m,n}(x, y) := \sum_{k=0}^{m} \sum_{\ell=0}^{n} c_{k,\ell} x^k y^\ell.$$

It is clear that if $(x, y) = (0, 0)$, then for any choice of the coefficients $c_{k,\ell}$, the double power series $\sum\sum_{(k,\ell)} c_{k,\ell} x^k y^\ell$ is convergent and its double sum is equal to $c_{0,0}$. Also, if $x \in \mathbb{R}$ and $y = 0$, then the double power series is convergent if and only if the (single) power series $\sum_{k=0}^{\infty} c_{k,0} x^k$ is convergent, and likewise, if $x = 0$ and $y \in \mathbb{R}$, then the double power series is convergent if and only if the (single) power series $\sum_{\ell=0}^{\infty} c_{0,\ell}\, y^\ell$ is convergent. On the other hand, if there is $(k_0, \ell_0) \in \mathbb{N}^2$ such that $c_{k,\ell} = 0$ whenever either $k > k_0$ or $\ell > \ell_0$, then the double power series is convergent for any $(x, y) \in \mathbb{R}^2$, and its double sum is equal to

$$\sum_{k=0}^{k_0} \sum_{\ell=0}^{\ell_0} c_{k,\ell} x^k y^\ell.$$

More generally, if $(x_0, y_0) \in \mathbb{R}^2$, then the double series

$$\sum\sum_{(k,\ell)\geq(0,0)} c_{k,\ell}(x - x_0)^k (y - y_0)^\ell$$

is called a **double power series** around (x_0, y_0). Its convergence can be discussed by letting $\tilde{x} = x - x_0$ and $\tilde{y} = y - y_0$, and considering the double power series $\sum\sum_{(k,\ell)} c_{k,\ell} \tilde{x}^k \tilde{y}^\ell$.

Typical sets of points (x, y) in $\mathbb{R}^2$ for which a double power series is convergent are illustrated by the following examples.

Examples 7.43. (i) Let $c_{k,\ell} := k^k \ell^\ell$ for $(k, \ell) \geq (0, 0)$, and let $(x, y) \in \mathbb{R}^2$. If $x \neq 0$ and $y \neq 0$, then $|c_{k,\ell} x^k y^\ell| > 1$ for all $(k, \ell) \in \mathbb{N}^2$ satisfying $k > 1/|x|$ and $\ell > 1/|y|$, and so by the (k, ℓ)th Term Test (Proposition 7.8), the double power series $\sum\sum_{(k,\ell)} c_{k,\ell} x^k y^\ell$ is divergent. Similarly, if $x \neq 0$ and $y = 0$, then the series $\sum_{k=0}^{\infty} c_{k,0} x^k$ is divergent, and if $x = 0$ and $y \neq 0$, then the series $\sum_{\ell=0}^{\infty} c_{0,\ell}\, y^\ell$ is divergent. Thus we see that the double power series $\sum\sum_{(k,\ell)} c_{k,\ell} x^k y^\ell$ is convergent if and only if $(x, y) = (0, 0)$.

(ii) Let $c_{k,\ell} := 1/k!\ell!$ for $(k, \ell) \geq (0, 0)$. It follows from Example 7.10 (ii) that the double power series $\sum\sum_{(k,\ell)} c_{k,\ell} x^k y^\ell$ is convergent for all $(x, y) \in \mathbb{R}^2$.

(iii) Let a and b be nonzero real numbers, and let $c_{k,\ell} := a^k b^\ell$ for $(k, \ell) \geq (0, 0)$. It follows from Example 7.10 (i) that the double power series $\sum\sum_{(k,\ell)} c_{k,\ell} x^k y^\ell$ is convergent if and only if $|ax| < 1$ and $|by| < 1$, that is, $|x| < 1/|a|$ and $|y| < 1/|b|$.

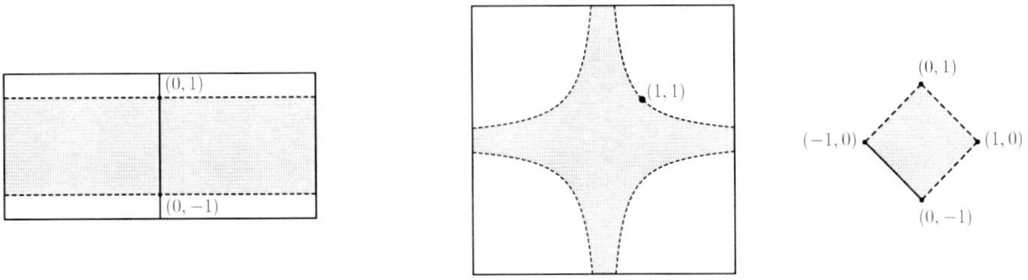

Fig. 7.2. Illustration of sets of convergence: The horizontal strip and the y-axis, the region bounded by rectangular hyperbolas, and the diamond-shaped region on which the double power series in Examples 7.43 (iv), (v), and (vi) converge, respectively.

(iv) For $(k, \ell) \geq (0, 0)$, let

$$c_{k,\ell} := \begin{cases} 1 & \text{if } k = 1, \\ 0 & \text{if } k \neq 1. \end{cases}$$

Then for $(x, y) \in \mathbb{R}^2$, the partial double sums of the double power series $\sum\sum_{(k,\ell)} c_{k\ell} x^k y^\ell$ are $A_{0,n}(x, y) := 0$ for $n \geq 0$, and

$$A_{m,n}(x, y) := x \sum_{\ell=0}^{n} y^\ell \quad \text{for } (m, n) \geq (1, 0).$$

Consequently, the double power series converges absolutely if $x = 0$ or $|y| < 1$, while it diverges if $x \neq 0$ and $|y| \geq 1$. It follows that the set of $(x, y) \in \mathbb{R}^2$ for which this double power series converges is the horizontal strip $\mathbb{R} \times (-1, 1)$ together with the y-axis, as shown in Figure 7.2. On this set, the convergence is absolute.

(v) For $(k, \ell) \geq (0, 0)$, let

$$c_{k,\ell} := \begin{cases} 1 & \text{if } k = \ell, \\ 0 & \text{if } k \neq \ell. \end{cases}$$

Then for $(x, y) \in \mathbb{R}^2$, the partial double sums of the double power series $\sum\sum_{(k,\ell)} c_{k,\ell} x^k y^\ell$ are

$$A_{m,n}(x, y) := \sum_{p=0}^{\min\{m,n\}} (xy)^p \quad \text{for } (m, n) \geq (0, 0).$$

Using the fact that the geometric series $\sum_p a^p$ converges absolutely if $|a| < 1$, while it diverges if $|a| \geq 1$, we see that the double power series converges absolutely if $|xy| < 1$, while it diverges if $|xy| \geq 1$. Thus the subset of $\mathbb{R}^2$ on which this double power series converges is precisely the region $\{(x, y) \in \mathbb{R}^2 : -1 < xy < 1\}$ bounded by the rectangular hyperbolas $xy = 1$ and $xy = -1$, as shown in Figure 7.2. On this set, the convergence is absolute.

(vi) Let $c_{k,\ell} := (k+\ell)!/k!\ell!$ for $(k,\ell) \geq (0,0)$, and let $(x,y) \in \mathbb{R}^2$. As in the proof of part (iii) of Proposition 7.16,

$$\sum_{k=0}^{m}\sum_{\ell=0}^{n} |c_{k,\ell}|\,|x|^k|y|^\ell \leq \sum_{j=0}^{m+n}\sum_{k=0}^{j} \frac{j!|x|^k|y|^{j-k}}{k!(j-k)!} = \sum_{j=0}^{m+n}(|x|+|y|)^j$$

for $(m,n) \geq (0,0)$, whereas

$$\sum_{k=0}^{n}\sum_{\ell=0}^{n} |c_{k,\ell}|\,|x|^k|y|^\ell \geq \sum_{j=0}^{n}\sum_{k=0}^{j} \frac{j!|x|^k|y|^{j-k}}{k!(j-k)!} = \sum_{j=0}^{n}(|x|+|y|)^j$$

for $n \geq 0$. Thus, in view of Example 7.10 (i), we see that the double power series $\sum\sum_{(k,\ell)} c_{k,\ell}x^k y^\ell$ converges absolutely if and only if $|x|+|y| < 1$. The subset of $\mathbb{R}^2$ on which this double power series converges absolutely is the diamond-shaped region $\{(x,y) \in \mathbb{R}^2 : |x|+|y| < 1\}$. It turns out that the set on which the double series converges is this diamond-shaped region together with the open line segment joining $(-1,0)$ and $(0,-1)$, as shown in Figure 7.2. (Compare Example 7.50 (ii). See Exercise 57.) $\diamond$

The above examples show that the set of all $(x,y) \in \mathbb{R}^2$ for which a double power series $\sum\sum_{(k,\ell)} c_{k,\ell}x^k y^\ell$ converges (absolutely) can be of a varied nature. This is in contrast to the convergence of a (single) power series for which the corresponding subset of $\mathbb{R}$ is always an interval. In this connection, we recall following result (Lemma 9.25 of ACICARA) for (single) power series.

Fact 7.44 (Abel's Lemma). *Let $x_0 \in \mathbb{R}$ and let $c_k \in \mathbb{R}$ for $k \geq 0$. If the set $\{c_k x_0^k : k \geq 0\}$ is bounded, then the power series $\sum_{k=0}^{\infty} c_k x^k$ is absolutely convergent for every $x \in \mathbb{R}$ with $|x| < |x_0|$.*

This leads, as in Proposition 9.26 of ACICARA, to the following fundamental result about the (absolute) convergence of a (single) power series.

Fact 7.45. *Either a power series $\sum_k c_k x^k$ converges absolutely for all $x \in \mathbb{R}$, or there is a nonnegative real number r such that it converges absolutely for all $x \in \mathbb{R}$ with $|x| < r$ and diverges for all $x \in \mathbb{R}$ with $|x| > r$.*

The **radius of convergence** of the power series is defined to be ∞ in the former case, and it is defined to be the unique nonnegative real number r with the stated properties in the latter case. We shall now attempt to obtain analogues of the above results for double power series.

Lemma 7.46 (Abel's Lemma for Double Power Series). *Let (x_0,y_0) be in $\mathbb{R}^2$ and let $c_{k,\ell} \in \mathbb{R}$ for $(k,\ell) \geq (0,0)$. If the set $\{c_{k,\ell}x_0^k y_0^\ell : (k,\ell) \geq (0,0)\}$ is bounded, then the double power series $\sum\sum_{(k,\ell)} c_{k,\ell}x^k y^\ell$ is absolutely convergent for every $(x,y) \in \mathbb{R}^2$ with $|x| < |x_0|$ and $|y| < |y_0|$.*

Proof. If $x_0 = 0$ or $y_0 = 0$, then there is nothing to prove. Suppose $x_0 \neq 0$ and $y_0 \neq 0$. Let $\alpha \in \mathbb{R}$ be such that $|c_{k,\ell}x_0^k y_0^\ell| \leq \alpha$ for all $(k,\ell) \geq (0,0)$. Given any $(x,y) \in \mathbb{R}^2$ with $|x| < |x_0|$ and $|y| < |y_0|$, let $\beta := |x|/|x_0|$ and $\gamma := |y|/|y_0|$. Then

$$|c_{k,\ell}x^k y^\ell| = |c_{k,\ell}x_0^k y_0^\ell|\beta^k\gamma^\ell \leq \alpha\beta^k\gamma^\ell \quad \text{for all } (k,\ell) \geq (0,0).$$

Since $\beta < 1$ and $\gamma < 1$, the geometric double series $\sum\sum_{(k,\ell)} \beta^k\gamma^\ell$ is convergent. (See Example 7.10(i).) By the Comparison Test for Double Series, it follows that $\sum\sum_{(k,\ell)} c_{k,\ell}x^k y^\ell$ is absolutely convergent. □

Proposition 7.47. *Either a double power series $\sum\sum_{(k,\ell)} c_{k,\ell}x^k y^\ell$ converges absolutely for all $(x,y) \in \mathbb{R}^2$, or there are nonnegative real numbers r and s such that it converges absolutely for all $(x,y) \in \mathbb{R}^2$ with $|x| < r$ and $|y| < s$, while the set $\{c_{k,\ell}x^k y^\ell : (k,\ell) \geq (0,0)\}$ is unbounded for all $(x,y) \in \mathbb{R}^2$ with $|x| > r$ and $|y| > s$.*

Proof. For $(x,y) \in \mathbb{R}^2$, let $C_{x,y} := \{c_{k,\ell}x^k y^\ell : (k,\ell) \geq (0,0)\}$. Consider $E := \{(x,y) \in \mathbb{R}^2 : C_{x,y} \text{ is bounded}\}$. For $(x,y) \in \mathbb{R}^2$, note that $(x,y) \in E$ if and only if $(|x|,|y|) \in E$. If $E = \mathbb{R}^2$, then given any $(x,y) \in \mathbb{R}^2$, we can find $(x_0,y_0) \in E$ such that $|x| < |x_0|$ and $|y| < |y_0|$. Since the set C_{x_0,y_0} is bounded, by Lemma 7.46, the double series $\sum\sum_{(k,\ell)} c_{k,\ell}x^k y^\ell$ is absolutely convergent. Next, suppose $E \neq \mathbb{R}^2$. The set E is nonempty since $(0,0) \in E$. By Proposition 2.8, E has a boundary point $(x^*,y^*) \in \mathbb{R}^2$. Define $r := |x^*|$ and $s := |y^*|$. Let $(x,y) \in \mathbb{R}^2$ with $|x| < r$ and $|y| < s$. By the definition of a boundary point, there is a sequence in E converging to (x^*,y^*), and so we can find $(x_0,y_0) \in E$ such that $|x| < |x_0|$ and $|y| < |y_0|$. Hence by Lemma 7.46, $\sum\sum_{(k,\ell)} c_{k,\ell}x^k y^\ell$ is absolutely convergent. On the other hand, let $(x,y) \in \mathbb{R}^2$ with $|x| > r$ and $|y| > s$. By the definition of a boundary point, there is a sequence in $\mathbb{R}^2 \setminus E$ converging to (x^*,y^*), and so we may find $(x_1,y_1) \in \mathbb{R}^2 \setminus E$ such that $|x_1| < |x|$ and $|y_1| < |y|$. Now since the set C_{x_1,y_1} is unbounded, it follows that the set $C_{x,y}$ is also unbounded. This proves the existence of nonnegative real numbers r and s with the desired properties. □

If a double power series $\sum\sum_{(k,\ell)} c_{k,\ell}x^k y^\ell$ is absolutely convergent for all $(x,y) \in \mathbb{R}^2$, then we say that its **biradius of convergence** is (∞,∞); otherwise, a pair (r,s) of nonnegative real numbers is said to be a **biradius of convergence** of the double power series, provided the double series converges absolutely for all $(x,y) \in \mathbb{R}^2$ with $|x| < r$ and $|y| < s$, while the set $C_{x,y} := \{c_{k,\ell}x^k y^\ell : (k,\ell) \geq (0,0)\}$ is unbounded for all $(x,y) \in \mathbb{R}^2$ with $|x| > r$ and $|y| > s$. This phenomenon is illustrated in Figure 7.3. Proposition 7.47 says that every double power series has a biradius of convergence.

Remarks 7.48. (i) It is interesting to observe that if r is the radius of convergence of a (single) power series, then the power series *diverges* for all $x \in \mathbb{R}$ with $|x| > r$, whereas if (r,s) is a biradius of convergence of a double power

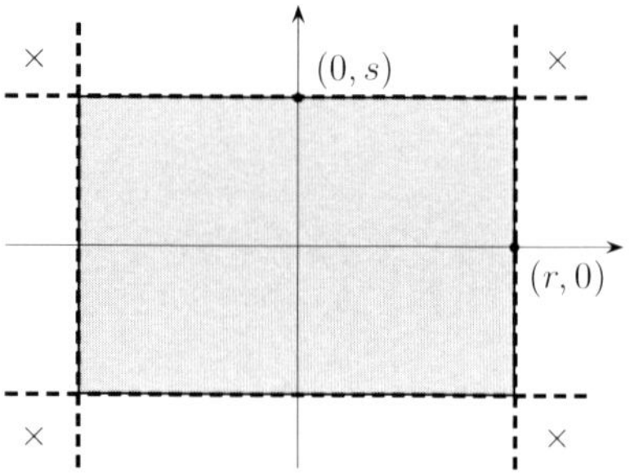

Fig. 7.3. When (r, s) is a biradius of convergence of a double power series, it converges absolutely in the shaded rectangle, while the set of its terms is unbounded in the four quadrangles marked by $\times$.

series, then the set $C_{x,y} := \{c_{k,\ell}x^k y^\ell : (k,\ell) \geq (0,0)\}$ *is unbounded* for all $(x,y) \in \mathbb{R}^2$ with $|x| > r$ and $|y| > s$. The unboundedness of the set $C_{x,y}$ cannot be replaced by the divergence of the double power series at (x,y), as the following example shows. Let $c_{0,0} := 1$, $c_{k,0} = c_{0,\ell} := 1$ for all $k, \ell \in \mathbb{N}$, $c_{1,1} := -1$, $c_{k,1} = c_{1,\ell} := -1/2$ for all $k, \ell \geq 2$, and $c_{k,\ell} := 0$ for all $(k,\ell) \geq (2,2)$. If $A_{m,n}(x,y)$ denotes the (m,n)th partial double sum of $\sum\sum_{(k,\ell)} c_{k,\ell}x^k y^\ell$, then $A_{0,0}(x,y) = 1$ and for $(m,n) \in \mathbb{N}^2$, we have

$$A_{m,0}(x,y) = \sum_{k=0}^{m} x^k, \quad A_{0,n}(x,y) = \sum_{\ell=0}^{n} y^\ell,$$

and

$$A_{m,n}(x,y) = 1 + \left(1 - \frac{y}{2}\right)\sum_{k=1}^{m} x^k + \left(1 - \frac{x}{2}\right)\sum_{\ell=1}^{n} y^k.$$

It is easy to see that the double power series converges absolutely for all $(x,y) \in \mathbb{R}^2$ with $|x| < 1$ and $|y| < 1$, and it diverges to ∞ for all (x,y) with $x \geq 1$ and $y \geq 1$ except for $(x,y) = (2,2)$. At $(2,2)$, a peculiar phenomenon occurs: Since $c_{k,0}2^k 2^0 = 2^k$ and $c_{0,\ell}2^0 2^\ell = 2^\ell$ for all $(k,\ell) \in \mathbb{N}^2$, we see that the set $C_{2,2}$ is unbounded, but since $A_{m,n}(2,2) = 1$ for all $(m,n) \in \mathbb{N}^2$, we see that the double power series converges to 1 at $(2,2)$. It follows that there are no nonnegative numbers r and s such that the double power series converges absolutely for all $(x,y) \in \mathbb{R}^2$ with $|x| < r$ and $|y| < s$, and it diverges for all $(x,y) \in \mathbb{R}^2$ with $|x| > r$ and $|y| > s$.

(ii) The radius of convergence of a (single) power series is unique. However, a double power series may have several biradii of convergence. For example, let $c_{k,\ell} := 1$ if $k = \ell$ and $c_{k,\ell} := 0$ if $k \neq \ell$ for $(k,\ell) \geq (0,0)$. Then the double power series $\sum\sum_{(k,\ell)} c_{k,\ell}x^k y^\ell = \sum_{k=0}^{\infty} x^k y^k$ converges absolutely if $|xy| < 1$. On the other hand, if $|xy| > 1$, then the set $C_{x,y} := \{x^k y^k : k \geq 0\}$ is unbounded. It follows that $(t, 1/t)$ is a biradius of convergence for each positive real number t.

It is therefore important to find all biradii of convergence, or failing this, as many biradii of convergence as possible, in order to obtain a fuller picture of the convergence behavior of a double power series. ◇

If r is the radius of convergence of a (single) power series, then the set $(-r, r)$ is known as the **interval of convergence** of the power series. It is the largest open subset of $\mathbb{R}$ in which the power series is absolutely convergent. Analogously, the **domain of convergence** of a double power series is defined to be the set of all $(x, y) \in \mathbb{R}^2$ such that the double power series converges absolutely at every point in some open square centered at (x, y). Note that if D is the domain of convergence of a double power series, then D is an open subset of $\mathbb{R}^2$ and moreover, $(x, y) \in D$ if and only if $(|x|, |y|) \in D$. It follows from the Comparison Test and Lemma 7.46 that $(x_0, y_0) \in \mathbb{R}^2$ belongs to the domain of convergence of $\sum\sum_{(k,\ell)} c_{k,\ell} x^k y^\ell$ if and only if the set $\{c_{k,\ell} x^k y^\ell : (k, \ell) \geq (0, 0)\}$ is bounded for every (x, y) in some open square centered at (x_0, y_0). It also follows that the domain of convergence of a double power series is empty if and only if $(0, 0)$ is a biradius of convergence of that double power series.

Let $\sum\sum_{(k,\ell)} c_{k,\ell} x^k y^\ell$ be a double power series, and let D be its domain of convergence. Assume that $(0, 0) \in D$, but $D \neq \mathbb{R}^2$. We show how to demarcate the subset D of $\mathbb{R}^2$. For $(x, y) \in \mathbb{R}^2$, let $C_{x,y} := \{c_{k,\ell} x^k y^\ell : (k, \ell) \geq (0, 0)\}$. By Lemma 7.46, for each $\theta \in (0, \pi/2)$, there is a unique point $(x(\theta), y(\theta))$ on the ray $L_\theta := \{(x, y) \in \mathbb{R}^2 : x > 0,\ y > 0, \text{ and } x \sin\theta = y \cos\theta\}$ such that the following two conditions hold: (i) double power series converges absolutely at every (x, y) on the open line segment between $(0, 0)$ and $(x(\theta), y(\theta))$, and (ii) the set $C_{x,y}$ is unbounded for each $(x, y) \in L_\theta$ with $x > x(\theta)$ and $y > y(\theta)$.

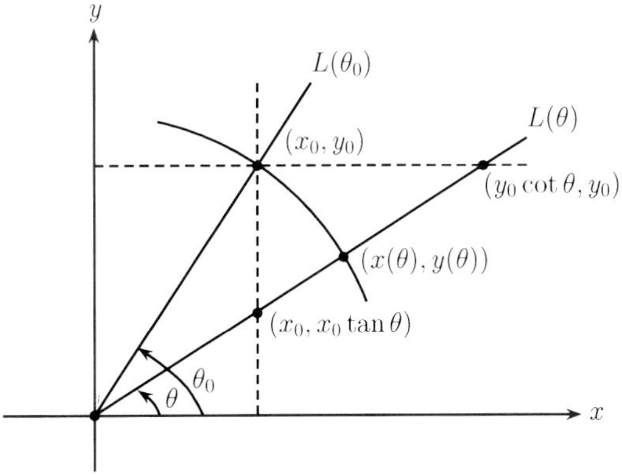

Fig. 7.4. Demarcation of a domain of convergence.

We show that the functions given by $\theta \longmapsto x(\theta)$ and $\theta \longmapsto y(\theta)$ from $(0, \pi/2)$ to $\mathbb{R}$ are continuous. Let $\theta_0 \in (0, \pi/2)$ and $(x_0, y_0) := (x(\theta_0), y(\theta_0))$.

Now the double power series converges absolutely at each $(x, y) \in \mathbb{R}^2$ satisfying $0 \le x < x_0$ and $0 \le y < y_0$, whereas the set $C_{x,y}$ is unbounded for each $(x, y) \in \mathbb{R}^2$ satisfying $x > x_0$ and $y > y_0$. Hence for any $\theta \in (0, \pi/2)$, the point $(x(\theta), y(\theta))$ lies on the closed line segment between $(x_0, x_0 \tan\theta)$ and $(y_0 \cot\theta, y_0)$. Since the functions tan and cot are continuous at θ_0, and since $x_0 \tan\theta_0 = y_0$ and $y_0 \cot\theta_0 = x_0$, it follows that $x(\theta) \to x_0$ and $y(\theta) \to y_0$ as $\theta \to \theta_0$. This proves the continuity at θ_0. We thus obtain a continuous curve demarcating the domain of convergence of the double power series in the first quadrant. (See Figure 7.4.) By symmetry, we obtain similar curves in the remaining three quadrants (excluding the x-axis and the y-axis).

In the following table we give the domains of convergence and biradii of convergence of the double power series considered in Example 7.43.

Double Power Series	Domain of Convergence	Biradii of Convergence
$\sum\sum_{(k,\ell)} k^k \ell^k x^k y^k$	$\emptyset$	$(0,0)$
$\sum\sum_{(k,\ell)} \frac{1}{k!\ell!} x^k y^\ell$	$\mathbb{R}^2$	(∞, ∞)
$\sum\sum_{(k,\ell)} a^k b^\ell x^k y^\ell$, $a \ne 0,\ b \ne 0$	$\left\{(x,y) \in \mathbb{R}^2 : \lvert x\rvert < \frac{1}{\lvert a\rvert} \text{ and } \lvert y\rvert < \frac{1}{\lvert b\rvert}\right\}$	$\left(r, \frac{1}{\lvert b\rvert}\right)$ for $0 \le r \le \frac{1}{\lvert a\rvert}$, $\left(\frac{1}{\lvert a\rvert}, s\right)$ for $0 \le s \le \frac{1}{\lvert b\rvert}$
$x\sum_{\ell=0}^{\infty} y^\ell$	$\{(x,y) \in \mathbb{R}^2 : \lvert y\rvert < 1\}$	$(r, 1)$ for $0 \le r < \infty$,
$\sum_{k=0}^{\infty} x^k y^k$	$\{(x,y) \in \mathbb{R}^2 : \lvert xy\rvert < 1\}$	$(t, 1/t)$ for $0 < t < \infty$
$\sum\sum_{(k,\ell)} \frac{(k+\ell)!}{k!\ell!} x^k y^\ell$	$\{(x,y) \in \mathbb{R}^2 : \lvert x\rvert + \lvert y\rvert < 1\}$	$(t, 1-t)$ for $0 \le t \le 1$

The above examples are typical and exhibit the variety of shapes that a domain of convergence of a double power series can have. The example in the penultimate row of the above table shows that such a domain D need not be a convex subset of $\mathbb{R}^2$. However, according to a result of Fabry (1902), the domain of convergence of every double power series is **log-convex**, that is, it is an open subset D of $\mathbb{R}^2$ such that $\{(\ln\lvert x\rvert, \ln\lvert y\rvert) : (x,y) \in D \text{ and } xy \ne 0\}$ is a convex subset of $\mathbb{R}^2$. (See Exercise 59.)

Taylor Double Series and Taylor Series

Let $D \subseteq \mathbb{R}^2$, (x_0, y_0) be an interior point of D and let $f : D \to \mathbb{R}$ be such that all partial derivatives of f of all orders exist and are continuous on a square neighborhood of (x_0, y_0). In analogy with the Taylor series of a function of one variable, the double power series

$$\sum_{(k,\ell)} \sum \frac{\partial^{k+\ell} f}{\partial x^k \partial y^\ell}(x_0, y_0) \frac{(x - x_0)^k}{k!} \frac{(y - y_0)^\ell}{\ell!}$$

is called the **Taylor double series** of f around (x_0, y_0). Note that the coefficients of this double power series are

$$c_{k,\ell} := \frac{1}{k!\ell!} \frac{\partial^{k+\ell} f}{\partial x^k \partial y^\ell}(x_0, y_0) \quad \text{for } (k, \ell) \geq (0, 0).$$

We observe that for $n = 0, 1, 2, \ldots$, the nth partial sum of the diagonal series $\sum_j c_j(x, y)$ corresponding to the above double series is

$$\begin{aligned} \sum_{j=0}^{n} c_j(x, y) &= \sum_{j=0}^{n} \sum_{\substack{k \geq 0 \\ k+\ell=j}} \sum_{\ell \geq 0} \frac{\partial^{k+\ell} f}{\partial x^k \partial y^\ell}(x_0, y_0) \frac{(x - x_0)^k}{k!} \frac{(y - y_0)^\ell}{\ell!} \\ &= \sum_{j=0}^{n} \sum_{k=0}^{j} \frac{\partial^j f}{\partial x^k \partial y^{j-k}}(x_0, y_0) \frac{(x - x_0)^k}{k!} \frac{(y - y_0)^{j-k}}{(j - k)!}, \end{aligned}$$

which is in fact the nth bivariate Taylor polynomial $P_n(x, y)$ of f around (x_0, y_0). (See Remark 3.48 (iii).) With this in view, the (single) series

$$\sum_{j=0}^{\infty} c_j(x, y), \quad \text{where } c_j(x, y) := \sum_{\substack{k \geq 0 \\ k+\ell=j}} \sum_{\ell \geq 0} c_{k,\ell}(x - x_0)^k (y - y_0)^\ell \ \text{ for } j \geq 0,$$

is called the **Taylor series** of f around (x_0, y_0). Thus the Taylor series of a function of two variables is the diagonal series corresponding to its Taylor double series.

An important question one would like to consider is whether the Taylor double series and/or the Taylor series of f around (x_0, y_0) converges (absolutely) at a given point $(x, y) \in \mathbb{R}^2$, and if so, then whether the corresponding double sum and/or the corresponding sum is equal to $f(x, y)$, provided $(x, y) \in D$. If $(x, y) := (x_0, y_0)$, then each partial double sum of the Taylor double series of f around (x_0, y_0) as well as each partial sum of the Taylor series of f around (x_0, y_0) is obviously equal to $f(x_0, y_0)$, and so our question has an affirmative answer if $(x, y) = (x_0, y_0)$. It is, however, possible that for each $(x, y) \in D \setminus \{(x_0, y_0)\}$, both the Taylor double series and the Taylor series of f around (x_0, y_0) converge but not to $f(x, y)$. For instance, let $f : \mathbb{R}^2 \to \mathbb{R}$ be defined by

$$f(x,y) := \begin{cases} e^{-1/(x^2+y^2)} & \text{if } (x,y) \neq (0,0), \\ 0 & \text{if } (x,y) = (0,0). \end{cases}$$

By considering the function $g : \mathbb{R} \to \mathbb{R}$ given by $g(0) := 0$ and $g(t) := e^{-1/t^2}$, and noting that $g^{(k)}(0) = 0$ for all $k \in \mathbb{N}$, it can be seen that

$$\frac{\partial^{k+\ell} f}{\partial x^k \partial y^\ell}(0,0) = 0 \quad \text{for all } (k,\ell) \geq (0,0).$$

Thus the Taylor double series of f around $(0, 0)$ as well as the Taylor series of f around $(0, 0)$ is identically zero, and neither converges to $f(x, y)$ at any $(x, y) \neq (0, 0)$.

If the Taylor double series of f around (x_0, y_0) converges absolutely at $(x, y) \in \mathbb{R}^2$, then by part (iii) of Proposition 7.21, the Taylor series of f around (x_0, y_0) also converges absolutely at (x, y). But the converse is not true, as we shall see in Example 7.50 (ii). (See also Exercises 32, 63, and 64.)

For $(x, y) \in D$ and $n = 0, 1, 2, \ldots$, let $R_n(x, y) := f(x, y) - P_n(x, y)$ and note that the Taylor series of f around (x_0, y_0) converges to $f(x, y)$ if and only if $R_n(x, y) \to 0$ as $n \to \infty$. The following results give sufficient conditions for the absolute convergence on $\mathbb{R}^2$ of the Taylor double series of a function and for deciding whether it converges to the function itself.

Proposition 7.49. *Let D be an open subset of $\mathbb{R}^2$, and let $(x_0, y_0) \in D$. Suppose $f : D \to \mathbb{R}$ has continuous partial derivatives of all orders on D, and there are positive real numbers M_0, α_0, and β_0 such that*

$$\left| \frac{\partial^{k+\ell} f}{\partial x^k \partial y^\ell}(x_0, y_0) \right| \leq M_0 \alpha_0^k \beta_0^\ell \quad \textit{for all } (k,\ell) \geq (0,0).$$

Then the Taylor double series of f and the Taylor series of f around (x_0, y_0) converge absolutely for all $(x, y) \in \mathbb{R}^2$. Moreover, both of these converge to $f(x, y)$, provided the line L joining (x_0, y_0) and (x, y) lies in D and there are positive real numbers M, α, and β such that

$$\left| \frac{\partial^{k+\ell} f}{\partial x^k \partial y^\ell}(\tilde{x}, \tilde{y}) \right| \leq M \alpha^k \beta^\ell \quad \textit{for all } (\tilde{x}, \tilde{y}) \in L \textit{ and all } (k,\ell) \geq (0,0).$$

Proof. Since the exponential double series

$$\sum_{(k,\ell)}\sum \frac{[\alpha_0(x - x_0)]^k}{k!} \frac{[\beta_0(y - y_0)]^\ell}{\ell!}$$

converges absolutely for all $(x, y) \in \mathbb{R}^2$, the Comparison Test for Double Series shows that the Taylor double series of f around (x_0, y_0) converges absolutely for all $(x,y) \in \mathbb{R}^2$. Consequently, by part (iii) of Proposition 7.21, the corresponding diagonal series, namely the Taylor series of f around (x_0, y_0), also converges absolutely for all $(x,y) \in \mathbb{R}^2$.

Next, let $(x,y) \in D$ be such that the line L joining (x_0, y_0) and (x,y) lies in D and there are positive real numbers M, α, and β such that

$$\left| \frac{\partial^{k+\ell} f}{\partial x^k \partial y^\ell}(\tilde{x}, \tilde{y}) \right| \leq M \alpha^k \beta^\ell \quad \text{for all } (\tilde{x}, \tilde{y}) \in L \text{ and all } (k,\ell) \geq (0,0).$$

Then by the Classical Version of the Bivariate Taylor Theorem (Proposition 3.47), there is $(c,d) \in L$ such that

$$R_n(x,y) := f(x,y) - P_n(x,y) = \sum_{\substack{k \geq 0 \\ k+\ell=n+1}} \sum_{\ell \geq 0} \frac{\partial^{n+1} f}{\partial x^k \partial y^\ell}(c,d) \frac{(x-x_0)^k}{k!} \frac{(y-y_0)^\ell}{\ell!},$$

and consequently,

$$\begin{aligned} |R_n(x,y)| &\leq \sum_{\substack{k \geq 0 \\ k+\ell=n+1}} \sum_{\ell \geq 0} M \frac{(\alpha|x-x_0|)^k}{k!} \frac{(\beta|y-y_0|)^\ell}{\ell!} \\ &= M \sum_{k=0}^{n+1} \frac{(\alpha|x-x_0|)^k}{k!} \frac{(\beta|y-y_0|)^{n+1-k}}{(n+1-k)!} \\ &= \frac{M(\alpha|x-x_0| + \beta|y-y_0|)^{n+1}}{(n+1)!}. \end{aligned}$$

This implies that $R_n(x,y) \to 0$ as $n \to \infty$. Hence the Taylor series of f about (x_0, y_0) converges to $f(x,y)$ at (x,y). Finally, the absolute convergence of the Taylor double series of f around (x_0, y_0) at (x,y) implies that its double sum is also equal to $f(x,y)$, thanks to part (iii) of Proposition 7.21. □

Examples 7.50. (i) Let $D := \{(x,y) \in \mathbb{R}^2 : x < 1 \text{ and } y < 1\}$ and let $f : D \to \mathbb{R}$ be defined by $f(x,y) := 1/(1-x)(1-y)$. It is easy to see that

$$\frac{\partial^{k+\ell} f}{\partial x^k \partial y^\ell}(0,0) = k!\,\ell! \quad \text{for all } (k,\ell) \geq (0,0).$$

Hence the Taylor double series of f around $(0,0)$ is the geometric double series $\sum\sum_{(k,\ell)} x^k y^\ell$. As we have seen in Example 7.10 (i), it converges absolutely if $|x| < 1$ and $|y| < 1$, while it diverges otherwise; moreover, if $|x| < 1$ and $|y| < 1$, then the double sum is $1/(1-x)(1-y) = f(x,y)$. The Taylor series of f around $(0,0)$ is

$$\sum_{j=0}^{\infty} c_j(x,y), \quad \text{where } c_j(x,y) := \sum_{k=0}^{j} x^k y^{j-k} \quad \text{for } (x,y) \in \mathbb{R}^2.$$

By part (iii) of Proposition 7.21, it converges absolutely if $|x| < 1$ and $|y| < 1$, and then its sum is equal to $f(x, y)$. We show that it diverges if $|x| \geq 1$ or $|y| \geq 1$. Assume that $|x| \geq 1$, and let $u := y/x$. Then

$$c_j(x, y) = x^j \sum_{k=0}^{j} u^{j-k} = x^j(1 + u + \cdots + u^j) \quad \text{for } j \geq 0.$$

If $u = 1$, then $|c_j(x, y)| = |x|^j(j + 1) \geq j + 1$, and if $u \neq 1$, then

$$|c_j(x, y)| = \frac{|x|^j|u^{j+1} - 1|}{|u - 1|} \geq \frac{|u^{j+1} - 1|}{|u - 1|} \quad \text{for } j \geq 0.$$

It follows that $c_j(x, y) \not\to 0$ as $j \to \infty$. Hence the Taylor series of f around $(0, 0)$ diverges if $|x| \geq 1$. Similarly, we see that it diverges if $|y| \geq 1$.

(ii) Let $D := \{(x, y) \in \mathbb{R}^2 : x + y < 1\}$ and let $f : D \to \mathbb{R}$ be defined by $f(x, y) := 1/(1 - x - y)$. It is easy to see that

$$\frac{\partial^{k+1} f}{\partial x^k \partial y^\ell}(0, 0) = (k + \ell)! \quad \text{for } k, \ell = 0, 1, 2, \ldots.$$

Hence the Taylor double series of f around $(0, 0)$ is

$$\sum\sum_{(k,\ell)} \frac{(k + \ell)!}{k!\ell!} x^k y^\ell.$$

As shown in Example 7.43 (vi), this double series converges absolutely if and only if $|x| + |y| < 1$. The Taylor series of f around $(0, 0)$ is

$$\sum_{j=0}^{\infty} \left(\sum_{k=0}^{j} \frac{j!}{k!(j - k)!} x^k y^{j-k} \right) = \sum_{j=0}^{\infty} (x + y)^j.$$

Clearly, this geometric series converges if and only if $|x + y| < 1$, and in this case, the convergence is absolute and the sum of the series at (x, y) is equal to $1/[1-(x+y)] = f(x, y)$. Thus if $(x, y) \in \mathbb{R}^2$ satisfies $|x+y| < 1 \leq |x| + |y|$, then the Taylor series of f around $(0, 0)$ converges absolutely at (x, y), but the Taylor double series of f around $(0, 0)$ does not. Since the Taylor series of f around $(0, 0)$ is the diagonal series corresponding to the Taylor double series of f around $(0, 0)$, it follows from Proposition 7.16 that if $(x, y) \in \mathbb{R}^2$ and $|x| + |y| < 1$, then the double sum of the Taylor double series of f around $(0, 0)$ at (x, y) is equal to $f(x, y)$. It can be shown that this Taylor double series converges conditionally at $(x, y) \in \mathbb{R}^2$ if and only if $x \in (-1, 0)$ and $x + y = -1$, and then its double sum is equal to $1/2$. (See Exercise 57.)

(iii) Let $D := \mathbb{R}^2$ and let $f : D \to \mathbb{R}$ be defined by $f(x,y) := \sin(x+y)$. Letting $g(u) := \sin u$ for $u \in \mathbb{R}$, it is easy to see that for $k, \ell = 0, 1, 2, \ldots,$

$$\frac{\partial^{k+\ell} f}{\partial x^k \partial y^\ell}(0,0) = g^{(k+\ell)}(0) = \begin{cases} 0 & \text{if } k+\ell \text{ is even,} \\ (-1)^{(k+\ell-1)/2} & \text{if } k+\ell \text{ is odd.} \end{cases}.$$

Hence the Taylor double series of f around $(0,0)$ is

$$\sum_{(k,\ell)}\sum c_{k,\ell}x^k y^\ell, \quad \text{where } c_{k,\ell} := \begin{cases} 0 & \text{if } k+\ell \text{ is even,} \\ \dfrac{(-1)^{(k+\ell-1)/2}}{k!\ell!} & \text{if } k+\ell \text{ is odd.} \end{cases}$$

The Taylor series of f around $(0,0)$ is

$$\sum_{j=0}^{\infty} c_j(x,y), \quad \text{where } c_j(x,y) := \sum_{k=0}^{j} g^{(j)}(0)\frac{x^k}{k!}\frac{y^{j-k}}{(j-k)!} = \frac{g^{(j)}(0)}{j!}(x+y)^j,$$

that is, by $\sum_{j=0}^{\infty}(-1)^j(x+y)^{2j+1}/(2j+1)!$. It follows from Proposition 7.49 that both the Taylor double series and the Taylor series of f around $(0,0)$ converge absolutely to $f(x,y)$ at all $(x,y) \in \mathbb{R}^2$.

(iv) Let $D := \mathbb{R}^2$ and let $f : D \to \mathbb{R}$ be defined by $f(x,y) := e^{x+y}$. Proceeding as in (iii) above, we see that both the Taylor double series $\sum\sum_{(k,\ell)} x^k y^\ell/k!\ell!$ of f around $(0,0)$ and the Taylor series $\sum_{j=0}^{\infty}(x+y)^j/j!$ of f around $(0,0)$ converge absolutely to $f(x,y)$ at all $(x,y) \in \mathbb{R}^2$. ◇

For additional examples about the convergence of Taylor double series and of Taylor series of functions of two variables, see Exercises 33, 63, 64, and 66.

Remark 7.51. Let D be an open subset of $\mathbb{R}^2$ and let $f : D \to \mathbb{R}$ be such that all partial derivatives of f of all orders exist and are continuous on D. If for every $(x_0, y_0) \in D$, there are $r > 0$ and $s > 0$ such that the Taylor double series of f around (x_0, y_0) converges absolutely to $f(x,y)$ for all $(x,y) \in D$ with $|x - x_0| < r$ and $|y - y_0| < s$, then f is said to be **real analytic** on D. In this case, by part (iii) of Proposition 7.21, the Taylor series of f around (x_0, y_0) also converges absolutely to $f(x,y)$ for all $(x,y) \in D$ with $|x - x_0| < r$ and $|y - y_0| < s$. Clearly, polynomial functions in two variables are real analytic on $\mathbb{R}^2$. Also, using Proposition 7.49, it can be seen that the functions defined by $f_1(x,y) := \sin(x+y)$ and $f_2(x,y) := e^{x+y}$ for $(x,y) \in \mathbb{R}^2$ are real analytic on $\mathbb{R}^2$. In fact, if D is the domain of convergence of a double power series and if its double sum is denoted by $f(x,y)$ for $(x,y) \in D$, then the function f is real analytic on D. (See, for example, 9.2.2 and 9.3.1 of [15].) On the other hand, a function having continuous partial derivatives of all orders on an open subset of $\mathbb{R}^2$ need not be real analytic there. Indeed, as noted earlier, it suffices to consider $f : \mathbb{R}^2 \to \mathbb{R}$ defined by $f(0,0) := 0$ and $f(x,y) := e^{-1/(x^2+y^2)}$ for $(x,y) \neq (0,0)$. ◇

7.5 Convergence of Improper Double Integrals

In Chapter 5, we considered integration of a bounded function defined on a bounded subset of $\mathbb{R}^2$. In this section, we shall extend the process of integration to functions defined on unbounded subsets of $\mathbb{R}^2$ of the form $[a,\infty) \times [c,\infty)$, where $a, c \in \mathbb{R}$, provided that the functions are bounded on $[a,x] \times [c,y]$ for every $(x,y) \geq (a,c)$. Our treatment will run parallel to that of infinite double series given in Sections 7.2 and 7.3. It will also be similar to the treatment of improper (single) integrals of functions defined on $[a,\infty)$, where $a \in \mathbb{R}$, that are bounded on $[a,x]$ for every $x \geq a$ (given, for example, in Section 9.4 of ACICARA).

We shall first give a formal (and pedantic) definition of an improper double integral and then adopt suitable conventions in order to simplify it.

Let $a, c \in \mathbb{R}$. An **improper double integral** on $[a,\infty) \times [c,\infty)$ is an ordered pair (f, F) of real-valued functions f and F defined on $[a,\infty)\times[c,\infty)$ such that f is integrable on $[a,x] \times [c,y]$ for every $(x,y) \geq (a,c)$ and

$$F(x,y) = \iint_{[a,x]\times[c,y]} f(s,t)d(s,t) \quad \text{for all } (x,y) \in [a,\infty) \times [c,\infty).$$

For simplicity and brevity, we shall use the informal but suggestive notation

$$\iint_{[a,\infty)\times[c,\infty)} f(s,t)d(s,t)$$

for the improper double integral (f, F) on $[a,\infty) \times [c,\infty)$. The function F is called the **partial double integral** of this improper double integral. Note that $F(x,c) = 0 = F(a,y)$ for all $(x,y) \in [a,\infty) \times [c,\infty)$, thanks to the convention stated in Remark 5.11. In view of Proposition 5.19, under suitable conditions on f and F such as the continuity of f and the vanishing of $F(x,c)$ and $F(a,y)$ for $(x,y) \in [a,\infty) \times [c,\infty)$, we see that

$$(f, F) \text{ is an improper double integral} \iff f = F_{xy}.$$

In particular, F is uniquely determined by f, and if f is continuous, then f is uniquely determined by F.

Let $a, c \in \mathbb{R}$ and let $f : [a,\infty) \times [c,\infty) \to \mathbb{R}$ be such that f is integrable on $[a,x] \times [c,y]$ for every $(x,y) \geq (a,c)$. We say that the improper double integral $\iint_{[a,\infty)\times[c,\infty)} f(s,t)d(s,t)$ is **convergent** if the limit

$$\lim_{(x,y)\to(\infty,\infty)} \iint_{[a,x]\times[c,y]} f(s,t)d(s,t)$$

exists. It is clear that if this limit exists, then it is unique, and we may denote it by the same symbol $\iint_{[a,\infty)\times[c,\infty)} f(s,t)d(s,t)$ used to denote the improper double integral. Usually, when we write

$$\iint_{[a,\infty)\times[c,\infty)} f(s,t)d(s,t) = I,$$

we mean that I is a real number and the improper double integral on the left side of the above equation is convergent and the limiting value is I. In this case, we may also say that $\iint_{[a,\infty)\times[c,\infty)} f(s,t)d(s,t)$ **converges** to I, or that I is the improper double integral of f on $[a,\infty)\times[c,\infty)$. An improper double integral that is not convergent is said to be **divergent**. In particular, we say that the improper integral **diverges** to ∞ or to $-\infty$ according as $\iint_{[a,x]\times[c,y]} f(s,t)d(s,t)$ tends to ∞ or to $-\infty$ as $(x,y)\to(\infty,\infty)$.

The convergence of an improper double integral $\iint_{[a,\infty)\times[c,\infty)} f(s,t)d(s,t)$ is not affected if we change the function f on a bounded subset of $[a,\infty)\times[c,\infty)$, although its limiting value may be altered by doing so. On the other hand, if we change f on an unbounded subset of $[a,\infty)\times[c,\infty)$, such as an infinite strip, we may affect the convergence of $\iint_{[a,\infty)\times[c,\infty)} f(s,t)d(s,t)$. For example, if $f(s,t) := 0$ for all $(s,t)\in[0,\infty)\times[0,\infty)$, then clearly $\iint_{[0,\infty)\times[0,\infty)} f(s,t)d(s,t)$ is convergent. But if we let $g(s,t) := 1$ for all $(s,t)\in[0,\infty)\times[0,1]$, and $g(s,t) := 0$ for all $(s,t)\in[0,\infty)\times(1,\infty)$, then $\iint_{[0,\infty)\times[0,\infty)} g(s,t)d(s,t)$ is divergent.

Examples 7.52. (i) Let α,β be positive real numbers, and let us consider $\iint_{[a,\infty)\times[c,\infty)}\alpha^s\beta^t d(s,t)$. For any (x,y) in $[a,\infty)\times[c,\infty)$,

$$\iint_{[a,x]\times[c,y]} \alpha^s\beta^t d(s,t) = \begin{cases} \left(\dfrac{\alpha^x-\alpha^a}{\ln\alpha}\right)\left(\dfrac{\beta^y-\beta^c}{\ln\beta}\right) & \text{if } \alpha\neq 1 \text{ and } \beta\neq 1,\\ (x-a)\left(\dfrac{\beta^y-\beta^c}{\ln\beta}\right) & \text{if } \alpha = 1 \text{ and } \beta\neq 1,\\ \left(\dfrac{\alpha^x-\alpha^a}{\ln\alpha}\right)(y-c) & \text{if } \alpha\neq 1 \text{ and } \beta = 1,\\ (x-a)(y-c) & \text{if } \alpha = 1 = \beta. \end{cases}$$

It follows that $\iint_{[a,\infty)\times[c,\infty)}\alpha^s\beta^t d(s,t)$ converges to $\alpha^a\beta^c/(\ln\alpha)(\ln\beta)$ if $\alpha<1$ and $\beta<1$, and diverges to ∞ otherwise.

(ii) Let $p,q\in\mathbb{R}$, and let us consider $\iint_{[1,\infty)\times[1,\infty)}(1/s^pt^q)d(s,t)$. For any (x,y) in $[1,\infty)\times[1,\infty)$,

$$\iint_{[1,x]\times[1,y]} \frac{1}{s^pt^q}d(s,t) = \begin{cases} \dfrac{(x^{1-p}-1)(y^{1-q}-1)}{(1-p)(1-q)} & \text{if } p\neq 1 \text{ and } q\neq 1,\\ \dfrac{(\ln x)(y^{1-q}-1)}{1-q} & \text{if } p=1 \text{ and } q\neq 1,\\ \dfrac{(x^{1-p}-1)(\ln y)}{1-p} & \text{if } p\neq 1 \text{ and } q=1,\\ (\ln x)(\ln y) & \text{if } p=1=q. \end{cases}$$

It follows that $\iint_{[1,\infty)\times[1,\infty)}(1/s^pt^q)d(s,t)$ converges to $1/(p-1)(q-1)$ if $p>1$ and $q>1$, and diverges to ∞ otherwise. ◇

It may be observed that there is a remarkable analogy between the definition of an infinite double series $\sum\sum_{(k,\ell)} a_{k,\ell}$ and the definition of an improper double integral $\iint_{[a,\infty)\times[c,\infty)} f(s,t)d(s,t)$. The double sequence of terms $(a_{k,\ell})$ corresponds to the function $f : [a,\infty) \times [c,\infty) \to \mathbb{R}$, and a partial double sum $A_{m,n} := \sum_{k=1}^{m}\sum_{\ell=1}^{n} a_{k,\ell}$, where $(m,n) \in \mathbb{N}^2$, corresponds to a partial double integral $F(x,y) = \iint_{[a,x]\times[c,y]} f(s,t)d(s,t)$, where $(x,y) \in [a,\infty)\times[c,\infty)$. The conventions $A_{k,0} = 0 = A_{0,\ell}$ for all $(k,\ell) \geq (0,0)$ correspond to the initial conditions $F(x,c) = 0 = F(a,y)$ for all $(x,y) \in [a,\infty)\times[c,\infty)$. Further, the equation involving the difference quotient of partial double sums, namely

$$a_{k,\ell} = \frac{A_{k,\ell} - A_{k,\ell-1} - A_{k-1,\ell} + A_{k-1,\ell-1}}{[k-(k-1)][\ell-(\ell-1)]},$$

corresponds to the equation involving the mixed partial derivative of the partial double integral F, namely

$$f(s,t) = \lim_{u\to s} \frac{1}{s-u}\left(\lim_{v\to t} \frac{F(s,t) - F(s,v) - F(u,t) + F(u,v)}{t-v}\right).$$

This analogy will become even more apparent as we develop the theory of improper double integrals further. However, this analogy may break down occasionally. For instance, we shall show in Section 7.6 that a straightforward analogue of the (k,ℓ)th Term Test for a double series fails to be true for improper double integrals.

The following results follow from the corresponding results for limits of functions of two real variables, just as similar results in the case of double series followed from the corresponding results for limits of double sequences. In what follows, we have let $a, c \in \mathbb{R}$ and f, g, h denote real-valued functions on $[a,\infty)\times[c,\infty)$.

1. **(Limit Theorem)** Let $\iint_{[a,\infty)\times[c,\infty)} f = I$ and $\iint_{[a,\infty)\times[c,\infty)} g = J$. Then $\iint_{[a,\infty)\times[c,\infty)} (f+g) = I + J$, and for any $r \in \mathbb{R}$, $\iint_{[a,\infty)\times[c,\infty)} (rf) = rI$. Further, if $f(s,t) \leq g(s,t)$ for all $(s,t) \in [a,\infty)\times[c,\infty)$, then $I \leq J$.
2. **(Sandwich Theorem)** If $f(s,t) \leq h(s,t) \leq g(s,t)$ for all $(s,t) \in [a,\infty)\times[c,\infty)$, and the improper double integrals of both f and g converge to I, then so does the improper double integral of h.
3. **(Cauchy Criterion)** An improper double integral (f,F) is convergent if and only if for every $\epsilon > 0$, there is $(x_0,y_0) \in [a,\infty)\times[c,\infty)$ such that

$$|F(x,y) - F(u,v)| < \epsilon \quad \text{for all } (x,y) \geq (u,v) \geq (x_0,y_0).$$

To see this, note that by Corollary 5.10, $F(x,y) - F(u,v)$ is the sum of the double integrals of f on $[a,u]\times[v,y]$, $[u,x]\times[c,v]$, and $[u,x]\times[v,y]$, and use the analogue of Proposition 2.54 for the case $(x,y) \to (\infty,\infty)$.

Remark 7.53. Our treatment of improper double integrals of real-valued functions on subsets of $\mathbb{R}^2$ of the form $[a,\infty) \times [c,\infty)$, where $a, c \in \mathbb{R}$, can be readily used to discuss the convergence of improper double integrals of functions on some other unbounded subsets of $\mathbb{R}^2$. This is outlined below.

First, suppose $b, \; c \in \mathbb{R}$ and $f : (-\infty, b] \times [c,\infty) \to \mathbb{R}$ is integrable on $[x,b] \times [c,y]$ for all $(x,y) \in (-\infty,b] \times [c,\infty)$. Define $\tilde{f} : [-b,\infty) \times [c,\infty) \to \mathbb{R}$ by $\tilde{f}(u,y) := f(-u,y)$. Then for every $(x,y) \in (-\infty,b] \times [c,\infty)$, we may apply Proposition 5.59 to the transformation $\Phi : [-b,-x] \times [c,y] \to [x,b] \times [c,y]$ defined by $\Phi(u,t) := (-u,t)$. Noting that $|J(\Phi)| = |-1| = 1$, we obtain

$$\iint_{[x,b]\times[c,y]} f(s,t)d(s,t) = \iint_{[-b,-x]\times[c,y]} \tilde{f}(u,t)d(u,t).$$

With this in view, we say that $\iint_{(-\infty,b]\times[c,\infty)} f(s,t)d(s,t)$ is **convergent** if the improper double integral $\iint_{[-b,\infty)\times[c,\infty)} \tilde{f}(u,t)d(u,t)$ is convergent, that is, if the limit

$$\lim_{(\xi,y)\to(\infty,\infty)} \iint_{[-b,\xi]\times[c,y]} \tilde{f}(u,t)d(u,t) = \lim_{(x,y)\to(-\infty,\infty)} \iint_{[x,b]\times[c,y]} f(s,t)d(s,t)$$

exists. In this case, this limit will be denoted by $\iint_{(-\infty,b]\times[c,\infty)} f(s,t)d(s,t)$ itself. Otherwise, we say that $\iint_{(-\infty,b]\times[c,\infty)} f(s,t)d(s,t)$ is **divergent**.

Next, suppose $c \in \mathbb{R}$ and $f : \mathbb{R} \times [c,\infty) \to \mathbb{R}$ is integrable on $[a,b] \times [c,y]$ for all $a, b, y \in \mathbb{R}$ with $a \le b$ and $c \le y$. We say that $\iint_{\mathbb{R}\times[c,\infty)} f(s,t)d(s,t)$ is **convergent** if both $\iint_{[0,\infty)\times[c,\infty)} f(s,t)d(s,t)$ and $\iint_{(-\infty,0]\times[c,\infty)} f(s,t)d(s,t)$ are convergent; in this case, their sum is denoted by $\iint_{\mathbb{R}\times[c,\infty)} f(s,t)d(s,t)$ itself. If any one of these is divergent, then we say that $\iint_{\mathbb{R}\times[c,\infty)} f(s,t)d(s,t)$ is **divergent**.

Similar definitions can be given for the convergence and divergence of

$$\iint_{(-\infty,b]\times(-\infty,d]} f(s,t)d(s,t), \quad \iint_{[a,\infty)\times(-\infty,d]} f(s,t)d(s,t)$$

and of

$$\iint_{(-\infty,b]\times\mathbb{R}} f(s,t)d(s,t), \quad \iint_{\mathbb{R}\times(-\infty,d]} f(s,t)d(s,t), \quad \iint_{[a,\infty)\times\mathbb{R}} f(s,t)d(s,t).$$

Finally, suppose $f : \mathbb{R}^2 \to \mathbb{R}$ is integrable on $[a,b] \times [c,d]$ for all $a, b, c, d \in \mathbb{R}$ with $a \le b$ and $c \le d$. We say that $\iint_{\mathbb{R}^2} f(s,t)d(s,t)$ is **convergent** if both $\iint_{\mathbb{R}\times[0,\infty)} f(s,t)d(s,t)$ and $\iint_{\mathbb{R}\times(-\infty,0]} f(s,t)d(s,t)$ are convergent; in this case, their sum is denoted by $\iint_{\mathbb{R}^2} f(s,t)d(s,t)$ itself. If any one of these is divergent, then we say that $\iint_{\mathbb{R}^2} f(s,t)d(s,t)$ is **divergent**.

In view of the above, we shall restrict ourselves to improper double integrals of functions on subsets of $\mathbb{R}^2$ of the form $[a,\infty) \times [c,\infty)$, where $a, c \in \mathbb{R}$, in this and the next section. Improper double integrals of functions on more general unbounded subsets of $\mathbb{R}^2$ are discussed in Section 7.7. ◇

Improper Double Integrals of Mixed Partials

The following result is an analogue of the result about the convergence of a telescoping double series (Proposition 7.13).

Proposition 7.54. *Let $g : [a, \infty) \times [c, \infty) \to \mathbb{R}$ be such that g_x and g_{xy} exist on $[a, \infty) \times [c, \infty)$, g_x is continuous on $[a, \infty) \times [c, \infty)$, and g_{xy} is integrable on $[a, b] \times [c, d]$ for every $(b, d) \geq (a, c)$. Then $\iint_{[a,\infty)\times[c,\infty)} g_{xy}(s, t)d(s, t)$ is convergent if and only if $\lim_{(b,d)\to(\infty,\infty)}[g(b, c) + g(a, d) - g(b, d)]$ exists, and in this case,*

$$\iint_{[a,\infty)\times[c,\infty)} g_{xy}(s,t)d(s,t) = g(a,c) - \lim_{(b,d)\to(\infty,\infty)} [g(b,c) + g(a,d) - g(b,d)].$$

Proof. By part (i) of Proposition 5.20, for all $(b, d) \geq (a, c)$, we have

$$\iint_{[a,b]\times[c,d]} g_{xy}(s,t)d(s,t) = g(b,d) - g(b,c) - g(a,d) + g(a,c).$$

Letting $(b, d) \to (\infty, \infty)$, we obtain the desired result. □

It may be noted that if a function $f : [a, \infty) \times [c, \infty) \to \mathbb{R}$ is continuous, then the improper double integral $\iint_{[a,\infty)\times[c,\infty)} f(s, t)d(s, t)$ can be written as $\iint_{[a,\infty)\times[c,\infty)} g_{xy}(s, t)d(s, t)$ for a suitable function $g : [a, \infty) \times [c, \infty) \to \mathbb{R}$. In fact, if we define $g : [a, \infty) \times [c, \infty) \to \mathbb{R}$ by

$$g(x,y) := \iint_{[a,x]\times[c,y]} f(s,t)d(s,t),$$

that is, if g is the partial double integral of $\iint_{[a,\infty)\times[c,\infty)} f(s, t)d(s, t)$, then by part (ii) of Proposition 5.20, $g_{xy} = f$. But then determining the existence of $\lim_{(x,y)\to(\infty,\infty)}[g(x, y) - g(x, c) - g(a, y)]$ is the same as determining the convergence of the given improper double integral $\iint_{[a,\infty)\times[c,\infty)} f(s, t)d(s, t)$. In some special cases, however, it is possible to find a function g satisfying the above conditions without involving any double integral. In these cases, we can easily determine the convergence of the improper double integral $\iint_{[a,\infty)\times[c,\infty)} f(s, t)d(s, t)$ using Proposition 7.54. For example, consider the improper double integral

$$\iint_{[0,\infty)\times[0,\infty)} s\,t\,e^{-(s^2+t^2)} d(s,t).$$

If we let $g(s, t) := e^{-(s^2+t^2)}/4$ for $(s, t) \in [0, \infty) \times [0, \infty)$, then it is easy to see that $g_{xy}(s, t) = s\,t\,e^{-(s^2+t^2)}$ for all $(s, t) \in [0, \infty) \times [0, \infty)$. Further, since

$$\lim_{(b,d)\to(\infty,\infty)} [g(b,0)+g(0,d)-g(b,d)] = \lim_{(b,d)\to(\infty,\infty)} \frac{e^{-b^2} + e^{-d^2} - e^{-(b^2+d^2)}}{4} = 0,$$

it follows from Proposition 7.54 that the given improper double integral converges to $g(0, 0) = 1/4$. Examples 7.52 (i)–(ii) also illustrate this technique.

Improper Double Integrals of Nonnegative Functions

The following result regarding the convergence of an improper double integral of a nonnegative function is an analogue of the result about the convergence of a double series of nonnegative terms (Proposition 7.14).

Proposition 7.55. *Let $f : [a, \infty) \times [c, \infty) \to \mathbb{R}$ be a nonnegative function that is integrable on $[a, x] \times [c, y]$ for every $(x, y) \geq (a, c)$. Then the improper double integral $\iint_{[a,\infty)\times[c,\infty)} f(s, t)d(s, t)$ is convergent if and only if its partial double integral F is bounded above, and in this case,*

$$\iint_{[a,\infty)\times[c,\infty)} f(s,t)d(s,t) = \sup\{F(x,y) : (x,y) \in [a,\infty) \times [c,\infty)\}.$$

If F is not bounded above, then $\iint_{[a,\infty)\times[c,\infty)} f(s, t)d(s, t)$ diverges to ∞.

Proof. Let $(x_2, y_2) \geq (x_1, y_1) \geq (a, c)$. By Corollary 5.10, we have

$$\begin{aligned} F(x_2, y_2) &= \iint_{[a,x_1]\times[c,y_1]} f(s,t)d(s,t) + \iint_{[x_1,x_2]\times[y_1,y_2]} f(s,t)d(s,t) \\ &\quad + \iint_{[a,x_1]\times[y_1,y_2]} f(s,t)d(s,t) + \iint_{[x_1,x_2]\times[c,y_1]} f(s,t)d(s,t) \\ &\geq \iint_{[a,x_1]\times[c,y_1]} f(s,t)d(s,t) \\ &= F(x_1, y_1), \end{aligned}$$

since $f(s, t) \geq 0$ for all $(s, t) \geq (a, c)$. Thus the function F is monotonically increasing. Hence by part (i) of Proposition 2.59 with $b = d = \infty$, we obtain the desired results. □

A result similar to Proposition 7.55 holds if $f(s, t) \leq 0$ for all (s, t) in $[a, \infty) \times [c, \infty)$. More generally, if there is a bounded subset E of $[a, \infty) \times [c, \infty)$ such that $f(s, t)$ has the same sign for all $(s, t) \in [a, \infty) \times [c, \infty)$ that are outside E, then $\iint_{[a,\infty)\times[c,\infty)} f(s, t)d(s, t)$ is convergent if and only if the function F is bounded. However, if there is no such bounded subset, then the improper double integral $\iint_{[a,\infty)\times[c,\infty)} f(s, t)d(s, t)$ may diverge, even though the function F is bounded. Moreover, the function F may be unbounded, even though the improper double integral $\iint_{[a,\infty)\times[c,\infty)} f(s, t)d(s, t)$ is convergent. These results are illustrated by the following examples.

Examples 7.56. (i) Let $f : [1, \infty) \times [1, \infty) \to \mathbb{R}$ be defined by

$$f(s,t) := \begin{cases} (-1)^{[s]} & \text{if } 1 \leq t \leq 2, \\ 0 & \text{if } t > 2, \end{cases}$$

where $[s]$ denotes, as usual, the integral part of $s \in [1, \infty)$. For (x, y) in $[1, \infty) \times [1, \infty)$, it can be easily checked that $F(x, y) = g(x)h(y)$, where

$$g(x) := \begin{cases} -1 + x - [x] & \text{if } [x] \text{ is even,} \\ -x + [x] & \text{if } [x] \text{ is odd,} \end{cases} \quad \text{and} \quad h(y) := \min\{1, y-1\}.$$

Clearly, F is bounded on $[1, \infty) \times [1, \infty)$, but since $F(2m-1, y) = 0$ and $F(2m, y) = -1$ for all $m \in \mathbb{N}$ and $y \geq 2$, the limit of $F(x, y)$ as (x, y) tends to (∞, ∞) does not exist, that is, $\iint_{[1,\infty)\times[1,\infty)} f(s,t)d(s,t)$ is divergent.

(ii) Let $f : [0, \infty) \times [0, \infty) \to \mathbb{R}$ be defined by

$$f(s,t) := \begin{cases} 1 & \text{if } 0 \leq t \leq 1, \\ -1 & \text{if } 1 < t \leq 2. \\ 0 & \text{if } t > 2. \end{cases}$$

It can be easily checked that for $(x, y) \in [0, \infty) \times [0, \infty)$,

$$F(x,y) = \begin{cases} xy & \text{if } 0 \leq y \leq 1, \\ x(2-y) & \text{if } 1 < y \leq 2, \\ 0 & \text{if } y > 2. \end{cases}$$

Clearly, F is unbounded on $[0, \infty) \times [0, \infty)$, but since $F(x, y) = 0$ for all $(x, y) \geq (0, 2)$, we see that $\iint_{[0,\infty)\times[0,\infty)} f(s,t)d(s,t)$ converges to 0. ◇

Analogues of the double series results "Summing by Squares" and "Summing by Diagonals" (Proposition 7.16) for improper double integrals are given in Exercise 67.

We now attempt to relate the convergence of an improper double integral $\iint_{[1,\infty)\times[1,\infty)} f(s,t)d(s,t)$ of a nonnegative function f to the convergence of the double series $\sum\sum_{(k,\ell)} f(k,\ell)$. Let us first consider $f : [1, \infty) \times [1, \infty) \to \mathbb{R}$ given by $f(s,t) := 1$ if $(s,t) \in \mathbb{N}^2$ and $f(s,t) := 0$ if $(s,t) \notin \mathbb{N}^2$. Then it is easy to see that $\iint_{[1,\infty)\times[1,\infty)} f(s,t)d(s,t)$ converges to 0, but $\sum\sum_{(k,\ell)} f(k,\ell)$ diverges to ∞. On the other hand, if we let $g := 1-f$, then it is easily seen that $\iint_{[1,\infty)\times[1,\infty)} g(s,t)d(s,t)$ diverges to ∞, but $\sum\sum_{(k,\ell)} g(k,\ell)$ converges to 0. Thus, in general, the convergence of $\iint_{[1,\infty)\times[1,\infty)} f(s,t)d(s,t)$ is independent of the convergence of $\sum\sum_{(k,\ell)} f(k,\ell)$ for nonnegative functions. In view of this, the following result is noteworthy.

Proposition 7.57 (Integral Test). *Let $f : [1, \infty) \times [1, \infty) \to \mathbb{R}$ be a nonnegative monotonically decreasing function. Then the improper double integral $\iint_{[1,\infty)\times[1,\infty)} f(s,t)d(s,t)$ is convergent if and only if the double series $\sum\sum_{(k,\ell)} f(k,\ell)$ is convergent, and in this case,*

$$\sum_{k=2}^{\infty}\sum_{\ell=2}^{\infty} f(k,\ell) \leq \iint_{[1,\infty)\times[1,\infty)} f(s,t)d(s,t) \leq \sum_{k=1}^{\infty}\sum_{\ell=1}^{\infty} f(k,\ell),$$

or, equivalently,

$$\iint_{[1,\infty)\times[1,\infty)} f(s,t)d(s,t) \le \sum_{k=1}^{\infty}\sum_{\ell=1}^{\infty} f(k,\ell) \le f(1,1) + \sum_{k=2}^{\infty} f(k,1) + \sum_{\ell=2}^{\infty} f(1,\ell) + \iint_{[1,\infty)\times[1,\infty)} f(s,t)d(s,t).$$

Also, the improper double integral $\iint_{[1,\infty)\times[1,\infty)} f(s,t)d(s,t)$ *diverges to* ∞ *if and only if the double series* $\sum\sum_{(k,\ell)} f(k,\ell)$ *diverges to* ∞.

Proof. Since f is monotonic, by part (i) of Proposition 5.12, f is integrable on $[1,x]\times[1,y]$ for every $(x,y) \ge (1,1)$. Define $F : [1,\infty)\times[1,\infty) \to \mathbb{R}$ by $F(x,y) := \iint_{[1,x)\times[1,y)} f(s,t)d(s,t)$. Since f is nonnegative, by Corollary 5.10, the function F is monotonically increasing. Hence Proposition 7.55 implies that the improper double integral $\iint_{[1,\infty)\times[1,\infty)} f(s,t)d(s,t)$ is convergent if and only if the set $\{F(m,n) : (m,n) \in \mathbb{N}^2\}$ is bounded above, and in this case

$$\begin{aligned}\iint_{[1,\infty)\times[1,\infty)} f(s,t)d(s,t) &= \sup\{F(x,y) : (x,y) \in [1,\infty)\times[1,\infty)\}\\ &= \sup\left\{F(m,n) : (m,n) \in \mathbb{N}^2\right\}\\ &= \lim_{(m,n)\to(\infty,\infty)} F(m,n).\end{aligned}$$

The penultimate equality follows since F is a monotonically increasing function, and the last equality follows from part (i) of Proposition 7.4. Similarly,

$$\iint_{[1,\infty)\times[1,\infty)} f(s,t)d(s,t) \text{ diverges to } \infty \iff F(m,n) \to \infty \text{ as } m,n \to \infty.$$

Define

$$a_{k,\ell} := \iint_{[k,k+1]\times[\ell,\ell+1]} f(s,t)d(s,t) \quad \text{for } (k,\ell) \in \mathbb{N}^2$$

and

$$A_{m,n} := \sum_{k=1}^{m}\sum_{\ell=1}^{n} a_{k,\ell} \quad \text{for } (m,n) \in \mathbb{N}^2.$$

Then by Domain Additivity of Double Integrals on Rectangles (Proposition 5.9), we have $A_{m,n} = F(m+1,n+1)$ for all $(m,n) \in \mathbb{N}^2$. Further, since $a_{k,\ell} \ge 0$ for all $(k,\ell) \in \mathbb{N}^2$, it follows from Proposition 7.14 that the double series $\sum\sum_{(k,\ell)} a_{k,\ell}$ is convergent if and only if the double sequence $(F(m,n))$ is bounded above, that is, the improper double integral $\iint_{[1,\infty)\times[1,\infty)} f(s,t)d(s,t)$ is convergent, and in this case, the sum of the double series is equal to the improper double integral. Similarly, $\sum\sum_{(k,\ell)} a_{k,\ell}$ diverges to ∞ if and only if the double sequence $(F(m,n))$ is not bounded above, that is, the improper double integral $\iint_{[1,\infty)\times[1,\infty)} f(s,t)d(s,t)$ diverges to ∞.

Now since f is monotonically decreasing,

$$f(k+1,\ell+1) \le a_{k,\ell} \le f(k,\ell) \quad \text{for all } (k,\ell) \in \mathbb{N}^2.$$

Hence $\sum\sum_{(k,\ell)} a_{k,\ell}$ is convergent if and only if $\sum\sum_{(k,\ell)} f(k,\ell)$ is convergent, and $\sum\sum_{(k,\ell)} a_{k,\ell}$ diverges to ∞ if and only if $\sum\sum_{(k,\ell)} f(k,\ell)$ diverges to ∞ by the Comparison Test for Double Series (Proposition 7.25).

Finally, since

$$\sum_{k=2}^{m+1}\sum_{\ell=2}^{n+1} f(k,\ell) = \sum_{k=1}^{m}\sum_{\ell=1}^{n} f(k+1,\ell+1) \le \sum_{k=1}^{m}\sum_{\ell=1}^{n} a_{k,\ell} \le \sum_{k=1}^{m}\sum_{\ell=1}^{n} f(k,\ell)$$

for all $(m,n) \in \mathbb{N}^2$, and

$$\lim_{(m,n)\to(\infty,\infty)} \sum_{k=1}^{m}\sum_{\ell=1}^{n} a_{k,\ell} = \lim_{(m,n)\to(\infty,\infty)} A_{m,n} = \iint_{[1,\infty)\times[1,\infty)} f(s,t)d(s,t),$$

we see that

$$\sum_{k=2}^{\infty}\sum_{\ell=2}^{\infty} f(k,\ell) \le \iint_{[1,\infty)\times[1,\infty)} f(s,t)d(s,t) \le \sum_{k=1}^{\infty}\sum_{\ell=1}^{\infty} f(k,\ell),$$

whenever $\iint_{[1,\infty)\times[1,\infty)} f(s,t)d(s,t)$ is convergent. □

The above result can be useful in determining whether a double series or an improper double integral is convergent, and in that case, to obtain lower bounds and upper bounds for them. This is illustrated in the example below.

Example 7.58. Let $f(s,t) := 1/(s+t)^p$ for $(s,t) \in [1,\infty)\times[1,\infty)$, where $p \in \mathbb{R}$ with $p > 0$. Then f is a nonnegative monotonically decreasing function. We have seen in Example 7.17 (i) that the double series $\sum\sum_{(k,\ell)} f(k,\ell)$ is convergent if and only if $p > 2$. Hence by the Integral Test, the improper double integral $\iint_{[1,\infty)\times[1,\infty)} f(s,t)d(s,t)$ is convergent if and only if $p > 2$.

Alternatively, we can directly show that $\iint_{[1,\infty)\times[1,\infty)} f(s,t)d(s,t)$ is convergent if and only if $p > 2$, and deduce that $\sum\sum_{(k,\ell)} f(k,\ell)$ is convergent if and only if $p > 2$. Indeed, let $(x,y) \ge (1,1)$, and let $F(x,y) := \iint_{[1,x]\times[1,y]} d(s,t)/(s+t)^p$. Suppose $p \le 2$. Then

$$\begin{aligned} F(x,y) &\ge \iint_{[1,x]\times[1,y]} \frac{d(s,t)}{(s+t)^2} = \int_1^x \Big(\int_1^y \frac{dt}{(s+t)^2}\Big) ds \\ &= \int_1^x \Big(\frac{1}{s+1} - \frac{1}{s+y}\Big) ds = \ln(x+1) - \ln 2 - \ln(x+y) + \ln(1+y) \\ &= \ln\frac{(x+1)(y+1)}{x+y} - \ln 2 \ge -\ln\Big(\frac{1}{x+1} + \frac{1}{y+1}\Big) - \ln 2. \end{aligned}$$

Hence $\iint_{[1,\infty)\times[1,\infty)} 1/(s+t)^p d(s,t)$ diverges to ∞. Next, suppose $p > 2$. Then

$$F(x,y) = \frac{1}{p-1}\int_1^x \left[\frac{1}{(s+1)^{p-1}} - \frac{1}{(s+y)^{p-1}}\right] ds$$
$$= \frac{1}{(p-1)(p-2)}\left[\frac{1}{2^{p-2}} - \frac{1}{(x+1)^{p-2}} - \frac{1}{(1+y)^{p-2}} + \frac{1}{(x+y)^{p-2}}\right].$$

Hence

$$\iint_{[1,\infty)\times[1,\infty)} \frac{1}{(s+t)^p} d(s,t) = \frac{1}{(p-1)(p-2)2^{p-2}} \quad \text{if } p > 2.$$

When $p > 2$, Proposition 7.57 also allows us to estimate the double sum $\sum\sum_{(k,\ell)} 1/(k+\ell)^p$ as follows:

$$\frac{1}{(p-1)(p-2)2^{p-2}} \le \sum\sum_{(k,\ell)\ge(1,1)} \frac{1}{(k+\ell)^p}$$
$$\le \frac{1}{2^p} + \sum_{k=2}^{\infty} \frac{1}{(k+1)^p} + \sum_{\ell=2}^{\infty} \frac{1}{(1+\ell)^p} + \frac{1}{(p-1)(p-2)2^{p-2}}$$
$$= \frac{p^2-3p+6}{2^p(p-1)(p-2)} + 2\sum_{k=2}^{\infty} \frac{1}{(k+1)^p}.$$

For $p = 4$, this gives

$$\frac{1}{24} \le \sum\sum_{(k,\ell)\ge(1,1)} \frac{1}{(k+\ell)^4} \le \frac{5}{48} + 2\left(\frac{\pi^4}{90} - 1 - \frac{1}{2^4}\right) = \frac{\pi^4}{45} - \frac{97}{48} < \frac{3}{20}.$$

The upper bound is obtained using the formula $\sum_{k=1}^{\infty} 1/k^4 = \pi^4/90$. (See, for example, Theorem 5.6.3 of Hijab's book [30] for a proof of this formula and in fact, a proof of the general formula for $\sum_{k=1}^{\infty} 1/k^{2n}$.) ◇

Absolute Convergence and Conditional Convergence

Recall that if a function f is integrable on a rectangle, then so is $|f|$. An improper double integral $\iint_{[a,\infty)\times[c,\infty)} f(s,t)d(s,t)$ is said to be **absolutely convergent** if the improper double integral $\iint_{[a,\infty)\times[c,\infty)} |f(s,t)|d(s,t)$ is convergent. The following result is an analogue of Proposition 7.18.

Proposition 7.59. *An absolutely convergent improper double integral is convergent.*

Proof. Let $\iint_{[a,\infty)\times[c,\infty)} f(s,t)d(s,t)$ be an absolutely convergent improper double integral. Consider $f^+, f^- : [a,\infty)\times[c,\infty) \to \mathbb{R}$ defined by

$$f^+(s,t) := \frac{|f(s,t)| + f(s,t)}{2} \quad \text{and} \quad f^-(s,t) = \frac{|f(s,t)| - f(s,t)}{2}.$$

For each $(x, y) \in [a, \infty) \times [c, \infty)$, the functions f^+ and f^- are integrable on $[a, x] \times [c, y]$. For $(x, y) \in [a, \infty) \times [c, \infty)$, let $F(x, y)$, $F^+(x, y)$, $F^-(x, y)$, and $\widetilde{F}(x, y)$ denote the partial double integrals of the improper double integrals of f, f^+, f^-, and $|f|$, respectively. Since the improper double integral $\iint_{[a,\infty)\times[c,\infty)} |f(s,t)|d(s,t)$ is convergent, the function $\widetilde{F}$ is bounded. Also, $0 \le F^+(x, y) \le \widetilde{F}(x, y)$ and $0 \le F^-(x, y) \le \widetilde{F}(x, y)$ for all (x, y) in $[a, \infty) \times [c, \infty)$. So by Proposition 7.55, both $\iint_{[a,\infty)\times[c,\infty)} f^+(s,t)d(s,t)$ and $\iint_{[a,\infty)\times[c,\infty)} f^-(s,t)d(s,t)$ are convergent. Since $f = f^+ - f^-$, we see that $\iint_{[a,\infty)\times[c,\infty)} f(s,t)d(s,t)$ is convergent. □

A convergent improper double integral that is not absolutely convergent is said to be **conditionally convergent**.

Example 7.60. We give a general method for constructing conditionally convergent improper double integrals. Let $a, c \in \mathbb{R}$, and consider $\phi : [a, \infty) \to \mathbb{R}$ and $\psi : [c, \infty) \to \mathbb{R}$ such that ϕ is Riemann integrable on $[a, x]$ for every $x \ge a$, and ψ is Riemann integrable on $[c, y]$ for every $y \ge c$, and moreover, both the improper integrals $\int_a^\infty \phi(s)ds$ and $\int_c^\infty \psi(t)dt$ are conditionally convergent. Define $f : [a, \infty) \times [c, \infty) \to \mathbb{R}$ by $f(s, t) := \phi(s)\psi(t)$. Let Φ and Ψ denote the partial integrals of the improper integrals $\int_a^\infty \phi(s)ds$ and $\int_c^\infty \psi(t)dt$ respectively, and let F denote the partial double integral of the improper double integral $\iint_{[a,\infty)\times[c,\infty)} f(s,t)d(s,t)$. Then it follows from Fubini's Theorem on Rectangles (Proposition 5.28) that $F(x, y) = \Phi(x)\Psi(y)$ for all $(x, y) \in [a, \infty) \times [c, \infty)$. Since $\int_a^\infty \phi(s)ds$ and $\int_c^\infty \psi(t)dt$ are convergent, we see that the improper double integral $\iint_{[a,\infty)\times[c,\infty)} f(s,t)d(s,t)$ is convergent. Similarly, let $\widetilde{\Phi}$ and $\widetilde{\Psi}$ denote the partial integrals of the improper integrals $\int_a^\infty |\phi(s)|ds$ and $\int_c^\infty |\psi(t)|dt$ respectively, and let $\widetilde{F}$ denote the partial double integral of the improper double integral $\iint_{[a,\infty)\times[c,\infty)} |f(s,t)|d(s,t)$. As before, we have $\widetilde{F}(x, y) = \widetilde{\Phi}(x)\widetilde{\Psi}(y)$ for all $(x, y) \in [a, \infty) \times [c, \infty)$. Since $\int_a^\infty |\phi(s)|ds$ and $\int_c^\infty |\psi(t)|dt$ diverge to ∞, it follows that the improper double integral $\iint_{[a,\infty)\times[c,\infty)} |f(s,t)|d(s,t)$ diverges to ∞. Thus $\iint_{[a,\infty)\times[c,\infty)} f(s,t)d(s,t)$ is conditionally convergent.

As a concrete example, we observe that the improper double integral

$$\iint_{[1,\infty)\times[1,\infty)} \frac{(\cos s)(\cos t)}{st} d(s,t)$$

is conditionally convergent. This follows since the improper integral $\int_1^\infty \frac{\cos t}{t} dt$ is conditionally convergent. (See, for instance, Example 9.38 of ACICARA.) ◇

We shall now give a characterization for the absolute convergence of an improper double integral. It may be compared with a similar characterization given in the case of a double series (Proposition 7.19).

Proposition 7.61. *Let $f : [a,\infty) \times [c,\infty) \to \mathbb{R}$ be such that f is integrable on $[a,x] \times [c,y]$ for each fixed $(x,y) \geq (a,c)$, the Riemann integral $\int_a^x |f(s,t)|ds$ exists for each fixed $(x,t) \geq (a,c)$, and the Riemann integral $\int_c^y |f(s,t)|dt$ exists for each fixed $(s,y) \geq (a,c)$. Then the improper double integral $\iint_{[a,\infty)\times[c,\infty)} f(s,t)d(s,t)$ is absolutely convergent if and only if the following conditions hold:*

(i) *There are $(s_0,t_0) \geq (a,c)$ and $\alpha_0 > 0$ such that*

$$\iint_{[s_0,x]\times[t_0,y]} |f(s,t)|d(s,t) \leq \alpha_0 \quad \text{for all } (x,y) \geq (s_0,t_0).$$

(ii) *For each fixed $x \geq a$, the improper integral $\int_c^\infty \big(\int_a^x |f(s,t)|ds\big)dt$ is convergent, and for each fixed $y \geq c$, the improper integral $\int_a^\infty \big(\int_c^y |f(s,t)|dt\big)ds$ is convergent.*

Proof. Since the function $|f|$ is integrable on $[a,x] \times [c,y]$, we may define

$$\widetilde{F}(x,y) := \iint_{[a,x]\times[c,y]} |f(s,t)|d(s,t) \quad \text{for } (x,y) \geq (a,c).$$

By Fubini's Theorem on Rectangles (Proposition 5.28),

$$\widetilde{F}(x,y) = \int_c^y \Big(\int_a^x |f(s,t)|ds\Big)dt = \int_a^x \Big(\int_c^y |f(s,t)|dt\Big)ds \quad \text{for } (x,y) \geq (a,c).$$

Suppose $\iint_{[a,\infty)\times[c,\infty)} f(s,t)d(s,t)$ is absolutely convergent. Since $|f(s,t)| \geq 0$ for all $(s,t) \geq (a,c)$, Proposition 7.55 shows that the function $\widetilde{F}$ is bounded above. Hence condition (i) holds with $(s_0,t_0) := (a,c)$ and $\alpha_0 := \sup\{\widetilde{F}(x,y) : (x,y) \geq (a,c)\}$. Also, in view of the first equality displayed above, for each fixed $x \geq a$, we have $\sup\big\{\int_c^y \big(\int_a^x |f(s,t)|ds\big)dt : y \geq c\big\} \leq \alpha_0$, so that the improper integral $\int_c^\infty \big(\int_a^x |f(s,t)|ds\big)dt$ is convergent. Similarly, for each fixed $y \geq c$, the improper integral $\int_a^\infty \big(\int_c^y |f(s,t)|dt\big)ds$ is convergent.

Conversely, assume that conditions (i) and (ii) hold. Let $(s_0,t_0) \geq (a,c)$ and $\alpha_0 > 0$ be such that $\iint_{[a,x]\times[c,y]} |f(s,t)|d(s,t) \leq \alpha_0$ for all $(x,y) \geq (s_0,t_0)$. By Domain Additivity (Proposition 5.9),

$$\widetilde{F}(x,y) = \iint_{[s_0,x]\times[t_0,y]} |f(s,t)|d(s,t) + \iint_{[a,s_0]\times[c,y]} |f(s,t)|d(s,t)$$
$$+ \iint_{[s_0,x]\times[c,t_0]} |f(s,t)|d(s,t) \quad \text{for all } (x,y) \geq (s_0,t_0).$$

Now since the improper integral $\int_c^\infty \big(\int_a^{s_0} |f(s,t)|ds\big)dt$ is convergent, there is $\beta_0 > 0$ such that

$$\iint_{[a,s_0]\times[c,y]} |f(s,t)|d(s,t) = \int_c^y \Big(\int_a^{s_0} |f(s,t)|ds\Big)dt \leq \beta_0 \quad \text{for all } y \geq c.$$

Similarly, since the improper integral $\int_a^\infty \big(\int_c^{t_0} |f(s,t)|dt\big)ds$ is convergent, there is $\gamma_0 > 0$ such that

$$\iint_{[s_0,x]\times[c,t_0]} |f(s,t)|d(s,t) = \int_{s_0}^{x}\Big(\int_c^{t_0} |f(s,t)|dt\Big)ds \le \gamma_0 \quad \text{for all } x \ge a.$$

Hence $\widetilde{F}(x,y) \le \alpha_0 + \beta_0 + \gamma_0$ for all $(x,y) \ge (s_0,t_0)$. This shows that the monotonically increasing function $\widetilde{F}$ is bounded on $[a,\infty)\times[c,\infty)$. Thus, by Proposition 7.55, $\iint_{[a,\infty)\times[c,\infty)} f(s,t)d(s,t)$ is absolutely convergent. □

7.6 Convergence Tests for Improper Double Integrals

In this section we shall consider several tests that enable us to conclude the convergence or divergence of improper double integrals. For most part, these are analogous to tests for double series. However, as remarked earlier, there is no straightforward analogue of the (k,ℓ)th Term Test for a double series. For example, consider $f : [0,\infty)\times[0,\infty) \to \mathbb{R}$ defined by $f(s,t) := 1$ if $(s,t) = (k,k)$ for some $k \in \mathbb{N}$, and $f(s,t) := 0$ otherwise. Clearly, f is bounded and the improper double integral of f on $[0,\infty)\times[0,\infty)$ converges to 0, but $f(s,t) \not\to 0$ as $(s,t) \to (\infty,\infty)$. By modifying this function, it is possible to find a continuous function of this kind, as the following example shows.

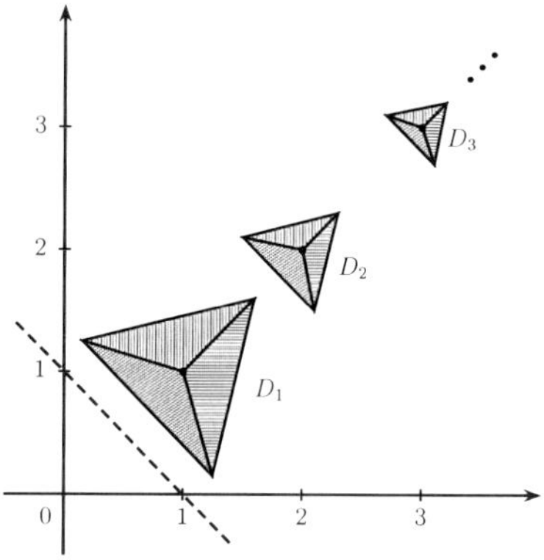

Fig. 7.5. Triangular regions $D_1, D_2, D_3, \ldots$ in Example 7.62 having one side parallel to the line $x + y = 1$, centroids at $(1,1), (2,2), (3,3), \ldots$ and areas $1, \frac{1}{4}, \frac{1}{9}, \ldots$.

Example 7.62. For $k \in \mathbb{N}$, let T_k denote the equilateral triangle having one of its sides parallel to the line given by $x + y = 1$ such that the centroid of the triangular region D_k enclosed by T_k is at (k,k), and the area of D_k is $1/k^2$.

(See Figure 7.5.) Fix any $k \in \mathbb{N}$. Let (a_k, b_k), (c_k, d_k), and (p_k, q_k) denote the vertices of T_k. Then the bivariate linear polynomial

$$\frac{1}{\Delta_k} \det \begin{bmatrix} x & y & 1 \\ a_k & b_k & 1 \\ c_k & d_k & 1 \end{bmatrix}, \quad \text{where } \Delta_k = \det \begin{bmatrix} k & k & 1 \\ a_k & b_k & 1 \\ c_k & d_k & 1 \end{bmatrix},$$

defines a polynomial function in two variables of degree 1 on the triangular subregion of D_k with vertices at (a_k, b_k), (c_k, d_k), and (k, k) such that its value at (a_k, b_k) and (c_k, d_k) is 0, while its value at (k, k) is 1. In a similar way, we obtain polynomial functions in two variables of degree 1 on the other two triangular subregions of D_k. Observe that the values of these functions lie between 0 and 1, and they coincide on the lines joining the vertices of T_k to its centroid. Hence, by piecing together these functions, we obtain a nonnegative, piecewise linear, continuous function $f_k : D_k \to \mathbb{R}$ such that $f_k(k, k) = 1$, whereas $f_k(a_k, b_k) = f_k(c_k, d_k) = f_k(p_k, q_k) = 0$. By Domain Additivity (Corollary 5.52), the double integral of f_k on D_k is the sum of the double integrals of f_k on the three triangular subregions of D_k. If E_k denotes one such subregion, say with vertices at (a_k, b_k), (c_k, d_k), and (k, k), then $\text{Area}(E_k) = \frac{1}{3}\text{Area}(D_k) = 1/3k^2$, and by part (i) of Proposition 6.29,

$$\iint_{E_k} f_k = \frac{\text{Area}(E_k)}{3} \Big(f_k(a_k, b_k) + f_k(c_k, d_k) + f_k(k, k) \Big) = \frac{1}{9k^2}.$$

Consequently, $\iint_{D_k} f_k = 3(1/9k^2) = 1/3k^2$. Now let us vary k and consider $f : [0, \infty) \times [0, \infty) \to \mathbb{R}$ given by

$$f(s, t) := \begin{cases} f_k(s, t) & \text{if } (s, t) \in D_k \text{ for some } k \in \mathbb{N}, \\ 0 & \text{otherwise.} \end{cases}$$

Since f_k vanishes on the vertices of T_k and hence on the sides of T_k, we see that f is a continuous function on $[0, \infty) \times [0, \infty)$. Moreover, since $\sum_{k=1}^{\infty} (1/k^2)$ is convergent, and for any $(x, y) \in \mathbb{R}^2$,

$$0 \le \iint_{[0,x]\times[0,y]} f \le \sum_{k=1}^{\infty} \iint_{D_k} f = \frac{1}{3} \sum_{k=1}^{\infty} \frac{1}{k^2},$$

we see that the improper double integral $\iint_{[0,\infty)\times[0,\infty)} f$ is convergent. But since $f(k, k) = 1$ for $k \in \mathbb{N}$, it is clear that $f(s, t) \not\to 0$ as $(s, t) \to (\infty, \infty)$. ◇

By modifying the function in the above example, one can obtain a continuous function $\tilde{f} : [0, \infty) \times [0, \infty) \to \mathbb{R}$ such that $\iint_{[0,\infty)\times[0,\infty)} \tilde{f}$ is convergent, but $\tilde{f}(k, k) \to \infty$ as $k \to \infty$. Indeed, it suffices to replace D_k by a triangular region $\widetilde{D}_k$ with $\text{Area}(\widetilde{D}_k) = 1/k^3$ and f_k by the function $\tilde{f}_k := k f_k$ for $k \in \mathbb{N}$.

Observe that the (k, ℓ)th Term Test for double series (Proposition 7.8) can be restated as follows. If a double series $\big((a_{k,\ell}), (A_{m,n})\big)$ is convergent, then $A_{k,\ell} - A_{k,\ell-1} - A_{k-1,\ell} + A_{k-1,\ell-1} \to 0$ as $(k, \ell) \to (\infty, \infty)$. This formulation has the following analogue for improper double integrals.

Proposition 7.63. *If an improper double integral (f, F) is convergent, then*

$$\iint_{[x-1,x]\times[y-1,y]} f(s,t)d(s,t) \to 0 \quad as\ (x,y) \to (\infty,\infty).$$

Proof. Consider $F(x,y) := \iint_{[a,x]\times[c,y]} f(s,t)d(s,t)$ for $(x,y) \geq (a,c)$. By Domain Additivity (Proposition 5.9), for all $(x,y) \geq (a+1, c+1)$, we obtain

$$\iint_{[x-1,x]\times[y-1,y]} f(s,t)d(s,t) = F(x,y) - F(x,y-1) - F(x-1,y) + F(x-1,y-1).$$

If (f, F) is convergent, then there is $I \in \mathbb{R}$ such that $F(x,y) \to I$, and so the right side of the above equality tends to 0 as $(x,y) \to (\infty,\infty)$. □

Tests for Absolute Convergence

The following test is an analogue of the Comparison Test for double series.

Proposition 7.64 (Comparison Test for Improper Double Integrals). *Suppose $a, c \in \mathbb{R}$ and $f, g : [a,\infty) \times [c,\infty) \to \mathbb{R}$ are such that both f and g are integrable on $[a,x] \times [c,y]$ for every $(x,y) \geq (a,c)$ and $|f| \leq g$. If $\iint_{[a,\infty)\times[c,\infty)} g(s,t)d(s,t)$ is convergent, then $\iint_{[a,\infty)\times[c,\infty)} f(s,t)d(s,t)$ is absolutely convergent and*

$$\left|\iint_{[a,\infty)\times[c,\infty)} f(s,t)d(s,t)\right| \leq \iint_{[a,\infty)\times[c,\infty)} g(s,t)d(s,t).$$

Proof. For $(x,y) \in [a,\infty) \times [c,\infty)$, let $F(x,y) := \iint_{[a,x]\times[c,y]} f(s,t)d(s,t)$, $G(x,y) := \iint_{[a,x]\times[c,y]} g(s,t)d(s,t)$, and $\widetilde{F}(x,y) := \iint_{[a,x]\times[c,y]} |f(s,t)|d(s,t)$. Assume that $\iint_{[a,\infty)\times[c,\infty)} g(s,t)d(s,t)$ is convergent. Then the function G is bounded above. Since $|f| \leq g$, we see that $\widetilde{F} \leq G$, and hence the function $\widetilde{F}$ is bounded above. Also, since $|f| \geq 0$, it follows from Proposition 7.55 that $\iint_{[a,\infty)\times[c,\infty)} |f(s,t)|d(s,t)$ is convergent, that is, $\iint_{[a,\infty)\times[c,\infty)} f(s,t)d(s,t)$ is absolutely convergent. Further, since $-f \leq |f| \leq g$ and $f \leq |f| \leq g$, we have $-F(x,y) \leq \widetilde{F}(x,y) \leq G(x,y)$ and $F(x,y) \leq \widetilde{F}(x,y) \leq G(x,y)$ for all $(x,y) \geq (a,c)$. Letting $(x,y) \to (\infty,\infty)$, we obtain

$$\left|\iint_{[a,\infty)\times[c,\infty)} f(s,t)d(s,t)\right| \leq \iint_{[a,\infty)\times[c,\infty)} g(s,t)d(s,t),$$

as desired. □

The improper double integrals given in Examples 7.52 are useful in employing the Comparison Test for Improper Double Integrals.

Examples 7.65. (i) Let $f : [1, \infty) \times [1, \infty) \to \mathbb{R}$ be defined by

$$f(s,t) := \frac{2^s 5^t + st^2}{3^s 7^t + s^3 + t^4}.$$

To see if $\iint_{[1,\infty)\times[1,\infty)} f$ is convergent, consider $g : [1, \infty) \times [1, \infty) \to \mathbb{R}$ defined by $g(s,t) := (2/3)^s(5/7)^t$. Then $\iint_{[1,\infty)\times[1,\infty)} g(s,t)d(s,t)$ is convergent since $0 < \frac{2}{3} < 1$ and $0 < \frac{5}{7} < 1$. Also, since $st^2 < 2^s 5^t$ and $s^3 + t^4 > 0$ for all $s, t \geq 1$,

$$|f(s,t)| < \frac{2^s 5^t + 2^s 5^t}{3^s 7^t} = 2\left(\frac{2}{3}\right)^s \left(\frac{5}{7}\right)^t \quad \text{for all } (s,t) \in [1,\infty) \times [1,\infty).$$

Hence $\iint_{[1,\infty)\times[1,\infty)} f(s,t)d(s,t)$ is (absolutely) convergent by the Comparison Test for Improper Double Integrals.

(ii) Let $f : [1, \infty) \times [1, \infty) \to \mathbb{R}$ be defined by

$$f(s,t) := \frac{1}{(1 + s + t + st + s^3t^4)^{1/2}}.$$

Consider $g : [1, \infty) \times [1, \infty) \to \mathbb{R}$ defined by $g(s,t) := 1/s^{3/2}t^2$. Then $\iint_{[1,\infty)\times[1,\infty)} g(s,t)d(s,t)$ is convergent since $\frac{3}{2} > 1$ and $2 > 1$. Also, $|f(s,t)| \leq |g(s,t)|$ for all $(s,t) \in [1,\infty) \times [1,\infty)$. Hence the improper double integral $\iint_{[1,\infty)\times[1,\infty)} f(s,t)d(s,t)$ is (absolutely) convergent by the Comparison Test for Improper Double Integrals. ◇

One can derive Limit Comparison Test and Root Test for improper double integrals from Proposition 7.61. These tests involve the concept of uniform convergence, which we have not introduced in this book. Hence we refrain from discussing them here.

Tests for Conditional Convergence

We shall now consider some tests that give conditional convergence of an improper double integral. They are based on the following result, which is analogous to the Partial Double Summation Formula (Proposition 7.37).

Proposition 7.66 (Partial Double Integration Formula). *Consider a rectangle $R := [a,b] \times [c,d]$ in $\mathbb{R}^2$ and let $f, g : R \to \mathbb{R}$ be such that f_x, f_y, and f_{xy} exist and are continuous on R, and that g is continuous on R. Define $G(x,y) := \iint_{[a,x]\times[c,y]} g(s,t)d(s,t)$ for (x,y) in R. Then*

$$\iint_R f(s,t)g(s,t)d(s,t) = f(b,d)G(b,d) + \iint_R f_{xy}(s,t)G(s,t)d(s,t) - \int_a^b f_x(s,d)G(s,d)ds - \int_c^d f_y(b,t)G(b,t)dt.$$

Proof. Since f_x and f_y are continuous on R, part (iii) of Proposition 3.3 shows that f is continuous on R. Also, by Corollary 5.23, we see that G_x, G_y, and G_{xy} exist and are continuous, and in fact, $G_{xy} = g$ on R. Carrying out double integration by parts (Proposition 5.25), we obtain

$$\iint_R fg = \triangle_{(a,c)}^{(b,d)}(fG) - \iint_R (f_x G_y + f_y G_x + f_{xy} G).$$

Moreover, $\triangle_{(a,c)}^{(b,d)}(fG) = f(b,d)G(b,d)$, since $G(a,c) = G(b,c) = G(a,d) = 0$. Next, by Fubini's Theorem on Rectangles (Proposition 5.28), we see that

$$\iint_R f_x G_y = \int_a^b \Big(\int_c^d f_x(s,t) G_y(s,t) dt \Big) ds.$$

For each fixed $s \in [a,b]$, we may use the one-variable formula for integration by parts (given, for example, in Proposition 6.25 of ACICARA) to obtain

$$\int_c^d f_x(s,t) G_y(s,t) dt = (f_x G)(s,d) - (f_x G)(s,c) - \int_c^d (f_x)_y(s,t) G(s,t) dt.$$

Since $G(s,c) = 0$ for all $s \in [a,b]$, it follows that

$$\iint_R f_x G_y = \int_a^b f_x(s,d) G(s,d) ds - \int_a^b \Big(\int_c^d f_{xy}(s,t) G(s,t) dt \Big) ds.$$

In a similar manner, we have

$$\iint_R f_y G_x = \int_c^d f_y(b,t) G(b,t) dt - \int_c^d \Big(\int_a^b f_{yx}(s,t) G(s,t) ds \Big) dt.$$

By the Mixed Partials Theorem (Proposition 3.14), $f_{yx} = f_{xy}$ on R. Now Fubini's Theorem on Rectangles yields

$$\int_a^b \Big(\int_c^d f_{xy}(s,t) G(s,t) dt \Big) ds = \iint_R f_{xy} G = \int_c^d \Big(\int_a^b f_{yx}(s,t) G(s,t) ds \Big) dt.$$

The desired result follows by adding appropriate equations stated above. □

Proposition 7.67 (Dirichlet's Test for Improper Double Integrals). *Let $f,\ g : [a,\infty) \times [c,\infty) \to \mathbb{R}$ be functions such that f_x, f_y, and f_{xy} exist and are continuous, and g is continuous. Assume that*

(i) *f is bimonotonic,*
(ii) *for each fixed $t \geq c$, the function given by $s \longmapsto f(s,t)$ is monotonic on $[a,\infty)$ and for each fixed $s \geq a$, the function given by $t \longmapsto f(s,t)$ is monotonic on $[c,\infty)$,*
(iii) *$\lim_{s\to\infty} f(s,s)$, $\lim_{s\to\infty} f(s,c)$, and $\lim_{t\to\infty} f(a,t)$ exist and each equals 0,*
(iv) *the partial double integral of $\iint_{[a,\infty]\times[c,\infty]} g(s,t) d(s,t)$ is bounded.*

Then the improper double integral $\iint_{[a,\infty)\times[c,\infty)} f(s,t)g(s,t)d(s,t)$ *is convergent and its partial double integral is bounded.*

Proof. As in the proof of Proposition 7.66, f is continuous on $[a,\infty)\times[c,\infty)$. First we show that $f(s,t)\to 0$ as $(s,t)\to(\infty,\infty)$ and that f is bounded. Let $\epsilon>0$ be given. By hypothesis (iii), there is $s_0\in[a,\infty)$ such that

$$(s,t)\geq(s_0,s_0)\Longrightarrow |f(s,s)|<\epsilon,\quad |f(s,c)|<\epsilon,\quad\text{and}\quad |f(a,t)|<\epsilon.$$

Consider $(s,t)\geq(a,c)$ with $s\geq s_0$ and $t\leq s$. By hypothesis (ii), either $f(s,c)\leq f(s,t)\leq f(s,s)$ or $f(s,c)\geq f(s,t)\geq f(s,s)$. Since both $f(s,c)$ and $f(s,s)$ are in $(-\epsilon,\epsilon)$, we see that $f(s,t)\in(-\epsilon,\epsilon)$. Similarly, if $(s,t)\geq(a,c)$ with $t\geq s_0$ and $s\leq t$, then $f(s,t)\in(-\epsilon,\epsilon)$. Thus for all $(s,t)\geq(a,c)$ with either $s\geq s_0$ or $t\geq s_0$, we have $|f(s,t)|<\epsilon$. Since $\epsilon>0$ is arbitrary, this implies that $f(s,t)\to 0$ as $(s,t)\to(\infty,\infty)$. Also, considering $\epsilon:=1$ and $\alpha:=\sup\{|f(s,t)| : a\leq s\leq s_0 \text{ and } c\leq t\leq s_0\}$, we obtain $|f(s,t)|\leq\max\{1,\alpha\}$, which proves that f is bounded.

We now examine each term on the right side of the Partial Double Integration Formula (Proposition 7.66). Let G denote the partial double integral of $\iint_{[a,\infty]\times[c,\infty]} g(s,t)d(s,t)$. Then G is bounded by hypothesis (iv), and so there is $\beta>0$ such that $|G(b,d)|\leq\beta$ for all $(b,d)\geq(a,c)$. Since $f(b,d)\to 0$ as $(b,d)\to(\infty,\infty)$, it follows that $f(b,d)G(b,d)\to 0$ as $(b,d)\to(\infty,\infty)$.

Also, by hypothesis (i), the function f is bimonotonic, and so Proposition 3.55 shows that f_{xy} does not change sign on $[a,\infty)\times[c,\infty)$. Hence for every $(b,d)\geq(a,c)$, we have

$$\begin{aligned}\iint_{[a,b]\times[c,d]} |f_{xy}(s,t)G(s,t)|d(s,t) &\leq \beta\left|\iint_{[a,b]\times[c,d]} f_{xy}(s,t)d(s,t)\right|\\ &= \beta\left|\int_a^b [f_x(s,d)-f_x(s,c)]ds\right|\\ &= \beta|f(b,d)-f(b,c)-f(a,d)+f(a,c)|.\end{aligned}$$

Since the function f is bounded, it follows by Proposition 7.55 that the improper double integral $\iint_{[a,\infty)\times[c,\infty)} f_{xy}(s,t)G(s,t)d(s,t)$ is absolutely convergent. By Proposition 7.55, its partial double integral is bounded, and by Proposition 7.59, it converges to a real number J.

Next, since for each fixed $t\in[c,\infty)$, the function $s\longmapsto f(s,t)$ is monotonic on $[a,\infty)$, it follows that f_x is of the same sign and hence for all $(b,d)\geq(a,c)$,

$$\left|\int_a^b f_x(s,d)G(s,d)ds\right|\leq\beta\left|\int_a^b f_x(s,d)ds\right| = \beta|f(b,d)-f(a,d)|.$$

Since $f(a,d)\to 0$ as $d\to\infty$, and $f(b,d)\to 0$ as $(b,d)\to(\infty,\infty)$, we see that $|f(b,d)-f(a,d)|\to 0$, and so $\int_a^b f_x(s,d)G(s,d)ds\to 0$ as $(b,d)\to(\infty,\infty)$. Similarly, it follows that $\int_c^d f_y(b,t)G(b,t)dt\to 0$ as $(b,d)\to(\infty,\infty)$.

By the Partial Double Integration Formula (Proposition 7.66), we obtain

$$\iint_{[a,b]\times[c,d]} f(s,t)g(s,t)d(s,t) \to 0 + J + 0 + 0 = J \quad \text{as } (b,d) \to (\infty,\infty).$$

Thus the improper double integral $\iint_{[a,\infty)\times[c,\infty)} f(s,t)g(s,t)d(s,t)$ is convergent. Also, its partial double integral is bounded, since each of the four terms on the right side of the Partial Double Integration Formula is bounded. □

One can derive analogues of Abel's Test and Dedekind's Tests (Exercises 48 and 49) for improper double integrals. Since their proofs would involve the concept of uniform convergence, we choose not to deal with these results. The Leibniz Test for Double Series given in Corollary 7.39 has no straightforward analogue for improper double integrals, essentially since the function $g : [a,\infty) \times [c,\infty) \to \mathbb{R}$ defined by

$$g(s,t) := \begin{cases} (-1)^{s+t} & \text{if } (s,t) \in \mathbb{N}^2, \\ 0 & \text{otherwise,} \end{cases}$$

is not continuous. The Convergence Test for Trigonometric Double Series given in Corollary 7.40 admits the following analogue for the so-called **Fourier double integrals**. The two improper double integrals in the corollary below are sometimes called the **Fourier sine double integral** and the **Fourier cosine double integral**, respectively.

Corollary 7.68 (Convergence Test for Fourier Double Integrals). *Let $a, c \in \mathbb{R}$ and let $f : [a,\infty) \times [c,\infty) \to \mathbb{R}$ be a function satisfying conditions* (i), (ii), *and* (iii) *of Proposition 7.67. Let θ and φ be nonzero real numbers. Then the improper double integrals*

$$\iint_{[a,\infty)\times[c,\infty)} f(s,t)\sin(s\theta + t\varphi)d(s,t) \text{ and } \iint_{[a,\infty)\times[c,\infty)} f(s,t)\cos(s\theta + t\varphi)d(s,t)$$

are convergent.

Proof. Define $g : [a,\infty) \times [c,\infty) \to \mathbb{R}$ by $g(s,t) := \sin(s\theta + t\varphi)$. Clearly, g is continuous. Consider the partial double integral $G : [a,\infty) \times [c,\infty) \to \mathbb{R}$ defined by $G(x,y) := \iint_{[a,x]\times[c,y]} g(s,t)d(s,t)$. Using Fubini's Theorem on Rectangles (Proposition 5.28) and noting that $\varphi \neq 0$ and $\theta \neq 0$, we obtain

$$\begin{aligned} |G(x,y)| &= \frac{1}{|\varphi|}\left|\int_a^x [\cos(s\theta + c\varphi) - \cos(s\theta + y\varphi)]ds\right| \\ &= \frac{1}{|\varphi|\,|\theta|}\left|\left|\sin(x\theta + c\varphi) - \sin(a\theta + c\varphi) + \sin(x\theta + y\varphi) - \sin(a\theta + y\varphi)\right|\right| \\ &\le \frac{4}{|\varphi|\,|\theta|}. \end{aligned}$$

Thus the function G is bounded. Hence by Proposition 7.67, the improper double integral $\iint_{[a,\infty)\times[c,\infty)} f(s,t)\sin(s\theta + t\varphi)d(s,t)$ is convergent. Similarly, $\iint_{[a,\infty)\times[c,\infty)} f(s,t)\cos(s\theta + t\varphi)d(s,t)$ is also convergent. □

Remark 7.69. If $\theta = \varphi = 0$, then $\sin(s\theta + t\varphi) = 0$ and $\cos(s\theta + t\varphi) = 1$ for all $(s,t) \in \mathbb{R}^2$, and so the integrand of the Fourier sine double integral $\iint_{[a,\infty)\times[c,\infty)} f(s,t)\sin(s\theta + t\varphi)d(s,t)$ is equal to zero, while the Fourier cosine double integral $\iint_{[a,\infty)\times[c,\infty)} f(s,t)\cos(s\theta + t\varphi)d(s,t)$ is just the improper double integral $\iint_{[a,\infty)\times[c,\infty)} f(s,t)d(s,t)$, which may or may not converge. If one of θ and φ is equal to 0, while the other is not equal to 0, then depending upon the choice of the function f (satisfying conditions (i), (ii), and (iii) given in Proposition 7.67), the corresponding Fourier double integrals may converge absolutely, or may converge conditionally, or may diverge. Exercise 36 illustrates each of these cases. ◇

Example 7.70. Let $f(s,t) := 1/(s+t)^p$ for $(s,t) \in [1,\infty) \times [1,\infty)$, where $p \in \mathbb{R}$ with $p > 0$. Then the function f clearly satisfies conditions (i) and (ii) of Proposition 7.67. Further, we have seen in Example 1.8 (ii) that f is bimonotonically increasing. Also, $f(s,s) \to 0$ and $f(s,1) \to 0$ as $s \to \infty$, and $f(1,t) \to 0$ as $t \to \infty$. Thus condition (iii) of Proposition 7.67 is also satisfied. Hence if θ and φ are nonzero real numbers, then by Corollary 7.68, we see that the improper double integrals

$$\iint_{[1,\infty)\times[1,\infty)} \frac{\sin(\theta s + \varphi t)}{(s+t)^p} d(s,t) \quad \text{and} \quad \iint_{[1,\infty)\times[1,\infty)} \frac{\cos(\theta s + \varphi t)}{(s+t)^p} d(s,t)$$

are convergent. In fact, in view of Exercise 7.58, both these improper double integrals are absolutely convergent if $p > 2$. ◇

7.7 Unconditional Convergence of Improper Double Integrals

In Chapter 5, we developed the theory of double integrals of bounded functions on bounded subsets of $\mathbb{R}^2$. When the function or the subset of $\mathbb{R}^2$ on which it is defined is unbounded, we are led to improper double integrals. In Sections 7.5 and 7.6, we discussed the theory of improper double integrals of functions defined on an unbounded subset of $\mathbb{R}^2$ of the form $[a,\infty) \times [c,\infty)$, where $a, c \in \mathbb{R}$. We have also noted in Remark 7.53 that this can be used to suitably define improper double integrals of functions defined on some other unbounded subsets of $\mathbb{R}^2$ such as $(-\infty,b]\times[c,\infty)$, $(-\infty,b]\times(-\infty,d]$, etc., where $b,c,d \in \mathbb{R}$. But what about functions defined on an arbitrary unbounded subset of $\mathbb{R}^2$ or unbounded functions defined on a bounded subset of $\mathbb{R}^2$? In these cases, there is no straightforward analogue of partial double integrals whose limit can be defined as the improper double integral. However, for a function $f : D \to \mathbb{R}$, where $D \subseteq \mathbb{R}^2$, it seems most natural to consider a suitable sequence (D_n) of bounded subsets of D such that f is integrable on each D_n, and then define $\iint_D f$ as the limit of $\iint_{D_n} f$ as $n \to \infty$. But of course this limit should exist and should be independent of the choice of the sequence (D_n). We shall see that this requirement leads to a more stringent notion of convergence even in the case of improper double integrals of functions on familiar unbounded subsets of $\mathbb{R}^2$ such as $[a,\infty)\times[c,\infty)$. This notion, called unconditional convergence, is discussed in this section. First, we shall consider the so-called improper double integrals of the **first kind**, which correspond to the case of functions on unbounded subsets of $\mathbb{R}^2$, and later we consider improper double integrals of the **second kind**, which correspond to the case of unbounded functions on bounded subsets of $\mathbb{R}^2$. Throughout, we shall restrict ourselves to continuous functions for the sake of simplicity.

Functions on Unbounded Subsets

We begin with an example of an unbounded subset D of $\mathbb{R}^2$ and a continuous function $f : D \to \mathbb{R}$ for which there are natural sequences (D_n) and (E_n) of bounded subsets of D such that f is integrable on D_n as well as E_n for each $n \in \mathbb{N}$, and $\lim_{n\to\infty} \iint_{D_n} f$ exists but $\lim_{n\to\infty} \iint_{E_n} f$ does not.

Example 7.71. Let $D := [0,\infty)\times[0,\infty)$. Consider $f : D \to \mathbb{R}$ defined by $f(s,t) := \sin(s^2+t^2)$. Then f is continuous on D, and for $(x,y)\in D$,

$$\begin{aligned} F(x,y) &:= \iint_{[0,x]\times[0,y]} f(s,t)d(s,t) \\ &= \int_0^x \left[\int_0^y \left(\sin s^2 \cos t^2 + \cos s^2 \sin t^2\right) dt\right] ds \\ &= \left(\int_0^x \sin s^2 ds\right)\left(\int_0^y \cos t^2 dt\right) + \left(\int_0^x \cos s^2 ds\right)\left(\int_0^y \sin t^2 dt\right). \end{aligned}$$

The substitution $u := s^2$ shows that for $x \ge 1$, we have

$$\int_1^x \sin s^2 ds = \frac{1}{2}\int_1^{x^2} \frac{\sin u}{\sqrt{u}} du.$$

By the Convergence Test for Fourier Integrals based on Dirichlet's Test (given, for example, in Corollary 9.52 of ACICARA), we see that the improper integral $\int_1^\infty (\sin u/\sqrt{u})\, du$ is convergent. Since $\int_0^x \sin s^2 ds = \int_0^1 \sin s^2 ds + \int_1^x \sin s^2 ds$ for all $x \ge 1$, it follows that the improper integral $\int_0^\infty \sin s^2 ds$ is convergent. Similarly, the improper integral $\int_0^\infty \cos s^2 ds$ is convergent.[1] Consequently, we conclude that $\lim_{(x,y)\to(\infty,\infty)} F(x,y)$ exists. In particular, if for $n\in\mathbb{N}$, we let $D_n := \{(s,t)\in D : s \le n \text{ and } t \le n\}$, then $\lim_{n\to\infty}\iint_{D_n} f(s,t)d(s,t) = \lim_{n\to\infty} F(n,n)$ exists (and is equal to $\pi/4$).

[1] The improper integrals $\int_0^\infty \sin s^2 ds$ and $\int_0^\infty \cos s^2 ds$ are known as **Fresnel integrals**. Each is equal to $\sqrt{\pi/8}$. See, for example, page 473 of [12, vol. II].

On the other hand, suppose $E_n := \{(s,t) \in D : s^2 + t^2 \le n\pi\}$ for $n \in \mathbb{N}$. Switching to polar coordinates $s := r\cos\theta$, $t := r\sin\theta$, we see that E_n is transformed to $G_n := [0, \sqrt{n\pi}] \times [0, \pi/2]$, and so by Proposition 5.65,

$$\begin{aligned}\iint_{E_n} f(s,t)d(s,t) &= \iint_{G_n} f(r\cos\theta, r\sin\theta)\, r\, d(r,\theta)\\ &= \left(\int_0^{\sqrt{n\pi}} r\,\sin r^2 dr\right)\left(\int_0^{\pi/2} d\theta\right)\\ &= \frac{\pi}{4}(1-\cos n\pi) = \frac{\pi}{4}\big(1-(-1)^n\big).\end{aligned}$$

It follows that $\lim_{n\to\infty} \iint_{E_n} f(s,t)d(s,t)$ does not exist. ◇

In view of the above example, and in analogy with our discussion in Section 7.2 of unconditionally convergent double series, we first define an appropriate class of sequences of bounded subsets of an unbounded set, and then we introduce the notion of unconditional convergence.

Let D be an unbounded subset of $\mathbb{R}^2$. A sequence (D_n) of subsets of D is said to be **exhausting** if it satisfies the following three conditions:

(i) D_n is bounded and ∂D_n is of content zero for each $n \in \mathbb{N}$,
(ii) $D_n \subseteq D_{n+1}$ for each $n \in \mathbb{N}$, and
(iii) each bounded subset of D is contained in D_n for some $n \in \mathbb{N}$.

Observe that the sequences (D_n) and (E_n) of subsets of $[0,\infty)\times[0,\infty)$ considered in Example 7.71 are exhausting. In general, if D is an unbounded subset of $\mathbb{R}^2$, and if for $n \in \mathbb{N}$, we let

$$D_n := \{(s,t) \in D : |s| \le n \text{ and } |t| \le n\} \quad\text{and}\quad E_n := \{(s,t) \in D : s^2+t^2 \le n^2\},$$

then (D_n) is exhausting, provided ∂D_n is of content zero for each $n \in \mathbb{N}$, and (E_n) is exhausting, provided ∂E_n is of content zero for each $n \in \mathbb{N}$. On the other hand, if $D := \mathbb{Q}^2$, then D does not admit any exhausting sequence of subsets. To see this, observe that if a subset E of D contains the bounded subset $([-1,1]\times[-1,1]) \cap \mathbb{Q}^2$ of D, then $\partial E \supseteq [-1,1]\times[-1,1]$, and therefore ∂E cannot be of content zero.

Let D be an unbounded subset of $\mathbb{R}^2$ that admits an exhausting sequence of subsets and let $f : D \to \mathbb{R}$ be a continuous function that is bounded on each bounded subset of D. Then by Proposition 5.43, f is integrable on each term of an exhausting sequence of subsets of D. We say that the improper double integral $\iint_D f(s,t)d(s,t)$ is **unconditionally convergent** if there is $I \in \mathbb{R}$ such that for every exhausting sequence (D_n) of subsets of D, the limit

$$\lim_{n\to\infty} \iint_{D_n} f(s,t)d(s,t)$$

exists and is equal to I. The hypothesis that D admits an exhausting sequence of subsets ensures that the real number I, when it exists, is unique. In this case, we write $\iint_D f(s,t)d(s,t) = I$, or simply $\iint_D f = I$.

It is easily seen that if the improper double integrals $\iint_D f$ and $\iint_D g$ are unconditionally convergent, then so are $\iint_D (f+g)$ and $\iint_D (rf)$ for any $r \in \mathbb{R}$.

We give below a necessary and sufficient condition for the unconditional convergence of an improper double integral of a nonnegative continuous function on an unbounded subset of $\mathbb{R}^2$. Its proof shows the importance of conditions (ii) and (iii) in the definition of an exhausting sequence.

Proposition 7.72. *Let D be an unbounded subset of $\mathbb{R}^2$ that admits an exhausting sequence of subsets, say (D_n), and let $f : D \to \mathbb{R}$ be a nonnegative continuous function that is bounded on each bounded subset of D. Then the improper double integral $\iint_D f$ is unconditionally convergent if and only if $\alpha \in \mathbb{R}$ such that $\iint_{D_n} f \le \alpha$ for all $n \in \mathbb{N}$, and in this case, $\iint_D f = \lim_{n\to\infty} \iint_{D_n} f$.*

Proof. Suppose $\iint_D f$ is unconditionally convergent. Then $\lim_{n\to\infty} \iint_{D_n} f$ exists. Since a convergent sequence of real numbers is bounded, there is $\alpha \in \mathbb{R}$ such that $\iint_{D_n} f \le \alpha$ for all $n \in \mathbb{N}$.

Conversely, suppose there is $\alpha \in \mathbb{R}$ such that $\iint_{D_n} f \le \alpha$ for all $n \in \mathbb{N}$. For $n \in \mathbb{N}$, let $I_n := \iint_{D_n} f$. Since f is nonnegative, (I_n) is a monotonically increasing sequence of real numbers. Also, (I_n) is bounded above by α. Hence (I_n) converges to $I := \sup\{I_n : n \in \mathbb{N}\}$. Next, let (E_n) be any other exhausting sequence of subsets of D, and let $J_n := \iint_{E_n} f$ for $n \in \mathbb{N}$. Then (J_n) is also a monotonically increasing sequence of real numbers. Now fix $m \in \mathbb{N}$. Since E_m is a bounded subset of D, there is $n_0 \in \mathbb{N}$ such that $E_m \subseteq D_{n_0}$, and so $J_m \le I_{n_0} \le I$. This shows that the sequence (J_n) is bounded and $J := \sup\{J_n : n \in \mathbb{N}\} \le I$. Interchanging the roles of (D_n) and (E_n), we see that $I \le J$. Hence the sequence (J_n) also converges to I. This proves that $\iint_D f$ is unconditionally convergent. □

Example 7.71 shows that the nonnegativity of the function f cannot be dropped from the above proposition. To obtain an analogue of the characterization in Proposition 7.72 for functions that may change sign, we require the following auxiliary results. Let us recall that if S is a subset of $\mathbb{R}^2$ and $f : S \to \mathbb{R}$ is a function, then $f^+, f^- : S \to \mathbb{R}$ are defined by $f^+ := (|f|+f)/2$ and $f^- := (|f|-f)/2$. Observe that f^+ and f^- are nonnegative, $f = f^+ - f^-$, and $|f| = f = f^+ + f^-$. Further, f is continuous on S if and only if both f^+ and f^- are continuous on S. Moreover, if S is bounded, then f is integrable on S if and only if both f^+ and f^- are integrable on S.

Lemma 7.73. *Let S be a bounded subset of $\mathbb{R}^2$ and $f : S \to \mathbb{R}$ an integrable function. Given any $\gamma \in \mathbb{R}$ with $\gamma > 2$, there is a subset T of S such that ∂T is of content zero, f is integrable on T, and*

$$\iint_S |f| \le \gamma \left| \iint_T f \right|.$$

Proof. Let $\delta := \iint_S |f|$. If $\delta = 0$, we let $T := \emptyset$. Suppose now that $\delta \ne 0$. Since $\delta = \iint_S f^+ + \iint_S f^-$, we see that either $\iint_S f^+ \ge \delta/2$ or $\iint_S f^- \ge \delta/2$. We assume without loss of generality that $\iint_S f^+ \geq \delta/2$. Let R be a rectangle containing S and let $g : R \to \mathbb{R}$ denote the function obtained by extending the function f^+ to R as usual, that is, by setting it to be zero on $R \setminus S$. Note that g is nonnegative, and if $g(x, y) > 0$ for some $(x, y) \in R$, then $(x, y) \in S$. Since $\gamma > 2$ and $\delta \neq 0$,

$$\sup \{L(P, g) : P \text{ is a partition of } R\} = L(g) = \iint_R g \geq \frac{\delta}{2} > \frac{\delta}{\gamma}.$$

Hence there is a partition $P := \{(x_i, y_j) : i = 0, 1, \ldots, n \text{ and } j = 0, 1, \ldots, k\}$ of R such that $L(P, g) = \sum_{i=1}^{n} \sum_{j=1}^{k} m_{i,j}(g)(x_i - x_{i-1})(y_j - y_{j-1}) > \delta/\gamma$, where $m_{i,j}(g) := \inf\{g(x, y) : (x, y) \in [x_{i-1}, x_i] \times [y_{j-1}, y_j]\}$ for $i = 0, 1, \ldots, n$ and $j = 0, 1, \ldots, k$. Let T denote the union of those subrectangles $[x_{i-1}, x_i] \times [y_{j-1}, y_j]$ of R for which $m_{i,j}(g) > 0$. Since $g(x, y) > 0$ for $(x, y) \in T$, we see that $T \subseteq S$. Also, ∂T is of content zero, since it consists of finitely many line segments. Further, by Corollary 5.50, f is integrable on T. Now for $(x, y) \in T$, $f^+(x, y) = g(x, y) > 0$ and so $f(x, y)$ is positive and is equal to $g(x, y)$. Thus

$$\left|\iint_T f\right| = \iint_T f = \iint_T g.$$

Next, let $h : R \to \mathbb{R}$ be the extension of $g|_T$ to R defined by $h(x, y) := g(x, y)$ if $(x, y) \in T$ and $h(x, y) := 0$ if $(x, y) \in R \setminus T$. Note that $0 \leq h \leq g$ and $m_{i,j}(h) = m_{i,j}(g)$ for all $i = 0, 1, \ldots, n$ and $j = 0, 1, \ldots, k$. Consequently,

$$\iint_T g = \iint_R h \geq L(P, h) = L(P, g) > \frac{\delta}{\gamma} = \frac{1}{\gamma} \iint_S |f|.$$

This yields $\gamma \left|\iint_T f\right| \geq \iint_S |f|$, as desired. □

The above result is interesting because of the reversal of the inequality sign in the basic inequality $\left|\iint_S f\right| \leq \iint_S |f|$. If the boundaries of the sets $S^+ := \{(x, y) \in S : f(x, y) \geq 0\}$ and $S^- := \{(x, y) \in S : f(x, y) \leq 0\}$ are of content zero, then we can let $T := S^+$ or $T := S^-$ and replace γ by 2 in the above proposition. (See Exercise 13 of Chapter 5.)

Lemma 7.74. *Let D be an unbounded subset of $\mathbb{R}^2$ that admits an exhausting sequence of subsets, and let $f : D \to \mathbb{R}$ be a continuous function that is bounded on each bounded subset of D. Let $\iint_D f$ be unconditionally convergent. Then there is $\alpha \in \mathbb{R}$ such that $\iint_S |f| \leq \alpha$ for every bounded subset S of D such that ∂S is of content zero.*

Proof. First we show that there is $\beta \in \mathbb{R}$ satisfying $\left|\iint_S f\right| \leq \beta$ for every bounded subset S of D such that ∂S is of content zero. Assume for a moment that this is not the case. Let (D_n) be an exhausting sequence of subsets of D, and set $U_1 := D_1$. Then there is a bounded subset T_1 of D such that ∂T_1 is of content zero and $\left|\iint_{T_1} f\right| \geq 1 + \iint_{U_1} |f|$, and for each $n \geq 2$, there is a bounded subset T_n of D such that ∂T_n is of content zero and

$$\left|\iint_{T_n} f\right| \geq n + \iint_{U_n} |f|, \quad \text{where } U_n := D_n \cup T_1 \cup \cdots \cup T_{n-1}.$$

Define $S_n := T_n \cup U_n$ for $n \in \mathbb{N}$. Note that (S_n) is an exhausting sequence of subsets of D. If for $n \in \mathbb{N}$, we let $V_n := S_n \setminus T_n$, then $V_n \subseteq U_n$ and ∂V_n is of content zero. Thus, by Domain Additivity (Corollary 5.52), we see that

$$\left|\iint_{S_n} f\right| = \left|\iint_{T_n} f + \iint_{V_n} f\right| \geq \left|\iint_{T_n} f\right| - \iint_{U_n} |f| \geq n.$$

Hence $\lim_{n\to\infty} \iint_{S_n} f$ cannot exist, which is a contradiction. This proves that there is $\beta \in \mathbb{R}$ satisfying the inequality stated at the beginning of the proof.

Now, given any bounded subset S of D such that ∂S is of content zero, by Lemma 7.73 with $\gamma := 3$, there is a subset T of S such that ∂T is of content zero, f is integrable on T, and $\iint_S |f| \leq 3\left|\iint_T f\right|$. Since $\left|\iint_T f\right| \leq \beta$, we obtain the desired result upon letting $\alpha := 3\beta$. □

Proposition 7.75. *Let D be an unbounded subset of $\mathbb{R}^2$ that admits an exhausting sequence of subsets, say (D_n), and let $f : D \to \mathbb{R}$ be a continuous function that is bounded on each bounded subset of D. Then $\iint_D f$ is unconditionally convergent if and only if there is $\alpha \in \mathbb{R}$ such that $\iint_{D_n} |f| \leq \alpha$ for all $n \in \mathbb{N}$, and in this case, $\iint_D f = \lim_{n\to\infty} \iint_{D_n} f$. Equivalently, $\iint_D f$ is unconditionally convergent if and only if $\iint_D |f|$ is unconditionally convergent.*

Proof. Assume that $\iint_D f$ is unconditionally convergent. By Lemma 7.74, there is $\alpha \in \mathbb{R}$ such that $\iint_{D_n} |f| \leq \alpha$ for all $n \in \mathbb{N}$.

Conversely, suppose there is $\alpha \in \mathbb{R}$ such that $\iint_{D_n} |f| \leq \alpha$ for all $n \in \mathbb{N}$. Since $0 \leq f^+ \leq |f|$ and $0 \leq f^- \leq |f|$, we obtain $\iint_{D_n} f^+ \leq \alpha$ and $\iint_{D_n} f^- \leq \alpha$ for all $n \in \mathbb{N}$. By Proposition 7.72 applied to the functions f^+ and f^-, we see that $\iint_D f^+$ and $\iint_D f^-$ are unconditionally convergent. Since $f = f^+ - f^-$, it follows that $\iint_D f$ is unconditionally convergent. Moreover, $\iint_D f = \lim_{n\to\infty} \iint_{D_n} f$, since the same holds with f replaced by f^+ and f^-.

Finally, the last assertion follows from Proposition 7.72. □

Corollary 7.76 (Comparison Test for Improper Double Integrals of First Kind). *Let D be an unbounded subset of $\mathbb{R}^2$ that admits an exhausting sequence of subsets. Let $f, g : D \to \mathbb{R}$ be continuous functions that are bounded on each bounded subset of D and satisfy $|f| \leq g$. If $\iint_D g$ is unconditionally convergent, then so is $\iint_D f$.*

Proof. Follows from Propositions 7.72 and 7.75. □

The corollary below may be compared with Proposition 7.24 concerning an analogous result for double series.

Corollary 7.77. *Let $D := [a, \infty) \times [c, \infty)$, where $a, c \in \mathbb{R}$, and let $f : D \to \mathbb{R}$ be a continuous function. Then $\iint_D f$ is unconditionally convergent if and only if it is absolutely convergent.*

Proof. For $n \in \mathbb{N}$, let $D_n := [a, a+n] \times [c, c+n]$. Then (D_n) is an exhausting sequence of subsets of D. Since D is a closed set, the closure of every bounded subset of D is contained in D. Hence Proposition 2.25 shows that f is bounded on each bounded subset of D. Also, f is integrable on $[a, x] \times [c, y]$ for every $(x, y) \in \mathbb{R}^2$ with $x \geq a$ and $y \geq c$. Further, given $(x, y) \in \mathbb{R}^2$, if we let $n := \max\{[x]+1, [y]+1\}$, then $[a, x] \times [c, y] \subset D_n$. Hence it follows from Propositions 7.72 and 7.55 that $\iint_D |f|$ is unconditionally convergent if and only if $\iint_D f$ is absolutely convergent. Thus, by Proposition 7.75, $\iint_D f$ is unconditionally convergent if and only if it is absolutely convergent. □

Examples 7.78. (i) Let $D := \{(s, t) \in \mathbb{R}^2 : s^2 + t^2 \geq 1\}$ and let $p \in \mathbb{R}$ with $p > 1$. Consider $f : D \to \mathbb{R}$ defined by $f(s, t) := 1/(s^2 + t^2)^p$. For $n \in \mathbb{N}$, let $D_n := \{(s, t) \in D : s^2 + t^2 \leq n^2\}$; switching to polar coordinates, D_n is transformed to $G_n := [1, n] \times [-\pi, \pi]$, and so by Proposition 5.65,

$$\iint_{D_n} f = \iint_{G_n} r^{-2p} r \, d(r, \theta) = 2\pi \int_1^n r^{1-2p} dr = \frac{\pi}{p-1}\left(1 - \frac{1}{n^{2p-2}}\right).$$

Hence $\lim_{n\to\infty} \iint_{D_n} f = \pi/(p-1)$. Since f is nonnegative, by Proposition 7.72, $\iint_D f$ is unconditionally convergent and is equal to $\pi/(p-1)$. Next, consider $g : D \to \mathbb{R}$ defined by $g(s, t) := \sin(s^2 + t^2)/(s^2 + t^2)^p$. Then $|g| \leq f$, and so by Corollary 7.76, $\iint_D g$ is unconditionally convergent. Note that the function g assumes both positive and negative values.

(ii) Let $D := [0, \infty) \times [0, \infty)$. Define $f : D \to \mathbb{R}$ by $f(s, t) := e^{-(s^2+t^2)}$. For $n \in \mathbb{N}$, let $D_n := \{(s, t) \in D : s^2 + t^2 \leq n^2\}$. As in Example 5.66 (iii),

$$\iint_{D_n} f = \frac{\pi}{4}(1 - e^{-n^2}) \quad \text{for } n \in \mathbb{N}.$$

Hence $\lim_{n\to\infty} \iint_{D_n} f = \pi/4$. Since f is nonnegative, Proposition 7.72 shows that $\iint_D f$ is unconditionally (and hence by Corollary 7.77, absolutely) convergent and is equal to $\pi/4$. Note that it is not possible to evaluate the partial double integrals $\iint_{[0,x]\times[0,y]} f$ in terms of an elementary function of x and y. ◊

Concept of Area of an Unbounded Subset of $\mathbb{R}^2$

We extend the concept of area to certain unbounded subsets of $\mathbb{R}^2$. Let D be an unbounded subset of $\mathbb{R}^2$ that admits an exhausting sequence (D_n) of subsets of D. Then Area(D_n) is well defined for each $n \in \mathbb{N}$. If the sequence (Area(D_n)) is bounded, then applying Proposition 7.72 to the function 1_D, we see that $\lim_{n\to\infty}$ Area(D_n) exists and is independent of the choice of an exhausting sequence (D_n) of bounded subsets of D. We define the **area** of the unbounded subset D of $\mathbb{R}^2$ to be equal to this limit, and we denote it by $A(D)$. On the other hand, if the sequence (Area(D_n)) is unbounded, then for every exhausting sequence (E_n) of subsets of D, the sequence (Area(E_n)) would be unbounded; in this case, we define $A(D) := \infty$.

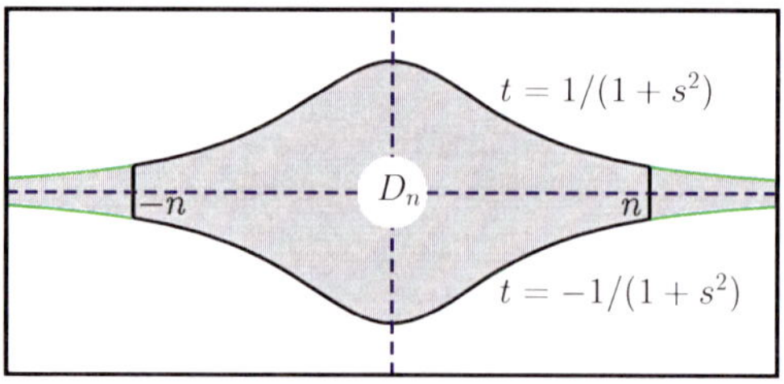

Fig. 7.6. Illustration of the bounded subsets D_n of D in Example 7.79 (i).

Examples 7.79. (i) Let $D := \{(s,t) \in \mathbb{R}^2 : |t| \le 1/(1+s^2)\}$. Define $D_n := \{(s,t) \in D : |s| \le n\}$ for $n \in \mathbb{N}$. (See Figure 7.6.) It is clear that (D_n) is an exhausting sequence of subsets of D. Also,

$$\text{Area}(D_n) := \iint_{D_n} 1_{D_n} = \int_{-n}^{n} \left(\int_{-1/(1+s^2)}^{1/(1+s^2)} dt \right) ds = 4 \int_0^n \frac{ds}{1+s^2} = 4 \arctan n.$$

Thus $A(D) := \lim_{n\to\infty} \text{Area}(D_n) = 2\pi$.

(ii) Let $D := \{(s,t) \in \mathbb{R}^2 : t \ge 0 \text{ and } |s^2 - t^2| \le 1\}$. Define $D_n := \{(s,t) \in D : |s| \le n\}$ for $n \in \mathbb{N}$. (See Figure 7.7.) It is clear that (D_n) is an exhausting sequence of subsets of D. Also,

$$\begin{aligned}\text{Area}(D_n) := \iint_{D_n} 1_{D_n} &= 4 \int_0^1 \left(\int_0^s dt \right) ds + 4 \int_1^n \left(\int_{\sqrt{s^2-1}}^{s} dt \right) ds \\ &= 2 + 4 \int_1^n \left(s - \sqrt{s^2-1} \right) ds \ge 2 + 4 \int_1^n \left(\frac{1}{2s} \right) ds = 2 + 2 \ln n.\end{aligned}$$

Thus the sequence $(\text{Area}(D_n))$ is unbounded, and so $A(D) = \infty$. ◇

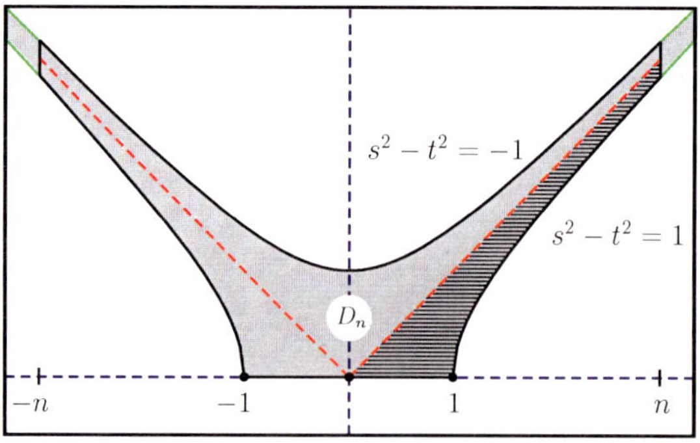

Fig. 7.7. Illustration of the bounded subsets D_n of D in Example 7.79 (ii).

Unbounded Functions on Bounded Subsets

We begin with an example of a bounded subset D of $\mathbb{R}^2$ and an unbounded continuous function $f : D \to \mathbb{R}$ for which there are natural sequences (D_n) and (E_n) of subsets of D such that f is integrable on D_n as well as E_n for each $n \in \mathbb{N}$, but the limits $\lim_{n\to\infty} \iint_{D_n} f$ and $\lim_{n\to\infty} \iint_{E_n} f$ are different.

Example 7.80. Let $D := [0,1] \times [0,1] \setminus \{(0,0)\}$. Consider $f : D \to \mathbb{R}$ defined by $f(s,t) := (s^2-t^2)/(s^2+t^2)^2$. Then ∂D is of content zero and f is continuous on D. For $(x,y) \neq (0,0)$ in D, let

$$F(x,y) := \iint_{[x,1]\times[y,1]} f(s,t)d(s,t).$$

By Fubini's Theorem on Rectangles (Proposition 5.28),

$$\begin{aligned} F(x,y) &= \int_x^1 \left[\int_y^1 \frac{s^2-t^2}{(s^2+t^2)^2}\,dt\right] ds \\ &= \int_x^1 \left[\frac{t}{s^2+t^2}\right]_{t=y}^{t=1} ds = \int_x^1 \frac{1}{s^2+1}ds - \int_x^1 \frac{y}{s^2+y^2}ds. \end{aligned}$$

If $x \in (0,1]$, then substituting $s := x/u$ in the second integral above,

$$F(x,x) = \int_x^1 \frac{1}{s^2+1}ds + \int_1^x \frac{1}{1+u^2}du = 0 \quad \text{and} \quad F(x,0) = \int_x^1 \frac{1}{s^2+1}ds.$$

Let $D_n := [1/n, 1] \times [1/n, 1]$ and $E_n := [1/n, 1] \times [0,1]$ for $n \in \mathbb{N}$. Then f is bounded on D_n as well as E_n for each $n \in \mathbb{N}$, and

$$\iint_{D_n} f(s,t)d(s,t) = F\Big(\frac{1}{n}, \frac{1}{n}\Big) = 0 \quad \text{for each } n \in \mathbb{N},$$

whereas

$$\iint_{E_n} f(s,t)d(s,t) = F\Big(\frac{1}{n}, 0\Big) \to \int_0^1 \frac{1}{s^2+1}ds = \frac{\pi}{4} \quad \text{as } n \to \infty. \qquad \diamond$$

In view of the above example, and in analogy with the notion of an exhausting sequence, we now define an appropriate class of sequences of subsets of a bounded set. Note that if the given set is bounded, then condition (iii) in the definition of an exhausting sequence implies that all except finitely many terms of the sequence coincide with the given set. With this in view, we will require instead that the areas of the subsets that constitute the sequence approach the area of the given set.

Let D be a bounded subset of $\mathbb{R}^2$ such that ∂D is of content zero. A sequence (D_n) of subsets of D is said to be **expanding** if it satisfies the following three conditions:

(i) ∂D_n is of content zero for each $n \in \mathbb{N}$,
(ii) $D_n \subseteq D_{n+1}$ for each $n \in \mathbb{N}$, and
(iii) $\text{Area}(D_n) \to \text{Area}(D)$ as $n \to \infty$.

Observe that the sequences (D_n) and (E_n) of subsets of $[0,1] \times [0,1] \setminus \{(0,0)\}$ considered in Example 7.80 are expanding. In contrast to exhausting sequences of subsets of an unbounded set, every bounded subset D of $\mathbb{R}^2$ whose boundary is of content zero admits an expanding sequence; indeed, we can simply consider the sequence (D_n) given by $D_n := D$ for all $n \in \mathbb{N}$.

Let $f : D \to \mathbb{R}$ be an unbounded continuous function that is bounded (and hence integrable) on each term of some expanding sequence of subsets of D. We say that the improper double integral $\iint_D f(s,t)d(s,t)$ is **unconditionally convergent** if there is $I \in \mathbb{R}$ such that for every expanding sequence (D_n) of subsets of D with the property that f is bounded on each D_n, the limit

$$\lim_{n\to\infty} \iint_{D_n} f(s,t)d(s,t)$$

exists and is equal to I. The hypothesis that f is bounded on each term of an expanding sequence of subsets of D ensures that the real number I, when it exists, is unique. In this case, we write $\iint_D f(s,t)d(s,t) = I$, or simply $\iint_D f = I$.

It is easily seen that if the improper double integrals $\iint_D f$ and $\iint_D g$ are unconditionally convergent, then so are $\iint_D (f+g)$ and $\iint_D (rf)$ for any $r \in \mathbb{R}$.

We give below a necessary and sufficient condition for the unconditional convergence of an improper double integral of a nonnegative unbounded continuous function on a bounded subset of $\mathbb{R}^2$. Its proof shows the importance of conditions (ii) and (iii) in the definition of an expanding sequence.

Proposition 7.81. *Let D be a bounded subset of $\mathbb{R}^2$ such that ∂D is of content zero and let $f : D \to \mathbb{R}$ be a nonnegative unbounded continuous function. Suppose D admits an expanding sequence (D_n) of subsets of D such that f is bounded on D_n for each $n \in \mathbb{N}$. Then $\iint_D f$ is unconditionally convergent if and only if there is $\alpha \in \mathbb{R}$ such that $\iint_{D_n} f \le \alpha$ for all $n \in \mathbb{N}$, and in this case, $\iint_D f = \lim_{n\to\infty} \iint_{D_n} f$.*

Proof. Suppose $\iint_D f$ is unconditionally convergent. Then $\lim_{n\to\infty} \iint_{D_n} f$ exists. Since a convergent sequence of real numbers is bounded, there is $\alpha \in \mathbb{R}$ such that $\iint_{D_n} f \le \alpha$ for all $n \in \mathbb{N}$.

Conversely, suppose there is $\alpha \in \mathbb{R}$ such that $\iint_{D_n} f \le \alpha$ for all $n \in \mathbb{N}$. For $n \in \mathbb{N}$, let $I_n := \iint_{D_n} f$. Since f is nonnegative, (I_n) is a monotonically increasing sequence of real numbers. Also, it is bounded above by α. Hence (I_n) converges to $I := \sup\{I_n : n \in \mathbb{N}\}$. Next, let (E_n) be any other expanding sequence of subsets of D such that f is bounded on E_n for each $n \in \mathbb{N}$, and let $J_n := \iint_{E_n} f$ for $n \in \mathbb{N}$. Then (J_n) is also a monotonically increasing sequence of real numbers. Now fix $m \in \mathbb{N}$ and let $\beta_m > 0$ be such that $f(s,t) \le \beta_m$ for all $(s,t) \in E_m$. Let $\epsilon > 0$ be given. Since $\text{Area}(D_n) \to \text{Area}(D)$ as $n \to \infty$, there is $n_0 \in \mathbb{N}$ such that $\text{Area}(D) - \text{Area}(D_{n_0}) < \epsilon/\beta_m$. By Corollary 5.38, $\partial(D \setminus D_{n_0})$, $\partial(E_m \setminus D_{n_0})$ and $\partial(E_m \cap D_{n_0})$ are of content zero. Also, Corollary 5.50 shows that f is integrable on $\partial(E_m \setminus D_{n_0})$ and on $\partial(E_m \cap D_{n_0})$. Hence by Domain Additivity (Corollary 5.52), we obtain

$$\iint_{E_m} f = \iint_{E_m \cap D_{n_0}} f + \iint_{E_m \setminus D_{n_0}} f.$$

But since $\text{Area}(D \setminus D_{n_0}) = \text{Area}(D) - \text{Area}(D_{n_0}) < \epsilon/\beta_m$, the Basic Inequality (Corollary 5.49) shows that

$$\iint_{E_m \setminus D_{n_0}} f \leq \beta_m \text{Area}(E_m \setminus D_{n_0}) \leq \beta_m \text{Area}(D \setminus D_{n_0}) < \beta_m \frac{\epsilon}{\beta_m} = \epsilon.$$

Hence
$$\iint_{E_m} f < \iint_{E_m \cap D_{n_0}} f + \epsilon \leq \iint_{D_{n_0}} f + \epsilon \leq I + \epsilon.$$

Since the inequality $\iint_{E_m} f < I + \epsilon$ holds for every $\epsilon > 0$, it follows that $J_m = \iint_{E_m} f \leq I$. This shows that the sequence (J_n) is bounded and $J := \sup\{J_n : n \in \mathbb{N}\} \leq I$. Interchanging the roles of (D_n) and (E_n), we see that $I \leq J$. Hence the sequence (J_n) also converges to I. This proves that $\iint_D f$ is unconditionally convergent. □

Example 7.80 shows that the nonnegativity of the function f cannot be dropped from the above proposition. To obtain an analogue of the characterization in Proposition 7.81 for functions that may change sign, we proceed exactly as we did in the case of functions defined on unbounded subsets of $\mathbb{R}^2$.

Lemma 7.82. *Let D be a bounded subset of $\mathbb{R}^2$ such that ∂D is of content zero and let $f : D \to \mathbb{R}$ be an unbounded continuous function. Suppose D admits an expanding sequence (D_n) of subsets of D such that f is bounded on D_n for each $n \in \mathbb{N}$. Let $\iint_D f$ be unconditionally convergent. Then there is $\alpha \in \mathbb{R}$ such that*

$$\iint_S |f| \leq \alpha$$

for every subset S of D such that ∂S is of content zero and f is bounded on S.

Proof. Similar to the proof of Lemma 7.74. □

Proposition 7.83. *Let D be a bounded subset of $\mathbb{R}^2$ such that ∂D is of content zero and let $f : D \to \mathbb{R}$ be an unbounded continuous function. Suppose D admits an expanding sequence (D_n) of subsets of D such that f is bounded on D_n for each $n \in \mathbb{N}$. Then $\iint_D f$ is unconditionally convergent if and only if there is $\alpha \in \mathbb{R}$ such that $\iint_{D_n} |f| \leq \alpha$ for all $n \in \mathbb{N}$, and in this case, $\iint_D f = \lim_{n\to\infty} \iint_{D_n} f$. Equivalently, $\iint_D f$ is unconditionally convergent if and only if $\iint_D |f|$ is unconditionally convergent.*

Proof. Similar to the proof of Proposition 7.75. □

Corollary 7.84 (Comparison Test for Improper Double Integrals of Second Kind). *Let D be a bounded subset of $\mathbb{R}^2$ such that ∂D is of content zero and let $f, g : D \to \mathbb{R}$ be unbounded continuous functions that satisfy $|f| \le g$. Suppose D admits an expanding sequence (D_n) of subsets of D such that both f and g are bounded on D_n for each $n \in \mathbb{N}$. If $\iint_D g$ is unconditionally convergent, then so is $\iint_D f$.*

Proof. Follows from Propositions 7.81 and 7.83. □

Examples 7.85. Let $D := \{(s,t) \in \mathbb{R}^2 : 0 < s^2 + t^2 \le 1\}$ and for $n \in \mathbb{N}$, let $D_n := \{(s,t) \in D : (1/n^2) \le s^2 + t^2 \le 1\}$. Then D is a bounded subset of $\mathbb{R}^2$, ∂D is of content zero, and (D_n) is an expanding sequence of subsets of D.

(i) Let $p \in \mathbb{R}$ with $0 < p < 1$. Consider $f : D \to \mathbb{R}$ defined by $f(s,t) := 1/(s^2+t^2)^p$. Then f is a nonnegative unbounded continuous function on D, and f is bounded on D_n for each $n \in \mathbb{N}$. Using polar coordinates and Proposition 5.65, we see that

$$\iint_{D_n} f = \iint_{[1/n,1]\times[-\pi,\pi]} r^{-2p} r \, d(r,\theta)$$
$$= 2\pi \int_{1/n}^{1} r^{1-2p} dr = \frac{\pi}{1-p}\left(1 - \frac{1}{n^{2-2p}}\right).$$

Hence $\lim_{n\to\infty} \iint_{D_n} f = \pi/(1-p)$. Since f is nonnegative, Proposition 7.81 shows that $\iint_D f$ is unconditionally convergent and is equal to $\pi/(1-p)$.

(ii) Define $g : D \to \mathbb{R}$ by $g(s,t) := -\ln(s^2+t^2)$. Then g is a nonnegative unbounded continuous function on D, and it is bounded on D_n for each $n \in \mathbb{N}$. Using polar coordinates and Proposition 5.65, we see that

$$\iint_{D_n} g = -2 \iint_{[1/n,1]\times[-\pi,\pi]} (\ln r) r \, d(r,\theta)$$
$$= -4\pi \int_{1/n}^{1} r(\ln r) dr$$
$$= \pi \left[r^2 - 2r^2(\ln r)\right]_{r=1/n}^{r=1} = \pi\left(1 - \frac{1}{n^2} - \frac{2\ln n}{n^2}\right).$$

Hence $\lim_{n\to\infty} \iint_{D_n} g = \pi$. Since f is nonnegative, Proposition 7.81 shows that $\iint_D g$ is unconditionally convergent and is equal to π. ◇

Before we conclude this chapter, we remark that "triple integrals" of functions defined on an unbounded subset of $\mathbb{R}^3$ and of unbounded functions defined on a bounded subset of $\mathbb{R}^3$ can be treated on the lines of "double integrals" treated in this section. For typical examples of such "triple integrals," see Exercise 71.

Notes and Comments

The topics of double series and improper double integrals are treated rather cursorily in most books on calculus. This is partly because these topics do not appear in applications as frequently as the corresponding one-variable topics of series and improper integrals, and partly because there is no universally accepted way of treating these two-variable topics. We have, however, given an extensive treatment of these topics. In this attempt, we have been guided by the analogy between one-variable topics and the corresponding two-variable topics on the one hand, and also the analogy between the discrete case (such as sequences, series, and double series) and the continuous case (such as functions defined on an interval, improper integrals, and improper double integrals) on the other hand.

If the limit of a double sequence $(a_{m,n})$ *is equal to* a*, the real numbers* $(a_{m,n})$ *come close to the real number* a *when both* m *and* n *are independently large. As a result, an interesting phenomenon occurs: a convergent double sequence need not be bounded, in contrast to the one-variable fact that a convergent sequence is always bounded. We have used the product order, that is, the componentwise partial order on* $\mathbb{N}^2$ *to introduce monotonic and bimonotonic double sequences of real numbers. These come in handy in discussing the convergence of a double series of nonnegative terms and in describing Dirichlet's test for conditional convergence of a double series.*

There are several ways of defining the convergence of a double series $\sum\sum_{(k,\ell)} a_{k,\ell}$ *depending on which finite sums of the terms* $a_{k,\ell}$ *are considered. We have chosen to define the convergence of a double series in terms of the convergence of the double sequence* $(A_{m,n})$*, where* $A_{m,n}$ *is the sum of* all *the terms* $a_{k,\ell}$ *with* $1 \leq k \leq m$ *and* $1 \leq \ell \leq n$*. This definition was first given by Pringsheim [42] in 1897 and has been adopted in the books of Hobson [32], Bromwich [7], and Buck [8]. For a double series of nonnegative terms, various ways of summing ("by squares," "by diagonals," etc.) coincide. There is yet another approach to the convergence of* $\sum\sum_{(k,\ell)} a_{k,\ell}$ *based on the concept of "unordered sums" given, for example, in the books of Protter and Morrey [44] and Ruiz [49]. We have given a version of this approach at the end of Section 7.2 under the nomenclature "unconditional convergence" and shown its equivalence to absolute convergence.*

In analogy with a telescoping series, we have considered the concept of a telescoping double series and shown how to find its double sum. This chapter contains analogues of the limit comparison test, the root test, and the ratio test for a double series. The analogue of the root test given here is sometimes attributed to Pringsheim [42] and Daniell [13]. For the analogues of the limit comparison test and the ratio test given in the text as well as for the analogues of Cauchy's condensation test, Abel's kth term test, the ratio comparison test, and Raabe's test for double series given in the exercises, we refer to [36]. This paper also contains results about the convergence of the Cauchy product of two double series introduced in Exercise 51. Another version of the ratio test and several other tests for the absolute convergence of a double series can be found in [5]. An analogue of the partial summation formula leads to a test for the conditional convergence of a double series that is analogous to Dirichlet's test for the conditional convergence of a single series. This result is essentially in the paper [27] of Hardy; it can also be found on page 97 of the book [7] of Bromwich. A variant of this test that is analogous to Abel's test, and a generalization of this test that is analogous to Dedekind's test for the conditional convergence of a single series are given in the exercises. Dedekind's test was considered by Hardy in [28]. A useful reference for similar considerations is [26].

We have treated double power series as special cases of double series. It is interesting to note that a double power series may have many biradii of convergence, in contrast to the uniqueness of the radius of convergence of a (single) power series. Also, the domain of convergence of a double power series can have a variety of shapes, and it need not even be a convex subset of $\mathbb{R}^2$. *A result of Fabry on the log-convexity of the domain of convergence is given as an exercise. We discuss the convergence of the so-called Taylor double series of a function having continuous partial derivatives of all orders on a square neighborhood of a point. We make a distinction between this double series and the corresponding diagonal series, called the Taylor series, whose* n*th partial sum is the* n*th bivariate Taylor polynomial of the function.*

We define the improper double integral of a real-valued function f *on a subset of* $\mathbb{R}^2$ *of the form* $[a, \infty) \times [c, \infty)$ *in analogy with the definition of a double series of real numbers, and then develop the concepts of absolute and conditional convergence. Several tests for absolute convergence and conditional convergence are discussed. It may be worthwhile to note that the Integral Test establishes a strong connection between the convergence of a double series of nonnegative terms and that of an improper double integral of a nonnegative function. Also, Dirichlet's Test for improper double integrals of bimonotonic functions is useful in establishing conditional convergence.*

In the last section of this chapter, we consider improper double integrals of continuous functions defined on an unbounded subset D *of* $\mathbb{R}^2$ *that are bounded on each bounded subset of* D*, and of unbounded continuous functions defined on a bounded subset of* $\mathbb{R}^2$. *The relevant notion here is that of unconditional convergence. Our development of this topic is partly based on the treatment given in Section 4.5 of Buck [8] and Section 4.7 of Courant and John [12, vol. II]. In particular, we prove a result stated by Buck [8, p. 223] that the unconditional convergence of the improper double integral of a continuous function and that of its absolute value are equivalent. This brings out the distinction between the notions of conditional and unconditional convergence of improper integrals of functions on subsets of* $\mathbb{R}^2$ *of the form* $[a, \infty) \times [c, \infty)$. *In fact, for improper integrals of such functions, unconditional convergence turns out to be equivalent to absolute convergence.*

Exercises

Part A

1. Show that each of the the following double sequences converges to 1.
(i) $\big((k+\ell)^{1/(k+\ell)}\big)$, (ii) $\big((k\ell)^{1/k\ell}\big)$, (iii) $\big((k+\ell)^{1/k\ell}\big)$, (iv) $\big((k\ell)^{1/(k+\ell)}\big)$.
2. Let (b_m) and (c_n) be sequences in $\mathbb{R}$. Define a double sequence $(a_{m,n})$ by $a_{m,n} := b_m + c_n$ for $(m,n) \in \mathbb{N}^2$. Show that $(a_{m,n})$ is convergent if and only if both (b_m) and (c_n) are convergent. (Hint: Cauchy Criterion)
3. Let $b, c, \beta, \gamma \in \mathbb{R}$ and define $a_{m,n} := \beta\, b^m + \gamma\, c^n$ for (m,n) in $\mathbb{N}^2$. Show that $(a_{m,n})$ is convergent if and only if one of the following conditions holds: (i) $\beta = 0 = \gamma$, (ii) $\beta = 0$ and $c \in (-1,1]$, (iii) $\gamma = 0$ and $b \in (-1,1]$, (iv) $b \in (-1,1]$ and $c \in (-1,1]$. (Hint: Exercise 2)
4. Let (b_m) and (c_n) be sequences in $\mathbb{R}$. Define a double sequence $(a_{m,n})$ by $a_{m,n} := b_m c_n$ for $(m,n) \in \mathbb{N}^2$. Show that $(a_{m,n})$ is convergent if and only if one of the following conditions holds: (i) Both (b_m) and (c_n) are convergent. (ii) One of (b_m) and (c_n) converges to zero and the other is bounded. (iii) All but finitely many terms of either (b_m) or (c_n) are equal to zero. (Hint: In case $(a_{k,\ell})$ is convergent, use the Cauchy Criterion to prove that if (b_m) has infinitely many nonzero terms, then (c_n) is bounded, and if (b_m) does not converge to zero, then (c_n) is convergent.)
5. Let $b, c \in \mathbb{R}$ and define $a_{m,n} := b^m c^n$ for (m,n) in $\mathbb{N}^2$. Show that $(a_{m,n})$ is convergent if and only if one of the following conditions holds: (i) $b = 0$, (ii) $c = 0$, (iii) $|b| < 1$ and $|c| \le 1$, (iv) $|b| \le 1$ and $|c| < 1$, (v) $b \in (-1,1]$ and $c \in (-1,1]$. (Hint: Exercise 4)
6. Let (b_k) and (c_ℓ) be sequences in $\mathbb{R}$. Show that $\sum\sum_{(k,\ell)}(b_k + c_\ell)$ is convergent if and only if there are $a \in \mathbb{R}$ and $n_0 \in \mathbb{N}$ such that $\sum_{k=1}^{n_0} b_k = n_0 a$ and $b_k = a$ for all $k > n_0$, and moreover, $\sum_{\ell=1}^{n_0} c_\ell = -n_0 a$ and $c_\ell = -a$ for all $\ell > n_0$.
7. Let (a_k) be a sequence of real numbers. Show that the double series $\sum\sum_{(k,\ell)} a_k a_\ell$ is convergent if and only if the series $\sum_k a_k$ is convergent. (Hint: Exercise 4.)
8. Let (b_k) and (c_ℓ) be sequences in $\mathbb{R}$. Show that $\sum\sum_{(k,\ell)} b_k c_\ell$ is convergent if and only if one of the following conditions holds: (i) Both the series $\sum_k b_k$ and $\sum_\ell c_\ell$ are convergent. (ii) One of the series $\sum_k b_k$ and $\sum_\ell c_\ell$ converges to zero and the sequence of partial sums of the other series is bounded. (iii) All but finitely many partial sums of either the series $\sum_k b_k$ or the series $\sum_\ell c_\ell$ are equal to zero. (Hint: Exercise 4)
9. Let $p \in \mathbb{R}$ and $a_{m,n} := \big(\ln(m+n)^p\big)$ for $(m,n) \in \mathbb{N}^2$. Show that the double sequence $(a_{m,n})$ is monotonically decreasing and bimonotonically increasing if $p \le 0$, and monotonically increasing and bimonotonically decreasing if $0 \le p \le 1 + \ln 2$. Also, show that $(a_{m,n})$ is bimonotonically decreasing if $p = 2$, but it is not bimonotonic if $p = 3$. (Compare Exercise 17 of Chapter 1. Hint: Example 7.7 (ii))

10. **(Abel's (k,ℓ)th Term Test)** Let $(a_{k,\ell})$ be a monotonically decreasing double sequence of nonnegative real numbers. If $\sum\sum_{(k,\ell)} a_{k,\ell}$ is convergent, then show that $k\ell\, a_{k,\ell} \to 0$ as $(k,\ell) \to (\infty,\infty)$. Further, show that the converse of this result does not hold. (Hint: Use Exercise 17. For the converse, consider $a_{k,\ell} := 1/k\ell\,(\ln k)(\ln \ell)$ for $(k,\ell) \in \mathbb{N}^2$.)
11. If p and q are positive real numbers such that $(1/p)+(1/q) \geq 1$, then show that the double series $\sum\sum_{(k,\ell)} 1/(k^p + \ell^q)$ is divergent. (Hint: Exercise 10 with $\ell := [k^{p/q}]$ for $k \in \mathbb{N}$)
12. Let $\sum\sum_{(k,\ell)} a_{k,\ell}$ be a double series whose terms are schematically given by

$$\begin{array}{ccccc} 1 & 2 & 4 & 8 & \cdots \\ -\frac{1}{2} & -1 & -2 & -4 & \cdots \\ -\frac{1}{4} & -\frac{1}{2} & -1 & -2 & \cdots \\ -\frac{1}{8} & -\frac{1}{4} & -\frac{1}{2} & -1 & \cdots \\ \vdots & \vdots & \vdots & \vdots & \end{array}$$

and let $A_{m,n}$ denotes its (m,n)th partial double sum. Show that each row-series is divergent, but each column-series converges to 0. Also, show that $A_{m,m} \to 2$ as $m \to \infty$. Is $\sum\sum_{(k,\ell)} a_{k,\ell}$ convergent?
13. For $(k,\ell) \in \mathbb{N}^2$, let $a_{k,\ell} := 1$ if $\ell = k$ and $a_{k,\ell} := (1-2^k)/2^k$ if $\ell = k+1$, while $a_{k,\ell} := 0$ if $\ell \neq k$ and $\ell \neq k+1$. Show that $\sum_{\ell=1}^{\infty} a_{k,\ell} = 1/2^k$ for each fixed $k \in \mathbb{N}$ and $\sum_{k=1}^{\infty} a_{k,\ell} = 1/2^{\ell-1}$ for each fixed $\ell \in \mathbb{N}$. Deduce that $\sum_k \left(\sum_\ell a_{k,\ell}\right) = 1$, whereas $\sum_\ell \left(\sum_k a_{k,\ell}\right) = 2$. Is $\sum\sum_{(k,\ell)} a_{k,\ell}$ convergent?
14. Let (a_k) and (b_ℓ) be monotonically decreasing sequences in $\mathbb{R}$ such that $a_k \to 0$ and $b_\ell \to 0$. Show that $\sum\sum_{(k,\ell)}(-1)^{k+\ell} a_k b_\ell$ is convergent and the double sum is equal to the sum of each of the two iterated series.
15. Let (a_j) be a sequence of nonnegative real numbers, and let $r, s \in [0,\infty)$. Show that the double series $\sum\sum_{(k,\ell)} a_{k+\ell}\, r^k s^\ell / k!\,\ell!$ is convergent if and only if the series $\sum_{j=0}^{\infty} a_j (r+s)^j/j!$ is convergent. (Hint: Proposition 7.16)
16. Let p and q be real numbers. Test the double series $\sum\sum_{(k,\ell)} a_{k,\ell}$ for convergence if the (k,ℓ)th term $a_{k,\ell}$ is equal to
(i) $\dfrac{\ln(k+\ell)}{(k+\ell)^p}$, (ii) $\dfrac{(\ln k\ell)}{k^p\ell^q}$, (iii) $\dfrac{1}{(\ln k\ell)^p}$ for $(k,\ell) \neq (1,1)$.
17. **(Cauchy's Condensation Test)** Let $(a_{k,\ell})$ be a monotonically decreasing double sequence of nonnegative real numbers. Show that the double series $\sum\sum_{(k,\ell)\geq(1,1)} a_{k,\ell}$ is convergent if and only if the double series $\sum\sum_{(k,\ell)\geq(0,0)} 2^{k+\ell} a_{2^k,2^\ell}$ is convergent. Deduce that for $p \in \mathbb{R}$, the double series $\sum\sum_{(k,\ell)\geq(1,1)} 1/(k+\ell)^p$ is convergent if and only if $p > 2$. (Hint: Proposition 7.14 and the A.M.-G.M. inequality)

18. **(Ratio Comparison Test for Double Series)** Let $(a_{k,\ell})$ and $(b_{k,\ell})$ be double sequences with $b_{k,\ell} > 0$ for all $(k,\ell) \in \mathbb{N}^2$. Prove the following:
    (i) Suppose $|a_{k+1,\ell}|b_{k,\ell} \le |a_{k,\ell}|b_{k+1,\ell}$ and $|a_{k,\ell+1}|b_{k,\ell} \le |a_{k,\ell}|b_{k,\ell+1}$ whenever both k and ℓ are large. If $\sum\sum_{(k,\ell)} b_{k,\ell}$ is convergent, and each row-series and each column-series corresponding to $\sum\sum_{(k,\ell)} |a_{k,\ell}|$ is convergent, then $\sum\sum_{(k,\ell)} |a_{k,\ell}|$ is convergent.
    (ii) Suppose $|a_{k+1,\ell}|b_{k,\ell} \ge |a_{k,\ell}|b_{k+1,\ell} > 0$ whenever k is large and $\ell \in \mathbb{N}$, and $|a_{k,\ell+1}|b_{k,\ell} \ge |a_{k,\ell}|b_{k,\ell+1} > 0$ whenever ℓ is large and $k \in \mathbb{N}$. If $\sum\sum_{(k,\ell)} b_{k,\ell}$ is divergent, then $\sum\sum_{(k,\ell)} |a_{k,\ell}|$ is divergent.
19. Let $(a_{k,\ell})$ be a double sequence of nonnegative real numbers.
    (i) If there is $p \in \mathbb{R}$ with $p > 1$ such that

$$a_{k,\ell+1} \le \left(1 - \frac{p}{\ell}\right) a_{k,\ell} \quad \text{and} \quad a_{k+1,\ell} \le \left(1 - \frac{p}{k}\right) a_{k,\ell},$$

    whenever both k and ℓ are large, and further, if each row-series and each column-series corresponding to $\sum\sum_{(k,\ell)} a_{k,\ell}$ are convergent, then show that $\sum\sum_{(k,\ell)} a_{k,\ell}$ is convergent.
    (ii) If there is $k \in \mathbb{N}$ such that

$$a_{k,\ell+1} \ge \left(1 - \frac{1}{\ell}\right) a_{k,\ell} > 0 \quad \text{for all large } \ell \in \mathbb{N},$$

    or if there is $\ell \in \mathbb{N}$ such that

$$a_{k+1,\ell} \ge \left(1 - \frac{1}{k}\right) a_{k,\ell} > 0 \quad \text{for all large } k \in \mathbb{N},$$

    then show that $\sum\sum_{(k,\ell)} a_{k,\ell}$ is divergent.

    (Hint: If $p > 1$ and $x \in [0,1]$, then $1 - px \le (1-x)^p$. Use Exercise 18 with $b_{k,\ell} := 1/(k\ell)^p$ and Raabe's test for single series stated, for example, in Exercise 13 of Chapter 9 of ACICARA.)
20. (i) If $a_{1,1} := 1$, $a_{k+1,1} := (2k-1)a_{k,1}/(2k+2)$ for $k \in \mathbb{N}$, and $a_{k,\ell+1} := (2\ell-1)a_{k,\ell}/(2\ell+2)$ for all $(k,\ell) \in \mathbb{N}^2$, then show that $\sum\sum_{(k,\ell)} a_{k,\ell}$ converges.
    (ii) If $a_{1,1} := 1$, $a_{k+1,1} := ka_{k,1}/(k+1)$ for $k \in \mathbb{N}$, and $a_{k,\ell+1} := \ell\, a_{k,\ell}/(\ell+1)$ for all $(k,\ell) \in \mathbb{N}^2$, then show that $\sum\sum_{(k,\ell)} a_{k,\ell}$ diverges.

    (Hint: Exercise 19)
21. **(Raabe's Test for Double Series)** Let $(a_{k,\ell})$ be a double sequence of nonzero real numbers. Use Exercise 19 to prove the following:
    (i) Suppose each row-series and each column-series corresponding to $\sum\sum_{(k,\ell)} a_{k,\ell}$ are absolutely convergent. If $\ell(1 - |a_{k,\ell+1}|/|a_{k,\ell}|) \to a$ and $k(1 - |a_{k+1,\ell}|/|a_{k,\ell}|) \to \tilde{a}$ as $(k,\ell) \to (\infty,\infty)$, where $a > 1$ and $\tilde{a} > 1$, then $\sum\sum_{(k,\ell)} a_{k,\ell}$ is absolutely convergent.
    (ii) If for some $k \in \mathbb{N}$, $\lim_{\ell\to\infty} \ell(1 - |a_{k,\ell+1}|/|a_{k,\ell}|)$ exists and is less than 1, or if for some $\ell \in \mathbb{N}$, $\lim_{k\to\infty} k(1 - |a_{k+1,\ell}|/|a_{k,\ell}|)$ exists and is less than 1, then $\sum\sum_{(k,\ell)} |a_{k,\ell}|$ is divergent.

22. Let $a_{k,\ell} := 1/\max\{k,\ell\}$ and $b_{k,\ell} := 1/\min\{k,\ell\}$ for all $(k,\ell) \in \mathbb{N}^2$. Show that $(a_{k,\ell})$ and $(b_{k,\ell})$ are bimonotonic double sequences such that $a_{k,\ell} \to 0$ and $b_{k,\ell} \to 0$ as $(k,\ell) \to (\infty,\infty)$. Further, show that the double series $\sum\sum_{(k,\ell)}(-1)^{k+\ell}a_{k,\ell}$ is convergent, but the double series $\sum\sum_{(k,\ell)}(-1)^{k+\ell}b_{k,\ell}$ is divergent. (Hint: Use Corollary 7.39 and the fact that if $\sum\sum_{(k,\ell)} c_{k,\ell}$ is convergent, then $c_{m,1}+\cdots+c_{m,m} \to 0$ as $m \to \infty$.)
23. Let $(a_{k,\ell})$ be a double sequence satisfying conditions (i), (ii), and (iii) of Dirichlet's Test for Double Series (Proposition 7.38), and let θ and φ be real numbers. Consider the two double series

$$\sum\sum_{(k,\ell)} a_{k,\ell}\sin(k\theta+\ell\varphi) \quad \text{and} \quad \sum\sum_{(k,\ell)} a_{k,\ell}\cos(k\theta+\ell\varphi).$$

Assume that θ is an integral multiple of 2π.
   (i) Suppose $\varphi = \pi$. Show that the first double series converges absolutely. Also, show that the second double series converges absolutely if $a_{k,\ell} := 1/k^2\ell^2$, it converges conditionally if $a_{k,\ell} := 1/k^2\ell$, and it diverges if $a_{k,\ell} := 1/k\ell$ for $(k,\ell) \in \mathbb{N}^2$.
   (ii) Suppose $\varphi = \pi/2$. Show that both the double series converge absolutely if $a_{k,\ell} := 1/k^2\ell^2$, they converge conditionally if $a_{k,\ell} := 1/k^2\ell$, and they diverge if $a_{k,\ell} := 1/k\ell$ for $(k,\ell) \in \mathbb{N}^2$.

(Hint: Dirichlet's test for single/double series.)
24. Let p be a positive real number, and let θ and φ be real numbers neither of which is an integral multiple of 2π. Show that the double series

$$\sum\sum_{(k,\ell)} \frac{\sin(k\theta+\ell\varphi)}{[\ln(k+\ell)]^p} \quad \text{and} \quad \sum\sum_{(k,\ell)} \frac{\cos(k\theta+\ell\varphi)}{[\ln(k+\ell)]^p}$$

are convergent. Deduce that the double series

$$\sum\sum_{(k,\ell)} \frac{(-1)^{k+\ell}}{[\ln(k+\ell)]^p}$$

is convergent. (Hint: Corollary 7.40 and Exercise 9)
25. Let $\sum\sum_{(k,\ell)} a_{k,\ell}$ be a convergent double series of nonnegative terms. Show that $a_{k,\ell} \to 0$ as $k+\ell \to \infty$, that is, for every $\epsilon > 0$, there is $n_0 \in \mathbb{N}$ such that $a_{k,\ell} < \epsilon$ for all $(k,\ell) \in \mathbb{N}^2$ with $k+\ell \geq n_0$.
26. In each of the following, determine all $(x,y) \in \mathbb{R}^2$ for which (a) the double power series $\sum\sum_{(k,\ell)} c_{k,\ell}x^ky^\ell$ is absolutely convergent, (b) the set $C_{x,y} := \{c_{k,\ell}x^ky^\ell : (k,\ell) \geq (0,0)\}$ is bounded, and (c) the double power series $\sum\sum_{(k,\ell)} c_{k,\ell}x^ky^\ell$ is convergent.
   (i) $c_{0,\ell} := 0$ for all $\ell \geq 0$ and $c_{k,\ell} := 1$ for all $(k,\ell) \geq (1,0)$,
   (ii) $c_{k,0} = c_{0,\ell} := 0$ for all $(k,\ell) \geq (0,0)$ and $c_{k,\ell} := 1$ for all $(k,\ell) \geq (1,1)$,
   (iii) $c_{k,\ell} := 0$ if $0 \leq k < \ell$, while $c_{k,\ell} := 1$ if $k \geq \ell \geq 0$.

(iv) $c_{0,0} := 1$, $c_{k,0} = c_{0,\ell} := 1$ for all $(k,\ell) \geq (1,1)$, $c_{1,1} := -1$, $c_{k,1} = c_{1,\ell} := -1/2$ for all $(k,\ell) \geq (2,2)$, and $c_{k,\ell} := 0$ for all $(k,\ell) \geq (2,2)$,
(v) $c_{k,0} := 1$ for all $k \geq 0$, $c_{k,1} := -1/2$ for all $k \geq 0$, and $c_{k,\ell} := 0$ for all $(k,\ell) \geq (0,2)$.

27. Let (r_1, s_1) and (r_2, s_2) be biradii of convergence of a double power series. Show that if $r_1 < r_2$, then $s_1 \geq s_2$, and if $s_1 < s_2$, then $r_1 \geq r_2$. Further, show that if $s_1 = 0$, then $(r_2, 0)$ is also a biradius of convergence for every $r_2 > r_1$, and if $r_1 = 0$, then $(0, s_2)$ is also a biradius of convergence for every $s_2 > s_1$.
28. Find the domain of convergence and all biradii of convergence of the double power series $\sum\sum_{(k,\ell)} c_{k,\ell} x^k y^\ell$, where
(i) $c_{k,\ell} := 1$ if $\ell = 1$ and $c_{k,\ell} := 0$ if $\ell \neq 1$, (ii) $c_{k,\ell} := k^k$, (iii) $c_{k,\ell} := \ell^\ell$, (iv) $c_{k,\ell} := 1/k!$, (v) $c_{k,\ell} := 1/\ell!$, (vi) $c_{k,\ell} := k^k/\ell!$, (vii) $c_{k,\ell} := \ell^\ell/k!$.
29. Let $c_{0,0} := 1$, $c_{k,0} = c_{0,\ell} := 1$ for all $(k,\ell) \geq (1,1)$, and $c_{k,\ell} := k^k \ell^\ell$ for all $(k,\ell) \geq (1,1)$. Show that the double power series $\sum\sum_{(k,\ell)} c_{k,\ell} x^k y^\ell$ is convergent if and only if $x = 0$ and $|y| < 1$, or $y = 0$ and $|x| < 1$. Find all biradii of convergence of this double power series.
30. Let $f, g : \mathbb{R}^2 \to \mathbb{R}$ be defined by

$$f(x,y) := \begin{cases} e^{-1/x^2} + e^{-1/y^2} & \text{if } x \neq 0 \text{ and } y \neq 0, \\ e^{-1/x^2} & \text{if } x \neq 0 \text{ and } y = 0, \\ e^{-1/y^2} & \text{if } x = 0 \text{ and } y \neq 0, \\ 0 & \text{if } x = 0 = y, \end{cases}$$

and

$$g(x,y) := \begin{cases} e^{-1/x^2} e^{-1/y^2} & \text{if } x \neq 0 \text{ and } y \neq 0, \\ 0 & \text{if } x = 0 \text{ or } y = 0. \end{cases}$$

Find the Taylor double series and the Taylor series of f as well as of g around $(0,0)$. Find all $(x,y) \in \mathbb{R}^2$ at which they converge to the corresponding functional values.
31. Let I and J be nonempty open intervals in $\mathbb{R}$, and let $\phi : I \to \mathbb{R}$ and $\psi : J \to \mathbb{R}$ be infinitely differentiable functions of one variable. Consider $f, g : I \times J \to \mathbb{R}$ defined by $f(x,y) := \phi(x) + \psi(y)$ and $g(x,y) := \phi(x)\psi(y)$. Let $x_0 \in I$ and $y_0 \in J$. Find the Taylor double series and the Taylor series of f as well as of g around (x_0, y_0) in terms of the coefficients of the Taylor series of ϕ around x_0 and of ψ around y_0. Also, determine whether the Taylor double series and the Taylor series of f as well as of g converge absolutely, and whether they converge to the corresponding functional values. (Hint: Example 3.50 (i))
32. Let $D \subseteq \mathbb{R}^2$ and let (x_0, y_0) be an interior point of D. Suppose $E \subseteq \mathbb{R}$ is such that $x + y \in E$ for all $(x,y) \in D$. Let $g : E \to \mathbb{R}$ be infinitely differentiable at $x_0 + y_0$. If $f : D \to \mathbb{R}$ is defined by $f(x,y) := g(x+y)$, then show that the Taylor double series of f around (x_0, y_0) is $\sum\sum_{(k,\ell)} g^{(k+\ell)}(x_0 + y_0)(x - x_0)^k (y - y_0)^\ell / k!\, \ell!$ and the Taylor series of

f around (x_0, y_0) is $\sum_{j=0}^{\infty} g^{(j)}(x_0+y_0)(x-x_0+y-y_0)^j/j!$. Further, if r is the radius of convergence of the Taylor series of g around $u_0 := x_0+y_0$, then prove the following statements.

(i) The Taylor double series of f around (x_0, y_0) converges absolutely at all $(x, y) \in \mathbb{R}^2$ with $|x-x_0|+|y-y_0| < r$, while it does not converge absolutely at all $(x, y) \in \mathbb{R}^2$ with $|x-x_0|+|y-y_0| > r$. Also, if $(x, y) \in D$ with $|x-x_0|+|y-y_0| < r$, and further, if the Taylor series of g around u_0 at $u := x+y$ converges to $g(u)$, then the Taylor double series of f around (x_0, y_0) at (x, y) converges to $f(x, y)$.

(ii) The Taylor series of f around (x_0, y_0) converges absolutely at all $(x, y) \in \mathbb{R}^2$ with $|u-u_0| < r$, while it diverges at all $(x, y) \in \mathbb{R}^2$ with $|u-u_0| > r$, where $u := x+y$. Also, if $(x, y) \in D$ with $|u-u_0| < r$, and further, if the Taylor series of g around u_0 at u converges to $g(u)$, then the Taylor series of f around (x_0, y_0) at (x, y) converges to $f(x, y)$.

(Hint: Example 3.17 (iii), Example 3.50 (ii), and Exercise 15)

33. Let $f : \mathbb{R}^2 \to \mathbb{R}$ be defined by $f(x, y) := \cos(x+y)$. Show that the Taylor double series of f around $(0, 0)$ is

$$\sum_{(k,\ell)}\sum c_{k,\ell}x^k y^\ell, \quad \text{where } c_{k,\ell} := \begin{cases} 0 & \text{if } k+\ell \text{ is odd,} \\ (-1)^{(k+\ell+1)/2} & \text{if } k+\ell \text{ is even,} \end{cases}$$

and the Taylor series of f around $(0, 0)$ is

$$\sum_{j=0}^{\infty}(-1)^j \frac{(x+y)^{2j}}{(2j)!}.$$

Further, show that both converge absolutely to $f(x, y)$ for all $(x, y) \in \mathbb{R}^2$. (Hint: Exercise 32)

34. Show that $\iint_{[0,\infty)\times[0,\infty)} d(s,t)/(1+s^2)(1+t^2)$ converges to $\pi^2/4$, while $\iint_{[0,\infty)\times[0,\infty)} d(s,t)/(1+s^2+t^2)$ diverges.

35. Consider $f : [1,\infty)\times[1,\infty) \to \mathbb{R}$ defined by (i), (ii), or (iii) below.
(i) $f(s,t) := (\cos s)(\sin t)/st$, (ii) $f(s,t) := (\sin s)(\sin t)/st$,
(iii) $f(s,t) := (-1)^{k+\ell}/k\ell$ for $s \in [k, k+1)$ and $t \in [\ell, \ell+1)$ with $k, \ell \in \mathbb{N}$.
Show that $\iint_{[1,\infty)\times[1,\infty)} f(s,t)d(s,t)$ is conditionally convergent. Further, show that $\iint_{[2,\infty)\times[2,\infty)} g(s,t)d(s,t)$ is conditionally convergent, where $g : [2,\infty)\times[2,\infty) \to \mathbb{R}$ is defined by $g(s,t) := (\cos s)(\sin t)/(\ln s)(\ln t)$.

36. Let $\theta, \varphi \in \mathbb{R}$ and let $f : [1,\infty)\times[1,\infty) \to \mathbb{R}$ be integrable on $[1,x]\times[1,y]$ for every $(x,y) \geq (1,1)$. Consider the Fourier sine double integral

$$\iint_{[1,\infty)\times[1,\infty)} f(s,t)\sin(s\theta+t\varphi)d(s,t)$$

and the Fourier cosine double integral

$$\iint_{[1,\infty)\times[1,\infty)} f(s,t)\cos(s\theta+t\varphi)d(s,t).$$

Prove the following statements.

(i) If $f(s,t) := 1/s^2t^2$ for $(s,t) \in [1,\infty) \times [1,\infty)$, then both the Fourier double integrals are absolutely convergent.

(ii) If $\theta = 0, \varphi \neq 0$, and if $f(s,t) := 1/s^2t$ for $(s,t) \in [1,\infty) \times [1,\infty)$, then both the Fourier double integrals are conditionally convergent.

(iii) If $\theta \neq 0, \varphi = 0$, and if $f(s,t) := 1/st^2$ for $(s,t) \in [1,\infty) \times [1,\infty)$, then both the Fourier double integrals are conditionally convergent.

(iv) If one of θ and φ is equal to 0 and the other is equal to π, and if $f(s,t) := 1/st$ for $(s,t) \in [1,\infty) \times [1,\infty)$, then the Fourier sine double integrals is divergent.

(v) If one of θ and φ is equal to 0 and the other is equal to $\pi/2$, and if $f(s,t) := 1/st$ for $(s,t) \in [1,\infty) \times [1,\infty)$, then the Fourier cosine double integrals is divergent.

37. Let $D := \{(s,t) \in \mathbb{R}^2 : t \geq 0 \text{ and } |s^4 - t^4| \leq 1\}$. Show that D is unbounded but $A(D)$ is well defined and is at most 4. (Compare Example 7.79 (ii).)

38. Define $f : [0,1] \times [0,1] \setminus \{(0,0)\} \to \mathbb{R}$ by $f(s,t) := (s^2 - t^2)/(s^2 + t^2)^2$. Show that f is integrable on $[0,1] \times [1/n, 1]$ for each $n \in \mathbb{N}$ and

$$\lim_{n\to\infty} \iint_{[0,1]\times[1/n,1]} f = -\frac{\pi}{4}.$$

(Compare Example 7.80.)

39. Let $D := (0,1] \times (0,1]$. Define $f,\ g : D \to \mathbb{R}$ by $f(s,t) := s/\sqrt{t}$ and $g(s,t) := t/\sqrt{s}$. Show that the improper double integrals $\iint_D f$ and $\iint_D g$ are unconditionally convergent, and each is equal to 1.

40. Let $D := (0,1] \times (0,1]$ and $p \in \mathbb{R}$ with $p > 0$. Define $f : D \to \mathbb{R}$ by $f(s,t) := 1/(s+t)^p$. Show that the improper double integral $\iint_D f$ is unconditionally convergent if and only if $p < 2$. Further, show that it is equal to $2 \ln 2$ if $p = 1$, while it is equal to $2(2^{1-p} - 1)/(1-p)(2-p)$ if $0 < |p - 1| < 1$. (Compare Exercise 68.)

Part B

41. Let $(a_{m,n})$ be a double sequence of real numbers and let $\ell \in \mathbb{R}$. Show that $a_{m,n} \to \ell$ as $(m,n) \to (\infty,\infty)$ if and only if $a_{x_n,y_n} \to \ell$ as $n \to \infty$ whenever (x_n) and (y_n) are sequences of real numbers such that $x_n \to \infty$ and $y_n \to \infty$ as $n \to \infty$. (Compare the definition of $f(x,y) \to \ell$ as $(x,y) \to (\infty,\infty)$ given in Section 2.3.)

42. Let $(a_{k,\ell})$ be a double sequence of nonnegative real numbers. For $j \in \mathbb{N}$, let d_j denote the sum of $a_{k,\ell}$, where (k,ℓ) varies over elements of $\mathbb{N}^2$ satisfying $j^2 < k^2 + \ell^2 \leq (j+1)^2$. Show that $\sum\sum_{(k,\ell)} a_{k,\ell}$ is convergent if and only if $\sum_{j=1}^{\infty} d_j$ is convergent, and in this case, $\sum\sum_{(k,\ell)} a_{k,\ell} = \sum_{j=1}^{\infty} d_j$.

43. Let (r_j) be a sequence of positive real numbers. Show that the double series $\sum\sum_{(k,\ell)} r_{k+\ell}/(k+\ell)$ is convergent if and only if the series $\sum_j r_j$ is convergent. (Hint: For $n \in \mathbb{N}$, the nth partial sum of the diagonal series is $\sum_{j=1}^{n} j\, r_{j+1}/(j+1)$. Use part (iii) of Proposition 7.16.)

44. Let $p \in \mathbb{R}$. Show that both

$$\sum_{k\geq 1}\sum_{\ell\geq 1}\frac{1}{(k+\ell)^2[\ln(k+\ell)]^p} \quad\text{and}\quad \iint_{[1,\infty)\times[1,\infty)}\frac{1}{(s+t)^2[\ln(s+t)]^p}d(s,t)$$

diverge to ∞. (Hint: Divergence of $\sum_{j=2}^{\infty} 1/j(\ln j)^p$, as shown in Example 9.40 (ii) of ACICARA, Exercise 43, and Proposition 7.57)

45. Let $p, \alpha, \beta, \gamma \in \mathbb{R}$ with $\alpha > 0$, $\gamma > 0$, and $\sqrt{\alpha\gamma} + \beta > 0$. Show that the double series $\sum\sum_{(k,\ell)} 1/(\alpha k^2 + 2\beta k\ell + \gamma\ell^2)^p$ is convergent if and only if $p > 1$. (Hint: If $M := \max\{\alpha, \beta, \gamma\}$, then $2(\sqrt{\alpha\gamma} + \beta)k\ell \leq \alpha k^2 + 2\beta k\ell + \gamma\ell^2 \leq M(k+\ell)^2$ for all $(k,\ell) \in \mathbb{N}^2$. Use Examples 7.10 (iii) and 7.17 (i).)

46. **(Regularly Convergent Double Sequence)** A double sequence $(a_{m,n})$ is said to be **regularly convergent** if it satisfies the following three conditions: (i) $(a_{m,n})$ is convergent, (ii) for each fixed $m \in \mathbb{N}$, the sequence $(a_{m,n})$ is convergent, and (iii) for each fixed $n \in \mathbb{N}$, the sequence $(a_{m,n})$ is convergent. Let $(a_{m,n})$ be regularly convergent, and let $a := \lim_{(m,n)\to(\infty,\infty)} a_{m,n}$, $a_m := \lim_{n\to\infty} a_{m,n}$ for each fixed $m \in \mathbb{N}$, and $\tilde{a}_n := \lim_{m\to\infty} a_{m,n}$ for each fixed n. Prove the following statements.
    (i) The double sequence $(a_{m,n})$ is bounded, and $\lim_{m\to\infty} a_m = a = \lim_{n\to\infty} \tilde{a}_n$. (Compare Part (iii) of Proposition 7.2.)
    (ii) For every $\epsilon > 0$, there is $j_0 \in \mathbb{N}$ such that if either $m \geq j_0$ or $n \geq j_0$, then $|a_{p,q} - a_{p,n} - a_{m,q} + a_{m,n}| < \epsilon$ for all $p \geq m$ and $q \geq n$.

47. **(Regularly Convergent Double Series)** A double series is said to be **regularly convergent** if the double sequence of its partial double sums is regularly convergent. Prove the following statements.
    (i) A double series is regularly convergent if and only if it is convergent, each row-series is convergent, and each column-series is convergent.
    (ii) If a double series is regularly convergent, then the double sequence of its partial double sums is bounded, both the iterated series are convergent, and their sums are equal to the double sum. (Compare part (iii) of Proposition 7.11.)
    (iii) If a double series $\sum\sum_{(k,\ell)} a_{k,\ell}$ is regularly convergent, then for every $\epsilon > 0$, there is $j_0 \in \mathbb{N}$ such that if either $m > j_0$ or $n > j_0$, then $\left|\sum_{k=m}^{p}\sum_{\ell=n}^{q} a_{k,\ell}\right| < \epsilon$ for all $p \geq m$ and $q \geq n$. (Hint: For all $p \geq m$ and $q \geq n$, we have $\sum_{k=m}^{p}\sum_{\ell=n}^{q} a_{k,\ell} = A_{p,q} - A_{p,n-1} - A_{m-1,q} + A_{m-1,n-1}$. Use Exercise 46 (ii).)
    (iv) An absolutely convergent double series is regularly convergent, but a regularly convergent double series need not be absolutely convergent.
    (v) If the hypotheses of Dirichlet's Test (Proposition 7.38) is satisfied, then the double series $\sum\sum_{(k,\ell)} a_{k,\ell}b_{k,\ell}$ is regularly convergent. (Hint: Proof of Proposition 7.38 and Dirichlet's Test for single series)
    (vi) If a double power series $\sum\sum_{(k,\ell)} c_{k,\ell}x^k y^\ell$ is regularly convergent at $(x_0, y_0) \in \mathbb{R}^2$, then it is absolutely convergent at all $(x, y) \in \mathbb{R}^2$ with $|x| < |x_0|$ and $|y| < |y_0|$. (Hint: Part (ii) above and Lemma 7.46)

48. **(Abel's Test for Double Series)** Let $(a_{k,\ell})$ and $(b_{k,\ell})$ be double sequences of real numbers satisfying the following four conditions.
   (i) $(a_{k,\ell})$ is bimonotonic,
   (ii) for each fixed $\ell \in \mathbb{N}$, the sequence given by $k \longmapsto a_{k,\ell}$ is monotonic, and for each fixed $k \in \mathbb{N}$, the sequence given by $\ell \longmapsto a_{k,\ell}$ is monotonic,
   (iii) the sequences $(a_{k,k})$, $(a_{k,1})$, and $(a_{1,\ell})$ are bounded, and
   (iv) the double series $\sum\sum_{(k,\ell)} b_{k,\ell}$ is regularly convergent.

   Show that the double series $\sum\sum_{(k,\ell)} a_{k,\ell}b_{k,\ell}$ is regularly convergent. (Hint: Proof of Proposition 7.38, Exercise 47, the equality $a_{k,n} - a_{k+1,n} = a_{k,1} - a_{k+1,1} - \sum_{\ell=1}^{n-1}(a_{k,\ell} - a_{k+1,\ell} - a_{k,\ell+1} + a_{k+1,\ell+1})$ for $(k,n) \geq (1,2)$, and the convergence of the third summand in the Partial Double Summation Formula given in Proposition 7.37 to $\sum_k (a_k - a_{k+1})B_k$ as $(k,n) \to (\infty,\infty)$, where $a_k := \lim_{n\to\infty} a_{k,n}$ and $B_k := \lim_{n\to\infty} \sum_{j=1}^{k}\sum_{\ell=1}^{n} b_{j,\ell}$)
49. **(Dedekind's Test for Double Series)** Let $(a_{k,\ell})$ be a double sequence of real numbers satisfying the following two conditions.
   (i) $\sum\sum_{(k,\ell)} |a_{k,\ell} - a_{k+1,\ell} - a_{k,\ell+1} + a_{k+1,\ell+1}|$ is convergent,
   (ii) both $\sum_k |a_{k,1} - a_{k+1,1}|$ and $\sum_\ell |a_{1,\ell} - a_{1,\ell+1}|$ are convergent.

   Show that the double series $\sum\sum_{(k,\ell)} a_{k,\ell}b_{k,\ell}$ is regularly convergent whenever the double series $\sum\sum_{(k,\ell)} b_{k,\ell}$ is regularly convergent. If, in addition,
   (iii) $\lim_{k\to\infty} a_{k,\ell} = 0$ for each fixed $\ell \in \mathbb{N}$ and $\lim_{\ell\to\infty} a_{k,\ell} = 0$ for each fixed $k \in \mathbb{N}$,

   is also satisfied, then show that the double series $\sum\sum_{(k,\ell)} a_{k,\ell}b_{k,\ell}$ is regularly convergent whenever the partial double sums of $\sum\sum_{(k,\ell)} b_{k,\ell}$ are bounded. (Hint: Proof of Proposition 7.38 and Exercise 48)
   [Note: As for the converse, Hardy [28, Theorem 12] has shown that the following results hold: (1) If $\sum\sum_{(k,\ell)} a_{k,\ell}b_{k,\ell}$ is regularly convergent whenever $\sum\sum_{(k,\ell)} b_{k,\ell}$ is regularly convergent, then (i) and (ii) above hold. (2) If $\sum\sum_{(k,\ell)} a_{k,\ell}b_{k,\ell}$ is regularly convergent whenever the partial double sums of $\sum\sum_{(k,\ell)} b_{k,\ell}$ are bounded, then (i), (ii), and (iii) above hold.]
50. Let $(a_{k,\ell})$ be a double sequence of real numbers satisfying the following three conditions.
   (i) $\sum\sum_{(k,\ell)} |a_{k,\ell} - a_{k+1,\ell} - a_{k,\ell+1} + a_{k+1,\ell+1}|$ is convergent,
   (ii) both $\sum_k |a_{k,1} - a_{k+1,1}|$ and $\sum_\ell |a_{1,\ell} - a_{1,\ell+1}|$ are convergent,
   (iii) there are $a, \tilde{a} \in \mathbb{R}$ such that $\lim_{k\to\infty} a_{k,\ell} = a$ for each fixed $\ell \in \mathbb{N}$ and $\lim_{\ell\to\infty} a_{k,\ell} = \tilde{a}$ for each fixed $k \in \mathbb{N}$.

   Show that the double series $\sum\sum_{(k,\ell)} a_{k,\ell}b_{k,\ell}$ is convergent and its partial double sums are bounded whenever the double series $\sum\sum_{(k,\ell)} b_{k,\ell}$ is convergent and its partial double sums are bounded. (Hint: Exercise 49)
   [Note: As for the converse, Hamilton [26, p. 283] has shown that the following result holds. If $\sum\sum_{(k,\ell)} a_{k,\ell}b_{k,\ell}$ is convergent and its partial double sums are bounded whenever $\sum\sum_{(k,\ell)} b_{k,\ell}$ is convergent and its partial double sums are bounded, then (i), (ii), and (iii) above hold.]

51. **(Cauchy Product)** Given any double sequences $(a_{k,\ell})$ and $(b_{k,\ell})$, let

$$a_{k,\ell} * b_{k,\ell} = \sum_{i=0}^{k} \sum_{j=0}^{\ell} a_{i,j} b_{k-i,\ell-j} \quad \text{for } (k,\ell) \geq (0,0).$$

The double series $\sum\sum_{(k,\ell)} a_{k,\ell} * b_{k,\ell}$ is known as the **Cauchy product** of the double series $\sum\sum_{(k,\ell)} a_{k,\ell}$ and $\sum\sum_{(k,\ell)} b_{k,\ell}$. If $\sum\sum_{(k,\ell)} a_{k,\ell}$ is absolutely convergent and $\sum\sum_{(k,\ell)} b_{k,\ell}$ is regularly convergent, then show that $\sum\sum_{(k,\ell)} a_{k,\ell} * b_{k,\ell}$ is regularly convergent and its double sum is equal to the product of the double sums of $\sum\sum_{(k,\ell)} a_{k,\ell}$ and $\sum\sum_{(k,\ell)} b_{k,\ell}$. Further, in contrast to the case of (single) series, show that there exist an absolutely convergent double series and a convergent double series whose Cauchy product is divergent. (Hint: Consider $a_{0,\ell} := 2^{-\ell}$ and $a_{k,\ell} := 0$ for $(k,\ell) \geq (1,0)$, and $b_{k,0} := 1$, $b_{k,1} := -1$, and $b_{k,\ell} := 0$ for $(k,\ell) \geq (0,2)$.) [Note: As for the converse, it is shown in [36, Theorem 3.8 and Remark 3.9 (i)] that the following result holds. If the double series $\sum\sum_{(k,\ell)} a_{k,\ell} * b_{k,\ell}$ is convergent and its partial double sums are bounded for every regularly convergent double series $\sum\sum_{(k,\ell)} b_{k,\ell}$, then the double series $\sum\sum_{(k,\ell)} a_{k,\ell}$ is absolutely convergent.]

52. Let $(a_{k,\ell})$ be a double sequence, and consider the set S of all finite sums $\sum_{i=1}^{n-1} |a_{p_i,q_i} - a_{p_{i+1},q_{i+1}}|$, where n varies over $\mathbb{N}$ and $(p_1,q_1), \ldots, (p_n,q_n)$ vary over elements of $\mathbb{N}^2$ satisfying $(p_1,q_1) \leq \cdots \leq (p_n,q_n)$. The double sequence $(a_{k,\ell})$ is said to be of **bounded variation** if the set S is bounded above. (Compare the definition of a function of bounded variation on $[a,b] \times [c,d]$ given in Section 1.2.) Prove the following statements.
   (i) If $(a_{k,\ell})$ is of bounded variation, then it is bounded.
   (ii) If both $(a_{k,\ell})$ and $(b_{k,\ell})$ are of bounded variation and $r \in \mathbb{R}$, then $(a_{k,\ell} + b_{k,\ell})$, $(ra_{k,\ell})$, and $(a_{k,\ell}\, b_{k,\ell})$ are of bounded variation.
   (iii) If $(a_{k,\ell})$ is bounded and monotonic, then $(a_{k,\ell})$ is of bounded variation. In particular, if $(a_{k,\ell})$ and $(b_{k,\ell})$ are bounded and monotonically increasing, then $(a_{k,\ell} - b_{k,\ell})$ is of bounded variation.
   (iv) If $(a_{k,\ell})$ is of bounded variation, then there are bounded and monotonically increasing double sequences $(b_{k,\ell})$ and $(c_{k,\ell})$ such that $a_{k,\ell} = b_{k,\ell} - c_{k,\ell}$ for all $(k,\ell) \in \mathbb{N}^2$. (Hint: For $(k,\ell) \in \mathbb{N}^2$, let $v_{k,\ell}$ denote the supremum of the set of all finite sums $\sum_{i=1}^{n-1} |a_{p_i,q_i} - a_{p_{i+1},q_{i+1}}|$, where n varies over $\mathbb{N}$ and $(p_1,q_1), \ldots, (p_n,q_n)$ vary over elements of $\mathbb{N}^2$ satisfying $(p_1,q_1) \leq \cdots \leq (p_n,q_n) = (k,\ell)$. Define $b_{k,\ell} := (v_{k,\ell} + a_{k,\ell})/2$ and $c_{k,\ell} := (v_{k,\ell} - a_{k,\ell})/2$.)

53. A double sequence $(a_{k,\ell})$ is said to be **bibounded** if the double sequence $(a'_{k,\ell})$ defined by $a'_{k,\ell} := a_{k,\ell} - a_{k,1} - a_{1,\ell} + a_{1,1}$ for $(k,\ell) \in \mathbb{N}^2$ is bounded. Further, a double sequence $(a_{k,\ell})$ is said to be of **bounded bivariation** if the double series

$$\sum\sum_{(k,\ell)} |a_{k,\ell} + a_{k+1,\ell+1} - a_{k,\ell+1} - a_{k+1,\ell}|$$

is convergent. (Compare the definition of a function of bounded bivariation on $[a, b] \times [c, d]$ given in Section 1.2.) Prove the following statements.

(i) If $(a_{k,\ell})$ is bounded, then it is bibounded, but the converse does not hold. In fact, $(a_{k,\ell})$ is bounded if and only if it is bibounded and the sequences $(a_{k,1})$ and $(a_{1,\ell})$ are bounded.

(ii) If $(a_{k,\ell})$ is of bounded bivariation, then it is bibounded, and further, the double sequence $(a'_{k,\ell})$ is convergent. (Hint: Use telescoping summation as in the proof of Proposition 7.13.)

(iii) If both $(a_{k,\ell})$ and $(b_{k,\ell})$ are of bounded bivariation and $r \in \mathbb{R}$, then $(a_{k,\ell}+b_{k,\ell})$ and $(ra_{k,\ell})$ are of bounded bivariation, but $(a_{k,\ell}\, b_{k,\ell})$ need not be of bounded bivariation.

(iv) If $(a_{k,\ell})$ is bibounded and bimonotonic, then $(a_{k,\ell})$ is of bounded bivariation. In particular, if $(a_{k,\ell})$ and $(b_{k,\ell})$ are bibounded and bimonotonically increasing, then $(a_{k,\ell} - b_{k,\ell})$ is of bounded bivariation.

(v) If $(a_{k,\ell})$ is of bounded bivariation, then there are bibounded and bimonotonically increasing double sequences $(b_{k,\ell})$ and $(c_{k,\ell})$ such that $a_{k,\ell} = b_{k,\ell} - c_{k,\ell}$ for all $(k, \ell) \in \mathbb{N}^2$. (Hint: For $(k, \ell) \in \mathbb{N}^2$, consider $w_{k,\ell} := \sum_{i=1}^{k} \sum_{j=1}^{\ell} |a_{i-1,j-1} + a_{i,j} - a_{i-1,j} - a_{i,j-1}|$, where $a_{0,0} := 0$, $a_{k,0} := 0$ for $k \in \mathbb{N}$ and $a_{0,\ell} := 0$ for $\ell \in \mathbb{N}$. and Define $b_{k,\ell} := (w_{k,\ell} + a_{k,\ell})/2$ and $c_{k,\ell} := (w_{k,\ell} - a_{k,\ell})/2$.)

54. Suppose a double sequence $(a_{k,\ell})$ is of bounded bivariation and the sequences $(a_{k,1})$ and $(a_{1,\ell})$ are of bounded variation. Then show that $(a_{k,\ell})$ is of bounded variation.

55. Find a double sequence that is of bounded variation, but not of bounded bivariation. Also, find a double sequence that is of bounded bivariation, but not of bounded variation.

56. Let $c_{k,\ell} \in \mathbb{R}$ for $(k, \ell) \geq (0, 0)$, and for $(x, y) \in \mathbb{R}^2$, consider $C_{x,y} := \{c_{k,\ell}x^k y^\ell : (k, \ell) \geq (0, 0)\}$ and $E := \{(x, y) \in \mathbb{R}^2 : C_{x,y}$ is a bounded set$\}$. If (x^*, y^*) is a boundary point of E, then show that $(|x^*|, |y^*|)$ is a biradius of convergence of the double power series $\sum\sum_{(k,\ell)} c_{k,\ell}x^k y^\ell$. Conversely, if $(r, s) \in \mathbb{R}^2$ is a biradius of convergence of the double power series $\sum\sum_{(k,\ell)} c_{k,\ell}x^k y^\ell$, then show that $(r, s), (-r, -s), (r, -s)$, and $(-r, s)$ are boundary points of E.

57. Let $c_{k,\ell} := (k + \ell)!/k!\ell!$ for $(k, \ell) \geq (0, 0)$, and let $(x, y) \in \mathbb{R}^2$. Show that the double series $\sum\sum_{(k,\ell)} c_{k,\ell}x^k y^\ell$ is convergent if and only if either $|x|+|y| < 1$, or $-1 < x < 0$ and $x+y = -1$, and in that event, the double sum is equal to $1/(1-x-y)$. Show that the set $\{c_{k,\ell}x^k y^\ell : (k, \ell) \geq (0, 0)\}$ is bounded if and only if $|x| + |y| \leq 1$. (Compare Example 7.43 (vi).)

58. **(Hadamard's Formula for Biradius of Convergence)** Let (r, s) be a biradius of convergence of a double power series $\sum\sum_{(k,\ell)} c_{k,\ell}x^k y^\ell$, where r and s are positive real numbers. Show that $\inf\{M_n : n \in \mathbb{N}\} = 1$, where

$$M_n := \sup\left\{ \left(|c_{k,\ell}|\, r^k s^\ell\right)^{1/(k+\ell)} : (k, \ell) \geq (0, 0) \text{ and } k + \ell \geq n \right\} \text{ for } n \in \mathbb{N}.$$

[Note: The above conclusion is sometimes written as

$$\limsup_{k+\ell\to\infty}(|c_{k,\ell}|r^k s^\ell)^{1/(k+\ell)} = 1.$$

This may be compared with the formula $\limsup_{k\to\infty}|c_k|^{1/k} = 1/r$ for the radius of convergence r of a (single) power series $\sum_{k=0}^{\infty} c_k x^k$, where r is a positive real number. See, for example, Proposition 9.27 of ACICARA.]

59. **(Fabry's Theorem)** Let D denote the domain of convergence of a double power series $\sum\sum_{(k,\ell)} c_{k,\ell}\, x^k y^\ell$. Show that for all $(x_1, y_1), (x_2, y_2) \in D$ and $t \in \mathbb{R}$ with $0 < t < 1$, we have $\left(|x_1|^t |x_2|^{1-t},\, |y_1|^t |y_2|^{1-t}\right) \in D$. Deduce that D is log-convex.
60. If D is the domain of convergence of a double power series, then show that the set $E := \left\{(u, v) \in \mathbb{R}^2 : (e^u, e^v) \in D\right\}$ is convex.
61. Sketch the subset $\{(\ln|x|, \ln|y|) : (x, y) \in D \text{ and } xy \neq 0\}$ of $\mathbb{R}^2$, where D is the domain of convergence of the double power series given in each of Examples 7.43 (iii), (iv), (v), and (vi). (Hint: Exercise 60)
62. If D denotes the domain of convergence of a double power series and D contains the set $\{(x, y) \in \mathbb{R}^2 : |x| < 1 \text{ or } |y| < 1\}$, then show that $D = \mathbb{R}^2$. (Hint: Exercise 60)
63. **(Logarithmic Double Series)** Let $D := \{(x, y) \in \mathbb{R}^2 : x + y > -1\}$ and let $f : D \to \mathbb{R}$ be defined by $f(x, y) := \ln(1 + x + y)$. Show that the Taylor double series of f around $(0, 0)$ is

$$\sum\sum_{(k,\ell)\neq(0,0)} (-1)^{k+\ell+1} \frac{(k+\ell-1)!}{k!\ell!} x^k y^\ell.$$

Also, show that this double series is absolutely convergent if and only if $|x| + |y| < 1$, and in that event, its double sum is equal to $f(x, y)$. Further, show that the Taylor series of f around $(0, 0)$ is

$$\sum_{j=1}^{\infty} \frac{(-1)^{j-1}}{j} (x + y)^j.$$

Also, show that this series converges absolutely to $f(x, y)$ if $|x + y| < 1$, it converges conditionally to $f(x, y)$ if $x + y = 1$, and it diverges if either $x + y \leq -1$ or $x + y > 1$. (Hint: Exercises 15 and 32)

64. **(Binomial Double Series)** Let $D := \{(x, y) \in \mathbb{R}^2 : x + y > -1\}$ and let $t \in \mathbb{R} \setminus \{0, 1, 2, \ldots\}$. Consider $f : D \to \mathbb{R}$ defined by $f(x, y) := (1 + x + y)^t$. Show that the Taylor double series of f around (0,0) is

$$1 + \sum\sum_{(k,\ell)\neq(0,0)} t(t-1)\cdots(t-k-\ell+1)\frac{x^k}{k!}\frac{y^\ell}{\ell!}.$$

Also, show that it converges absolutely if and only if $|x| + |y| \leq 1$, provided $t > 0$, and it converges absolutely if and only if $|x| + |y| < 1$, provided $t < 0$, and whenever it is absolutely convergent, its double sum is equal to $f(x, y)$. Further show that the Taylor series of f around (0,0) is

$$1+\sum_{j=1}^{\infty} t(t-1)\cdots(t-j+1)\frac{(x+y)^j}{j!}.$$

Also, show that this series converges absolutely if and only if $|x+y| \le 1$, provided $t > 0$, and it converges absolutely if and only if $|x+y| < 1$, provided $t < 0$, and whenever it is absolutely convergent, its sum is equal to $f(x,y)$. (Hint: Exercises 15 and 32)

65. Let $D \subseteq \mathbb{R}^2$ be such that $(0,0)$ is an interior point of D. Suppose $E \subseteq \mathbb{R}$ is such that $xy \in E$ for all $(x,y) \in D$. Let $g : E \to \mathbb{R}$ be infinitely differentiable at 0. If $f : D \to \mathbb{R}$ is defined by $f(x,y) := g(xy)$, then show that the Taylor double series of f around $(0,0)$ is

$$\sum\sum_{(k,\ell)} c_{k,\ell}\, x^k y^\ell, \quad \text{where } c_{k,\ell} := \begin{cases} k!g^{(k)}(0) & \text{if } k = \ell, \\ 0 & \text{if } k \ne \ell, \end{cases}$$

and the Taylor series of f around $(0,0)$ is

$$\sum_{j=0}^{\infty} c_j(x,y), \quad \text{where } c_j(x,y) := \begin{cases} \dfrac{g^{(j/2)}(0)}{(j/2)!}(xy)^{j/2} & \text{if } j \text{ is even}, \\ 0 & \text{if } j \text{ is odd}. \end{cases}$$

Further, if r is the radius of convergence of the Taylor series of g around 0, then prove the following statements.

(i) If $(x,y) \in \mathbb{R}^2$ with $|xy| < r$, then the Taylor double series of f and the Taylor series of f around $(0,0)$ both converge absolutely, while if $|xy| > r$, then both diverge.

(ii) If $(x,y) \in D$ with $|xy| < r$, and further, if the Taylor series of g around 0 at $u := xy$ converges to $g(u)$, then both the Taylor double series of f and the Taylor series of f around $(0,0)$ at (x,y) converge to $f(x,y)$.

(Hint: Exercise 40 of Chapter 3.)

66. Find the Taylor double series and the Taylor series around $(0,0)$ of the following functions. In each case, find r such that both these converge absolutely if $|xy| < r$ and diverge if $|xy| > r$. Also, state whether these converge to the corresponding functional values.

(i) $f(x,y) := \sin xy$ for $(x,y) \in \mathbb{R}^2$,

(ii) $f(x,y) := e^{xy}$ for $(x,y) \in \mathbb{R}^2$,

(iii) $f(x,y) := \ln(1+xy)$ for $(x,y) \in \mathbb{R}^2$ with $xy > -1$,

(iv) $f(x,y) := (1+xy)^t$ for $(x,y) \in \mathbb{R}^2$ with $xy > -1$ and $t \ne 0, 1, \ldots$,

(v) $f(x,y) := 1/(1-xy)$ for $(x,y) \in \mathbb{R}^2$ satisfying $xy \ne -1$. (Compare Example 7.43 (v).)

(Hint: Exercise 65)

67. **(Integrating over Squares and Triangles)** Let $a, c \in \mathbb{R}$ and let $\iint_{[a,\infty)\times[c,\infty)} f(s,t)d(s,t)$ be an improper double integral with $f(s,t) \ge 0$ for all $(s,t) \in [a,\infty) \times [c,\infty)$. For $r \ge 0$, let $D_r := [a, a+r] \times [c, c+r]$, $E_r := \{(s,t) \in \mathbb{R}^2 : a \le s,\ c \le t \text{ and } s+t \le a+c+r\}$, and define

$$G(r) := \iint_{D_r} f(s,t)d(s,t) \quad \text{and} \quad H(r) := \iint_{E_r} f(s,t)d(s,t).$$

Show that $\iint_{[a,\infty)\times[c,\infty)} f(s,t)d(s,t)$ is convergent if and only if either of the two limits $\lim_{r\to\infty} G(r)$ and $\lim_{r\to\infty} H(r)$ exists, and in this case

$$\iint_{[a,\infty)\times[c,\infty)} f(s,t)d(s,t) = \lim_{r\to\infty} G(r) = \lim_{r\to\infty} H(r).$$

(Compare Proposition 7.16. Hint: Proposition 7.55)

68. Given any $\alpha, \beta \in \mathbb{R}$ with $\alpha > 0$ and $\beta > 1$, consider

$$\iint_{[1,\infty)\times[1,\infty)} \frac{d(s,t)}{(s+t^\alpha)^\beta} \quad \text{and} \quad \iint_{(0,1]\times(0,1]} \frac{d(s,t)}{(s+t^\alpha)^\beta}.$$

Prove that the former is unconditionally convergent if and only if $\alpha(\beta-1) > 1$, whereas the latter is unconditionally convergent if and only if $\alpha(\beta-1) < 1$. (Hint: Use Proposition 5.28 and integrate with respect to s first. Compare Example 7.58.)

69. Given any $p, q \in \mathbb{R}$ with $p > 0$ and $q > 0$, consider

$$\iint_{[1,\infty)\times[1,\infty)} \frac{d(s,t)}{s^p + t^q} \quad \text{and} \quad \iint_{(0,1]\times(0,1]} \frac{d(s,t)}{s^p + t^q}.$$

Prove that the former is unconditionally convergent if and only if $(1/p) + (1/q) < 1$, whereas the latter is unconditionally convergent if and only if $(1/p) + (1/q) > 1$. (Hint: Note that for all $(s,t) \geq (0,0)$,

$$\frac{1}{2^p}\left(s + t^{q/p}\right)^p \leq s^p + t^q \leq 2\left(s + t^{q/p}\right)^p.$$

Use Exercise 68 with $\alpha := q/p$ and $\beta := p$ if $p > 1$.)

70. Given any $p, q \in \mathbb{R}$ with $p > 0$ and $q > 0$, show that the double series $\sum\sum_{(k,\ell)} 1/\left(k^p + \ell^q\right)$ is convergent if and only if $(1/p) + (1/q) < 1$. (Hint: Exercise 69 and Proposition 7.57)

71. Let $p \in \mathbb{R}$ with $p > 0$ and let $f : \mathbb{R}^3 \setminus \{(0,0,0)\} \to \mathbb{R}$ be defined by $f(s,t,u) := 1/(s^2+t^2+u^2)^p$. If $D := \{(s,t,u) \in \mathbb{R}^3 : s^2+t^2+u^2 \geq 1\}$, then show that $\iiint_D f(s,t,u)d(s,t,u)$ is unconditionally convergent if and only if $p > 3/2$, and in this case, it is equal to $4\pi/(2p-3)$. Further, if $E := \{(s,t,u) \in \mathbb{R}^3 : 0 < s^2+t^2+u^2 \leq 1\}$, then show that $\iiint_E f(s,t,u)d(s,t,u)$ is unconditionally convergent if and only if $p < 3/2$, and in this case, it is equal to $4\pi/(3-2p)$. (Hint: Consider $D_n := \{(s,t,u) \in \mathbb{R}^3 : n^2 \geq s^2+t^2+u^2 \geq 1\}$ and $E_n := \{(s,t,u) \in \mathbb{R}^3 : (1/n^2) \leq s^2+t^2+u^2 \leq 1\}$ for $n \in \mathbb{N}$; use spherical coordinates and part (ii) of Proposition 5.72. Compare Examples 7.78 (i) and 7.85 (i).)

References

1. T. Apostol, *Mathematical Analysis*, first ed. and second ed., Addison-Wesley, Reading, Mass., 1957 and 1974.
2. T. Apostol, *Calculus*, second ed., vols. I and II, John Wiley, New York, 1967 and 1969.
3. K. E. Atkinson, *The Numerical Solution of Integral Equations of the Second Kind*, Cambridge University Press, Cambridge, 1997.
4. J. C. Baez, The octonions, *Bull. Amer. Math. Soc.* **39** (2002), pp. 145–205.
5. S. A. Baron, Deriving convergence tests for double series (in Russian), *Tartu Riikliku Ülikooli Toimetised* (*Acta Comment. Univ. Tartu. Math.*) **55** (1958), pp. 9–19.
6. E. F. Beckenbach and R. Bellman, *Inequalities*, Springer-Verlag, New York, 1965.
7. T. J. Bromwich, *An Introduction to the Theory of Infinite Series*, third ed., Chelsea, New York, 1991.
8. R. C. Buck, *Advanced Calculus*, third ed., McGraw-Hill, New York, 1978.
9. J. A. Clarkson and C. R. Adams, On definitions of bounded variation for functions of two variables, *Trans. Amer. Math. Soc.* **35** (1933), pp. 824–854.
10. J. A. Clarkson and C. R. Adams, Properties of functions $f(x, y)$ of bounded variation, *Trans. Amer. Math. Soc.* **36** (1934), pp. 711–730.
11. L. J. Corwin and R. H. Szczarba, *Calculus in Vector Spaces*, second ed., Marcel Dekker, New York, 1995.
12. R. Courant and F. John, *Introduction to Calculus and Analysis*, vols. I and II, Springer-Verlag, New York, 1989.
13. P. J. Daniell, Ratio tests for double power series, *Quart. J. Math.* **11** (1940), pp. 183–192.
14. G. de Rham, *Differentiable Manifolds*, Springer-Verlag, Berlin, 1984.
15. J. Dieudonné, *Foundations of Modern Analysis*, Academic Press, New York and London, 1969.
16. H.-D. Ebbinghaus, H. Hermes, F. Hirzebruch, M. Koecher, K. Mainzer, A. Prestel, and R. Remmert, *Numbers*, Springer-Verlag, New York, 1991.
17. H. Engles, *Numerical Quadrature and Cubature*, Academic Press, New York, 1980.

18. K. Eriksson, D. Estep, and C. Johnson, *Applied Mathematics: Body and Soul*, vols. 1, 2, and 3, Springer, Berlin, 2004.
19. W. H. Fleming, *Functions of Several Variables*, second ed., Springer-Verlag, New York, 1977.
20. L. Fuchs, *Partially Ordered Algebraic Systems*, Pergamon Press, Oxford, 1963.
21. C. F. Gauss, Disquisitiones Arithmeticae, revised ed. of an English translation, Springer-Verlag, New York, 1986.
22. S. R. Ghorpade and B. V. Limaye, *A Course in Calculus and Real Analysis*, Springer, New York, 2006.
23. S. R. Ghorpade and B. V. Limaye, Sylvester's minorant criterion, Lagrange-Beltrami identity, and nonnegative definiteness, *Math. Student*, Spec. Centenary Vol. (2007), pp. 123–130.
24. S. R. Ghorpade and B. V. Limaye, A geometric approach to saddle points, *Austral. Math. Soc. Gaz.* **36** (2009), pp. 127–136.
25. H. Goldstein, *Classical Mechanics*, second ed., Addison-Wesley, Reading, Mass., 1980.
26. H. J. Hamilton, On transformations of double series, *Bull. Amer. Math. Soc.* **42** (1936), pp. 275–283.
27. G. H. Hardy, On the convergence of certain multiple series, *Proc. London Math. Soc. (2)* **1** (1903), pp. 124–128.
28. G. H. Hardy, On the convergence of certain multiple series, *Proc. Cambridge Philosoph. Soc.* **19** (1916–1919), pp. 86–95.
29. G. H. Hardy, *A Course of Pure Mathematics*, reprint of the tenth ed. (1952), Cambridge University Press, Cambridge, 1992.
30. O. Hijab, *Introduction to Calculus and Classical Analysis*, second ed., Springer, New York, 2007.
31. T. H. Hildebrandt, *Introduction to the Theory of Integration*, Academic Press, New York and London, 1963.
32. E. W. Hobson, *The Theory of Functions of a Real Variable and the Theory of Fourier Series*, vols. I and II, third ed., Dover, New York, 1927.
33. I. L. Kantor and A. S. Solodovnikov, *Hypercomplex Numbers - An Elementary Introduction to Algebras*, Springer-Verlag, Berlin, 1989.
34. P. D. Lax, Change of variables in multiple integrals, *Amer. Math. Monthly* **106** (1999), pp. 497–501.
35. P. D. Lax, Change of variables in multiple integrals II, *Amer. Math. Monthly* **108** (2001), pp. 115–119.
36. B. V. Limaye and M. Zeltser, On the Pringsheim convergence of double series, *Proc. Estonian Acad. Sci.* **58** (2009), pp. 108–121.
37. J. Lynnes and D. Jesperson, Moderate degree symmetric triangulation rules for the triangle, *J. Inst. Math. Appl.* **15** (1975), pp. 19–32.
38. J. E. Marsden and M. Hoffman, *Elementary Classical Analysis*, second ed., W. H. Freeman, San Francisco, 1993.
39. J. R. Munkres, *Analysis on Manifolds*, Addison-Wesley, Redwood City, CA, 1991.
40. J. R. Munkres, *Topology*, second ed., Prentice Hall, Upper Saddle River, NJ, 2000.
41. R. M. Murray, Z. Li, and S. S. Sastry, *A Mathematical Introduction to Robotic Manipulation*, CRC Press, Boca Raton, 1994.

42. A. Pringsheim, Elementare theorie der unendlichen doppelreihen, *Münch. Ber.* **27** (1897), pp. 101–152.
43. A. Pringsheim, *Vorlesungen über Zahlen- und Funktionenlehre*, vols. I and II, B. G. Teubner, Leipzig/Berlin, 1916 and 1925.
44. M. H. Protter and C. B. Morrey, *A First Course in Real Analysis*, second ed., Springer, New York, 1991.
45. C. C. Pugh, *Real Mathematical Analysis*, Springer-Verlag, New York, 2002.
46. R. Remmert, *Theory of Complex Functions*, Springer-Verlag, New York, 1991.
47. A. W. Roberts and D. E. Varberg, *Convex Functions*, Academic Press, New York, 1973.
48. W. Rudin, *Principles of Mathematical Analysis*, third ed., McGraw-Hill, New York–Auckland–Düsseldorf, 1976.
49. J. M. Ruiz, *The Basic Theory of Power Series*, Vieweg Verlag, Braunschweig/Wiesbaden, Germany, 1993.
50. H. Sagan, *Space-Filling Curves*, Springer, New York, 1994.
51. J. Schwartz, The formula for change of variables in a multiple integral, *Amer. Math. Monthly* **61** (1954), pp. 81–85.
52. S. L. Sobolev, *Theory of Approximations of Integrals of Functions of Several Variables*, Amer. Math. Soc. Translations Ser. 2, vol. 70, American Mathematical Society, Providence, 1968.
53. S. L. Sobolev and V. L. Vaskevich, *The Theory of Cubature Formulas*, Kluwer Academic, Dordrecht, 1997.
54. M. Spivak, *Calculus on Manifolds*, Benjamin, New York, 1965.
55. A. H. Stroud, *Approximate Calculation of Multiple Integrals*, Prentice-Hall, Englewood Cliffs, NJ, 1971.
56. J. van Tiel, *Convex Analysis: An Introductory Text*, John Wiley, New York, 1984.
57. J. R. L. Webb, *Functions of Several Real Variables*, Ellis Horwood, Chichester, 1991.

List of Symbols and Abbreviations

	Definition/Description	Page
ACICARA	Reference [22]: A Course in Calculus and Real Analysis	VI
$\mathbb{R}$	set of all real numbers	1
$\mathbb{R}^n$	n-dimensional Euclidean space	1
$\mathbb{N}$	set of all positive integers	1
$\mathbf{x} = (x_1, \ldots, x_n)$	vector in $\mathbb{R}^n$ with coordinates $x_1, \ldots, x_n$	1
$\mathbf{0}$	zero vector	1
$\mathbf{x} \cdot \mathbf{y}$	dot product of $\mathbf{x}$ and $\mathbf{y}$	3
$\lvert\mathbf{x}\rvert$	norm $\sqrt{x_1^2 + \cdots + x_n^2}$ of the vector $\mathbf{x} = (x_1, \ldots, x_n) \in \mathbb{R}^n$	3
$\mathbf{x} \leq \mathbf{y}$	$\mathbf{x}$ is less than or equal to $\mathbf{y}$ in the product order on $\mathbb{R}^n$	5
$I_{a,b}$	closed interval between real numbers a and b	6
$I_{\mathbf{a},\mathbf{b}}$	$I_{a_1,b_1} \times \cdots \times I_{a_n,b_n}$, where $\mathbf{a} = (a_1, \ldots, a_n)$ and $\mathbf{b} = (b_1, \ldots, b_n)$	6
$I \times J$	$\{(x, y) \in \mathbb{R}^2 : x \in I \text{ and } y \in J\}$	7
$\mathbb{S}_r(\mathbf{c})$	$\{\mathbf{x} \in \mathbb{R}^n : \lvert x_i - c_i\rvert < r \text{ for } i = 1, \ldots, n\}$	7
$\mathbb{B}_r(\mathbf{c})$	$\{\mathbf{x} \in \mathbb{R}^n : \lvert\mathbf{x} - \mathbf{c}\rvert < r\}$	7
$\mathrm{diam}(D)$	diameter of a nonempty bounded subset D of $\mathbb{R}^n$	8
$\mathbb{Q}$	set of all rational numbers	9
$I + J$	$\{x + y : x \in I \text{ and } y \in J\}$, where I, J are intervals in $\mathbb{R}$	15
$V(f)$	total variation of a function f of bounded variation	17
v_f	total variation function corresponding to f	18
$W(f)$	total bivariation of a function f of bounded bivariation	20
w_f	total bivariation function corresponding to f	22
$\triangle_{\mathbf{a}}^{\mathbf{b}} f$	alternating difference of $f(c_1, \ldots, c_n)$, where $c_i \in \{a_i, b_i\}$	25
$\triangle_{(a_1,a_2)}^{(b_1,b_2)} f$	$f(b_1, b_2) + f(a_1, a_2) - f(b_1, a_2) - f(a_1, b_2)$	25
$\triangle_{(a_1,a_2,a_3)}^{(b_1,b_2,b_3)} f$	$f(b_1, b_2, b_3) + f(b_1, a_2, a_3) + f(a_1, b_2, a_3) + f(a_1, a_2, b_3)$ $-f(b_1, b_2, a_3) - f(a_1, b_2, b_3) - f(b_1, a_2, b_3) - f(a_1, a_2, a_3)$	25
IVP	Intermediate Value Property	29
$[x]$	integer part of a real number x	29

	Definition/Description	Page
$\mathbf{x} \preceq \mathbf{y}$	$\mathbf{x}$ is less than or equal to $\mathbf{y}$ in the lexicographic order on $\mathbb{R}^n$	35
$\mathbb{M}_r(\mathbf{c})$	$\{\mathbf{x} \in \mathbb{R}^n : \lvert x_1 - c_1 \rvert + \cdots + \lvert x_n - c_n \rvert < r\}$	35
$\lVert \mathbf{x} \rVert_p$	p-norm $\left(\lvert x_1\rvert^p + \cdots + \lvert x_n\rvert^p\right)^{1/p}$ of $\mathbf{x} = (x_1, \ldots, x_n) \in \mathbb{R}^n$	38
$\lVert \mathbf{x} \rVert_\infty$	∞-norm $\max\{\lvert x_1\rvert, \ldots, \lvert x_n\rvert\}$ of $\mathbf{x} = (x_1, \ldots, x_n) \in \mathbb{R}^n$	38
SIVP	Strong Intermediate Value Property	42
$((x_n, y_n))$	sequence in $\mathbb{R}^2$ whose nth term is (x_n, y_n)	43
$(x_n, y_n) \to (x_0, y_0)$	sequence $((x_n, y_n))$ converges to (x_0, y_0)	44
$((x_{n_k}, y_{n_k}))$	subsequence of $((x_n, y_n))$	45
$\overline{D}$	closure of a subset D of $\mathbb{R}^n$	46, 48
∂D	boundary of a subset D of $\mathbb{R}^n$	46, 48
$\lim\limits_{(x,y)\to(x_0,y_0)} f(x,y)$	limit of f as (x, y) tends to (x_0, y_0)	67
$\lim\limits_{(x,y)\to(x_0^+,y_0^+)} f(x,y)$	limit of f from first quadrant as (x, y) tends to (x_0, y_0)	71
$\lim\limits_{(x,y)\to(\infty,\infty)} f(x,y)$	limit of f as (x, y) tends to (∞, ∞)	73
$f_x, \dfrac{\partial f}{\partial x}$	partial derivative of f w.r.t. x	84 138
∇f	gradient of f	85, 139
$(f_x)_-$	left(-hand) partial derivative of f w.r.t. x	85
$(f_x)_+$	right(-hand) partial derivative of f w.r.t. x	85
MVT	Mean Value Theorem	87
$\mathbf{D}_{\mathbf{u}} f$	directional derivative of f along $\mathbf{u}$	88, 138
$f_{xx}, \dfrac{\partial^2 f}{\partial x^2}$	partial derivative of f_x w.r.t. x	91, 138
$f_{xy}, \dfrac{\partial^2 f}{\partial y \partial x}$	partial derivative of f_x w.r.t. y	91, 138
$\dfrac{\partial^n f}{\partial x^{n-m} \partial y^m}$	nth-order partial derivative of f $(m = 0, 1, \ldots, n)$	96
$\mathcal{D}_{h,k}$	partial differential operator $h\dfrac{\partial}{\partial x} + k\dfrac{\partial}{\partial y}$	97
$\mathcal{D}_{h,k}^n$	$\left(h\dfrac{\partial}{\partial x} + k\dfrac{\partial}{\partial y}\right)^n = \sum\limits_{m=0}^{n} \dbinom{n}{m} h^{n-m} k^m \dfrac{\partial^n}{\partial x^{n-m} \partial y^m}$	98
$\mathbf{D}_{\mathbf{u}}^n f$	nth-order directional derivative of f along $\mathbf{u}$	99
$D_{\mathbf{uv}}^2 f$	directional derivative of $\mathbf{D}_{\mathbf{u}} f$ along $\mathbf{v}$	99
$J(\Phi)$	Jacobian of the transformation Φ	123
$\dfrac{\partial(x,y)}{\partial(u,v)}$	Jacobian of the functions x and y w.r.t. u and v	123
Δf	discriminant of f	136
$\mu(P)$	mesh of partition P	187
$\mathsf{P}_{n,k}$	partition into $n \times k$ equal parts	187

Index

MIX
Papier aus verantwortungsvollen Quellen
Paper from responsible sources
FSC® C105338

Printed by Books on Demand, Germany